Comprehensive
PHYSICS for AIEEE and Other Competitive Examinations

Comprehensive
PHYSICS
for
AIEEE
and
Other Competitive Examinations

By

Prof. Narinder Kumar
Ex-Proffessor
Deptt. of Physics
S.D. Govt. College
Ludhiana

Dr. J.K. Juneja
Deptt. of Physics
Hindu College
Sonepat

GOLDEN BELLS

(*AN IMPRINT OF LAXMI PUBLICATIONS PVT. LTD.*)

BANGALORE • CHENNAI • COCHIN • GUWAHATI • HYDERABAD
JALANDHAR • KOLKATA • LUCKNOW • MUMBAI • RANCHI

NEW DELHI

Published by :
GOLDEN BELLS
[*A Unit of Laxmi Publications*]
94-B, Hindi Park, Daryaganj,
New Delhi-110002.

Phones : 011-2326 01 90
011-5179 16 84

Faxes : 011-2326 02 48
011-2326 22 79

Branches :

- 129/1, IIIrd Main Road, IX Cross Chamrajpet, **Bangalore** (*Phone* : 080-26 61 15 61)
- 26, Damodaran Street, T. Nagar, **Chennai** (*Phone* : 044-24 34 47 26)
- St. Benedict's Road, **Cochin** (*Phone* : 0484-239 70 04)
- Pan Bazar, Rani Bari, **Guwahati** (*Phones* : 0361-254 36 69, 251 38 81)
- 4-2-453, Ist Floor, Ramkote, **Hyderabad** (*Phone* : 040-24 75 02 47)
- Adda Tanda Chowk, N.D. 365, **Jalandhar City** (*Phone* : 0181-222 12 72)
- 106/A, Ist Floor, S.N. Banerjee Road, **Kolkata** (*Phones* : 033-22 27 37 73, 22 27 52 47)
- 18, Madan Mohan Malviya Marg, **Lucknow** (*Phone* : 0522-220 95 78)
- 128A, Block 3, First Floor, Noorani Building, L.J. Road, **Mumbai** (*Phone* : 022-24 46 39 98)
- Radha Govind Street, Tharpagna, **Ranchi** (*Phone* : 0651-230 77 64)

EMAIL : colaxmi@hotmail.com

WEBSITE : www.laxmipublications.com

GBC-5020-285-PHYSICS FOR AIEEE

4th Edition

Price Rs: 150.00

C—115 |4

Laser Typesetted at : Goswami Printers, Delhi-110053.

Printed at : Odin Press (P) Ltd, 04023770271

TREND ANALYSIS OF AIEEE PAPERS

(No. of Questions asked Chapterwise in Previous Years)

Chapters / Topics	*2002*	*2003*	*2004*	*2005*
1. Introduction and Measurement	1	—	—	—
2. Units and Dimensions	—	2	1	1
3. Vectors	—	—	1	—
4. Kinematics and Dynamics	11	8	8	10
5. Work, Power and Energy	2	4	4	5
6. Rotational Mechanics	4	4	2	4
7. Gravitation	3	3	4	3
8. Properties of Matter	1	1	4	2
9. Heat and Thermodynamics	11	6	5	5
10. Oscillations and Waves	10	8	6	6
11. Electrostatics	5	5	4	5
12. Current Electricity, Chemical, Thermal and Thermoelectric Effects	4	8	9	7
13. Magnetic Effects of Current, Magnetism, EMI and AC	10	10	12	11
14. Optics	4	5	4	6
15. Modern Physics	5	9	11	7
16. Solids and Semiconductor Devices	4	2	—	3
Total	75	75	75	75

SYLLABUS FOR AIEEE (AET)

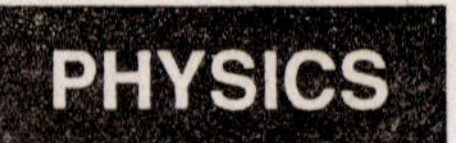

PART A : MECHANICS

Unit 1 **Units and Measurement**

Units for measurement, system of units—S.I., fundamental and derived units. Dimensions and their applications.

Unit 2 **Description of Motion in One Dimension**

Motion in a straight line, uniform and non-uniform motion, their graphical representation. Uniformly accelerated motion, and its applications.

Unit 3 **Description of Motion in Two and Three Dimensions**

Scalars and vectors, vector addition, a real number, zero-vector and its properties. Resolution of vectors. Scalar and vector products, uniform circular motion and its applications, projectile motion.

Unit 4 **Laws of Motion**

Force and inertia—Newton's Laws of Motion. Conservation of linear momentum and its applications, rocket propulsion, friction—laws of friction.

Unit 5 **Work, Energy and Power**

Concept of work, energy and power. Energy—kinetic and potential. Conservation of energy and its applications, Elastic collisions in one and two dimensions. Different forms of energy.

Unit 6 **Rotational Motion and Moment of Inertia**

Centre of mass of a two-particle system. Centre of mass of a rigid body, general motion of a rigid body, nature of rotational motion, torque, angular momentum, its conservation and applications. Moment of Inertia, parallel and perpendicular axes theorem, expression of moment of inertia for ring, disc and sphere.

Unit 7 **Gravitation**

Acceleration due to gravity, one and two-dimensional motion under gravity. Universal law of gravitation, variation in the acceleration due to gravity of the earth. Planetary motion, Kepler's laws, artificial satellite—geostationary satellite, gravitational potential energy near the surface of earth, gravitational potential and escape velocity.

Unit 8 **Solids and Fluids**

Inter-atomic and inter-molecular forces, states of matter.

A. ***Solids*** : Elastic properties, Hooke's law, Young's modulus, bulk modulus, modulus of rigidity.

B. ***Liquids*** : Cohesion and adhesion. Surface energy and surface tension. Flow of fluids, Bernoulli's theorem and its applications. Viscosity, Stoke's Law, terminal velocity.

PART B : OSCILLATIONS, WAVES, HEAT & THERMODYNAMICS

Unit 9 **Oscillations**

Periodic motion, simple harmonic motion and its equation of motion, energy in S.H.M., Oscillations of a spring and simple pendulum.

Unit 10 **Waves**

Wave motion, speed of a wave, longitudinal and transverse waves, superposition of waves, progressive and standing waves, free and forced Oscillations, resonance, vibration of strings and air-columns, beats, Doppler effect.

Unit 11 **Heat and Thermodynamics**

Thermal expansion of solids, liquids and gases and their specific heats, Relationship between C_p and C_v for gases, first law of thermodynamics, thermodynamic processes. Second law of thermodynamics, Carnot cycle, efficiency of heat engines.

Unit 12 **Transference of Heat**

Modes of transference of heat. Thermal conductivity. Black body radiations, Kirchoff's law, Wien's law, Stefan's law of radiation and Newton's law of cooling.

PART C : ELECTRICITY AND MAGNETISM

Unit 13 **Electrostatics**

Electric charge—its unit and conservation, Coulomb's law, dielectric constant, electric field, lines of force, field due to dipole and its behaviour in a uniform electric field, electric flux, Gauss's theorem and its applications. Electric potential, potential due to a point charge. Conductors and insulators, distribution of charge on conductors. Capacitance, parallel plate capacitor, combination of capacitors, energy of capacitor.

Unit 14 **Current Electricity**

Electric current and its unit, sources of energy, cells—primary and secondary, grouping of cells resistance of different materials, temperature dependence, specific resistivity, Ohm's law, Kirchoff's law, series and parallel circuits. Wheatstone Bridge with their applications and potentiometer with their applications.

Unit 15 **Thermal and Chemical Effects of Currents**

Heating effects of current, electric power, simple concept of thermoelectricity—Seebeck effect and thermocouple, Chemical effect of current—Faraday's laws of electrolysis.

Unit 16 **Magnetic Effects of Currents**

Oersted's experiment, Biot-Savart's law, magnetic field due to straight wire, circular loop and solenoid, force on a moving charge in a uniform magnetic field (Lorentz force), forces and torques on currents in a magnetic field, force between two current carrying wires, moving coil galvanometer and conversion to ammeter ar voltmeter.

Unit 17 **Magnetostatics**

Bar magnet, magnetic field, lines of force, torque on a bar magnet in a magnetic field, earth's magnetic f para, dia and ferro magnetism, magnetic induction, magnetic susceptibility.

Unit 18 **Electromagnetic Induction and Alternating Currents**

Induced e.m.f., Faraday's Law, Lenz's Law, self and mutual Induction, alternating currents, impedance and reactance, power in a.c. Circuits with L.C. and R Series Combination, resonant circuits. Transformer and A.C. generator.

PART D : OPTICS

Unit 19 **Ray Optics**

Reflection and refraction of light at plane and curved surfaces, total internal reflection, optical fibre ; deviation and dispersion of light by a prism ; Lens formula, magnification and resolving power ; microscope and telescope.

Unit 20 **Wave Optics**

Wave nature of light ; Interference—Young's double slit experiment. Diffraction—diffraction due to a single slit. Elementary idea of polarization.

Unit 21 **Electromagnetic Waves**

Electromagnetic waves and their characteristics, Electromagnetic wave spectrum from gamma to radio waves—propagation of EM waves in atmosphere.

PART E : MODERN PHYSICS

Unit 22 **Electrons and Photons**

Charge on an electron, e/m for an electron, photoelectric effect and Einstein's equation of photoelectric effect.

Unit 23 **Atoms, Molecules and Nuclei**

Alpha-particles scattering experiment, Atomic masses, size of the nucleus ; Radioactivity ; Alpha, beta and gamma particles/rays and their properties, radioactive decay law, half life and mean life of radio-active nuclei, binding energy, mass-energy relationship, nuclear fission and nuclear fusion.

Unit 24 **Solids and Semi-Conductors Devices**

Energy bands in solids, conductors, insulators and semi-conductors, PN junction, diodes, diode as rectifier, transistor action, transistor as an amplifier.

Contents

SOLVED TEST PAPERS—

AIEEE (AET) 2005—Solved

1. Out of the following pair, which one does NOT have identical dimensions is
   (*a*) moment of inertia and moment of a force
   (*b*) Work and torque
   (*c*) angular momentum and Planck's constant
   (*d*) impulse and momentum.
2. The relation between time t and distance x is $t = ax^2 + bx$ where a and b are constants. The acceleration is
   (*a*) $-2av^3$ (*b*) $2av^2$
   (*c*) $-2abv^2$ (*d*) $2bv^3$.
3. A car, starting from rest, accelerates at the rate f through a distance S, then continues at constant speed for time t and then decelerates at the rate $\frac{f}{2}$ to come to rest. If the total distance traversed is 15 S, then
   (*a*) $S = \frac{1}{2} ft^2$ (*b*) $S = \frac{1}{4} ft^2$
   (*c*) $S = ft$ (*d*) $S = \frac{1}{6} ft^2$.
4. A particle is moving eastwards with a velocity of 5 m s^{-1}. In 10 second, the velocity changes to 5 m s^{-1} northwards. The average acceleration in this time is
   (*a*) zero
   (*b*) $\frac{1}{\sqrt{2}}$ m s^{-2} towards north-west
   (*c*) $\frac{1}{\sqrt{2}}$ m s^{-2} towards north-east
   (*d*) $\frac{1}{2}$ m s^{-2} towards north.
5. A projectile can have the same range 'R' for two angles of projection. If 't_1' and 't_2' be the times of flights in the two cases, then the product of the two times of flight is proportional to
   (*a*) $\frac{1}{R}$ (*b*) R
   (*c*) R^2 (*d*) $\frac{1}{R^2}$.
6. An annular ring withs inner and outer radii R_1 and R_2 is rolling without slipping with a uniform angular speed. The ratio of the forces experienced by the two particles situated on the inner and outer parts of the ring, $\frac{F_1}{F_2}$ is
   (*a*) 1 (*b*) $\frac{R_1}{R_2}$
   (*c*) $\frac{R_2}{R_1}$ (*d*) $\left(\frac{R_1}{R_2}\right)^2$.
7. A smooth block is released at rest on a 45° incline and then slides a distance 'd'. The time taken to slide is 'n' times as much to slide on rough incline than on a smooth incline. The coefficient of friction is
   (*a*) $\mu_s = 1 - \frac{1}{n^2}$ (*b*) $\mu_s = \sqrt{1-\frac{1}{n^2}}$
   (*c*) $\mu_k = 1 - \frac{1}{n^2}$ (*d*) $\mu_k = \sqrt{1-\frac{1}{n^2}}$.
8. The upper half of an inclined plane with inclination ϕ is perfectly smooth while the lower half is rough. A body starting from rest at the top will again come to rest at the bottom if the coefficient of friction for the lower half is given by
   (*a*) $2 \tan \phi$ (*b*) $\tan \phi$
   (*c*) $2 \sin \phi$ (*d*) $2 \cos \phi$.
9. A bullet fired into a fixed target loses half of its velocity after penetrating 3 cm. How much further it will penetrate before coming to rest assuming that it faces constant resistance to motion ?
   (*a*) 1.5 cm (*b*) 1.0 cm
   (*c*) 3.0 cm (*d*) 2.0 cm.
10. A parachutist, after bailing out, falls 50 m without friction. When parachute opens, it decelerates at 2 m/s^2. He reaches the ground with a speed of 3 m/s. At what height, did he bail out ?
    (*a*) 293 m (*b*) 111 m
    (*c*) 91 m (*d*) 182 m.
11. A particle of mass 0.3 kg is subjected to a force $F = -kx$ with $k = 15$ N/m. What will be its initial acceleration if it is released from a point 20 cm away from the origin ?
    (*a*) 5 m/s^2 (*b*) 10 m/s^2
    (*c*) 3 m/s^2 (*d*) 15 m/s^2.
12. The block of mass M moving on the frictionless horizontal surface collides with the spring of spring constant K and compresses it by length L. The maximum momentum of the block after collision is

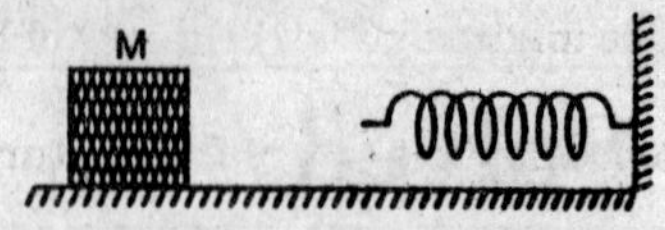

(a) zero (b) $\frac{ML^2}{K}$

(c) $\sqrt{MK}$ L (d) $\frac{KL^2}{2M}$.

13. A mass '*m*' moves with a velocity '*v*' collides inelastically with another identical mass. After collision the 1st mass moves with velocity $\frac{v}{\sqrt{3}}$ in a direction perpendicular to the initial direction of motion. Find the speed of the 2nd mass after collision.

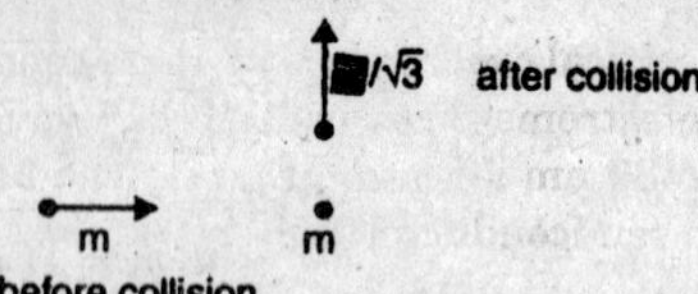

(a) $\frac{2}{\sqrt{3}}v$ (b) $\frac{v}{\sqrt{3}}$

(c) v (d) $\sqrt{3}\,v$.

14. A spherical ball of mass 20 kg is stationary at the top of a hill of height 100 m. It rolls down a smooth surface to the ground, then climbs up another hill of height 30 m and finally rolls down to a horizontal base at a height of 20 m above the ground. The velocity attained by the ball is

(a) 10 m/s (b) $10\sqrt{30}$ m/s

(c) 40 m/s (d) 20 m/s.

15. A block is kept on a frictionless inclined surface with angle of inclination 'α'. The incline is given an acceleration '*a*' to keep the block stationary. Then *a* is equal to

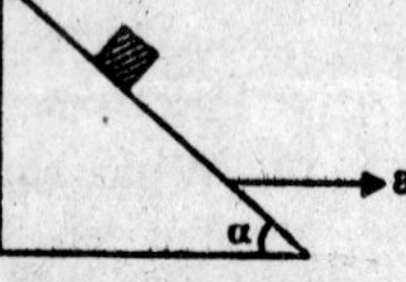

(a) g (b) $g \tan \alpha$

(c) $g/\tan \alpha$ (d) g cosec α.

16. The moment of inertia of a uniform semicircular disc of mass M and radius *r* about a line perpendicular to the plane of the disc through the centre is

(a) Mr^2 (b) $\frac{1}{2}Mr^2$

(c) $\frac{1}{4}Mr^2$ (d) $\frac{2}{5}Mr^2$.

17. A body A of mass M while falling vertically downwards under gravity breaks into two parts ; a body B of mass $\frac{1}{3}$ M and a body C of mass $\frac{2}{3}$ M. The centre of mass of bodies B and C taken together shifts compared to that of body A towards

(a) body C (b) body B

(c) depends on height of breaking

(d) does not shift.

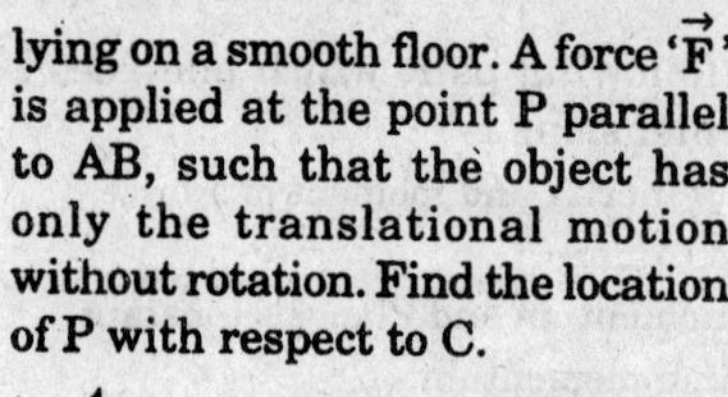

18. A 'T' shaped object with dimensions shown in the figure, is lying on a smooth floor. A force '$\vec{F}$' is applied at the point P parallel to AB, such that the object has only the translational motion without rotation. Find the location of P with respect to C.

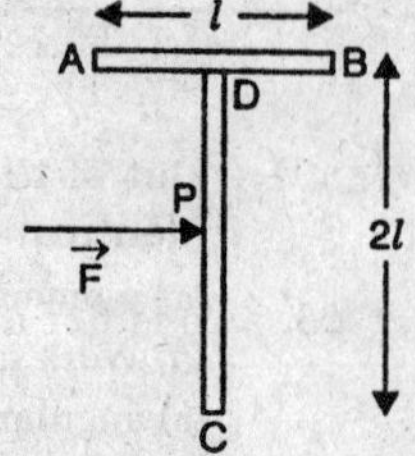

(a) $\frac{4}{3}l$ (b) l

(c) $\frac{2}{3}l$ (d) $\frac{3}{2}l$.

19. The change in the value of '*g*' at a height '*h*' above the surface of the earth is the same as at a depth '*d*' below the surface of earth. When both '*d*' and '*h*' are much smaller than the radius of earth, then which one of the following is correct ?

(a) $d = 2h$ (b) $d = h$

(c) $d = \frac{h}{2}$ (d) $d = \frac{3h}{2}$.

20. A particle of mass 10 g is kept on the surface of a uniform sphere of mass 100 kg and radius 10 cm. Find the work to be done against the gravitational force between them to take the particle far away from the sphere.

(you may take G = 6.67×10^{-11} Nm^2/kg^2)

(a) 6.67×10^{-9} J (b) 6.67×10^{-10} J

(c) 13.34×10^{-10} J (d) 3.33×10^{-10} J.

21. A 20 cm long capillary tube is dipped in water. The water rises up to 8 cm. If the entire arrangement is put in a freely falling elevator, the length of water column in the capillary tube will be

(a) 4 cm (b) 20 cm

(c) 8 cm (d) 10 cm.

22. If 'S' is stress and 'Y' is Young's modulus of material of a wire, the energy stored in the wire per unit volume is

(a) $\frac{2Y}{S^2}$ (b) $\frac{S}{2Y}$

(c) $2S^2$ Y (d) $\frac{S^2}{2Y}$.

23. Average density of the earth

(a) is directly proportional to *g*

(b) is inversely proportional to *g*

(c) does not depend on *g*

(d) is a complex function of *g*.

24. A body of mass m is accelerated uniformly from rest to a speed v in a time T. The instantaneous power delivered to the body as a function of time is given by

(a) $\frac{1}{2}\frac{mv^2}{T^2}.t$ (b) $\frac{1}{2}\frac{mv^2}{T^2}.t^2$

(c) $\frac{mv^2}{T^2}.t$ (d) $\frac{mv^2}{T^2}.t^2$.

25. Consider a car moving on a straight road with a speed of 100 m/s. The distance at which car can be stopped is $[\mu_k = 0.5]$

(a) 100 m (b) 400 m

(c) 800 m (d) 1000 m.

26. Which of the following is **incorrect** regarding the first law of thermodynamics ?

(a) It introduces the concept of the internal energy.

(b) It introduces the concept of entropy.

(c) It is not applicable to any cyclic process.

(d) It is a restatement of the principle of conservation of energy.

27. The temperature-entropy diagram of a reversible engine cycle is given in the figure. Its efficiency is

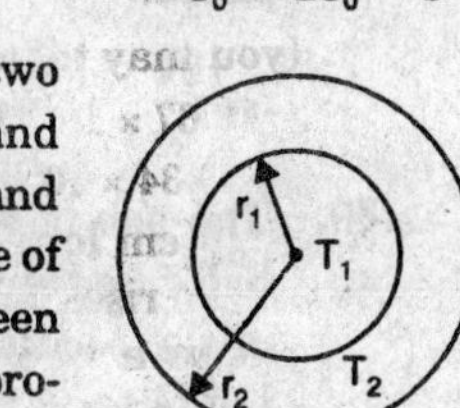

(a) 1/3 (b) 2/3

(c) 1/2 (d) 1/4.

28. The figure shows a system of two concentric spheres of radii r_1 and r_2 and kept at temperatures T_1 and T_2, respectively. The radial rate of flow of heat in a substance between the two concentric spheres is proportional to

(a) $\frac{r_1 r_2}{(r_2 - r_1)}$ (b) $(r_2 - r_1)$

(c) $(r_2 - r_1)/(r_1 r_2)$ (d) $\ln\left(\frac{r_2}{r_1}\right)$.

29. A system goes from A to B via two processes I and II as shown in figure. If ΔU_1 and ΔU_2 are the changes in internal energies in the processes I and II respectively, then

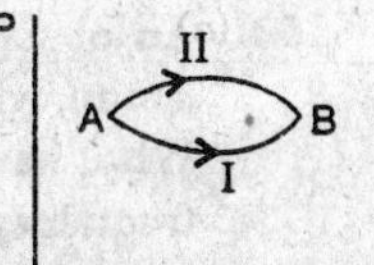

(a) $\Delta U_2 > \Delta U_1$

(b) $\Delta U_2 < \Delta U_1$

(c) $\Delta U_1 = \Delta U_2$

(d) relation between ΔU_1 and ΔU_2 cannot be determined.

30. A gaseous mixture consists of 16 g of helium and 16 g of oxygen. The ratio $\frac{C_p}{C_v}$ of the mixture is

(a) 1.4 (b) 1.54

(c) 1.59 (d) 1.62.

31. The intensity of gamma radiation from a given source is I. On passing through 36 mm of lead, it is reduced to $\frac{I}{8}$. The thickness of lead which will reduce the intensity to $\frac{I}{2}$ will be

(a) 18 mm (b) 12 mm

(c) 6 mm (d) 9 mm.

32. The electrical conductivity of a semiconductor increases when electromagnetic radiation of wavelength shorter than 2480 nm is incident on it. The band gap in (eV) for the semiconductor is

(a) 0.5 eV (b) 0.7 eV

(c) 1.1 eV (d) 2.5 eV.

33. A photocell is illuminated by a small bright source placed 1 m away. When the same source of light is placed $\frac{1}{2}$ m away, the number of electrons emitted by photocathode would

(a) decrease by a factor of 2 (b) increase by a factor of 2

(c) decrease by a factor of 4 (d) increase by a factor of 4.

34. Starting with a sample of pure ^{66}Cu, $\frac{7}{8}$ of it decays into Zn in 15 minutes. The corresponding half-life is

(a) 5 minutes (b) $7\frac{1}{2}$ minutes

(c) 10 minutes (d) 15 minutes.

35. If radius of the $^{27}_{13}$Al nucleus is estimated to be 3.6 Fermi then the radius of $^{125}_{52}$Te nucleus be nearly

(a) 4 fermi (b) 5 fermi

(c) 6 fermi (d) 8 fermi.

36. The diagram shows the energy levels for an electron in a certain atom. Which transition shown represents the emission of a photon with the most energy ?

n = 4, n = 3, n = 2, n = 1; transitions I, II, III, IV

(a) I (b) II

(c) III (d) IV.

37. If the kinetic energy of a free electron doubles, its de Broglie wavelength changes by the factor

(a) $\frac{1}{\sqrt{2}}$ (b) $\sqrt{2}$

(c) $\frac{1}{2}$ (d) 2.

38. In a common base amplifier, the phase difference between the input signal voltage and output voltage is

(a) 0 (b) $\frac{\pi}{2}$

(c) $\frac{\pi}{4}$ (d) π.

39. In a full wave rectifier circuit operating from 50 Hz mains frequency, the fundamental frequency in the ripple would be

(a) 100 Hz (b) 70.7 Hz

(c) 50 Hz (d) 25 Hz.

40. A nuclear transformation is denoted by $X(n, \alpha)\,^7_3Li$. Which of the following is the nucleus of element X ?

(a) ^9_5B (b) $^{11}_4Be$

(c) $^{12}C_6$ (d) $^{10}_5B$.

41. The function $\sin^2(\omega t)$ represents

(a) a simple harmonic motion with a period $2\pi/\omega$

(b) a simple harmonic motion with a period π/ω

(c) a periodic, but not simple harmonic, motion with a period $2\pi/\omega$

(d) a periodic, but not simple harmonic, motion with a period π/ω.

42. A Young's double slit experiment uses a monochromatic source. The shape of the interference fringes formed on a screen is

(a) straight line (b) parabola

(c) hyperbola (d) circle.

43. Two simple harmonic motions are represented by the equations $y_1 = 0.1 \sin\left(100\pi t + \frac{\pi}{3}\right)$ and $y_2 = 0.1 \cos \pi t$.

The phase difference of the velocity of particle 1 with respect to the velocity of particle 2 is

(a) $\frac{-\pi}{3}$ (b) $\frac{\pi}{6}$

(c) $\frac{-\pi}{6}$ (d) $\frac{\pi}{3}$.

44. A fish looking up through the water sees the outside world contained in a circular horizon.

If the refractive index of water is $\frac{4}{3}$ and the fish is 12 cm below the surface, the radius of this circle in cm is

(a) $36\sqrt{5}$ (b) $4\sqrt{5}$

(c) $36\sqrt{7}$ (d) $36/\sqrt{7}$.

45. Two point white dots are 1 mm apart on a black paper. They are viewed by eye of pupil diameter 3 mm. Approximately, what is the maximum distance at which these dots can be resolved by the eye ? [Take wavelength of light = 500 nm]

(a) 6 m (b) 3 m

(c) 5 m (d) 1 m.

46. A thin glass (refractive index 1.5) lens has optical power of – 5 D in air. Its optical power in a liquid medium with refractive index 1.6 will be

(a) 25 D (b) – 25 D

(c) 1 D (d) – 1 D.

47. When two tuning forks (fork 1 and fork 2) are sounded simultaneously, 4 beats per second are heard. Now, some tape is attached on the prong of the fork 2. When the tuning forks are sounded again, 6 beats per second are heard. If the frequency of fork 1 is 200 Hz, then what was the original frequency of fork 2 ?

(a) 196 Hz (b) 204 Hz

(c) 200 Hz (d) 202 Hz.

48. If a simple harmonic motion is represented by $\frac{d^2x}{dt^2} + \alpha x = 0$, its time period is

(a) $2\pi\alpha$ (b) $2\pi\sqrt{\alpha}$

(c) $\frac{2\pi}{\alpha}$ (d) $\frac{2\pi}{\sqrt{\alpha}}$.

49. The bob of a simple pendulum is a spherical hollow ball filled with water. A plugged hole near the bottom of the oscillating bob gets suddenly unplugged. During observation, till water is coming out, the time period of oscillation would

(a) remain unchanged

(b) increase towards a saturation value

(c) first increase and then decrease to the original value

(d) first decrease and then increase to the original value.

50. An observer moves towards a stationary source of sound, with a velocity one-fifth of the velocity of sound. What is the percentage increase in the apparent frequency ?

(a) 5% (b) 20%

(c) zero (d) 0.5%.

51. If I_0 is the intensity of the principal maximum in the single slit diffraction pattern, then what will be its intensity when the slit width is doubled ?

(a) I_0 (b) $\frac{I_0}{2}$

(c) $2I_0$ (d) $4I_0$.

52. When an unpolarised light of intensity I_0 is incident on a polarising sheet, the intensity of the light which does not get transmitted is

(a) zero (b) I_0

(c) $\frac{1}{2}I_0$ (d) $\frac{1}{4}I_0$.

53. A charged ball B hangs from a silk thread S, which makes an angle θ with a large charged conducting sheet P, as shown in the figure. The surface charge density σ of the sheet is proportional to

(a) sin θ (b) tan θ

(c) cos θ (d) cot θ.

54. Two point charges + 8q and – 2q are located at x = 0 and x = L respectively. The location of a point on the x-axis at which the net electric field due to these two point charges is zero is

(a) 8 L (b) 4 L

(c) 2 L (d) $\frac{L}{4}$.

55. Two thin wire rings each having a radius R are placed at a distance d apart with their axes coinciding. The charges on the two rings are +q and –q. The potential difference between the centres of the two rings is

(a) zero (b) $\frac{q}{4\pi\varepsilon_0}\left[\frac{1}{R}-\frac{1}{\sqrt{R^2+d^2}}\right]$

(c) $qR/4\pi\varepsilon_0 d^2$ (d) $\frac{q}{2\pi\varepsilon_0}\left[\frac{1}{R}-\frac{1}{\sqrt{R^2+d^2}}\right]$.

56. A parallel plate capacitor is made by stacking n equally spaced plates connected alternatively. If the capacitance between any two adjacent plates is 'C' then the resultant capacitance is

(a) C (b) nC

(c) (n – 1) C (d) (n + 1) C.

57. A fully charged capacitor has a capacitance 'C'. It is discharged through a small coil of resistance wire embedded in a thermally insulated block of specific heat capacity 's' and mass 'm'. If the temperature of the block is raised by 'ΔT', the potential difference 'V' across the capacitance is

(a) $\frac{ms\,\Delta T}{C}$ (b) $\sqrt{\frac{2ms\,\Delta T}{C}}$

(c) $\sqrt{\frac{2mC\,\Delta T}{s}}$ (d) $\frac{mC\,\Delta T}{s}$.

58. One conducting U tube can slide inside another as shown in figure, maintaining electrical contacts between the tubes. The magnetic field B is perpendicular to the plane of the figure. If each tube moves towards the other at a constant speed V, then the emf induced in the circuit in terms of B, l and V where l is the width of each tube, will be

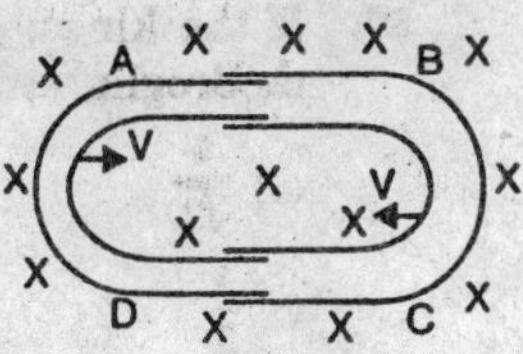

(a) zero (b) 2 BlV

(c) BlV (d) – BlV.

59. A heater coil is cut into two equal parts and only one parts is now used in the heater. The heat generated will now be

(a) one fourth (b) halved

(c) doubled (d) four times.

60. Two thin, long parallel wires, separated by a distance 'd' carry a current of 'i' A in the same direction. They will

(a) attract each other with a force of $\mu_0 i^2/(2\pi d^2)$

(b) repel each other with a force of $\mu_0 i^2/(2\pi d^2)$

(c) attract each other with a force of $\mu_0 i^2/(2\pi d)$

(d) repel each other with a force of $\mu_0 i^2/(2\pi d)$.

61. A moving coil galvanometer has 150 equal divisions. Its current sensitivity is 10 divisions per milliampere and voltage sensitivity is 2 divisions per millivolt. In order that each division reads 1 volt, the resistance in ohms needed to be connected in series with the coil will be

(a) 99995 (b) 9995

(c) 10^3 (d) 10^5.

62. Two voltmeters, one of copper and another of silver, are joined in parallel. When a total charge q flows through the voltmeters, equal amount of metals are deposited. If the electrochemical equivalents of copper and silver are z_1 and z_2 respectively, the charge which flows through the silver voltmeter is

(a) $q\frac{z_1}{z_2}$ (b) $q\frac{z_2}{z_1}$

(c) $\frac{q}{1+\frac{z_1}{z_2}}$ (d) $\frac{q}{1+\frac{z_2}{z_1}}$.

63. In the circuit, the galvanometer G shows zero deflection. If the batteries A and B have negligible internal resistance, the value of the resistor R will be

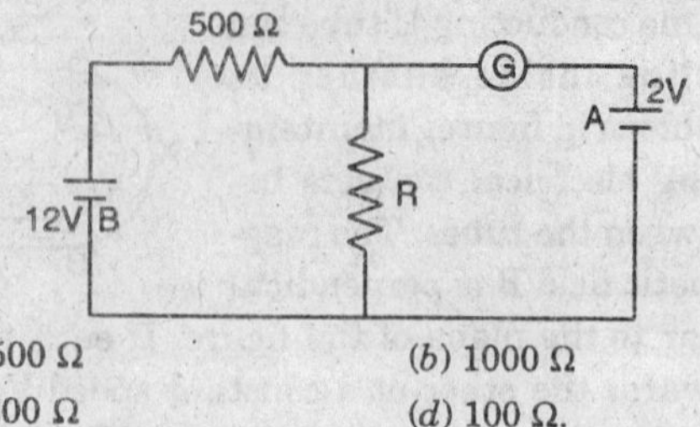

(a) 500 Ω (b) 1000 Ω

(c) 200 Ω (d) 100 Ω.

64. Two sources of equal emf are connected to an external resistance R. The internal resistances of the two sources are R_1 and R_2 ($R_2 > R_1$). If the potential difference across the source having internal resistance R_2 is zero, then

(a) $R = R_1 R_2/(R_1 + R_2)$ (b) $R = R_1 R_2/(R_2 - R_1)$

(c) $R = R_2 \times (R_1 + R_2)/(R_2 - R_1)$

(d) $R = R_2 - R_1$.

65. Two concentric coils, each of radius equal to 2π cm, are placed at right angles to each other. 3 ampere and 4 ampere are the currents flowing in each coil respectively. The magnetic induction in weber/m^2 at the centre of the coils will be

($\mu_0 = 4\pi \times 10^{-7}$ Wb/A . m)

(a) 5×10^{-5} (b) 7×10^{-5}

(c) 12×10^{-5} (d) 10^{-5}.

66. A charged particle of mass m and charge q travels on a circular path of radius r that is perpendicular to a magnetic field B. The time taken by the particle to complete one revolution is

(a) $\dfrac{2\pi qB}{m}$ (b) $\dfrac{2\pi m}{qB}$

(c) $\dfrac{2\pi mq}{B}$ (d) $\dfrac{2\pi q^2 B}{m}$.

67. In a potentiometer experiment the balancing with a cell is at length 240 cm. On shunting the cell with a resistance of 2 Ω, the balancing length becomes 120 cm. The internal resistance of the cell is

(a) 4 Ω (2) 2 Ω

(c) 1 Ω (4) 0.5 Ω.

68. The resistance of hot tungsten filament is about 10 times the cold resistance. What will be the resistance of 100 W and 200 V lamp when not in use ?

(a) 400 Ω (b) 200 Ω

(c) 40 Ω (d) 20 Ω.

69. A magnetic needle is kept in a non-uniform magnetic field. It experiences

(a) a force and a torque (b) a force but not a torque

(c) a torque but not a force (d) neither a force nor a torque.

70. A uniform electric field and a uniform magnetic field are acting along the same direction in a certain region. If an electron is projected along the direction of the fields with a certain velocity then

(a) it will turn towards right of direction of motion

(b) it will turn towards left of direction of motion

(c) its velocity will decrease

(d) its velocity will increase.

71. A coil of inductance 300 mH and resistance 2 Ω is connected to a source of voltage 2 V. The current reaches half of its steady state value in

(a) 0.15 s (b) 0.3 s

(c) 0.05 s (d) 0.1 s.

72. The self inductance of the motor of an electric fan is 10 H. In order to impart maximum power at 50 Hz, it should be connected to a capacitance of

(a) 1 μ F (b) 2 μ F

(c) 4 μ F (d) 8 μ F.

73. An energy source will supply a constant current into the load if its internal resistance is

(a) zero

(b) non-zero but less than the resistance of the load

(c) equal to the resistance of the load

(d) very large as compared to the load resistance.

74. A circuit has a resistance of 12 ohm and an impedance of 15 ohm. The power factor of the circuit will be

(a) 1.25 (b) 0.125

(c) 0.8 (d) 0.4.

75. The phase difference between the alternating current and emf is $\dfrac{\pi}{2}$. Which of the following cannot be the constituent of the circuit ?

(a) LC (b) L alone

(c) C alone (d) R, L.

ANSWERS

1. (a)	2. (a)	3. None	4. (b)	5. (b)	6. (b)	7. (c)	8. (a)
9. (b)	10. (a)	11. (b)	12. (c)	13. (a)	14. (c)	15. (b)	16. (b)
17. (d)	18. (a)	19. (a)	20. (b)	21. (b)	22. (d)	23. (a)	24. (c)

25. (d)	26. (b)	27. (a)	28. (a)	29. (c)	30. (d)	31. (b)	32. (a)
33. (d)	34. (a)	35. (c)	36. (c)	37. (a)	38. (a)	39. (a)	40. (d)
41. (d)	42. (c)	43. (c)	44. (d)	45. (c)	46. None	47. (a)	48. (d)
49. (c)	50. (b)	51. (d)	52. (c)	53. (b)	54. (c)	55. (d)	56. (c)
57. (b)	58. (b)	59. (c)	60. (c)	61. (b)	62. (d)	63. (d)	64. (d)
65. (a)	66. (b)	67. (b)	68. (c)	69. (a)	70. (c)	71. (d)	72. (a)
73. (d)	74. (c)	75. (d).					

Solutions

1. (*a*) [Work, torque] $= [MLT^{-2}]\,[L]$
$= [ML^2T^2]$

[Angular momentum, Planck's constant]
$= [M]\,[LT^{-1}]\,[L]$
$= [ML^2T^{-1}]$

[Impulse, momentum] $= [M]\,[LT^{-1}]$
$= [MLT^{-1}]$

[Moment of inertia] $= [ML^2]$

[Moment of force] $= [MLT^{-2}]\,[L]$
$= [ML^2T^{-2}]$

Clearly, the dimensions of moment of inertia are different from the dimensions of moment of force.

2. (*a*) $t = ax^2 + bx$

$$\frac{dt}{dx} = 2ax + b$$

or $$\frac{dx}{dt} = (2ax + b)^{-1}$$

or $$v = (2ax + b)^{-1}$$

$$\text{acceleration} = \frac{dv}{dt} = \frac{dv}{dx}\cdot\frac{dx}{dt}$$
$$= (-1)\,(2ax + b)^{-2}\,(2a)\,(2ax + b)^{-1}$$
$$= -2a(2ax + b)^{-3} = -2av^3.$$

3. All given options are **Wrong**.

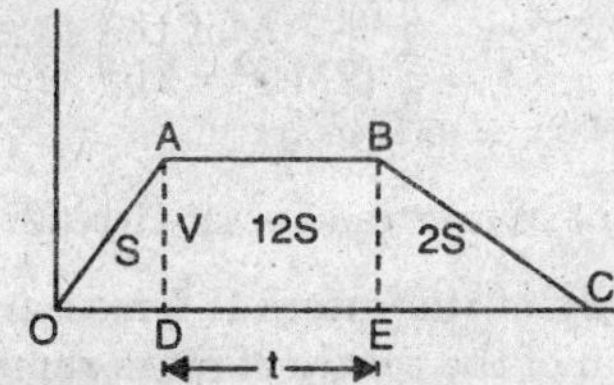

Retardation is given to be half of acceleration. So, distance covered has to be 2S. Clearly, the distance covered during uniform motion is 12S.

Now, $12S = Vt = \sqrt{2fs}\;t$

or $144S^2 = 2fSt^2$

or $S = \frac{1}{72}\,ft^2.$

4. (*b*) $\Delta\vec{v} = \vec{v}_2 + (-\vec{v}_1)$
$= \vec{v}_2 - \vec{v}_1$

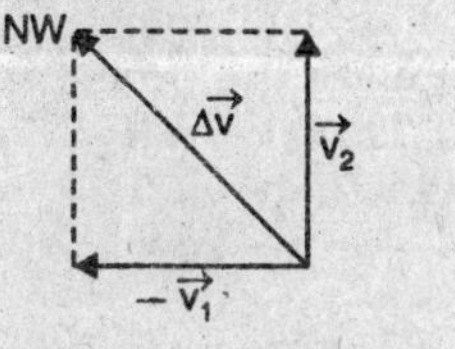

$$\Delta v = \sqrt{v_1^2 + v_2^2} = \sqrt{5^2 + 5^2}\ \text{m s}^{-1} = 5\sqrt{2}\ \text{m s}^{-1}$$

$$\text{Acceleration} = \frac{5\sqrt{2}}{10}\ \text{m s}^{-2} = \frac{1}{\sqrt{2}}\ \text{m s}^{-2}$$

The direction is north-west.

5. (*b*) $$t_1 = \frac{2v\sin\theta}{g}$$

$$t_2 = \frac{2v\sin(90° - \theta)}{g} = \frac{2v\cos\theta}{g}$$

$$t_1 t_2 = \frac{(2v\sin\theta)(2v\cos\theta)}{g^2}$$

$$= \frac{2}{g}\left[\frac{v^2\,(2\sin\theta\cos\theta)}{g}\right]$$

$$= \frac{2}{g}\,\frac{v^2\sin 2\theta}{g} = \frac{2}{g}R$$

$\therefore\quad t_1 t_2 \propto R.$

6. (*b*) $$\frac{F_1}{F_2} = \frac{m\omega^2 R_1}{m\omega^2 R_2} \quad\text{or}\quad \frac{F_1}{F_2} = \frac{R_1}{R_2}.$$

7. (*c*) *For the case of smooth inclined plane*

Using $S = ut + \frac{1}{2}at^2$, we get

$$d = \frac{1}{2}(g\sin\theta)\,t^2 \qquad ...(i)$$

For the case of rough inclined plane

$$d = \frac{1}{2}(g \sin\theta - \mu_k g \cos\theta)(nt)^2 \qquad ...(ii)$$

$$\sin\theta = (\sin\theta - \mu_k \cos\theta)n^2$$

But $\theta = 45°$, $\therefore$ $\sin\theta = \cos\theta$

$\therefore$ $1 = (1 - \mu_k)n^2$

or $1 - \mu_k = \frac{1}{n^2}$ or $\mu_k = 1 - \frac{1}{n^2}$.

8. **(a)** *For the smooth upper half*

Using $v^2 - u^2 = 2aS$, we get

$v^2 - 0^2 = 2g \sin\phi l$

or $v = \sqrt{2g \sin\phi l}$

For the lower half

Using $v^2 - u^2 = 2aS$, we get

$0^2 - 2g \sin\phi l = -2(\mu_k \cos\phi - \sin\phi)gl$

or $\sin\phi = \mu_k \cos\phi - \sin\phi$

or $\mu_k \cos\phi = 2 \sin\phi$

or $\mu_k = 2 \tan\phi$.

9. **(b)** $\left(\frac{v}{2}\right)^2 - v^2 = -2a \times 3$

Again, $0^2 - \left(\frac{v}{2}\right)^2 = -2ax$

Dividing, $\frac{x}{3} = \frac{-v^2/4}{-3v^2/4} = \frac{1}{3}$

or $x = 1$ cm.

10. **(a)** *For the first 50 m*

Velocity attained at the end of first 50 m

$= \sqrt{2 \times 9.8 \times 50}$ m s^{-1} $= \sqrt{980}$ m s^{-1}

After the parachute opens

'u' $= \sqrt{980}$ m s^{-1}

$v = 3$ m s^{-1}

$a = -2$ m s^{-2}

$h = ?$

Using $v^2 - u^2 = 2aS$, we get

$9 - 980 = -2 \times 2 \times h$

or $h = \frac{971}{4}$ m ≈ 243 m

Total height $= (243 + 50)$ m $= 293$ m.

11. **(b)** $|F| = 15 \times 0.2$ N $= 3$ N

acceleration $= \frac{3 \text{ N}}{0.3 \text{ kg}} = 10$ m s^{-2}.

12. **(c)** $\frac{1}{2}Mv^2 = \frac{1}{2}KL^2$ or $v = \sqrt{\frac{K}{M}}L$

Maximum momentum

$= Mv = M\sqrt{\frac{K}{M}}L = \sqrt{MK}L$.

13. **(a)**

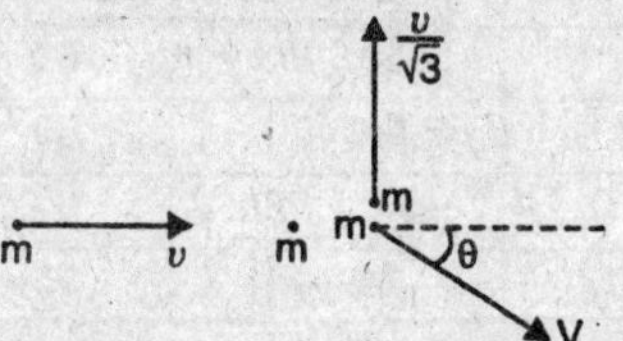

Conserving horizontal momentum, we get

$mv = mV \cos\theta$

or $v = V\cos\theta \qquad ...(i)$

Conserving vertical momentum,

$\frac{mv}{\sqrt{3}} = mV \sin\theta$

or $\frac{v}{\sqrt{3}} = V \sin\theta \qquad ...(ii)$

Squaring and adding (i) and (ii), we get

$V^2 = v^2 + \frac{v^2}{3} = \frac{4v^2}{3}$ or $V = \frac{2}{\sqrt{3}}v$.

14. **(c)** $\frac{1}{2}mv^2 = mg(h_1 - h_2)$

or $v^2 = 2g(h_1 - h_2)$

or $v = \sqrt{2g(h_1 - h_2)}$

$= \sqrt{2 \times 10(100 - 20)}$ m s^{-1}

$= \sqrt{1600}$ m s^{-1} $= 40$ m s^{-1}.

15. **(b)** $ma \cos\alpha = mg \sin\alpha$

or $a = g \tan\alpha$.

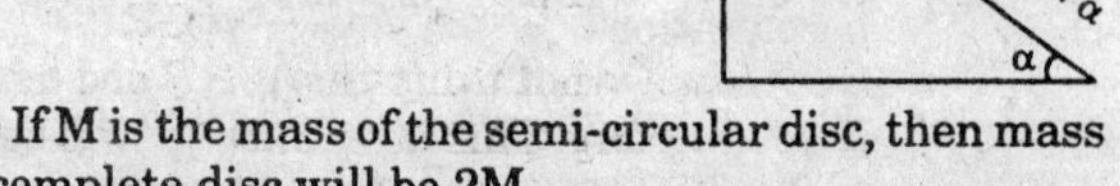

16. **(b)** If M is the mass of the semi-circular disc, then mass of complete disc will be 2M.

Moment of inertia of complete disc about the given axis

$= \frac{1}{2}(2M)r^2 = Mr^2$

Moment of inertia of semi-circular disc $= \frac{1}{2}Mr^2$.

17. **(d)** Only of the internal forces are involved. So, the position of the centre of mass remains unaffected.

18. **(a)** Since the object has only the translational motion without rotation therefore the centre of mass of the object is the point where the force has been applied. So, we have to find the centre of mass of the object.

Let us take C as the origin and CD to be along Y-axis. If m be the mass of AB, then the mass of the CD is 2 m.

The centre of mass of AB is at a distance $2l$ from C. The centre of mass of CD is at a distance l from C.

Distance of centre of mass of the object from C

$$= \frac{2m \times l + m \times 2l}{2m + m} = \frac{4ml}{3m} = \frac{4l}{3}.$$

19. **(*a*)** $g_d = g_h$

$$g\left(1 - \frac{d}{R}\right) = g\left(1 - \frac{2h}{R}\right) \Rightarrow d = 2h.$$

20. **(*b*)** The work to be done against the gravitational force between the particle and the sphere so that they are separated by infinite distance is numerically equal to the gravitational potential energy U.

Now, $U = -\frac{GMm}{R}$

$$= -\frac{6.67 \times 10^{-11} \times 100 \times 10 \times 10^{-3} \times 100}{10} \text{ J}$$

$$= -6.67 \times 10^{-10} \text{ J}$$

$\therefore$ Work done $= 6.67 \times 10^{-10}$ J.

21. **(*b*)** In condition of weightlessness, water rises to the whole of the available length.

22. **(*d*)** Energy stored/volume

$$= \frac{1}{2} \times \text{stress} \times \text{strain}$$

$$= \frac{1}{2} \times \text{stress} \times \frac{\text{strain}}{Y} \qquad \left(\because\ Y = \frac{\text{stress}}{\text{strain}}\right)$$

$$= \frac{(\text{stress})^2}{2Y} = \frac{S^2}{2Y}.$$

23. **(*a*)** $g = \frac{GM}{R^2} = \frac{G}{R^2} \times \frac{4}{3}\pi R^3 \rho$

or $g = \frac{4}{3}\pi G R \rho$ or $\rho = \frac{3g}{4\pi G R} \Rightarrow \rho \propto g$

Caution. Do not think that $g \propto R$ and hence ρ does not depend upon g. This is because "g" will be proportional to R only if it is given that ρ is constant. In the given problem, ρ has to be treated as a VARIABLE at the start of our reasoning.

24. **(*c*)** $P = \vec{F} . \vec{v} = Fv \cos 0° = Fv$

or $P = (ma)\, at = ma^2 t$

or $P = m\left[\frac{v - 0}{T}\right]^2 t$ or $P = \frac{mv^2}{T^2}\, t.$

25. **(*d*)** 'u' $= 100$ m s^{-1}, $v = 0$, $a = -\mu g$

Using $v^2 - u^2 = 2aS$,

$0^2 - 100 \times 100 = -2 \times 0.5 \times 10 \times S$

or $S = \frac{100 \times 100}{10}$ m $= 1000$ m.

26. **(*b*)** The following arguments shall help us to arrive at the right choice.

(*i*) The zeroth law of thermodynamics is related to the concept of temperature. The first law is related to the concept of internal energy. The second law of thermodynamics is related to entropy.

(*ii*) Statement of first law of thermodynamics.

"In any thermodynamic process *between equilibrium states i and f*, the equantity Q + W has the same value the for any path between *i* and *f*. This quantity is equal to the change in the value of a state function called internal energy"

(*iii*) The first law of thermodynamics is a restatement of the law of conservation of energy.

> **Note 1.** In a cyclical process, using first law of thermodynamics,
>
> $$Q + W = 0.$$
>
> **Note 2.** Strictly speaking, the options (*b*) and (*c*) are both incorrect.

27. **(*a*)** $\eta = \frac{W}{Q_1} = \frac{\frac{1}{2} S_0 T_0}{2T_0 S_0 - \frac{1}{2} T_0 S_0} = \frac{1}{3}.$

28. **(*a*)** For radial flow of heat in a spherical shell

$$Q = 4\pi K (T_1 - T_2)\, r_1 r_2 / r_2 - r_1.$$

29. **(*c*)** Change in internal energy does not depend upon path.

30. **(*d*)** $n_1 = 4, n_2 = \frac{1}{2},$

$$\gamma_1 = \frac{5}{3} \text{ (for monoatomic gas)}$$

$$\gamma_2 = \frac{7}{5} \text{ (for diatomic gas)}$$

$$\frac{n_1 + n_2}{\gamma - 1} = \frac{n_1}{\gamma_1 - 1} + \frac{n_2}{\gamma_2 - 1}$$

$$\frac{4 + 0.5}{\gamma - 1} = \frac{4}{\frac{5}{3} - 1} + \frac{1/2}{\frac{7}{5} - 1}$$

or $\frac{4.5}{\gamma - 1} = \frac{12}{2} + \frac{5}{4} = \frac{29}{4}$

or $\frac{\gamma - 1}{4.5} = \frac{4}{29}$ or $\gamma - 1 = \frac{4 \times 4.5}{29} = 0.62$

or $\gamma = 1.62.$

31. **(*b*)** $I' = Ie^{-\mu x}$

or $e^{\mu x} = \frac{I}{I'}$ or $\mu x = \log_e \frac{I}{I'}$

Now, $36\mu = \log_e \dfrac{I \times 8}{I} = 3 \log_e 2$

$x\mu = \log_e \dfrac{I \times 2}{I}$

or $\dfrac{x}{36} = \dfrac{\log_e 2}{3 \log_e 2} = \dfrac{1}{3}$

or $x = \dfrac{36}{3}$ mm = 12 mm.

32. (*a*) $E = \dfrac{12375}{\lambda (\text{in Å})}$ eV

$= \dfrac{12375}{24800}$ eV = 0.5 eV.

33. (*d*) Intensity $\propto \dfrac{1}{(\text{distance})^2}$

Since distance is halved therefore intensity is increased by a factor of 4. Moreover, photoelectric current and hence number of photoelectrons is proportional to intensity.

34. (*a*) $N = \dfrac{N_0}{2^{t/T_{1/2}}}$

$\dfrac{N}{N_0} = \dfrac{1}{2^{\frac{15}{T_{1/2}}}}$

$1 - \dfrac{7}{8} = \dfrac{1}{2^{15/T_{1/2}}}$

$\dfrac{1}{2^3} = \dfrac{1}{2^{15/T_{1/2}}}$

$\therefore \quad T_{1/2} = \dfrac{15}{3} = 5$ min

35. (*c*) $R = R_0 A^{1/3}$

$\dfrac{R_2}{R_1} = \left[\dfrac{125}{27}\right]^{1/3} = \dfrac{5}{3}$

or $R_2 = \dfrac{5}{3} R_1 = \dfrac{5}{3} \times 3.6$ fm = 6 fm.

36. (*c*) $E_n = -\dfrac{13.6 Z^2}{n^2}$ eV

or $E_n = -\dfrac{k}{n^2}$ eV

$E_1 = -\dfrac{k}{1}$ eV, $E_2 = -\dfrac{k}{4}$, $E_4 = -\dfrac{k}{16}$

Now, $E_2 - E_1 = -\dfrac{k}{4} - \left(\dfrac{-k}{1}\right)$

$= k - \dfrac{k}{4} = \dfrac{3k}{4}$

$E_4 - E_2 = -\dfrac{k}{16} - \left(\dfrac{-k}{4}\right)$

$= \dfrac{k}{4} - \dfrac{k}{16} = \dfrac{3k}{16}$

Clearly, $(E_2 - E_1) > (E_4 - E_2)$.

37. (*a*) $\lambda = \dfrac{h}{mv} = \dfrac{h}{p} = \dfrac{h}{\sqrt{2mE_k}}$

$\lambda \propto \dfrac{1}{\sqrt{E_k}} \quad \lambda' \propto \dfrac{1}{\sqrt{2E_k}}$

$\dfrac{\lambda'}{\lambda} = \dfrac{1}{\sqrt{2}}$.

38. (*a*) Factual information.

39. (*a*) Factual information.

Input → $\sin \omega t$

Output → $\sin^2 \omega t$ → which involves $\cos 2 \omega t$.

40. (*d*) $X + {}_0n^1 \longrightarrow {}_3Li^7 + {}_2He^4 \Rightarrow$ X is ${}_5B^{10}$.

41. (*d*) $\cos 2\omega t = 1 - 2 \sin^2 \omega t$

or $\sin^2 \omega t = \dfrac{1}{2}(1 - \cos 2\omega t)$

$2\omega = \dfrac{2\pi}{T}$ or $T = \dfrac{\pi}{\omega}$.

42. (*c*) $S_2P - S_1P$ = constant

This is the equation of a hyperbola with focii S_1 and S_2.

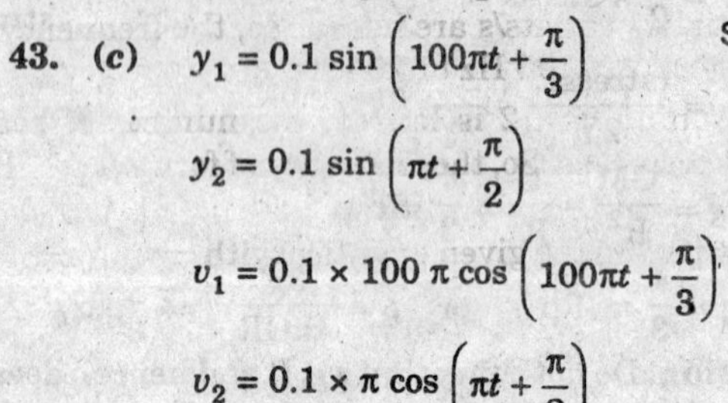

43. (*c*) $y_1 = 0.1 \sin \left(100\pi t + \dfrac{\pi}{3}\right)$

$y_2 = 0.1 \sin \left(\pi t + \dfrac{\pi}{2}\right)$

$v_1 = 0.1 \times 100\,\pi \cos \left(100\pi t + \dfrac{\pi}{3}\right)$

$v_2 = 0.1 \times \pi \cos \left(\pi t + \dfrac{\pi}{2}\right)$

Initial phase of v_1 w.r.t. v_2

$= \dfrac{\pi}{3} - \dfrac{\pi}{2} = \dfrac{2\pi - 3\pi}{6} = -\dfrac{\pi}{6}$.

44. (*d*) $\tan i_c = \dfrac{r}{h}$ or $r = h \tan i_c$

But $\sin i_c = \dfrac{1}{\mu}$

$\tan i_c = \dfrac{1}{\sqrt{\mu^2 - 1}}$

$\therefore \quad \dfrac{r}{h} = \dfrac{1}{\sqrt{\mu^2 - 1}}$

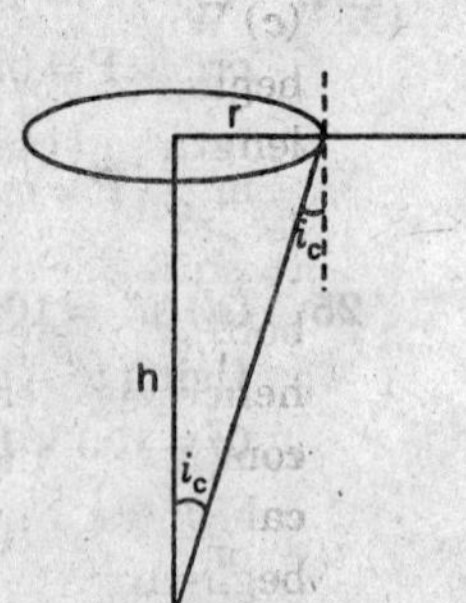

or $r = \dfrac{h}{\sqrt{\mu^2 - 1}} = \dfrac{12}{\sqrt{\dfrac{16}{9} - 1}} = \dfrac{36}{\sqrt{7}}$ cm.

45. (*c*) $\dfrac{1.22\,\lambda}{a} = \dfrac{x}{d}$ or $d = \dfrac{xa}{1.22\,\lambda}$

$$= \frac{1 \times 10^{-3} \times 3 \times 10^{-3}}{1.22 \times 500 \times 10^{-9}} \text{ m} \approx 5 \text{ m}.$$

46. All options are **Wrong.**

$$P = \left(\frac{\mu_2}{\mu_1} - 1\right)\left(\frac{1}{R_1} - \frac{1}{R_2}\right)$$

$$-5 = \left(\frac{1.5}{1} - 1\right)\left(\frac{1}{R_1} - \frac{1}{R_2}\right)$$

$$\frac{-5}{0.5} = \frac{1}{R_1} - \frac{1}{R_2}$$

or $\dfrac{1}{R_1} - \dfrac{1}{R_2} = -10$

Again, $P = \left(\dfrac{1.5}{1.6} - 1\right)(-10)$

$$= \frac{(-0.1)(-10)}{1.6} = \frac{5}{8} \text{D}.$$

47. (*a*) Initially, 4 beats/s are heard. So, the frequency of tuning fork 2 is 196 Hz or 204 Hz.

When tuning fork 2 is loaded, the number of beats heard/s increases. So, the frequency of fork 2 is 196 Hz.

48. (*d*) Comparing the given equation with $\dfrac{d^2x}{dt^2} + \omega^2 x = 0$, we get

$$\omega^2 = \alpha$$

or $\omega = \sqrt{\alpha}$ or $\dfrac{2\pi}{T} = \sqrt{\alpha}$

or $\dfrac{T}{2\pi} = \dfrac{1}{\sqrt{\alpha}}$ or $T = \dfrac{2\pi}{\sqrt{\alpha}}$.

49. (*c*) When water begins to come out, the centre of mass begins to lower. There is an increase in the effective length and hence an increase in the time period. When water flows out completely, the centre of mass, begins to shift back to the geometrical centre of the hollow bob. So, there is a decrease in the effective length and hence a decrease in time period. When water flows out completely, the centre of mass is again at the geometrical centre. Now, the time period is the same as in the beginning.

50. (*b*) $v_0 = \dfrac{1}{5} v$

$$\text{v}' = \frac{v + v_0}{v}\text{v} \quad \text{v}' = \frac{v + \frac{v}{5}}{v}\text{v}$$

$$\text{v}' = 1.2\,\text{v}, \quad \frac{\text{v}'}{\text{v}} = 1.2$$

$$\frac{\text{v}' - \text{v}}{\text{v}} = 1.2 - 1 = 0.2$$

$$\frac{\text{v}' - \text{v}}{\text{v}} \times 100 = 20.$$

51. (*d*) When the slit width is doubled, the half angular width of the central maximum which is $\dfrac{\lambda}{a}$ reduces of half. The area of the central diffraction band would become $\dfrac{1}{4}$ th. So the intensity of the central maximum would become 4 times.

52. (*c*) 50% of light is not transmitted.

53. (*b*) $T \sin\theta = qE$

$T \cos\theta = mg$

$$\tan\theta = \frac{qE}{mg}$$

$$\tan\theta = \frac{q}{mg}\left[\frac{\sigma}{2\varepsilon_0}\right]$$

or $\sigma \propto \tan\theta$.

54. (*c*)

Electric field cannot be zero at any point between A and B. This is because both the fields will act in the same direction.

Electric field cannot be zero at a point to the left of A because the electric field of charge $8q$ shall always dominate the electric field of charge "$-2q$". So, electric field can be zero only at a point to the right of charge "$-2q$". Suppose electric field is zero at a distance l to the right of point B.

Now, $\dfrac{1}{4\pi\varepsilon_0}\dfrac{2q}{l^2} = \dfrac{1}{4\pi\varepsilon_0}\dfrac{8q}{(L + l)^2}$

or $4l^2 = (L + l)^2$

or $2l = L + l$ or $l = L$

Distance from A is 2L.

55. (*d*)

$\sqrt{R^2+d^2}$, R, R, C_1, C_2, +q, −q, d

Potential at C_1 $= \dfrac{1}{4\pi\varepsilon_0}\dfrac{q}{R} - \dfrac{1}{4\pi\varepsilon_0}\dfrac{q}{\sqrt{R^2+d^2}}$

$= \dfrac{q}{4\pi\varepsilon_0}\left[\dfrac{1}{R} - \dfrac{1}{\sqrt{R^2+d^2}}\right]$

Potential at C_2 $= \dfrac{1}{4\pi\varepsilon_0}\dfrac{-q}{R} + \dfrac{1}{4\pi\varepsilon_0}\dfrac{q}{\sqrt{R^2+d^2}}$

$= \dfrac{q}{4\pi\varepsilon_0}\left[\dfrac{-1}{R} + \dfrac{q}{\sqrt{R^2+d^2}}\right]$

Potential difference

$= \dfrac{q}{4\pi\varepsilon_0}\left[\dfrac{2}{R} - \dfrac{2}{\sqrt{R^2+d^2}}\right]$

$= \dfrac{q}{2\pi\varepsilon_0}\left[\dfrac{1}{R} - \dfrac{1}{\sqrt{R^2+d^2}}\right].$

56. (*c*) Factual information.

57. (*b*) $\dfrac{1}{2}CV^2 = ms\,\Delta T$ or $V = \sqrt{\dfrac{2ms\,\Delta T}{C}}$.

58. (*b*) Two emfs, each of value B*l*V, are produced. Both the emfs tend to send current in the same direction. So, they are added up.

59. (*c*) $Q = \dfrac{V^2}{R}t$

When R is halved, then Q is doubled.

60. (*c*) $\dfrac{F}{l} = \dfrac{\mu_0}{4\pi}\dfrac{2I_1I_2}{d} = \dfrac{\mu_0 i^2}{2\pi d}$

Since like currents attract therefore force is attractive.

61. (*b*) Current required for full scale deflection,

$I_g = \dfrac{150}{10}$ mA = 15 mA

Voltage corresponding to full scale deflection

$= \dfrac{150}{2}$ mV = 75 mV

Galvanometer resistance, $G = \dfrac{75\text{ mV}}{15\text{ mA}} = 5\,\Omega$

Voltage to be measured = 150 V

$V = I_g(R+G)$

or $R = \dfrac{V}{I_g} - G = \dfrac{150}{15\times10^{-3}} - 5$

$= 10000 - 5 = 9995\,\Omega.$

62. (*d*) $m = zq,\quad z \propto \dfrac{1}{q}$

$\dfrac{z_1}{z_2} = \dfrac{q_2}{q_1}$

Now, $q = q_1 + q_2$

$\dfrac{q}{q_1} = \dfrac{q_1}{q_2} + 1$

$\dfrac{q}{q_2} = \dfrac{z_2}{z_1} + 1$ or $q_2 = \dfrac{q}{\dfrac{z_2}{z_1}+1}.$

63. (*d*) Current through $R = \dfrac{12}{500+R}$

Voltage across $R = \dfrac{12}{500+R}$

Since galvanometer shows zero deflection,

$\dfrac{12}{500+R} = 2$

or $12R = 1000 + 2R$

or $10R = 1000$ or $R = 100\,\Omega.$

64. (*d*) $V = E - IR_2$

$V = 0$

$E = IR_2$

$E = \dfrac{2ER_2}{R+R_1+R_2}$

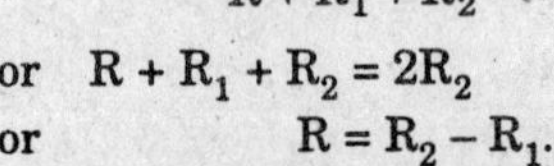

or $R + R_1 + R_2 = 2R_2$

or $R = R_2 - R_1.$

R, E, R_1, E, R_2

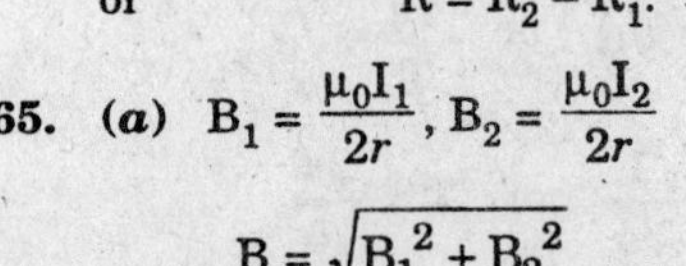

65. (*a*) $B_1 = \dfrac{\mu_0 I_1}{2r}$, $B_2 = \dfrac{\mu_0 I_2}{2r}$

$B = \sqrt{B_1^2 + B_2^2}$

$B = \dfrac{\mu_0}{2r}\sqrt{I_1^2 + I_2^2}$

$B = \dfrac{4\pi\times10^{-7}}{2\times2\pi\times10^{-2}}\sqrt{9+16}$

$= 5\times10^{-5}$ Wb m^{-2}.

66. (*b*) $mr\omega^2 = Bqv = Bqr\omega$

or $\omega = \dfrac{Bq}{m}$ or $\dfrac{2\pi}{T} = \dfrac{Bq}{m}$ or $T = \dfrac{2\pi m}{Bq}.$

67. *(b)* $r = \frac{l_1 - l_2}{l_2} R$

$= \frac{240 - 120}{120} \times 2\,\Omega = 2\,\Omega.$

68. *(c)* $R_{hot} = 10\,R_{cold}$

$R_{hot} = \frac{V^2}{P} = \frac{200 \times 200}{100} = 400\,\Omega$

$\therefore \quad R_{cold} = \frac{400}{10}\,\Omega = 40\,\Omega.$

69. *(a)* Factual information.

70. *(c)* Magnetic field shall not exert any force. However, the electric field shall exert a force in a direction opposite to the direction of motion of the electron. So, there will be decrease in the velocity of the electron.

71. *(d)* $L = 300 \times 10^{-3}$ H

$R = 2\,\Omega$

$I = \frac{I_0}{2}$

Using $I = I_0\left(1 - e^{\frac{-R}{L}t}\right)$

$\frac{I_0}{2} = I_0\left(1 - e^{-\frac{R}{L}t}\right)$

or $\frac{1}{2} = 1 - e^{-\frac{R}{L}t}$ or $e^{-\frac{R}{L}t} = \frac{1}{2}$

or $e^{\frac{R}{L}t} = 2$ or $\frac{R}{L}t = \log_e 2 = 0.693$

or $t = \frac{0.693 \times 300 \times 10^{-3}}{2}$ s $= 0.1$ s.

72. *(a)* $\nu = \frac{1}{2\pi\sqrt{LC}}$

$50 = \frac{1}{2\pi\sqrt{10C}}$ or $2500 = \frac{1}{40\pi^2 C}$

or $C = \frac{1}{2500 \times 40 \times 10^{-1}}$ F $= 10^{-6}$ F $= 1\,\mu$F.

73. *(a)* Factual information.

74. *(c)* $\cos\phi = \frac{R}{Z} = \frac{12}{15} = 0.8.$

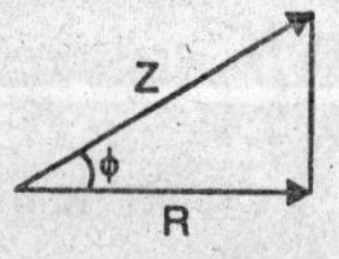

75. *(d)* The presence of R forbids a phase difference of $\frac{\pi}{2}$.

AIEEE (AET)2004—Solved

1. Which one of the following represents the correct dimensions of the coefficient of viscosity ?

(*a*) $ML^{-1}T^{-2}$ (*b*) MLT^{-1}
(*c*) $ML^{-1}T^{-1}$ (*d*) $ML^{-2}T^{-2}$.

2. A particle moves in a straight line with retardation proportional to its displacement. Its loss of kinetic energy for any displacement x is proportional to

(*a*) x^2 (*b*) e^x
(*c*) x (*d*) $\log_e x$.

3. A ball is released from the top of a tower of height h metres. It takes T seconds to reach the ground. What is the position of the ball in T/3 seconds ?

(*a*) $h/9$ metres from the ground
(*b*) $7h/9$ metres from the ground
(*c*) $8h/9$ metres from the ground
(*d*) $17h/18$ metres from the ground.

4. If $\vec{A} \times \vec{B} = \vec{B} \times \vec{A}$, then the angle between $\vec{A}$ and $\vec{B}$ is

(*a*) π (*b*) $\pi/3$
(*c*) $\pi/2$ (*d*) $\pi/4$.

5. A projectile can have the same range R for two angles of projection. If T_1 and T_2 be the time of flights in the two cases, then the product of the two time of flights is directly proportional to

(*a*) $1/R^2$ (*b*) $1/R$
(*c*) R (*d*) R^2.

6. Which of the following statements is false for a particle moving in a circle with a constant angular speed ?

(*a*) The velocity vector is tangent to the circle.
(*b*) The acceleration vector is tangent to the circle.
(*c*) The acceleration vector points to the centre of the circle.
(*d*) The velocity and acceleration vectors are perpendicular to each other.

7. An automobile travelling with a speed of 60 km/h, can brake to stop within a distance of 20 m. If the car is going twice as fast,, *i.e.* 120 km/h, the stopping distance will be

(*a*) 20 m (*b*) 40 m
(*c*) 60 m (*d*) 80 m.

8. A machine gun fires a bullet of mass 40 g with a velocity 1200 m s^{-1}. The man holding it can exert a maximum force of 144 N on the gun. How many bullets can he fire per second at the most ?

(*a*) one (*b*) four
(*c*) two (*d*) three.

9. Two masses $m_1 = 5$ kg and $m_2 = 4.8$ kg tied to a string are hanging over a light frictionless pulley. What is the acceleration of the masses when left free to move ? ($g = 9.8$ m/s^2)

(*a*) 0.2 m/s^2
(*b*) 9.8 m/s^2
(*c*) 5 m/s^2
(*d*) 4.8 m/s^2.

10. A uniform chain of length 2 m is kept on a table such that a length of 60 cm hangs freely from the edge of the table. The total mass of the chain is 4 kg. What is the work done in pulling the entire chain on the table ?

(*a*) 7.2 J (*b*) 3.6 J
(*c*) 120 J (*d*) 1200 J.

11. A block rests on a rough inclined plane making an angle of 30° with the horizontal. The coefficient of static friction between the block and the plane is 0.8. If the frictional force on the block is 10 N, the mass of the block (in kg) is (take $g = 10$ m/s^2)

(*a*) 2.0 (*b*) 4.0
(*c*) 1.6 (*d*) 2.5.

12. A force $\vec{F} = (5\hat{i} + 3\hat{j} + 2\hat{k})$ N is applied over a particle which displaces it from its origin to the point $\vec{r} = (2\hat{i} - \hat{j})$ m. The work done on the particle in joules is

(*a*) − 7 (*b*) + 7
(*c*) + 10 (*d*) + 13.

13. A body of mass m, accelerates uniformly from rest to velocity v_1 in time t_1. The instantaneous power delivered to the body as a function of time t is

(*a*) $\frac{mv_1 t}{t_1}$ (*b*) $\frac{mv_1^2 t}{t_1^2}$
(*c*) $\frac{mv_1 t^2}{t_1}$ (*d*) $\frac{mv_1^2 t}{t_1}$.

14. A particle is acted upon by a force of constant magnitude which is always perpendicular to the velocity of the particle. The motion of the particle takes place in a plane. It follows that

(*a*) its velocity is constant
(*b*) its acceleration is constant
(*c*) its kinetic energy is constant
(*d*) it moves in a straight line.

15. A solid sphere is rotating in free space. If the radius of the sphere is increased keeping mass same, which one of the following will not be affected ?

(a) moment of inertia (b) angular momentum
(c) angular velocity (d) rotational kinetic energy.

16. A ball is thrown from a point with a speed v_0 at an angle of projection θ. From the same point and at the same instant, a person starts running with a constant speed $v_0/2$ to catch the ball. Will the person be able to catch the ball ? If yes, what should be the angle of projection ?

(a) yes, 60° (b) yes, 30°
(c) no (d) yes, 45°.

17. One solid sphere A and another hollow sphere B are of same mass and same outer radii. Their moments of inertia about their diameters are respectively I_A and I_B such that

(a) $I_A = I_B$ (b) $I_A > I_B$
(c) $I_A < I_B$ (d) $I_A/I_B = d_A/d_B$.

where d_A and d_B are their densities.

18. A satellite of mass m revolves around the earth of radius R at a height x from its surface. If g is the acceleration due to gravity on the surface of the earth, the orbital speed of the satellite is

(a) gx (b) $\frac{gR}{R-x}$
(c) $\frac{gR^2}{R+x}$ (d) $\left(\frac{gR^2}{R+x}\right)^{1/2}$

19. The time period of an earth satellite in circular orbit is independent of

(a) the mass of the satellite
(b) radius of its orbit
(c) both the mass and radius of the orbit
(d) neither the mass of the satellite nor the radius of its orbit.

20. If g is the acceleration due to gravity on the earth's surface, the gain in the potential energy of an object of mass m raised from the surface of the earth to a height equal to the radius R of the earth is

(a) 2 mgR (b) $\frac{1}{2}$ mgR
(c) $\frac{1}{4}$ mgR (d) mgR.

21. Suppose the gravitational force varies inversely as the nth power of distance. Then the time period of a planet in circular orbit of radius R around the sun will be proportional to

(a) $R^{\left(\frac{n+1}{2}\right)}$ (b) $R^{\left(\frac{n-1}{2}\right)}$
(c) R^n (d) $R^{\left(\frac{n-2}{2}\right)}$.

22. A wire fixed at the upper end stretches by length l by by applying a force F. The work done in stretching is

(a) $F/2l$ (b) Fl
(c) $2Fl$ (d) $Fl/2$.

23. Spherical balls of radius R are falling in a viscous fluid of viscosity η with a velocity v. The retarding viscous force acting on the spherical ball is

(a) directly proportional to R but inversely proportional to v
(b) directly proportional to both radius R and velocity v
(c) inversely proportional to both radius R and velocity v
(d) inversely proportional to R but directly proportional to velocity v.

24. If two soap bubbles of different radii are connected by a tube,

(a) air flows from the bigger bubble to the smaller bubble till the sizes become equal
(b) air flows from bigger bubble to the smaller bubble till the sizes are interchanged
(c) air flows from the smaller bubble to the bigger
(d) there is no flow of air.

25. The bob of a simple pendulum executes simple harmonic motion in water with a period t, while the period of oscillation of the bob is t_0 in air. Neglecting frictional force of water and given that the density of the bob is (4/3) × 1000 kg/m³. What relationship between t and t_0 is true ?

(a) $t = t_0$ (b) $t = t_0/2$
(c) $t = 2t_0$ (d) $t = 4t_0$.

26. A particle at the end of a spring executes simple harmonic motion with a period t_1, while the corresponding period for another spring is t_2. If the period of oscillation with the two springs in series is T, then

(a) $T = t_1 + t_2$ (b) $T^2 = t_1^2 + t_2^2$
(c) $T^{-1} = t_1^{-1} + t_2^{-1}$ (d) $T^{-2} = t_1^{-2} + t_2^{-2}$.

27. The total energy of a particle, executing simple harmonic motion, is

(a) $\propto x$ (b) $\propto x^2$
(c) independent of x (d) $\propto x^{1/2}$.

where x is the displacement from the mean position.

28. The displacement y of a particle in a medium can be expressed as : $y = 10^{-6} \sin(100t + 20x + \pi/4)$ m, where t is in second and x in metre. The speed of the wave is

(a) 2000 m/s (b) 5 m/s
(c) 20 m/s (d) 5π m/s.

29. A particle of mass m is attached to a spring (of spring constant k) and has a natural angular frequency ω_0. An external force $F(t)$ proportional to $\cos \omega t$ $(\omega \neq \omega_0)$ is applied to the oscillator. The time-displacement of the oscillator will be proportional to

(a) $\frac{m}{\omega_0^2 - \omega^2}$ (b) $\frac{1}{m(\omega_0^2 - \omega^2)}$
(c) $\frac{1}{m(\omega_0^2 + \omega^2)}$ (d) $\frac{m}{\omega_0^2 + \omega^2}$.

30. In forced oscillation of a particle, the amplitude is maximum for a frequency ω_1 of the force, while the energy is maximum for a frequency ω_2 of the force, then

(*a*) $\omega_1 = \omega_2$ (*b*) $\omega_1 > \omega_2$

(*c*) $\omega_1 < \omega_2$ when damping is small and $\omega_1 > \omega_2$ when damping is large

(*d*) $\omega_1 < \omega_2$.

31. One mole of ideal monoatomic gas ($\gamma = 5/3$) is mixed with one mole of diatomic gas ($\gamma = 7/5$). What is γ for the mixture ? γ denotes the ratio of specific he at constant pressure, to that at constant volume.

(*a*) 3/2 (*b*) 23/15

(*c*) 35/23 (*d*) 4/3.

32. If the temperature of the sun were to increase from T to 2T and its radius from R to 2R, then the ratio of the radiant energy received on earth to what it was previously will be

(*a*) 4 (*b*) 16

(*c*) 32 (*d*) 64.

33. Which of the following statements is correct for any thermodynamic system ?

(*a*) The internal energy changes in all processes.

(*b*) Internal energy and entropy are state functions.

(*c*) The change in entropy can never be zero.

(*d*) The work done in an adiabatic process is always zero.

34. Two thermally insulated vessels 1 and 2 are filled with air at temperatures (T_1, T_2), volumes (V_1, V_2) and pressures (P_1, P_2) respectively. If the valve joining the two vessels is opened, the temperature inside the vessel at equilibrium will be

(*a*) $T_1 + T_2$ (*b*) $(T_1 + T_2)/2$

(*c*) $\dfrac{T_1T_2(P_1V_1 + P_2V_2)}{P_1V_1T_2 + P_2V_2T_1}$ (*d*) $\dfrac{T_1T_2(P_1V_1 + P_2V_2)}{P_1V_1T_1 + P_2V_2T_2}$.

35. A radiation of energy E falls normally on a perfectly reflecting surface. The momentum transferred to the surface is

(*a*) E/*c* (*b*) 2E/*c*

(*c*) Ec (*d*) E/c^2.

36. The temperature of the two outer surfaces of a composite slab, consisting of two materials having coefficients of thermal conductivity K and 2K and thickness x and $4x$ respectively are T_2 and T_1 ($T_2 > T_1$). The rate of heat transfer through the slab, in a steady state is $\left(\dfrac{A(T_2 - T_1)K}{x}\right) f$, with f equal to

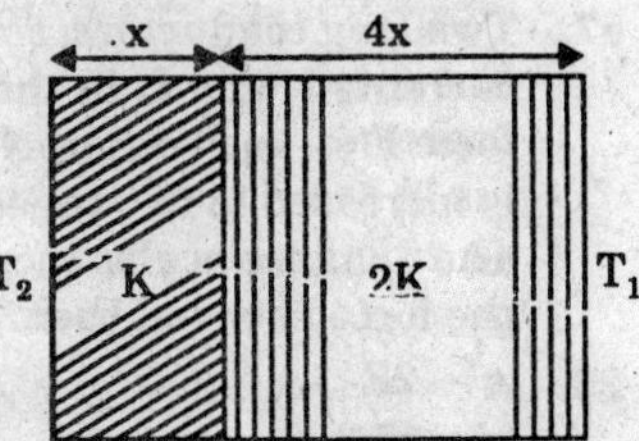

(*a*) 1 (*b*) 1/2

(*c*) 2/3 (*d*) 1/3.

37. A light ray is incident perpendicular to one face of a 90° prism and is totally internally reflected at the glass-air interface. If the angle of reflection is 45°, we conclude that the refractive index n

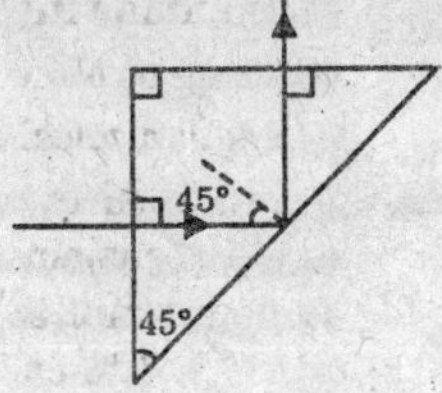

(*a*) $n < \dfrac{1}{\sqrt{2}}$ (b) $n > \sqrt{2}$

(c) $n > \dfrac{1}{\sqrt{2}}$ (*d*) $n < \sqrt{2}$.

38. A plano-convex lens of refractive index 1.5 and radius of curvature 30 cm is silvered at the curved surface. Now this lens has been used to form the image of an object. At what distance from this lens an object be placed in order to have a real image of the size of the object ?

(*a*) 20 cm (*b*) 30 cm

(*c*) 60 cm (*d*) 80 cm.

39. The angle of incidence at which reflected light is totally polarised for reflection from air to glass (refractive index n), is

(*a*) $\sin^{-1}(n)$ (*b*) $\sin^{-1}(1/n)$

(*c*) $\tan^{-1}(1/n)$ (*d*) $\tan^{-1}(n)$.

40. The maximum number of possible interference maxima for slit-separation equal to twice the wavelength in Young's double-slit experiment is

(*a*) infinite (*b*) five

(*c*) three (*d*) zero.

41. An electromagnetic wave of frequency $\nu = 3.0$ MHz passes from vacuum into a dielectric medium with permitivity $\varepsilon = 4.0$. Then

(*a*) wavelength is doubled and the frequency remains unchanged

(*b*) wavelength is doubled and frequency becomes half

(*c*) wavelength is halved and frequency remains unchanged

(*d*) wavelength and frequency both remain unchanged.

42. Two spherical conductors B and C having equal radii and carrying equal charges in them repel each other with a force F when kept apart at some distance. A third spherical conductor having same radius as that of B but uncharged is brought in contact with B, then brought in contact with C and finally removed away from both. The new force of repulsion between B and C is

(*a*) F/4 (*b*) 3F/4

(*c*) F/8 (*d*) 3F/8.

43. A charged particle q is shot towards another charged particle Q which is fixed, with a

speed v. It approaches Q upto a closest distance r and then returns. If q were given a speed $2v$, the closest distances of approach would be

(a) r (b) $2r$

(c) $r/2$ (d) $r/4$.

44. Four charges, equal to – Q, are placed at the four corners of a square and a charge q is at its centre. If the system is in equilibrium the value of q is

(a) $-\frac{Q}{4}(1+2\sqrt{2})$ (b) $\frac{Q}{4}(1+2\sqrt{2})$

(c) $-\frac{Q}{2}(1+2\sqrt{2})$ (d) $\frac{Q}{2}(1+2\sqrt{2})$.

45. Alternating current cannot be measured by D.C. ammeter because

(a) A.C. cannot pass through D.C. ammeter

(b) A.C. changes direction

(c) Average value of current for complete cycle is zero

(d) D.C. ammeter will get damaged.

46. The total current supplied to the circuit by the battery is

6V, 2 Ω, 6 Ω, 3 Ω, 1.5 Ω

(a) 1 A

(b) 2 A

(c) 4 A

(d) 6 A.

47. The resistance of the series combination of two resistances is S. When they are joined in parallel, the total resistance is P. If S = nP, then the minimum possible value of n is

(a) 4 (b) 3

(c) 2 (d) 1.

48. An electric current is passed through a circuit containing two wires of the same material, connected in parallel. If the lengths and radii of the wires are in the ratio of 4/3 and 2/3, then the ratio of the currents passing through the wires will be

(a) 3 (b) 1/3

(c) 8/9 (d) 2.

49. In a metre bridge experiment, null point is obtained at 20 cm from one end of the wire when resistance X is balanced against another resistance Y. If X < Y, then where will be the new position of the null point from the same end, if one decides to balance a resistance of 4X against Y ?

(a) 50 cm (b) 80 cm

(c) 40 cm (d) 70 cm.

50. The thermistors are usually made of

(a) metals with low temperature coefficient of resistivity

(b) metals with high temperature coefficient of resistivity

(c) metal oxides with high temperature coefficient of resistivity

(d) semiconducting materials having low temperature coefficient of resistivity.

51. Time taken by a 836 W heater to heat one litre of water from 10°C to 40°C is

(a) 50 s (b) 100 s

(c) 150 s (d) 200 s.

52. The thermo emf of a thermocouple varies with the temperature θ of the hot junction as $E = a\theta + b\theta^2$ in volts where the ratio a/b is 700°C. If the cold junction is kept at 0°C, then the neutral temperature is

(a) 700°C (b) 350°C

(c) 1400°C

(d) no neutral temperature is possible for this thermocouple.

53. The electrochemical equivalent of a metal is 3.3×10^{-7} kg per coulomb. The mass of the metal liberated at the cathode when a 3A current is passed for 2 second will be

(a) 19.8×10^{-7} kg (b) 9.9×10^{-7} kg

(c) 6.6×10^{-7} kg (d) 1.1×10^{-7} kg.

54. A current i ampere flows along an infinitely long straight thin walled tube. Then the magnetic induction at any point inside the tube is

(a) infinite (b) zero

(c) $\frac{\mu_0}{4\pi}\cdot\frac{2i}{r}$ tesla (d) $\frac{2i}{r}$ tesla.

55. A long wire carries a steady current. It is bent into a circle of one turn and the magnetic field at the centre of the coil is B. It is then bent into a circular loop of n turns. The magnetic field at the centre of the coil will be

(a) nB (b) n^2B

(c) $2n$B (d) $2n^2$B.

56. The magnetic field due to a current-carrying circular loop of radius 3 cm at a point on the axis at a distance of 4 cm from the centre is 54 μT. What will be its value at the centre of the loop ?

(a) 250 μT (b) 100 μT

(c) 125 μT (d) 75 μT.

57. Two long conductors, separated by a distance d, carry currents I_1 and I_2 in the same direction. They exert a force F on each other. Now the current in one of them is increased to two times and its direction is reversed. The distance is also increased to $3d$. The new value of the force between them is

(a) – 2F (b) F/3

(c) – 2F/3 (d) – F/3.

58. The length of a magnet is large compared to its width and breadth. The time period of its oscillation in a vibration magnetometer is 2 s. The magnet is cut along its length into three equal parts and three parts are then placed on each other with their like poles together. The time period of this combination will be

(*a*) 2 s (*b*) 2/3 s

(c) $2\sqrt{3}$ s (*d*) $2/\sqrt{3}$ s.

59. The materials suitable for making electromagnets should have

(*a*) high retentivity and high coercivity

(*b*) low retentivity and low coercivity

(*c*) high retentivity and low coercivity

(*d*) low retentivity and high coercivity.

60. In an LCR series a.c. circuit, the voltage across each of the components, L, C and R is 50 V. The voltage across the LC combination will be

(*a*) 50 V (*b*) $50\sqrt{2}$ V

(*c*) 100 V (*d*) 0 V (zero).

61. A coil having *n* turns and resistance R Ω is connected with a galvanometer of resistance 4 R Ω. This combination is moved in time *t* seconds from a magnetic field W_1 weber to W_2 weber. The induced current in the circuit is

(*a*) $-\dfrac{W_2 - W_1}{5Rnt}$ (*b*) $-\dfrac{n(W_2 - W_1)}{5Rt}$

(*c*) $-\dfrac{(W_2 - W_1)}{Rnt}$ (*d*) $-\dfrac{n(W_2 - W_1)}{Rt}$.

62. In a uniform magnetic field of induction B, a wire in the form of a semicircle of radius *r* rotates about the diameter of the circle with angular frequency ω. The axis of rotation is perpendicular to the field. If the total resistance of the circuit is R, the mean power generated per period of rotation is

(*a*) $\dfrac{B\pi r^2\omega}{2R}$ (*b*) $\dfrac{(B\pi r^2\omega)^2}{8R}$

(*c*) $\dfrac{(B\pi r\omega)^2}{2R}$ (*d*) $\dfrac{(B\pi r\omega^2)^2}{8R}$.

63. In a LCR circuit, capacitance is changed from C to 2C. For the resonant frequency to remain unchanged, the inductance should be changed from L to

(*a*) 4L (*b*) 2L

(*c*) L/2 (*d*) L/4.

64. A metal conductor of length 1 m rotates vertically about one of its ends at angular velocity 5 radians per second. If the horizontal component of earth's magnetic field is 0.2×10^{-4} T, then the emf developed between the two ends of the conductor is

(*a*) 5 μV (*b*) 50 μV

(*c*) 5 mV (*d*) 50 mV.

65. According to Einstein's photoelectric equation, the plot of the kinetic energy of the emitted photo electrons from a metal vs the frequency of the incident radiation gives a straight line whose slope

(*a*) depends on the nature of the metal used

(*b*) depends on the intensity of the radiation

(*c*) depends both on the intensity of the radiation and the metal used

(*d*) is the same for all metals and independent of the intensity of the radiation.

66. The work function of a substance is 4.0 eV. The longest wavelength of light that can cause photoelectron emission from this substance is approximately

(*a*) 540 nm (*b*) 400 nm

(*c*) 310 nm (*d*) 220 nm.

67. A charged oil drop is suspended in a uniform field of 3×10^4 V/m so that it neither falls nor rises. The charge on the drop will be (take the mass of the charge = 9.9×10^{-15} kg and $g = 10$ m/s²)

(*a*) 3.3×10^{-18} C (*b*) 3.2×10^{-18} C

(*c*) 1.6×10^{-18} C (*d*) 4.8×10^{-18} C.

68. A nucleus disintegrates into two nuclear parts which have their velocities in the ratio 2 : 1. The ratio of their nuclear sizes will be

(*a*) $2^{1/3} : 1$ (*b*) $1 : 3^{1/2}$

(*c*) $3^{1/2} : 1$ (*d*) $1 : 2^{1/3}$.

69. The binding energy per nucleon of deuteron ($^2_1$H) and helium nucleus ($^4_2$He) is 1.1 MeV and 7 MeV respectively. If two deuteron nuclei react to form a single helium nucleus, then the energy released is

(*a*) 13.9 MeV (*b*) 26.9 MeV

(*c*) 23.6 MeV (*d*) 19.2 MeV.

70. An α-particle of energy 5 MeV is scattered through 180° by a fixed uranium nucleus. The distance of the closest approach is of the order of

(*a*) 1 Å (*b*) 10^{-10} cm

(*c*) 10^{-12} cm (*d*) 10^{-15} cm.

71. When *npn* transistor is used as an amplifier

(*a*) electrons move from base to collector

(*b*) holes move from emitter to base

(*c*) electrons move from collector to base

(*d*) holes move from base to emitter.

72. For a transistor amplifier in common emitter configuration for load impedance of 1 kΩ (h_{fe} = 50 and h_{oe} = 25) the current gain is

(*a*) – 5.2 (*b*) – 15.7

(*c*) – 24.8 (*d*) – 48.78.

73. A piece of copper and another of germanium are cooled from room temperature to 77 K, the resistance of

(*a*) each of them increases

(*b*) each of them decreases

(*c*) copper decreases and germanium increases

(*d*) copper increases and germanium decreases.

74. The manifestation of band structure in solids is due to

(*a*) Heisenberg's uncertainty principle
(*b*) Pauli's exclusion principle
(*c*) Bohr's correspondence principle
(*d*) Boltzmann's law.

75. When *p-n* junction diode is forward biased, then

(*a*) the depletion region is reduced and barrier height is increased
(*b*) the depletion region is widened and barrier height is reduced.
(*c*) both the depletion region and barrier height are reduced
(*d*) both the depletion region and barrier height are increased.

ANSWERS

1. (*c*)	**2.** (*a*)	**3.** (*c*)	**4.** (*a*)	**5.** (*c*)	**6.** (*b*)	**7.** (*d*)	**8.** (*d*)
9. (*a*)	**10.** (*b*)	**11.** (*a*)	**12.** (*b*)	**13.** (*b*)	**14.** (*c*)	**15.** (*b*)	**16.** (*a*)
17. (*c*)	**18.** (*d*)	**19.** (*a*)	**20.** (*b*)	**21.** (*a*)	**22.** (*d*)	**23.** (*b*)	**24.** (*c*)
25. (*c*)	**26.** (*b*)	**27.** (*c*)	**28.** (*b*)	**29.** (*b*)	**30.** (*a*)	**31.** (*a*)	**32.** (*d*)
33. (*b*)	**34.** (*c*)	**35.** (*b*)	**36.** (*d*)	**37.** (*b*)	**38.** (*a*)	**39.** (*d*)	**40.** (*b*)
41. (*c*)	**42.** (*d*)	**43.** (*d*)	**44.** (*b*)	**45.** (*c*)	**46.** (*c*)	**47.** (*a*)	**48.** (*a*)
49. (*a*)	**50.** (*c*)	**51.** (*c*)	**52.** (*d*)	**53.** (*a*)	**54.** (*b*)	**55.** (*b*)	**56.** (*a*)
57. (*c*)	**58.** (*a*)	**59.** (*b*)	**60.** (*d*)	**61.** (*b*)	**62.** (*b*)	**63.** (*c*)	**64.** (*b*)
65. (*d*)	**66.** (*c*)	**67.** (*a*)	**68.** (*b*)	**69.** (*c*)	**70.** (*c*)	**71.** (*a*)	**72.** (*d*)
73. (*c*)	**74.** (*b*)	**75.** (*c*).					

Solutions

1. (*c*) $F = 6\pi\eta rv$

$$[\eta] = \frac{[F]}{[r][v]} = \frac{[MLT^{-2}]}{[L][LT^{-1}]} = [ML^{-1}T^{-1}]$$

2. (*a*) $-a \propto x$

or $a \propto -x$ or $a = -kx$

or $\frac{dv}{dt} = -kx$ or $\frac{dv}{dx}.\frac{dx}{dt} = -kx$

or $\frac{vdv}{dx} = -kx$ or $vdv = -kxdx$

or $\int_u^v vdv = -k\int_0^x xdx$

or $\frac{v^2}{2} - \frac{u^2}{2} = -kx^2$ or $\frac{1}{2}mv^2 - \frac{1}{2}mu^2 = -mkx^2$

or $\frac{1}{2}mu^2 - \frac{1}{2}mv^2 = mkx^2$

$\therefore$ loss of kinetic energy $\propto x^2$

3. (*c*) $h = \frac{1}{2}gT^2$ or $T = \sqrt{\frac{2h}{g}}$

Now, $S = \frac{1}{2}g\left(\frac{T}{3}\right)^2 = \frac{g}{18} \times \frac{2h}{g} = \frac{h}{9}$

Height above the ground $= h - \frac{h}{9} = \frac{8h}{9}$

4. (*a*) $\vec{A} \times \vec{B} = \vec{B} \times \vec{A}$

or $\vec{A} \times \vec{B} - \vec{B} \times \vec{A} = 0$

or $\vec{A} \times \vec{B} + \vec{A} \times \vec{B} = 0$ or $2(\vec{A} \times \vec{B}) = 0$

or $\vec{A} \times \vec{B} = 0$

Clearly, the angle between $\vec{A}$ and $\vec{B}$ is either 0° or π. So, 'π' is the correct choice in the given problem.

5. (*c*) Range is same for two angles (θ and $90° - \theta$) of projection.

$$T_1 = \frac{2v\sin\theta}{g}$$

$$T_2 = \frac{2v\sin(90° - \theta)}{g} = \frac{2v\cos\theta}{g}$$

$$T_1T_2 = \frac{4v^2\sin\theta\cos\theta}{g^2}$$

$$T_1T_2 = \frac{2v^2(2\sin\theta\cos\theta)}{g^2} = \frac{2v^2\sin 2\theta}{g^2}$$

or $$T_1T_2 = \frac{2}{g}\left[\frac{v^2 \sin 2\theta}{g}\right]$$

or $$T_1T_2 = \frac{2R}{g} \quad \text{or} \quad T_1T_2 \propto R$$

6. (*b*) The acceleration vector is along the radius of the circle and not along the tangent to the circle.

7. (*d*) Loss of kinetic energy ∝ stopping distance. When velocity is doubled, kinetic enregy is quadrupled. Loss of kinetic energy is four times. So, stopping distance is also increased by a factor of 4. Thus, the new stopping distance is 4 × 20 m = 80 m.

8. (*d*) $$F = mnv$$

or $$n = \frac{F}{mv} = \frac{144}{\frac{40}{1000}\times 1200}$$

$$= \frac{144}{40 \times 1.2} = \frac{144}{48} = 3$$

9. (*a*) $$a = \frac{m_1 - m_2}{m_1 + m_2}\, g$$

$$= \frac{5 - 4.8}{5 + 4.8} \times 9.8 \text{ m s}^{-2} = 0.2 \text{ m s}^{-2}$$

10. (*b*) Mass of hanging chain, $m = \frac{4}{2} \times \frac{60}{100} = \frac{6}{5}$ kg

Height of centre of gravity, $h = 0.3$ m

Work done, $W = mgh = \frac{6}{5} \times 10 \times 0.3 \text{ J} = 3.6 \text{ J}$

11. (*a*) Since the block is at rest,

∴ $mg \sin\theta = 10$

or $$m = \frac{10}{g \sin\theta} = \frac{10}{10 \sin 30^\circ} = 2 \text{ kg}$$

12. (*b*) $$\vec{S} = \vec{r} - \vec{0} = 2\hat{i} - \hat{j}$$

$$\vec{F} = 5\hat{i} + 3\hat{j} + 2\hat{k}$$

$$W = \vec{F} \cdot \vec{S}$$

$$W = (5\hat{i} + 3\hat{j} + 2\hat{k}) \cdot (2\hat{i} - \hat{j})$$
$$= (10 - 3) \text{ J} = 7 \text{ J}$$

13. (*b*) If a be the acceleration, then

$$v_1 = 0 + at_1 \quad \text{or} \quad a = \frac{v_1}{t_1}$$

If v is the velocity at any instant of time t, then

$$v = at = \frac{v_1}{t_1} t$$

∴ Power $P = \vec{F} \cdot \vec{v} = mav$

$$= m\left(\frac{v_1}{t_1}\right)\frac{v_1}{t_1} t = \frac{m{v_1}^2 t}{{t_1}^2}$$

14. (*c*) When force and velocity vectors are perpendicular, the work done is zero. So, there is no change of kinetic energy.

15. (*b*) The given space is a free space. Clearly, there is no external torque. Thus, angular momentum is conserved.

16. (*a*) The person will be able to catch the ball if the horizontal component of the velocity of the ball is equal to the speed of the person.

∴ $$u_0 \cos\theta = \frac{v_0}{2}$$

or $$\cos\theta = \frac{1}{2} \quad \text{or} \quad \theta = 60^\circ$$

17. (*c*) $$I_A = \frac{2}{5} MR^2$$

$$I_B = \frac{2}{3} MR^2$$

Clearly $I_A < I_B$

18. (*d*) The gravitational force on the satellite provides the necessary centripetal force.

∴ $$\frac{GM_e m}{(R+x)^2} = \frac{m{v_0}^2}{R+x} \quad \text{or} \quad \frac{GM_e}{R+x} = {v_0}^2$$

or $$v_0 = \left[\frac{GM_e}{R+x}\right]^{1/2} = \left[\frac{gR^2}{R+x}\right]^{1/2}$$

19. (*a*) Time period of satellite, $T = 2\pi\sqrt{\frac{(R+h)^3}{GM}}$

Clearly, the time period does not depend upon the mass of the satelite.

20. (*b*) Potential energy on the surface of earth $= -\frac{GMm}{R}$

Potential energy at height h above the surface of earth

i.e. at a distance 2R from the centre of earth $= -\frac{GMm}{2R}$

Gain in potential energy

$$= \left[-\frac{GMm}{2R}\right] - \left[-\frac{GMm}{R}\right]$$

$$= \frac{GMm}{R} - \frac{GMm}{2R} = \frac{GMm}{2R}$$

$$= \frac{GMm}{2R} \times \frac{R}{R} = \frac{GMmR}{2R^2}$$

$$= \frac{mgR}{2} \qquad \left[\because g = \frac{GM}{R^2} \text{ or } GM = gR^2\right]$$

21. **(a)** The required centripetal force is obtained from gravitational force.

$$\frac{mv^2}{R} = \frac{GMm}{R^n} \text{ or } v = \sqrt{\frac{GM}{R^{n-1}}}$$

Again, $$T = \frac{2\pi R}{v} = 2\pi R\sqrt{\frac{R^{n-1}}{GM}}$$

$$= 2\pi\sqrt{\frac{R^{n+1}}{GM}} = 2\pi\frac{R^{\frac{n+1}{2}}}{\sqrt{GM}}$$

or $$T \propto R^{\frac{n+1}{2}}$$

22. **(d)** Work done $= \frac{1}{2}$ stress × strain × volume

$$= \frac{1}{2}\frac{F}{A}\frac{l}{L}(AL) = \frac{1}{2}Fl$$

23. **(b)** Viscous force, $F = 6\pi\eta rv$

24. **(c)** $$p = \frac{2\sigma}{R}$$

R is less for smaller bubble. p is more. For larger bubble, R is more. p is less. Since air flows from higher pressure to lower pressure therefore the air shall flow from smaller bubble to larger bubble.

25. **(c)** $$t_0 = 2\pi\sqrt{\frac{l}{g}}$$

Effective weight inside water, mg'

$$= \rho Vg - \rho' Vg = \rho Vg\left(1 - \frac{\rho'}{\rho}\right)$$

$$= mg\left(1 - \frac{\rho'}{\rho}\right)$$

$\therefore$ $$g' = g\left(1 - \frac{\rho'}{\rho}\right)$$

$$t = 2\pi\sqrt{\frac{l}{g\left(1 - \frac{\rho'}{\rho}\right)}} = 2\pi\sqrt{\frac{l}{g\left(1 - \frac{3}{4}\right)}}$$

$$= 2 \times 2\pi\sqrt{\frac{l}{g}} = 2\,t_0$$

26. **(b)** $$t_1 = 2\pi\sqrt{\frac{m}{k_1}} \qquad \text{...(1)}$$

$$t_2 = 2\pi\sqrt{\frac{m}{k_2}} \qquad \text{...(2)}$$

When springs are in series, then

$$T = 2\pi\sqrt{\frac{m}{\frac{k_1k_2}{k_1+k_2}}}$$

or $$T = 2\pi\sqrt{\frac{m(k_1+k_2)}{k_1k_2}}$$

squaring and adding (1) and (2), we get

$$t_1^2 + t_2^2 = 4\pi^2\frac{m}{k_1} + 4\pi^2\frac{m}{k_2}$$

or $$t_1^2 + t_2^2 = 4\pi^2 m\left(\frac{1}{k_1} + \frac{1}{k_2}\right)$$

or $$t_1^2 + t_2^2 = 4\pi^2 m\left(\frac{k_1+k_2}{k_1k_2}\right)$$

or $$t_1^2 + t_2^2 = T^2$$

27. **(c)** The total energy remains constant at all locations.

28. **(b)** $$y = 10^{-6}\sin\left(100\,t + 20x + \frac{\pi}{4}\right)$$

Comparing with

$$y = a\sin(\omega t + kx + \phi),$$

we get

$$\omega = 100 \text{ rad s}^{-1} \text{ and } k = 20 \text{ m}^{-1}$$

$$v = \frac{\omega}{k} = \frac{100}{20} = 5 \text{ m s}^{-1}$$

29. **(b)** $$x = A\sin(\omega t + \phi)$$

with $$A = \frac{F_0/m}{\omega_0^2 - \omega^2}$$

30. **(a)** In the condition of resonance, the frequency is equal to the natural frequency. Both amplitude and energy get maximised.

$\therefore$ $$\omega_1 = \omega_2$$

31. **(a)** Total number of degrees of freedom of

2 moles = 3N + 5N = 8N

Total number of degrees of freedom/molecule, n

$$= \frac{8N}{2N} = 4$$

Now, $$\gamma = 1 + \frac{2}{4} = \frac{3}{2}$$

32. (*d*) Surface area is increased by a factor of 4. The temperature is increased by a factor of 16. So, the radiant energy is increased by a factor of 64. [Using Stefan's law]

33. (*b*) Factual information.

34. (*c*) The guiding principle in this problem is that the total number of moles of the system remain the same.

$$\therefore \quad \frac{P_1V_1}{RT_1} + \frac{P_2V_2}{RT_2} = \frac{P(V_1+V_2)}{RT}$$

$$\text{or} \quad T = \frac{P(V_1+V_2)T_1T_2}{P_1V_1T_2 + P_2V_2T_1}$$

Using Boyle's law,

$$P(V_1+V_2) = P_1V_1 + P_2V_2$$

$$\text{or} \quad P = \frac{P_1V_1 + P_2V_2}{V_1+V_2}$$

$$\therefore \quad T = \frac{(P_1V_1 + P_2V_2)T_1T_2}{P_1V_1T_2 + P_2V_2T_1}$$

35. (*b*) $P_i = \frac{E}{c}$

If the surface is perfectly reflecting, then

$$p_f = \frac{E}{c}$$

$$\text{Change} = \frac{2E}{c}$$

36. (*d*) $Q = \frac{KA(\theta_1-\theta_2)t}{d}$

$$\Rightarrow \frac{K(\theta_1-\theta_2)}{d} = \text{constant}$$

$$\therefore \quad \frac{K(T_2-T_0)}{x} = \frac{2k(T_0-T_1)}{4x}$$

$$\text{or} \quad T_2 - T_0 = \frac{T_0-T_1}{2}$$

$$\text{or} \quad 2T_2 - 2T_0 = T_0 - T_1$$

$$\text{or} \quad 3T_0 = 2T_2 + T_1$$

$$\text{or} \quad T_0 = \frac{2T_2+T_1}{3}$$

$$\text{Rate of heat transfer} = \frac{Q}{T} = \frac{KA(T_2-T_0)}{x}$$

$$= \frac{KA}{x}\left[T_2 - \frac{2T_2+T_1}{3}\right] = \frac{KA}{x}\left[\frac{3T_2-2T_2-T_1}{3}\right]$$

$$= \frac{KA}{x}\left[\frac{T_2-T_1}{3}\right] = \frac{KA}{3x}(T_2-T_1)$$

Now, calculate $\frac{Q}{t}$.

37. (*b*)

$$i > i_C$$

$$45° > i_C$$

$$\text{or} \quad \sin 45° > \sin i_C$$

$$\text{or} \quad \frac{1}{\sqrt{2}} > \sin i_C$$

$$\text{or} \quad \frac{1}{\sqrt{2}} > \frac{1}{\mu} \quad \text{or} \quad \sqrt{2} < \mu$$

$$\text{or} \quad \mu > \sqrt{2}$$

38. (*a*) The effective focal length is given by

$$\frac{1}{F} = \frac{2}{f_l} + \frac{1}{f_m}$$

$$\text{But} \quad \frac{1}{f_l} = (1.5-1)\left(\frac{1}{\infty} - \frac{1}{30}\right) = \frac{1}{60}$$

$$\text{or} \quad \frac{2}{f_l} = \frac{1}{30}$$

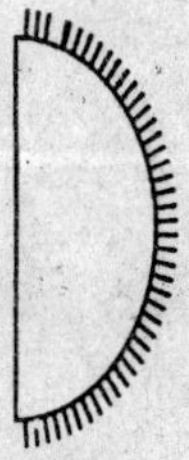

Again, $R = 30$ cm

$$f_m = \frac{R}{2} = 15 \text{ cm}$$

$$\text{Now,} \quad \frac{1}{F} = \frac{1}{30} + \frac{1}{15}$$

$$\text{or} \quad \frac{1}{F} = \frac{1+2}{30} = \frac{3}{30} = \frac{1}{10} \quad \text{or} \quad F = 10 \text{ cm}$$

To have a real image of the size of the object, the object must be placed at the centre of curvature of the equivalent mirror.

So, the required distance is 2 × 10 cm *i.e.* 20 cm.

39. (*d*) According to Brewester's law,

$$n = \tan i_p \quad \text{or} \quad i_p = \tan^{-1}(n)$$

40. (*b*) The condition for interference maximum is

$$d \sin\theta = n\lambda$$

But $d = 2\lambda$ (given)

$$\therefore \quad 2\lambda \sin\theta = n\lambda \quad \text{or} \quad \sin\theta = \frac{n}{2}$$

Clearly, n can have only integral values – 2, – 1, 0, + 1, + 2. [Note that the value of sin θ can vary between – 1 and + 1 only.]

So, the maximum number of possible interference maxima is 5.

41. (*c*) Energy ($h\nu$) is conserved in refraction.

So, ν remains unchanged.

Speed halved.

$$v = \nu\lambda$$

$$v \propto \lambda.$$

Since v is halved therefore λ is halved.

42. **(*d*)** In the first case, the charge on each sphere is q.

$$F = \frac{1}{4\pi\varepsilon_0}\frac{q^2}{d^2}$$

When A is brought in contact with B, the charge on each becomes $\frac{q}{2}$. Now, when A is brought in contact with C, the charge on each is $\frac{\frac{q}{2}+q}{2}$

or $\frac{3q}{4}$. If F′ is the new force of repulsion, then

$$F' = \frac{1}{4\pi\varepsilon_0}\frac{\left(\frac{q}{2}\right)\left(\frac{3q}{4}\right)}{d^2}$$

or $F' = \frac{1}{4\pi\varepsilon_0}\frac{3q^2}{8d^2}$ or $F' = \frac{3F}{8}$

43. **(*d*)** $\frac{1}{2}mv^2 = \frac{1}{4\pi\varepsilon_0}\frac{qQ}{r}$

or $v^2 \propto \frac{1}{r}$ or $r \propto \frac{1}{v^2}$

When v is doubled, r becomes one-fourth.

44. **(*b*)**

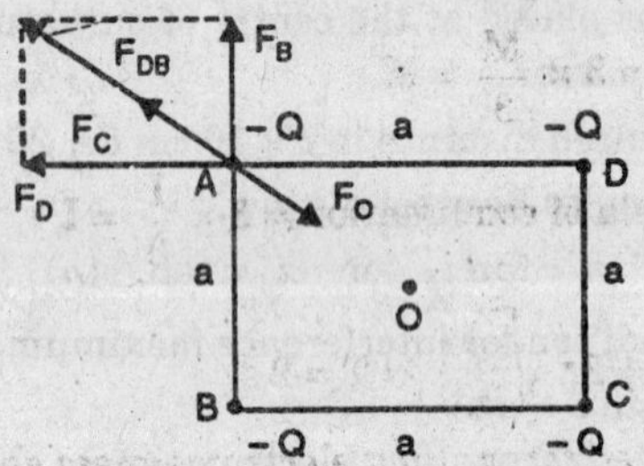

Consider the equilibrium of charge "– Q" at A.

$$F_B = \frac{1}{4\pi\varepsilon_0}\frac{Q^2}{a^2}\ ;\ F_D = \frac{1}{4\pi\varepsilon_0}\frac{Q^2}{a^2}$$

$$F_{BD} = \sqrt{2}\,F_B = \frac{1}{4\pi\varepsilon_0}\frac{\sqrt{2}\,Q^2}{a^2}$$

$$F_C = \frac{1}{4\pi\varepsilon_0}\frac{Q^2}{(\sqrt{2}a)^2} \text{ or } F_C = \frac{1}{4\pi\varepsilon_0}\frac{Q^2}{2a^2}$$

$$F_O = \frac{1}{4\pi\varepsilon_0}\frac{qQ}{\left(\frac{\sqrt{2}a}{2}\right)^2} = \frac{1}{4\pi\varepsilon_0}\frac{2qQ}{a^2}$$

For equilibrium,

$$F_{BD} + F_C = F_O$$

$$\frac{1}{4\pi\varepsilon_0}\frac{\sqrt{2}\,Q^2}{a^2} + \frac{1}{4\pi\varepsilon_0}\frac{Q^2}{2a^2} = \frac{1}{4\pi\varepsilon_0}\frac{2qQ}{a^2}$$

or $\sqrt{2}\,Q^2 + \frac{Q^2}{2} = 2qQ$ or $\sqrt{2}\,Q + \frac{Q}{2} = 2q$

or $q = \frac{\sqrt{2}Q}{2} + \frac{Q}{4}$ or $q = \frac{Q}{4}\left[\frac{4\sqrt{2}}{2} + \frac{4}{4}\right]$

or $q = \frac{Q}{4}(2\sqrt{2}+1)$

45. **(*c*)** Factual information.

AC measuring instruments are based on "heating effect of current".

46. **(*c*)** 2 Ω and 6 Ω are in parallel.

This gives $\frac{2\times 6}{2+6}\,\Omega$ or $\frac{12}{8}\,\Omega$

or $\frac{3}{2}\,\Omega$ or 1.5 Ω. This is in series with 1.5 Ω. This gives 3 Ω. This is in parallel with 3 Ω. This gives 1.5 Ω.

Now $I = \frac{6\text{ V}}{1.5\,\Omega} = 4\text{ A}$

47. **(*a*)** $S = R_1 + R_2$

$$P = \frac{R_1R_2}{R_1+R_2}$$

$\because$ $S = nP$,

$\therefore$ $R_1 + R_2 = \frac{nR_1R_2}{R_1+R_2}$

or $(R_1+R_2)^2 = nR_1R_2$

For minimum value of n, $R_1 = R_2$.

$\therefore$ $(2R_1)^2 = nR_1^2$

or $n = 4$

48. **(*a*)** $\frac{I_1}{I_2} = \frac{R_2}{R_1}$

$$\frac{I_1}{I_2} = \frac{\rho L_2}{A_2}\times\frac{A_1}{\rho L_1}$$

or $\frac{I_1}{I_2} = \left(\frac{L_2}{L_1}\right)\left(\frac{A_1}{A_2}\right)$ or $\frac{I_1}{I_2} = \left(\frac{L_2}{L_1}\right)\left(\frac{\pi r_1^2}{\pi r_2^2}\right)$

$$= \frac{L_2r_1^2}{L_1r_2^2} = \frac{3}{4}\left(\frac{2}{3}\right)^2 = \frac{3}{4}\times\frac{4}{9} = \frac{1}{3}$$

49. **(a)** $\frac{X}{Y} = \frac{20}{80} = \frac{1}{4}$ or $Y = 4X$

$\frac{4X}{Y} = \frac{l}{100 - l}$ or $\frac{4X}{4X} = \frac{l}{100 - l}$

or $l = 100 - l$ or $2l = 100$ cm

or $l = 50$ cm

50. **(c)** Thermistors ("thermal resistors") are devices having high temperature coefficient of resistivity. They can have either negative temperature coefficient of resistance or positive temperature coefficient of resistance. The positive temperature coefficient types are based on barium titanate. The negative temperature coefficient types are made from a mixture of oxides of iron, nickel and cobalt with small amounts of other substances.

51. **(c)** $Pt = mS\theta$

or $t = \frac{mS\theta}{P} = \frac{1 \times 4180 \times 30}{836}$ s = 150 s

52. **(d)** $E = a\theta + b\theta^2$

$\frac{dE}{d\theta} = a + 2b\theta$

$0 = a + 2b\theta_n$

or $\theta_n = -\frac{a}{2b}$

$= -\frac{1}{2}\left(\frac{a}{b}\right) = -350°C$

But temperature of cold function is given to be 0°C. So, θ_n is not possible.

53. **(a)** $m = ZIt = 3.3 \times 10^{-7} \times 3 \times 2$

$= 19.8 \times 10^{-7}$ kg

54. **(b)** No current is enclosed. Using Ampere's circuital law, B = 0

55. **(b)** $B = \frac{\mu_0 NI}{2r}$

But $l = 2\pi r \times N$

or $r = \frac{l}{2\pi N}$

∴ $B = \frac{\mu_0 N^2 I \times 2\pi}{2l}$

or $B \propto N^2$

This leads us to the right choice.

56. **(a)** $B = \frac{\mu_0 I r^2}{2(r^2 + x^2)^{3/2}}$

$B' = \frac{\mu_0 I}{2r}$

$\frac{B'}{B} = \frac{(r^2 + x^2)^{3/2}}{r^3}$

$\frac{B'}{B} = \left(1 + \frac{x^2}{r^2}\right)^{3/2} = \left(1 + \frac{16}{9}\right)^{3/2}$

$= \left(\frac{25}{9}\right)^{3/2} = \left(\frac{5}{3}\right)^3 = \frac{125}{27}$

or $B' = \frac{125}{27} B = \frac{125}{27} \times 54\ \mu T = 250\ \mu T$

57. **(c)** $F \propto \frac{I_1 I_2}{d}$

$F' \propto -\frac{2I_1 I_2}{3d}$

$\frac{F'}{F} = -\frac{2}{3}$

or $F' = -\frac{2F}{3}$

58. **(a)** Note that the magnet is cut along length and not perpendicular to length.

Magnetic moment of each part = $\frac{M}{3}$ moment of inertia of each part = $\frac{I}{3}$

Magnetic moment of given combination

$= 3 \times \frac{M}{3} = M$

Moment of inertia of combination = $3 \times \frac{I}{3} = I$

$T' = 2\pi \sqrt{\frac{I}{MB}} = T = 2$ s

59. **(b)** Materials used for making electromagnets should have the following properties :

(*i*) low retentivity

(*ii*) high value of saturation magnetisation

(*iii*) low coercivity

(*iv*) small hysteresis loss.

60. **(d)** The voltages across L and C are in opposite phase. So, they cancel out.

61. **(b)** $I = -\frac{n d\phi}{R' dt} = -\frac{n(W_2 - W_1)}{5Rt}$

62. **(b)** $\phi = BA \cos\theta$

$\phi = B\left(\frac{\pi r^2}{2}\right) \cos \omega t$

$$E = -\frac{d}{dt}\phi = \frac{B\pi r^2}{2}\omega \sin \omega t$$

$$P = \frac{E^2}{R}$$

$$P = \frac{B^2\pi^2 r^4 \omega^2}{4R}\sin^2 \omega t$$

But $\langle \sin^2 \omega t \rangle = \frac{1}{2}$

$$\therefore \quad \langle P \rangle = \frac{(B\pi r^2 \omega)^2}{8R}$$

63. (*c*) $\omega_0 = \frac{1}{\sqrt{LC}}$

Now, $L'(2C) = LC$ or $L' = \frac{L}{2}$

64. (*b*) $E = \frac{1}{2}B\omega R^2$

$$= \frac{1}{2} \times 0.2 \times 10^{-4} \times 5 \times 1 \times 1$$

$$= 0.5 \times 10^{-4}\ V = 5 \times 10^{-5}\ V$$

$$= 50 \times 10^{-6}\ V = 50\ \mu V$$

65. (*d*)

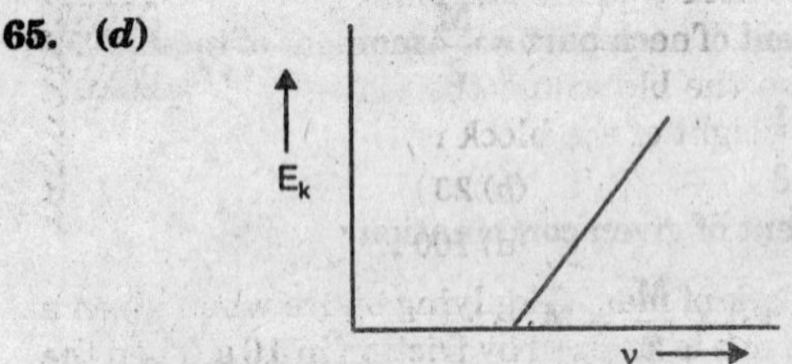

According to Einstein's photoelectric equation, $(K.E)_{max.} = h\nu - \phi_0$. Comparing with the straight line equation $y = mx + c$, we find that the slope of the graph is h. Clearly, the slope is same for all metals and is independent of the intensity of the radiation.

66. (*c*) $\lambda_{max.} = \frac{hc}{\phi_0}$

$$= \frac{6.6 \times 10^{-34} \times 3 \times 10^8}{4 \times 1.6 \times 10^{-19}} = 310\text{ nm}$$

67. (*a*) $qE = mg$

$$q = \frac{mg}{E} = \frac{9.9 \times 10^{-15} \times 10}{3 \times 10^4}\ C$$

$$= 3.3 \times 10^{-18}\ C$$

68. (*b*) Using principle of conservation of momentum, $m_1 v_1 = m_2 v_2$

or $\frac{m_1}{m_2} = \frac{v_2}{v_1}$ or $\frac{m_1}{m_2} = \frac{1}{2}$

or $m_1 = \frac{m_2}{2}$

We know that $R = R_0 A^{1/3}$ where R is the radius of nucleus and A is the mass number.

$$\frac{R_1}{R_2} = \frac{A_1^{1/3}}{A_2^{1/3}} = \frac{m_1^{1/3}}{m_2^{1/3}}$$

$$= \left(\frac{m_1}{m_2}\right)^{1/3} = \left(\frac{m_2}{2m_2}\right)^{1/3}$$

$$= \left(\frac{1}{2}\right)^{1/3} = 1 : 2^{1/3}$$

69. (*c*) Binding energy per nucleon for $_1^2H$ is 1.1.

$\therefore$ Binding energy = 2.2 MeV

Binding energy per nucleon for $_2^4He$ is 7.

$\therefore$ Binding energy = 28 MeV

Energy released = total binding energy of product – total binding energy of reactants

$$= 28 - (2 \times 2.2) = 28 - 4.4$$

$$= 23.6\text{ MeV}$$

70. (*c*) $5\text{MeV} = \frac{1}{4\pi\varepsilon_0}\frac{q_1 q_2}{r}$

$$5 \times 10^6 \times e = 9 \times 10^9 \frac{Z_1 \times Z_2 \rho^2}{r}$$

$$r = \frac{9 \times 10^9 \times 92 \times 2 \times (1.6 \times 10^{-19})^2}{5 \times 10^6}\ m$$

$$= 5.3 \times 10^{-14}\ m = 5.3 \times 10^{-12}\ cm$$

71. (*a*) When *npn* transistor is used, electrons of *n*-type emitter move from base to collector.

72. (*d*) In CE configuration,

$$A_i = \frac{-h_{fe}}{1 + h_{oe} R_L}$$

$$= \frac{-50}{1 + 25 \times 10^{-6} \times 1 \times 10^3} = -48.78$$

73. (*d*) Copper has positive temperature coefficient of resistance but germanium has negative temperature coefficient of resistance.

74 and 75. (*b*) Factual information.

75. (*c*) Factual information.

AIEEE (AET) 2003—Solved

1. The physical quantities *not* having same dimensions are

(*a*) speed and $(\mu_0 \varepsilon_0)^{-1/2}$ (*b*) torque and work

(*c*) momentum and Planck's constant

(*d*) stress and Young's modulus.

2. Three forces start acting simultaneously on a particle moving with velocity, $\vec{v}$. These forces are represented in magnitude and direction by the three sides of a triangle ABC (as shown). The particle will now move with velocity

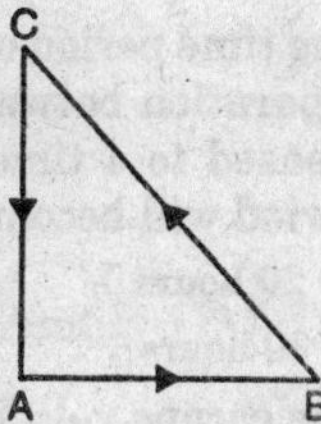

(*a*) $\vec{v}$, remaining unchanged

(*b*) less than $\vec{v}$

(*c*) greater than $\vec{v}$

(*d*) $|v|$ in the direction of the largest force BC.

3. The coordinates of a moving particle at any time '*t*' are given by $x = \alpha t^3$ and $y = \beta t^3$. The speed of the particle at time '*t*' is given by

(*a*) $\sqrt{\alpha^2+\beta^2}$ (*b*) $3t\sqrt{\alpha^2+\beta^2}$

(*c*) $3t^2\sqrt{\alpha^2+\beta^2}$ (*d*) $t^2\sqrt{\alpha^2+\beta^2}$.

4. A car, moving with a speed of 50 km/hr, can be stopped by brakes after at least 6 m. If the same car is moving at a speed of 100 km/hr, the minimum stopping distance is

(*a*) 6 m (*b*) 12 m

(*c*) 18 m (*d*) 24 m.

5. A boy playing on the roof of a 10 m high building throws a ball with a speed of 10 m/s at an angle of 30° with the horizontal. How far from the throwing point will the ball be at the height of 10 m from the ground?

$\left[g = 10 \text{ m/s}^2, \sin 30° = \frac{1}{2}, \cos 30° = \frac{\sqrt{3}}{2}\right]$

(*a*) 8.66 m (*b*) 5.20 m

(*c*) 4.33 m (*d*) 2.60 m.

6. The displacement of a particle varies according to the relation $x = 4(\cos \pi t + \sin \pi t)$. The amplitude of the particle is

(*a*) 8 (*b*) – 4

(*c*) 4 (*d*) $4\sqrt{2}$.

7. Consider the following two statements :

A. Linear momentum of a system of particles is zero.

B. Kinetic energy of a system of particles is zero.

Then

(*a*) A implies B and B implies A

(*b*) A does not imply B and B does not imply A

(*c*) A implies B but B does not imply A

(*d*) A does not imply B but B implies A.

8. A horizontal force of 10 N is necessary to just hold a block stationary against a wall. The coefficient of friction between the block and the wall is 0.2. The weight of the block is

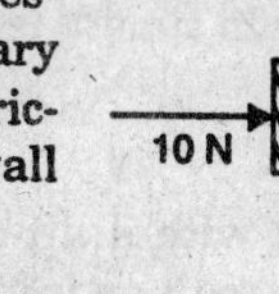

(*a*) 2 N (*b*) 20 N

(*c*) 50 N (*d*) 100 N.

9. A marble block of mass 2 kg lying on ice when given a velocity of 6 m/s is stopped by friction in 10 s. Then the coefficient of friction is

(*a*) 0.01 (*b*) 0.02

(*c*) 0.03 (*d*) 0.06.

10. A light spring balance hangs from the hook of the other light spring balance and a block of mass M kg hangs from the former one. Then the *true* statement about the scale reading is

(*a*) Both the scales read M/2 kg each

(*b*) Both the scales read M kg each

(*c*) The scale of the lower one reads M kg and of the upper one zero

(*d*) The readings of the two scales can be anything but the sum of the readings will be M kg.

11. A block of mass M is pulled along a horizontal frictionless surface by a rope of mass *m*. If a force P is applied at the free end of the rope, the force exerted by the rope on the block is

(a) $\frac{PM}{M+m}$ (b) $\frac{Pm}{M+m}$

(c) $\frac{Pm}{M-m}$ (d) P.

12. A spring balance is attached to the ceiling of a lift. A man hangs his bag on the spring and the spring reads 49 N, when the lift is stationary. If the lift moves downward with an acceleration of 5 m/s^2, the reading of the spring balance will be

(a) 49 N (b) 24 N

(c) 74 N (d) 15 N.

13. When a U^{238} nucleus originally at rest, decays by emitting an alpha particle having a speed 'u', the recoil speed of the residual nucleus is

(a) $-\frac{4u}{238}$ (b) $\frac{4u}{238}$

(c) $-\frac{4u}{234}$ (d) $\frac{4u}{234}$.

14. A body is moved along a straight line by a machine delivering a constant power. The distance moved by the body in time 't' is proportional to

(a) $t^{1/2}$ (b) $t^{3/4}$

(c) $t^{3/2}$ (d) $t^{1/4}$.

15. A rocket with a lift-off mass 3.5×10^4 kg is blasted upwards with an initial acceleration of 10 m/s^2. Then the initial thrust of the blast is nearly

(a) 1.75×10^5 N (b) 3.5×10^5 N

(c) 7.0×10^5 N (d) 14.0×10^5 N.

16. Two spherical bodies of mass M and 5 M and radii R and 2 R respectively are released in free space with initial separation between their centres equal to 12 R. If they attract each other due to gravitational force only, then the distance covered by the smaller body just before collision is

(a) 1.5 R (b) 2.5 R

(c) 4.5 R (d) 7.5 R.

17. A particle performing uniform circular motion has angular momentum L. If its angular frequency is doubled and its kinetic energy halved, then the new angular momentum is

(a) $\frac{L}{2}$ (b) $\frac{L}{4}$

(c) 2 L (d) 4 L.

18. Let $\vec{F}$ be the force acting on a particle having position vector $\vec{r}$, and $\vec{\tau}$ be the torque of this force about the origin. Then

(a) $\vec{r} \cdot \vec{\tau} = 0$ and $\vec{F} \cdot \vec{\tau} = 0$ (b) $\vec{r} \cdot \vec{\tau} = 0$ and $\vec{F} \cdot \vec{\tau} \neq 0$

(c) $\vec{r} \cdot \vec{\tau} \neq 0$ and $\vec{F} \cdot \vec{\tau} = 0$ (d) $\vec{r} \cdot \vec{\tau} \neq 0$ and $\vec{F} \cdot \vec{\tau} \neq 0$.

19. A circular disc X of radius R is made from an iron plate of thickness t, and another disc Y of radius 4 R is made from an iron plate of thickness $\frac{t}{4}$. Then the relation between the moment of inertia I_X and I_Y is

(a) $I_Y = 64\, I_X$ (b) $I_Y = 32\, I_X$

(c) $I_Y = 16\, I_X$ (d) $I_Y = I_X$.

20. The time period of a satellite of earth is 5 hours. If the separation between the earth and the satellite is increased to 4 times the previous value, the new time period will become

(a) 20 hours (b) 10 hours

(c) 80 hours (d) 40 hours.

21. The escape velocity for a body projected vertically upwards from the surface of earth is 11 km/s. If the body is projected at an angle of 45° with the vertical, the escape velocity will be

(a) $\frac{11}{\sqrt{2}}$ km/s (b) $11\sqrt{2}$ km/s

(c) 22 km/s (d) 11 km/s.

22. A mass M is suspended from a spring of negligible mass. The spring is pulled a little and then released so that the mass executes SHM of time period T. If the mass is increased by m, the time period becomes $\frac{5T}{3}$. Then the ratio of $\frac{m}{M}$ is

(a) $\frac{5}{3}$ (b) $\frac{3}{5}$

(c) $\frac{25}{9}$ (d) $\frac{16}{9}$.

23. A spring of spring constant 5×10^3 N/m is stretched initially by 5 cm from the unstretched position. Then the work required to stretch it further by another 5 cm is

(a) 6.25 Nm (b) 12.50 Nm

(c) 18.75 Nm (d) 25.00 Nm.

24. A wire suspended vertically from one of its ends is stretched by attaching a weight of 200 N to the lower end. The weight stretches the wire by 1 mm. Then the elastic energy stored in the wire is

(a) 0.1 J (b) 0.2 J

(c) 10 J (d) 20 J.

25. A body executes simple harmonic motion. The potential energy (P.E.), the kinetic energy (K.E.) and total energy (T.E.) are measured as a function of displacement x. Which of the following statements is *true* ?

(a) P.E. is maximum when $x = 0$

(b) K.E. is maximum when $x = 0$

(c) T.E. is zero when $x = 0$

(d) K.E. is maximum when x is maximum.

26. The length of a simple pendulum executing simple harmonic motion is increased by 21%. The percentage increase in the time period of the pendulum of increased length is

(a) 10% (b) 11%

(c) 21% (d) 42%.

27. Two particles A and B of equal masses are suspended from two massless springs of spring constants k_1 and k_2, respectively. If the maximum velocities, during oscillation, are equal, the ratio of amplitudes of A and B is

(a) $\frac{k_1}{k_2}$ (b) $\sqrt{\frac{k_1}{k_2}}$

(c) $\frac{k_2}{k_1}$ (d) $\sqrt{\frac{k_2}{k_1}}$.

28. A metal wire of linear mass density of 9.8 g/m is stretched with a tension of 10 kg-wt between two rigid supports 1 metre apart. The wire passes at its middle point between the poles of a permanent magnet, and it vibrates in resonance when carrying an alternating current of frequency n. The frequency n of the alternating source is

(a) 25 Hz (b) 50 Hz

(c) 100 Hz (d) 200 Hz.

29. The displacement y of a wave travelling in the x-direction is given by

$$y = 10^{-4} \sin\left(600t - 2x + \frac{\pi}{3}\right) \text{ metres,}$$

where x is expressed in metres and t in seconds. The speed of the wave-motion in ms^{-1} is

(a) 200 (b) 300

(c) 600 (d) 1200.

30. A tuning fork of known frequency 256 Hz makes 5 beats per second with the vibrating string of a piano. The beat frequency decreases to 2 beats per second when the tension in the piano string is slightly increased. The frequency of the piano string before increasing the tension was

(a) 256 + 5 Hz (b) 256 + 2 Hz

(c) 256 – 2 Hz (d) 256 – 5 Hz.

31. A Carnot engine takes 3×10^6 cal of heat from a reservoir at 627°C, and gives it to a sink at 27°C. The work done by the engine is

(a) zero (b) 4.2×10^6 J

(c) 8.4×10^6 J (d) 16.8×10^6 J.

32. "Heat cannot by itself flow from a body at lower temperature to a body at higher temperature" is a statement or consequence of

(a) first law of thermodynamics

(b) second law of thermodynamics

(c) conservation of momentum

(d) conservation of mass.

33. During an adiabatic process, the pressure of a gas is found to be proportional to the cube of its absolute temperature. The ratio C_p/C_v for the gas is

(a) $\frac{3}{2}$ (b) $\frac{4}{3}$

(c) 2 (d) $\frac{5}{3}$.

34. Which of the following parameters does not characterize the thermodynamic state of matter ?

(a) Volume (b) Temperature

(c) Pressure (d) Work.

35. According to Newton's law of cooling, the rate of cooling of a body is proportional to $(\Delta\theta)^n$, where $\Delta\theta$ is the difference of the temperature of the body and the surroundings, and n is equal to

(a) one (b) two

(c) three (d) four.

36. The earth radiates in the infra-red region of the spectrum. The spectrum is correctly given by

(a) Wien's law (b) Rayleigh Jeans law

(c) Planck's law of radiation (d) Stefan's law of radiation.

37. To get three images of a single object, one should have two plane mirrors at an angle of

(a) 30° (b) 60°

(c) 90° (d) 120°.

38. Consider telecommunication through optical fibres. Which of the following statements is *not* true ?

(a) Optical fibres may have homogeneous core with a suitable cladding

(b) Optical fibres can be of graded refractive index

(c) Optical fibres are subject to electromagnetic interference from outside

(d) Optical fibres have extremely low transmission loss.

39. The image formed by an objective of a compound microscope is

(a) virtual and enlarged (b) virtual and diminished

(c) real and diminished (d) real and enlarged.

40. To demonstrate the phenomenon of interference, we require two sources which emit radiation

(*a*) of the same frequency and having a definite phase relationship

(*b*) of nearly the same frequency

(*c*) of the same frequency

(*d*) of different wavelengths.

41. Dimensions of $\frac{1}{\mu_0 \varepsilon_0}$, where symbols have their usual meaning, are

(*a*) $[LT^{-1}]$ (*b*) $[L^{-1}T]$

(*c*) $[L^{-2}T^2]$ (*d*) $[L^2T^{-2}]$.

42. Three charges $-q_1$, $+q_2$ and $-q_3$ are placed as shown in the figure. The *x*-component of the force on $-q_1$ is proportional to

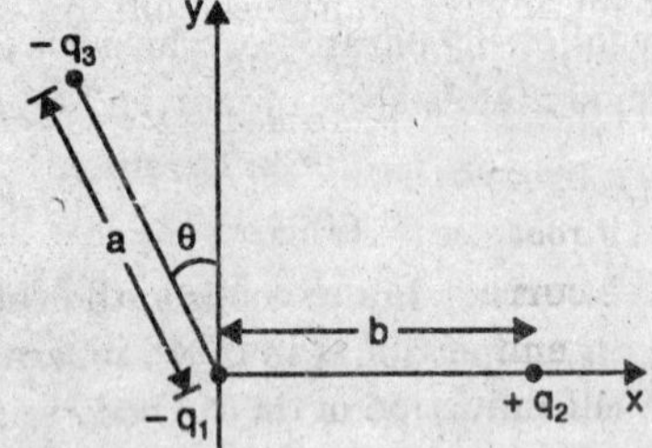

(*a*) $\frac{q_2}{b^2} - \frac{q_3}{a^2} \sin \theta$ (*b*) $\frac{q_2}{b^2} - \frac{q_3}{a^2} \cos \theta$

(*c*) $\frac{q_2}{b^2} + \frac{q_3}{a^2} \sin \theta$ (*d*) $\frac{q_2}{b^2} + \frac{q_3}{a^2} \cos \theta$.

43. A thin spherical conducting shell of radius R has a charge *q*. Another charge Q is placed at the centre of the shell. The electrostatic potential at a point P at a distance $\frac{R}{2}$ from the centre of the shell is

(*a*) $\frac{(q+Q)}{4\pi\varepsilon_0} \frac{2}{R}$ (*b*) $\frac{2Q}{4\pi\varepsilon_0 R}$

(*c*) $\frac{2Q}{4\pi\varepsilon_0 R} - \frac{2q}{4\pi\varepsilon_0 R}$ (*d*) $\frac{2Q}{4\pi\varepsilon_0 R} + \frac{q}{4\pi\varepsilon_0 R}$.

44. If the electric flux entering and leaving an enclosed surface respectively is ϕ_1 and ϕ_2, the electric charge inside the surface will be

(*a*) $(\phi_1 + \phi_2)\,\varepsilon_0$ (*b*) $(\phi_2 - \phi_1)\,\varepsilon_0$

(*c*) $(\phi_1 + \phi_2) / \varepsilon_0$ (*d*) $(\phi_2 - \phi_1) / \varepsilon_0$.

45. The work done in placing a charge of 8×10^{-18} coulomb on a condenser of capacity 100 micro-farad is

(*a*) 32×10^{-32} joule (*b*) 16×10^{-32} joule

(*c*) 31×10^{-26} joule (*d*) 4×10^{-10} joule.

46. A sheet of aluminium foil of negligible thickness is introduced between the plates of a capacitor. The capacitance of the capacitor

(*a*) increases (*b*) decreases

(*c*) remains unchanged (*d*) becomes infinite.

47. A 3 volt battery with negligible internal resistance is connected in a circuit as shown in the figure. The current, I, in the circuit will be

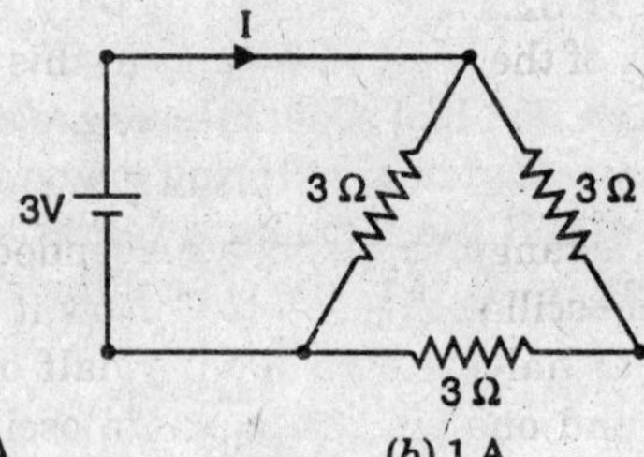

(*a*) 1/3 A (*b*) 1 A

(*c*) 1.5 A (*d*) 2 A.

48. An ammeter reads upto 1 ampere. Its internal resistance is 0.81 ohm. To increase the range to 10 A, the value of the required shunt is

(*a*) 0.09 Ω (*b*) 0.03 Ω

(*c*) 0.3 Ω (*d*) 0.9 Ω.

49. The length of a wire of a potentiometer is 100 cm, and the emf of its standard cell is E volt. It is employed to measure the emf of a battery whose internal resistance is 0.5 Ω. If the balance point is obtained at l = 30 cm from the positive end, emf of the battery is

(*a*) $\frac{30\,E}{100}$ (*b*) $\frac{30\,E}{100.5}$

(*c*) $\frac{30\,E}{(100 - 0.5)}$ (*d*) $\frac{30\,(E - 0.5i)}{100}$.

where *i* is the current in the potentiometer wire.

50. The length of a given cylindrical wire is increased by 100%. Due to the consequent decrease in diameter the change in the resistance of the wire will be

(*a*) 300% (*b*) 200%

(*c*) 100% (*d*) 50%.

51. A strip of copper and another of germanium are cooled from room temperature to 80 K. The resistance of

(*a*) each of these increases

(*b*) each of these decreases

(*c*) copper strip increases and that of germanium decreases

(*d*) copper strip decreases and that of germanium increases.

52. A 220 volt, 1000 watt bulb is connected across a 110 volt mains supply. The power consumed will be

(*a*) 1000 watt (*b*) 750 watt

(*c*) 500 watt (*d*) 250 watt.

53. The thermo emf of a thermo-couple is 25 μV/°C at room temperature. A galvanometer of 40 ohm resistance, capable of detecting current as low as 10^{-5} A, is connected with the thermocouple. The smallest temperature difference that can be detected by this system is

(a) 20° C (b) 16° C

(c) 12° C (d) 8° C.

54. The negative Zn pole of Daniell cell sending a constant current through a circuit, decreases in mass by 0.13 g in 30 minutes. If the electrochemical equivalent of Zn and Cu are 32.5 and 31.5 respectively, the increase the mass of the positive Cu pole in this time is

(a) 0.242 g (b) 0.180 g

(c) 0.141 g (d) 0.126 g.

55. A thin rectangular magnet suspended freely has a period of oscillation equal to T. Now it is broken into two equal halves (each having half of the original length) and one piece is made to oscillate freely in the same field. If its period of oscillation is T′, the ratio $\frac{T'}{T}$ is

(a) $\frac{1}{4}$ (b) $\frac{1}{2\sqrt{2}}$

(c) $\frac{1}{2}$ (d) 2.

56. A particle of mass M and charge Q moving with velocity $\vec{v}$ describes a circular path of radius R when subjected to a uniform transverse magnetic field of induction B. The work done by the field when the particle completes one full circle is

(a) BQ v 2 π R (b) $\left(\frac{Mv^2}{R}\right)2\pi R$

(c) zero (d) BQ 2 π R.

57. A particle of charge – 16 × 10^{-18} coulomb moving with velocity 10 ms^{-1} along the x-axis enters a region where a magnetic field of induction B is along the y-axis, and an electric field of magnitude 10^4 V/m is along the negative z-axis. If the charged particle continues moving along the x-axis, the magnitude of B is

(a) 10^{-3} Wb/m² (b) 10^3 Wb/m²

(c) 10^5 Wb/m² (d) 10^{16} Wb/m².

58. Curie temperature is the temperature above which

(a) a paramagnetic matetrial becomes ferromagnetic

(b) a ferromagnetic material becomes paramagnetic

(c) a paramagnetic material becomes diamagnetic

(d) a ferromagnetic material becomes diamagnetic.

59. A magnetic needle lying parallel to a magnetic field requires W units of work to turn it through 60°. The torque needed to maintain the needle in this position will be

(a) 2 W (b) $\sqrt{3}$ W

(c) W (d) $\frac{\sqrt{3}}{2}$ W.

60. The magnetic lines of force inside a bar magnet

(a) are from south-pole to north-pole of the magnet

(b) are from north-pole to south-pole of the magnet

(c) do not exist

(d) depend upon the area of cross-section of the bar magnet.

61. Two coils are placed close to each other. The mutual inductance of the pair of coils depends upon

(a) the currents in the two coils

(b) the rates at which currents are changing in the two coils

(c) relative position and orientation of the two coils

(d) the material of the wire of the coils.

62. The core of any transformer is laminated so as to

(a) increase the secondary voltage

(b) reduce the energy loss due to eddy currents

(c) make it light weight

(d) make it robust and strong.

63. When the current changes from + 2 A to – 2 A in 0.05 second, an emf of 8 V is induced in a coil. The coefficient of self-induction of the coil is

(a) 0.1 H (b) 0.2 H

(c) 0.4 H (d) 0.8 H.

64. In an oscillating LC circuit, the maximum charge on the capacitor is Q. The charge on the capacitor when the energy is stored equally between the electric and magnetic field is

(a) Q (b) $\frac{Q}{2}$

(c) $\frac{Q}{\sqrt{3}}$ (d) $\frac{Q}{\sqrt{2}}$.

65. Which of the following radiations has the least wavelength ?

(a) X-rays (b) γ-rays

(c) β-rays (d) α-rays.

66. Two identical photocathodes receive light of frequencies f_1 and f_2. If the velocities of the photo electrons (of mass m) coming out are respectively v_1 and v_2, then

(a) $v_1 - v_2 = \left[\frac{2h}{m}(f_1 - f_2)\right]^{1/2}$

(b) $v_1^2 - v_2^2 = \frac{2h}{m}(f_1 - f_2)$

(c) $v_1 + v_2 = \left[\frac{2h}{m}(f_1 + f_2)\right]^{1/2}$

(d) $v_1^2 + v_2^2 = \frac{2h}{m}(f_1 + f_2)$.

67. Which of the following *cannot* be emitted by radioactive substances during their decay ?

(*a*) Electrons (*b*) Protons

(*c*) Neutrinoes (*d*) Helium nuclei.

68. A radioactive sample at any instant has its disintegration rate 5000 disintegrations per minute. After 5 minutes, the rate is 1250 disintegrations per minute. Then, the decay constant (per minute) is

(*a*) 0.8 ln 2 (*b*) 0.4 ln 2

(*c*) 0.2 ln 2 (*d*) 0.1 ln 2.

69. A nucleus with Z = 92 emits the following in a sequence :

$\alpha, \beta^-, \beta^-, \alpha, \alpha, \alpha, \alpha, \alpha, \beta^-, \beta^-, \alpha, \beta^+, \beta^+, \alpha$.

The Z of the resulting nucleus is

(*a*) 74 (*b*) 76

(*c*) 78 (*d*) 82.

70. Which of the following atoms has the lowest ionization potential ?

(*a*) ${}^{16}_{8}O$ (*b*) ${}^{14}_{7}N$

(*c*) ${}^{133}_{55}Cs$ (*d*) ${}^{40}_{18}Ar$.

71. The wavelengths involved in the spectrum of deuterium (${}^{2}_{1}D$) are slightly different from that of hydrogen spectrum, because

(*a*) the attraction between the electron and the nucleus is different in the two cases

(*b*) The size of the two nuclei are different

(*c*) the nuclear forces are different in the two cases

(*d*) the masses of the two nuclei are different.

72. In the nuclear fusion reaction ${}^{2}_{1}H + {}^{3}_{1}H \rightarrow {}^{4}_{2}He + n$, given that the repulsive potential energy between the two nuclei is ~ 7.7×10^{-14} J, the temperature at which the gases must be heated to initiate the reaction is nearly [Boltzmann's constant $k = 1.38 \times 10^{-23}$ J/K]

(*a*) 10^9 K (*b*) 10^7 K

(*c*) 10^5 K (*d*) 10^3 K.

73. If the binding energy of the electron in hydrogen atom is 13.6 eV, the energy required to remove the electron from the first excited state of Li^{++} is

(*a*) 122.4 eV (*b*) 30.6 eV

(*c*) 13.6 eV (*d*) 3.4 eV.

74. The difference in the variation of resistance with temperature in a metal and a semiconductor arises essentially due to the difference in the

(*a*) variation of scattering mechanism with temperature

(*b*) crystal structure

(*c*) variation of the number of charge carriers with temperature

(*d*) type of bonding.

75. In the middle of the depletion layer of a reverse-biased p-n junction, the

(*a*) potential is zero (*b*) electric field is zero

(*c*) potential is maximum (*d*) electric field is maximum.

ANSWERS

1. (c)	2. (a)	3. (c)	4. (d)	5. (a)	6. (d)	7. (d)	8. (a)
9. (d)	10. (b)	11. (a)	12. (b)	13. (c)	14. (c)	15. (c)	16. (d)
17. (b)	18. (a)	19. (a)	20. (d)	21. (d)	22. (d)	23. (c)	24. (a)
25. (b)	26. (a)	27. (d)	28. (b)	29. (b)	30. (d)	31. (c)	32. (b)
33. (b)	34. (d)	35. (a)	36. (a)	37. (c)	38. (c)	39. (d)	40. (a)
41. (c)	42. (c)	43. (d)	44. (b)	45. (a)	46. (c)	47. (c)	48. (a)
49. (a)	50. (a)	51. (d)	52. (d)	53. (b)	54. (d)	55. (c)	56. (c)
57. (b)	58. (b)	59. (b)	60. (a)	61. (d)	62. (b)	63. (a)	64. (d)
65. (b)	66. (b)	67. (b)	68. (b)	69. (c)	70. (c)	71. (d)	72. (a)
73. (b)	74. (c)	75. (a).					

Solutions

1. (*c*) [momentum] = [MLT^{-1}]
[Planck's constant] = [$ML^2 T^{-1}$]. Clearly, momentum and Planck's constant do not have the same dimensions.

2. (*a*)Using triangle law of vectors, we can conclude that if three forces can be represented by the three sides a triangle taken in the same order, then the vect sum of the three forces is zero. So, the acceleration is zero. Thus, the velocity $\vec{v}$ remains unchanged.

3. (*c*) $x = \alpha t^3$

$$v_x = \frac{d}{dt}(x) \quad \text{or} \quad v_x = \frac{d}{dt}(\alpha t^3)$$

or $v_x = 3\alpha t^2$

Again, $y = \beta t^3$

$$v_y = \frac{d}{dt}(y) \quad \text{or} \quad v_y = \frac{d}{dt}(\beta t^3)$$

or $v_y = 3\beta t^2$

Now, $v^2 = v_x^2 + v_y^2$ or $v^2 = (3\alpha t^2)^2 + (3\beta t^2)^2$

or $v = 3t^2\sqrt{\alpha^2 + \beta^2}$.

4. (*d*) $\frac{1}{2}mv^2 = f \times S$ or $S \propto v^2$

Since v is doubled therefore S is quadrupled.

5. (*a*) $R = \frac{v^2 \sin 2\theta}{g}$

$$= \frac{10 \times 10 \sin 60^\circ}{10} = 10 \times \frac{\sqrt{3}}{2} \text{ m}$$

$$= 5 \times \sqrt{3} \text{ m} = 5 \times 1.732 \text{ m} = 8.66 \text{ m}.$$

6. (*d*) $x = 4(\cos \pi t + \sin \pi t)$

or $x = 4\sqrt{2}\left(\frac{1}{\sqrt{2}}\cos \pi t + \frac{1}{\sqrt{2}}\sin \pi t\right)$

or $x = 4\sqrt{2}\left(\sin\frac{\pi}{4}\cos \pi t + \cos\frac{\pi}{4}\sin \pi t\right)$

or $x = 4\sqrt{2}\sin\left(\frac{\pi}{4} + \pi t\right)$

or $x = 4\sqrt{2}\sin\left(\pi t + \frac{\pi}{4}\right)$

Comparing with

$x = A\sin(\omega t + \phi_0)$, we find that $A = 4\sqrt{2}$.

7. (*d*) Consider two particles having equal and opposite linear momenta. So, the net momentum is zero. However, kinetic energy is not zero. If kinetic energy is zero, then momentum is surely zero.

8. (*a*) $W = f = \mu R = 0.2 \times 10 \text{ N} = 2 \text{ N}$.

9. (*d*) $v = u + at$.

$v = u - \mu g t$ [$\because$ $\mu mg = ma$]

$0 = 6 - \mu \times 10 \times 10$

or $100\mu = 6$

or $\mu = \frac{6}{100} = 0.06$.

10. (*b*) The correct choice follows from Newton's third law of motion.

11. (*a*) Acceleration of system $= \frac{P}{M+m}$

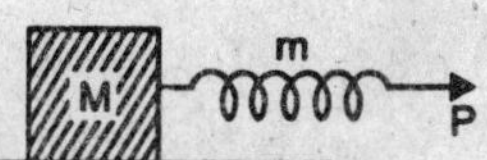

Force exerted by rope on block = Mass of block × Acceleration of block

$$= \frac{MP}{M+m} = \frac{PM}{M+m}.$$

12. (*b*) $mg = 49$

or $m = \frac{49}{9.8}\text{ kg} = \frac{490}{98}\text{ kg} = \frac{10}{2}\text{ kg} = 5\text{ kg}$

Again, $g' = g - a$

$= (9.8 - 5)\text{ ms}^{-2} = 4.8\text{ ms}^{-2}$

Weight $= mg'$

$= 5 \times 4.8 \text{ N} = 24 \text{ N}$.

13. (*c*) The momentum of the system remains the same before and after decay. This is in accordance with the principle of conservation of momentum.

Now, $4u + 234v = 238 \times 0$

or $234v = -4u$

or $v = -\frac{4u}{234}$.

14. (*c*) $P = Fv = mav$

$$P = \frac{mS}{t^2} \times \frac{S}{t}$$

Both P and m are constants.

$\therefore$ $S^2 \propto t^3$ or $S \propto t^{3/2}$

15. (*c*) Initial thrust $= mg + ma = m(g + a)$

$= 3.5 \times 10^4 (10 + 10) = 7.0 \times 10^5$ N.

16. (*d*) Ratio of masses is 1 : 5.

Ratio of accelerations is 5 : 1.

Ratio of distances covered is 5 : 1.

Distance to be covered

$= (12\,R - R - 2\,R) = 9\,R$

Distance moved by smaller sphere

$$= \frac{5}{6} \times 9R = \frac{45R}{6} = 7.5\,R.$$

17. (*b*) $L = I\omega$

But $E_k = \frac{1}{2} I\omega^2$ or $I = \frac{2\,E_k}{\omega^2}$

$\therefore \quad L = \frac{2\,E_k}{\omega^2}\,\omega$ or $L = \frac{2\,E_k}{\omega}$

$$L' = \frac{2\,(E_k/2)}{2\omega}$$

$$L' = \frac{L}{4}.$$

18. (*a*) $\vec{\tau} = \vec{r} \times \vec{F}$

$\vec{r} \perp \vec{\tau}$. So, $\vec{r} \cdot \vec{\tau} = 0$

$\vec{\tau} \perp \vec{F}$. So, $\vec{F} \cdot \vec{\tau} = 0$.

19. (*a*) Moment of inertia of disc about central axis

$$= \frac{1}{2} MR^2 = \frac{1}{2}\,[\pi R^2 t\rho]\,R^2 = \frac{1}{2}\,\pi t\,\rho R^4$$

For the given problem,

$I_X \propto tR^4$

$I_Y \propto \frac{t}{4}(4R)^4$ or $I_Y \propto 64\,tR^4$

None, $\frac{I_Y}{I_X} = 64$ or $I_Y = 64\,I_X$.

20. (*d*) $T^2 \propto r^3$

$T'^2 \propto (4r)^3$

or $\frac{T'^2}{T^2} = 64$ or $\frac{T'}{T} = 8$

or $T' = 8 \times 5$ h $= 40$ h.

21. (*d*) The escape velocity does not depend upon the angle of projection.

22. (*d*) $T = 2\pi\sqrt{\frac{M}{K}}$

$T' = 2\pi\sqrt{\frac{M+m}{K}}$ or $\frac{T'}{T} = \sqrt{\frac{M+m}{M}}$

or $\frac{5T}{3T} = \sqrt{\frac{M+m}{M}}$ or $\frac{5}{3} = \sqrt{\frac{M+m}{M}}$

or $\frac{25}{9} - 1 = \frac{m}{M}$ or $\frac{m}{M} = \frac{16}{9}$.

23. (*c*) $W = \frac{1}{2} \times 5 \times 10^3 \left[\left(\frac{10}{100}\right)^2 - \left(\frac{5}{100}\right)^2\right]$

$$= \frac{5 \times 10^3}{2 \times 10^4}\,[100 - 25] = \frac{5 \times 75}{20}\ \text{J}$$

$= 18.75$ J $= 18.75$ Nm.

24. (*a*) Energy stored

$$= \frac{1}{2}\,F\,\Delta l = \frac{1}{2} \times 200 \times 1 \times 10^{-3}\ \text{J}$$

$= 100 \times 10^{-3}$ J $= 0.1$ J.

25. (*b*) At the mean position, kinetic energy is maximum and potential energy is minimum.

26. (*a*) $T = 2\pi\sqrt{\frac{l}{g}}$

$T \propto \sqrt{l}$

$T' \propto \sqrt{l + \frac{21}{100}\,l}$

$\therefore \quad \frac{T'}{T} = \sqrt{\frac{121}{100}} = \frac{11}{10}$

$$\left(\frac{T'}{T} - 1\right) \times 100 = \left(\frac{11}{10} - 1\right) \times 100 = \frac{1}{10} \times 100 = 10.$$

27. (*d*) Maximum velocity

$$= a\omega = a\sqrt{\frac{k}{m}}$$

If maximum velocities are equal, then

$a\sqrt{\frac{k}{m}}$ = constant or $a\sqrt{k}$ = constant

or $a \propto \frac{1}{\sqrt{k}}$ or $\frac{a_1}{a_2} = \sqrt{\frac{k_2}{k_1}}$

28. (*b*) $n = \frac{1}{2l}\sqrt{\frac{T}{m}}$

$$= \frac{1}{2 \times 1}\sqrt{\frac{10 \times 9.8}{9.8 \times 10^{-3}}} = \frac{100}{2}\ \text{Hz} = 50\ \text{Hz}.$$

29. **(*b*)** Comparing the given equation with
$y = a \sin(\omega t - kx + \phi)$, we get
$\omega = 600\ s^{-1}$ and $k = 2\ m^{-1}$

$$v = \frac{\omega}{k} = \frac{600}{2} = 300\ m\ s^{-1}.$$

30. **(*d*)** The following arguments shall lead us to the right choice.

(*i*) Since the tuning fork is making 5 beats/s with the vibrating string therefore the frequency of the string can be either 251Hz or 261 Hz.

(*ii*) When the tension in the string is increased, the frequency of the string increases.

(*iii*) If the original frequency was 261 Hz, then the number of beats/s should increase and not decrease. So, 261 Hz is rejected.

(*iv*) Obviously, the correct frequency is 251 Hz which must have increased to 254 Hz giving 2 beats/s.

31. **(*c*)** $\frac{W}{Q_1} = 1 - \frac{T_2}{T_1}$

$$\frac{W}{3 \times 10^6 \times 4.2} = 1 - \frac{273 + 27}{273 + 627}.$$

or $$\frac{W}{12.6 \times 10^6} = 1 - \frac{300}{900} = 1 - \frac{1}{3} = \frac{2}{3}$$

or $$W = \frac{2}{3} \times 12.6 \times 10^6\ J = 8.4 \times 10^6\ J.$$

32. **(*b*)** Factual statement. For more information, refer to Comprehensive Physics for class XI.

33. **(*b*)** In an adiabatic process, $\frac{T^\gamma}{P^{\gamma - 1}} = C$

But $P \propto KT^3$ (given)

$\therefore$ $$\frac{T^\gamma}{K^{\gamma - 1} T^{3\gamma - 3}} = C$$

or $$\frac{T^\gamma}{T^{3\gamma - 3}} = CK^{\gamma - 1}$$

or $$T^{-2\gamma + 3} = CK^{\gamma - 1}$$

Using principle of homogeneity of dimensions,

$$-2\gamma + 3 = \gamma - 1$$

or $$-3\gamma = -1 - 3 = -4$$

or $$\gamma = \frac{4}{3} \text{ or } \frac{C_p}{C_v} = \frac{4}{3}.$$

34. **(*d*)** Work gives a relation between two different thermodynamic states. It does not characterise the thermodynamic state of matter.

35. **(*a*)** According to Newton's law of cooling, the rate of cooling of a body is directly proportional to the temperature difference between the body and the surroundings.

36. **(*a*)** Factual information.

37. **(*c*)** $$n = \frac{360}{\theta} - 1$$

$$3 = \frac{360}{\theta} - 1$$

or $$4 = \frac{360}{\theta} \quad \text{or} \quad \theta = 90°.$$

38. **(*c*)** Factual information.

39. **(*d*)** Note that the objective of a compound microscope is a convex lens.

40. **(*a*)** Coherent sources are those sources which have the same frequency and which have a constant phase relationship.

41. **(*c*)** $$c = \frac{1}{\sqrt{\mu_0 \varepsilon_0}}$$

or $$\mu_0 \varepsilon_0 = \frac{1}{c^2} = \frac{1}{(LT^{-1})^2} = L^{-2}T^2.$$

42. **(*c*)**

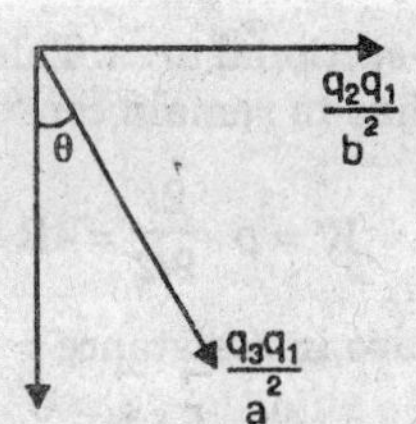

The x-component of $\frac{q_3 q_1}{a^2}$ is $\frac{q_3 q_1}{a^2} \sin\theta$.

Clearly, the x-component of total force is proportional to $\frac{q_2}{b^2} + \frac{q_3}{a^2} \sin\theta$.

43. **(*d*)** Total potential at P = Potential at P due to charge q on the surface of the sphere + Potential at P due to charge Q at the centre of the shell.

$$= \frac{1}{4\pi\varepsilon_0} \frac{q}{R} + \frac{1}{4\pi\varepsilon_0} \frac{Q}{R/2} = \frac{q}{4\pi\varepsilon_0 R} + \frac{2Q}{4\pi\varepsilon_0 R}.$$

44. **(*b*)** Using Gauss's law, $\phi_2 - \phi_1 = \frac{q}{\varepsilon_0}$

or $q = (\phi_2 - \phi_1)\varepsilon_0$.

45. **(*a*)** $$W = \frac{Q^2}{2C}$$

$$= \frac{8 \times 10^{-18} \times 8 \times 10^{-18}}{2 \times 100 \times 10^{-6}}\ J = 32 \times 10^{-32}\ J.$$

46. **(*c*)** $$C = \frac{\varepsilon_0 A}{d - t}$$

In the given problem, $t \to 0$.

47. (*c*) Resistance, $R = \frac{6 \times 3}{6+3}\,\Omega = 2\,\Omega$

$$I = \frac{3}{2}\text{ A} = 1.5\text{ A}.$$

48. (*a*) $S = \frac{GI_g}{I - I_g}$

$$= \frac{0.81 \times 1}{10-1}\,\Omega = \frac{0.81}{9}\,\Omega = 0.09\,\Omega.$$

49. (*a*) Using the principle of potentiometer, $V \propto l$

$$\therefore \quad \frac{V}{E} = \frac{l}{L}$$

or $V = \frac{l}{L} E = \frac{30}{100} E = \frac{30\,E}{100}$.

50. (*a*) New length is $2l$, if the original length is l. Clearly, the new cross-sectional area is $\frac{a}{2}$, if a is the initial cross-sectional area. This is because the volume of the wire has to remain constant.

Now, $R' = \rho \frac{2l}{9/2} = 4R$

Increase in resistance = 4R – R = 3R

Percentage increase in resistance = $\frac{3R}{R} \times 100 = 300$.

51. (*d*) The temperature coefficient of resistance of copper is positive. On the other hand, the temperature coefficient of resistance of germanium is negative.

52. (*d*) $P = \frac{V^2}{R}$. If V is halved, P is reduced by a factor of 4.

So, new power is $\frac{1000}{4}$ W *i.e.* 250 W.

53. (*b*) If θ is the small temperature difference, then

$$E = 25\,\theta \times 10^{-6}\text{ volt}$$

or $IR = 25\,\theta \times 10^{-6}$

or $10^{-5} \times 40 = 25\,\theta \times 10^{-6}$

or $\theta = \frac{10^{-5} \times 40}{25 \times 10^{-6}} = \frac{400}{25}\,°\text{C} = 16°\text{ C}$.

54. (*d*) Using Faraday's law of electrolysis,

$$\frac{m_{Zn}}{m_{Cu}} = \frac{Z_{Zn}}{Z_{Cu}}$$

$$\frac{0.13}{m_{Cu}} = \frac{32.5}{31.5}$$

or $m_{Cu} = \frac{31.5 \times 0.13}{32.5}\text{ g} = 0.126\text{ g}$.

55. (*c*) Moment of inertia = $\frac{\text{mass} \times (\text{length})^2}{12}$

When magnet is divided into two equal halves, mass is reduced by a factor of 2 and length is also reduced by a factor of 2. So, new moment of inertia is $\frac{1}{8}$th of the initial moment of inertia.

Also, magnetic moment = pole strength × length

Pole strength is unchanged and the length is halved. So, new magnetic moment is one-half of the initial magnetic moment.

Now $T' = 2\pi\sqrt{\frac{I'}{M'B}}$

$$= 2\pi\sqrt{\frac{I/8}{\frac{M}{2}B}} = \frac{T}{\sqrt{4}} = \frac{T}{2}$$

$$\frac{T'}{T} = \frac{1}{2}.$$

56. (*c*) Work done by magnetic field is zero.

57. (*b*) $v = \frac{E}{B}$

or $B = \frac{E}{v} = \frac{10^4}{10}\text{ Wb m}^{-2} = 10^3\text{ Wb m}^{-2}$.

58. (*b*) Curie temperature is that temperature above which a ferromagnetic material becomes paramagnetic.

59. (*b*) $W = MB(1 - \cos 60°)$

$$W = \frac{MB}{2} \text{ or } MB = 2\,W$$

Again, $\tau = MB \sin 60°$

$$= MB \times \frac{\sqrt{3}}{2} = 2\,W \times \frac{\sqrt{3}}{2} = \sqrt{3}\,W.$$

60. (*a*) Factual information.

61. (*d*) Factual information.

62. (*b*) Factual information.

63. (*a*) $8 = L\frac{4}{0.05}$

or $L = 2 \times 0.05\text{ H} = 0.1\text{ H}$.

64. (*d*) $\frac{1}{2}\frac{Q^2}{2C} = \frac{q^2}{2C}$ or $q = \frac{Q}{\sqrt{2}}$.

65. (*b*) Factual information.

66. (*b*) $\frac{1}{2} m v_1^2 = hf_1 - \phi_0$

and $\frac{1}{2} m v_2^2 = hf_2 - \phi_0$

Subtracting,

$$\frac{1}{2} m (v_1^2 - v_2^2) = h (f_1 - f_2)]$$

or $$v_1^2 - v_2^2 = \frac{2h}{m} (f_1 - f_2)$$

67. (*b*) Factual information.

68. (*b*) $N = N_0 e^{-\lambda t}$

$$\frac{dN}{dt} = -\lambda N_0 e^{-\lambda t}$$

$$-5000 = -\lambda N_0 e^{-\lambda \times 0}$$

$$-1250 = -\lambda N_0 e^{-\lambda \times 5}$$

or $$e^{-5\lambda} = \frac{1250}{5000}$$

or $$e^{5\lambda} = \frac{5000}{1250} = 4$$

or $$5\lambda = \log_e 4 = 2 \log_e 2$$

or $$\lambda = \frac{2}{5} \log_e 2 = 0.4 \log_e 2.$$

69. (*c*) Decrease in Z due to emission of 8α-particles = 16
Increase in Z due to emission of 4β-particles = 4
Decrease in Z due to emission of 2 positrons = 2

∴ Z of resultant nucleus

$$= 92 - 16 + 4 - 2 = 78.$$

70. (*c*) Largest size of the atom. Electrons in outermost orbit farthest from nucleus. Minimum coulomb force of attraction with the nucleus.

71. (*d*) Factual information.

72. (*a*) To initiate the reaction,

$$\frac{3}{2} kT = 7.7 \times 10^{-14} \quad \text{or} \quad \frac{3}{2} \times 1.38 \times 10^{-23} T = 7.7 \times 10^{-14}$$

or $$T = \frac{7.7 \times 10^{-14} \times 2}{3 \times 1.38 \times 10^{-23}} K = 3.7 \times 10^9 K.$$

73. (*b*) Energy of first excited state of helium

$$= -\frac{3^2 \times 13.6}{2^2} \text{ eV} = -30.6 \text{ eV}$$

Energy required to remove electron

$$= -(-30.6) \text{ eV}$$
$$= 30.6 \text{ eV}.$$

74. (*c*) Factual information.

75. (*a*) Factual information.

AIEEE (AET)2002—Solved

1. If a body looses half of its velocity on penetrating 3 cm in a wooden block, then how much will it penetrate more before coming to rest ?
(*a*) 1 cm (*b*) 2 cm
(*c*) 3 cm (*d*) 4 cm.

2. If two mirrors are kept at 60° to each other, then the number of images formed by them is
(*a*) 5 (*b*) 6
(*c*) 7 (*d*) 8.

3. The inductance between A and D is
(*a*) 3.66 H
(*b*) 9 H
(*c*) 0.66 H
(*d*) 1 H.

4. Tube A has both ends open while tube B has one end closed, otherwise they are identical. The ratio of fundamental frequency of tube A and B is
(*a*) 1 : 2 (*b*) 1 : 4
(*c*) 2 : 1 (*d*) 4 : 1.

5. A charged particle q is placed at the centre O of cube of length L (ABCDEFGH). Another same charge q is placed at a distance L from O. Then the electron flux emerging from q at O is

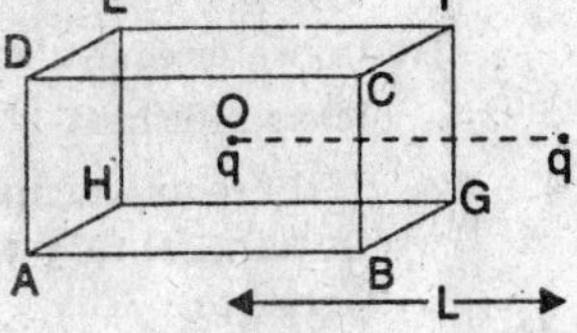

(*a*) q/ε_0 (*b*) zero
(*c*) $q/2\pi\varepsilon_0 L$ (*d*) $q/3\pi\varepsilon_0 L$.

6. A ball whose kinetic energy is E, is projected at an angle of 45° to the horizontal. The kinetic energy of the ball at the highest point of its flight will be
(*a*) E (*b*) $E/\sqrt{2}$
(*c*) E/2 (*d*) zero.

7. If 13.6 eV energy is required to ionize the hydrogen atom, then the energy required to remove an electron from $n = 2$ is
(*a*) 10.2 eV (*b*) 0 eV
(*c*) 3.4 eV (*d*) 6.8 eV.

8. On moving a charge of 20 coulomb by 2 cm, 2 J of work is done, then the potential difference between the points is
(*a*) 0.1 V (*b*) 8 V
(*c*) 2 V (*d*) 0.5 V.

9. If a current is passed through a spring then the spring will
(*a*) expand (*b*) compress
(*c*) remains same (*d*) none of these.

10. A tuning fork arrangement (pair) produces 4 beats/sec with one fork of frequency 288 cps. A little wax is placed on the unknown fork and it then produces 2 beats/sec. The frequency of the unknown fork is
(*a*) 286 cps (*b*) 292 cps
(*c*) 294 cps (*d*) 288 cps.

11. If an ammeter is to be used in place of a voltmeter, then we must connect with the ammeter a
(*a*) low resistance in parallel
(*b*) high resistance in parallel
(*c*) high resistance in series
(*d*) low resistance in series.

12. Wavelength of light used is an optical instrument are $\lambda_1 = 4000$ Å and $\lambda_2 = 5000$ Å, then ratio of their respective resolving powers (corresponding to λ_1 and λ_2) is
(*a*) 16 : 25 (*b*) 9 : 1
(*c*) 4 : 5 (*d*) 5 : 4.

13. A wave $y = a \sin(\omega t - kx)$ on a string meets with another wave producing a node at $x = 0$. Then the equation of the unknown wave is
(*a*) $y = a \sin(\omega t + kx)$ (*b*) $y = -a \sin(\omega t + kx)$
(*c*) $y = a \sin(\omega t - kx)$ (*d*) $y = -a \sin(\omega t - kx)$.

14. From a building two balls A and B are thrown such that A is thrown upwards and B downwards (both vertically with same speed). If v_A and v_B are their respective velocities on reaching the ground, then
(*a*) $v_B > v_A$ (*b*) $v_A = v_B$
(*c*) $v_A > v_B$
(*d*) their velocities depends on their masses.

15. If suddenly the gravitational force of attraction between earth and a satellite revolving around it becomes zero, then the satellite will
(*a*) continue to move in its orbit with same velocity
(*b*) move tangentially to the original orbit in the same velocity
(*c*) become stationary in its orbit
(*d*) move towards the earth.

16. If an electron and a proton having same momenta enter perpendicular to a magnetic field, then
(a) curved path of electron and proton will be same (ignoring the sense of revolution)
(b) they will more undeflected
(c) curved path of electron is more curved than that of proton
(d) path of proton is more curved.

17. Electromagnetic waves are transverse in nature is evident by
(a) polarisation (b) interference
(c) reflection (d) diffraction.

18. Which of the following is used in optical fibres ?
(a) Total internal reflection (b) Scattering
(c) Diffraction (d) Refraction.

19. Which of the following is more close to a black body ?
(a) Black board paint (b) Green leaves
(c) Black holes (d) Red roses.

20. Which of the following are not electromagnetic waves ?
(a) Cosmic-rays (b) γ-rays
(c) β-rays (d) X-rays.

21. If in a circular coil A of radius R, current i is flowing and in another coil B of radius 2R a current $2i$ is flowing, then the ratio of the magnetic fields, B_A and B_B produced by them will be
(a) 1 (b) 2
(c) 1/2 (d) 4.

22. If in the circuit, power dissipation is 150 W, then R is

R
2 Ω
15 V

(a) 2 Ω
(b) 6 Ω
(c) 5 Ω
(d) 4 Ω.

23. Energy required to move a body of mass m from an orbit of radius 2R to 3R is
(a) $GMm/12R^2$ (b) $GMm/3R^2$
(c) $GMm/8R$ (d) $GMm/6R$.

24. If θ_i is the inversion temperature, θ_n is the neutral temperature, θ_c is the temperature of the cold junction then
(a) $\theta_i + \theta_c = \theta_n$ (b) $\theta_i - \theta_c = 2\theta_n$
(c) $\dfrac{\theta_i + \theta_c}{2} = \theta_n$ (d) $\theta_c - \theta_i = 2\theta_n$.

25. Which statement is incorrect ?
(a) All reversible cycles have same efficiency
(b) Reversible cycle has more efficiency than an irreversible one
(c) Carnot cycle is a reversible one
(d) Carnot cycle has the maximum efficiency in all cycles.

26. Two identical particles move towards each other with velocity 2v and v respectively. The velocity of centre of mass is
(a) v (b) v/3
(c) v/2 (d) zero.

27. By increasing the temperature, the specific resistance of a conductor and a semiconductor
(a) increases for both (b) decreases for both
(c) increases, decreases (d) decreases, increases.

28. Cooking gas containers are kept in a lorry moving with uniform speed. The temperature of the gas molecules inside will
(a) increase (b) decrease
(c) remain same
(d) decrease for some, while increase for other.

29. Identify the pair whose dimensions are equal
(a) torque and work (b) stress and energy
(c) force and stress (d) force and work.

30. The kinetic energy needed to project a body of mass m from the earth surface (radius R) to infinity is
(a) $mgR/2$ (b) $2mgR$
(c) mgR (d) $mgR/4$.

31. Heat given to a body which raises its temperature by 1°C is
(a) water equivalent (b) thermal capacity
(c) specific heat (d) temperature gradient.

32. If there are n capacitors, each of capacity C, connected in parallel to V volt source, then the energy stored is equal to
(a) CV (b) $\frac{1}{2}nCV^2$
(c) CV^2 (d) $\frac{1}{2n}CV^2$.

33. When temperature increases, the frequency of a tuning fork
(a) increases (b) decreases
(c) remains same
(d) increases or decreases depending on the material.

34. If a spring has time period T, and is cut into n equal parts, then the time period of each part will be
(a) $T\sqrt{n}$ (b) $T/\sqrt{n}$
(c) nT (d) T.

35. At absolute zero, Si acts as
(a) n-metal (b) metal
(c) insulator (d) none of these.

36. Infrared radiation is detected by
(a) spectrometer (b) pyrometer
(c) nanometer (d) photometer.

37. Length of a string tied to two rigid supports is 40 cm. Maximum length (wavelength in cm) of a stationary wave produced on it is

(a) 20 (b) 80
(c) 40 (d) 120.

38. If N_0 is the original mass of the substance of half-life period $t_{1/2}$ = 5 years, then the amount of substance left after 15 years is

(a) $N_0/8$ (b) $N_0/16$
(c) $N_0/2$ (d) $N_0/4$.

39. The escape velocity of a body depends upon mass as

(a) m^0 (b) m^1
(c) m^2 (d) m^3.

40. If mass-energy equivalence is taken into account, when water is cooled to form ice, the mass of ice obtained is

(a) more (b) unchanged
(c) less (d) may increase or decrease.

41. In a simple harmonic oscillator, at the mean position

(a) kinetic energy is minimum, potential energy is maximum
(b) both kinetic and potential energies are maximum
(c) kinetic energy is maximum, potential energy is minimum
(d) both kinetic and potential energies are minimum.

42. The power factor of an A.C. circuit having resistance (R) and inductance (L) connected in series and an angular velocity ω is

(a) $R/\omega L$ (b) $R/(R^2 + \omega^2L^2)^{1/2}$
(c) $\omega L/R$ (d) $R/(R^2 - \omega^2L^2)^{1/2}$.

43. An astronomical telescope has a large aperture to

(a) reduce spherical aberration
(b) have high resolution
(c) increase span of observation
(d) have low dispersion.

44. A spring of force constant 800 N/m has an extension of 5 cm. The work done in extending it from 5 cm to 15 cm is

(a) 16 J (b) 8 J
(c) 32 J (d) 24 J.

45. The part of a transistor which is most heavily doped to produce large number of majority carriers is

(a) emitter (b) base
(c) collector
(d) can be any of the above three.

46. At what temperature is the r.m.s. velocity of a hydrogen molecule equal to that of an oxygen molecule at 47°C

(a) 80 K (b) – 73 K
(c) 3 K (d) 20 K.

47. A lift is moving down with acceleration a. A man in the lift drops a ball inside the lift. The acceleration of the ball as observed by the man in the lift and a man standing stationary on the ground are respectively

(a) g, g (b) $g - a, g - a$
(c) $g - a, g$ (d) a, g.

48. The energy band gap is maximum in

(a) metals (b) superconductors
(c) insulators (d) semiconductors.

49. A solid sphere, a hollow sphere and a ring are released from top of an inclined plane (frictionless) so that they slide down the plane. Then maximum acceleration down the plane is for (no rolling)

(a) solid sphere (b) hollow sphere
(c) ring (d) all same.

50. Even Carnot engine cannot give 100% efficiency because we cannot

(a) prevent radiation (b) find ideal sources
(c) reach absolute zero temperature
(d) eliminate friction.

51. A light string passing over a smooth light pulley connects two blocks of masses m_1 and m_2 (vertically). If the acceleration of the system is $g/8$, then the ratio of the masses is

(a) 8 : 1 (b) 9 : 7
(c) 4 : 3 (d) 5 : 3.

52. Capacitance (in F) of a spherical conductor with radius 1 m is

(a) 1.1×10^{-10} (b) 10^{-6}
(c) 9×10^{-9} (d) 10^{-3}.

53. Two forces are such that the sum of their magnitudes is 18 N and their resultant is perpendicular to the smaller force. Then the magnitudes of the forces are

(a) 12 N, 6 N (b) 13 N, 5 N
(c) 10 N, 8 N (d) 16 N, 2 N.

54. A cylinder of height 20 m is completely filled with water. The velocity of efflux of water (in ms^{-1}) through a small hole on the side wall of the cylinder near its bottom is

(a) 10 m/s (b) 20 m/s
(c) 25.5 m/s (d) 5 m/s.

55. Initial angular velocity of a circular disc of mass M is ω_1. Then two small spheres of mass m are attached gently to two diametrically opposite points on the edge of the disc. What is the final angular velocity of the disc ?

(a) $\left(\frac{M+m}{M}\right)\omega_1$ (b) $\left(\frac{M+m}{m}\right)\omega_1$
(c) $\left(\frac{M}{M+4m}\right)\omega_1$ (d) $\left(\frac{M}{M+2m}\right)\omega_1$.

56. Moment of inetia of a circular wire of mass M and radius R about is diameter is

(a) $MR^2/2$ (b) MR^2
(c) $2MR^2$ (d) $MR^2/4$.

57. 1 mole of a gas with $\gamma = 7/5$ is mixed with 1 mole of a gas with $\gamma = 5/3$; then the value of γ for the resulting mixture is

(a) 3/2 (b) 2/5
(c) 24/16 (d) 12/7.

58. Two spheres of the same material have radii 1 m and 4 m and temperatures 4000 K and 2000 K respectively. The ratio of the energy radiated per second by the first sphere to that by the second is

(a) 1 : 1 (b) 16 : 1
(c) 4 : 1 (d) 1 : 9.

59. Three identical blocks of masses $m = 2$ kg are drawn by a force F = 10.2 N on a frictionless surface, then what is the tension (in N) in the string between the blocks B and C ?

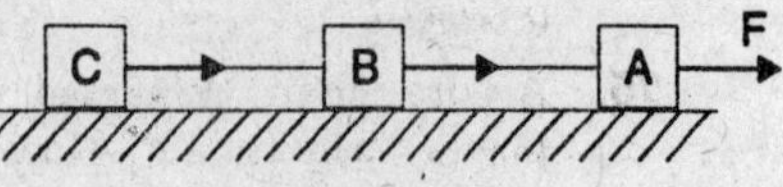

(a) 9.2 (b) 8
(c) 3.4 (d) 9.8.

60. When forces F_1, F_2, F_3 are acting on a particle of mass m such that F_2 and F_3 are mutually perpendicular, then the particle remains stationary. If the force F_1 is now removed then the acceleration of the particle is

(a) F_1/m (b) F_2F_3/mF_1
(c) $(F_2 - F_3)/m$ (d) F_2/m.

61. A child swinging on a swing in sitting position, stands up, then the time period of the swing will

(a) increase (b) decrease
(c) remain same
(d) increases if the child is long and decreases if the child is short.

62. One end of massless rope, which passes over a massless and frictionless pulley P is tied to a hook C while the other end is free. Maximum tension that the rope can bear is 360 N. With what value of maximum safe acceleration (in ms^{-2}) can a man of 60 kg climb down the rope ?

(a) 16 (b) 6
(c) 4 (d) 8.

63. Speeds of two identical cars are u and $4u$ at a specific instant. The ratio of the respective distances in which the two cars are stopped from that instant is

(a) 1 : 1 (b) 1 : 4
(c) 1 : 8 (d) 1 : 16.

64. The mass of a product liberated on anode in an electrochemical cell depends on

(a) $(It)^{1/2}$ (b) It
(c) I/t (d) I^2t.

(where t is the time period for which the current is passed)

65. The minimum velocity (in ms^{-1}) with which a car driver must traverse a flat curve of radius 150 m and coefficient of friction 0.6 to avoid skidding is

(a) 60 (b) 30
(c) 15 (d) 25.

66. A particle of mass m moves along line PC with velocity v as shown. What is the angular momentum of the particle about P ?

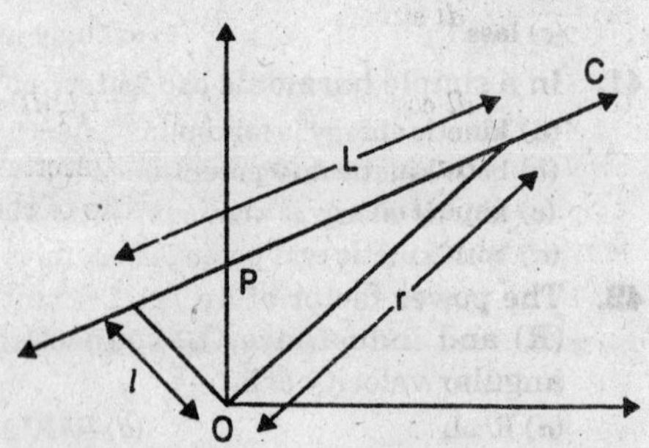

(a) mvL (b) mvl
(c) mvr (d) zero.

67. At a specific instant emission of radioactive compound is deflected in a magnetic field. The compound can emit

(i) electrons (ii) protons
(iii) He^{2+} (iv) neutrons.

The emission at the instant can be

(a) (i), (ii) (iii) (b) (i), (ii), (iii), (iv)
(c) (iv) (d) (ii), (iii).

68. A conducting square loop of side L and resistance R moves in its place with a uniform velocity v perpendicular to one of its sides. A magnetic induction B constant in time and space, pointing perpendicular and into the plane at the loop exists everywhere with half the loop outside the field, as shown in figure. The induced e.m.f. is

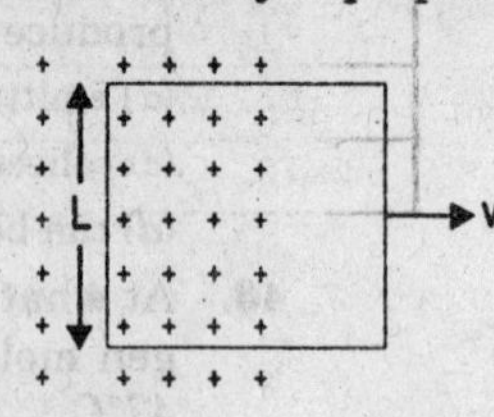

(a) zero (b) RvB
(c) vBl/R (d) vBl.

69. If a charge q is placed at the centre of the line joining two equal charges Q such that the system is in equilibrium then the value of q is

(a) Q/2 (b) – Q/2

(c) Q/4 (d) – Q/4.

70. Wires 1 and 2 carrying currents i_1 and i_2 respectively are inclined at an angle θ to each other. What is the force on a small element dl of wire 2 at a distance of r from wire 1 (as shown in figure) due to the magnetic field of wire 1 ?

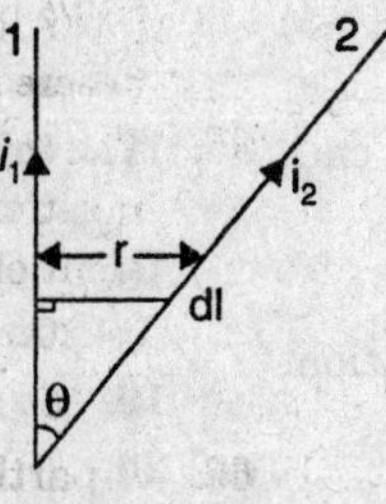

(a) $\frac{\mu_0}{2\pi r} i_1 i_2 \, dl \tan \theta$

(b) $\frac{\mu_0}{2\pi r} i_1 i_2 \, dl \sin \theta$

(c) $\frac{\mu_0}{2\pi r} i_1 i_2 \, dl \cos \theta$ (d) $\frac{\mu_0}{4\pi r} i_1 i_2 \, dl \sin \theta$.

71. Sodium and copper have work functions 2.3 eV and 4.5 eV respectively. Then the ratio of the wavelengths is nearest to

(a) 1 : 2 (b) 4 : 1

(c) 2 : 1 (d) 1 : 4.

72. Formation of covalent bonds in compounds exhibits

(a) wave nature of electron

(b) particle nature of electron

(c) both wave and particle nature of electron

(d) none of these.

73. In a transformer, number of turns in the primary are 140 and that in the secondary are 280. If current in primary is 4 A, then that in the secondary is

(a) 4 A (b) 2 A

(c) 6 A (d) 10 A.

74. The time period of a charged particle undergoing a circular motion in a uniform magnetic field is independent of its

(a) speed (b) mass

(c) charge (d) magnetic induction.

75. A wire when connected to 220 V mains supply has power dissipation P_1. Now the wire is cut into two equal pieces which are connected in parallel to the same supply. Power dissipation in this case is P. Then $P_2 : P_1$ is

(a) 1 (b) 4

(c) 2 (d) 3.

ANSWERS

1. (a)	2. (a)	3. (d)	4. (c)	5. (a)	6. (c)	7. (c)	8. (a)
9. (b)	10. (b)	11. (c)	12. (d)	13. (b)	14. (b)	15. (b)	16. (a)
17. (a)	18. (a)	19. (a)	20. (c)	21. (a)	22. (b)	23. (d)	24. (c)
25. (a)	26. (c)	27. (c)	28. (c)	29. (a)	30. (c)	31. (b)	32. (b)
33. (b)	34. (b)	35. (c)	36. (b)	37. (b)	38. (a)	39. (a)	40. (c)
41. (c)	42. (b)	43. (b)	44. (b)	45. (a)	46. (d)	47. (c)	48. (c)
49. (a)	50. (c)	51. (b)	52. (a)	53. (b)	54. (b)	55. (c)	56. (a)
57. (a)	58. (a)	59. (c)	60. (a)	61. (b)	62. (c)	63. (d)	64. (b)
65. (b)	66. (d)	67. (a)	68. (d)	69. (d)	70. (c)	71. (c)	72. (c)
73. (b)	74. (a)	75. (b).					

Solutions

1. (*a*) Loss of kinetic energy $= \left(1-\frac{1}{4}\right) E = \frac{3}{4} E$

Now, $\frac{3}{4}E = f \times 3$

For complete loss, $E = f \times S$ or $\frac{S}{3} = \frac{4}{3}$ or S = 4 cm.

2. (*a*) $n = \frac{360}{60} - 1 = 5$.

3. (*d*) The given arrangement is a parallel combination of inductances.

$\therefore \quad \frac{1}{L} = \frac{1}{3} + \frac{1}{3} + \frac{1}{3} = \frac{3}{3} = 1$

or $\quad L = 1$ henry.

4. (*c*) $\nu_A = \frac{v}{2l}$; $\nu_B = \frac{v}{4l}$

$\frac{\nu_A}{\nu_B} = \frac{v}{2l} \times \frac{4l}{v} = \frac{2}{1}$.

5. (*a*) Use Gauss's law.

6. (*c*) At the highest point. velocity is $v \cos \theta$. Kinetic energy is $\frac{1}{2} mv^2 \cos^2\theta$ or $E \cos^2\theta$, where E is the initial kinetic energy.
When $\theta = 45°$, kinetic energy at the highest point is $E \cos^2 45°$ or $\frac{E}{2}$.

7. (*c*) $E_2 = \frac{13.6}{4}$ eV = 3.4 eV.

8. (*a*) $V = \frac{2\text{ J}}{20\text{ C}} = 0.1$ V.

9. (*b*) Like currents attract.

10. (*b*) From the data of 4 beats/s, we find that ν is either 292 cps or 284 cps.
When the wax is attached, the number of beats/s is reduced.
Clearly, the correct frequency is 292 cps.

11. (*c*) Voltmeter is a high resistance instrument.

12. (*d*) Resolving power $\propto \frac{1}{\lambda}$.

13. (*b*) Following two arguments lead us to the right choice.
(*i*) If one wave is travelling from right to left, then the other wave is travelling from left to right. So, $\omega t - kx$ is replaced by $\omega t + kx$.
(*ii*) The sign of amplitude changes. So, '*a*' becomes '$-a$'.

14. (*b*) $v^2 - u^2 = 2gh$
Clearly, the sign of initial velocity is not important.

15. (*b*) Use newton's first law of motion.

16. (*a*) $Bqv = \frac{mv^2}{r}$ or $r = \frac{mv}{Bq}$ or $r = \frac{p}{Bq}$
All quantities on right hand side remain unchanged. So, r remains unchanged.

17. (*a*) Polarisation is a property of transverse wave motion.

18. (*a*) The working of optical fibres is based on total internal reflection.

19. (*a*) Based on simple facts.

20. (*c*) β-rays are made up of electrons.

21. (*a*) $B = \frac{\mu_0 I}{2R}$

In the given problem, $\frac{I}{R}$ is unchanged.

22. (*b*) $150 = \frac{15 \times 15}{R}$ or $R = \frac{15 \times 15}{150}\ \Omega$

or $R = \frac{15}{10}\ \Omega = \frac{3}{2}\ \Omega$ Now $\frac{2R}{2+R} = \frac{3}{2}$

$4R - 3R = 6$ or $R = 6\ \Omega$.

23. (*d*) $W = \left(-\frac{GMm}{3R}\right) - \left(-\frac{GMm}{2R}\right)$

$= \frac{GMm}{2R} - \frac{GMm}{3R} = \frac{GMm}{6R}$.

24. (*c*) Neutral temperature is the arithmetic mean of temperature of cold junction and temperature of inversion.

25. (*a*) Based on simple facts.

26. (*c*) $(2m) V = m \times 2v - m \times v$

or $2 mV = 2 mv - mv$ or $V = \frac{mv}{2m} = \frac{v}{2}$.

27. (*c*) Metallic conductors have positive value of temperature coefficient of resistance. On the other hand, semiconductors have negative value of temperature coefficient of resistance.

28. (*c*) The velocity of centre of mass of container has no effect on the temperature of the gas in the container.

29. (*a*) Both torque and work have the same dimensional formula $[ML^2 T^{-2}]$.

30. (*c*) $\frac{1}{2} m v_e^2 = \frac{1}{2} m \frac{2GM}{R} = \frac{GMm}{R} = \frac{mgR^2}{R} = mgR$.

31. (*b*) $Q = mS\theta$
If $\theta = 1$, then $Q = mS$ = thermal capacity.

32. (*b*) When n capacitors, each of capacity C, are connected in parallel, then the combined capacity is nC. So, the energy stored is $\frac{1}{2}(nC) V^2$.

33. (*b*) With increase in temperature, l increases.

Again, $\quad v = \frac{m^2 k}{4\sqrt{3}\,\pi l^2}\sqrt{\frac{Y}{\rho}} \quad$ or $\quad v \propto \frac{1}{l^2}$

Due to increase in l, v decreases.

34. (*b*) $\quad K' = nK$

$$T' = 2\pi\sqrt{\frac{m}{K'}} \quad \text{or} \quad T' = 2\pi\sqrt{\frac{m}{nK}} \quad \text{or} \quad T' = \frac{T}{\sqrt{n}}.$$

35. (*c*) No charge carriers are available. All the covalent bonds are stable.

36. (*b*) Pyrometer is a device used for the detection of infrared radiation.

37. (*b*) $\frac{\lambda}{2} = 40$ cm.

38. (*a*) $N = \frac{N_0}{2^{\frac{15}{5}}}$ or $N = \frac{N_0}{2^3} = \frac{N_0}{8}$.

39. (*a*) Escape velocity does not depend upon mass.

40. (*c*) If we assume that the energy lost by water is due to conversion of mass, then the mass if ice obtained is less than the mass of water.

41. (*c*) Based on facts.

42. (*b*) Power factor, $\cos\phi = \frac{R}{Z} = \frac{R}{\sqrt{R^2 + \omega^2 L^2}}$.

43. (*b*) Larger the aparture of the telescope, more is the resolving power.

44. (*b*) $W = \frac{1}{2}k[3\xi]^2 - \frac{1}{2}k\xi^2 = 8 \times \frac{1}{2}k\xi^2 = 4\,k\xi^2$

$$= 4 \times 800 \times 0.05 \times 0.05 \text{ J} = 8 \text{ J}.$$

45. (*a*) Based on facts.

46. (*d*) We know that $\quad C = \sqrt{\frac{3RT}{M}}$

In the given problem, C and R constants.

$\therefore \quad \frac{T}{M} = \text{constant}$

$$\frac{T}{2} = \frac{273 + 47}{32} \quad \text{or} \quad \frac{T}{2} = \frac{320}{32} = 10 \quad \text{or} \quad T = 20 \text{ K}.$$

47. (*c*) Note that the man inside the lift is in non-inertial frame. On the other hand, the man standing stationary on the ground is in inertial frame.

48. (*c*) Based on facts.

49. (*a*) $$a = \frac{g\sin\theta}{1 + \frac{K^2}{R^2}}$$

For solid sphere, $\quad MK^2 = \frac{2}{5}MR^2 \quad$ or $\quad \frac{K^2}{R^2} = \frac{2}{5}$

$$\therefore \quad a = \frac{g\sin\theta}{1 + \frac{2}{5}} = \frac{5}{7}g\sin\theta = 0.71\,g\sin\theta$$

For hollow sphere, $\quad MK^2 = \frac{2}{3}MR^2 \quad$ or $\quad \frac{K^2}{R^2} = \frac{2}{3}$

$$\therefore \quad a = \frac{g\sin\theta}{1 + \frac{2}{3}} = \frac{3}{5}g\sin\theta = 0.6\,g\sin\theta$$

For ring, $\quad MK^2 = MR^2 \quad$ or $\quad \frac{K^2}{R^2} = 1$

$$\therefore \quad a = \frac{g\sin\theta}{1 + 1} = \frac{1}{2}g\sin\theta = 0.5\,g\sin\theta$$

Clearly, solid sphere has maximum acceleration.

50. (*c*) $$\eta = 1 - \frac{T_2}{T_1} = 1 - \frac{Q_2}{Q_1}$$

For 100% efficiency, $\quad T_2 = 0, Q_2 = 0$

So, we need a sink at absolute zero of temperature. In this situation, the whole of the heat will be converted into work. But we cannot reach absolute zero of temperature. So, 100% efficiency is not possible.

51. (*b*) $$a = \frac{m_1 - m_2}{m_1 + m_2}g$$

Let $\quad \frac{m_1}{m_2} = x \quad$ or $\quad m_1 = m_2 x$

$$\therefore \quad a = \frac{m_2 x - m_2}{m_2 x + m_2}g \quad \text{or} \quad a = \frac{x - 1}{x + 1}g$$

But $\quad a = \frac{g}{8} \quad \therefore \quad \frac{g}{8} = \frac{x - 1}{x + 1}g$

or $\quad \frac{1}{8} = \frac{x - 1}{x + 1} \quad$ or $\quad x + 1 = 8x - 8$

or $\quad 9 = 7x \quad$ or $\quad x = \frac{9}{7}$.

52. (*a*) $C = 4\pi\epsilon_0 R$ or $C = \frac{1}{9 \times 10^9}$ F $= 0.11 \times 10^{-9}$ F

$$= 1.1 \times 10^{-10} \text{ F}.$$

53. (*b*) $\quad P + Q = 18$

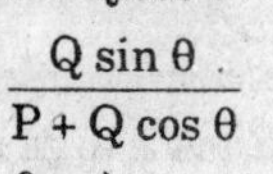

$$\tan 90° = \frac{Q\sin\theta}{P + Q\cos\theta}$$

$$\infty = \frac{Q\sin\theta}{P + Q\cos\theta}$$

$\Rightarrow \quad P + Q\cos\theta = 0$

Again, $\quad 12^2 = P^2 + Q^2 + 2PQ\cos\theta$

Also, $\quad 18^2 = P^2 + Q^2 + 2PQ$

Simplifying, we get $\quad P = 5N$ and $Q = 13$ N.

54. (*b*) $v = \sqrt{2gh} = \sqrt{2 \times 10 \times 20}$ m s^{-1} = 20 m s^{-1}.

55. (*c*) Applying principle of conservation of angular momentum,

$$\left[\frac{1}{2}MR^2\right]\omega_1 = \left[\frac{1}{2}MR^2 + mR^2 + mR^2\right]\omega$$

or $$\omega = \frac{M}{M + 4m}\omega_1.$$

56. (*a*) Factual information.

57. (*a*) $\frac{n_1 + n_2}{r - 1} = \frac{n_1}{r_1 - 1} + \frac{n_2}{r_2 - 1}$ or $\frac{1+1}{r-1} = \frac{1}{\frac{5}{3} - 1} + \frac{1}{\frac{7}{5} - 1}$

$\frac{2}{r-1} = \frac{3}{2} + \frac{5}{2} = 4$ or $r - 1 = \frac{2}{4} = \frac{1}{2}$

or $r = 1 + \frac{1}{2} = \frac{3}{2}$.

58. (*a*) $\frac{E_1}{E_2} = \frac{A_1 T_1^4}{A_2 T_2^4} = \frac{R_1^2 T_1^4}{R_2^2 T_2^4} = \frac{1^2 \times 4000^4}{4^2 \times 2000^4} = \frac{16}{16} = 1.$

59. (*c*) $a = \frac{10.2}{3 \times 2}$ m s^{-2} = $\frac{10.2}{6}$ m s^{-2} = 1.7 m s^{-2}

T = 2 × 1.7 N = 3.4 N.

60. (*a*) F_1 is equilibrant force.

61. (*b*) When the child stands up, the 'effective length of simple pendulum' decreases.

$$T = 2\pi\sqrt{\frac{l}{g}}$$

Clearly, T decreases.

62. (*c*) $mg - T = ma$ or $600 - 360 = 60\,a$

or $a = \frac{240}{60}$ m s^{-2} = 4 m s^{-2}.

63. (*d*) $f \times S = \frac{1}{2} mv^2$ or $S \propto v^2$ or $\frac{S_1}{S_2} = \left(\frac{1}{4}\right)^2 = \frac{1}{16}$.

64. (*b*) Refer to Faraday's first law of electrolysis.

65. (*b*) $v = \sqrt{\mu_s rg} = \sqrt{0.6 \times 150 \times 10} = \sqrt{900}$ m s^{-1}

= 30 m s^{-1}.

66. (*d*) The line of action passes through the observation point. So, angular momentum is zero.

67. (*a*) Charged particles can be deflected by magnetic field.

68. (*d*) Refer to Comprehensive Physics for Class XII.

Q q Q

r r

2r

69. (*d*) Considering the equilibrium of corner charge,

$$\frac{qQ}{r^2} = -\frac{Q^2}{4r^2} \quad \text{or} \quad q = -\frac{Q}{4}.$$

70. (*c*) $B = \frac{\mu_0 i_1}{2\pi r}$

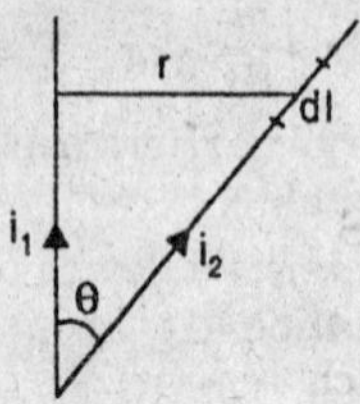

$$F = BIl \cos\theta = \frac{\mu_0 i_1}{2\pi r} i_2\, dl \cos\theta.$$

71. (*c*) $\phi_0 = \frac{hc}{\lambda}$ ∴ $\frac{\lambda_1}{\lambda_2} = \frac{4.5}{2.3} \approx \frac{2}{1}$

72. (*c*) Factual information.

73. (*b*) $\frac{n_s}{n_p} = \frac{280}{140} = 2$

$\frac{I_s}{I_p} = \frac{n_p}{n_s} = \frac{1}{2}$ or $I_s = \frac{I_p}{2}$ A = $\frac{4\,A}{2}$ = 2 A.

74. (*a*) $T = \frac{2\pi m}{Bq}$

Clearly, T is independent of speed.

75. (*b*) $P = \frac{V^2}{R}$. R is reduced by a factor of 4. So, P is increased by a factor of 4.

1.1 UNITS, MEASUREMENTS AND DIMENSIONS

Important Terms, Facts and Formulae

BASIC CONCEPTS

1. The magnitude of a physical quantity is the product of unit (u) in which the quantity is measured and the number (n) of times that unit is contained in the given quantity.
2. Units are classified into two categories—basic or fundamental units and derived units.
3. *The basic or fundamental units are the units of fundamental quantities.* These units are so named because they can neither be derived from one another nor can be further resolved into other more simpler units.
4. *Derived units are those units which are derived from the basic units.* As an example, the unit of speed is cm s^{-1} and the unit of acceleration is cm s^{-2}. These are derived units.
5. Following are the *seven fundamental units* of 'International System of Units':
metre, kilogram, second, ampere, kelvin, candela, mole.
6. In addition to seven fundamental units, there are following **two supplementary units :**
radian, steradian.
7. *Coherent system of units is that system of units which is based upon a set of fundamental units from which all other units can be derived by either multiplying or dividing them without introducing numerical factors.*
8. A system of units which uses only one unit for a base quantity is called rationalised system of units. Examples : MKS and SI.
9. ☞ ***The dimensional formula of a physical quantity does not depend upon the system of units used to represent that quantity.***
10. The value (magnitude) of a physical quantity remains the same in all systems of measurement. However, the numerical value changes.
In general, $n_1u_1 = n_2u_2 = n_3u_3 = ...$.

IMPORTANT FORMULAE

1. **Formula for plane angle**
The angle subtended at the centre of a circle of radius r by an arc of length l is given by
$$\theta = \frac{l}{r} \text{ radian}$$
2. **Formula for solid angle**
The solid angle subtended at the centre of a sphere of radius r by an area ΔS on the surface of the sphere is given by
$$\Delta\Omega = \frac{\Delta S}{r^2} \text{ steradian}$$
3. **Formula for conversion from one system of units to another**
Consider a physical quantity whose dimensional formula is $[M^aL^bT^c]$. Let n_1 be its numerical value in a system of fundamental units M_1, L_1 and T_1. Then the magnitude of the physical quantity in this system is $n_1[M_1{}^aL_1{}^bT_1{}^c]$.
Let n_2 be the numerical value in another system of fundamental units M_2, L_2 and T_2. The magnitude of the physical quantity in this system is $n_2[M_2{}^aL_2{}^bT_2{}^c]$.
Since the value of the physical quantity is the same in all systems,
$$\therefore \quad n_2[M_2{}^aL_2{}^bT_2{}^c] = n_1[M_1{}^aL_1{}^bT_1{}^c]$$
$$n_2 = n_1\left[\frac{M_1}{M_2}\right]^a\left[\frac{L_1}{L_2}\right]^b\left[\frac{T_1}{T_2}\right]^c$$
This equation is to be used only after expressing the given quantity in absolute units.

☞ **Do not use this formula**
if the given quantity is in gravitational units.

IMPORTANT POINTS TO REMEMBER

1. All of the following have the same dimensional formula $[M^0L^0T^{-1}]$.
   Frequency, angular frequency, angular velocity and velocity gradient.
2. All of the following are dimensionless.
   Angle, Solid angle, T-ratios, Strain, Poisson's ratio, Relative density, Relative permittivity, Refractive index and Relative permeability.
3. Following three quantities have the same dimensional formula $[M^0L^2T^{-2}]$.
   Square of velocity, gravitational potential, latent heat.
4. Following quantities have the same dimensional formula $[ML^2T^{-2}]$.
   Work, energy, torque, heat.
5. Following have the same dimensional formula $[MLT^{-1}]$.
   Momentum, impulse.
6. Both acceleration and gravitational field intensity have the same dimensional formula $[M^0LT^{-2}]$.
7. Force, weight, thrust and energy gradient have the same dimensional formula $[MLT^{-2}]$.
8. Entropy, gas constant, Boltzmann constant and thermal capacity have the same dimensions in mass, length and time.
9. Light year, radius of gyration and wavelength have the same dimensional formula $[M^0LT^0]$.
10. Surface tension and spring constant have the same dimensional formula $[ML^0T^{-2}]$.
11. Planck's constant and angular momentum have the same dimensional formula $[ML^2T^{-1}]$.
12. Rydberg constant and propagation constant have the same dimensional formula $[M^0L^{-1}T^0]$.
13. Following have the same dimensional formula $[ML^{-1}T^{-2}]$.
    Pressure, Stress, Modulii of elasticity, Energy density.

QUESTION BANK

Level I

Choose the correct answer from the four alternatives given in each of the following questions :

1. The density of water is equal to

(*a*) 10^{-3} kg m^{-3} (*b*) 10 kg m^{-3}
(*c*) 10^3 kg m^{-3} (*d*) 10^6 kg m^{-3}.

2. One atmospheric pressure is equal to

(*a*) 10^8 N m^{-2} (*b*) 10^7 N m^{-2}
(*c*) 10^6 N m^{-2} (*d*) 10^5 N m^{-2}.

3. If C is the capacity and R is the resistance, then the dimensional formula of $\frac{1}{CR}$ is

(*a*) $[M^0L^0T^2]$ (*b*) $[M^0L^0T^{-1}]$
(*c*) $[M^0L^1T^{-1}]$ (*d*) $[ML^2T^{-1}A^{-1}]$.

4. Which of the following sets cannot enter into the list of fundamental quantities in any system of units ?

(*a*) length, time and mass (*b*) mass, time and velocity
(*c*) length, time and velocity (*d*) length, mass and velocity.

5. Which of the following is a dimensional constant ?

(*a*) refractive index (*b*) dielectric constant
(*c*) relative density (*d*) gravitational constant.

6. If the size of a unit be represented by *k* and its numerical value as *n*, then

(*a*) $n \propto k$ (*b*) $n \propto \frac{1}{k}$
(*c*) $n \propto k^2$ (*d*) $n \propto \frac{1}{k^2}$.

7. The SI unit of electrochemical equivalent is

(*a*) kg C (*b*) C kg^{-1}
(*c*) kg C^{-1} (*d*) kg^2 C^{-1}.

8. With the usual notations, the equation $\tan\theta = \frac{rg}{v^2}$ said to give the angle of banking θ is

(*a*) dimensionally correct only
(*b*) numerically correct only
(*c*) both numerically and dimensionally correct
(*d*) neither numerically nor dimensionally correct.

9. What is the power of a 100 W bulb in cgs units ?

(*a*) 10^6 erg/s (*b*) 10^7 erg/s
(*c*) 10^9 erg/s (*d*) 10^{11} erg/s.

10. The dimensional formula for the modulus of rigidity is

(*a*) $[ML^2T^{-2}]$ (*b*) $[ML^{-1}T^{-3}]$
(*c*) $[ML^{-2}T^{-2}]$ (*d*) $[ML^{-1}T^{-2}]$.

11. The expression $[ML^{-1}T^{-1}]$ represents

(*a*) momentum (*b*) force
(*c*) pressure (*d*) coefficient of viscosity.

12. The unit of momentum is

(*a*) N s (*b*) N s^{-1}
(*c*) N m (*d*) N m^{-1}.

13. Density of liquid is 15.7 g cm^{-3}. Its value in the International System of Units is

(*a*) 15.7 kg m^{-3} (*b*) 157 kg m^{-3}
(*c*) 1570 kg m^{-3} (*d*) 15700 kg m^{-3}.

14. N kg^{-1} is the unit of

(*a*) velocity (*b*) acceleration
(*c*) momentum (*d*) angular momentum.

15. A weber is equivalent to

(*a*) A m^{-2} (*b*) A m^{-1}
(*c*) A m^2 (*d*) T m^2.

16. A sextant is used to measure

(*a*) area of hill (*b*) height of an object
(*c*) breadth of a tower (*d*) volume of the building.

17. Which of the following quantities has unit newton per metre (N m^{-1}) ?

(*a*) force (*b*) power
(*c*) energy (*d*) surface tension.

18. Which of the following is dimensional constant ?

(*a*) refractive index (*b*) relative density
(*c*) Poisson's ratio (*d*) gravitational constant.

19. Order of magnitude of human life is

(*a*) 10^{12} s (*b*) 10^9 s
(*c*) 10^8 s (*d*) 10^6 s.

20. What is the dimensional formula of coefficient of linear expansion ?

(*a*) $[ML^2T^{-2}K^{-1}]$ (*b*) $[MLT^{-2}K^{-1}]$
(*c*) $[M^0L^0TK^{-1}]$ (*d*) $[M^0L^0T^0K^{-1}]$.

21. The dimensional formula of magnetic permeability is

(*a*) $[M^0L^{-1}T]$ (*b*) $[M^0L^2T^{-1}]$
(*c*) $[M^0L^2T^{-1}A^2]$ (*d*) $[MLT^{-2}A^{-2}]$.

22. The time taken by sunlight to travel from sun to earth is

(a) 10^{-1} s (b) 1 s
(c) 10 s (d) 5×10^2 s.

23. If the velocity of light be taken as unit of velocity and a year as unit of time, then the unit of length is called
(a) AU (b) parsec
(c) fermi (d) light year.

24. The dimensional formula for areal velocity is
(a) $[M^0L^{-2}T^1]$ (b) $[M^0L^{-2}T^{-1}]$
(c) $[M^0L^2T^{-1}]$ (d) $[M^0L^2T^1]$.

25. The radius of curvature of a spherical surface is measured using
(a) a spectrometer (b) vernier calliper
(c) screw gauge (d) a spherometer.

26. Energy per unit volume represents
(a) Pressure (b) Force
(c) Thrust (d) Work.

27. Which of the following pairs has the same dimensions ?
(a) Impulse and Momentum
(b) Specific heat and Latent heat
(c) Moment of inertia and Moment of momentum
(d) Surface tension and tension (force).

28. A pressure of 10^6 dyne cm^{-2} is equivalent to
(a) 10^5 N m^{-2} (b) 10^4 N m^{-2}
(c) 10^6 N m^{-2} (d) 10^7 N m^{-2}.

29. Let L denote the self-inductance of a coil which is in series with a capacitor of capacitance C. Which of the following has the unit second ?
(a) $\sqrt{LC}$ (b) C/L
(c) CL (d) L^2/C^2.

30. Which of the following is not a unit of energy ?
(a) Ws (b) kg m s^{-1}
(c) N m (d) joule.

31. The radio-active decay constant has the same dimensional formula as
(a) mole (b) frequency
(c) time (d) mass.

32. The velocity of a body is 10 mm/nano second. It is also
(a) 3.6×10^3 km h^{-1} (b) 36×10^5 km h^{-1}
(c) 3.6×10^7 km h^{-1} (d) 3.6 km h^{-1}.

33. The surface tension of mercury is 32 dyne cm^{-1}. Its value in SI units is
(a) 0.032 (b) 0.32
(c) 3200 (d) 32000.

34. If L denotes the inductance and R the resistance, then the dimensional formula of L/R is
(a) $[M^1L^0T^1]$ (b) $[M^0L^0T^1]$
(c) $[M^1L^1T^1]$ (d) $[M^0L^1T^1]$.

35. Which of the following expressions for energy can be ruled out on the basis of dimensional considerations ?
(a) mgh (b) $\frac{1}{2}mv^2$
(c) $\frac{1}{2}K\xi^2$ (d) None of these.

36. $[M^1L^{-2}T^{-2}]$ represents dimensional formula of which of the following physical quantities ?
(a) energy (b) pressure
(c) torque (d) pressure gradient.

37. The density of wood is 0.5 g cm^{-3} in cgs system of units. The corresponding value in SI units is
(a) 5000 (b) 500
(c) 5 (d) 0.5.

38. If L is the inductance and C is the capacitance, then the dimensional representation of expression $\frac{1}{2\pi\sqrt{LC}}$ will be
(a) $[M^0L^1T^{-1}]$ (b) $[M^0L^0T^{-1}]$
(c) $[M^0L^0T^1]$
(d) cannot be expressed in terms of M L T.

39. The dimensional formula of action (= energy × time) is
(a) $[ML^2T^{-2}]$ (b) $[MLT^{-2}]$
(c) $[ML^3T^{-2}]$ (d) $[ML^2T^{-1}]$.

40. What is the SI unit of Stefan-Boltzmann's constant σ ?
(a) W m^{-2} K^{-4} (b) W m^2 K^4
(c) W K^{-4} (d) erg s^{-2} K^{-4}.

41. The unit of electrical permittivity of free space ε_0 is
(a) C N^{-1} m^{-1} (b) N m^2 C^{-2}
(c) $C^2 \cdot N^{-2}$ m^{-2} (d) C^2 N^{-1} m^{-2}.

42. The dimensional formula for root mean square velocity is
(a) $[M^0LT^{-1}]$ (b) $[M^0L^0T^{-2}]$
(c) $[M^0L^0T^{-1}]$ (d) $[MLT^{-3}]$.

43. The dimensional formula of 8 calorie is
(a) $[ML^2T^{-2}]$ (b) $[ML^2T^{-2}]$
(c) $4[ML^2T^{-2}]$ (d) $2[ML^2T^{-2}]$.

44. What is the SI unit of permeability ?
(a) henry per metre (b) tesla metre per ampere
(c) weber per ampere metre (d) All of these.

45. The velocity of a body is given by the equation :

$$v = \frac{b}{t} + ct^2 + dt^3$$

The dimensional formula of b is

(a) $[M^0LT^0]$ (b) $[ML^0T^0]$
(c) $[M^0L^0T]$ (d) $[MLT^{-1}]$.

46. If $y = a + bt + ct^2$, where y is in metre and t in second, then the unit of c is

(a) m (b) s^{-2}
(c) $m\ s^{-1}$ (d) $m\ s^{-2}$.

47. $g\ cm\ s^{-2}$ stands for the unit of

(a) Energy (b) Force
(c) Momentum (d) Acceleration.

48. The dimensional formula of PV, where P is pressure and V is volume is the same as that of

(a) Work (b) Power
(c) Elastic modulus (d) Pressure.

49. One poise is equal to

(a) $1\ N\ s\ m^{-2}$ (b) $0.1\ N\ s\ m^{-2}$
(c) $0.01\ N\ s\ m^{-2}$ (d) $10\ N\ s\ m^{-2}$.

50. The quantity having dimensions – 2 in the time is

(a) force (b) pressure
(c) gravitational constant (d) all of these.

51. In the equation $\left(P + \frac{a}{V^2}\right)(V - b) = RT$, the SI unit of a is

(a) $N\ m^2$ (b) $N\ m^4$
(c) $N\ m^{-3}$ (d) $N\ m^{-2}$.

52. The dimensional formula of latent heat is

(a) $[M^0L^2T^{-2}]$ (b) $[MLT^2]$
(c) $[ML^2T^{-2}]$ (d) $[MLT^{-1}]$.

53. A spherometer has 250 equal divisions marked along the peripheri of its disc and one full rotation of the disc advances on the main scale by 0.0625 cm. The least count of the system is

(a) 2.5×10^{-4} cm (b) 25×10^{-3} cm
(c) 2.5×10^{-2} cm (d) none of these.

54. In the equation $\xi = a \sin(\omega t - kx)$, the dimensional formula of ω is

(a) $[M^0L^0T^{-1}]$ (b) $[M^0L^1T^{-1}]$
(c) $[ML^0T^0]$ (d) $[M^0L^{-1}T^0]$.

55. In Q. 54, the dimensional formula of k is

(a) $[M^0L^0T]$ (b) $[M^0LT]$
(c) $[MLT]$ (d) $[M^0L^{-1}T^0]$.

56. In Q. 54, the dimensional formula of ω/k is

(a) $[M^0LT^{-1}]$ (b) $[M^0L^0T^{-1}]$
(c) $[M^{-1}LT^0]$ (d) $[M^0L^0T^0]$.

57. The dimensional formula of $[ML^{-1}T^{-2}]$ does not represent the following

(a) stress (b) power
(c) pressure (d) Young's modulus.

58. Linear momentum and Angular momentum have the same dimensions in

(a) Mass and length (b) Length and time
(c) Mass and time (d) Mass, length and time.

59. The velocity of a particle is given by

$$v = a + \frac{b}{t} + ct^2$$

The unit of b will be

(a) m (b) $m\ s^2$
(c) $m\ s^{-1}$ (d) $m\ s^{-2}$.

60. The SI unit of potential energy is

(a) $kg\ (m\ s^{-1})^2$ (b) $kg\ m\ (s^{-1})^3$
(c) $(kg\ s)^2\ m$ (d) $kg^{-1}\ m^{-1}\ s^{-1}$.

61. Universal time is based on

(a) rotation of earth on its axis
(b) oscillations of quartz crystal
(c) vibrations of cesium atom
(d) earth's orbital motion around the sun.

62. A vernier calliper has its main scale of 10 cm equally divided into 200 equal parts. Its vernier scale of 25 divisions coincides with 12 mm on the main scale. The least count of the instrument is

(a) 0.020 cm (b) 0.002 cm
(c) 0.010 cm (d) 0.001 cm.

63. The physical quantity which has the dimensional formula $[M^1T^{-3}]$ is

(a) surface tension (b) density
(c) solar constant (d) compressibility.

64. The equation of a wave is given by

$$y = A \sin \omega \left[\frac{x}{v} - k\right]$$

where ω is the angular velocity and v is the linear velocity. The dimensional formula of k is

(a) $[LT]$ (b) $[T]$
(c) $[T^{-1}]$ (d) $[T^2]$.

65. $2l\sqrt{\frac{m}{T}}$, where l is the length of a string of linear density m under tension T has the same dimensional formula as that of

(a) mass (b) time
(c) length (d) mole.

66. Which of the following cannot be regarded as an essential characteristic of a unit of measurement ?

(a) Inaccessibility (b) Indestructibility
(c) Invariability (d) Reproducibility.

67. Force constant has the same dimensions as

(a) Coefficient of viscosity (b) Surface Tension
(c) Frequency (d) Impulse.

68. If h is height and g is acceleration due to gravity, then the dimensional formula of $\sqrt{\frac{2h}{g}}$ is the same as that of

(a) time (b) mass
(c) volume (d) velocity.

69. The electrostatic force between two point charges q_1 and q_2 separated by distance r is given by

$$F = k\frac{q_1 q_2}{r^2}$$

The constant k

(a) depends upon medium between charges only
(b) depends upon system of units only
(c) depends upon both (a) and (b)
(d) is independent of both (a) and (b).

70. The dimensional formula of the ratio of angular to linear momentum is

(a) $[M^0LT^0]$ (b) $[MLT]$
(c) $[ML^2T^{-1}]$ (d) $[M^{-1}L^{-1}T^{-1}]$.

71. The dimensional formula of pressure gradient is

(a) $[MLT^{-2}]$ (b) $[ML^2T^{-2}]$
(c) $[ML^{-2}T^{-2}]$ (d) $[MLT^{-1}]$.

72. In the equation : $S_{nth} = u + \frac{a}{2}(2n-1)$, the letters have their usual meanings. The dimensional formula of S_{nth} is

(a) $[M^1L^0T^1]$ (b) $[M^1L^{-1}T^{-1}]$
(c) $[M^0L^1T^{-1}]$ (d) $[M^0L^1T^0]$.

73. An athletic coach told his team that 'muscle time speed equals power'. What dimension does he view for muscle ?

(a) $[MLT^{-2}]$ (b) $[ML^2T^{-2}]$
(c) $[MLT^2]$ (d) $[MLT]$.

74. Two quantities A and B are related by the relation $\frac{A}{B} = m$ where m is linear mass density and A is force. The dimensions of B will be

(a) same as that of latent heat
(b) same as that of pressure
(c) same as that of work
(d) same as that of momentum.

75. The vernier of a circular scale is divided into 30 divisions which coincide against 29 main scale divisions. Each main scale division is $\frac{1°}{2}$. The least count of the instrument is

(a) 30′ (b) 10′
(c) 1′ (d) 0.1′.

76. To measure the radius of curvature with a spherometer, we use the formula

(a) $R = \frac{h^2}{6} + \frac{1}{l}$ (b) $R = \frac{l^2}{6h} + \frac{h}{2}$
(c) $R = \frac{h^2}{2l} + \frac{l}{h}$ (d) $R = \frac{2l^2}{h} + \frac{6}{l}$.

77. If L denotes the inductance of an inductor through which a current I is flowing, then the dimensional formula of LI^2 is

(a) $[MLT^{-2}]$ (b) $[ML^2T^{-2}]$
(c) $[M^2L^2T^{-2}]$
(d) not expressible in terms of M, L, T.

78. The physical quantity that has a ratio of 10^3 between its SI unit and cgs unit is

(a) Young's modulus (b) Boltzmann constant
(c) Planck's constant
(d) Universal gravitational constant.

79. What is the dimensional formula of mc^2, where the letters have their usual meanings ?

(a) $[MLT^{-1}]$ (b) $[ML^0L^0]$
(c) $[ML^2T^{-2}]$ (d) $[M^{-1}L^3T^6]$.

80. Which of the following does not have the dimensions of frequency ?

(a) $\frac{1}{CR}$ (b) $\frac{R}{L}$
(c) $\frac{1}{\sqrt{LC}}$ (d) $\frac{C}{L}$.

81. The dimensional formula for specific heat is

(a) $[M^0L^2T^{-2}K^{-1}]$ (b) $[M^0LT^{-1}K^{-1}]$
(c) $[M^0LTK]$ (d) $[M^0LT^{-2}K]$.

82. The dimensional formula of Wien's constant is

(a) $[M^0LT^0K]$ (b) $[ML]$
(c) $[LTK]$ (d) $[MK]$.

83. The dimensional formula for gas constant is

(a) $[MT^{-2}K]$ (b) $[MLT^{-2}]$
(c) $[LK]$ (d) $[ML^2T^{-2}K^{-1}]$.

84. The ampere-hour is the unit of

(*a*) energy (*b*) power

(*c*) current strength

(*d*) quantitative measure of electricity.

85. If $S = \frac{1}{3} ft^3$, '*f*' has the dimensions of

(*a*) $[M^0L^{-1}T^3]$ (*b*) $[M^1L^1T^{-3}]$

(*c*) $[M^0L^1T^{-3}]$ (*d*) $[M^0L^{-1}T^{-3}]$.

86. The dimensional formula of $\sqrt{\frac{I}{BM}}$ is

(*a*) $[M^0L^0T^0A^0]$ (*b*) $[M^0L^0T^1A^0]$

(*c*) $[M^1L^2A^1T^1]$ (*d*) $[M^0L^2A^1T^1]$.

87. A dimensionless quantity

(*a*) does not exist (*b*) always has a unit

(*c*) never has a unit (*d*) may have a unit.

88. The damping force on an oscillating particle is observed to be proportional to velocity. The constant of proportionality can be measured in

(*a*) kg s^{-1} (*b*) kg s

(*c*) kg m s^{-1} (*d*) kg m^{-1} s^{-1}.

89. The SI unit of electrical conductivity is

(*a*) S (*b*) S m

(*c*) S m^{-1} (*d*) S^{-1} m^{-1}.

90. If the units of M and L are increased three times, the unit of energy will be increased by

(*a*) 3 times (*b*) 6 times

(*c*) 27 times (*d*) 81 times.

91. What is the dimensional formula of

$$\frac{\text{Planck's constant}}{\text{Linear momentum}}?$$

(*a*) $[M^0L^0T^0]$ (*b*) $[M^0L^0T]$

(*c*) $[M^0LT^0]$ (*d*) $[MLT^{-1}]$.

92. $[ML^2T^{-3}A^{-2}]$ is the dimensional formula of

(*a*) Electric resistance (*b*) Capacity

(*c*) Electric potential (*d*) Specific resistance.

93. The unit of inductance is

(*a*) V A^{-1} (*b*) J A^{-1}

(*c*) V s A^{-1} (*d*) V A s^{-1}.

94. The equation of stationary wave is

$$\xi = 2a \sin \frac{2\pi ct}{\lambda} \cos \frac{2\pi x}{\lambda}.$$

Which of the following is correct ?

(*a*) The units of *ct* and λ are the same.

(*b*) The units of *x* and λ are the same.

(*c*) The unit of ξ is the same as that of *a*.

(*d*) All of these.

95. The force F is given in terms of time *t* and displacement *x* by the equation F = A cos B*x* + C sin D*t*. The dimensional formula of $\frac{D}{B}$ is

(*a*) $[M^0L^0T^0]$ (*b*) $[M^0L^0T^{-1}]$

(*c*) $[M^0L^{-1}T^0]$ (*d*) $[M^0L^1T^{-1}]$.

96. In a vernier callipers, ten smallest divisions of the vernier scale are equal to nine smallest division on the main scale. If the smallest division on the main scale is half millimetre, then the vernier constant is

(*a*) 0.5 mm (*b*) 0.1 mm

(*c*) 0.05 mm (*d*) 0.005 mm.

97. The pitch of a screw gauge is 0.5 mm. Its head scale contains 50 divisions. The least count of the screw gauge is

(*a*) 0.001 mm (*b*) 0.01 mm

(*c*) 0.02 mm (*d*) 0.025 mm.

98. The dimensional formula of current density is

(*a*) $[M^0L^{-2}T^{-1}Q]$ (*b*) $[M^0L^2T^1Q^{-1}]$

(*c*) $[MLT^{-1}Q]$ (*d*) $[ML^{-2}T^{-1}Q^2]$.

QUESTION BANK

Level II

Choose the correct answer from the four alternatives given in each of the following questions :

1. Suppose a quantity *y* can be dimensionally represented in terms M, L and T, that is $[y] = [M^a L^b T^c]$. The quantity mass

(*a*) may be represented in terms of L, T and *y* if $a = 0$

(*b*) may be represented in terms of L, T and *y* if $a \neq 0$

(*c*) can always be dimensionally represented in terms of L, T and *y*

(*d*) can never be dimensionally represented in terms of L, T and *y*.

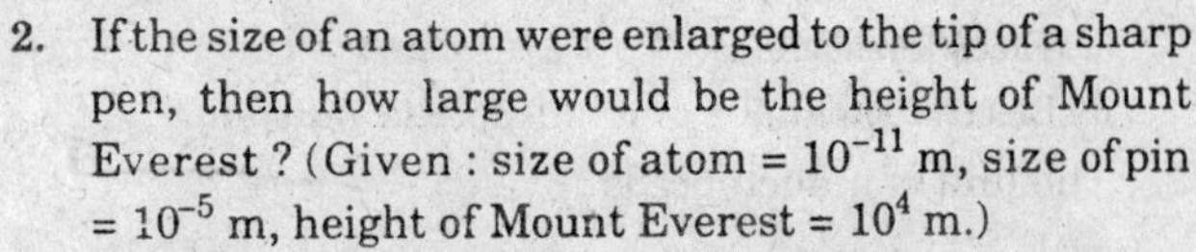

2. If the size of an atom were enlarged to the tip of a sharp pen, then how large would be the height of Mount Everest ? (Given : size of atom = 10^{-11} m, size of pin = 10^{-5} m, height of Mount Everest = 10^4 m.)

(a) 10^{-2} m (b) 10^6 m
(c) 10^{10} m (d) 10^{11} m.

3. In the equation $X = 3YZ^2$, X and Z have dimensions of capacitance and magnetic induction respectively. In MKSQ system, the dimensional formula of Y is

(a) $[M^{-3}L^{-2}T^{-2}Q^{-4}]$ (b) $[ML^{-2}]$
(c) $[M^{-3}L^{-2}Q^4T^8]$ (d) $[M^{-3}L^{-2}Q^4T^4]$.

4. If I is the moment of inertia and ω the angular velocity, what is the dimensional formula of rotational kinetic energy $\frac{1}{2}I\omega^2$?

(a) $[ML^2T^{-1}]$ (b) $[M^2L^{-1}T^{-2}]$
(c) $[ML^2T^{-2}]$ (d) $[M^2L^{-1}T^{-2}]$.

5. The dimensional formula for Rydberg constant is

(a) $[M^0LT^{-1}]$ (b) $[M^0L^{-1}T^0]$
(c) $[M^0L^0T^0]$ (d) [MLT].

6. The fundamental unit, which has the same power in the dimensional formulae of surface tension and viscosity is

(a) mass (b) length
(c) time (d) none of these.

7. Given that force (F) is given $F = Pt^{-1} + Qt$. Here t is time. The unit of P is same as that of

(a) displacement (b) velocity
(c) acceleration (d) momentum.

8. Given that $r = m^2 \sin pt$, where t represents time. If the unit of m is N, then the unit of r is

(a) N (b) N^2
(c) N s (d) N^2 s.

9. The e.m.f. of a cell is defined as work done per unit charge. The dimensional formula of e.m.f. in terms of charge Q is

(a) $[MT^{-2}Q^{-2}]$ (b) $[ML^2T^{-2}Q^2]$
(c) $[ML^2T^{-2}Q^{-2}]$ (d) $[ML^2T^{-2}Q^{-1}]$.

10. Coefficient of thermal conductivity has the dimensions

(a) $[MLT^{-3}K^{-1}]$ (b) $[ML^3T^3K^2]$
(c) $[ML^3T^{-3}K^{-2}]$ (d) $[M^2L^3T^{-3}K^2]$.

11. Consider a new system of units in which c (speed of light in vacuum), h (Planck's constant) and G (gravitational constant) are taken as fundamental units. Which of the following would correctly represent mass in this new system ?

(a) $\sqrt{\frac{hC}{G}}$ (b) $\sqrt{\frac{GC}{h}}$
(c) $\sqrt{\frac{hG}{c}}$ (d) $\sqrt{hGC}$.

12. A rectangular beam which is supported at its two ends and loaded in the middle with weight W sags by an amount δ such that $\delta = \frac{Wl^3}{4Yd^3\ ?}$, where l, d and Y represent length, depth and elasticity respectively. Guess the unknown factor using dimensional considerations

(a) breadth (b) $(\text{breadth})^2$
(c) $(\text{breadth})^3$ (d) mass.

13. The dimensional formula of self-inductance is

(a) $[MLT^{-2}]$ (b) $[ML^2T^{-1}A^{-2}]$
(c) $[ML^2T^{-2}A^{-2}]$ (d) $[ML^2T^{-2}A^{-1}]$.

14. The main scale of a spectrometer is divided into 720 divisions in all. If the vernier scale consists of 30 divisions, the least count of the instrument is

(Given : 30 vernier divisions coincide with 29 main scale divisions.)

(a) 0.1° (b) 1″
(c) 1′ (d) 0.1″.

15. The dimensions of $(\text{velocity})^2 \div$ radius are the same as that of

(a) Planck's constant (b) gravitational constant
(c) dielectric constant (d) none of these.

16. The dimensional formula of magnetic induction B is

(a) $[M^0ALT^0]$ (b) $[M^0AL^{-1}T^0]$
(c) $[M^0AL^2T^0]$ (d) $[ML^0T^{-2}A^{-1}]$.

17. If time T, acceleration A and force F are regarded as base units, then the dimensional formula of work is

(a) [FA] (b) [FAT]
(c) $[FAT^2]$ (d) $[FA^2T]$.

18. The unit of amplification factor is

(a) ohm (b) mho
(c) AV^{-1} (d) a non-dimensional constant.

19. The dimensional formula for entropy is

(a) $[MLT^{-2}K^{-1}]$ (b) $[ML^2T^{-2}]$
(c) $[ML^2T^{-2}K^{-1}]$ (d) $[ML^{-2}T^{-2}K^{-1}]$.

20. The dimensional formula of farad is

(a) $[M^{-1}L^{-2}TQ]$ (b) $[M^{-1}L^{-2}T^2Q^2]$
(c) $[M^{-1}L^{-2}TQ^2]$ (d) $[M^{-1}L^{-2}T^2Q]$.

21. The dimensional formula of magnetic moment M is

(a) $[M^0ALT^0]$ (b) $[M^0AL^{-1}T^0]$
(c) $[M^0AL^2T^0]$ (d) $[ML^0T^{-2}A^{-1}]$.

22. Let Q denote the charge on the plates of a capacitor of capacitance C. The dimensional formula for Q^2/C is

(a) $[M^2L^2T^2]$ (b) $[ML^2T^{-2}]$
(c) $[MLT^2]$ (d) $[M^2L^2T]$.

23. If the acceleration due to gravity is 10 m s^{-2} and the units of length and time are changed to kilometre and hour respectively, the numerical value of the acceleration is

(a) 360000 (b) 72000
(c) 36000 (d) 129600.

24. Consider the following statements about dimensions

P : Force = Acceleration due to gravity × mass

Q : Electric charge = current × time

R : Magnetic flux = Electric voltage × time

The correct statements are

(a) R, P only (b) P, Q only
(c) Q, R only (d) P, Q and R.

25. A highly rigid cubical block A of small mass M and side L is fixed rigidly on to another cubical block of same dimensions and of low modulus of rigidity η such that the lower face of A completely covers the upper face of B. The lower face of B is rigidly held on a horizontal surface. A small force F is applied perpendicular to one of the side faces of A. After the force is withdrawn, block A executes small oscillations, the time period of which is given by

(a) $2\pi\sqrt{M\eta L}$ (b) $2\pi\sqrt{\frac{M\eta}{L}}$

(c) $2\pi\sqrt{\frac{ML}{\eta}}$ (d) $2\pi\sqrt{\frac{M}{\eta L}}$.

26. If C is the restoring couple per unit radian twist and I is the moment of inertia, then the dimensional representation of $2\pi\sqrt{\frac{I}{C}}$ will be

(a) $[M^0L^0T^{-1}]$ (b) $[M^0L^0T^1]$
(c) $[M^0L^1T^{-1}]$ (d) $[M^1L^2T^{-2}]$.

27. The dimensional formula of capacitance in terms of M, L, T and I is

(a) $[ML^2T^2I^2]$ (b) $[ML^{-2}T^4I^2]$
(c) $[M^{-1}L^3T^3I]$ (d) $[M^{-1}L^{-2}T^4I^2]$.

28. SI unit of intensity of wave is

(a) J m^{-2} s^{-2} (b) J m^{-1} s^{-2}
(c) W m^{-2} (d) J m^{-2}.

29. In a new system of units, unit of mass is 10 kg, unit of length is 1 km and unit of time is 1 minute. The value of 1 joule in this new hypothetical system is

(a) 3.6×10^{-4} new units (b) 6×10^7 new units
(c) 10^{11} new units (d) 1.67×10^4 new units.

30. The time dependence of physical quantity P is given by $P = P_0 e^{-\alpha t^2}$. The constant α will

(a) be dimensionless
(b) have dimensions T^{-2}
(c) have dimensions same as that of P
(d) have dimensions equal to the dimension of P multiplied by T^{-2}.

31. $[ML^3T^{-1}Q^{-2}]$ is the dimensional formula of

(a) resistance (b) resistivity
(c) conductance (d) conductivity.

32. μ_0 and ε_0 denote the permeability and permittivity respectively of free space. The dimensional formula of $\mu_0\varepsilon_0$ is

(a) $[L^{-2}T^2]$ (b) $[L^{-1}T]$
(c) $[LT^{-1}]$ (d) $[L^2T^{-2}]$.

33. The number of particles given by

$$n = -D\frac{n_2 - n_1}{x_2 - x_1}$$

are crossing a unit area perpendicular to x-axis in unit time, where n_1 and n_2 are the number of particles per unit volume for the values x_1 and x_2 of x respectively. Then the dimensional formula of diffusion constant D is

(a) $[M^0LT^2]$ (b) $[M^0L^2T^{-4}]$
(c) $[M^0LT^{-3}]$ (d) $[M^0L^2T^{-1}]$.

34. The dimensional formula of $\frac{1}{\varepsilon_0}\frac{e^2}{hc}$ is

(a) $[M^0L^0T^0A^0]$ (b) $[M^{-1}L^3T^2A]$
(c) $[ML^3T^{-4}A^{-2}]$ (d) $[M^{-1}L^{-3}T^4]$.

35. The dimensional formula of electrical conductivity is

(a) $[M^{-1}L^{-3}T^3A^2]$ (b) $[ML^3T^3A^2]$
(c) $[M^2L^3T^{-3}A^2]$ (d) $[ML^3T^{-3}A^{-2}]$.

36. If energy E, velocity V and time T are chosen as fundamental quantities, then the dimensional formula of surface tension is

(a) $[EV^{-2}T^{-2}]$ (b) $[EV^{-1}T^{-2}]$
(c) $[EV^{-2}T^{-1}]$ (d) $[E^2V^{-1}T^{-2}]$.

37. The radius of the proton is about 10^{-15} m. The radius of the observable universe is 10^{26} m. Identify the distance which is half-way between these two extremes on a logarithmic scale.

(*a*) 10^{21} m (*b*) 10^6 m
(*c*) 10^{-6} m (*d*) 10^0 m.

38. What will be the unit of time in that system in which unit of length is *m*, unit of mass is kg and unit of force is kg wt ?
(*a*) 9.8 s (*b*) $\sqrt{9.8}$ s
(*c*) $(9.8)^2$ s (*d*) $\frac{1}{\sqrt{9.8}}$ s.

39. If the unit of time is taken as one second and the acceleration due to gravity is also taken as unity in a particular system, then the unit of length in that system is
(*a*) 9.8 m (*b*) $(9.8)^{-1}$ m
(*c*) $(9.8)^{-2}$ m (*d*) zero.

40. The velocity v of water waves may depend on their wavelength (λ), the density of water (ρ) and the acceleration due to gravity (g). The method of dimensions gives the relation between these quantities as
(*a*) $v^2 \propto \lambda^{-1} \rho^{-1}$ (*b*) $v^2 \propto g\lambda$
(*c*) $v^2 \propto g\lambda\rho$ (*d*) $g^{-1} \propto \lambda^3$.

41. The frequency of vibration f of a mass m suspended from a spring of spring constant k is given by relation of the type $f = cm^x k^y$, where c is a dimensionless constant. The values of x and y are
(*a*) 1/2, 1/2 (*b*) – 1/2, – 1/2
(*c*) 1/2, – 1/2 (*d*) – 1/2, 1/2.

42. Given : $X = (Gh/c^3)^{1/2}$, where G, h and c are gravitational constant, Planck's constant and the velocity of light respectively. Dimensions of X are the same as those of
(*a*) mass (*b*) time
(*c*) length (*d*) acceleration.

43. If R is Rydberg constant, h is Planck's constant and c is the velocity of light, then Rhc has the same dimensional formula as that of
(*a*) Force (*b*) Power
(*c*) Energy (*d*) Angular momentum.

44. The dimensional formula of coefficient of permittivity for free space (ε_0) in the equation

$$F = \frac{1}{4\pi\varepsilon_0}\frac{q_1 q_2}{r^2},$$

where symbols have their usual meanings, is
(*a*) $[M^1L^3A^{-2}T^{-4}]$ (*b*) $[M^{-1}L^{-3}T^4A^2]$
(*c*) $[M^{-1}L^{-3}A^{-2}T^{-4}]$ (*d*) $[M^1L^3A^2T^{-4}]$.

45. The thrust developed by a rocket-motor is given by $F = mv + A(P_1 - P_2)$, where m is the mass of the gas ejected per unit time, v is velocity of the gas, A is area of cross-section of the nozzle, P_1, P_2 are the pressures of the exhaust gas and surrounding atmosphere. The formula is dimensionally
(*a*) correct (*b*) wrong
(*c*) sometimes wrong, sometimes correct
(*d*) Data is not adequate.

46. Which of the following quantities can be written in the international system of units kg m^2 A^{-2} s^{-3} ?
(*a*) resistance (*b*) inductance
(*c*) capacitance (*d*) magnetic flux.

47. Dimensional formula of the Hall coefficient is
(*a*) $[M^0L^0A^{-2}T^{-1}]$ (*b*) $[M^0L^3AT]$
(*c*) $[M^0L^0AT]$ (*d*) $[M^0L^3A^{-1}T^{-1}]$.

48. What is the unit of k in the relation

$$U = \frac{ky}{y^2 + a^2}$$

where U represents the potential energy, y represents the displacement and a represents the maximum displacement *i.e.*, amplitude ?
(*a*) m s^{-1} (*b*) m s
(*c*) J m (*d*) J s^{-1}.

49. If E, m, J and G represent energy, mass, angular momentum and gravitational constant respectively, then the dimensional formula of EJ^2/m^5G^2 is
(*a*) angle (*b*) length
(*c*) mass (*d*) time.

50. A vernier calliper has 20 divisions on the vernier scale, which coincide with 19 on the main scale. The least count of the instrument is 0.1 mm. The main scale divisions are of
(*a*) 0.5 mm (*b*) 1 mm
(*c*) 2 mm (*d*) $\frac{1}{4}$ mm.

51. The order of magnitude of the number of nitrogen molecules in an air bubble of diameter 2 mm under ordinary conditions is
(*a*) 10^5 (*b*) 10^9
(*c*) 10^{13} (*d*) 10^{17}.

52. The dimensional formula for hole mobility in a semi-conductor is
(*a*) $[M^{-1}L^0T^2A]$ (*b*) $[M^{-1}L^2T^{-2}A]$
(*c*) $[M^0L^0T^0A^0]$ (*d*) $[ML^{-2}T^{-2}A]$.

53. The unit of electric field is not equivalent to
(*a*) N C^{-1} (*b*) J C^{-1}
(*c*) V m^{-1} (*d*) J C^{-1} m^{-1}.

54. A calorie is a unit of heat and equals 4.2 J. Suppose we employ a system of units in which the unit of mass is α kg, the unit of length is β metre and the unit of time is γ second. In this new system, 1 calorie =

(a) $\alpha^{-1}\beta^{-2}\gamma^2$ (b) $4.2\,\alpha\beta^2\gamma^{-2}$
(c) $\alpha\beta^2\gamma^{-2}$ (d) $4.2\,\alpha^{-1}\beta^{-2}\gamma^2$.

55. A unit of area, often used in measuring land areas, is the hectare defined as 10^4 m^2. An open-pit coal mine consumes 75 hectares of land, down to a depth of 26 m, each year. What volume of earth, in cubic kilometre, is removed in this time ?

(a) 0.01 (b) 0.02
(c) 0.03 (d) 0.04.

56. The velocity of a body which has fallen freely under gravity varies as $g^p h^q$, where g is the acceleration due to gravity and 'h' is the height through which it has fallen. The values of p and q are

(a) $-\frac{1}{2}, \frac{1}{2}$ (b) $\frac{1}{2}, -\frac{1}{2}$
(c) $\frac{1}{2}, \frac{1}{2}$ (d) $-\frac{1}{2}, -\frac{1}{2}$.

57. If frequency ν, velocity v and density d are considered as fundamental units, then the dimensional formula for momentum will be

(a) $dv^4\nu^{-3}$ (b) $dv^2\nu^{-1}$
(c) $dv\,\nu^2$ (d) $d^2 v^2 \nu^2$.

58. The wavelength associated with a moving particle depends upon power p of its mass m, qth power of its velocity v and rth power of Planck's constant h. Then the correct set of values of p, q and r is

(a) $p = 1, q = -1, r = 1$ (b) $p = 1, q = 1, r = 1$
(c) $p = -1, q = -1, r = -1$ (d) $p = -1, q = -1, r = 1$.

59. Let us choose a new unit of length such that the velocity of light in vacuum is unity. If light takes 8 minute and 20 second to cover the distance between sun and earth, this distance in terms of the new unit is

(a) 5 (b) 50
(c) 500 (d) 3×10^8
(e) 1.5×10^{11}.

60. For the equation $F \propto A^a v^b d^c$, where F is the force, A is the area, v is the velocity and d is the density, the values of a, b and c are respectively

(a) 1, 2, 1 (b) 2, 1, 1
(c) 1, 1, 2 (d) 0, 1, 1.

61. Length is measured in metre and time in second as usual. But a new unit of mass is so chosen that G = 1. This new unit of mass is equal to

(a) 1.5×10^7 kg (b) 1.5×10^{10} kg
(c) 6.67×10^{-11} kg (d) 6.67×10^{-8} kg.

62. If gravitational constant G, velocity of light c and Planck's constant h are taken as the fundamental quantities, then the dimensional formula for length is

(a) $[G^{-1/2} c^{-3/2} h^{1/2}]$ (b) $[G^{1/2} c^{-3/2} h^{1/2}]$
(c) $[G^{-1/2} c^{3/2} h^{-1/2}]$ (d) $[G^{1/2} c^{-3/2} h^{-1/2}]$.

63. Express T in terms of G, h and c

(a) G h c (b) $G^{1/2} h^{1/2} c^{-3/2}$
(c) $G^{1/2} h^{1/2} c^{-5/2}$ (d) $G^{-1/2} h^{1/2} c^{1/2}$.

64. If the unit of force is 1 kilonewton, the length is 1 km and time is 100 second, what will be the unit of mass ?

(a) 1 kg (b) 100 kg
(c) 1000 kg (d) 10000 kg.

65. If 1 g cm s^{-1} = x newton-second, then the number x is equal to

(a) 1×10^{-1} (b) 3.6×10^{-3}
(c) 1×10^{-5} (d) 6×10^{-4}.

66. The SI unit of length is the metre. Suppose we adopt a new unit of length which equals x metre. The area of 1 m^2 expressed in terms of the new unit has a magnitude

(a) x (b) x^2
(c) x^{-1} (d) x^{-2}.

67. An important milestone in the evolution of the universe just after the Big Bang is the Planck time t_p, the value of which depends on three fundamental constants-speed c of light in vacuum, gravitational constant G and Planck's constant h. Then, $t_p \propto$

(a) Ghc^5 (b) $\frac{c^5}{Gh}$
(c) $\frac{Gh}{c^5}$ (d) $\left(\frac{Gh}{c^5}\right)^{1/2}$.

68. If force F, acceleration A and time T be taken as fundamental physical quantities, then the dimensions of length in this system of units are

(a) FT (b) FAT
(c) AT^2 (d) FAT^2.

69. The ratio of 1 kWh to 1 MeV is

(a) 2.25×10^{17} (b) 2.25×10^{19}
(c) 2.25×10^{23} (d) $2.25 \times 4.4 \times 10^9$.

70. In a particular system, the units of length, mass and time are chosen to be 10 cm, 10 g and 0.1 s respectively. The unit of force in this system will be

(*a*) 0.1 N (*b*) 1 N
(*c*) 10 N (*d*) 100 N.

71. Let us redefine newton as the force of attraction between two particles, each of mass 1 kg, separated by 1 m. Then the value of universal constant of gravitation is

(*a*) 1 N m^2 kg^{-2} (*b*) 10 N m^2 kg^{-2}
(*c*) 100 N m^2 kg^{-2} (*d*) $\frac{1}{99}$ N m^2 kg^{-2}.

72. In a hypothetical new system of measurement, the gravitational force between two particles, each of mass 1 kg, separated by 1 km is taken as a unit of force. Let us call this new unit of force* 'notwen'. How many newton will be there in one 'notwen'? Given ; G = 6.67 × 10^{-11} N m^2 kg^{-2}

(*a*) 6.67×10^{-11} (*b*) 6.67×10^{-17}
(*c*) $\frac{1}{6.67 \times 10^{-11}}$ (*d*) $\frac{1}{6.67 \times 10^{-17}}$.

73. Let us consider a new hypothetical system of measurement in which the unit of energy is called** eluoj. Suppose, in this system, the gravitational force of attraction between two particles, each of mass 1 kg, separated by 1 km is taken as a unit of force. Then, how many joule are contained in one eluoj ? Given : G = 6.67 × 10^{-11} N m^2 kg^{-2}.

(*a*) 6.67×10^{-7} (*b*) 6.67×10^{-8}
(*c*) 6.67×10^{-14} (*d*) 6.67×10^{-17}.

74. If K represents kinetic energy, V represents velocity and T represents time, and these are chosen as the fundamental units, then the dimensional formula of surface tension is

(*a*) $[KV^{-2}T^{-2}]$ (*b*) $[KV^{-1}T^{-2}]$
(*c*) $[K^2V^{-1}T^{-3}]$ (*d*) $[KV^{-2}T^{-1}]$.

75. If P represents radiation pressure, *c* represents speed of light and Q represents radiation striking unit area per second, then non-zero integers *x*, *y* and *z* such that $P^x Q^y C^z$ is dimensionless are

(*a*) $x = 1, y = 1, z = -1$ (*b*) $x = 1, y = -1, z = 1$
(*c*) $x = -1, y = 1, z = 1$ (*d*) $x = 1, y = 1, z = 1$.

76. The mass *m* of the largest stone that can be moved by flowing water depends upon the density ρ of water, acceleration due to gravity *g* and velocity of flow *v*, then $m \propto$

(*a*) v^2 (*b*) v^4
(*c*) v^6 (*d*) v^8.

ANSWERS

QUESTION BANK (Level I)

1. (c) 2. (d) 3. (b) 4. (c) 5. (d) 6. (b) 7. (c) 8. (a)
9. (c) 10. (d) 11. (d) 12. (a) 13. (d) 14. (b) 15. (d) 16. (b)
17. (d) 18. (d) 19. (b) 20. (d) 21. (d) 22. (d) 23. (d) 24. (c)
25. (d) 26. (a) 27. (a) 28. (a) 29. (a) 30. (b) 31. (b) 32. (c)
33. (a) 34. (b) 35. (d) 36. (d) 37. (b) 38. (b) 39. (d) 40. (a)
41. (d) 42. (a) 43. (a) 44. (d) 45. (a) 46. (d) 47. (b) 48. (a)
49. (b) 50. (d) 51. (b) 52. (a) 53. (a) 54. (a) 55. (d) 56. (a)
57. (b) 58. (c) 59. (a) 60. (a) 61. (a) 62. (b) 63. (c) 64. (b)
65. (b) 66. (a) 67. (b) 68. (a) 69. (c) 70. (a) 71. (c) 72. (c)
73. (a) 74. (a) 75. (c) 76. (b) 77. (b) 78. (d) 79. (c) 80. (d)
81. (a) 82. (a) 83. (d) 84. (d) 85. (c) 86. (b) 87. (d) 88. (a)
89. (c) 90. (c) 91. (c) 92. (a) 93. (c) 94. (d) 95. (d) 96. (c)
97. (b) 98. (a).

QUESTION BANK (Level II)

1. (b) 2. (c) 3. (d) 4. (c) 5. (b) 6. (a) 7. (d) 8. (b)
9. (d) 10. (a) 11. (a) 12. (a) 13. (c) 14. (c) 15. (d) 16. (d)
17. (c) 18. (d) 19. (c) 20. (b) 21. (c) 22. (b) 23. (d) 24. (b)

25. (d)	26. (b)	27. (d)	28. (c)	29. (a)	30. (b)	31. (b)	32. (a)
33. (d)	34. (a)	35. (a)	36. (a)	37. (b)	38. (d)	39. (a)	40. (b)
41. (d)	42. (c)	43. (c)	44. (b)	45. (a)	46. (a)	47. (d)	48. (c)
49. (a)	50. (c)	51. (d)	52. (a)	53. (b)	54. (d)	55. (b)	56. (c)
57. (a)	58. (d)	59. (c)	60. (a)	61. (b)	62. (d)	63. (c)	64. (d)
65. (c)	66. (d)	67. (d)	68. (c)	69. (b)	70. (a)	71. (a)	72. (b)
73. (c)	74. (a)	75. (b)	76. (c).				

HINTS/SOLUTIONS

QUESTION BANK (Level I)

1. (c) Ideally speaking, the examiner should specify the temperature in this question. This is because the density of water varies with temperature. It is maximum (10^3 kg m^{-3}) at 4°C.

> ☞ **Author's suggestion**
> *In questions relating to the density of water, if no temperature is specified, assume the temperature to be 4°C.*

2. (d) 1 atmospheric pressure = 76 cm of Hg

$= 76 \times 13.6 \times 981$ dyne cm^{-2}

$= 1.01 \times 10^6$ dyne cm^{-2} = 1.01×10^5 N m^{-2}

> ☞ **Commit to memory (CTM)**
> $1\ N\ m^{-2} = 10\ dyne\ cm^{-2}$

3. (b) CR is time constant of CR circuit.

4. (c) Since velocity is derivable from length and time therefore it cannot be grouped with length and time as fundamental quantity.

5. (d) Refractive index, Dielectric constant and Relative density are all ratios of like quantities. So, they are dimensionless.

6. (b) Value = nk. Since value is fixed therefore nk = constant.

7. (c) According to Faraday's first law of electrolysis, $m = ZQ$ or $Z = \frac{m}{Q}$. So, SI unit of Z is kg C^{-1}.

8. (a) The equation $\tan\theta = \frac{v^2}{rg}$ is both dimensionally and numerically correct.

9. (c) 100 W = 100 J s^{-1} = 100×10^7 erg s^{-1} = 10^9 erg s^{-1}.

10. (d) $[\eta] = \frac{[F]}{[A]} = \frac{[MLT^{-2}]}{[L^2]} = [ML^{-1}T^{-2}]$.

11. (d) Use Newton's law of viscous flow $F = -\eta A \frac{dv}{dx}$

12. (a) $F = \frac{dp}{dt}$.

13. (d) $15.7\ g\ cm^{-3} = 15.7 \times 10^{-3}\ kg\ (10^{-2}\ m)^{-3} = 15700\ kg\ m^{-3}$.

14. (b) Acceleration is force per unit mass.

15. (d) 1 T = 1 W b m^{-2}.

16. (b) The height of a tree, building, tower, hill etc. can be determined with the help of a sextant.

17. (d) Surface tension is force per unit length.

18. (d) Refractive index, Relative density and Poisson's ratio are all dimensionless.

19. (b) Think of average life span in years and then proceed.

20. (d) $l_t = l_0(1 + \alpha t)$ or $\alpha = \frac{l_t - l_0}{l_0 t}$.

21. (d) $\frac{F}{l} = \frac{\mu_0 I_1 I_2}{2\pi r}$ or $[\mu_0] = \frac{[F]}{[I_1 I_2]} = \frac{[MLT^{-2}]}{[A^2]} = [MLT^{-2}A^{-2}]$

22. (d) The distance of sun from earth is 8 light minutes.

23. (d) Speed = $\frac{\text{distance}}{\text{time}}$

24. (c) Areal velocity is area covered per unit time.

25. (d) For knowing more details, consult your book of Practical Physics.

26. (a) $\frac{[\text{Energy}]}{[\text{Volume}]} = \frac{[ML^2T^{-2}]}{[L^3]} = [ML^{-1}T^{-2}]$

$[\text{Pressure}] = \frac{[MLT^{-2}]}{[L^2]} = [ML^{-1}T^{-2}]$.

27. (a) Impulse is change of momentum.

28. (a) Remember the conversion factor of 10.

29. (a) Based on $T = 2\pi\sqrt{LC}$.

30. (*b*) Pressure, surface tension and $\frac{\text{stress}}{\text{strain}}$ heave appropriate units. Energy is incorrectly expressed in units of momentum.

31. (*b*) Radioactive decay constant has the dimensions of 'reciprocal of time'.

32. (*c*) $\frac{10 \times 10^{-3} \text{m}}{10^{-9} \text{s}} = 10^7 \text{ m s}^{-1}$

$= 10^7 \times \frac{18}{5} \text{ km h}^{-1} = 3.6 \times 10^7 \text{ km h}^{-1}.$

33. (*a*) $\frac{32 \times 10^{-5} \text{ N}}{(10)^{-2} \text{ m}} = 32 \times 10^{-3} \text{ N m}^{-1} = 0.032 \text{ N m}^{-1}.$

34. (*b*) $\frac{L}{R}$ is time-constant of LR circuit.

35. (*d*) mgh is potential energy. $\frac{1}{2}mv^2$ is kinetic energy. $\frac{1}{2}k\xi^2$ is elastic potential energy.

36. (*d*) $[ML^{-2}T^{-2}] = \frac{[MLT^{-2}]}{[L][L^2]} = \frac{\text{Force}}{\text{distance} \times \text{area}} = \frac{\text{Pressure}}{\text{distance}}$

= Pressure gradient.

37. (*b*) $0.5 \text{ g cm}^{-3} = 0.5 \times 10^{-3} \text{ kg } (10^{-2} \text{ m})^{-3} = 500 \text{ kg m}^{-3}.$

> ☞ **CTM**
> *$1 \text{ g cm}^{-3} = 1000 \text{ kg m}^{-3}$*
> *Conversion factor for density is 1000.*

38. (*b*) Resonance frequency, $\upsilon = \frac{1}{2\pi\sqrt{LC}}$

39. (*d*) [energy][time] = $[ML^2T^{-2}][T] = [ML^2T^{-1}]$.

40. (*a*) Energy/area/time = σT^4.

41. (*d*) $\varepsilon_0 = \frac{q_1 q_2}{Fr^2}$

42. (*a*) Root mean square velocity is basically velocity.

43. (*a*) 8 is dimensionless constant.

44. (*d*)

(*a*) Self-inductance of a solenoid, $L = \frac{\mu_0 N^2 A}{l}$

(*b*) $B = \frac{\mu_0 I}{2\pi r}$ (*c*) Use $B = \frac{\phi}{A}$ in (*b*).

45. (*a*) $\left[\frac{b}{t}\right] = [v]$ or $[b] = [vt] = [L]$.

> ☞ ***Note that each term in the right-hand side of the given equation has the dimensions of velocity. It follows from the principle of homogeneity of dimensions.***

46. (*d*) $[ct^2] = [y]$ or $[c] = \frac{[y]}{t^2}$

47. (*b*) Force is the product of mass and acceleration.

48. (*a*) $PV = \frac{\text{force}}{\text{area}} \times \text{volume} = \text{force} \times \text{length}.$

49. (*b*) In cgs system, the unit of coefficient of viscosity is poise.

In SI, the unit is decapoise (1 N s m^{-2})

$1 \text{ poise} = \frac{1}{10}$ decapoise.

> ☞ **An interesting variation of Q. 49**
> *The ratio of SI and cgs units of coefficient of viscosity is*
> *(a) 1* *(b) 10*
> *(c) $\frac{1}{10}$* *(d) 100.*

50. (*d*) Refer to the dimensional formulae of force, pressure and gravitational constant.

> ☞ **A new technique**
> *You know that [force] = $[MLT^{-2}]$; pressure = $\frac{force}{area}$ and area is not related to time, $G = \frac{Fr^2}{m_1 m_2}$ and $\frac{r^2}{m_1 m_2}$ is not related to time.*

51. (*b*) $\frac{a}{V^2}$ has the units of pressure. So, a has the units of PV^2. Thus, SI unit of a is $\text{N m}^{-2} \text{ m}^6$ or N m^4.

> ☞ **A basic concept of physics**
> *Quantities of only like nature can be added.*
> *Can you add 7 donkeys and 3 dogs ?*

52. (*a*) Energy per unit mass is latent heat.

53. (*a*) Least count = $\frac{\text{Pitch}}{\text{Total number of divisions}}$

$= \frac{0.0625}{250} \text{ cm} = 2.5 \times 10^{-4} \text{ cm}.$

> ☞ **CAUTION**
> *Do not ignore questions on (i) vernier callipers (ii) screw gauge (iii) spherometer (iv) spectrometer.*

54-56. $\sin(\omega t - kx)$ is dimensionless. ωt is dimension- less. ω has dimensions of inverse time. kx is dimensionless. k has the dimensions of inverse x. $\frac{\omega}{k} = \frac{x}{t}$ = velocity.

> ☞ **CTM** $\frac{\omega}{k}$ *represents wave velocity.*

57. (*b*) $[ML^{-1}T^{-2}] = \frac{[MLT^{-2}]}{[L^2]} = \frac{\text{Force}}{\text{Area}}$

= Pressure or stress or Young's modulus.

58. (*c*) [Linear momentum] = $[MLT^{-1}]$

[Angular momentum] = $[ML^2T^{-1}]$.

59. (*a*) $\frac{b}{t}$ has the unit of velocity.

60. (*a*) Think in terms of kinetic energy.

61. (*a*) Time defined in terms of the rotation of the earth is called universal time (UT).

62. (*b*) 1 SD = $\frac{10}{200}$ cm = 0.050 cm ; 1 VD = $\frac{1.2}{25}$ cm = 0.048 cm.

> ☞ *Vernier constant = 1 SD – 1 VD.*

63. (*c*) $[MT^{-3}] = \frac{[ML^2T^{-2}]}{[L^2][T]}$ = energy/area/time.

64. (*b*) ωk is dimensionless. So, k has the dimensions of inverse ω *i.e.* k has the dimensions of time.

65. (*b*) $v = \frac{1}{2l}\sqrt{\frac{T}{m}}$, $2l\sqrt{\frac{m}{T}}$ has the dimensions of time.

66. (*a*) Indestructibility, Invariability and Reproducibility are essential characteristics of a unit of measurement.

67. (*b*) Both force constant and surface tension represent force per unit length.

68. (*a*) Think in terms of the time taken by a body in falling freely from rest through a height h.

69. (*c*) Refer to the mathematical form of Coulomb's law.

70. (*a*) $L = pr$.

71. (*c*) Pressure gradient is pressure per unit distance.

72. (*c*) S_{nth} represents the distance covered in nth second.

73. (*a*) [muscle] = $\frac{[\text{Power}]}{[\text{Speed}]}$ = [Force].

74. (*a*) $[B] = \left[\frac{\text{force} \times \text{length}}{\text{mass}}\right] = \left[\frac{\text{energy}}{\text{mass}}\right]$ = [Latent heat].

75. (*c*) 30 vernier divisions coincide with 29 main scale divisions.

Least count = 1 SD – 1 VD

$= 1\text{ SD} - \frac{29}{30}\text{ SD} = \frac{1}{30}\text{ SD}$

One main scale division corresponds to $\frac{1°}{2}$ *i.e.* 30′

$\therefore$ Least count $= \frac{1}{30} \times 30' = 1'$.

76. (*b*) A spherometer (Fig. 1) is used to determine the radius of curvature of a spherical surface. The theory of the method is briefly described below :

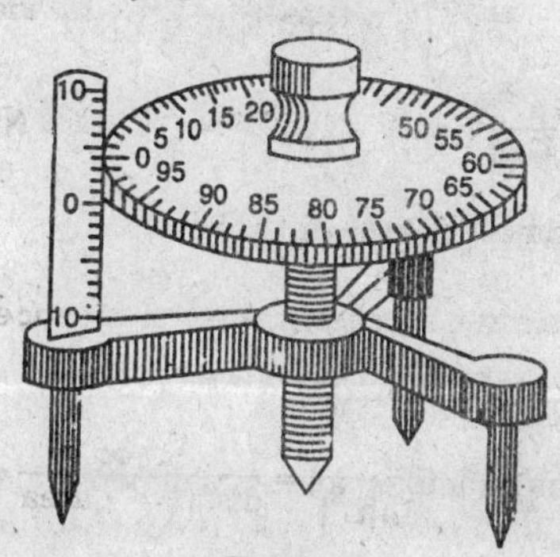

Fig. 1

In the Fig. 2, $r^2 = h(2R - h)$

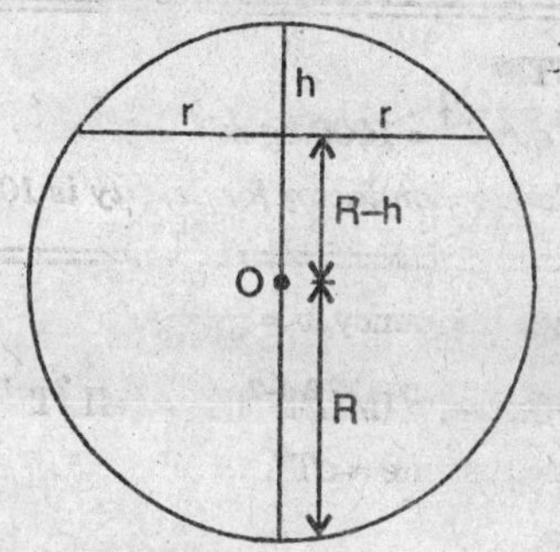

Fig. 2

On simplification,

$$R = \frac{r^2}{2h} + \frac{h}{2}$$

But $r = \frac{l}{\sqrt{3}}$

[Think of an equilateral triangle of side l.]

$$\therefore \quad R = \frac{l^2}{6h} + \frac{h}{2}.$$

77. (*b*) Both $\frac{1}{2}LI^2$ and $\frac{1}{2}CV^2$ represent energy.

78. (*d*) $\frac{1\text{ Nm}^2}{1\text{ kg}^2} = \frac{10^5\text{ dyne} \times 10^4\text{ cm}^2}{10^6\text{ g}^2} = 10^3\text{ dyne cm}^2\text{ g}^{-2}$

79. (*c*) Einstein's mass-energy equivalence is $E = mc^2$.

80. (*d*)

(*a*) CR is time constant, (*b*) $\frac{L}{R}$ is time constant,

(*c*) $\nu = \frac{1}{2\pi\sqrt{LC}}$.

81. (*a*) $[C] = \left[\frac{Q}{m\theta}\right] = \frac{[ML^2T^{-2}]}{[M][K]} = [M^0L^2T^{-2}K^{-1}]$.

82. (*a*) $[b] = [\lambda_m T] = [M^0LT^0K]$.

83. (*d*) $[R] = \left[\frac{PV}{\mu T}\right] = [ML^2T^{-2}][K^{-1}] = [ML^2T^{-2}K^{-1}]$.

84. (*d*) $I = \frac{q}{t}$, $q = It$.

85. (*c*) $[f] = \frac{[S]}{[t^3]} = [M^0LT^{-3}]$.

86. (*b*) $T = 2\pi\sqrt{\frac{I}{MB}}$ (formula for time period).

87. (*d*) As an example, angle is a dimensionless quantity. It has a unit. Specific gravity is dimensionless and has no unit.

88. (*a*) $F = kv$, $[k] = \frac{[F]}{[v]} = \frac{[MLT^{-2}]}{[LT^{-1}]} = [MT^{-1}]$.

So, unit is kg s^{-1}.

89. (*c*) Conductivity $\left(\frac{1}{\rho}\right) = \frac{l}{aR}$. The result follows.

90. (*c*) [Energy] = $[ML^2T^{-2}]$. Increase M and L by a factor of 3 and get the required result.

91. (*c*) Planck's constant and angular momentum are dimensionally same. Moreover, angular momentum is the product of momentum and lever arm of momentum.

92. (*a*) $V = IR$, $[R] = \left[\frac{V}{I}\right] = \left[\frac{\text{Work done}}{\text{Charge} \times I}\right]$

$$= \frac{[ML^2T^{-2}]}{[A^2T]} = [ML^2T^{-3}A^{-2}].$$

93. (*c*) $E = L\frac{dI}{dt}$ or $L = \frac{Edt}{dI}$. Clearly, SI unit is Vs A^{-1}.

94. (*d*) (*a*) and (*b*) are correct because angle is dimensionless. (*c*) is correct because T-ratio is dimensionless.

95. (*d*) $[D] = [T^{-1}]$, $[B] = [x^{-1}]$, $\left[\frac{D}{B}\right] = \left[\frac{x}{T}\right]$ = [velocity]

$$= [M^0LT^{-1}].$$

96. (*c*) 10 VD = 9 SD, 1 VD = $\frac{9}{10}$ SD,

Vernier constant = 1 SD – 1 VD

$$= \left(1 - \frac{9}{10}\right)\text{SD} = \frac{1}{10}\text{ SD} = \frac{1}{10} \times \frac{1}{2}\text{ mm}$$

$$= \frac{1}{20}\text{ mm} = 0.05\text{ mm}.$$

> ☞ *Vernier constant of vernier callipers*
> *= 1 SD – 1 VD.*

97. (*b*) Least count = $\frac{0.5\text{ mm}}{50}$ = 0.01 mm.

98. (*a*) Current density = $\frac{\text{Current}}{\text{area}} = \frac{Q}{\text{area} \times t}$.

QUESTION BANK (Level II)

1. (*b*) It is a brain teaser.

> ☞ In Q. 1, ***the focus of your thought process should be on power a.***

2. (*c*) We are scaling up to $\frac{10^{-5}}{10^{-11}}$ *i.e.* 10^6. In the new scale, the approximate height of mount everest is $10^6 \times 10^4$ m *i.e.* 10^{10} m.

3. (*d*) [Capacitance X] = $[M^{-1}L^{-2}T^2Q^2]$

[Magnetic induction Z] = $[MT^{-1}Q^{-1}]$

$$[Z]^2 = [M^2T^{-2}Q^{-2}]$$

Given : $X = 3YZ^2$ or $Y = \frac{X}{3Z^2}$ or $[Y] = \frac{[X]}{[Z^2]}$

(3 has no dimensions)

$$\therefore\ [Y] = \frac{[M^{-1}L^{-2}T^2Q^2]}{[M^2T^{-2}Q^{-2}]} = [M^{-3}L^{-2}T^4Q^4]$$

4. (*c*) Do not think in terms of I and ω. Remember : kinetic energy is fundamentally 'work'.

5. (*b*) Rydberg constant has dimensions of inverse length.

6. (*a*) [surface tension] = $[ML^0T^{-2}]$, [viscosity] = $[ML^{-1}T^{-1}]$. Clearly, mass has the same power in the two physical quantities.

7. (*d*) $\frac{P}{t}$ has the units of force. P has the units of force × time *i.e.* momentum.

8. (*b*) T-ratios are dimensionless.

9. (*d*) $[e] = \frac{[ML^2T^{-2}]}{[Q]} = [ML^2T^{-2}Q^{-1}]$

10. (*a*) $Q = \frac{KA(\theta_1 - \theta_2)t}{d}$ or $K = \frac{Qd}{(\theta_1 - \theta_2)t}$

SI unit of K is $J m\, m^{-2} K^{-1} s^{-1}$ or $J\, m^{-1} s^{-1} K^{-1}$.

11. (*a*) After having known the answer, you would be tempted to proceed as follows :

$$\left|\sqrt{\frac{hc}{G}}\right| = \frac{[ML^2T^{-1}LT^{-1}]^{1/2}}{[M^{-1}L^3T^{-2}]^{1/2}} = [M^2]^{1/2} = [M]$$

But this is not a correct approach.

Correct approach $[h] = [ML^2T^{-1}]$

$[c] = [LT^{-1}]$

$[G] = [M^{-1}L^3T^{-2}]$

Focus on L and T. You would soon realise that $\frac{hc}{G}$ is required. But $\frac{hc}{G}$ gives M^2. No problem. Take the square root of $\frac{hc}{G}$ and you have done it !

12. (*a*) $[?] = \frac{[Wl^3]}{[4Yd^3\delta]} = \frac{[MLT^{-2}][L^3][L^2]}{[MLT^{-2}][L^3][L]} = [L]$.

13. (*c*) $E = \frac{1}{2}LI^2$, $[L] = \frac{[E]}{[I^2]}$.

14. (*c*) 30 vernier divisions of a spectrometer coincide with 29 main scale divisions.

Least count = 1 SD – 1 VD = 1 SD – $\frac{29}{30}$ SD = $\frac{1}{30}$ SD

720 main scale divisions correspond to 360°. So, one main scale division corresponds to $\frac{1°}{2}$ *i.e.* 30′.

∴ Least count = $\frac{1}{30} \times 30' = 1'$.

> ☞ **Useful information for solving questions on screw gauge, spherometer and spectrometer**
>
> 1. *The principle of a spherometer is the same as that of a screw gauge. So, the pitch and least count of a spherometer are determined in the same way as in the case of screw gauge.*
> 2. *Pitch of a screw is the linear distance through which the screw moves when one complete rotation is given to the screw.*
> 3. *Pitch is the perpendicular distance between two consecutive threads on the screw.*
> 4. *Least count* = $\frac{\textit{Pitch}}{\textit{Number of divisions on circular scale}}$
>
> 5. *The circular scale of spectrometer is divided into 360 equal parts. Each degree is further divided into two parts. So, the smallest scale division is equal to 0.5 degree or 30 minutes. The vernier scale has 30 equal divisions which coincide with 29 equal divisions of the main scale. In such a case, the vernier constant of the spectrometer is 1′.*

15. (*d*) $\left[\frac{v^2}{r}\right] = \left[\frac{L^2T^{-2}}{L}\right] = [LT^{-2}]$. The given quantity is acceleration which is different from (*a*), (*b*) and (*c*).

> ☞ **Think carefully !**
>
> *Centripetal acceleration* $= \frac{v^2}{r}$.

16. (*d*) $F_m = Bqv$ or $[B] = \left[\frac{F_m}{qv}\right] = \left[\frac{F_m}{Itv}\right]$

$= \frac{[MLT^{-2}]}{[A][T][LT^{-1}]} = [ML^0T^{-2}A^{-1}]$

> ☞ **Time-saving solution**
>
> $F = BIl$, $[B] = \frac{[F]}{[Il]} = [ML^0T^{-2}A^{-1}]$

17. (*c*) $[A] = [LT^{-2}]$ or $[L] = [AT^2]$

[Work] = [Force × Distance] = [FL] = $[FAT^2]$.

18. (*d*) Since amplification factor is the ratio of output voltage and input voltage therefore it is a dimensionless constant.

19. (*c*) $[\text{Entropy}] = \frac{[Q]}{[T]} = \frac{[ML^2T^{-2}]}{[K]} = [ML^2T^{-2}K^{-1}]$.

20. (*b*) $[C] = \left[\frac{Q}{V}\right] = \left[\frac{Q^2}{W}\right] = [M^{-1}L^{-2}T^2Q^2]$.

21. (*c*) Magnetic dipole moment of a current loop is the product of current and area of the loop.

22. (*b*) Electrostatic potential energy of a capacitor is $\frac{Q^2}{2C}$.

23. (*d*) $n = \frac{10 \text{ m s}^{-2}}{\text{km h}^{-2}} = \frac{10 \text{ mh}^2}{\text{km s}^2} = \frac{10\text{m}\,(3600)(3600)\text{s}^2}{1000 \text{ m s}^2} = 129600$.

24. (*b*) $W = mg$ and $I = \frac{q}{t}$.

25. (*d*) Do not fear long statements.

$[\eta] = [ML^{-1}T^{-2}]$ or $[T^2] = \left[\frac{M}{L\eta}\right]$.

☞ CAUTION

η is not coefficient of viscosity in this question.

26. (*b*) Think of formula for time period in angular SHM.

27. (*d*) $C = \frac{Q}{V} = \frac{QQ}{W} = \frac{Q^2}{W} = \frac{I^2t^2}{W}$,

$[C] = [M^{-1}L^{-2}T^2][I^2T^2] = [M^{-1}L^{-2}T^4I^2]$.

28. (*c*) Intensity is energy/area/time.

29. (*a*) Note that the dimensional formula of energy is $[ML^2T^{-2}]$

$$n_2 = 1\left[\frac{1\text{ kg}}{10\text{ kg}}\right]^1\left[\frac{1\text{ m}}{1\text{ km}}\right]^2\left[\frac{1\text{ s}}{1\text{ minute}}\right]^{-2}$$

$$= \frac{1}{10} \times \frac{1}{10^6} \times \frac{1}{(60)^{-2}} = \frac{3600}{10^7} = 3.6 \times 10^{-4}.$$

30. (*b*) Applying principle of homogeneity of dimensions, we find that $e^{-\alpha t^2}$ is dimensionless. So, αt^2 is dimensionless. Thus, α should have the dimensions of inverse square time.

31. (*b*) $R = \rho\frac{l}{a}, \frac{V}{I} = \rho\frac{l}{a}, \rho = \frac{Va}{Il} = \frac{[ML^2T^{-2}][T][L]}{[Q][Q]}$

$= [ML^3T^{-1}Q^{-2}]$.

32. (*a*) Velocity, $c = \frac{1}{\sqrt{\mu_0\varepsilon_0}}$.

33. (*d*) From the given relation, $D = -\frac{n(x_2 - x_1)}{n_2 - n_1}$

Here $[n] = \left[\frac{1}{\text{area} \times \text{time}}\right] = \frac{1}{[L^2T]} = L^{-2}T^{-1}$

$[x_2 - x_1] = [L]$ and $[n_2 - n_1] = \left[\frac{1}{\text{volume}}\right] = \left[\frac{1}{L^3}\right] = [L^{-3}]$

So, $[D] = \frac{[L^{-2}T^{-1}L]}{[L^{-3}]} = [L^2T^{-1}]$.

34. (*a*) The formula for fine structure constant is $\frac{e^2}{4\pi\varepsilon_0\left(\frac{h}{2\pi}\right)c}$

The result follows from here.

☞ *Do not deal with ε_0, e, h and c individually.*

35. (*a*) Conductivity $\left(\frac{1}{\rho}\right) = \frac{l}{aR}$.

The result follows.

36. (*a*) $[E] = [ML^2T^{-2}]$ ...(1)

$[V] = [LT^{-1}]$ ...(2)

$[\sigma] = [MT^{-2}]$ ...(3)

Eliminate L from (1) and (2). Then equation (3) would give the required result.

37. (*b*) $r_1 = 10^{-15}$ m, $r_2 = 10^{26}$ m,

$\log r = \frac{1}{2}[\log 10^{-15} + \log 10^{26}]$

$= \frac{1}{2}[-15 + 26] = 5.5 \approx 6, r = 10^6$ m.

38. (*d*) $T'^{-2} = 9.8T^{-2}, \frac{1}{T'^2} = \frac{9.8}{T^2}$

or $T'^2 = \frac{T^2}{9.8} = \frac{(1s)^2}{9.8}, T' = \frac{1}{\sqrt{9.8}}$ s.

39. (*a*) $9.8[LT^{-2}] = [L'T^{-2}] \Rightarrow L' = 9.8$ m.

40. (*b*) $v = k\lambda^a\rho^bg^c : [M^0LT^{-1}] = [L^a][M^bL^{-3b}][L^cT^{-2c}]$

or $[M^0LT^{-1}] = [M^b L^{a-3b+c} T^{-2c}]$

Equating powers of M, L and T, we get

$b = 0, a - 3b + c = 1$ and $-2c = -1$ or $c = \frac{1}{2}$

Again, $a - 3b + c$ or $a + \frac{1}{2} = 1$ or $a = \frac{1}{2}$

$\therefore\ v = k\lambda^{1/2}\rho^0 g^{1/2}$ or $v^2 \propto g\lambda$

41. (*d*) $[M^0L^0T^{-1}] = [M^x][M^yT^{-2y}] = [M^{x+y}T^{-2y}]$

Equating powers of M and T,

$x + y = 0, -2y = -1$

or $y = \frac{1}{2}, x + \frac{1}{2} = 0$ or $x = -\frac{1}{2}$.

42. (*c*) $[X] = \left[\frac{M^{-1}L^3T^{-2} \times ML^2T^{-1}}{L^3T^{-3}}\right]^{1/2} = [L]$.

43. (*c*) $[Rhc] = [L^{-1}ML^2T^{-1}LT^{-1}] = [ML^2T^{-2}]$.

44. (*b*) $[\varepsilon_0] = \frac{[A^2T^2]}{[ML^3T^{-2}]} = [M^{-1}L^{-3}T^4A^2]$.

45. (*a*) Each of the three terms in the given equation has the dimensional formula of force.

46. (*a*) Think in terms of dimensional formula.

47. (*d*) $\frac{1}{ne}$ is hall coefficient. $[n] = [L^{-3}]$, $[e] = [AT]$.

48. (*c*) The right hand side of the given relation is basically $\frac{k}{\text{metre}}$. But, since the left hand side is joule, therefore *h* should be J m.

49. (*a*) $\frac{[ML^2T^{-2}][ML^2T^{-1}]^2}{[M^5][M^{-1}L^3T^{-2}]^2} = [M^0L^0T^0]$.

50. (*c*) $\frac{0.1}{10} = \left(1 - \frac{19}{20}\right)$SD or $\frac{1}{100} = \frac{1}{20} \times 1$ SD

or $1\text{ SD} = \frac{1}{5}\text{ cm} = \frac{1}{5} \times 10\text{ mm} = 2\text{ mm}$.

51. (*d*) Molar volume is 2×10^4 cm^3 ; Bubble volume is 4×10^{-3} cm^3.
This gives 2×10^{-7} mole. With $N = 6 \times 10^{23}$, we reach 10^{17}.

52. (*a*) $[\mu_h] = \dfrac{\text{drift velocity } v}{\text{electric field E}} = \dfrac{[LT^{-1}][AT]}{[MLT^{-2}]} = [M^{-1}T^2A]$.

53. (*b*) JC^{-1} represents work done per unit charge which corresponds to electric potential.

54. (*d*) [Calorie] = $[M^1L^2T^{-2}]$. Comparing with general dimensional formula $[M^a\ L^b\ T^c]$, we get
$a = 1, b = 2, c = -2$.

$$n_2 = 4.2\left[\frac{1\text{ kg}}{\alpha\text{ kg}}\right]^1\left[\frac{1\text{ m}}{\beta\text{ m}}\right]^2\left[\frac{1\text{ s}}{\gamma\text{ s}}\right]^{-2} = 4.2\ \alpha^{-1}\beta^{-2}\gamma^2.$$

55. (*b*) Required volume $= \dfrac{75 \times 10^4 \times 26}{10^3 \times 10^3 \times 10^3}\ (\text{km})^3$
$= 0.0195\ (\text{km})^3 = 0.02\ (\text{km})^3$.

56. (*c*) $[M^0LT^{-1}] = [L^p\ T^{-2p}][L^q] = [L^{p+q}\ T^{-2p}]$
Comparing powers of L and T,

$p + q = 1$ and $-2p = -1$ or $p = \dfrac{1}{2}$.

Again, $\dfrac{1}{2} + q = 1$ or $q = 1 - \dfrac{1}{2} = \dfrac{1}{2}$.

57. (*a*) $[p] = [MLT^{-1}] = [ML^{-3}L^4T^{-4}T^3] = \dfrac{[ML^{-3}][LT^{-1}]^4}{[T^{-1}]^3}$

or $[p] = \left[\dfrac{dv^4}{\nu^3}\right]$.

58. (*d*) $[M^0LT^0] = [M^{p+r}\ L^{q+2r}\ T^{-q-r}]$. Now, compare powers of M, L and T. Then, you have three equations and three unknowns.

> ☞ **An important note**
>
> *If number of unknowns is greater than the number of equations, then something is wrong with your solution.*

59. (*c*) New unit of distance = Speed in new units × 500 s.

60. (*a*) $[MLT^{-2}] = [L^{2a}]\ [L^b\ T^{-b}]\ [M^c\ L^{-3c}] = [M^c\ L^{2a+c-3c}\ T^{-b}]$
Comparing powers of M, L and T, we get
$c = 1, 2a + b - 3c = 1, -b = -2$ or $b = 2$
$2a + 2 - 3(1) = 1 \Rightarrow 2a = 2$ or $a = 1$.

61. (*b*) New unit of mass is $\dfrac{1}{6.67 \times 10^{-11}}$ kg *i.e.* 1.5×10^{10} kg.

62. (*d*) In this case, $-p + q = 0$, $3p + 2q + r = 1$ and $-2p - q - r = 0$.

Solving, $p = \dfrac{1}{2}, q = -\dfrac{3}{2}, r = -\dfrac{1}{2}$.

63. (*c*) $-p + q = 0$, $3p + 2q + r = 0$, $-2p - q - r = 1$.

On solving, $p = \dfrac{1}{2}, q = \dfrac{1}{2}, r = -\dfrac{5}{2}$.

64. (*d*) $\dfrac{ML}{T^2} = 10^3$, $M = \dfrac{10^3 \times 100 \times 100}{10^3}$ kg $= 10^4$ kg

65. (*c*) $x = \dfrac{1\text{ g cm s}^{-1}}{1\text{ Ns}} = \dfrac{1\text{ g cm s}^{-1}}{1\text{ kg} \times 1\text{ ms}^{-2} \times 1\text{ s}}$

$= \dfrac{1\text{ g cm s}^{-1}}{10^3\text{ g} \times 10^2\text{ cm s}^{-2} \times 1\text{ s}} = 10^{-5}$.

66. (*d*) $n(x\text{ m})^2 = 1\text{ m}^2$ or $n = \dfrac{1}{x^2}$.

67. (*d*) *Note carefully that every alternative has 'Gh' and 'c^5'.*
$[Gh] = [M^{-1}L^3T^{-2}]\ [ML^2T^{-1}] = [M^0L^5T^{-3}]$
$[c^5] = [L^5T^{-5}]$. It is now clear that $\dfrac{Gh}{c^5}$ has the dimensions of (time)^2.

68. (*c*) Forget about F. Just think of acceleration.

69. (*b*) $\dfrac{1\text{ kWh}}{1\text{ MeV}} = \dfrac{10^3 \times 3600}{10^6 \times 1.6 \times 10^{-19}} = 2.25 \times 10^{19}$.

70. (*a*) $F = MLT^{-2} = \dfrac{10}{1000} \times \dfrac{10}{100} \times 10 \times 10\text{ N} = \dfrac{1}{10}\text{ N} = 0.1\text{ N}$.

71. (*a*) $1\text{ N} = G\dfrac{1\text{ kg} \times 1\text{ kg}}{1\text{ m}^2}$ or $G = 1\text{ N m}^2\text{ kg}^{-2}$.

72. (*b*) $1\text{ newton} = 6.67 \times 10^{-11}\dfrac{1\text{ kg} \times 1\text{ kg}}{(1\text{ km})^2}\text{ N} = 6.67 \times 10^{-17}\text{ N}$.

73. (*c*) 1 eluoj = 1 notwen × 1 km
$= 6.67 \times 10^{-17}\text{ N} \times 10^3\text{ joule} = 6.67 \times 10^{-14}$ joule.

74. (*a*) $[\sigma] = \left[\dfrac{F}{L}\right] = \left[\dfrac{F \times L}{L^2}\right] = \left[\dfrac{FLT^{-2}}{L^2T^{-2}}\right] = [KV^{-2}T^{-2}]$.

75. (*b*) $[M^0L^0T^0] = [ML^{-1}T^{-2}]^x\ [MT^{-3}]^y\ [LT^{-1}]^z$
or $[M^0L^0T^0] = [M^{x+y}\ L^{-x+z}\ T^{-2x-3y-z}]$
Comparing powers of M, L and T, we get
$x + y = 0, -x + z = 0, -2x - 3y - z = 0$
Solving, $x = 1, y = -1, z = 1$.

76. (*c*) $m = k\rho^a\ g^b\ v^c$
$[M] = [M^a\ L^{-3a}]\ [L^b\ T^{-2b}]\ [L^c\ T^{-c}]$
$[ML^0T^0] = [M^a\ L^{-3a+b+c}\ T^{-2b-c}]$
Comparing powers of M, L and T, $a = 1, -3a + b + c = 0$, $-2b - c = 0$; on solving, $a = 1, b = -3, c = 6$
$\therefore$ $m \propto v^6$

1.2 SIGNIFICANT FIGURES AND ERROR ANALYSIS

Important Terms, Facts and Formulae

SEVEN RULES FOR FINDING THE NUMBER OF SIGNIFICANT FIGURES

1. All non-zero digits are significant.
2. All zeros between two non-zero digits are significant.
3. (*a*) All zeros to the left of an understood decimal point but to the right of a non-zero digit are not significant.

   (*b*) All zeros to the left of an understood decimal point but to the right of a non-zero digit are significant if they come from a measurement.
4. All zeros to the left of an expressed decimal point and to the right of a non-zero digit are significant.
5. All zeros to the right of a decimal point but to the left of non-zero digit are not significant provided there is only a zero to the left of the decimal point.
6. All zeros to the right of a decimal point and to the right of a non-zero digit are significant.
7. The number of significant figures does not vary with the choice of different units.

PROPAGATION OF ERRORS

1. When the result involves the sum of two observed quantities.

$$Z = A + B$$

The maximum possible error in Z is given by

$$\Delta Z = \Delta A + \Delta B$$

2. When the result involves the difference of two observed quantities.

$$Z = A - B$$

The maximum possible error in Z is given by

$$\Delta Z = \Delta A + \Delta B$$

3. When the result involves the product of two observed quantities.

$$Z = AB$$

The maximum possible relative error in Z is given by

$$\frac{\Delta Z}{Z} = \frac{\Delta A}{A} + \frac{\Delta B}{B}$$

In terms of percentage error,

$$\frac{\Delta Z}{Z} \times 100 = \frac{\Delta A}{A} \times 100 + \frac{\Delta B}{B} \times 100$$

4. When the result involves the quotient of two observed quantities.

$$Z = \frac{A}{B}$$

The maximum possible relative error in Z is given by

$$\frac{\Delta Z}{Z} = \frac{\Delta A}{A} + \frac{\Delta B}{B}$$

The maximum percentage error in Z is given by

$$\frac{\Delta Z}{Z} \times 100 = \frac{\Delta A}{A} \times 100 + \frac{\Delta B}{B} \times 100$$

5. When the result involves the product of some powers of the measured values.

$$Z = A^n B^m$$

The maximum possible relative error in Z is given by

$$\frac{\Delta Z}{Z} = n\frac{\Delta A}{A} + m\frac{\Delta B}{B}$$

QUESTION BANK

Level I

Choose the correct answer from the four alternatives given in each of the following questions :

1. The decimal equivalent of $\frac{1}{20}$ upto three significant figures is

(a) 0.05 (b) 0.050
(c) 0.500 (d) 0.0500.

2. Which of the following measurements is most accurate ?

(a) 9×10^{-2} m (b) 90×10^{-3} m
(c) 900×10^{-4} m (d) 0.090 m.

3. When 999 mm is added to 100 m, we get

(a) 101 m (b) 100.999 m
(c) 100.99 m (d) 100.9 m.

4. If 1.76 cm^3 is the volume of one cube, then the volume of twenty-five such cubes is

(a) 44 cm^3 (b) 44.0 cm^3
(c) 44.00 cm^3 (d) 44.000 cm^3.

5. If X = A + B and ΔX, ΔA and ΔB are maximum errors in X, A and B respectively, then

(a) $\Delta X = \Delta A + \Delta B$ (b) $\Delta X = \Delta A - \Delta B$
(c) $\frac{\Delta X}{X} = \frac{\Delta A}{A} - \frac{\Delta B}{B}$ (d) $\frac{\Delta X}{X} = \frac{\Delta A}{A} + \frac{\Delta B}{B}$.

6. If $x = a - b$, then the maximum percentage error in the measurement of x will be

(a) $\left(\frac{\Delta a}{a} + \frac{\Delta b}{b}\right) \times 100\%$ (b) $\left(\frac{\Delta a}{a} - \frac{\Delta b}{b}\right) \times 100\%$
(c) $\left(\frac{\Delta a}{a-b} + \frac{\Delta b}{a-b}\right) \times 100\%$ (d) $\left(\frac{\Delta a}{a-b} - \frac{\Delta b}{a-b}\right) \times 100\%$.

7. If X = A × B and ΔX, ΔA and ΔB are maximum absolute errors in X, A and B respectively, then the maximum relative error in X is given by

(a) $\Delta X = \Delta A + \Delta B$ (b) $\Delta X = \Delta A - \Delta B$
(c) $\frac{\Delta X}{X} = \frac{\Delta A}{A} - \frac{\Delta B}{B}$ (d) $\frac{\Delta X}{X} = \frac{\Delta A}{A} + \frac{\Delta B}{B}$.

8. If $X = \frac{A}{B}$ and ΔX, ΔA and ΔB are the maximum absolute errors in X, A and B respectively, then the maximum fractional error in X is given by

(a) $\Delta X = \Delta A + \Delta B$ (b) $\Delta X = \Delta A - \Delta B$
(c) $\frac{\Delta X}{X} = \frac{\Delta A}{A} - \frac{\Delta B}{B}$ (d) $\frac{\Delta X}{X} = \frac{\Delta A}{A} + \frac{\Delta B}{B}$.

9. If the percentage errors of A, B and C are a, b and c respectively, then the total percentage error in the product ABC is

(a) abc (b) $a + b + c$
(c) $\frac{1}{a} + \frac{1}{b} + \frac{1}{c}$ (d) $ab + bc + ca$
(e) none of these.

10. If $f = x^2$, then the relative error in f is

(a) $\frac{2\Delta x}{x}$ (b) $\frac{(\Delta x)^2}{x}$
(c) $\frac{\Delta x}{x}$ (d) $(\Delta x)^2$.

11. With due regard for significant figures, $(12.5)^2$ =

(a) 156.250 (b) 156.25
(c) 156.2 (d) 156.

12. The initial temperature of a liquid is (80.0 ± 0.1)°C. After it has been cooled, its temperature is (10.0 ± 0.1)°C. The fall in temperature in degree centigrade is

(a) 70.0 (b) 70.0 ± 0.3
(c) 70.0 ± 0.2 (d) 70.0 ± 0.1.

13. Which of the following measurements is most significant ?

(a) 0.003 mm (b) 3.00 mm
(c) 30.00 mm (d) 3.0 mm.

14. The value of 0.99 – 0.989 is

(a) 0.001 (b) 0.010×10^{-1}
(c) 0.01×10^{-1} (d) 0.1×10^{-3}.

15. The length of a pencil is 10.4 cm. Out of it, 2.01 cm is cut out. The remaining length of the pencil is

(a) 8 cm (b) 8.3 cm
(c) 8.4 cm (d) 8.39 cm.

16. Two rods with lengths 20.123 cm and 18.1 cm are placed side by side. The difference in their lengths is

(a) 2 cm (b) 2.0 cm
(c) 2.02 cm (d) 2.023 cm.

17. In the previous question, the two rods are placed end to end. Then the combined length is

(a) 38 cm (b) 38.2 cm
(c) 38.72 cm (d) 38.223 cm.

18. The length of a cylinder is 2.346 m. What will be the new length of the cylinder if 1.24 m length is welded to the cylinder along the length ?

(*a*) 3.586 m (*b*) 3.59 m
(*c*) 3.6 m (*d*) 4 m.

19. The length of a rod as measured in an experiment is 2.56 m, 2.51 m, 2.49 m, 2.58 m, 2.48 m and 2.55 m respectively. The average length is

(*a*) $\frac{15.17}{6}$ m (*b*) 2.5283 m
(*c*) 2.53 m (*d*) 2.528 m.

20. A physical quantity is represented by $X = M^a L^b T^{-c}$. If percentage errors in the measurements of M, L and T are α%, β% and γ% respectively, then total percentage error is

(*a*) $(\alpha a + \beta b - \gamma c)$% (*b*) $(\alpha a + \beta b + \gamma c)$%
(*c*) $(\alpha a - \beta b - \gamma c)$% (*d*) 0%.

21. The heat generated in a circuit is given by $Q = I^2 Rt$, where I is current, R is resistance and *t* is time. If the percentage errors in measuring I, R and *t* are 2%, 1% and 1% respectively, then the maximum error in measuring heat will be

(*a*) 2% (*b*) 3%
(*c*) 4% (*d*) 6%.

22. The sides of a rectangle are 6.01 m and 12 m. Taking the significant figures into account, the area of the rectangle is

(*a*) 72 m^2 (*b*) 72.1 m^2
(*c*) 72.00 m^2 (*d*) 72.12 m^2.

23. What is the area of a disc of radius 1.1 cm ?

(*a*) 3.8028571 cm^2 (*b*) 3.8029 cm^2
(*c*) 3.803 cm^2 (*d*) 3.8 cm^2.

24. What is the volume of a sphere whose radius is 5 cm ?

(*a*) 523.80952 cm^3 (*b*) 523.8 cm^3
(*c*) 524 cm^3 (*d*) 5×10^2 cm^3.

25. Error in the measurement of radius of sphere is 2%. Then error in the measurement of volume is

(*a*) 1% (*b*) 5%
(*c*) 3% (*d*) 6%.

26. The percentage errors in the measurement of mass and speed are 2% and 3% respectively. How much will be the maximum error in the estimate of kinetic energy obtained by measuring mass and speed ?

(*a*) 11% (*b*) 8%
(*c*) 5% (*d*) 1%.

27. The internal and external diameters of a hollow cylinder are measured with the help of a vernier callipers. Their values are 4.23 ± 0.01 cm and 3.87 ± 0.01 cm respectively. The thickness of the wall of the cylinder is

(*a*) 0.36 ± 0.02 cm (*b*) 0.18 ± 0.02 cm
(*c*) 0.36 ± 0.01 cm (*d*) 0.18 ± 0.01 cm.

28. Given : Resistance, $R_1 = (8 \pm 0.4)\ \Omega$ and Resistance, $R_2 = (8 \pm 0.6)\ \Omega$. What is the net resistance when R_1 and R_2 are connected in series ?

(*a*) $(16 \pm 0.4)\ \Omega$ (*b*) $(16 \pm 0.6)\ \Omega$
(*c*) $(16 \pm 1.0)\ \Omega$ (*d*) 16 ± 0.2.

29. The length, breadth and thickness of a metal block is given by l = 9 cm, b = 8 cm, t = 2.45 cm. The volume of the block is

(*a*) 2×10^2 cm^3 (*b*) 1.8×10^2 cm^3
(*c*) 1.77×10^2 cm^3 (*d*) 1.764×10^2 cm^2.

30. The effective length of a simple pendulum is the sum of the following three :

length of string, radius of bob, length of hook

In a simple pendulum experiment, the length of the string, as measured by a metre scale is 92.0 cm. The radius of the bob combined with the length of the hook, as measured by a vernier callipers is 2.15 cm. The effective length of the pendulum is

(*a*) 92.0 cm (*b*) 94.2 cm
(*c*) 94.15 cm (*d*) 94 cm.

31. The pressure on a square plate is measured by measuring the force on the plate and the length of the sides of the plate by using the formula $P = \frac{F}{l^2}$. If the maximum errors in the measurement of force and length are 4% and 2% respectively, then the maximum error in the measurement of pressure is

(*a*) 1% (*b*) 2%
(*c*) 8% (*d*) 10%.

32. The focal length of a mirror is given by $\frac{1}{f} = \frac{1}{u} + \frac{1}{v}$ where u and v represent object and image distances respectively.

The maximum relative error in f is

(*a*) $\frac{\Delta f}{f} = \frac{\Delta u}{u} + \frac{\Delta v}{v}$ (*b*) $\frac{\Delta f}{f} = \frac{1}{\Delta u/u} + \frac{1}{\Delta v/v}$

(*c*) $\frac{\Delta f}{f} = \frac{\Delta u}{u} + \frac{\Delta v}{v} - \frac{\Delta (u+v)}{u+v}$ (*d*) $\frac{\Delta f}{f} = \frac{\Delta u}{u} + \frac{\Delta v}{v} + \frac{\Delta u}{u+v} + \frac{\Delta v}{u+v}$.

33. The density of the material of a cube is measured by measuring its mass and length of its side. If the maximum errors in the measurement of mass and the length are 3% and 2% respectively, the maximum error in the measurement of density is

(*a*) 1% (*b*) 5%
(*c*) 7% (*d*) 9%.

34. There are atomic clocks capable of measuring time with an accuracy of 1 part in 10^{11}. If two such clocks are operated with precision, then after running for 5000 years, these will record

(*a*) a difference of nearly 1 s (*b*) a difference of 1 day
(*c*) a difference of 10^{11} s (*d*) a difference of 1 year.

35. If there is a positive error of 50% in the measurement of speed of a body, then the error in the measurement of kinetic energy is

(*a*) 25% (*b*) 50%
(*c*) 100% (*d*) 125%.

36. The length of a cube is 2.1×10^{-2} m. The volume in significant figures will be

(*a*) 9.2×10^{-6} m^3 (*b*) 9.3×10^{-6} m^3
(*c*) 9.26×10^{-6} m^3 (*d*) 9.261×10^{-6} m^3.

37. The mass of a box is 2.3 g. Two gold pieces, each of mass 0.035 g, are placed in it. The total mass of the box and gold pieces is

(*a*) 2.3 g (*b*) 2.4 g
(*c*) 2.37 g (*d*) 2.370 g.

38. A certain body weighs 22.42 g and has a measured volume of 4.7 cm^3. The possible errors in the measurement of mass and volume are 0.01 g and 0.1 cm^3. Then maximum error in the density will be nearly

(*a*) 22% (*b*) 2%
(*c*) 0.2% (*d*) 0.02%.

39. The least count of a stop watch is 0.2 second. The time of 20 oscillations of a pendulum is measured to be 25 second. The percentage error in the measurement of time will be

(*a*) 8% (*b*) 1.8%
(*c*) 0.8% (*d*) 0.1%.

40. One side of a cubical block is measured with the help of a vernier callipers of vernier constant 0.01 cm. This side comes out to be 1.23 cm. What is the percentage error in the measurement of area ?

(*a*) $\frac{1.23}{0.01} \times 100$ (*b*) $\frac{0.01}{1.23} \times 100$
(*c*) $2 \times \frac{0.01}{1.23} \times 100$ (*d*) $3 \times \frac{0.01}{1.23} \times 100$.

41. If the value of the resistance is 10.845 Ω and the value of the current is 3.23 ampere, then the potential difference is 35.02935 volt. Its value in correct significant figures would be

(*a*) 35 V (*b*) 35.0 V
(*c*) 35.03 V (*d*) 35.029 V.

42. The radius of the sphere is (4.3 ± 0.1) cm. The percentage error in its volume is

(*a*) $\frac{0.1}{4.3} \times 100$ (*b*) $3 \times \frac{0.1 \times 100}{4.3}$
(*c*) $\frac{1}{3} \times \frac{0.1 \times 100}{4.3}$ (*d*) $3 + \frac{0.1 \times 100}{4.3}$.

43. In an experiment, we measure quantities *a*, *b* and *c*. Then *x* is calculated from the formula, $x = \frac{ab^2}{c^3}$

The percentage errors in *a*, *b*, *c* are ± 1%, ± 3%, and ± 2% respectively. The percentage error in *x* can be

(*a*) ± 1% (*b*) ± 4%
(*c*) 7% (*d*) ± 13%.

44. The density of a cube is measured by measuring its mass and the length of its sides. If the maximum errors in the measurement of mass and length are 3% and 2% respectively, then the maximum error in the measurement of the density of cube is

(*a*) 7% (*b*) 5%
(*c*) 10% (*d*) 9%.

45. A public park, in the form of a square, has an area of (100 ± 0.2) m^2. The side of park is

(*a*) (10 ± 0.01) m (*b*) (10 ± 0.1) m
(*c*) (10.0 ± 0.1) m (*d*) (10.0 ± 0.2) m.

46. If the error in the measurement of momentum of a particle is (+ 100%), then the error in the measurement of kinetic energy is

(*a*) 25% (*b*) 200%
(*c*) 300% (*d*) 400%.

QUESTION BANK

Level II

Choose the correct answer from the four alternatives given in each of the following questions :

1. The dimensions of a rectangular block measured with callipers having least count of 0.01 cm are 5 mm × 10 mm × 5 mm. The maximum percentage error in the measurement of the volume of the block is

(*a*) 5% (*b*) 10%
(*c*) 15% (*d*) 20%.

2. The length l, breadth b and thickness t of a block of wood were measured with the help of a measuring scale. The results with permissible errors are

$l = 15.12 \pm 0.01$ cm, $b = 10.15 \pm 0.01$ cm,
$t = 5.28 \pm 0.01$ cm.

The percentage error in volume upto proper significant figures is

(*a*) 0.28% (*b*) 0.36%
(*c*) 0.48% (*d*) 0.64%.

3. Given : Potential difference, V = (8 ± 0.5) V and current, I = (2 ± 0.2) A. The value of resistance R is

(*a*) 4 ± 16.25% (*b*) 4 ± 6.25%
(*c*) 4 ± 10% (*d*) 4 ± 8%.

4. The radius of a wire is 0.24 mm. Then its area of cross-section by taking significant figures into consideration is

(*a*) 0.1 mm^2 (*b*) 0.2 mm^2
(*c*) 0.18 mm^2 (*d*) 0.180 mm^2
(*e*) 0.1808 mm^2.

5. The measured mass and volume of a body are 23.42 g and 4.9 cm^3 respectively with possible error 0.01 g and 0.1 cm^3. The maximum error in density is nearly

(*a*) 0.2% (*b*) 2%
(*c*) 5% (*d*) 10%.

6. The velocity of transverse wave in a string is $v = \sqrt{\frac{T}{m}}$, where T is the tension in the string and m is mass per unit length. If T = 3.0 kgf, mass of string is 2.5 g and length of string is 1.000 m, then the percentage error in the measurement of velocity is

(*a*) 0.5 (*b*) 0.7
(*c*) 2.3 (*d*) 3.6.

7. A current of 2.34 ampere flows in a resistance of 11.111111 Ω. The potential difference across the given resistance with due regard for significant figures is

(*a*) 26.000 V (*b*) 26.00 V
(*c*) 26.0 V (*d*) 26 V.

8. A resistor of 10 kΩ having tolerance 10% is connected in series with another resistor of 20 kΩ having tolerance 20%. The tolerance of the combination will be approximately

(*a*) 10% (*b*) 13%
(*c*) 17% (*d*) 20%.

9. The length, breadth and thickness of a block is measured to be 50 cm, 2.0 cm and 1.00 cm. The percentage error in the measurement of volume is

(*a*) 0.8% (*b*) 8%
(*c*) 10% (*d*) 12.5%.

10. A resistor of 4 kΩ with tolerance 10% is connected in parallel with a resistor of 6 kΩ with tolerance 10%. The tolerance of the parallel combination is nearly

(*a*) 10% (*b*) 20%
(*c*) 30% (*d*) 40%.

11. The relative density of the material of a body is the ratio of its weight in air and the loss of its weight in water. By using a spring balance, the weight of the body in air is measured to be 5.00 ± 0.05 N. The weight of the body in water is measured to be 4.00 ± 0.05 N. Then the maximum possible percentage error in relative density is

(*a*) 11% (*b*) 10%
(*c*) 9% (*d*) 7%.

12. In Q. 11, the relative density is

(*a*) 1.25 ± 9% (*b*) 1.25 ± 11%
(*c*) 5.00 ± 11% (*d*) 5.00 ± 9%.

13. Given : π = 3.14. The value of π^2 with due regard for significant figures is

(*a*) 9.86 (*b*) 9.859
(*c*) 9.8596 (*d*) 9.85960.

14. The length l, breadth b and thickness t of a block are measured with the help of a metre scale. Given : $l = 15.12 \pm 0.01$ cm, $b = 10.15 \pm 0.01$ cm, $t = 5.28 \pm 0.01$ cm. The percentage error in volume is

(*a*) 0.64% (*b*) 0.28%
(*c*) 0.37% (*d*) 0.48%.

15. The specific resistance ρ of a circular wire of radius r, resistance R and length l is given by $\rho = \frac{\pi r^2 R}{l}$

Given : $r = 0.24 \pm 0.02$ cm, $R = 30 \pm 1\ \Omega$ and $l = 4.80 \pm 0.01$ cm. The percentage error in ρ is nearly

(*a*) 7% (*b*) 9%
(*c*) 13% (*d*) 20%.

16. The following observations were taken for determining surface tension of water by capillary tube method :

Diameter of capillary, $D = 1.25 \times 10^{-2}$ m and rise of water in capillary, $h = 1.45 \times 10^{-2}$ m.

Taking $g = 9.80$ m s^{-2} and using the relation $T = (rgh/2) \times 10^3$ N m^{-1}, what is the possible error in surface tension T ?

(*a*) 2.4% (*b*) 15%
(*c*) 1.6% (*d*) 0.15%.

ANSWERS

QUESTION BANK (Level I)

1. (*d*)	**2.** (*c*)	**3.** (*a*)	**4.** (*b*)	**5.** (*a*)	**6.** (*c*)	**7.** (*d*)	**8.** (*d*)
9. (*b*)	**10.** (*a*)	**11.** (*d*)	**12.** (*c*)	**13.** (*c*)	**14.** (*c*)	**15.** (*c*)	**16.** (*b*)
17. (*b*)	**18.** (*b*)	**19.** (*c*)	**20.** (*b*)	**21.** (*d*)	**22.** (*a*)	**23.** (*d*)	**24.** (*d*)
25. (*d*)	**26.** (*b*)	**27.** (*b*)	**28.** (*c*)	**29.** (*a*)	**30.** (*b*)	**31.** (*c*)	**32.** (*d*)
33. (*d*)	**34.** (*a*)	**35.** (*d*)	**36.** (*b*)	**37.** (*b*)	**38.** (*b*)	**39.** (*c*)	**40.** (*c*)
41. (*b*)	**42.** (*b*)	**43.** (*d*)	**44.** (*d*)	**45.** (*a*)	**46.** (*c*		

QUESTION BANK (Level II)

1. (*a*)	**2.** (*b*)	**3.** (*a*)	**4.** (*c*)	**5.** (*b*)	**6.** (*d*)	**7.** (*c*)	**8.** (*c*)
9. (*b*)	**10.** (*c*)	**11.** (*a*)	**12.** (*c*)	**13.** (*a*)	**14.** (*c*)	**15.** (*d*)	**16.** (*c*).

HINTS/SOLUTIONS

QUESTION BANK (Level I)

1. (*d*) Two zeros to the right of the decimal point and to the right of 5 are significant.

2. (*c*) In (*a*), there is an error of 1 in 9. In (*b*), there is an error of 1 in 90. In (*c*), there is an error of 1 in 900. In (*d*), there is an error of 0.01 in 0.09.

3. (*a*) 100 m + 0.999 m = 100.999 m = 101 m.

4. (*b*) Note that there are three significant figures in 1.76 cm^3.

> ☞ *In Q. 4, do not bother yourself about the significant figures of the multiplicative factor '25'.*

5. (*a*) When two quantities are added, their absolute errors are added up.

6. (*c*) Maximum absolute error is $\Delta a + \Delta b$. Now work out the relative error and finally the percentage error.

7. (*d*) When two quantities are multiplied, their maximum relative errors are added up.

8. (*d*) When two quantities are divided, their maximum fractional or relative errors are added up.

9. (*b*) In a product, percentage errors are added up.

10. (*a*) Required relative error = power × relative error in x.

11. (*d*) $(12.5)^2 = 156.25$. There are only three significant figures in 12.5. So, rounding off to three significant figures, we get 156.

> ☞ **CAUTION**
> *When a quantity is squared, the number of significant figures is not squared.*

12. (*a*) When quantities are subtracted, their maximum absolute errors are added up.

13. (*c*) 30.00 has four significant digits.

14. (*c*) The result should have two decimal places (same as in 0.99).

15-18. The number of decimal places in the result is the same as the smallest number of significant decimal places in any number used in the sum or difference.

19. (*c*) Average length

$$= \frac{2.56 + 2.51 + 2.49 + 2.58 + 2.48 + 2.55}{6} \text{ m} = 2.528 \text{ m}.$$

But only two decimal places are required. So, the average length is 2.53 m.

20. (*b*) Treat '$-c$' as '$+c$'.

> ☞ *In this type of questions, the negative sign in the power is not to be considered.*

21. (*d*) Required error is $2 \times 2\% + 1\% + 1\%$ *i.e.* 6%.

22. (*a*) Result should have only two significant numbers (same as in 12 m).

23. (*d*) Express the result in two significant figures.

24. (*d*) Volume of sphere $= \frac{4}{3}\pi r^3$. Use this formula and then express the result in one significant number.

25. (*d*) Volume $\alpha\ r^3$. So, error is $3 \times 2\%$ *i.e.* 6%.

26. (*b*) Remember : kinetic energy $= \frac{1}{2}\ mv^2$. Required percentage error is $2\% + 2 \times 3\%$ *i.e.* 8%.

27. (*b*) Subtract 3.87 from 4.23 and then divide by 2.

28. (*a*) $R = (8 + 8 \pm 0.4 \pm 0.6)\ \Omega = (16 \pm 1.0)\ \Omega$.

29. (*a*) The result has to be in one significant number only.

30. (*b*) (92.0 + 2.15) cm = 94.15 cm. Rounding off to first decimal place, we get 94.2 cm.

31. (*c*) Maximum percentage error in $P = 4\% + 2 \times 2\% = 8\%$.

32. (*d*) $f = \frac{uv}{u+v}, \frac{\Delta f}{f} = \frac{\Delta u}{u} + \frac{\Delta v}{v} + \frac{\Delta (u+v)}{u+v}$.

33. (*d*) Required error in density $= 3\% + 3 \times 2\% = 9\%$.

34. (*a*) Required time $= \frac{5000 \times 86400 \times 365.25}{10^{11}}$ s = 1.6 s.

35. (*d*) $\frac{E'}{E} = \frac{9}{4}$ or $\frac{E'}{E} - 1 = \frac{9}{4} - 1 = \frac{5}{4}$,

$$\frac{E' - E}{E} \times 100 = \frac{5}{4} \times 100 = 125$$

36. (*b*) Volume $= (2.1 \times 10^{-2})^3\ m^3 = 9.261 \times 10^{-6}\ m^3$. Rounding off to two significant figures, we get $9.3 \times 10^{-6}\ m^3$.

37. (*b*) (2.3 + 0.035 + 0.035) g = 2.37 g

But we have to retain only one decimal place.

So, the total mass is 2.4 g.

38. (*b*) Maximum error $= \frac{0.01}{22.42} \times 100 + \frac{0.1}{4.7} \times 100 \approx 2$.

39. (*c*) $\frac{0.2}{25} \times 100 = 0.8$.

40. (*c*) Percentage error in measurement of a side $= \frac{0.01}{1.23} \times 100$

Percentage error in measurement of area

$$= 2 \times \frac{0.01}{1.23} \times 100.$$

41. (*b*) We have to retain three significant figures in the result.

42. (*b*) Percentage error in radius is $\frac{0.1}{4.3} \times 100$. Again, $V \propto R^3$.

> ☞ *It clearly follows from Q. 42 that in an experiment based on the volume of a sphere, the radius of the sphere has to be measured very accurately. A better approach will be to measure the volume directly. As an illustration, take a graduated cylinder containing water and put a sphere into it.*

43. (*d*) Percentage error in $x = 1\% + 2 \times 3\% + 3 \times 2\% = 13\%$. The sign ± has been used because the words 'maximum percentage error' have not been used.

> ☞ **An Important Note**
>
> *Percentage error is* $\pm \frac{\Delta A}{A} \times 100$
>
> *Maximum percentage error is* $\frac{\Delta A}{A} \times 100$.

44. (*d*) Maximum error in density $= (3 + 3 \times 2)\% = 9\%$.

45. (*a*) Percentage error in side $= \frac{1}{2}\left[\frac{0.2}{100} \times 100\right] = 0.1$.

Absolute error in side $= \frac{0.1}{100} \times 10 = 0.01$.

46. (*c*) Relation between E_k and p is $E_k = \frac{p^2}{2m}$. When p is doubled, E_k becomes four times. So, E_k increases by 300.

> ☞ **You were invited to commit a mistake**
>
> *And you probably committed it and thought that (b) is the right choice !*
>
> *Note that the laws of error propagation are to be applied for small percentage only.*

QUESTION BANK (Level II)

1. (*a*) Required percentage error

$= 2 \times \frac{0.01}{0.5} \times 100 + \frac{0.01}{1.0} \times 100 = 4 + 1 = 5$

2. (*b*) Percentage error in volume

$= \frac{0.01}{15.12} \times 100 + \frac{0.01}{10.15} \times 100 + \frac{0.01}{5.28} \times 100$

$= 0.07 + 0.10 + 0.19 = 0.36$

3. (*a*) Percentage error in V is $\frac{0.5}{8} \times 100$ *i.e.* 6.25. Percentage error in I is $\frac{0.2}{2} \times 100$ *i.e.* 10. Now, add up the percentage errors.

4. (*c*) Area of cross-section $= \frac{22}{7} \times 0.24 \times 0.24\ \text{mm}^2 = 0.18\ \text{mm}^2$.

5. (*b*) Percentage error in mass $= \frac{0.01}{23.42} \times 100 = 0.04$,

Percentage error in volume $= \frac{0.1}{4.9} \times 100 = 2.04$. Adding up the percentage errors, we get nearly 2%.

6. (*d*) $v = \sqrt{\frac{T}{m}} = \left[\frac{m'g}{\frac{M}{l}}\right]^{1/2} = \left[\frac{m'lg}{M}\right]^{1/2}$

It follows from here, $\frac{\Delta v}{v} = \frac{1}{2}\left[\frac{\Delta m'}{m} + \frac{\Delta l}{l} + \frac{\Delta M}{M}\right]$

$= \frac{1}{2}\left[\frac{0.1}{3.0} + \frac{0.001}{1.000} + \frac{0.1}{2.5}\right] = \frac{1}{2}[0.03 + 0.001 + 0.04] = 0.036$

Percentage error in the measurement = 3.6.

7. (*c*) V = 2.34 × 11.111111 volt = 26.0 volt.

8. (*c*) $\Delta R_s = \Delta R_1 + \Delta R_2 = \left[\frac{10}{100} \times 10 + \frac{20}{100} \times 20\right] k\Omega = 5\ k\Omega$

$R_s = (10 + 20)\ k\Omega = 30\ k\Omega$,

$\frac{\Delta R_s}{R_s} \times 100 = \frac{5}{30} \times 100 = \frac{50}{3} \approx 17.$

9. (*b*) Percentage error in length $= \frac{1}{50} \times 100 = 2$,

Percentage error in breadth $= \frac{0.1}{2.0} \times 100 = 5$,

Percentage error in thickness $= \frac{0.01}{1.00} \times 100 = 1$,

Percentage error in volume = 2 + 5 + 1 = 8.

10. (*c*) $R_s = \frac{R_1 R_2}{R_1 + R_2}$, $\frac{\Delta R_s}{R_s} \times 100$

$= \frac{\Delta R_1}{R_1} \times 100 + \frac{\Delta R_2}{R_2} \times 100 + \frac{\Delta (R_1 + R_2)}{R_1 + R_2} \times 100$

Now, $\Delta R_1 = \frac{10}{100} \times 4\ k\Omega = 0.4\ k\Omega$,

$\Delta R_2 = \frac{10}{100} \times 6\ k\Omega = 0.6\ k\Omega$

Again, $\frac{\Delta R_s}{R_s} \times 100 = \frac{0.4}{4} \times 100 + \frac{0.6}{6} \times 100$

$+ \frac{0.4 + 0.6}{10} \times 100 = 10 + 10 + 10 = 30$

11. (*a*) $R.D = \frac{w_1}{w_1 - w_2}$, $\frac{\Delta (R.D)}{R.D} \times 100$

$= \frac{\Delta w_1}{w_1} \times 100 + \frac{\Delta (w_1 - w_2)}{w_1 - w_2} \times 100$

$= \frac{0.05}{5.00} \times 100 + \frac{0.05 + 0.05}{1.00} \times 100 = 1 + 10 = 11$

12. (*c*) $R.D = \frac{5.00}{1.00} \pm 11\% = 5.00 \pm 11\%$

14. (*c*) $\frac{\Delta l}{l} \times 100 = \frac{0.01}{15.12} \times 100 = 0.07$,

$\frac{\Delta b}{b} \times 100 = \frac{0.01}{10.15} \times 100 = 0.1$,

$\frac{\Delta t}{t} \times 100 = \frac{0.01}{5.28} \times 100 = 0.2$

Percentage error = 0.07 + 0.1 + 0.2 = 0.37.

15. (*d*) Required percentage $= \frac{0.02}{0.24} \times 100 \times 2 + \frac{1}{30} \times 100$

$+ \frac{0.01}{4.80} \times 100 = 16.7 + 3.3 + 0.2 \approx 20.$

16. (*c*) Percentage error in T $= \frac{0.01}{1.25} \times 100 + \frac{0.01}{9.80} \times 100$

$+ \frac{0.01}{1.45} \times 100 = 0.8 + 0.1 + 0.7 = 1.6.$

2 DESCRIPTION OF MOTION IN ONE DIMENSION

Important Terms, Facts and Formulae

EQUATIONS OF UNIFORMLY ACCELERATED MOTION

1. **Velocity-time equation**

$$v = u + at$$

This equation is basically a vector equation

$$\vec{v} = \vec{v} + \vec{at}$$

Here, u is initial velocity, v is final velocity, a is uniform acceleration and t is the time during which the velocity changes from u to v.

If $v(0)$ is taken as velocity at $t = 0$, $v(t)$ the velocity at time t and a is acceleration, then

$$v(t) = v(0) + at$$

2. **Displacement-time equation or position-time equation**

$$S = ut + \frac{1}{2}at^2$$

This equation is basically a vector equation

$$\vec{S} = \vec{ut} + \frac{1}{2}\vec{at}^2$$

Here, $\vec{S}$ represents the displacement of the particle in time t.

If $x(0)$ represents the position of particle at $t = 0$ and $x(t)$ represents the position of particle at time t, then

$$x(t) = x(0) + v(0)t + \frac{1}{2}at^2$$

3. **Displacement-velocity equation**

$$v^2 - u^2 = 2aS$$

This is a scalar equation. However it can be written in vector form as follows :

$$\vec{v}\,.\,\vec{v} - \vec{u}\,.\,\vec{u} = 2\vec{a}\,.\,\vec{S}$$

4. **Displacement in nth second**

$$S_{nth} = u + \frac{a}{2}(2n - 1)$$

Here, S_{nth} represents displacement during nth second.

> ☞ **CAUTION**
>
> *If the acceleration is constant both in magnitude and direction, only then the above equations are to be used.*
>
> *The above equations are not applicable to simple harmonic motion or circular motion.*

AN IN DEPTH DISCUSSION OF A SERIOUS CONCEPTUAL CONFUSION

It is generally thought that **the acceleration due to gravity is to be taken as positive in the case of freely falling bodies and negative in the case of bodies thrown upwards.**

This is a misconception. **The correct concept is as under :**

In any problem of rising or falling body, choose upward or downward direction as positive. Suppose a ball is thrown vertically upwards. Let us further suppose that the upward direction is to be taken as positive. The downward direction would then clearly be taken as negative. Now, both the initial velocity and displacement will be taken as positive. But acceleration due to gravity will be taken as negative. If the downward direction is taken as positive, then both the initial velocity and displacement will be taken as negative. But acceleration due to gravity will be taken as positive.

Conclusion. It is your personal choice whether to take upward direction as positive or negative in a given problem.

Two Important Notes

1. The acceleration due to gravity is always directed vertically downwards.
2. Never give sign to the unknown quantity. If your solution is correct, the proper sign shall be there in the answer.

BASIC CONCEPTS

1. In uniform motion, the velocity does not depend upon the choice of origin.

2. Both speed and velocity have the same units and dimensions.
3. A body can have average speed without having average velocity. Think of a ball thrown upwards and returning to the point of throw.
4. Algebraically, the velocity can be positive, negative or zero.
5. The speed is always positive or zero.
6. Generally, the positive velocity implies that the object is moving to the right of the origin.
7. There is no difference between instantaneous velocity and average velocity in the case of uniform motion.
8. If a body covers distance S_1 with velocity v_1, distance S_2 with velocity v_2 and so on in the same direction, then

$$\text{average speed, } v_{av} = \frac{S_1 + S_2 + \ldots\ldots}{\frac{S_1}{v_1} + \frac{S_2}{v_2} + \ldots\ldots}.$$

9. **The magnitude of displacement may or may not be equal to the distance covered by the body.**
10. In uniformly accelerated motion, the average acceleration over a given time interval is equal to the instantaneous acceleration during that time interval.
11. Apply the following kinematic equations to uniformly accelerated one-dimensional motion only

(*i*) $v = u + at$ (*ii*) $v^2 - u^2 = 2aS$

(*iii*) $S = ut + \frac{1}{2}at^2$ (*iv*) $S = \left(\frac{v+u}{2}\right)t$

(*v*) $S_{nth} = u + \frac{a}{2}(2n - 1)$

12. Facts about UNIFORM VELOCITY

- It is on a straight line path and always in the same direction.
- The magnitude of velocity is equal to speed.
- The average velocity is equal to instantaneous velocity.
- Acceleration is zero.
- Net force is zero.

13. In the case of accelerated motion, either the magnitude or the direction or both magnitude and direction of velocity change.

14. **What distinguish displacement from distance ?**

- Displacement has direction.
- Distance does not have direction.
- The magnitude of displacement can be both positive and negative.
- Distance is always positive. It never decreases with time.
- Distance ≥ | Displacement |.

IMPORTANT POINTS TO REMEMBER

1. **When a moving particle returns to its initial point**

   (*i*) Displacement is zero.

   (*ii*) Average velocity is zero.

   (*iii*) Distance is NOT zero.

   (*iv*) Average speed is NOT zero.
2. Unless otherwise specified, the motion of the body is understood to be taking place in vacuum.
3. For uniform motion along a straight line in a given direction, the magnitude of displacement is equal to the actual distance travelled by the body.
4. In uniformly accelerated motion, average velocity over any interval of time is not equal to the instantaneous velocity.
5. Speed may increase or decrease with time.

☞ CAUTION

Decreasing speed is NOT negative speed.

EFFECTIVE USE OF MATHEMATICAL TOOLS IN SOLVING PROBLEMS OF ONE-DIMENSIONAL MOTION

If displacement-time equation is given, we can get velocity-time equation with the help of differentiation. Again, we can get acceleration-time equation with the help of differentiation.

If acceleration-time equation is given, we can get velocity-time equation by integration. From velocity-time equation, we can get displacement-time equation by integration.

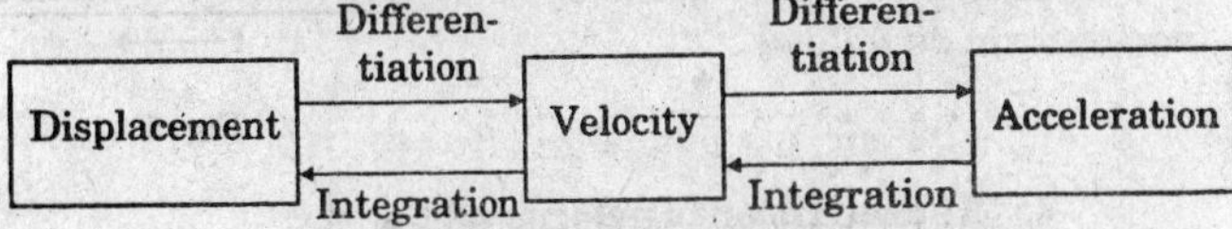

QUESTION BANK

Level I

Choose the correct answer from the four alternatives given in each of the following questions :

1. A car travels from A to B at a speed of 20 km h^{-1} and returns at a speed of 30 km h^{-1}. The average speed of the car for the whole journey is

   (*a*) 5 km h^{-1} (*b*) 24 km h^{-1}
   (*c*) 25 km h^{-1} (*d*) 50 km h^{-1}.

2. A body is thrown vertically upwards with a speed of 100 m s^{-1}. On the return journey, the speed in m s^{-1} at the starting point will be

   (*a*) 100 m s^{-1} (*b*) 9.8 m s^{-1}
   (*c*) 100×9.8 m s^{-1} (*d*) $\frac{100}{9.8}$ m s^{-1}.

3. If a body having initial velocity zero is moving with a uniform acceleration of 8 m s^{-2}, then the distance travelled by it in fifth second will be

   (*a*) zero (*b*) 36 m
   (*c*) 40 m (*d*) 100 m.

4. A pebble is dropped into a well of depth *h*. The splash is heard after time *t*. If *c* be the velocity of sound, then

   (*a*) $t = \sqrt{\frac{gc}{2h}}$ (*b*) $t = c + gh$
   (*c*) $t = c - v$ (*d*) $t = \sqrt{\frac{2h}{g}} + \frac{h}{c}$.

5. A ball released from a certain height falls 5 m in one second. In 4 s, it falls through

   (*a*) 80 m (*b*) 40 m
   (*c*) 1.25 m (*d*) 20 m.

6. The displacement-time graph of a moving particle is shown below. The instantaneous velocity of the particle is negative at the point

   (*a*) C
   (*b*) D
   (*c*) E
   (*d*) F.

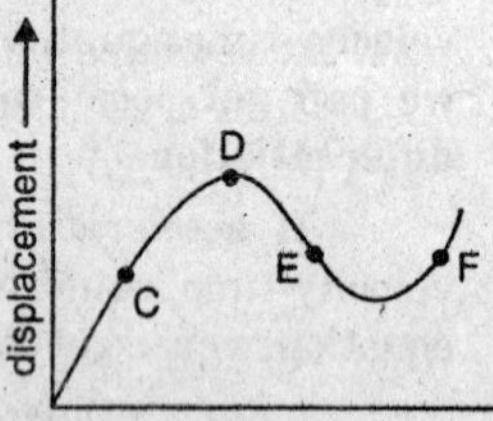

Fig. 1

7. The variation of velocity of a particle moving along a straight line is shown in the Fig. 2. The distance travelled by the particle in 4 s is

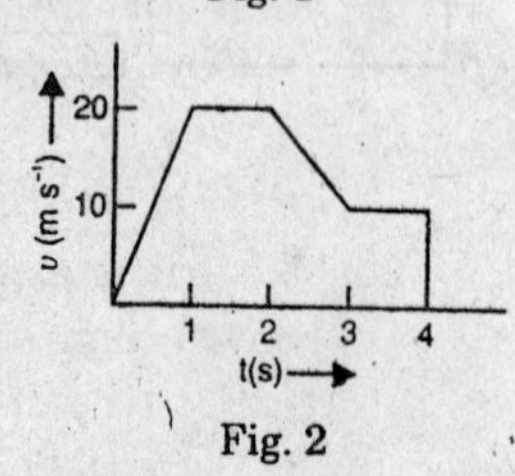

Fig. 2

   (*a*) 25 m (*b*) 30 m
   (*c*) 55 m (*d*) 60 m.

8. A particle starts from rest and moves along a straight line with constant acceleration. The variation of velocity *v* with displacement S is

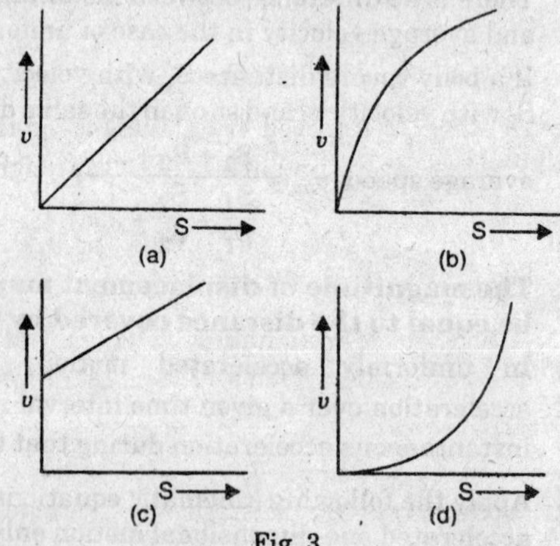

Fig. 3

9. A particle starts from rest at time $t = 0$ and moves on a straight line with acceleration as plotted in Fig. 4. The speed of the particle will be maximum after time

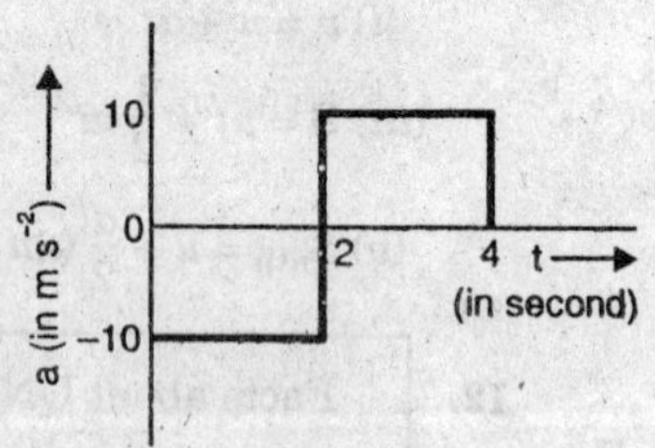

Fig. 4

   (*a*) 1 s (*b*) 2 s
   (*c*) 3 s (*d*) 4 s.

10. The displacement-time graphs of two moving particles make angles of 30° and 45° with the *x*-axis. The ratio of the two velocities is

    (*a*) $\sqrt{3} : 1$
    (*b*) 1 : 1
    (*c*) 1 : 2
    (*d*) $1 : \sqrt{3}$.

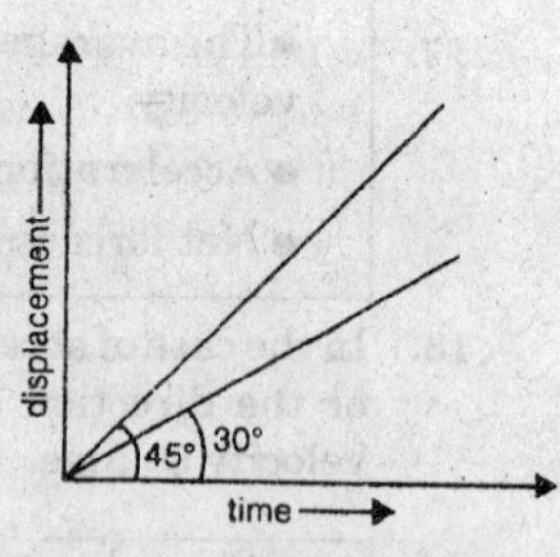

Fig. 5

11. The graph of displacement-time for a body travelling in a straight line is given. We can conclude that

(a) the velocity is constant.

(b) the velocity increases uniformly.

(c) the body is subjected to acceleration from O to A.

(d) the velocity of the body at A is zero.

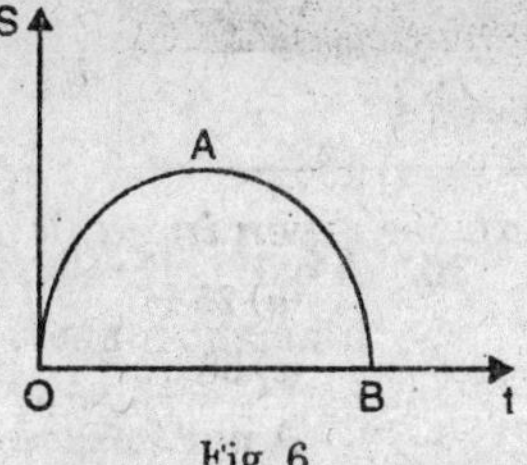

Fig. 6

12. A ball takes t second to fall from a height h_1 and $2t$ second to fall from a height h_2. Then h_1/h_2 is

(a) 2 (b) 4
(c) 0.5 (d) 0.25.

13. A stone is dropped from the top of the tower and reaches the ground in 3 s. Then the height of the tower is

(a) 18.6 m (b) 39.2 m
(c) 44.1 m (d) 98 m.

14. A body covers one half of its journey at 40 m s^{-1} and the next half at 50 m s^{-1}. Its average velocity is

(a) 44.4 m s^{-1} (b) 50 m s^{-1}
(c) 45 m s^{-1} (d) 40 m s^{-1}.

15. A car travels equal distances in the same direction with velocities 60 km h^{-1}, 20 km h^{-1} and 10 km h^{-1} respectively. The average velocity of the car over the whole journey of motion is

(a) 8 m s^{-1} (b) 7 m s^{-1}
(c) 6 m s^{-1} (d) 5 m s^{-1}.

16. A body starts from rest and is uniformly accelerated for 30 s. The distance travelled in the first 10 s is x_1, next 10 s is x_2 and the last 10 s is x_3. Then $x_1 : x_2 : x_3$ is the same as

(a) 1 : 2 : 4 (b) 1 : 2 : 5
(c) 1 : 3 : 5 (d) 1 : 3 : 9.

17. A ball is dropped from the top of a building. The ball takes 0.5 s to fall past the 3 m length of a window some distance from the top of the building. If the velocities of the ball at the top and at the bottom of the window are v_T and v_B respectively, then

(a) $v_T + v_B = 12$ m s^{-1} (b) $v_T - v_B = 4.9$ m s^{-1}
(c) $v_B v_T = 1$ m s^{-1} (d) $\frac{v_B}{v_T} = 1$ m s^{-1}.

18. In Q. 17, $\frac{v_T^2}{2g}$ gives the

(a) distance of the top of the building from the top of the window

(b) distance of the bottom of the window from the top of the building

(c) height of building

(d) distance of top of window from the ground.

19. Two cars are moving in the same direction with a speed of 30 km h^{-1}. They are separated from each other by 5 km. Third car moving in the opposite direction meets the two cars after an interval of 4 minutes. What is th speed of the third car ?

(a) 30 km h^{-1} (b) 35 km h^{-1}
(c) 40 km h^{-1} (d) 45 km h^{-1}.

20. The displacement-time graphs of two particles A and B are straight lines making angles of respectively 30° and 60° with the time- axis. If the velocity of A is v_A and that of B is v_B, then the value of $\frac{v_A}{v_B}$ is

(a) $\frac{1}{2}$ (b) $\frac{1}{\sqrt{3}}$
(c) $\sqrt{3}$ (d) $\frac{1}{3}$.

21. The position vector of a particle in SI units is given by

$$\vec{r} = 4t^2\,\hat{i} + 3t^2\,\hat{j} + 2\hat{k}$$

The acceleration of the particle is

(a) 4 m s^{-2} (b) 3 m s^{-2}
(c) 2 m s^{-2} (d) 10 m s^{-2}.

22. In Q. 21, the distance covered in the first 10 second is

(a) 200 m (b) 300 m
(c) 400 m (d) 500 m.

23. The velocity of a stone thrown vertically upwards is halved in 1.5 second. The maximum height attained by the stone is (Take g = 10 m s^{-2})

(a) 20 m (b) 25 m
(c) 30 m (d) 45 m.

24. A boy releases a ball from the top of a building. It will clear a window 2 m high at a distance 10 m below the top in nearly

(a) 1 s (b) 1.3 s
(c) 0.6 s (d) 0.13 s.

25. Two balls are dropped to the ground from different heights. One ball is dropped 2 s after the other but they both strike the ground at the same time, 5 s after the first was dropped. The difference in the heights at which they were dropped is (Given : g = 10 m s^{-2})

(a) 10 m (b) 20 m
(c) 40 m (d) 80 m.

26. A stone is allowed to fall from the top of a tower 100 metre high and at the same time another stone is projected vertically upwards from the ground with a velocity of 25 m s^{-1}. The two stones will meet after

(*a*) 4 s (*b*) 0.4 s
(*c*) 0.04 s (*d*) 40 s.

27. In Q. 26, the two stones meet at a height of
(*a*) 21.6 metre from the ground
(*b*) 50.0 metre from the ground
(*c*) 78.4 metre from the ground
(*d*) 19.6 metre from the ground.

28. A body starts from rest and moves along a straight line with uniform acceleration. It covers a distance of 150 m during 8th second of its motion. The acceleration of the body is
(*a*) 20 m s^{-2} (*b*) 10 m s^{-2}
(*c*) 30 m s^{-2} (*d*) 15 m s^{-2}.

29. The displacement of a body is given by $2s = gt^2$ where *g* is a constant. The velocity of the body at any time *t* is
(*a*) gt (*b*) $\frac{gt}{2}$
(*c*) $\frac{gt^2}{2}$ (*d*) $\frac{gt^3}{6}$.

30. A ball is thrown straight upward with a speed v from a height h above the ground. The time taken for the ball to strike the ground is given by
(*a*) $-h = vt - \frac{1}{2}gt^2$ (*b*) $h = vt - \frac{1}{2}gt^2$
(*c*) $\frac{1}{2}gt^2$ (*d*) $\sqrt{\frac{2g}{h}}$.

31. The velocity of a body at the end of 5 s is 30 m s^{-1}, at the end of 12 s is 58 m s^{-1} and at the end of 22 s, it is 98 m s^{-1}. The body is moving with
(*a*) uniform velocity (*b*) uniform acceleration
(*c*) uniform displacement (*d*) uniform retardation.

QUESTION BANK

Level II

Choose the correct answer from the four alternatives given in each of the following questions :

1. A body having uniform acceleration of 10 m s^{-2} has a velocity of 100 m s^{-1}. In what time, the velocity will be doubled ?
(*a*) 8 s (*b*) 10 s
(*c*) 12 s (*d*) 14 s.

2. The velocity of a body travelling with a uniform acceleration of 2 m s^{-2} is 10 m s^{-1}. What is its velocity after a time interval of 4 s ?
(*a*) 18 m s^{-1} (*b*) 16 m s^{-1}
(*c*) 14 m s^{-1} (*d*) 12 m s^{-1}.

3. The displacement-time graph of a moving object is shown in Fig. 7. Which of the velocity-time graphs shown in Fig. 8 could represent the motion of the same body ?

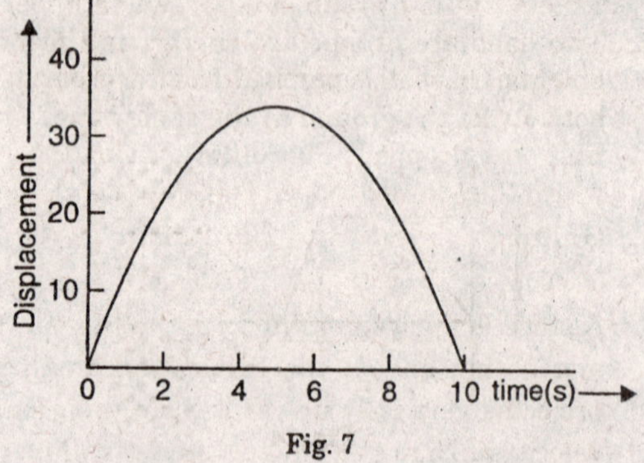

Fig. 7

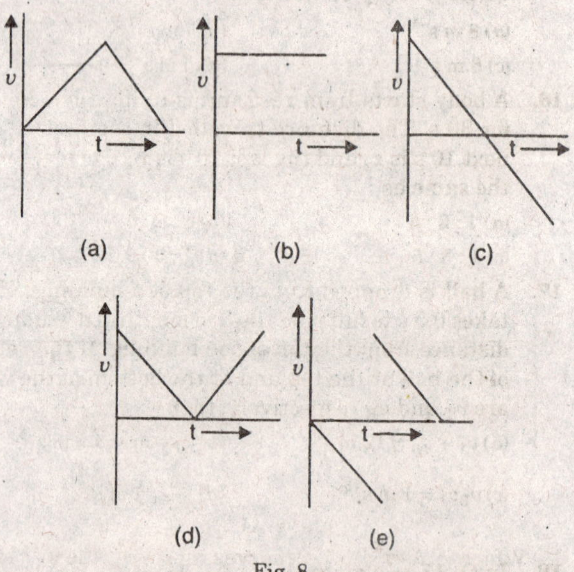

Fig. 8

4. A graph of acceleration versus time of a particle starting from rest at $t = 0$ is as shown in Fig. 9. The speed of the particle at $t = 14$ second is

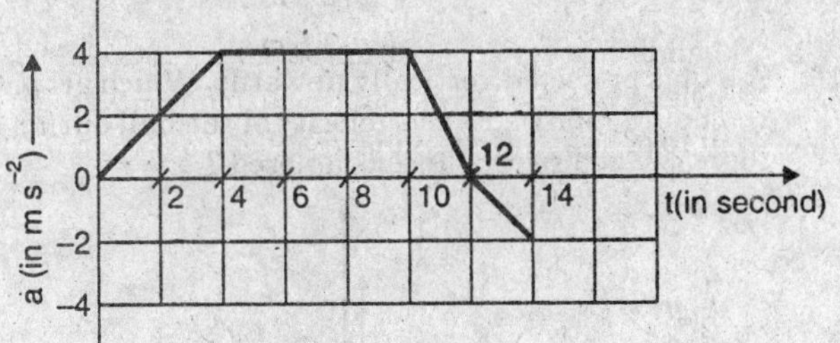

Fig. 9

(*a*) 2 m s^{-1} (*b*) 34 m s^{-1}
(*c*) 20 m s^{-1} (*d*) 42 m s^{-1}.

5. The graph shown in Fig. 10 shows the velocity v versus time t for a body. Which of the graphs shown in Fig. 11 represents the corresponding acceleration versus time graphs ?

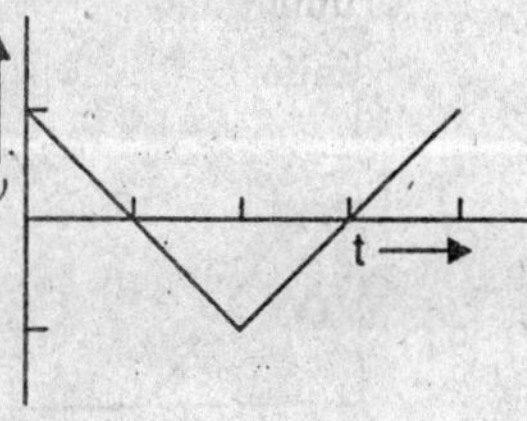

Fig. 10

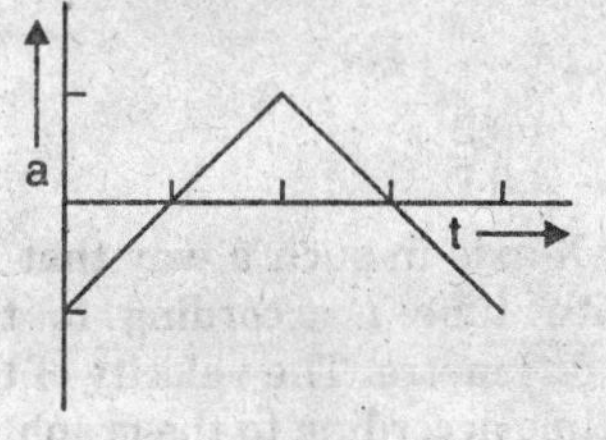

(a)

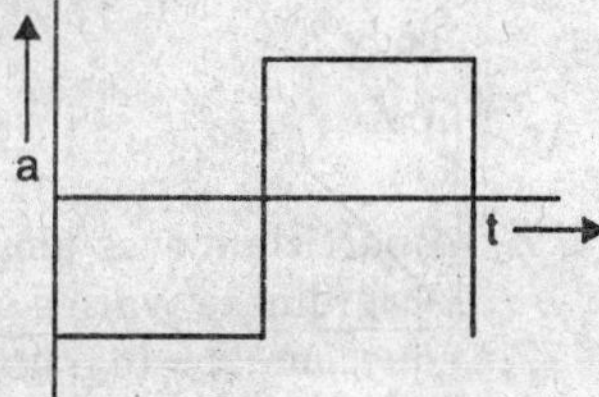

(b)

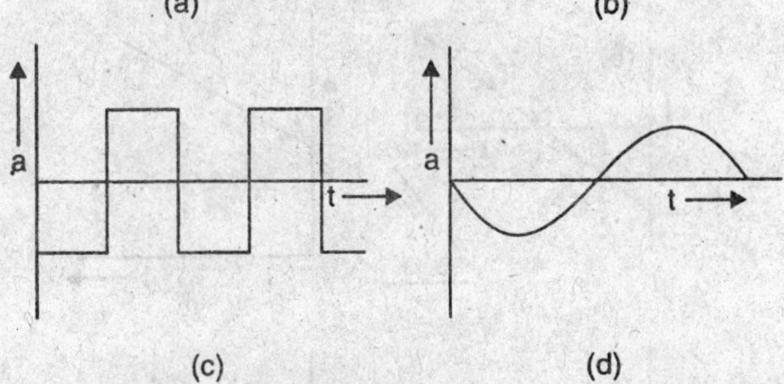

Fig. 11

6. Which of the following curves represent the v – t graph of an object falling on a metallic surface and bouncing back ?

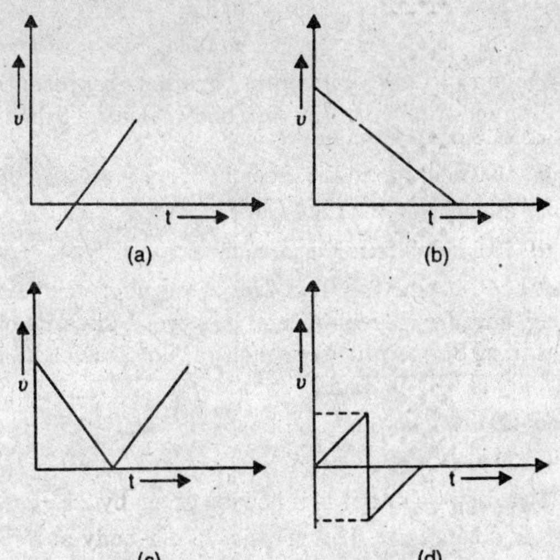

Fig. 12

7. In the given $v - t$ graph, the distance travelled by the body in 5 second will be

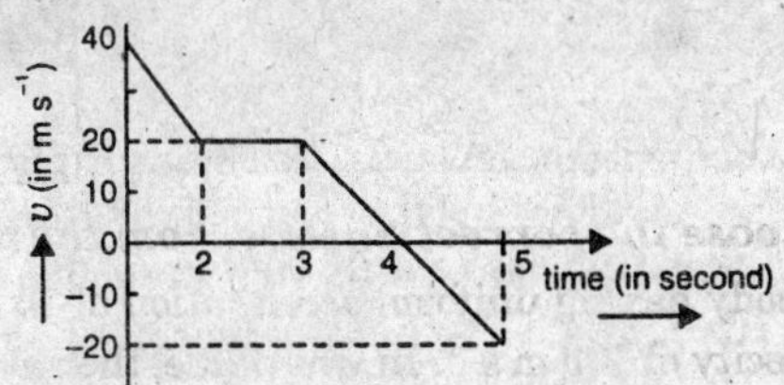

Fig. 13

(*a*) 20 m (*b*) 40 m
(*c*) 80 m (*d*) 100 m.

8. In Q. 7, the displacement of the body in 5 second will be

(*a*) 20 m (*b*) 40 m
(*c*) 80 m (*d*) 100 m.

9. In Q. 7, the average velocity of the body in 5 second is

(*a*) 4 m s^{-1} (*b*) 8 m s^{-1}
(*c*) 16 m s^{-1} (*d*) 20 m s^{-1}.

10. The velocity-time graph of a particle in linear motion is shown. Both v and t are in SI units. What is the displacement of the particle from the origin after 8 second ?

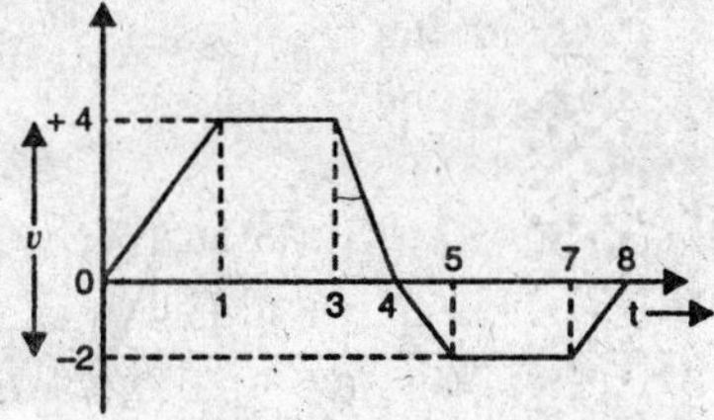

Fig. 14

(*a*) 6 m (*b*) 8 m
(*c*) 16 m (*d*) 18 m.

11. Which of the following graphs represents the distance-time variation of a body released from the top of a building ?

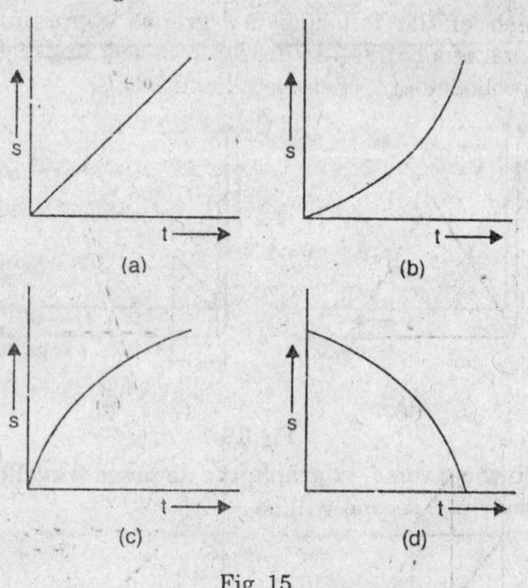

Fig. 15

12. A rocket is fired upwards. Its engine explodes fully in 12 second. The height reached by the rocket as calculated from its velocity-time graph is

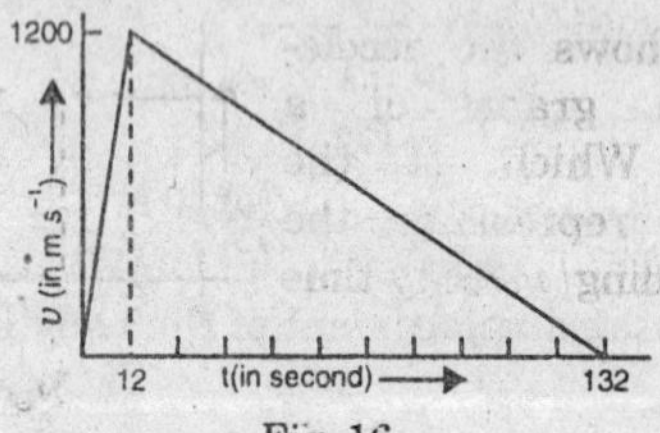

Fig. 16

(*a*) 1200×66 m (*b*) 1200×132 m
(*c*) $\frac{1200}{12}$ m (*d*) 1200×12^2 m.

13. An elevator is going up. The variation in the velocity of the elevator is as given in the graph. What is the height to which the elevator takes the passengers ?

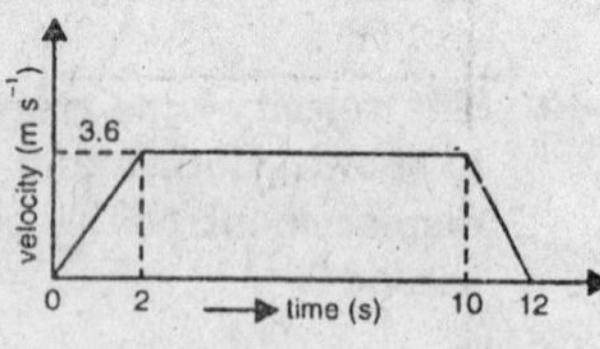

Fig. 17

(*a*) 3.6 m (*b*) 28.8 m
(*c*) 36.0 m (*d*) 72.0 m.

14. In Q. 13, what will be the average velocity of the elevator ?

(*a*) 1 m s^{-1} (*b*) 3 m s^{-1}
(*c*) 2.88 m s^{-1} (*d*) 3.24 m s^{-1}.

15. In Q. 13, the average acceleration of the elevator is
(*a*) zero (*b*) 0.3 m s^{-2}
(*c*) – 1.8 m s^{-1} (*d*) 1.8 m s^{-2}

16. A ball is projected vertically upwards. Which graph in the figure represents the velocity of the ball during its flight when air resistance is ignored ?

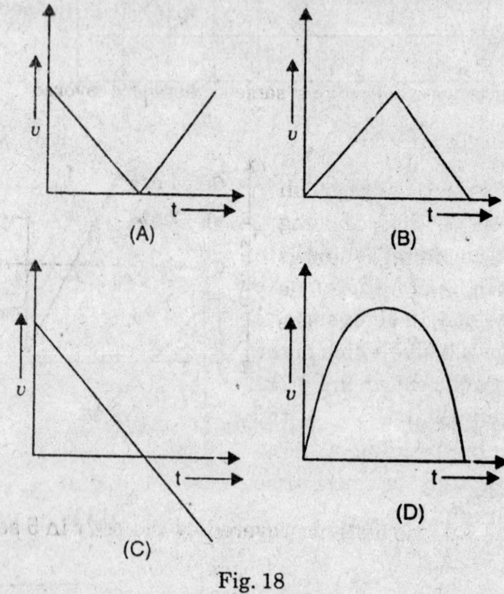

Fig. 18

(*a*) A (*b*) B
(*c*) C (*d*) D.

17. A particle moves along X-axis in such a way that its *x*-coordinate varies with time *t* according to the equation : $x = (6 - 4t + 6t^2)$ metre. The velocity of the particle will vary with time according to the graph

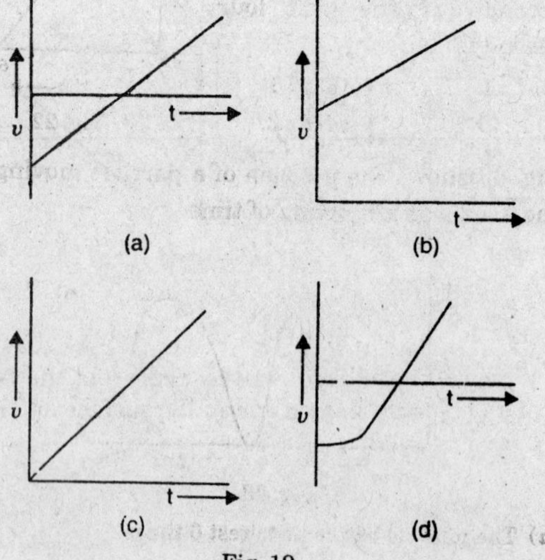

Fig. 19

18. The velocity-time graph of a body is shown in Fig. 20. The ratio of the ...... during the intervals OA and AB is ......

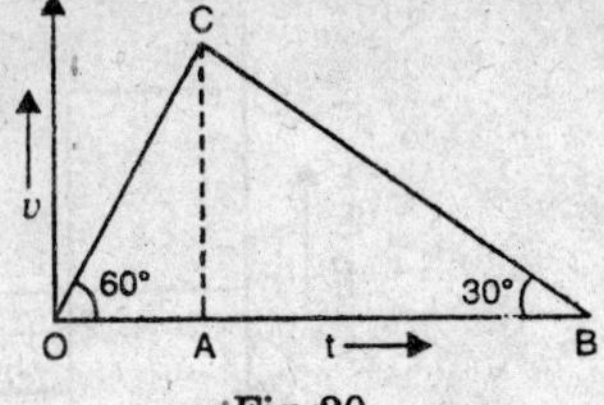

Fig. 20

(a) average velocities ; 1

(b) $\frac{OA}{AB}$; $\frac{1}{4}$

(c) average accelerations, same as distances covered

(d) distances covered ; $\frac{1}{2}$.

19. The velocity-time graph of a particle moving along a straight line is shown in Fig. 21. The displacement of the body in 5 second is

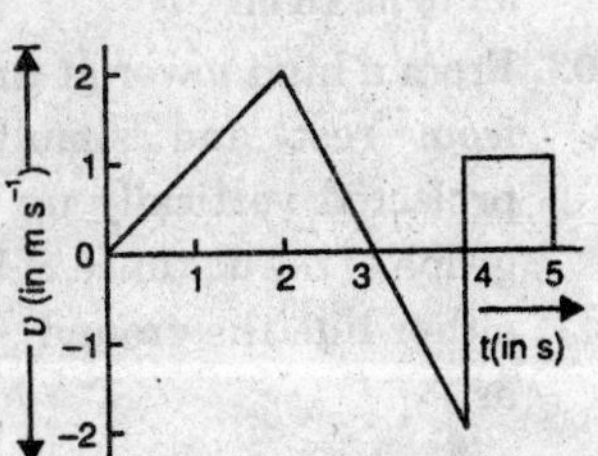

Fig. 21

(a) 0.5 m

(b) 1 m

(c) 2 m

(d) 3 m.

20. In Q. 19, the distance covered by the body in 5 second is

(a) 1 m (b) 2 m

(c) 3 m (d) 5 m.

21. For the displacement-time graph shown in Fig. 22, the ratio of the magnitudes of the speeds during the first two second and the next four second is

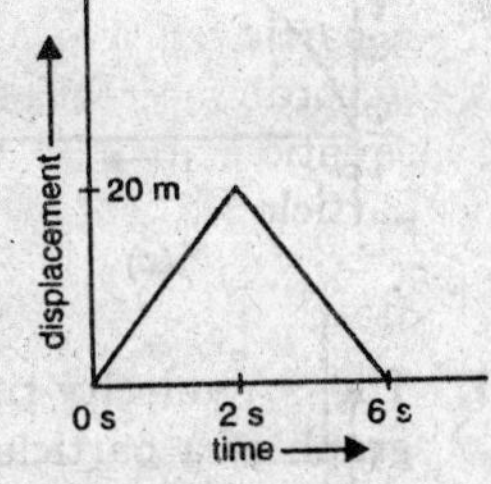

Fig. 22

(a) 1 : 1 (b) 2 : 1

(c) 1 : 2 (d) 3 : 2.

22. Fig. 23 shows the position of a particle moving along the X-axis as a function of time

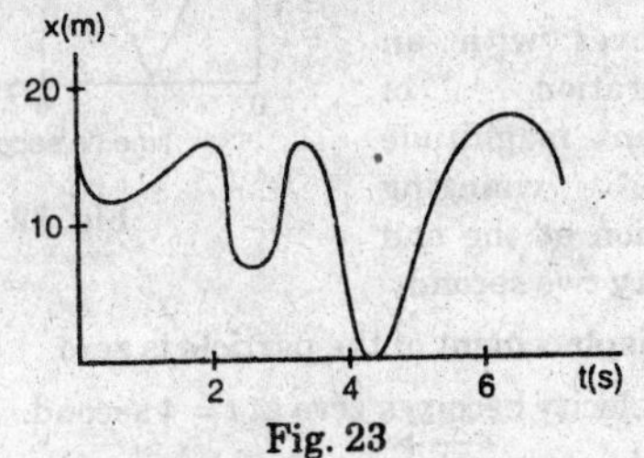

Fig. 23

(a) The particle has come to rest 6 times.

(b) The maximum speed is at $t = 6$ s.

(c) The velocity remains positive for $t = 0$ to $t = 6$ s.

(d) The average velocity for the total period shown is negative.

23. Which of the following v-t graphs represents the motion of a ball falling freely from rest under gravity and rebounding from a metallic surface ?

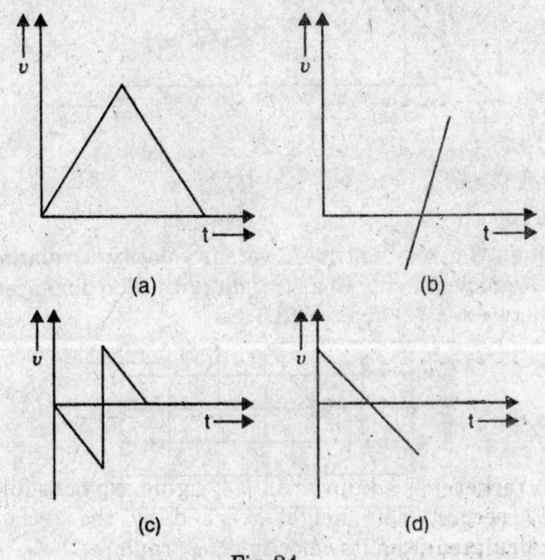

Fig. 24

24. Fig. 25 shows the acceleration-time graph of a particle. Which of the following represents the corresponding velocity-time graph ?

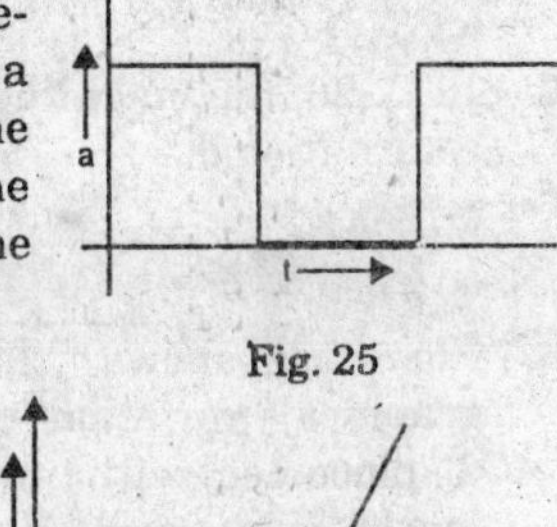

Fig. 25

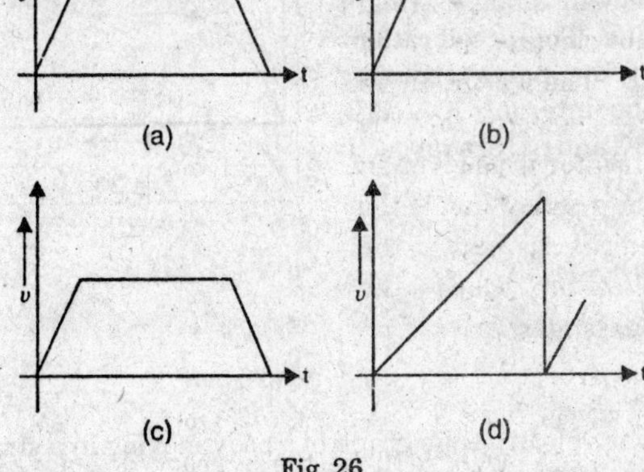

Fig. 26

25. Fig. 27 shows the path A to C traced by a particle in xy plane. B is a point on the path between $x = 10$ cm and

50 cm such that the instantaneous parti'e velocity at B is the same in direction as the ave.age particle velocity during the path A to B. The x co-ordinate of B is nearest to

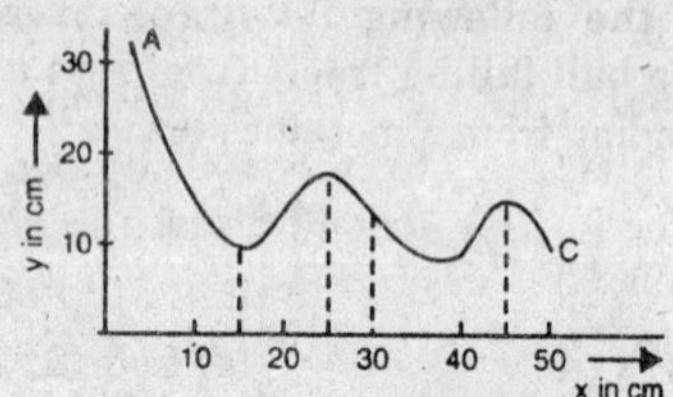

Fig. 27

(*a*) 16 cm (*b*) 22 cm

(*c*) 30 cm (*d*) 44 cm.

26. Fig. 28 gives the time 't' versus velocity 'v' relation for a vehicle moving in a straight path. The displacement between t = 0 and t = 12.5 s is

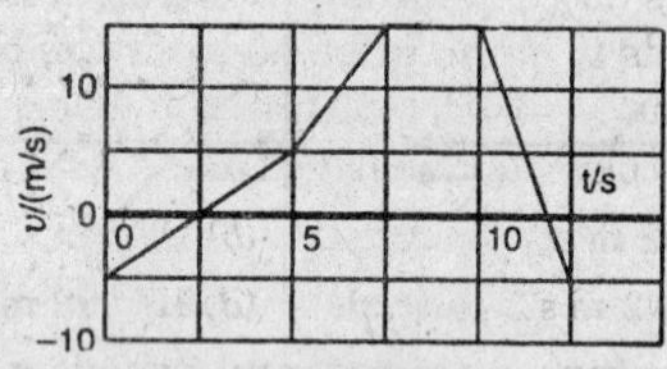

Fig. 28

(*a*) 30 m (*b*) 150 m

(*c*) 75 m (*d*) 125 m.

27. In Q. 26, the magnitude | a | of highest acceleration/retardation is

(*a*) 8 m s^{-2} (*b*) 4 m s^{-2}

(*c*) 2 m s^{-2} (*d*) 16 m s^{-2}.

28. Fig. 29 shows the graphical variation of displacement with time for the case of a particle moving along a straight line. The accelerations of the particle during the intervals OA, AB, BC and CD are respectively

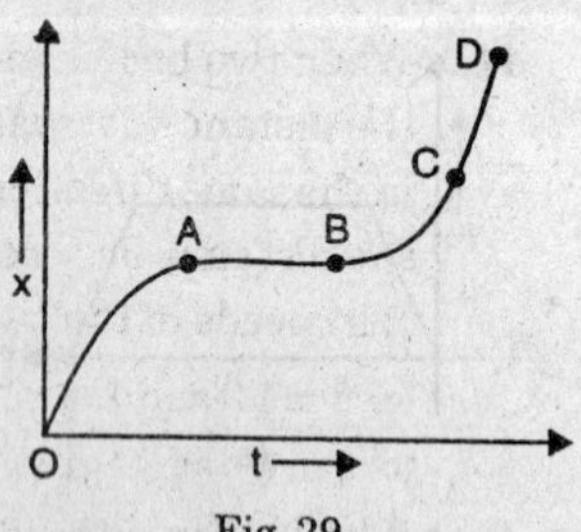

Fig. 29

	OA	AB	BC	CD
(*a*)	–	0	+	0
(*b*)	+	0	+	+
(*c*)	–	0	–	0
(*d*)	+	0	–	+

29. The velocity-time graph of a body moving in a straight line is shown in Fig. 30. The displacement and distance travelled by the body in 6 second are respectively

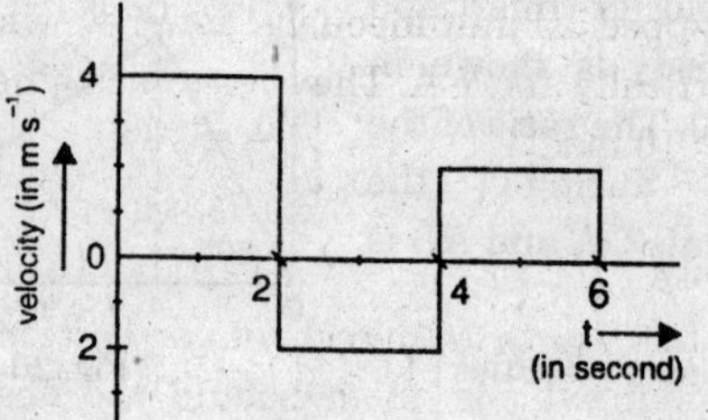

Fig. 30

(*a*) 8 m, 16 m (*b*) 16 m, 8 m

(*c*) 16 m, 16 m (*d*) 8 m, 8 m.

30. From a high tower at time t = 0, one stone is dropped from rest and simultaneously another stone is projected vertically up with an initial velocity. The graph of the distance 'S' between the two stones, before either hits the ground, plotted against time 't' will be as

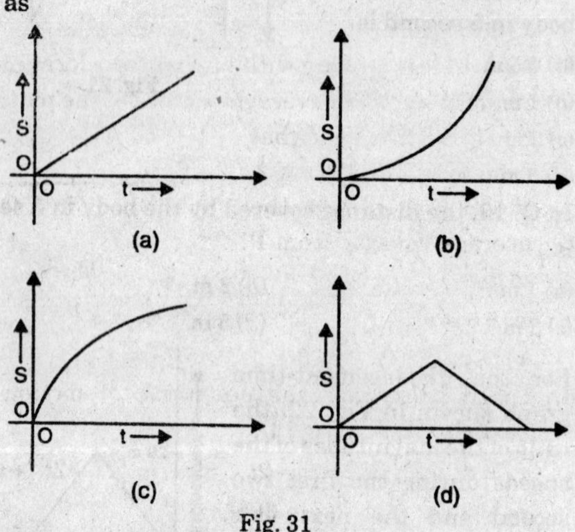

Fig. 31

31. The velocity-time graph of a particle is as shown in Fig. 32

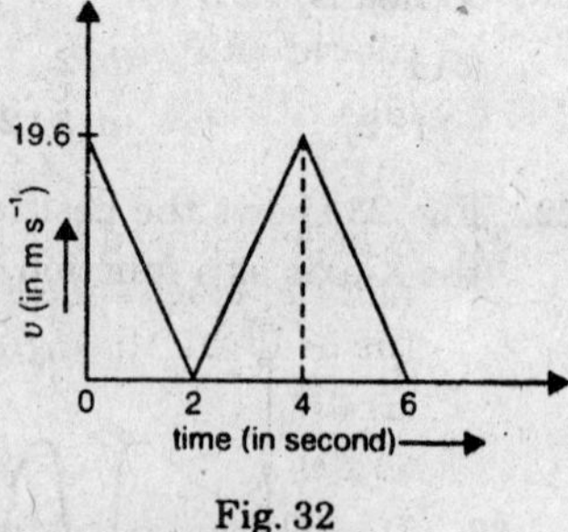

Fig. 32

(*a*) It moves with a constant acceleration throughout.

(*b*) It moves with an acceleration of constant magnitude but changing direction at the end of every two second.

(*c*) The displacement of the particle is zero.

(*d*) The velocity becomes zero at t = 4 second.

32. A particle is projected from a point A vertically upwards with a speed of 50 m s^{-1} and another is

dropped simultaneously from B which is 200 m vertically above A. They cross each other after [Given : $g = 10 \text{ m s}^{-2}$]

(*a*) 8 s (*b*) 6 s
(*c*) 4 s (*d*) 5 s.

33. A bus starts from rest with an acceleration of 1 m s^{-2}. A man who is 48 m behind the bus starts with a uniform velocity of 10 m s^{-1}. The minimum time after which the man will catch the bus is

(*a*) 4.8 s (*b*) 8 s
(*c*) 10 s (*d*) 12 s.

34. A body travelling with uniform acceleration crosses two points A and B with velocities 20 m s^{-1} and 30 m s^{-1} respectively. The speed of the body at the mid-point of A and B is

(*a*) 24 m s^{-1} (*b*) 25 m s^{-1}
(*c*) 25.5 m s^{-1} (*d*) $10\sqrt{6}$ m s^{-1}.

35. A particle is moving with uniform acceleration along a straight line. The average velocity of the particle from P to Q is 8 m s^{-1} and that from Q to S is 12 m s^{-1}. If QS = PQ, then the average velocity from P to S is

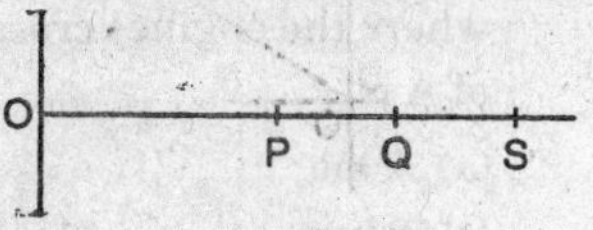

Fig. 33

(*a*) 9.6 m s^{-1} (*b*) 12.87 m s^{-1}
(*c*) 64 m s^{-1} (*d*) 327 m s^{-1}.

36. Given below are the equations of motion of four particles A, B, C and D.

$x_A = 6t - 3$; $x_B = 4t^2 - 2t + 3$; $x_C = 3t^3 - 2t^2 + t - 7$;
$x_D = 7 \cos 60° - 3 \sin 30°$

Which of these four particles move with uniform acceleration ?

(*a*) A (*b*) B
(*c*) C (*d*) D.

37. A body released from the top of a tower falls through half the height of the tower in 2 s. In what time shall the body fall through the height of the tower ?

(*a*) 4 s (*b*) 3.26 s
(*c*) 3.48 s (*d*) 2.828 s.

38. Drops of water fall from the roof of a building 9 m high at regular intervals of time, the first drop reaching the ground at the same instant the fourth drop starts its fall. What are the distances of the second and third drops from the roof ?

(*a*) 4 m and 1 m (*b*) 4 m and 2 m
(*c*) 6 m and 2 m (*d*) 6 m and 3 m.

39. A train is moving with a velocity of 30 m s^{-1}. When brakes are applied, it is found that the velocity reduces to 10 m s^{-1} in 240 m. When the velocity of the train is zero, the distance travelled is

(*a*) 220 m (*b*) 240 m
(*c*) 250 m (*d*) 270 m.

40. A ball falls freely from rest. The ratio of the distances travelled in first, second, third and fourth second is

(*a*) 1 : 1 : 1 : 2 (*b*) 1 : 1 : 1 : 3
(*c*) 1 : 2 : 3 : 4 (*d*) 1 : 3 : 5 : 7.

41. The equation of motion of a particle is

$$x = t^2 - t^3$$

The particle is moving with

(*a*) uniform velocity (*b*) uniform acceleration
(*c*) variable retardation (*d*) uniform retardation.

42. A ball is dropped on the floor from a height of 10 m. It rebounds to a height of 2.5 m. If the ball is in contact with the floor for 0.01 s, the average acceleration during contact is nearly [Take $g = 10 \text{ m s}^{-2}$]

(*a*) $500\sqrt{2}$ m s^{-2} upwards (*b*) 1800 m s^{-2} downwards
(*c*) $1500\sqrt{2}$ m s^{-2} upwards (*d*) $1500\sqrt{2}$ m s^{-2} downwards.

43. A body starts from rest and travels a distance x with uniform acceleration, then it travels a distance $2x$ with uniform speed, finally it travels a distance $3x$ with uniform retardation and comes to rest. If the complete motion of the particle is along a straight line, then the ratio of its average velocity to maximum velocity is

(*a*) 2/5 (*b*) 3/5
(*c*) 4/5 (*d*) 6/7.

44. When two bodies move uniformly towards each other, the distance decreases by 6 m s^{-1}. If both bodies move in the same directions with the same speeds (as above), the distance between them increases by 4 m s^{-1}. Then the speeds of the two bodies are

(*a*) 3 m s^{-1} and 3 m s^{-1} (*b*) 4 m s^{-1} and 2 m s^{-1}
(*c*) 5 m s^{-1} and 1 m s^{-1} (*d*) 7 m s^{-1} and 3 m s^{-1}.

45. A ball is thrown vertically upwards from the surface of earth with a speed of 18 km h^{-1}. If g = 10 m s^{-2}, then the maximum height attained by the ball is

(*a*) 1.25 m (*b*) 3 m
(*c*) 10 m (*d*) 180 m.

46. Two bodies begin a free fall from rest, from the same height, 1 second apart. How long after the first body begins to fall will the two bodies be 10 m apart ?

(*a*) 2 s (*b*) 3.5 s
(*c*) 0.5 s (*d*) 1.5 s.

47. A drunkard is walking along a straight road. He takes 5 steps forward and 3 steps backward, followed by 5 steps forward and 3 steps backward and so on. Each step is one metre long and takes one second. There is a pit on the road 11 metre away from the starting point. The drunkard will fall into the pit after

(*a*) 21 s (*b*) 29 s
(*c*) 31 s (*d*) 37 s.

48. A particle slides from rest from the topmost point of a vertical circle of radius r along a smooth chord making an angle θ with the vertical. The time of descent is

(*a*) least for $\theta = 0°$ (*b*) least for $\theta = 45°$
(*c*) maximum for $\theta = 0°$ (*d*) independent of θ.

49. A body travels 2 m in the first two second and 2.20 m in the next 4 second with uniform deceleration. The velocity of the body at the end of 9 second is

(*a*) – 10 m s^{-1} (*b*) – 0.20 m s^{-1}
(*c*) – 0.40 m s^{-1} (*d*) – 0.80 m s^{-1}.

50. A particle starting from rest travels a distance x in first 2 second and a distance y in next 2 second, then

(*a*) $y = 2x$ (*b*) $y = 3x$
(*c*) $y = 4x$ (*d*) $y = x$.

51. A stone thrown vertically upwards attains a maximum height of 45 m. In what time the velocity of stone become equal to one-half the velocity of throw? (Given : $g = 10$ m s^{-2}.)

(*a*) 2 s (*b*) 1.5 s
(*c*) 1 s (*d*) 0.5 s.

52. The position x of a body as a function of time t is given by the equation :

$$x = 2t^3 - 6t^2 + 12t + 6$$

The acceleration of the body is zero at time t is equal to

(*a*) 1 s (*b*) 2 s
(*c*) 3 s (*d*) 0.5 s.

53. A ball is thrown vertically upwards from the top of a tower at 4.9 m s^{-1}. It strikes the pond near the base of the tower after 3 second. The height of the tower is

(*a*) 29.4 m (*b*) 44.1 m
(*c*) 73.5 m (*d*) 490 m.

54. A body covers 26 m, 28 m, 30 m, 32 m in 10th s, 11th s, 12th s and 13th s respectively. The body starts

(*a*) from rest and moves with uniform acceleration
(*b*) from rest and moves with uniform velocity
(*c*) with an initial velocity and moves with uniform acceleration
(*d*) with an initial velocity and moves with uniform velocity.

55. A stone is dropped under gravity from rest from a height h ($g = 9.8$ m s^{-2}) and it travels a distance $\frac{9h}{25}$ in the last second. What is the height h ?

(*a*) 145 m (*b*) 100 m
(*c*) 122.5 m (*d*) 100 m.

56. A body is released from the top of a tower of height h metre. It takes T second to reach the ground. Where is the ball at the time $\frac{T}{2}$ second ?

(*a*) at $\frac{h}{4}$ metre from the ground
(*b*) at $\frac{h}{2}$ metre from the ground
(*c*) at $\frac{3h}{4}$ metre from the ground
(*d*) depends upon the mass and volume of the ball.

57. Two trains A and B 100 km apart are travelling towards each other with starting speed of 50 km h^{-1} for both. The train A is accelerating at 18 km h^{-2} and the train B is decelerating at 18 km h^{-2}. The distance where the engines cross each other from starting point of A is

(*a*) 32 km (*b*) 68 km
(*c*) 59 km (*d*) 50 km.

58. A body is thrown vertically upwards in air. When air resistance is taken into consideration, let the time of ascent be t_1 and time of descent be t_2. Then

(*a*) $t_1 = t_2$. (*b*) $t_1 < t_2$
(*c*) $t_1 > t_2$ (*d*) $t_1 >=< t_2$.

59. The velocity of a particle at an instant is 10 ms^{-1}. After 5 s, the velocity of the particle is 20 ms^{-1}. The velocity at 3 s before the given instant is

(*a*) 4 m s^{-1} (*b*) 6 m s^{-1}
(*c*) 7 m s^{-1} (*d*) 8 m s^{-1}.

60. A ball is dropped from a height of 20 m and rebounds with a velocity which is $\frac{3}{4}$ th of the velocity with which it hits the ground. What is the time interval between the first and second bounces ? ($g = 10$ m s^{-2})

(*a*) 3 s (*b*) 4 s
(*c*) 5 s (*d*) 6 s.

61. A particle is moving along x-axis. Its displacement x (in m) and time t (in s) are related by the equation $t = 5 + \sqrt{x}$. The displacement of the particle when its velocity becomes zero, is

(*a*) 2.5 m (*b*) zero
(*c*) 5 m (*d*) 10 m

62. With what speed must a ball be thrown down for it to bounce 10 m higher than its original level ? Neglect any loss of energy in striking the ground.

(*a*) 20 m s^{-1} (*b*) 14 m s^{-1}

(*c*) 5 m s^{-1} (*d*) 1 m s^{-1}.

63. A particle initially at rest moves along x-axis. It is subjected to an acceleration which varies with time according to the equation : $a = 2t + 5$. Its velocity after 2 second will be

(*a*) 9 m s^{-1} (*b*) 12 m s^{-1}

(*c*) 14 m s^{-1} (*d*) 18 m s^{-1}.

64. Two cars A and B are travelling in the same direction with velocities v_A and v_B ($v_A > v_B$). When the car A is at a distance S behind the car B, the driver of the car A applies the brakes producing a uniform retardation a. There will be no collision when

(*a*) $S < \frac{(v_A - v_B)^2}{2a}$ (*b*) $S = \frac{(v_A - v_B)^2}{2a}$ only

(*c*) $S \geq \frac{(v_A - v_B)^2}{2a}$ (*d*) $S \leq \frac{(v_A - v_B)^2}{2a}$.

65. A balloon is going upwards with a velocity 12 m s^{-1}. It releases a packet when it is at a height of 65 m from the ground. How much time the packet will take to reach the ground if $g = 10$ m s^{-2} ?

(*a*) 5 s (*b*) 6 s

(*c*) 7 s (*d*) 8 s.

66. A body sliding on a smooth inclined plane requires 4 second to reach the bottom, starting from rest at the top. How much time does it take to cover one-fourth the distance starting from rest at top ?

(*a*) 1 s (*b*) 2 s

(*c*) 4 s (*d*) 16 s.

67. A stone is released from the top of a tower. It covers 24.5 m distance in the last second of its journey. What is the height of the tower ?

(*a*) 44.1 m (*b*) 58.5 m

(*c*) 73.5 m (*d*) 98 m.

68. The displacement y of a particle along a straight line at time t is given by $y = a_0 + a_1 t + a_2 t^2$

The acceleration of the particle is

(*a*) a_0 (*b*) a_1

(*c*) a_2 (*d*) $2a_2$.

69. A particle moves with constant acceleration. If v_1, v_2 and v_3 are the average velocities in the three successive intervals t_1, t_2 and t_3 of time, then the correct relation is

(*a*) $\frac{v_1 - v_2}{v_2 - v_3} = \frac{t_1 - t_2}{t_2 + t_3}$ (*b*) $\frac{v_1 - v_2}{v_2 - v_3} = \frac{t_1 - t_2}{t_1 - t_3}$

(*c*) $\frac{v_1 - v_2}{v_2 - v_3} = \frac{t_1 - t_2}{t_2 - t_3}$ (*d*) $\frac{v_1 - v_2}{v_2 - v_3} = \frac{t_1 + t_2}{t_2 + t_3}$.

70. A, B, C and D are the points in a vertical line such that AB = BC = CD. If a body falls from rest at A, then the time of descent through AB, BC and CD are in the ratio

(*a*) $1 : \sqrt{2} - 1 : \sqrt{3} - \sqrt{2}$ (*b*) $1 : \sqrt{2} : \sqrt{3}$

(*c*) 1 : 2 : 3 (*d*) 1 : 4 : 9.

71. A ball is dropped from a bridge 125 m high. After the ball has been falling for 2 second, another ball is thrown downwards with such a velocity that the two balls strike the surface of water simultaneously. The velocity of the second ball is nearly

(*a*) 27 m s^{-1} (*b*) 30 m s^{-1}

(*c*) 40 m s^{-1} (*d*) infinite.

72. A body released from a great height falls freely towards the earth. Another body is released from the same height exactly one second later. The separation between the two bodies two second after the release of the second body is

(*a*) 9.8 m (*b*) 4.9 m

(*c*) 24.5 m (*d*) 19.6 m.

73. A ball is thrown vertically upward with a velocity 'u' from the balloon descending with velocity v. The ball will pass by the balloon after time

(*a*) $\frac{u - v}{2g}$ (*b*) $\frac{u + v}{2g}$

(*c*) $\frac{2(u - v)}{g}$ (*d*) $\frac{2(u + v)}{g}$.

74. A particle experiences constant acceleration for 20 second after starting from rest. If it travels a distance S_1 in the first 10 second and a distance S_2 in the next 10 second, then

(*a*) $S_2 = 2S_1$ (*b*) $S_2 = 3S_1$

(*c*) $S_2 = 4S_1$ (*d*) $S_2 = 5S_1$.

75. A steamer takes 12 days to reach from port A to B. Every day only one steamer sets out from both the ports. How many steamers does each boat meet in the open sea ?

(*a*) 12 (*b*) 13

(*c*) 23 (*d*) 24.

76. A student decides to test the law of gravity for himself. He walks off a skyscraper 320 m high with a stopwatch in hand and starts his free fall (zero initial velocity). Five second later, superman arrives at the scene and dives off the roof of the skyscraper to save the student. What must be superman's initial velocity in order that he catches the student just before the ground is reached ? (Given : $g = 10$ m s^{-2})

(a) 67 m s^{-1} (b) 91.7 m s^{-1}

(c) 103 m s^{-1} (d) 140 m s^{-1}.

77. One car moving on a straight road covers one-third of the distance with 20 km h^{-1} and the rest with 60 km h^{-1}. The average speed is

(a) 40 km h^{-1} (b) 80 km h^{-1}

(c) $46\frac{2}{3}$ km h^{-1} (d) 36 km h^{-1}.

78. A particle covers 4 m, 5 m, 6 m and 7 m in 3rd, 4th, 5th and 6th second respectively. The particle starts

(a) with an initial non-zero velocity and moves with uniform acceleration

(b) from rest and moves with uniform velocity

(c) with an initial velocity and moves with uniform velocity

(d) from rest and moves with uniform acceleration.

79. If a splash is heard 4.23 second after a stone is dropped into a well 78.4 m deep, then velocity of sound in air is

(a) 300 m s^{-1} (b) 309 m s^{-1}

(c) 330 m s^{-1} (d) 341 m s^{-1}.

80. A body moving with uniform retardation covers 3 km before its speed is reduced to half its initial value. It comes to rest in another distance of

(a) 1 km (b) 2 km

(c) 3 km (d) $\frac{1}{2}$ km.

81. The displacement x of a body varies with time t as

$$x = -\frac{2}{3}t^2 + 16t + 2$$

In what time the body comes to rest ? (x is measured in metre and t in second.)

(a) 6 s (b) 12 s

(c) 18 s (d) 20 s.

82. The two ends of a train moving with constant acceleration pass a certain point with velocities u and v. The velocity with which the middle point of the train passes the same point is

(a) $\frac{u+v}{2}$ (b) $\frac{u^2+v^2}{2}$

(c) $\sqrt{\frac{u^2+v^2}{2}}$ (d) $\sqrt{u+v}$.

83. A rifle bullet loses 1/20th of its velocity in passing through a plank. The least number of such planks required to just stop the bullet is

(a) 5 (b) 10

(c) 11 (d) 20.

84. A stone is dropped into water from a bridge 44.1 m above the water. Another stone is thrown vertically downward 1 s later. Both strike the water simultaneously. What was the initial speed of the second stone ?

(a) 12.25 m s^{-1} (b) 14.75 m s^{-1}

(c) 16.23 m s^{-1} (d) 17.15 m s^{-1}.

85. The driver of a train travelling at 115 km h^{-1} sees on the same track, 100 m infront of him, a slow train travelling in the same direction at 25 km h^{-1}. The least retardation that must be applied to faster train to avoid a collision is

(a) 25 m s^{-2} (b) 50 m s^{-2}

(c) 75 m s^{-2} (d) 3.125 m s^{-2}.

86. A ball is dropped from the top of a building 120 m high. At the same time, another ball is thrown upwards with a velocity of 40 m s^{-1} from the bottom of the building. The two balls will meet after

(a) 5.6 s (b) 3.5 s

(c) 4 s (d) 3 s.

87. A stone falls freely from rest and the total distance covered by it in the last second of its motion equals the distance covered by it in the first three second of its motion. The stone remains in air for

(a) 6 s (b) 5 s

(c) 7 s (d) 4 s.

88. A body starts from a point O with a velocity v and uniform acceleration a. The direction of acceleration is reversed when the velocity of the body becomes $5v$. The velocity of body at point O will be

(a) $-v$ (b) $-5v$

(c) $-7v$ (d) $-9v$.

89. A ball dropped from the 9th storey of a multi-storeyed building reaches the ground in 3 second. In the first second of its free fall, it passes through n storeys, where n is equal to (Take $g = 10$ m s^{-2}.)

(a) 1 (b) 2

(c) 3 (d) 4.

90. On applying brakes, an automobile stops after 5 second covering 25 m with uniformly retarded motion. The initial speed of the automobile is

(a) 5 m s^{-1} (b) 10 m s^{-1}

(c) 15 m s^{-1} (d) 20 m s^{-1}.

91. A balloon is at a height of 81 m and is ascending upwards with a velocity of 12 m s^{-1}. A body of 2 kg weight is dropped from it. If $g = 10$ m s^{-2}, the body will reach the surface of the earth in

(*a*) 1.5 s (*b*) 4.025 s
(*c*) 5.4 s (*d*) 6.75 s.

92. A boat takes 2 hours to travel 8 km and back in still water lake. With water velocity of 4 km h^{-1}, the time taken for going upstream of 8 km and coming back is
(*a*) 160 minute (*b*) 80 minute
(*c*) 100 minute (*d*) 120 minute.

93. A stone is dropped from the top of a tall tower and after 1 s another stone is dropped from a balcony 20 m below the top. If both the stones reach the ground at the same time, then the height of the tower is (Take $g = 10$ m s^{-2})
(*a*) 3.125 m (*b*) 11.2 m
(*c*) 31.25 m (*d*) 312.5 m.

94. A car accelerates from rest at a constant rate of 2 m s^{-2} for sometime. Then, it retards at a constant rate of 4 m s^{-2} and comes to rest. If the total time for which it remains in motion is 3 second, what is the total distance travelled ?
(*a*) 2 m (*b*) 3 m
(*c*) 4 m (*d*) 6 m.

95. A rocket is fired upward from the earth's surface such that it creates an acceleration of 19.6 m s^{-2}. If after 5 s, its engine is switched off, the maximum height of the rocket from earth's surface would be
(*a*) 980 m (*b*) 735 m
(*c*) 490 m (*d*) 245 m.

96. A wooden block of mass 10 g is dropped from the top of a cliff 100 m high. Simultaneously a bullet of mass 10 g is fired from the foot of the cliff upwards with a velocity of 100 m s^{-1}. The bullet and the wooden block will meet, each other, after a time
(*a*) 1 s (*b*) 7 s
(*c*) 0.5 s (*d*) 10 s.

97. A boy takes 90 second to walk up a stalled escalator. However if the escalator is moving and the boy stands on the escalator, he is carried up in one minute. How much time the boy would take to walk up the moving escalator ?
(*a*) 36 s (*b*) 40 s
(*c*) 41.25 s (*d*) 43.75 s.

98. A particle starts from rest with uniform acceleration a. Its velocity after n second is u. The displacement of the body in the last two second is
(*a*) $\frac{2v(n-1)}{n}$ (*b*) $\frac{v(n-1)}{n}$
(*c*) $\frac{v(n+1)}{n}$ (*d*) $\frac{2v(2n+1)}{n}$.

99. A stone thrown vertically upward flies past a window one second after it was thrown upward and after three second on its way downward. The height of the window above the ground is (Take $g = 10$ m s^{-2})
(*a*) 20 m (*b*) 15 m
(*c*) 10 m (*d*) 5 m.

100. Two cars leave with a 1 minute gap and move with an acceleration of 0.2 m s^{-2}. How long after the departure of the second car does the distance between them become equal to three times its initial value ?
(*a*) 2 minute (*b*) $\frac{3}{2}$ minute
(*c*) 1 minute (*d*) $\frac{1}{2}$ minute.

101. A particle travels half the distance with a velocity of 6 m s^{-1}. The remaining half distance is covered with a velocity of 4 m s^{-1} for half the time and with a velocity of 8 m s^{-1} for the rest of the half time. What is the velocity of the particle averaged over the whole time of motion ?
(*a*) 9 m s^{-1} (*b*) 6 m s^{-1}
(*c*) 5.35 m s^{-1} (*d*) 5 m s^{-1}.

102. A stone is dropped from a certain height which can reach the ground in 5 second. If the stone is stopped after 3 second of its fall and then allowed to fall again, then the time taken by the stone to reach the ground for the remaining distance is
(*a*) 4 s (*b*) 3 s
(*c*) 2 s (*d*) None of these.

103. The acceleration of a particle increases linearly with time t as '$6t$'. If the initial velocity of the particle is zero and the particle starts from the origin, then the distance travelled by the particle in time t will be
(*a*) t (*b*) t^2
(*c*) t^3 (*d*) t^4.

104. For a particle moving in a straight line, the displacement of the particle at time t is given by

$$S = t^3 - 6t^2 + 3t + 7$$

What is the velocity of the particle when its acceleration is zero ?
(*a*) – 9 m s^{-1} (*b*) – 12 m s^{-1}
(*c*) 3 m s^{-1} (*d*) 42 m s^{-1}.

105. The distance through which a body falls in the nth second is h. The distance through which it falls in the next second is
(*a*) h (*b*) $h + \frac{g}{2}$
(*c*) $h - g$ (*d*) $h + g$.

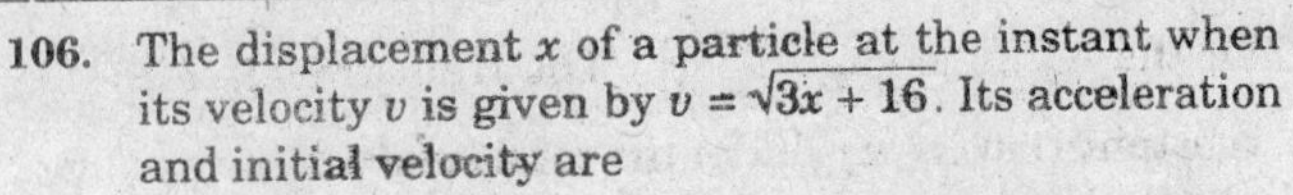

106. The displacement x of a particle at the instant when its velocity v is given by $v = \sqrt{3x + 16}$. Its acceleration and initial velocity are

(a) 1.5 units, 4 units (b) 3 units, 4 units
(c) 16 units, 1.6 units (d) 16 units, 3 units.

107. A sprinter can reach his maximum speed of 10 m s^{-1} by accelerating at 2.5 m s^{-2}. In order to cover a distance of 100 m, he will take

(a) 9 s (b) 10 s
(c) 11 s (d) 12 s.

108. Two bodies begin a free fall from the same height at a time interval of N s. If vertical separation between the two bodies is 1 after n second from the start of the first body, then n is equal to

(a) $\sqrt{nN}$ (b) $\frac{1}{gN}$
(c) $\frac{1}{gN} + \frac{N}{2}$ (d) $\frac{1}{gN} - \frac{N}{4}$.

109. An object fall from rest. In the last second of its fall, it covers half of the total distance. If g = 9.8 m s^{-2}, then the total time of its fall is

(a) $(2 - \sqrt{2})$ s (b) $(2 + \sqrt{2})$ s
(c) $(2 \pm \sqrt{2})$ s (d) 4 s.

110. In Q. 109, the total height of the fall is

(a) 19.6 m (b) 50 m
(c) 57 m (d) 78.4 m.

111. Two bodies begin a free fall from rest, from the same height 2 second apart. How long after the first body begins to fall, the two bodies will be 40 metre apart ? (Take g = 10 m s^{-2})

(a) 1 s (b) 2 s
(c) 3 s (d) 4 s.

112. Two balls are projected simultaneously with the same speed from the top of a tower, one vertically upwards and the other vertically downwards. They reach the ground in 9 second and 4 second respectively. The height of the tower is (Take g = 10 m s^{-2})

(a) 100 m (b) 120 m
(c) 180 m (d) 200 m.

113. A lift in which a man is standing, is moving upwards with a speed of 10 m s^{-1}. The man drops a coin from a height of 4.9 metre and if g = 9.8 m s^{-2}, then the coin reaches the floor of the lift after a time

(a) $\sqrt{2}$ s (b) 1 s
(c) $\frac{1}{2}$ (d) $\frac{1}{\sqrt{2}}$.

114. The greatest acceleration or deceleration that a train may have is a. The minimum time in which the train can get from one station to the next at a distance s is

(a) $\sqrt{\frac{s}{a}}$ (b) $\sqrt{\frac{2s}{a}}$
(c) $2\sqrt{\frac{s}{a}}$ (d) $\frac{1}{2}\sqrt{\frac{s}{a}}$

115. A ball is thrown vertically upwards from the top of a tower of height h with velocity v. The ball strikes the ground after

(a) $\frac{v}{g}\left[1 + \sqrt{1 + \frac{2gh}{v^2}}\right]$ (b) $\frac{v}{g}\left[1 - \sqrt{1 + \frac{2gh}{v^2}}\right]$
(c) $\frac{v}{g}\left(1 + \frac{2gh}{v^2}\right)^{1/2}$ (d) $\frac{v}{g}\left(1 - \frac{2gh}{v^2}\right)^{1/2}$

116. A body starts from rest and falls vertically from a height of 19.6 metre. If g = 9.8 m s^{-2}, then the time taken by the body to fall through the last metre of its fall, is

(a) 2.00 s (b) 0.05 s
(c) 0.45 s (d) 1.95 s.

117. A rubber ball is dropped from a height of 5 metre on a plane where the acceleration due to gravity is not known. On bouncing, it rises to a height of 1.8 m. On bouncing, the ball loses its velocity by a factor of

(a) $\frac{3}{5}$ (b) $\frac{9}{25}$
(c) $\frac{2}{5}$ (d) $\frac{16}{25}$.

118. A steam boat goes across a lake and comes back (i) on a quiet day when the water is still and (ii) on a rough day when there is a uniform current so as to help the journey onwards and to impede the journey back. If the speed of the launch on both days was same, the time required for complete journey on the rough day, as compared to the quiet day will be

(a) more (b) less
(c) same (d) None of these.

119. A train runs past a telegraph pole in 15 second, and through a tunnel 450 metre long in 45 s. The length of the train is

(a) 225 m (b) 325 m
(c) 425 m (d) 450 m.

120. A ship travels downstream from point A to point B in two hours and upstream in three hours. Then the time that it will take a log of wood to cover the distance from point A to point B is

(a) 12 hrs. (b) 7 hrs,
(c) 5 hrs. (d) 1 hrs.

121. Two trains travelling on the same track are approaching each other with equal speeds of 40 m s^{-1}. The drivers of the trains begin to decelerate simultaneously when they are just 2 km apart. If the decelerations are both uniform and equal, then the value of deceleration to barely avoid collision should be

(*a*) 0.8 m s^{-2} (*b*) 2.1 m s^{-2}
(*c*) 11.0 m s^{-2} (*d*) 13.2 m s^{-2}.

122. The distance x covered by a particle in one-dimensional motion varies with time t as : $x^2 = at^2 + 2bt + c$. The acceleration of the particle varies as

(*a*) x^{-3} (*b*) $x^{3/2}$
(*c*) x^2 (*d*) $x^{-2/3}$.

123. Two balls of equal masses are thrown upwards, along the same vertical direction, at an interval of 2 s, with the same initial velocity of 39.2 m s^{-1}. Then these collide at a height of

(*a*) 44.1 m (*b*) 73.5 m
(*c*) 117.6 m (*d*) 196.0 m.

124. A car is moving along a straight road with uniform acceleration. It passes through two points P and Q separated by a distance with velocities 30 km h^{-1} and 40 km h^{-1} respectively. The velocity of car midway between P and Q is

(*a*) 33.3 km h^{-1} (*b*) $20\sqrt{3}$ km h^{-1}
(*c*) $25\sqrt{2}$ km h^{-1} (*d*) 35.35 km h^{-1}.

125. A man in a balloon rising vertically with an acceleration of 4.9 m s^{-2} releases a ball 2 s after the balloon is let go from the ground. The greatest height above the ground reached by the ball is (g = 9.8 m s^{-2})

(*a*) 24.5 m (*b*) 19.6 m
(*c*) 14.7 m (*d*) 9.8 m.

126. A body, moving along a straight line with uniform acceleration, starts from point O and passes through points A, B, C, D at distances x_1, x_2, x_3, x_4 from O. It covers distances AB, BC and CD in equal intervals of time. Then $(x_4 - x_1)$ is equal to

(*a*) $(x_2 - x_3)$ (*b*) $2(x_2 - x_4)$
(*c*) $3(x_2 - x_3)$ (*d*) $3(x_3 - x_2)$.

127. The relation between time t and distance x is

$$t = \alpha x^2 + \beta x$$

where α and β are constants. The retardation is

(*a*) $2\alpha v^3$ (*b*) $2\beta v^3$
(*c*) $2\alpha\beta v^3$ (*d*) $2\beta^2 v^3$.

128. A train starting from rest travels the first part of its journey with constant acceleration a, second part with constant velocity v and third part with constant retardation a, being brought to rest. The average speed for the whole journey is $\frac{7v}{8}$. The train travels with constant velocity for ...... of the total time

(*a*) $\frac{3}{4}$ (*b*) $\frac{7}{8}$
(*c*) $\frac{5}{6}$ (*d*) $\frac{9}{7}$.

129. In Q. 128, the fraction of the distance travelled by body with uniform speed to the whole distance is

(*a*) $\frac{6}{7}$ (*b*) $\frac{7}{8}$
(*c*) $\frac{5}{6}$ (*d*) $\frac{3}{4}$.

130. A particle oscillates along the x-axis according to the equation $x = 0.05 \sin(5t - \pi/6)$, where x is in metre and t is in second. Its velocity at $t = 0$ is

(*a*) 0.25 m s^{-1} (*b*) 0.22 m s^{-1}
(*c*) 2.5 m s^{-1} (*d*) 0.13 m s^{-1}.

131. In Q. 130, the acceleration of the particle at $t = 0$ is

(*a*) 0.63 m s^{-2} (*b*) 0.36 m s^{-2}
(*c*) 0.86 m s^{-2} (*d*) 0.13 m s^{-2}.

132. A particle starts from the origin and moves along the X-axis such that the velocity at any instant is given by $4t^3 - 2t$, where t is in second and velocity is in m s^{-1}. What is the acceleration of the particle when it is 2 m from the origin ?

(*a*) 10 m s^{-2} (*b*) 12 m s^{-2}
(*c*) 22 m s^{-2} (*d*) 28 m s^{-2}.

133. A particle moving along a straight line has a velocity v m s^{-1}, when it cleared a distance of y metre. These two are connected by the relation $v = \sqrt{49 + y}$. When its velocity is 1 m s^{-1}, its acceleration (in m s^{-2}) is

(*a*) 1 (*b*) 2
(*c*) 7 (*d*) 0.5.

134. A particle moving in a straight line with uniform acceleration is observed to be at a distance 'a' from a fixed point initially. It is at distances b, c, d from the same point after n, $2n$, $3n$ second. The acceleration of the particle is

(*a*) $\frac{c - 2b + a}{n^2}$ (*b*) $\frac{c + b + a}{9n^2}$
(*c*) $\frac{c + 2b + a}{4n^2}$ (*d*) $\frac{c - b + a}{n^2}$.

135. If a, b, and c be the distances travelled by the body during xth, yth and zth second from start, then which of the following relations is true ?

(*a*) $a(y - z) + b(z - x) + c(x - y) = 0$

(*b*) $a(x - y) + b(y - z) + c(z - x) = 0$

(c) $a(z - x) + b(x - y) + c(y - z) = 0$

(d) $ax + by + cz = 0$.

136. A balloon is moving upwards with velocity 9.8 m s^{-1}. It releases a stone which comes down to the ground in 11 s. The height of the balloon from the ground at the moment when the stone was dropped is

(a) 494.9 m (b) 592.9 m
(c) 490 m (d) 485.1 m.

137. A body of mass 3 kg falls from the multi-storeyed building 100 m high and buries itself 2 m deep in the sand. The time of penetration will be

(a) 0.09 s (b) 0.9 s
(c) 9 s (d) 10 s.

138. A particle moves in a straight line so that its displacement x metre at time t second is given by :

$$t = \sqrt{x^2 - 1}$$

Its acceleration in ms^{-2} at time t second is

(a) $\frac{1}{x^3}$ (b) $\frac{t^2}{x^3}$
(c) $\frac{1}{x} - \frac{t^2}{x^3}$ (d) $\frac{t^2}{x^3} - \frac{1}{x^2}$.

139. A stone dropped from a tower reaches the ground after 4 s. If it is stopped 2 s after its fall, and then released again, it will reach the ground after how many more second ?

(a) 2 (b) 3
(c) $\sqrt{8}$ (d) $\sqrt{12}$.

140. A body thrown up from the ground vertically passes the height of 10.2 m twice in an interval of 10 second. Its initial velocity was (Given : $g = 10$ m s^{-2})

(a) 100 m s^{-1} (b) 98 m s^{-1}
(c) 52 m s^{-1} (d) 49 m s^{-1}.

141. A particle moving with uniform acceleration is found to travel 35 m in the 8th second and 51 m in the 12th second. Its velocity in m s^{-1} at the beginning of the 11th second is

(a) 49 (b) 45
(c) 47 (d) 51.

142. Three particles start from the origin at the same time, one with a velocity v_1 along x-axis, the second along the y-axis with a velocity v_2 and the third along $x = y$ line. The velocity of the third so that the three may always lie on the same line is

(a) $\frac{v_1 v_2}{v_1 + v_2}$ (b) $\frac{\sqrt{2}\, v_1 v_2}{v_1 + v_2}$
(c) $\frac{\sqrt{3}\, v_1 v_2}{v_1 + v_2}$ (d) zero.

143. A particle travels at a constant velocity of 3 m s^{-1} for 20 s. It then stops for 10 s before completing its journey in a further 30 s at a constant velocity of 2 m s^{-1}. Its average velocity for the whole journey is

(a) 2.0 m s^{-1} (b) 2.4 m s^{-1}
(c) 2.5 m s^{-1} (d) 3.0 m s^{-1}.

144. A body freely falling from rest has a velocity v after it falls through distance h. The distance it has to fall down further for its velocity to become double is

(a) h (b) $2h$
(c) $3h$ (d) $4h$.

145. One body is dropped, while a second body is thrown downward with an initial velocity of 1 m s^{-1} simultaneously. The separation between these is 1.8 m after a time

(a) 4.5 s (b) 9 s
(c) 1.8 s (d) 36 s.

146. Two balls A and B are thrown simultaneously from the top of a tower. A is thrown vertically up with a speed of 4 m s^{-1}. B is thrown vertically down with a speed of 4 m s^{-1}. The balls A and B hit the ground with speeds v_A and v_B respectively. Then

(a) $v_A < v_B$ (b) $v_A > v_B$
(c) $v_A \gtreqless v_B$ (d) $v_A = v_B$.

147. A car travelling at a velocity of 80 km h^{-1} slowed down to 44 km h^{-1} in 15 s. The retardation is

(a) 0.67 m s^{-2} (b) 1 m s^{-2}
(c) 1.25 m s^{-2} (d) 1.5 m s^{-2}.

148. Two balls A and B are thrown from the top of a tower. The ball A is thrown vertically upwards. The ball B is thrown vertically downwards with the same speed. Which of the following is true ?

(a) A reaches the ground earlier than B
(b) B reaches the ground earlier than A
(c) Ball A strikes the ground with higher velocity
(d) Both A and B reach the ground at the same time.

149. The motion of a particle along a straight line is shown in figure. The average velocity of the particle is

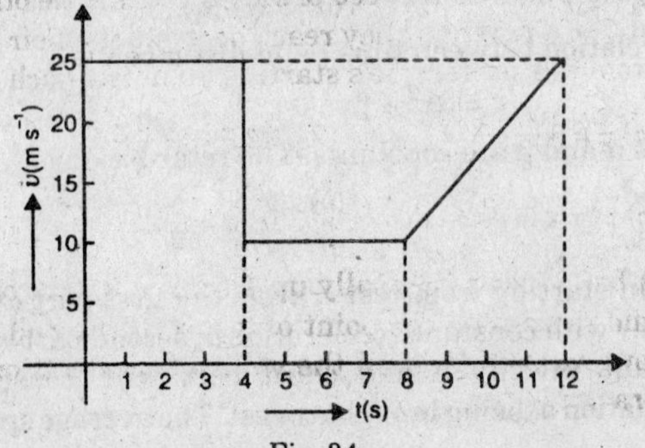

Fig. 34

(*a*) 12.5 m s^{-1} (*b*) 15 m s^{-1}
(*c*) 17.5 m s^{-1} (*d*) 20 m s^{-1}.

150. A railway train 200 m long passes over a bridge 600 m long. The time taken by the train to cross the bridge with uniform speed of 36 km h^{-1} is
(*a*) 60 s (*b*) 80 s
(*c*) 800 s (*d*) 20 s.

151. A boat can travel at a speed of 8 km h^{-1} in still water on a lake. In the flowing water of a stream, it can move at 8 km h^{-1} relative to the water in the stream. If the stream speed is 3 km h^{-1}, how fast can the boat move past a tree on the shore in travelling (*i*) upstream (*ii*) downstream ?
(*a*) 5 km h^{-1} and 11 km h^{-1} (*b*) 11 km h^{-1} and 5 km h^{-1}
(*c*) 8 km h^{-1} and 8 km h^{-1} (*d*) 5 km h^{-1} and 5 km h^{-1}.

152. A man runs at a speed of 4 m s^{-1} to overtake a standing bus. When he is 6 m behind the door at $t = 0$, the bus moves forward and continues with a constant acceleration of 1.2 m s^{-2}. The man reaches the door in time t. Then
(*a*) $4t = 6 + 0.6t^2$ (*b*) $1.2t^2 = 4t$
(*c*) $4t^2 = 1.2t$ (*d*) $6 + 4t = 0.6t^2$.

153. Water drops fall at regular intervals from a tap which is 5 m above the ground. The third drop is leaving the tap at the instant the first drop touches the ground. How far above the ground is the second drop at that instant ? (Take $g = 10$ m s^{-2})
(*a*) 1.25 m (*b*) 4.00 m
(*c*) 2.50 m (*d*) 3.75 m.

154. A particle moving in a straight line covers half the distance with a speed of 3 m s^{-1}. The other half of the distance is covered in two equal time intervals with speeds of 4.5 m s^{-1} and 7.5 m s^{-1}. The *average* speed of the particle during this motion is
(*a*) 4 m s^{-1} (*b*) 5 m s^{-1}
(*c*) 5.5 m s^{-1} (*d*) 4.8 m s^{-1}.

155. Two boys start running straight towards each other simultaneously from two points that are 100 m apart. One runs with a speed of 5 m s^{-1}, while the other runs at 7 m s^{-1}. When they reach each other, their distance from the slower one's starting point is x such that
(*a*) $\frac{x}{5} = \frac{100 - x}{7}$ (*b*) $\frac{x}{7} = \frac{100 - x}{5}$
(*c*) $\pm 7x = 5(100 - x)$ (*d*) $2 = \frac{5}{7}x$.

156. A ball kicked vertically up attains a height of 19.6 m and returns to the point of throw. If the ball is in air for four second, then the value of acceleration due to gravity is
(*a*) 4.9 m s^{-2} (*b*) 9.8 m s^{-2}
(*c*) 10 m s^{-2} (*d*) 2×9.8 m s^{-2}.

157. From the top of a tower, a particle is thrown vertically downwards with a velocity of 10 m s^{-1}. The ratio of the distances covered by it in the 3rd and 2nd seconds of its motion is (Given : $g = 10$ m s^{-2})
(*a*) 7 : 5 (*b*) 3 : 4
(*c*) 4 : 3 (*d*) 6 : 5.

158. Two particles P and Q get 4 m closer each second while travelling in opposite direction. They get 0.4 m closer every second while travelling in same direction. The speeds of P and Q are respectively
(*a*) 2.2 m s^{-1} and 0.4 m s^{-1} (*b*) 2.2 m s^{-1} and 1.8 m s^{-1}
(*c*) 4 m s^{-1} and 0.4 m s^{-1} (*d*) None of these.

159. Two particles held at different heights a and b above the ground are allowed to fall from rest. The ratio of their velocities on reaching the ground is
(*a*) $a : b$ (*b*) $\sqrt{a} : \sqrt{b}$
(*c*) $a^2 : b^2$ (*d*) $a^3 : b^3$.

160. A point mass initially at rest moves along x-axis. Its acceleration varies with time as $a = 6t + 5$ m s^{-2}. If it starts from origin, the distance covered by it in 1 s is
(*a*) 3 m (*b*) 5.5 m
(*c*) 3.5 m (*d*) 4 m.

161. A particle is moving with constant initial velocity 4 m s^{-1} till $t = 1.5$ s. Then it accelerates at 10 m s^{-2} till $t = 3$ s. The distance covered is (Take $g = 10$ m s^{-2})
(*a*) 17.25 m (*b*) 36.25 m
(*c*) 40 m (*d*) 23.25 m.

162. A particle starting from rest falls from a certain height. Assuming that the value of acceleration due to gravity remains the same throughout motion, its displacements in three successive half second intervals are S_1, S_2, S_3. Then
(*a*) $S_1 : S_2 : S_3 :: 1 : 5 : 9$ (*b*) $S_1 : S_2 : S_3 :: 1 : 2 : 3$
(*c*) $S_1 : S_2 : S_3 :: 1 : 1 : 1$ (*d*) $S_1 : S_2 : S_3 :: 1 : 3 : 5$.

163. A ball released from the top of a tower travels $\frac{11}{36}$ of the height of the tower in the last second of its journey. The height of the tower is (Take : $g = 10$ m s^{-2})
(*a*) 11 m (*b*) 36 m
(*c*) 47 m (*d*) 180 m.

164. Two cars leave one after the other and travel with an acceleration of 0.4 m s^{-2}. Two minutes after the departure of the first car, the distance between them becomes 1.90 km. The time interval between their departures is (*a*) 50 s (*b*) 60 s

(*c*) 70 s (*d*) 80 s.

165. The displacement of a body is given by $x = 4bt^3$, where a and b are constants. The velocity after 1 s from start is

(*a*) $12b$ (*b*) $a - 4b$

(*c*) $a + 4b$ (*d*) $1 + 4b/3$.

166. A stone thrown upward with a speed v from the top of the tower reaches the ground with a velocity $3v$. The height of the tower is

(*a*) $3v^2/g$ (*b*) $4v^2/g$

(*c*) $6v^2/g$ (*d*) $9v^2/g$.

167. The acceleration of a particle increasing linearly with time t is bt. The particle starts from the origin with an initial velocity, v_0. The distance travelled by the particle in time t will be

(*a*) $v_0t + \frac{1}{6}bt^3$ (*b*) $v_0t + \frac{1}{3}bt^2$

(*c*) $v_0t + \frac{1}{3}bt^3$ (*d*) $v_0t + \frac{1}{2}bt^2$.

168. A rocket is fired vertically upwards and moves with net vertical acceleration of 20 m s^{-2}. After 1 minute, the fuel is exhausted. The time taken by it to reach the highest point after the fuel is exhausted will be (Take $g = 10$ m s^{-2})

(*a*) 30 s (*b*) 60 s

(*c*) 120 s (*d*) 240 s.

169. A particle moving in a straight line with uniform acceleration is observed to be at a distance '*a*' from a fixed point initially. It is at distances *b*, *c*, *d* from the same point after *n*, 2*n*, 3*n* second. Then $(d - a) =$

(*a*) $(c - b)$ (*b*) $2(c - b)$

(*c*) $3(c - b)$ (*d*) $4(c - b)$.

170. In the one-dimensional motion of a particle, the relation between position x and time t is given by

$$x^2 + 2x = t$$

Choose the correct statement :

(*a*) The retardation of the particle is $\frac{1}{4(x+1)^3}$.

(*b*) The uniform acceleration of the particle is $\frac{1}{(x+1)^3}$.

(*c*) The uniform velocity of the particle is $\frac{1}{(x+1)^3}$.

(*d*) The particle has a variable acceleration of $4t + 6$.

171. A body starts from rest and falls vertically from a height of 19.6 metre. If $g = 9.8$ m s^{-2}, then the distance travelled by the body in the last 0.1 second of its motion is

(*a*) 1.9 metre (*b*) 0.049 metre

(*c*) 17.7 metre (*d*) 19.6 metre.

172. The retardation experienced by a moving motor boat, after its engine is cut off, is given by

$$\frac{dv}{dt} = -kv^3, \text{ where } k \text{ is a constant.}$$

If v_0 is the magnitude of the velocity at cut-off, the magnitude of the velocity at time t after the cut-off is

(*a*) v_0 (*b*) $\frac{v_0}{2}$

(*c*) v_0e^{-kt} (*d*) $\frac{v_0}{\sqrt{2v_0^2kt + 1}}$.

173. The engine of a train can impart a maximum acceleration of 1 m s^{-2} and the brakes can give a maximum retardation of 3 m s^{-2}. The least time during which a train can go from one place to the other place at a distance of 1.2 km is nearly

(*a*) 108 s (*b*) 191 s

(*c*) 56.6 s (*d*) time is fixed.

174. The driver of a train moving at 72 km h^{-1} sights another train moving at 4 m s^{-1} on the same track and in the same direction. He instantly applies brakes to produce a retardation of 1 m s^{-2}. The minimum distance between the trains so that no collision occurs is

(*a*) 32 m (*b*) 64 m

(*c*) 128 m (*d*) 256 m.

175. A car starts from rest and acquires a velocity v with uniform acceleration α. Then it comes to a stop with uniform retardation β. The average velocity of the car is

(*a*) v (*b*) $\frac{v}{2}$

(*c*) $\frac{\alpha\beta}{\alpha + \beta}$ (*d*) $\frac{\alpha + \beta}{\alpha\beta}$.

176. A stone falls from a balloon that is descending at a uniform rate of 12 m s^{-1}. The displacement of the stone from the point of release after 10 s is

(*a*) 725 m (*b*) 610 m

(*c*) 510 m (*d*) 490 m.

177. The driver of a car moving with a speed of 10 m s^{-1} sees a red light ahead, applies brakes and stops after covering 10 metre distance. If the same car were moving with a speed of 20 m s^{-1}, then the same driver would have stopped the car after covering 30 m distance. Within what distance the car can be stopped if travelling with a velocity of 15 m s^{-1} ? Assume the same reaction time and the same deceleration in each case.

(a) 18.75 m (b) 20.75 m
(c) 22.75 m (d) 25 m.

178. A tiger chases a deer 30 m ahead of it and gains 3 metre in 5 second after the chase began. The distance gained by the tiger in 10 second is

(a) 6 m (b) 12 m
(c) 18 m (d) 20 m.

179. A car starts from rest and accelerates at 5 m s^{-2} for sometime. Then, the car moves with uniform velocity for 15 second. Again, it decelerates at 5 m s^{-2} and comes to rest. If the total time for the journey is 25 second, then the average speed for the journey is

(a) 10 m s^{-1} (b) 20 m s^{-1}
(c) 30 m s^{-1} (d) 40 m s^{-1}.

180. A ball thrown upward from the top of a tower with speed v reaches the ground in t_1 second. If this ball is thrown downward from the top of the same tower with speed v_1 it reaches the ground in t_2 second. In what time the ball shall reach the ground if it is allowed to fall freely under gravity from the top of the tower ?

(a) $\frac{t_1 + t_2}{2}$ (b) $\frac{t_1 - t_2}{2}$
(c) $\sqrt{t_1 t_2}$ (d) $t_1 + t_2$.

181. Fig. 35 shows the variation of displacement of a body, moving along a straight line, with time. The velocity of the body during AB is

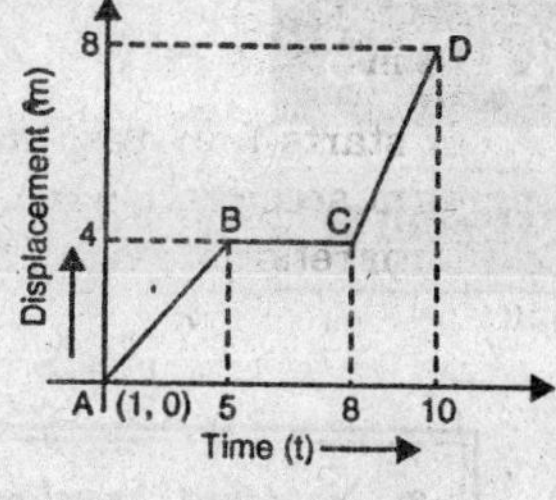

Fig. 35

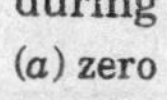

(a) zero
(b) 1 m s^{-1}
(c) 2 m s^{-1}
(d) 4 m s^{-1}.

182. In Q. 181, the velocity of the body during BC is

(a) zero (b) 1 m s^{-1}
(c) 2 m s^{-1} (d) 4 m s^{-1}.

183. As shown in figure the velocity of body as it moves from C to D is

(a) 0 m s^{-1} (b) 1 m s^{-1}
(c) 2 m s^{-1} (d) 4 m s^{-1}.

184. A scooter can produce a maximum acceleration of 5 m s^{-2}. Its brakes can produce a maximum retardation of 10 m s^{-2}. The minimum time in which it can cover a distance of 1.5 km is

(a) 30 s (b) 60 s
(c) 90 s (d) 120 s.

185. Given : $ax^2 + bx = t$. Here x and t represent displacement and time respectively. The acceleration of the particle at the origin of the co-ordinate axes is

(a) $\frac{a}{b^3}$ (b) $\frac{2a}{b^3}$
(c) $-\frac{2a}{b^3}$ (d) $-\frac{3a}{b^3}$.

186. A 0.05 kg ball thrown up with a velocity of 30 m s^{-1} reaches the highest point in 2.5 second. If g = 10 m s^{-2}, then the average air resistance experienced by the ball is (Given : g = 10 m s^{-2})

(a) 0.1 N (b) 0.2 N
(c) 0.3 N (d) 0.4 N.

187. The driver of an express train moving with a velocity v_1 finds that a goods train is moving with a velocity v_2 in the same direction on the same track. He applies the brakes and produces a retardation a. The minimum time required to avoid collision is

(a) $\frac{v_1 - v_2}{a}$ (b) $\frac{v_1}{a}$
(c) $\frac{v_2}{a}$ (d) $\frac{v_1 + v_2}{a}$.

188. If the velocity of a car is increased by 20%, then the minimum distance in which it can be stopped increases by

(a) 44% (b) 55%
(c) 66% (d) 88%.

ANSWERS

QUESTION BANK (Level I)

1. (b)	2. (a)	3. (b)	4. (d)	5. (a)	6. (c)	7. (c)	8. (b)
9. (b)	10. (d)	11. (d)	12. (d)	13. (c)	14. (a)	15. (d)	16. (c)
17. (a)	18. (a)	19. (d)	20. (d)	21. (d)	22. (d)	23. (d)	24. (d)
25. (d)	26. (a)	27. (a)	28. (a)	29. (a)	30. (a)	31. (b).	

QUESTION BANK (Level II)

1. (b)	2. (a)	3. (c)	4. (b)	5. (b)	6. (d)	7. (d)	8. (c)
9. (c)	10. (a)	11. (b)	12. (a)	13. (c)	14. (b)	15. (a)	16. (c)
17. (a)	18. (a)	19. (d)	20. (d)	21. (b)	22. (a)	23. (c)	24. (b)
25. (c)	26. (c)	27. (a)	28. (a)	29. (a)	30. (a)	31. (b)	32. (c)
33. (b)	34. (c)	35. (a)	36. (b)	37. (d)	38. (a)	39. (d)	40. (d)
41. (c)	42. (c)	43. (b)	44. (c)	45. (a)	46. (d)	47. (b)	48. (d)
49. (b)	50. (b)	51. (b)	52. (a)	53. (a)	54. (c)	55. (c)	56. (c)
57. (c)	58. (b)	59. (a)	60. (a)	61. (b)	62. (b)	63. (c)	64. (c)
65. (a)	66. (b)	67. (a)	68. (d)	69. (d)	70. (a)	71. (a)	72. (c)
73. (d)	74. (b)	75. (c)	76. (b)	77. (d)	78. (a)	79. (d)	80. (a)
81. (b)	82. (c)	83. (c)	84. (a)	85. (d)	86. (d)	87. (b)	88. (c)
89. (a)	90. (b)	91. (c)	92. (a)	93. (c)	94. (d)	95. (b)	96. (a)
97. (a)	98. (a)	99. (b)	100. (c)	101. (b)	102. (a)	103. (c)	104. (a)
105. (d)	106. (a)	107. (d)	108. (c)	109. (b)	110. (c)	111. (c)	112. (c)
113. (b)	114. (c)	115. (a)	116. (b)	117. (c)	118. (a)	119. (a)	120. (a)
121. (a)	122. (a)	123. (b)	124. (d)	125. (c)	126. (d)	127. (a)	128. (a)
129. (a)	130. (b)	131. (a)	132. (c)	133. (d)	134. (a)	135. (a)	136. (d)
137. (a)	138. (a)	139. (d)	140. (c)	141. (b)	142. (b)	143. (a)	144. (c)
145. (c)	146. (d)	147. (a)	148. (b)	149. (c)	150. (b)	151. (a)	152. (a)
153. (d)	154. (a)	155. (a)	156. (b)	157. (a)	158. (b)	159. (b)	160. (c)
161. (d)	162. (d)	163. (d)	164. (a)	165. (a)	166. (b)	167. (a)	168. (c)
169. (c)	170. (a)	171. (a)	172. (d)	173. (c)	174. (c)	175. (b)	176. (b)
177. (a)	178. (b)	179. (b)	180. (c)	181. (b)	182. (a)	183. (c)	184. (a)
185. (c)	186. (a)	187. (a)	188. (a).				

HINTS/SOLUTIONS

QUESTION BANK (Level I)

1. (*d*) $\bar{v} = \frac{2 \times 20 \times 30}{20 + 30}$ km h^{-1} = 24 km h^{-1}

2. (*a*) The speed with which a ball is thrown vertically upwards is the same as the speed with which the ball returns.

3. (*b*) $S_{5th} = 0 + \frac{8}{2}(2 \times 5 - 1)$ m = 36 m

4. (*d*) Add the times taken by pebble and the time taken by sound.

5. (*a*) $5 = \frac{1}{2} g \times 1 \times 1$ or $g = 10$ m s^{-2},

$S = \frac{1}{2} \times 10 \times 4 \times 4$ m = 80 m

6. (*a*) Slope is negative at the point E.

7. (*c*) $S = \frac{1}{2} \times 1 \times 20 + 1 \times 20 + \frac{1}{2} \times (20 + 10) \times 1 + 1 \times 10$

= 55 m

8. (*b*) Think of $v^2 - u^2 = 2aS$. In the given problem, $u = 0$ and a is constant.

$\therefore \quad v^2 \propto S.$

- *Note that the velocity increases at a uniform rate.*
- *Note that the 'displacement changes' in each second are in the ratio* 1 : 3 : 5 : ....

9. (*b*) There is an increase in negative velocity upto 2 second. After two second, acceleration becomes positive. Now, the speed that is being asked for starts decreasing.

10. (*d*) $\frac{\tan 30°}{\tan 45°} = \frac{1}{\sqrt{3}} + 1 = 1 : \sqrt{3}$

11. (*d*) At the point A, the tangent to the curve is parallel to time-axis. So, velocity at A is zero. But acceleration is not zero. Note that the displacement corresponding to the point A is not zero.

12. (*d*) $t = \sqrt{\frac{2h}{g}}$, $h \propto t^2$, $\frac{h_1}{h_2} = \frac{t^2}{4t^2} = \frac{1}{4} = 0.25.$

13. (*c*) $h = \frac{1}{2} \times 9.8 \times 3 \times 3 \text{ m} = 44.1 \text{ m}$

14. (*a*) $\bar{v} = \frac{x}{\frac{x}{80} + \frac{x}{100}} = \frac{80 \times 100}{180} = 44.44 \text{ m s}^{-1}$

15. (*d*) Average velocity $= \frac{3x}{\frac{x}{60} + \frac{x}{20} + \frac{x}{10}} = \frac{3x}{\frac{x + 3x + 6x}{60}}$

$= \frac{3x \times 60}{10x} = 18 \text{ km h}^{-1} = 18 \times \frac{5}{18} \text{ m s}^{-1} = 5 \text{ m s}^{-1}$

16. (*c*) $x_1 = \frac{1}{2} a \times 10 \times 10 = 50a,\ x_1 + x_2 = \frac{1}{2} a \times 20 \times 20 = 200a$

$x_1 + x_2 + x_3 = \frac{1}{2} a \times 30 \times 30 = 450a$

$x_2 = 200a - 50a = 150a,\ x_3 = 450a - 200a = 250a$

$x_1 : x_2 : x_3 :: 50a : 150a : 250a,\ x_1 : x_2 : x_3 :: 1 : 3 : 5$

17. (*a*) $\frac{v_T + v_B}{2} = \frac{3}{0.5} = 6$ or $v_T + v_B = 12 \text{ m s}^{-1}$.

18. (*a*) $v_T^2 - 0^2 = 2gh$ or $\frac{v_T^2}{2g} = h.$

19. (*d*) This problem is best solved by the use of concept of relative velocity.

$v + 30 = 75$ or $v = 45 \text{ km h}^{-1}$

20. (*d*) $\frac{v_A}{v_B} = \frac{\tan 30°}{\tan 60°} = \frac{1}{\sqrt{3}} \times \frac{1}{\sqrt{3}} = \frac{1}{3}$

21. (*d*) $\vec{v} = 8t\,\hat{i} + 6t\,\hat{j},\ \vec{a} = 8\,\hat{i} + 6\,\hat{j}$

$a = \sqrt{8^2 + 6^2} = \sqrt{64 + 36} = \sqrt{100} = 10 \text{ m s}^{-2}.$

22. (*d*) $\vec{\Delta r} = \vec{r_{10}} - \vec{r_0} = [4 \times 10 \times 10\hat{i} + 3 \times 10 \times 10\hat{j} + 2\hat{k}]$

$- [4 \times 0 \times 0\hat{i} + 3 \times 0 \times 0\hat{j} + 2\hat{k}]$

$= 400\hat{i} + 300\hat{j}$

$\Delta r = \sqrt{(400)^2 + (300)^2}$ or $\Delta r = 100\sqrt{16 + 9} = 500 \text{ m}$

23. (*d*) $\frac{v}{2} = v - 10 \times 1.5$ or $\frac{v}{2} = 15$ or $v = 30 \text{ m s}^{-1}$

$h_{max} = \frac{v^2}{2g} = \frac{30 \times 30}{2 \times 10} \text{ m} = 45 \text{ m}$

24. (*d*) $\Delta t = \sqrt{\frac{2 \times 12}{10}} - \sqrt{\frac{2 \times 10}{10}}$

$= 1.549 \text{ s} - 1.414 \text{ s} = 0.135 \text{ s}$

25. (*d*) $h_1 - h_2 = \frac{1}{2} g(t_1^2 - t_2^2) = \frac{1}{2} \times 10(5^2 - 3^2)$

$= 5 \times 8 \times 2 \text{ m} = 80 \text{ m}$

26. (*a*) $x = \frac{1}{2} gt^2,\ 100 - x = 25t - \frac{1}{2} gt^2$;

Adding $25t = 100$ or $t = 4 \text{ s}$

27. (*a*) $x = \frac{1}{2} \times 9.8 \times 4 \times 4 = 78.4 \text{ m}$

Required height = 21.6 m.

28. (*a*) $150 = 0 + \frac{a}{2}(2 \times 8 - 1)$ or $\frac{300}{15} = a$ or $a = 20 \text{ m s}^{-2}$

29. (*a*) $s = \frac{1}{2} gt^2,\ v = \frac{1}{2} g \times 2t = gt$

30. (*a*) Take vertically upward direction as positive and vertically downward direction as negative.

31. (*b*) $30 = u + 5a,\ 58 = u + 12a$. Subtracting, $28 = 7a,\ a = 4 \text{ m s}^{-2}$. So, the body is moving with constant acceleration.

QUESTION BANK (Level II)

1. (*b*) $200 = 100 + 10t$ or $t = 10 \text{ s}.$

2. (*a*) $v = 10 + 2 \times 4 = 18 \text{ m s}^{-1}$

3. (*c*) Think of the slope of the given displacement-time graph at different points and you would arrive at the correct answer.

Alternatively, look at the self-illustrative figure.

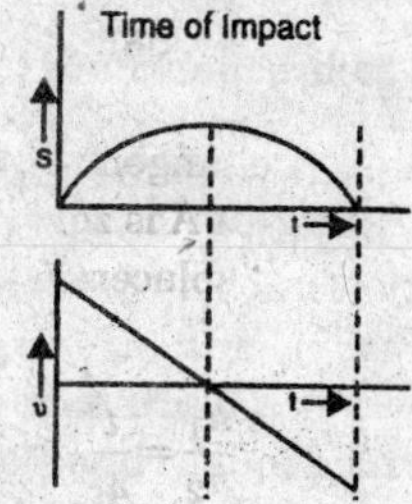

Fig. 36. Highest point of flight.

☞ **The above problem is based on the following physical situation :**

A ball thrown vertically upwards and returning to the point of throw.

As the ball rises, its velocity becomes less and less positive. It becomes zero at the maximum height. Then the velocity becomes more and more negative because it is now falling at greater speed until it returns to the ground.

The area above the time-axis gives +ve displacement. The area below the time-axis gives –ve displacement. There is as much area above the time-axis as there is below. So, the total displacement from launch to impact is zero.

4. (*b*) The problem is based on variable acceleration.

$$v = \int a\,dt = \frac{1}{2}(12+6)4 - \frac{1}{2}\times 2\times 2 = 36 - 2 = 34 \text{ m s}^{-1}$$

Since initial velocity is zero and the gain in velocity is 34 m s^{-1}, therefore, the velocity at 14 second is 34 m s^{-1}.

Problems based on variable acceleration

We know that $a = \frac{dv}{dt}$ *or* $dv = a\,dt$

Integrating, $v = \int a\,dt$

☞ **CAUTION.** *'a' is not to be taken out of the integral sign.*

5. (*b*) Acceleration is negative and constant for first half. It is positive and constant over next half.
6. (*d*) The straight vertical line in the graph represents change in the direction of velocity.
7. (*d*) Area between *v-t* graph and time-axis

$$= \frac{1}{2}\times 2\times 20 + 3\times 20 + \frac{1}{2}\times 1\times 20 + \frac{1}{2}\times 1\times 20 = 100 \text{ units.}$$

8. (*c*) Area between *v-t* graph and time-axis

$$= \frac{1}{2}\times 2\times 20 + 3\times 20 + \frac{1}{2}\times 1\times 20 - \frac{1}{2}\times 1\times 20 = 80 \text{ units.}$$

☞ **An important note**

While calculating displacement from v-t graph, give due regard to positive and negative signs. For calculation of distance, only positive signs are to be used.

9. (*c*) Average velocity $= \frac{80 \text{ m}}{5 \text{ s}} = 16 \text{ m s}^{-1}$
10. (*a*) Displacement = Area of upper trapezium − Area of lower trapezium

$$= \frac{1}{2}(2+4)4 + \frac{1}{2}(2+4)2 = 12 - 6 = 6 \text{ m.}$$

11. (*b*) When a body is released from the top of a building, the motion of the body is uniformly accelerated.
12. (*a*) Height reached $= \frac{1}{2}\times 132\times 1200 \text{ m} = 66\times 1200 \text{ m}$
13. (*c*) Height $= \frac{1}{2}(12+8)\,3.6 \text{ m} = 36 \text{ m}$
14. (*b*) Average velocity $= \frac{36 \text{ m}}{12 \text{ s}} = 3 \text{ m s}^{-1}$
15. (*a*) The change in velocity is zero.
16. (*c*) If upward velocity is taken as positive, then downward velocity is taken as negative.
17. (*a*)
   - $v = \frac{d}{dt}(x) = -4 + 12t$
   - When $t = 0$, $v = -4$ units
   - When $v = 0$, $12t = 4$ or $t = \frac{4}{12}$ units $= \frac{1}{3}$ units
   - Acc. = 12 (Positive slope).
18. (*a*)

(*a*) $\frac{\frac{1}{2}\times OA\times AC}{OA} + \frac{\frac{1}{2}\times AB\times AC}{AB} = 1$

(*b*) $\tan 60° = \frac{CA}{OA}$ and $\tan 30° = \frac{CA}{AB}$

$OA \tan 60° = AB \tan 30°$

or $\frac{OA}{AB} = \frac{\tan 30°}{\tan 60°} = \frac{1}{\sqrt{3}}\times\frac{1}{\sqrt{3}} = \frac{1}{3}$

19. (*d*) Displacement (in magnitude)

$$= \left(\frac{1}{2}\times 3\times 2 - \frac{1}{2}\times 1\times 2 + 1\times 1\right)\text{m}$$

$$= 3 \text{ m}$$

20. (*d*) Distance covered $= \left(\frac{1}{2}\times 3\times 2 + \frac{1}{2}\times 1\times 2 + 1\times 1\right)\text{m}$

$$= (3+1+1) \text{ m} = 5 \text{ m}$$

21. (*b*) Ratio is $\frac{20}{2} \div \frac{20}{4}$, *i.e.*, 2 : 1.
22. (*a*) There are six points in this graph where the tangents have zero slope.
23. (*c*) On rebound, there is an instantaneous change in the direction of velocity. This settles the answer.
24. (*b*) Since acceleration is constant, therefore there is a uniform increase in velocity. So, the *v-t* graph is a straight line sloping upward to the right. When acceleration becomes zero, velocity is constant. So, *v-t* graph is a straight line parallel to the time-axis.
25. (*c*) Select point B in such a way that AB is tangent to the curve at B.
26. (*c*) Rectangles under the curve = 6, each = 12.5 m.
27. (*a*) Slope having magnitudes 1, 2, 0, 4. One unit $= \frac{5}{2.5} \text{ m s}^{-2}$. Hence maximum $|a| = 8$.
28. (*a*) The portion OA of the graph is convex upward. It represents negative acceleration. The portion AB represents that x is not changing with time. Clearly, it is a case of zero acceleration. The portion BC of the graph is concave upward. It represents positive acceleration. The portion CD is a straight line sloping upward to the right.

It represents uniform velocity and hence zero acceleration.

29. (*a*) Displacement = $(2 \times 4 - 2 \times 2 + 2 \times 2)$ m = 8 m

Distance = $(2 \times 4 + 2 \times 2 + 2 \times 2)$ m = 16 m

30. (*a*) $S_1 = \frac{1}{2}gt^2$, $S_2 = -v_0t + \frac{1}{2}gt^2$, Clearly, $(S_1 - S_2) \propto t$

31. (*b*)

(*a*) is eliminated because if acceleration is constant, then both the magnitude and direction of acceleration should remain the same.

Area under velocity-time graph is all positive. So, (*c*) is ruled out.

The velocity at $t = 4$ second is 19.6 m s^{-1}. So, (*d*) is ruled out.

32. (*c*) $x = 50t - \frac{1}{2} \times 10 \times t^2$, $-(200 - x) = -\frac{1}{2} \times 10 \times t^2$

Subtracting $200 = 50t$ or $t = 4$ s

33. (*b*) $10t = 48 + \frac{1}{2} \times 1 \times t^2$ or $t^2 - 20t + 96 = 0$

or $t^2 - 8t - 12t + 96 = 0$ or $t(t-8) - 12(t-8) = 0$

or $(t-12)(t-8) = 0$ or $t = 8$ s or 12 s

But we are interested in minimum time.

34. (*c*) $30^2 - 20^2 = 2al$, $30^2 - v^2 = 2a\frac{l}{2}$;

$30^2 - 20^2 = 2 \times 30^2 - 2v^2$ or $2v^2 = 2 \times 30^2 - 30^2 + 20^2$

$= 900 + 400 = 1300$ or $v^2 = 650$

or $v = \sqrt{650}$ m s^{-1} = 25.5 m s^{-1}

35. (*a*) Average velocity = $\frac{2 \times 8 \times 12}{8 + 12}$ m s^{-1} = 9.6 m s^{-1}

36. (*b*) $x_A = 6t - 3$

$v_A = \frac{d}{dt}(6t - 3) = 6$

Particle A moves with **constant velocity.**

$v_B = \frac{d}{dt}(4t^2 - 2t + 3) = 8t - 2$, $a_B = \frac{d}{dt}(8t - 2) = 8$

Particle B moves with **constant acceleration.**

$v_C = \frac{d}{dt}(3t^3 - 2t^2 + t - 7) = 9t^2 - 4t + 1$

$a_C = 18t - 4$

So, the particle moves with **variable acceleration.**

$x_D = 7\cos 60° - 3\sin 30°$

Since x_D does not depend upon time therefore particle is at rest.

37. (*d*) $t = \sqrt{\frac{2h}{g}}$; $h \propto t^2$. If h is doubled, t becomes $\sqrt{2}$ times.

So, the new time is $2\sqrt{2}$ s *i.e.* 2×1.414 s or 2.828 s.

38. (*a*) For the fourth drop, 't' = 0, for the third drop, 't' = t ; For the second drop, 't' = $2t$; For the first drop, 't' = $3t$.

Now, $9 = \frac{1}{2}g(3t)^2$ [For the first drop]

or $\frac{1}{2}gt^2 = 1$

For the second drop,

$$h_2 = \frac{1}{2}g(2t)^2 = 4 \times \frac{1}{2}gt^2 = 4 \times 1 \text{ m} = 4 \text{ m}$$

For the third drop, $h_3 = \frac{1}{2}gt^2 = 1$ m

So, the distances of second and third drops are 4 m and 1 m respectively.

39. (*d*) $10^2 - 30^2 = 2 \times a \times 240$ or $a = -\frac{800}{480}$ m s^{-2} = $-\frac{5}{3}$ m s^{-2}

$0^2 - 10^2 = 2\left(-\frac{5}{3}\right)S$ or $S = \frac{100 \times 3}{10}$ m = 30 m

Total distance = (240 + 30) m = 270 m

40. (*d*) $h = \frac{1}{2}gt^2$, when t = 1 s, 2 s, 3 s and 4 s, then the ratio of distances is 1 : 4 : 9 : 16. The ratio of the distances travelled in the first, second, third and fourth second is 1 : 3 : 5 : 7.

41. (*c*) $v = \frac{d}{dt}(x) = \frac{d}{dt}(t^2) - \frac{d}{dt}(t^3) = 2t - 3t^2$

Acceleration = $\frac{d}{dt}(v) = \frac{d}{dt}(2t - 3t^2) = 2 - 6t$

$t > 0$. So, acceleration is negative. Moreover, it is variable.

42. (*c*) Average acceleration = $\frac{\Delta v}{\Delta t}$

$$= \frac{\sqrt{2gh'} - (-\sqrt{2gh})}{\Delta t} = \frac{\sqrt{2gh'} + \sqrt{2gh}}{\Delta t}$$

$$= \frac{\sqrt{2 \times 10 \times 2.5} + \sqrt{2 \times 10 \times 10}}{0.01} \text{ m s}^{-2}$$

$$= \frac{\sqrt{50} + \sqrt{200}}{0.01} \text{ m s}^{-2} = \frac{5\sqrt{2} + 10\sqrt{2}}{0.01} \text{ m s}^{-2}$$

$$= \frac{15\sqrt{2}}{0.01} \text{ m s}^{-2} = 1500\sqrt{2} \text{ m s}^{-2}$$

The upward velocity has been taken as positive. Since average acceleration is positive therefore its direction is vertically upward.

> ☞ **CAUTION.** *In this type of questions, the common mistake is to take Δv as $\sqrt{2gh'} - \sqrt{2gh}$.*

43. (*b*) $\frac{V}{2} = \frac{x}{t_1}$ or $t_1 = \frac{2x}{V}$, $V = \frac{2x}{t_2}$ or $t_2 = \frac{2x}{V}$,

$\frac{V}{2} = \frac{3x}{t_3}$ or $t_3 = \frac{6x}{V}$

$$v_{av} = \frac{x + 2x + 3x}{\frac{2x}{V} + \frac{2x}{V} + \frac{6x}{V}} = \frac{6x}{\frac{10x}{V}} = \frac{3V}{5}$$

or $\frac{v_{av}}{V} = \frac{3}{5}$.

> ☞ *In this solution, time has been saved by the use of the formulae :*
>
> (i) $v_{av} = \frac{u+v}{2}$ (ii) $v_{av} = \frac{S}{t}$

44. (c) $v_1 + v_2 = 6$ m s^{-1}, $v_1 - v_2 = 4$ m s^{-1} ; Adding $2v_1 = 10$ or $v_1 = 5$ m s^{-1} ; Subtracting, $2v_2 = 2$ or $v_2 = 1$ m s^{-1}.

> - *When two bodies move towards each other, | relative velocity | = sum of speeds of the two particles*
> - *When two bodies move in the same direction | relative velocity | = difference of speeds of the two particles.*

45. (a) $v = 18 \times \frac{5}{18}$ m s^{-1} = 5 m s^{-1},

$h_{max} = \frac{5 \times 5}{2 \times 10}$ m $= \frac{25}{20}$ m = 1.25 m.

> ☞ **Author's suggestion**
>
> ***In this type of questions, do not waste your time in using $v^2 - u^2 = 2aS$.***

46. (d) $S_1 - S_2 = \frac{1}{2}gt^2 - \frac{1}{2}g(t-1)^2$

$$10 = \frac{1}{2} \times g(1)(2t - 1),\ 2t = 3,\ t = 1.5 \text{ s.}$$

47. (b) Effective displacement is 2 m in 8 second. Time for (11 – 5) m *i.e.* 6 m will 3 × 8 s *i.e.* 24 s. Last 5 m will take 5 s. So, total time is 29 s.

48. (d) 'a' = $g \cos\theta$, S = $2r\cos\theta$, Using $S = \frac{1}{2}at^2$, $2r\cos\theta$

$= \frac{1}{2}g\cos\theta t^2$ or $4r = gt^2$; clearly, t is independent of θ.

49. (b) $2 = u \times 2 + \frac{1}{2} \times a \times 2 \times 2$ or $1 = u + a$,

$4.20 = u \times 6 + \frac{1}{2}a \times 6 \times 6$

or $0.7 = u + 3a$, $2a = -0.3$ or $a = -0.15$ m s^{-2},

$u = 1 - a = (1 + 0.15)$ m s^{-1} = 1.15 m s^{-1},

$v = 1.15 - 0.15 \times 9 = 1.15 - 1.35 = -0.2$ m s^{-1}

50. (b) $x = \frac{1}{2} \times a \times 2 \times 2 = 2a$, $x + y = \frac{1}{2} \times a \times 4 \times 4 = 8a$

or $x + y = 4(2a) = 4x$ or $y = 3x$

51. (b) $v = \sqrt{2gh} = \sqrt{2 \times 10 \times 45} = 30$ m s^{-1},

$15 = 30 - 10t$ or $10t = 15$ or $t = 1.5$ s

52. (a) $x = 2t^3 - 6t^2 + 12t + 61$; $v = 6t^2 - 12t + 12$

acc. $= \frac{d}{dt}(v) = 12t - 12$;

$0 = 12t - 12$ or $12t = 12$ or $t = 1$ s

53. (a) $h = -4.9 \times 3 + \frac{1}{2} \times 9.8 \times 3 \times 3 = 4.9 \times 3 \times 2$ m = 29.4 m

54. (c) $26 = u + \frac{a}{2}(2 \times 10 - 1) = u + \frac{19a}{2}$,

$28 = u + \frac{a}{2}(2 \times 11 - 1)$

$= u + \frac{21a}{2}$; $\frac{21a}{2} - \frac{19a}{2} = 28 - 26 = 2$ or $\frac{2a}{2} = 2$

or $a = 2$ m s^{-2}, $26 = u + \frac{2}{2}(19)$ or $u = 7$ m s^{-1}

Clearly, the body has a non-zero initial velocity and moves with uniform acceleration.

55. (c) $\frac{9h}{25} = \frac{g}{2}(2n - 1)$, $h = \frac{1}{2}gn^2$. Dividing, $\frac{9}{25} = \frac{2n-1}{n^2}$

$\Rightarrow n = 5$, $h = \frac{1}{2} \times 9.8 \times 5 \times 5$ m = 122.5 m.

56. (c) $h = \frac{1}{2}gT^2$, $h' = \frac{1}{2}g\left(\frac{T}{2}\right)^2$, $h' = \frac{h}{4}$,

Height above the ground $= h - \frac{h}{4} = \frac{3h}{4}$

57. (c) $x = 50t + \frac{1}{2} \times 18 \times t^2$, $100 - x = 50t - \frac{1}{2} \times 18 \times t^2$

Adding, $100 = 100t$ or $t = 1$ hour

$$x = 50 \times 1 + \frac{1}{2} \times 18 \times 1 \times 1 = 59 \text{ km}$$

58. (b) Let g_1 be the retardation due to gravity in the presence of air resistance. Clearly, $|g_1| > |g|$.

If v be the velocity of throw, then

time of ascent, $t_1 = \frac{v}{g_1}$ ...(1)

maximum height $= \frac{v^2}{2g_1}$

Let g_2 be the acceleration due to gravity in the presence of air resistance. Clearly $|g_2| < |g|$

Now, $\frac{v^2}{2g_1} = \frac{1}{2}g_2 t_2^2$, where t_2 is the time of descent.

Now, $t_2^2 = \frac{2v^2}{2g_1g_2} = \frac{v^2}{g_1g_2}$ or $t_2 = \frac{v}{\sqrt{g_1g_2}}$ ...(2)

$\frac{(2)}{(1)}$ gives $\frac{t_2}{t_1} = \frac{v}{\sqrt{g_1g_2}} \times \frac{g_1}{v} = \frac{g_1}{\sqrt{g_1g_2}} = \sqrt{\frac{g_1}{g_2}}$

But $|g_1| > |g|$ and $|g_2| < |g|$

$\therefore \sqrt{\frac{g_1}{g_2}} > 1$

$\therefore \frac{t_2}{t_1} > 1$ or $t_2 > t_1$ or $t_1 < t_2$

59. (*a*) $20 = 10 + 5a$ or $a = 2\text{ m s}^{-2}$. Now, let u be the required velocity. Let us now treat 10 m s^{-1} as the final velocity. $10 = u + 2 \times 3$ or $u = 4\text{ m s}^{-1}$.

60. (*a*) $v = \sqrt{2 \times 10 \times 20}\text{ m s}^{-1} = 20\text{ m s}^{-1}$,

$v' = \frac{3}{4} \times 20\text{ m s}^{-1} = 15\text{ m s}^{-1}$

$t = \frac{2 \times 15}{10}\text{ s} = \frac{30}{10}\text{ s} = 3\text{ s}$

61. (*b*) $\sqrt{x} = t - 5$, $x = (t-5)^2$, $v = \frac{dx}{dt} = 2(t-5)$,

When $v = 0$, $t - 5 = 0$ or $t = 5$

$\sqrt{x} = 5 - 5 = 0 \Rightarrow x = 0$

62. (*b*) Refer to Fig. 37

$\frac{v^2}{2g} = 10$

or $v^2 = 2 \times 9.8 \times 10$

or $v = \sqrt{196}\text{ m s}^{-1}$

$= 14\text{ m s}^{-1}$

10 m

Fig. 37

63. (*c*) $a = 2t + 5$, $\frac{dv}{dt} = 2t + 5$ or $dv = (2t+5)\,dt$

Integrating, $\int_0^v dv = \int_0^2 (2t+5)\,dt$

or $v = \left|\frac{2t^2}{2} + 5t\right|_0^2 = 4 + 5 \times 2 = 14\text{ m s}^{-1}$

64. **(*c*) The central idea in this problem is to reduce initial relative velocity ($v_A - v_B$) to zero in distance S′ such that S′ ≤ S**

Now, $0^2 - (v_A - v_B)^2 = -2aS'$ *or* $S' = \frac{(v_A - v_B)^2}{2a}$.

To avoid collision, $S' \le S$

$\therefore \frac{(v_A - v_B)^2}{2a} \le S$ *or* $S \ge \frac{(v_A - v_B)^2}{2a}$.

65. (*a*) $65 = -12t + \frac{1}{2} \times 10 \times t^2$ or $5t^2 - 12t - 65 = 0$

On simplification, $t = 5$ s.

66. (*b*) Using $t = \sqrt{\frac{2h}{a}}$, $t \propto \sqrt{l}$, $\frac{t_2}{t_1} = \sqrt{\frac{l/4}{l}} = \frac{1}{2}$

or $t_2 = \frac{t_1}{2} = \frac{4\text{ s}}{2} = 2\text{ s}$

67. (*a*) $24.5 = \frac{1}{2} \times 9.8\,(2n-1) = 4.9(2n-1)$

or $2n - 1 = 5$ or $n = 3$

$h = \frac{1}{2} \times 9.8 \times 3 \times 3\text{ m} = 4.9 \times 9\text{ m} = 44.1\text{ m}$

68. (*d*) $v = \frac{dy}{dt} = \frac{d}{dt}(a_0 + a_1t + a_2t^2) = a_1 + 2a_2t$

Acceleration $= \frac{d}{dt}(v) = \frac{d}{dt}(a_1 + 2a_2t) = 2a_2$

69. (*d*) If v_1', v_2' and v_3' be the velocities at the ends of time intervals t_1, $t_1 + t_2$ and $t_1 + t_2 + t_3$ respectively, then $v_1' = u + at_1$, $v_2' = u + a(t_1 + t_2)$, $v_3' = u + a(t_1 + t_2 + t_3)$

$v_1 = \frac{u + v_1'}{2} = \frac{u + u + at_1}{2} = u + \frac{1}{2}at_1$

$v_2 = \frac{v_1' + v_2'}{2} = \frac{(u + at_1) + [u + a(t_1 + t_2)]}{2}$

$= u + at_1 + \frac{1}{2}at_2$

$v_3 = \frac{v_2' + v_3'}{2} = \frac{u + at_1 + at_2 + u + a(t_1 + t_2 + t_3)}{2}$

or $v_3 = u + at_1 + at_2 + \frac{1}{2}at_3$

$v_1 - v_2 = u + \frac{1}{2}at_1 - u - at_1 - \frac{1}{2}at_2 = -\frac{1}{2}a(t_1 + t_2)$

$v_2 - v_3 = u + at_1 + \frac{1}{2}at_2 - u - at_1 - at_2 - \frac{1}{2}at_3$

$= -\frac{1}{2}at_2 - \frac{1}{2}at_3 = -\frac{1}{2}a(t_2 + t_3)$

$\frac{v_1 - v_2}{v_2 - v_3} = \frac{t_1 + t_2}{t_2 + t_3}$

70. (*a*) $x = \frac{1}{2}gt^2$, $2x = \frac{1}{2}gt_1^2$, $3x = \frac{1}{2}gt_2^2$

$t \propto \sqrt{x}$, $t_1 \propto \sqrt{2x}$, $t_2 \propto \sqrt{3x}$

The required ratio is $1 : \sqrt{2} - 1 : \sqrt{3} - \sqrt{2}$

71. (*a*) For first ball, $u = 0$, $h = 125$ m, $a = +10\text{ m s}^{-2}$ (say), $t = ?$

Applying $h = ut + \frac{1}{2}at^2$, $t = 5$ s

For second ball, $u = $ 'u', $h = 125$ m, $a = +10\text{ m s}^{-2}$, $t = 3$ s

Applying $h = ut + \frac{1}{2}at^2$, 'u' $\approx 27\text{ m s}^{-1}$

72. (*c*) $\Delta x = \frac{1}{2}gt^2 - \frac{1}{2}g(t-1)^2$

$$= \frac{1}{2}g[t^2 - (t-1)^2] = \frac{1}{2}g(2t-1)$$

$$= \frac{1}{2} \times 9.8 \times 5 \text{ m} = 24.5 \text{ m}$$

> **In Qs. such as 72 use**
>
> $$\Delta x = \frac{1}{2}g\,(t_1 - t_2)(t_1 + t_2)$$

73. (*d*) The discussion of solution of this question is **extremely important**

> **Solution I.** Rel. velocity of ball w.r.t. balloon = $u + v$
>
> Time taken by ball to reach highest point = t (say)
>
> $$0 = -(u+v) + gt \quad \text{or} \quad t = \frac{u+v}{g}$$
>
> Total time = $\frac{2(u+v)}{g}$.
>
> **Note that the balloon, has been stopped in the above solution.** In the above solution, we are considering velocity relative to balloon.
>
> **Solution II.** In this solution, we shall consider velocities with respect to earth.
>
> $$vt = -ut + \frac{1}{2}gt^2 \quad \text{or} \quad (v+u)t = \frac{1}{2}gt^2$$
>
> or $$t = \frac{2(u+v)}{g}$$
>
> **WRONG SOLUTION**
>
> $$vt = -(u+v)t + \frac{1}{2}gt^2$$
>
> $$v + u + v = \frac{1}{2}gt \quad \text{or} \quad t = \frac{2u+v}{g}$$

74. (*b*) $S_1 = \frac{1}{2}gt^2, S_1 + S_2 = \frac{1}{2}g(2t)^2$

$$S_2 = \frac{1}{2}g(4t)^2 - \frac{1}{2}gt^2 = \frac{1}{2}g \times 3t^2$$

$$S_2 = 3 \times \frac{1}{2}gt^2 = 3S_1$$

75. (*c*) When a steamer say S starts its journey from B, eleven steamers are present in the sea and moving towards B. The steamer S, in its journey towards A shall meet 12 steamers. So, there will be a total of 23 steamers.

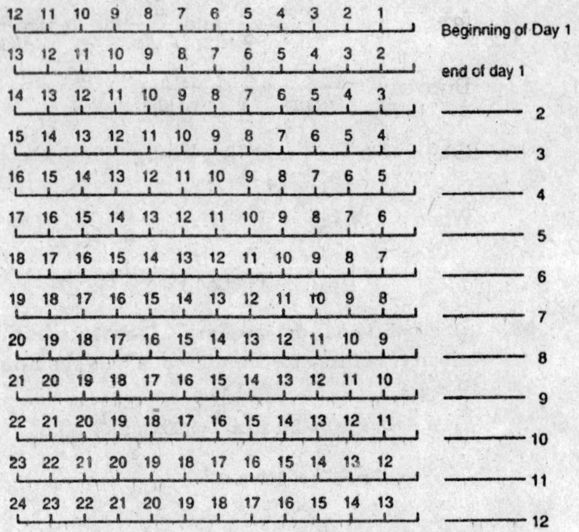

12 11 10 9 8 7 6 5 4 3 2 1 — Beginning of Day 1
13 12 11 10 9 8 7 6 5 4 3 2 — end of day 1
14 13 12 11 10 9 8 7 6 5 4 3 — 2
15 14 13 12 11 10 9 8 7 6 5 4 — 3
16 15 14 13 12 11 10 9 8 7 6 5 — 4
17 16 15 14 13 12 11 10 9 8 7 6 — 5
18 17 16 15 14 13 12 11 10 9 8 7 — 6
19 18 17 16 15 14 13 12 11 10 9 8 — 7
20 19 18 17 16 15 14 13 12 11 10 9 — 8
21 20 19 18 17 16 15 14 13 12 11 10 — 9
22 21 20 19 18 17 16 15 14 13 12 11 — 10
23 22 21 20 19 18 17 16 15 14 13 12 — 11
24 23 22 21 20 19 18 17 16 15 14 13 — 12

Fig. 38

76. (*b*) For the student, $320 = \frac{1}{2} \times 10 \times t^2$

or $$t = \sqrt{\frac{320 \times 2}{10}} \text{ s} \quad \text{or} \quad t = 8 \text{ s}.$$

Again, for the superman, $t = 3$ s.

$$\therefore\ 320 = v \times 3 + \frac{1}{2} \times 10 \times 3 \times 3 \quad \text{or} \quad v = \frac{320 - 45}{3} \text{ m s}^{-1}$$

$$= 91.7 \text{ m s}^{-1}$$

77. (*d*) $$v_{av.} = \frac{x}{\frac{x}{3 \times 20} + \frac{2x}{3 \times 60}} = \frac{x}{\frac{x}{60} + \frac{x}{90}}$$

or $$v_{av.} = \frac{60 \times 90}{90 + 60} = 36 \text{ km h}^{-1}$$

78. (*a*) $4 = u + \frac{a}{2}(2 \times 3 - 1)$ or $4 = u + \frac{5a}{2}$,

$$5 = u + \frac{a}{2}(2 \times 4 - 1) \quad \text{or} \quad 5 = u + \frac{7a}{2}$$

Subtracting, $1 = \frac{7a}{2} - \frac{5a}{2} = \frac{2a}{2} = a$

Again, $\quad 4 = u + \frac{5}{2}$ or $u = 4 - \frac{5}{2} = 1.5 \text{ m s}^{-1}$

So, the initial velocity is non-zero and acceleration is uniform.

79. (*d*) $t = \sqrt{\frac{2h}{g}} = \sqrt{\frac{2 \times 78.4}{9.8}} \text{ s} = 4 \text{ s}$

Time taken by sound = $0.23 = \frac{78.4}{v}$

or $$v = \frac{78.4}{0.23} \text{ m s}^{-1} = 340.9 \text{ m s}^{-1} \approx 341 \text{ m s}^{-1}$$

80. (*a*) $\left(\frac{v}{2}\right)^2 - v^2 = 2a \times 3$ and $0^2 - \left(\frac{v}{2}\right)^2 = 2aS$

Dividing, $\frac{S}{3} = \frac{1}{3}$ or S = 1 km.

81. (*b*) $v = \frac{d}{dt}(x) = \frac{d}{dt}\left(-\frac{2}{3}t^2 + 16t + 2\right)$ or $v = -\frac{2}{3} \times 2t + 16$

When $v = 0, -\frac{4t}{3} + 16 = 0$ or $-\frac{4t}{3} = -16$ or $t = 12$ s

82. (*c*) $v^2 - u^2 = 2al$ and $v'^2 - u^2 = 2a\frac{l}{2} = al$

or $2(v'^2 - u^2) = 2al$

Equating, $2(v'^2 - u^2) = v^2 - u^2$

or $v'^2 = u^2 + \frac{v^2 - u^2}{2} = \frac{v^2 + u^2}{2}$

or $v' = \sqrt{\frac{u^2 + v^2}{2}}$

83. (*c*) $\left(\frac{19v}{20}\right)^2 - v^2 = 2ax$

$0^2 - v^2 = 2anx$

Dividing, $n = \frac{-v^2}{\left(\frac{19}{20}v\right)^2 - v^2} = \frac{1}{1 - \left(\frac{19}{20}\right)^2}$

$= \frac{20 \times 20}{(20 + 19)(20 - 19)} = \frac{400}{39} = 10.3$

So, the bullet shall stop in 11th plank.

84. (*a*) $t = \sqrt{\frac{2 \times 44.1}{9.8}}$ s $= \sqrt{9}$ s = 3 s,

$44.1 = v \times 2 + \frac{1}{2} \times 9.8 \times 2 \times 2$

or $2v = 44.1 - 4.9 \times 4 = 24.5$

or $v = \frac{24.5}{2}$ m s^{-1} = 12.25 m s^{-1}

85. (*d*) Relative velocity = 90 km h^{-1} = $90 \times \frac{5}{18}$ m s^{-1} = 25 m s^{-1}

$0^2 - 25^2 = 2 \times a \times 100$

or $200a = -625a = -\frac{625}{200}$ m s^{-2} or $a = -3.125$ m s^{-2}

86. (*d*) $h = \frac{1}{2}gt^2$, $120 - h = 40t - \frac{1}{2}gt^2$,

Adding, $120 = 40t$ or $t = 3$ s

87. (*b*) $\frac{1}{2}g \times 3 \times 3 = \frac{g}{2}(2n - 1)$

or $2n - 1 = 9$ or $2n = 10$ or $n = 5$ s

☞ *When a body is falling freely from rest, the distances travelled in the first, second, third, fourth and fifth second of motion are in the ratio of 1 : 3 : 5 : 7 : 9.*

Let us now solve Q. 87 by a different method.

If x is the distance travelled in the first second, then the total distance travelled in the first three second is $x + 3x + 5x$ *i.e.* $9x$. This is clearly equal to the distance travelled in the fifth second. So, the body has fallen for 5 second.

88. (*c*) We know that $v^2 - u^2 = 2aS$

Note that 2*a*S is constant.

$\therefore \quad v^2 - u^2 = \text{constant}$

$\therefore \quad (5v)^2 - v^2 = v'^2 - (5v)^2$ or $v'^2 = 49v^2$

or $v' = -7v$ [+ ve sign ignored]

89. (*a*) $9y = \frac{1}{2} \times 10 \times 3 \times 3$ or $y = 5$ m

Again, $n \times 5 = \frac{1}{2} \times 10 \times 1 \times 1 = 5$ or $n = 1$.

90. (*b*) Bad solution

$0 = v + a \times 5;\quad 25 = v \times 5 + \frac{1}{2}a \times 5 \times 5$

Now, $25 = 5v - \frac{5v}{2} = \frac{5v}{2}$ or $5v = 50$ or $v = 10$ m s^{-1}

Good solution

$\frac{v + 0}{2} = \frac{25}{5} = 5$ or $v = 10$ m s^{-1}

91. (*c*) $81 = -12t + \frac{1}{2} \times 10 \times t^2$

or $5t^2 - 12t - 81 = 0$

or $t = \frac{12 \pm \sqrt{144 + 4 \times 5 \times 81}}{2 \times 5}$ s

$= \frac{12 \pm \sqrt{1764}}{10} = \frac{12 \pm 42}{10} = \frac{54}{10}$ s = 5.4 s

[Negative sign has been ignored]

92. (*a*) $v_b = \frac{8 + 8}{2}$ km h^{-1} = $\frac{16}{2}$ km h^{-1} = 8 km h^{-1}

$v_w = 4$ km h^{-1}

$t = \frac{8}{8 - 4} + \frac{8}{8 + 4} = \frac{8}{4} + \frac{8}{12} = \left(2 + \frac{2}{3}\right)h$

$= \frac{8}{3}h = \frac{8}{3} \times 60$ minute = 160 minute.

93. (*c*) $h = \frac{1}{2}gt^2$, $h - 20 = \frac{1}{2}g(t - 1)^2$,

Subtracting, $20 = \frac{1}{2}g[t^2 - (t-1)^2]$ or $\frac{40}{10} = (2t - 1)$

or $2t - 1 = 4$ or $2t = 5$ or $t = 2.5$ s

$$h = \frac{1}{2} \times 10 \times 2.5 \times 2.5 \text{ m} = 31.25 \text{ m}$$

94. (*d*) Using $v = u + at$ or $v - u = at$, we find that if $|\vec{a}|$ is doubled, t is halved.

If t is the time for acceleration, then $\frac{t}{2}$ is the time for retardation.

Now, $t + \frac{t}{2} = 3$ or $\frac{3t}{2} = 3$ or $t = 2$ s

$$S = \frac{1}{2} \times 2 \times 2 \times 2 + \frac{1}{2} \times 4 \times 1 \times 1 = (4 + 2) \text{ m} = 6 \text{ m}$$

95. (*b*) $v = 19.6 \times 5 \text{ m s}^{-1} = 98 \text{ m s}^{-1}$

$$h_{max} = \frac{1}{2} \times 19.6 \times 5 \times 5 + \frac{98 \times 98}{2 \times 9.8}$$

$$= 245 + 490 = 735 \text{ m}$$

96. (*a*) $h = \frac{1}{2}gt^2$, $100 - h = 100t - \frac{1}{2}gt^2$,

$100 = 100t$ or $t = 1$ s

97. (*a*) Let v_b be the velocity of the boy relative to the escalator. Let v_e be the velocity of the escalator. Then, the relative velocity of the boy with respect to the ground in the moving escalator is $(v_b + v_e)$.

If l is the length of the escalator, then

$$\frac{l}{v_b} = 90, \frac{l}{v_e} = 60, \frac{l}{v_b + v_e} = t$$

or $\frac{1}{t} = \frac{v_b + v_e}{l} = \frac{v_b}{l} + \frac{v_e}{l} = \frac{1}{90} + \frac{1}{60}$

or $\frac{1}{t} = \frac{60 + 90}{90 \times 60} = \frac{150}{90 \times 60} = \frac{1}{36}$ or $t = 36$ s

98. (*a*) Displacement in last 2 second $= \frac{1}{2}an^2 - \frac{1}{2}a(n-2)^2$

$$= \frac{a}{2}(2n - 2)2 = 2a(n - 1)$$

But $v = an$ or $a = \frac{v}{n}$

$\therefore$ Displacement in last 2 second $= \frac{2v}{n}(n - 1)$.

99. (*b*) $h = vt - \frac{1}{2}gt^2$ or $\frac{1}{2}gt^2 - vt + h = 0$

or $gt^2 - 2vt + 2h = 0 \Rightarrow t_1 t_2 = \frac{2h}{g}$

$1 \times 3 = \frac{2h}{10}$ or $2h = 30$ m or $h = 15$ m

> **Commit to memory**
>
> $ax^2 + bx + c = 0$
>
> *Sum of roots* $= -\frac{b}{a}$; *Product of roots* $= \frac{c}{a}$

100. (*c*) $\frac{1}{2} \times 0.2\,[(60 + t)^2 - t^2] = 3 \times \frac{1}{2} \times 0.2 \times 60 \times 60$

or $(60 + 2t)60 = 3 \times 60 \times 60$

or $60 + 2t = 180$ or $2t = 120$

or $t = 60$ s $= 1$ minute

101. (*b*) Average velocity for the second half of the distance is $\frac{4 + 8}{2}$ m s^{-1} *i.e.* 6 m s^{-1}

The first half distance is covered with a velocity of 6 m s^{-1}. So, the average velocity for the whole time of motion is 6 m s^{-1}.

102. (*a*) Distance to be covered $= \frac{1}{2}g(5^2 - 3^2) = 8$ g

$$t = \sqrt{\frac{2 \times 8g}{g}} \text{ s} = 4 \text{ s}$$

103. (*c*) $\frac{dv}{dt} = 6t$ or $dv = 6t\,dt$, $v = \frac{6t^2}{2} = 3t^2$,

$dx = 3t^2\,dt \Rightarrow x = 3\frac{t^3}{3} = t^3$

104. (*a*) $v = 3t^2 - 12t + 3$; $a = 6t - 12$. When $a = 0$, $6t - 12 = 0$ or $6t = 12$ or $t = 2$ s. When $t = 2$ s,

$$v = 3 \times 2 \times 2 - 12 \times 2 + 3 = -9 \text{ m s}^{-1}$$

105 (*d*) $y = \frac{1}{2}g(n+1)^2 - \frac{1}{2}gn^2 = \frac{g}{2}[(n+1)^2 - n^2] = \frac{g}{2}(2n + 1)$

Also, $h = \frac{g}{2}(2n - 1)$. $\therefore$ $y = \frac{g}{2}\left[\frac{2h}{g} + 1 + 1\right]$

or $y = h + g$

106. (*a*) $v = \sqrt{3x + 16}$ or $v^2 = 3x + 16$ or $v^2 - 16 = 3x$

Comparing with $v^2 - u^2 = 2aS$, we get, $u = 4$ units, $2a = 3$ or $a = 1.5$ units

107. (*d*) Let us first calculate the time required to attain a maximum speed of 10 m s^{-1}. Using $v = u + at$, we get $10 = 0 + 2.5t \Rightarrow t = 4$ s. Distance covered in 4 s $= \frac{1}{2} \times 2.5 \times 4 \times 4$ m $= 20$ m. The remaining 80 m distance is clearly covered with a constant velocity of 10 m s^{-1}. Time taken $= \frac{80}{10}$ s $= 8$ s. So, total time is $(4 + 8)$ s *i.e.* 12 s.

108. (*c*) $S_1 = \frac{1}{2}gn^2$, $S_2 = \frac{1}{2}g(n - N)^2$

$S_1 - S_2 = \frac{1}{2} g[n^2 - (n-N)^2]$ or $1 = \frac{g}{2}(2n-N)(N)$

or $2n - N = \frac{2}{gN}$ or $2n = \frac{2}{gN} + N$ or $n = \frac{1}{gN} + \frac{N}{2}$

109. (*b*) $\frac{h}{2} = \frac{g}{2}(2n-1)$ or $h = g(2n-1)$ or $\frac{1}{2}gn^2$

$= g(2n-1)$ or $n^2 = 4n - 2$ or $n^2 - 4n + 2 = 0$

or $n = \frac{4 \pm \sqrt{16-8}}{2} = 2 + \sqrt{2}$ [Ignoring –ve sign]

110. (*c*) $h = \frac{1}{2} \times 9.8(2+\sqrt{2})^2 = 4.9 \times 11.65$ m ≈ 57 m

111. (*c*) $\frac{1}{2}gt^2 - \frac{1}{2}g(t-2)^2 = 40$

or $\frac{1}{2} \times 10\,(2t-2)(2) = 40$ or $2t - 2 = 4$ or $t = 3$ s

112. (*c*) $h = -v \times 9 + \frac{1}{2} \times 10 \times 9 \times 9$

$h = v \times 4 + \frac{1}{2} \times 10 \times 4 \times 4$

Subtracting, $0 = 13v + \frac{10}{2}(16 - 81)$

or $13v = 325$ or $v = 25$ m s^{-1}

Again, $h = 25 \times 4 + 5 \times 4 \times 4$

$= (100 + 80)$ m $= 180$ m

113. (*b*) Let us solve the problem in terms of relative initial velocity, relative acceleration and relative displacement of the coin with respect to the floor of the lift.

$u = 10 - 10 = 0$ m s^{-1}, $a = 9.8$ m s^{-2}, S = 4.9 m, t = ?

$4.9 = 0 \times t + \frac{1}{2} \times 9.8 \times t^2$ or $4.9t^2 = 4.9$ or $t = 1$ s

114. (*c*) Clearly, if t is the required time, than acceleration and deceleration have taken place each for time $\frac{t}{2}$.

Now, $s = \frac{1}{2}\left(\frac{at}{2}\right)(t)$ or $\frac{at^2}{4} = $ s.

115. (*a*) $h = -vt + \frac{1}{2}gt^2$ or $gt^2 - 2vt - h = 0$

$$t = \frac{-(-2v) \pm \sqrt{4v^2 + 4gh}}{2g} = \frac{2v \pm 2\sqrt{v^2 + gh}}{2g}$$

$$= \frac{v}{g} \pm \frac{[v^2 + 2gh]^{1/2}}{g} = \frac{v}{g}\left[1 \pm \sqrt{1 + \frac{2gh}{v^2}}\right]$$

Now retain only the positive sign.

116. (*b*) $\Delta t = \sqrt{\frac{2 \times 19.6}{9.8}}$ s $- \sqrt{\frac{2 \times 18.6}{9.8}}$ s $= 2 - 1.95 = 0.05$ s.

117. (*c*) Fractional loss in velocity $= \frac{v_1 - v_2}{v_1}$

$$= 1 - \frac{v_2}{v_1} = 1 - \sqrt{\frac{2g \times 1.8}{2g \times 5}} = 1 - \frac{3}{5} = \frac{2}{5}.$$

118. (*c*) $t_1 = \frac{x}{v} + \frac{x}{v} = \frac{2x}{v}$

$$t_2 = \frac{x}{v+w} + \frac{x}{v-w} = \frac{2xv}{v^2 - w^2}$$

or $$t_2 = \frac{2xv}{v^2\left(1 - \frac{w^2}{v^2}\right)} = \frac{2x}{v\left(1 - \frac{w^2}{v^2}\right)}$$

or $t_2 = \frac{t_1}{1 - \frac{w^2}{v^2}}$ or $t_2 > t_1$

119. (*a*) $\frac{l}{v_1} = 15$ and $\frac{l + 450}{v_1} = 45$

or $\frac{l+450}{v_1} = \frac{3l}{v_1}$ or $2l = 450$ or $l = 225$ m

120. (*a*) $\frac{x}{v_s + v_w} = 2$, $\frac{x}{v_s - v_w} = 3$

Dividing, $\frac{x}{v_s + v_w} \div \frac{v_s - v_w}{x} = \frac{2}{3}$

or $3v_s - 3v_w = 2v_s + 2v_w$ or $v_s = 5v_w$

Now, $\frac{x}{5v_w + v_w} = 2$ or $\frac{x}{6v_w} = 2$ or $\frac{x}{v_w} = 12\,h$

121. (*a*) Let us calculate relative deceleration by considering relative velocity

Using $v^2 - u^2 = 2aS$, $0^2 - 80^2 = 2 \times a \times 2000$

or $a = -\frac{80 \times 80}{4000} = -\frac{64}{40}$ m s^{-2} $= -1.6$ m s^{-2}

Deceleration of each train is $\frac{1.6}{2}$ m s^{-2} *i.e.* 0.8 m s^{-2}

122. (*a*) $x^2 = at^2 + 2bt + c$

Differentiating w.r.t. time, we get

$$2x\frac{dx}{dt} = 2at + 2b \text{ or } xv = at + b \left(\text{or } v = \frac{at+b}{x}\right)$$

Again differentiating w.r.t. time, we get

$$x\frac{dv}{dt} + v\frac{dx}{dt} = a$$

or $xA + v^2 = a$ or $xA = a - v^2$

or $xA = a - \left(\frac{at+b}{x}\right)^2$

or $xA = \frac{ax^2 - a^2t^2 - b^2 - 2abt}{x^2}$

or $xA = \frac{a^2t^2 + 2abt + ac - a^2t^2 - b^2 - 2abt}{x^2}$

or $A = \frac{ac - b^2}{x^3}$ or $A \propto x^{-3}$

123. (*b*) $h = 39.2t - \frac{1}{2}gt^2 = 39.2(t-2) - \frac{1}{2}g(t-2)^2$

or $39.2(t - t + 2) = \frac{1}{2}g[t^2 - (t-2)^2]$

or $\frac{39.2 \times 2 \times 2}{9.8} = (2t - 2)(2)$ or $2t - 2 = \frac{39.2 \times 2}{9.8} = 8$

or $2t = 10$ or $t = 5$

Now, $h = 39.2 \times 5 - \frac{1}{2} \times 9.8 \times 5 \times 5$

$= (196 - 122.5)$ m $= 73.5$ m

124. (*d*) $40^2 - 30^2 = 2aS$, $v^2 - 30^2 = 2a\frac{S}{2}$ or $2(v^2 - 30^2) = 2aS$

Comparing, $2(v^2 - 900) = 1600 - 900 = 700$

or $v^2 = 900 + 350 = 1250$ or $v = 35.35$ km h^{-1}.

125. (*c*) $v = 4.9 \times 2$ m s^{-1} $= 9.8$ m s^{-1}, $0^2 - 9.8^2 = -2 \times 9.8 \times h_1$

or $h_1 = 4.9$ m. Again, if h_2 is the distance travelled by the balloon in 2 second, then

$h_2 = \frac{1}{2} \times 4.9 \times 2 \times 2$ m $= 9.8$ m

Total height, $h = h_1 + h_2 = (4.9 + 9.8)$ m $= 14.7$ m

126. (*d*) $x_1 = \frac{1}{2}gt_0^2$, $x_2 = \frac{1}{2}g(t_0 + t)^2$, $x_3 = \frac{1}{2}g(t_0 + 2t)^2$,

$x_4 = \frac{1}{2}g(t_0 + 3t)^2$; $x_4 - x_1 = \frac{1}{2}g[(t_0 + 3t)^2 - t_0^2]$

$= \frac{1}{2}g(2t_0 + 3t)(3t)$; $x_3 - x_2 = \frac{g}{2}(t)(2t_0 + 3t)$

Clearly, $x_4 - x_1 = 3(x_3 - x_2)$

127. (*a*) $t = \alpha x^2 + \beta x$

Differentiating, $1 = 2\alpha xv + \beta v$; Differentiating again,

$0 = 2\alpha xa + 2\alpha v^2 + \beta a$, $a = \frac{-2\alpha v^2}{2\alpha x + \beta}$

or $a = (-2\alpha v^2)\left(\frac{1}{2\alpha x + \beta}\right) = -2\alpha v^3$ or $-a = 2\alpha v^3$

128. (*a*) $\frac{7v}{8} = \frac{\frac{1}{2}vt + vt' + \frac{1}{2}vt}{t + t' + t}$ or $\frac{7}{8} = \frac{t + t'}{2t + t'}$

or $6t = t'$

Now $\frac{t'}{2t + t'} = \frac{6t}{8t} = \frac{3}{4}$.

129. (*a*) $\frac{vt'}{\frac{1}{2}vt + vt' + \frac{1}{2}vt} = \frac{v \times 6t}{vt + 6vt} = \frac{6}{7}$.

130. (*b*) $x = 0.05 \sin\left(5t - \frac{\pi}{6}\right)$; $v = \frac{dx}{dt} = 5 \times 0.05 \cos\left(5t - \frac{\pi}{6}\right)$

or $v = 0.25 \cos\left(5t - \frac{\pi}{6}\right)$

At $t = 0$, velocity $= 0.25 \cos\left(-\frac{\pi}{6}\right) = 0.25 \times \frac{\sqrt{3}}{2}$

$= \frac{\sqrt{3}}{8}$ m s^{-1} $= 0.2165$ m s^{-1} ≈ 0.22 m s^{-1}

131. (*a*) Acceleration $= \frac{d}{dt}\left[0.25 \cos\left(5t - \frac{\pi}{6}\right)\right]$

$= -0.25 \times 5 \sin\left(5t - \frac{\pi}{6}\right) = -1.25 \sin\left(5t - \frac{\pi}{6}\right)$

At $t = 0$, acceleration $= -1.25 \sin\left(-\frac{\pi}{6}\right)$

$= 1.25 \times \frac{1}{2}$ m s^{-2} $= 0.625$ m s^{-2} ≈ 0.63 m s^{-2}

132. (*c*) $\frac{dx}{dt} = 4t^3 - 2t$

or $dx = 4t^3\,dt - 2t\,dt$

Integrating, $x = \frac{4t^4}{4} - \frac{2t^2}{2} = t^4 - t^2$

When $x = 2$, $t^4 - t^2 - 2 = 0$,

$t^2 = \frac{-(-1) \pm \sqrt{1 + 8}}{2}$

or $t^2 = \frac{1 \pm 3}{2} = 2$ (Ignoring – ve sign)

Again, $\frac{d^2x}{dt^2} = 12t^2 - 2$

When $t^2 = 2$, acceleration $= 12 \times 2 - 2 = 22$ m s^{-2}

133. (*d*) $v = \sqrt{49 + y}$, $a = \frac{dv}{dy} \cdot \frac{dy}{dt} = v\frac{dv}{dy}$

$= (49 + y)^{1/2} \times \frac{1}{2}(49 + y)^{1/2 - 1} = \frac{1}{2}$ m s^{-2}

134. (*a*) $b - a = un + \frac{1}{2}An^2$

$2b - 2a = 2un + An^2$

Again,

$c - a = u(2n) + \frac{1}{2}A(2n)^2$

Subtracting, $c - a - 2b + 2a = An^2$

$A = \frac{c - 2b + a}{n^2}$

t = 0 t = n t = 2n
a
b
c

Fig. 39

135. (*a*) $a = u + \frac{A}{2}(2x - 1)$, $b = u + \frac{A}{2}(2y - 1)$, $c = u + \frac{A}{2}(2z - 1)$

$a - b = A(x - y)$, $b - c = A(y - z)$

Dividing, $\frac{a - b}{b - c} = \frac{x - y}{y - z}$ or $ay - az - by + bz$

$= bx - by - cx + cy$

or $ay - az + bz - bx + cx - cy = 0$

or $a(y - z) + b(z - x) + c(x - y) = 0$

136. (d) $h = \left(-9.8 \times 11 + \frac{1}{2} \times 9.8 \times 11 \times 11\right)$ metre

$= (-107.8 + 592.9)$ m $= 485.1$ m

137. (a) $v = \sqrt{2 \times 9.8 \times 100} = \sqrt{1960}$ m s^{-1}

$\frac{\sqrt{1960}+0}{2} = \frac{2}{t}$ or $t = \frac{4}{\sqrt{1960}}$ s $= \frac{4}{44.27}$ s $= 0.09$ s

138. (a) $t^2 = x^2 - 1$ or $x^2 = t^2 + 1$

or $2x\frac{dx}{dt} = 2t$ or $xv = t$

or $x\frac{dv}{dt} + v\frac{dx}{dt} = 1$ or $x\frac{dv}{dt} = 1 - v^2$

or $\frac{dv}{dt} = \frac{1-v^2}{x} = \frac{1-\frac{t^2}{x^2}}{x} = \frac{x^2-t^2}{x^3}$

But $x^2 - t^2 = 1$ $\therefore$ $\frac{dv}{dt} = \frac{1}{x^3}$

139. (d) **Step I.** $h = \frac{1}{2} \times 10 \times 4 \times 4$ m $= 80$ m

Step II. $h' = \frac{1}{2} \times 10 \times 2 \times 2$ m $= 20$ m

Step III. $t = \sqrt{\frac{2 \times 60}{10}}$ s $= \sqrt{12}$ s

140. (c) $10.2 = ut - \frac{1}{2} \times 10 \times t^2$ or $5t^2 - ut + 10.2 = 0$

$t_1 + t_2 = -\frac{-u}{5} = \frac{u}{5}$, $t_1t_2 = \frac{10.2}{5} = \frac{102}{50}$

$(t_1 - t_2)^2 = \left(\frac{u}{5}\right)^2 - \frac{4 \times 102}{50}$ or $100 + \frac{204}{25} = \left(\frac{u}{5}\right)^2$

or $\frac{u}{5} = \frac{\sqrt{2704}}{5}$ or $u = 52$ m s^{-1}

141. (b) $35 = u + \frac{a}{2} \times 15$; $51 = u + \frac{a}{2} \times 23$; Subtracting,

$16 = \frac{a}{2} \times 8$ or $4a = 16$ or $a = 4$ m s^{-2},

Again, $35 = u + \frac{4}{2}(2 \times 8 - 1) = u + 30$

or $u = 5$ m s^{-1} ; Again, $v = 5 + 4 \times 10 = 45$ m s^{-1}

142. (b) Let time interval be chosen as 1 second

$\frac{PA}{PB} = \frac{OA}{OB} = \frac{v_x}{v_y}$

So, P(x, y) divides AB in the ratio $v_x : v_y$.

Using section formula,

$x = \frac{v_x \times 0 + v_y \times v_x}{v_x + v_y} = \frac{v_xv_y}{v_x + v_y}$

$y = \frac{v_xv_y + v_y \times 0}{v_x + v_y}$

$= \frac{v_xv_y}{v_x + v_y}$

$v = \sqrt{x^2 + y^2}$

$= \sqrt{2}\,\frac{v_xv_y}{v_x + v_y}$.

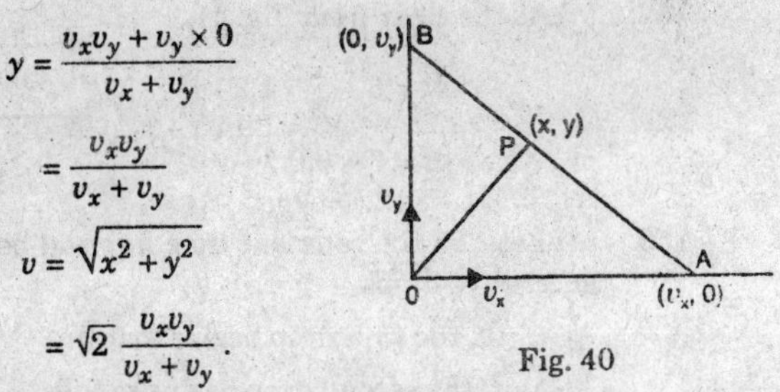

Fig. 40

Now, replace v_x by v_1 and v_y by v_2.

143. (a) Total distance $= 3$ m s$^{-1} \times 20$ s $+ 0 + 2$ m s$^{-1} \times 30$ s $= 120$ m

Total time $= 20$ s $+ 10$ s $+ 30$ s $= 60$ s

Average velocity $= \frac{120 \text{ m}}{60 \text{ s}} = 2$ m s^{-1}

144. (c) $(2v)^2 - v^2 = 2gh'$ or $4v^2 - v^2 = 2gh'$ or $3v^2 = 2gh'$

or $3 \times 2gh = 2gh'$ or $h' = 3h$

145. (c) For the dropped body, $h_1 = \frac{1}{2}gt^2$; For the thrown body,

$h_2 = 1 \times t + \frac{1}{2}gt^2 = t + \frac{1}{2}gt^2$; $h_2 - h_1 = t$;

So, $t = 1.8$ second.

146. (d) When A returns to the 'level' of top of tower, its downward velocity is 4 m s^{-1}. This velocity is the same as that of B. So, both A and B hit the ground with the same velocity.

147. (a) $u - v = (80 - 44)$ km h^{-1} $= 36$ km h^{-1}

$= 36 \times \frac{5}{18}$ m s^{-1} $= 10$ m s^{-1}

Retardation $= \frac{10 \text{ m s}^{-1}}{15 \text{ s}} = 0.67$ m s^{-2}

148. (b) Ball A shall have to waste its time in covering up and down distance above the top of the tower.

149. (c) Average velocity $= \frac{25 \times 4 + 10 \times 4 + \frac{1}{2}(10 + 25)4}{12}$ m s^{-1}

$= \frac{100 + 40 + 70}{12}$ m s^{-1} $= \frac{210}{12}$ m s^{-1} $= 17.5$ m s^{-1}

150. (b) $v = 36$ km h^{-1} $= 36 \times \frac{5}{18}$ m s^{-1} $= 10$ m s^{-1},

$t = \frac{600 + 200}{10}$ s $= 80$ s

151. (a) For upstream motion, $v = (8 - 3)$ km h^{-1} $= 5$ km h^{-1}

For downstream motion,

$v = (8 + 3)$ km h^{-1} $= 11$ km h^{-1}

152. (*a*) As the clear from Fig. 41,

$$4t = 6 + \frac{1}{2} \times 1.2 \times t^2$$

or $$4t = 6 + 0.6t^2$$

Fig. 41

153. (*d*) Let *t* be the constant time interval between any two successive drops.

At '*t*' = 0, the first drop begins to fall

At '*t* = *t*, the second drop begins to fall

At '*t*' = 2*t*, the third drop begins to fall.

It is given that **the third drop is leaving the tap at the instant the first drop touches the ground.**

So, the first drop has been in motion for time interval '2*t*'.

For the first drop,

$$5 = \frac{1}{2} \times 10 \times (2t)^2 = 20t^2$$

or $$t^2 = \frac{1}{4}\,s^2$$

For the second drop,

$$y = \frac{1}{2} \times 10t^2 = 5t^2 = \frac{5}{4}\,m$$

$$h = \left(5 - \frac{5}{4}\right) m = 3.75\ m$$

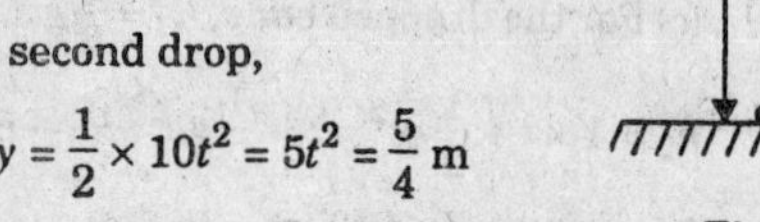

Fig. 42

154. (*a*) Average speed for other half of distance

$$= \frac{4.5 + 7.5}{2}\ m\ s^{-1} = 6\ m\ s^{-1}$$

Average speed during whole motion

$$= \frac{2 \times 3 \times 6}{3 + 6}\ m\ s^{-1} = 4\ m\ s^{-1}$$

155. (*a*) Equate times in the two cases. [Both start at the same time and stop at the same time]

156. (*b*) $2 = \sqrt{\frac{2 \times 19.6}{g}}$ or $g = \frac{2 \times 19.6}{4}\ m\ s^{-2}$

or $g = 9.8\ m\ s^{-2}$

157. (*a*) $S_{3rd} = 10 + \frac{10}{2}(2 \times 3 - 1) = 35$

$$S_{2nd} = 10 + \frac{10}{2}(2 \times 2 - 1) = 25.$$

Ratio is $\frac{35}{25}$ *i.e.* $= \frac{7}{5}$

158. (*b*) $v_1 + v_2 = 4\ m\ s^{-1}, v_1 - v_2 = 0.4\ m\ s^{-1}$

$$2v_1 = 4.4 \text{ or } v_1 = 2.2\ m\ s^{-1},$$

$$v_2 = (4 - 2.2)\ m\ s^{-1} = 1.8\ m\ s^{-1}$$

159. (*b*) $v \propto \sqrt{h}\ \therefore\ \frac{v_1}{v_2} = \sqrt{\frac{a}{b}}$

So, (*b*) is the correct choice.

The velocity acquired by a body in falling fr[illegible] from rest through height *h* is $\sqrt{2gh}$. [$u = 0$, [illegible] g, 'S' $= h$, $v^2 - u^2 = 2aS$]

160. (*c*) $a = 6t + 5 \Rightarrow dv = (6t + 5)\,dt$

$$v = \frac{6t^2}{2} + 5t = 3t^2 + 5t \Rightarrow dS = (3t^2 + 5t)\,dt$$

$$S = \frac{3t^3}{3} + \frac{5t^2}{2}$$

$$(S)_{t=1s} = \frac{3 \times 1 \times 1 \times 1}{3} + \frac{5}{2} 1 \times 1 = 3.5\ m.$$

161. (*d*) $S = \left(4 \times 1.5 + 4 \times 1.5 + \frac{1}{2} \times 10 \times 1.5 \times 1.5\right) m$

$$= (6 + 6 + 5 \times 2.25)\ m = 23.25\ m.$$

162. (*d*) $S' \propto t^2$. Now, $S'_1 : S'_2 : S'_3 :: \frac{1}{4}, 1, \frac{9}{4}$ or $1 : 4 : 9$

For successive intervals,

$$S_1 : S_2 : S_3 :: 1 : (4 - 1) : (9 - 4)$$

or $S_1 : S_2 : S_3 :: 1 : 3 : 5.$

163. (*d*) $\frac{11h}{36} = \frac{9.8}{2}(2n - 1)$ or $\frac{11}{36} \times \frac{1}{2} \times 9.8n^2 = \frac{9.8}{2}(2n - 1)$

or $2n - 1 = \frac{11}{36}n^2$ or $11n^2 = 72n - 36$

or $11n^2 - 72n + 36 = 0$ or $11n^2 - 66n - 6n + 36 = 0$

or $11n(n - 6) - 6(n - 6) = 0$ or $(11n - 6)(n - 6) = 0$

$\Rightarrow$ $n = 6$ [Rejecting fractional value]

$$h = \frac{1}{2} \times 10 \times 6 \times 6\ m = 180\ m$$

164. (*a*) $1900 = \frac{1}{2} \times 0.4[(120)^2 - (120 - t)^2]$

or $\frac{1900}{0.2} = (240 - t)t$ or $9500 = 240t - t^2$

or $t^2 - 240t + 9500 = 0$

or $t^2 - 190t - 50t + 9500 = 0$

or $t(t - 190) - 50(t - 190) = 0$

or $(t - 50)(t - 190) = 0 \Rightarrow t = 50\ s, 190\ s$

Rejecting $t = 190$ s, we get $t = 50$ s

165. (*a*) $v = \frac{d}{dt}(x) = \frac{d}{dt}(a + 4bt^3) = 4b \times 3t^2 = 12bt^2$

When $t = 1$ s, then $v = 12b$

166. (*b*) $(3v)^2 - (-v)^2 = 2gh$ or $8v^2 = 2gh$ or $h = \frac{4v^2}{g}$

167. (*a*) $\frac{dv}{dt} = bt$ or $dv = bt\,dt$

$$\int_{v_0}^{v} dv = \int_0^t bt\,dt \text{ or } v - v_0 = \frac{bt^2}{2}$$

or $v = v_0 + \frac{bt^2}{2}$

or $dx = v_0 dt + \frac{bt^2}{2} dt$

$$\int_0^x dx = \int_0^t v_0\, dt + \frac{b}{2}\int_0^t t^2\, dt$$

or $x = v_0 t + \frac{1}{2}\frac{bt^3}{3} = v_0 t + \frac{bt^3}{6}$

168. (*c*) $v = 20 \times 60 \text{ m s}^{-1} = 1200 \text{ m s}^{-1}$

$t = \frac{1200}{10} \text{ s} = 120 \text{ s}$

169. (*c*) $b - a = \frac{1}{2}\text{A}n^2, c - b = \frac{1}{2}\text{A}n^2, d - c = \frac{1}{2}\text{A}n^2$

Adding, $d - a = 3 \times \frac{1}{2}\text{A}n^2 = 3(c - b)$

170. (*a*) $x^2 + 2x = t$

$2xv + 2v = 1$ or $v = \frac{1}{2(x+1)}$

Again, $2x \times \text{acc.} + 2v^2 + 2 \times \text{acc.} = 0$

or $2 \times \text{acc.}\ (x + 1) = -2v^2$

or $\text{acc.} = -\frac{2v^2}{2(x+1)} = -\frac{v^2}{(x+1)} = -\frac{1}{4(x+1)^3}$

The negative sign indicates retardation.

171. (*a*) $t = \sqrt{\frac{2 \times 19.6}{9.8}} = 2 \text{ s}; 1.9 = \sqrt{\frac{2 \times h}{9.8}}$

or $h = \frac{1.9 \times 1.9 \times 9.8}{2} \text{ m} = 17.689 \text{ m}$,

Required distance = 19.6 – 17.689 m = 1.9 m.

172. (*d*) $\int_{v_0}^{v} \frac{dv}{-kv^3} = \int_0^t dt$

$-\frac{1}{k}\int_{v_0}^{v} v^{-3}\, dv = t$ or $-\frac{1}{k}\left|\frac{v^{-3+1}}{-3+1}\right|_{v_0}^{v} = t$

or $\frac{1}{2k}\left[\frac{1}{v^2} - \frac{1}{v_0^2}\right] = t$ or $\frac{1}{v^2} - \frac{1}{v_0^2} = 2kt$

or $\frac{1}{v^2} = \frac{1}{v_0^2} + 2kt$ or $\frac{1}{v^2} = \frac{1 + 2v_0^2 kt}{v_0^2}$

or $v = \frac{v_0}{\sqrt{2v_0^2 kt + 1}}$

173. (*c*) $1 = \frac{\text{V}}{t_1}, 3 = \frac{\text{V}}{t_2}, 1200 = \frac{1}{2}(t_1 + t_2)\text{V}$,

$1200 = \frac{1}{2}\left(\text{V} + \frac{\text{V}}{3}\right)\text{V} = \frac{1}{2}\frac{4\text{V}^2}{3} = \frac{2\text{V}^2}{3}$

or $\text{V}^2 = 1800, 1200 = \frac{1}{2} t \times \sqrt{1800}$,

$t = \frac{2400}{\sqrt{1800}} \text{ s} = \frac{2400}{42.43} \text{ s} = 56.6 \text{ s}$.

174. (*c*) Relative initial velocity $= 72 \times \frac{5}{18} - 4 = 16 \text{ m s}^{-1}$

Now, $0^2 - 16 \times 16 = 2(-1)\text{ S}$ or $\text{S} = 128 \text{ m}$

175. (*b*) $v^2 = 2\alpha\text{S}$. Also, $v^2 = 2\beta\text{S}_2$

$\therefore \text{S} = \text{S}_1 + \text{S}_2 = \frac{v^2}{2\alpha} + \frac{v^2}{2\beta}$

Again, $v = \alpha t_1$ and $v = \beta t_2$

$\therefore t = t_1 + t_2 = \frac{v}{\alpha} + \frac{v}{\beta}$

$\therefore v_{av.} = \frac{\frac{v^2}{2\alpha} + \frac{v^2}{2\beta}}{\frac{v}{\alpha} + \frac{v}{\beta}} = \frac{v}{2}$

176. (*b*) $u = 12 \text{ m s}^{-1}, g = 9.8 \text{ m s}^{-2}, t = 10 \text{ s}$

$\text{S} = \left(12 \times 10 + \frac{1}{2} \times 9.8 \times 10 \times 10\right) \text{metre}$

$= (120 + 4.9 \times 100) \text{ m} = 610 \text{ m}$

177. (*a*) If t_0 is the reaction time, then the distance covered during decelerated motion is $10 - 10t_0$.

Now, in the first case,

$10^2 = 2a(10 - 10t_0)$ ...(*i*)

Similarly, in the second case,

$20^2 = 2a(30 - 20t_0)$ ...(*ii*)

Again, in the third case,

$15^2 = 2a(x - 15t_0)$ ...(*iii*)

Dividing (*ii*) by (*i*), $\frac{20^2}{10^2} = \frac{30 - 20t_0}{10 - 10t_0}$

or $40 - 40t_0 = 30 - 20t_0$ or $20t_0 = 10$ or $t_0 = \frac{1}{2} \text{ s}$

Dividing (*iii*) by (*i*), we get

$\frac{225}{100} = \frac{x - 15t_0}{10 - 10t_0}$ or $\frac{9}{4} = \frac{x - 15 \times \frac{1}{2}}{10 - 10 \times \frac{1}{2}}$

$45 = 4x - 30$ or $4x = 75$ or $x = \frac{75}{4} \text{ m} = 18.75 \text{ m}$.

178. (*b*) If a is the relative acceleration, then

$3 = \frac{1}{2} \times a \times 5 \times 5$ or $a = \frac{6}{25} \text{ m s}^{-2}$

Again, $\text{S} = \frac{1}{2} \times \frac{6}{25} \times 10 \times 10 \text{ m} = 12 \text{ m}$

179. (*b*) Clearly, the time for accelerated motion is $\frac{25-15}{2}$ s or 5 s. Similarly, the time for decelerated motion is also 5 s.

Velocity at the end of accelerated motion

$$= 5 \times 5 = 25 \text{ m s}^{-1}$$

Total distance covered

$$= \left(\frac{1}{2} \times 5 \times 25 + 25 \times 15 + \frac{1}{2} \times 5 \times 25\right) \text{m}$$

$$= (125 + 375) \text{ m} = 500 \text{ m}$$

Average speed $= \frac{500 \text{ m}}{25 \text{ s}} = 20 \text{ m s}^{-1}$

180. (*c*) $h = -vt_1 + \frac{1}{2}gt_1^2, h = vt_2 + \frac{1}{2}gt_2^2,$

Now $\frac{h}{t_1} + \frac{h}{t_2} = \frac{1}{2}g(t_1 + t_2)$ or $h = \frac{1}{2}gt_1t_2$

Also, $h = \frac{1}{2}gt^2 \therefore t^2 = t_1t_2$ or $t = \sqrt{t_1t_2}$

181. (*b*) $v = \frac{4}{4} \text{ m s}^{-1} = 1 \text{ m s}^{-1}$

182. (*a*) There is no change of displacement with time. So, velocity is zero during BC.

183. (*c*) $v = \frac{4}{2} \text{ m s}^{-1} = 2 \text{ m s}^{-1}$

184. (*a*) If v is the maximum velocity attained, then $v^2 - 0^2 = 2 \times 5 \times S_1$. Also, $0^2 - v^2 = -2 \times 10 \times S_2$

$$S_1 = \frac{v^2}{10}, S_2 = \frac{v^2}{20}$$

$$S = S_1 + S_2 \Rightarrow 1500 = \frac{v^2}{10} + \frac{v^2}{20} = \frac{3v^2}{20}$$

or $v^2 = \frac{1500 \times 20}{3} = 10000$ or $v = 100 \text{ m s}^{-1}$

Again, $100 = 0 + 5t_1$ or $t_1 = 20$ s, $0 = 100 - 10t_2$

or $10t_2 = 100$ or $t_2 = 10$ s

Total time = (20 + 10) s = 30 s

185. (*c*) $\frac{dt}{dx} = 2ax + b$ or $\frac{dx}{dt} = \frac{1}{2ax+b}$ or $v = \frac{1}{2ax+b}$

$$\frac{dv}{dt} = \frac{dv}{dx} \cdot \frac{dx}{dt} = \frac{(2ax+b)\frac{d}{dx}(1) - (1)\frac{d}{dx}(2ax+b)}{(2ax+b)^2} \cdot v$$

$$= \frac{-2a}{(2ax+b)^2} \cdot \frac{1}{2ax+b} = -\frac{2a}{(2ax+b)^3}$$

At $x = 0$, $\frac{dv}{dt} = -\frac{2a}{b^3}$

186. (*a*) $a = \frac{0-30}{2.5} \text{ m s}^{-2} = -12 \text{ m s}^{-2}$; Retardation due to air resistance = 2 m s^{-2}; F = 0.05 × 2 N = 0.1 N

187. (*a*) $v_1 - v_2 = at$ or $t = \frac{v_1 - v_2}{a}$

188. (*a*) $0^2 - v^2 = -2aS$ or $v^2 \propto S$; $\left(v + \frac{20v}{100}\right)^2 \propto S'$

or $\frac{S'}{S} = \left(1 + \frac{20}{100}\right)^2 = \frac{36}{25}$ or $\left(\frac{S'}{S} - 1\right) \times 100$

$$= \left(\frac{36}{25} - 1\right) \times 100 = 44$$

3.1 VECTORS

Important Terms, Facts and Formulae

BASIC CONCEPTS

(*i*) A vector quantity possesses both magnitude and direction. But any quantity which has magnitude and direction is not necessarily a vector. This is because vectors obey different laws and this is just possible that a quantity which has both magnitude and direction may not obey the vector laws. This is of course true that we do not come across such cases very often. But we cannot overlook this fact.

(*ii*) In the case of position and displacement vectors, both the tail and the tip of the arrow convey an idea of position in space. This may not be true in the case of other vectors. As an example, let a particle be at P at any instant of time. At this instant of time, its velocity is shown to be $\vec{v}$ while the force it experiences is $\vec{F}$.

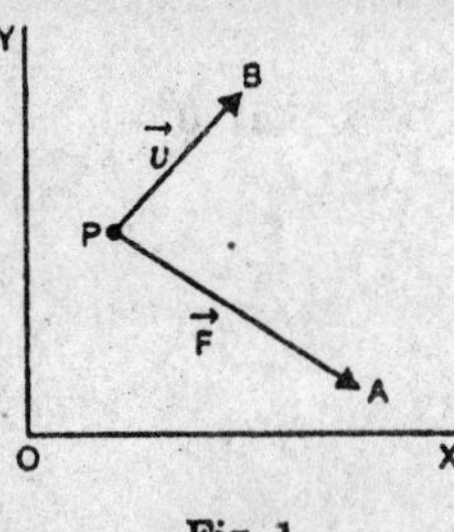

Fig. 1

So, the tail-tip representation of position and displacement vectors is not relevant in this case. In this case, the tip has no direct meaning as such. As an example in the present case it will be useless to write $\vec{PB}$ and $\vec{PA}$. Instead we shall simply write $\vec{v}$ and $\vec{F}$.

(*iii*) Any vector $\vec{a}$ can be expressed in terms of its unit vector $\hat{a}$ as follows :

$$\vec{a} = a\hat{a}$$

Here $\hat{a}$ is in the direction of $\vec{a}$. $\hat{a}$ is read as '*a* hat' or '*a* cap' or '*a* caret'.

$$\hat{a} = \frac{\vec{a}}{a} \text{ or } \frac{\vec{a}}{|\vec{a}|}$$

So, if a given vector is divided by its magnitude, we get a unit vector.

(*iv*) The three rectangular unit vectors $\hat{i}$, $\hat{j}$ and $\hat{k}$ are shown in Fig. 2. $\hat{i}$ denotes the direction of X-axis, $\hat{j}$ denotes the direction of Y-axis and $\hat{k}$ denotes the direction of Z-axis. The three unit vectors $\hat{i}$, $\hat{j}$ and $\hat{k}$ are collectively known as '*orthogonal triad of unit vectors*'. These are also known as *base vectors*.

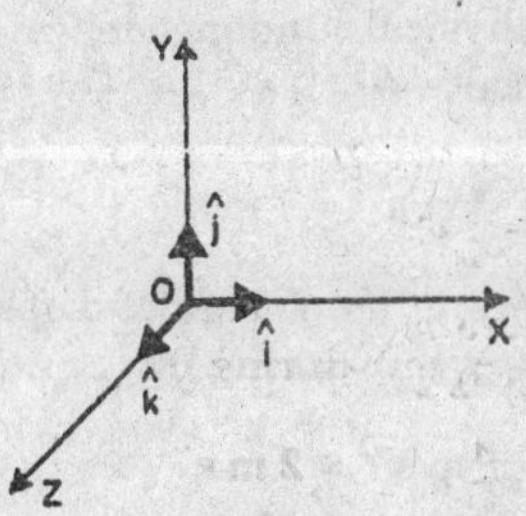

Fig. 2. Orthogonal triad of unit vectors.

(*v*) Vector addition or subtraction is possible only between vectors of same kind.

(*vi*) The product of vectors is possible between vectors of different kinds.

(*vii*) The negative sign reverses the direction of vectors. But the negative sign in the case of scalars has a different significance. As an example, an angle measured anticlockwise may be taken as +θ while an angle measured clockwise may be taken as –θ.

(*viii*) While finding angle between two vectors, we should ensure that the two vectors are either directed towards the point or away from the point.

(*ix*) The resultant of two vectors can be determined by parallelogram law of vector addition only if the vectors are either directed towards the point or away from the point.

(*x*) **It would be meaningless to compare two vectors with different physical dimensions.** As an example, there is no point in comparing velocity vector and force vector.

(*xi*) Unit vector has neither units nor dimensions.

(*xii*) Unit vector of velocity is not velocity.

(*xiii*) If $\vec{a} + \vec{b} = \vec{c}$ or $\vec{a} - \vec{b} = \vec{c}$ or $\vec{a} + \vec{b} + \vec{c} = 0$, then $\vec{a}$, $\vec{b}$ and $\vec{c}$ are in the same plane.

IMPORTANT POINTS TO REMEMBER

1. A unit vector $\hat{n}$ which is perpendicular to $\vec{A}$ and $\vec{B}$ is given by

$$\hat{n} = \frac{\vec{A} \times \vec{B}}{|\vec{A} \times \vec{B}|}$$

2. The minimum number of unequal coplanar vectors whose vector sum is zero is 3.
3. The minimum number of unequal non-coplanar vectors whose vector sum is zero is 4.
4. Component of a vector perpendicular to itself is zero. This fact ultimately leads us to the 'principle of physical independence of horizontal and vertical'.
5. The magnitude of the resultant of $\vec{A}$ and $\vec{B}$ varies from A + B to A – B.
6. Resultant of two vectors of equal magnitude bisects the angle between two vectors.
7. Resultant of two vectors of equal magnitude (say, P) acting at an angle of 60° is $\sqrt{3}$P.
8. Resultant of two vectors of equal magnitude (say, P) acting at an angle of 90° is $\sqrt{2}$P.
9. Resultant of two vectors of equal magnitude (say, P) acting at an angle of 120° is P.
10. The magnitude of the resultant of two vectors is maximum when the vectors act in the same direction.
11. The magnitude of the resultant of two vectors is minimum when they act in opposite directions.

QUESTION BANK

Level I

Choose the correct answer from the four alternatives given in each of the following questions :

1. If l, m and n are the direction cosines of a vector, then

(a) $l + m + n = 1$ (b) $l^2 + m^2 + n^2 = 1$

(c) $\frac{1}{l} + \frac{1}{m} + \frac{1}{n} = 1$ (d) $lmn = 1$.

2. Given : θ is the angle between $\vec{A}$ and $\vec{B}$. Then $|\hat{A} \times \hat{B}|$ is equal to

(a) sin θ (b) cos θ

(c) tan θ (d) cot θ.

3. Two vectors $\vec{A}$ and $\vec{B}$ are such that $\vec{A} + \vec{B} = \vec{C}$ and $A^2 + B^2 = C^2$. Which of the following is correct ?

(a) $\vec{A}$ is parallel to $\vec{B}$.

(b) $\vec{A}$ is anti-parallel to $\vec{B}$.

(c) $\vec{A}$ is perpendicular to $\vec{B}$.

(d) $\vec{A}$ and $\vec{B}$ are equal in magnitude.

4. The angle between $\vec{A}$ and $\vec{B}$ is θ. The value of the triple product $\vec{A} \cdot \vec{B} \times \vec{A}$ is

(a) A^2B (b) zero

(c) $A^2B \sin\theta$ (d) $A^2B \cos\theta$.

5. The minimum number of vectors in different planes which can be added to give zero resultant is

(a) 5 (b) 4

(c) 3 (d) 2.

6. If $\vec{A}$ is perpendicular to $\vec{B}$, then

(a) $\vec{A} \cdot (\vec{A} + \vec{B}) = A^2 + AB$ (b) $\vec{A} \cdot \vec{B} = 0$

(c) $\vec{A} \cdot (\vec{A} + \vec{B}) = A^2$ (d) $\vec{A} \cdot \vec{B} = AB$.

7. What is the numerical value of the vector $3\hat{i} + 4\hat{j} + 5\hat{k}$?

(a) $3\sqrt{2}$ (b) $5\sqrt{2}$

(c) $7\sqrt{2}$ (d) $9\sqrt{2}$.

8. $\vec{A}$ is directed vertically downwards and $\vec{B}$ is directed along the north. What is the direction of $\vec{B} \times \vec{A}$?

(a) east (b) west

(c) north east (d) north west.

9. The resultant of two forces, each P, acting at an angle θ is

(a) $2P \sin\frac{\theta}{2}$ (b) $2P \cos\frac{\theta}{2}$

(c) $2P \cos\theta$ (d) $P\sqrt{2}$.

10. Two forces, each equal to $\frac{P}{2}$, act at right angles. Their effect may be neutralised by a third force acting along their bisector in the opposite direction with a magnitude of

(a) P (b) $\frac{P}{2}$

(c) $\frac{P}{\sqrt{2}}$ (d) $\sqrt{2}$ P.

11. For any two vectors $\vec{A}$ and $\vec{B}$, if $\vec{A} \cdot \vec{B} = |\vec{A} \times \vec{B}|$, the magnitude of $\vec{C} = \vec{A} + \vec{B}$ is

(a) $\sqrt{A^2 + B^2}$ (b) A + B

(c) $\sqrt{A^2 + B^2 + \frac{AB}{\sqrt{2}}}$ (d) $\sqrt{A^2 + B^2 + \sqrt{2}\,AB}$.

12. The magnitude of the vector product of two vectors is $\sqrt{3}$ times their scalar product. The angle between vectors is

(a) $\frac{\pi}{2}$ (b) $\frac{\pi}{6}$

(c) $\frac{\pi}{3}$ (d) $\frac{\pi}{4}$.

13. The vector sum of the forces of 10 N and 6 N can be

(a) 2 N (b) 8 N

(c) 18 N (d) 20 N.

14. Given : $\vec{P} = 3\hat{j} + 4\hat{k}$ and $\vec{Q} = 2\hat{i} + 5\hat{k}$. The magnitude of the scalar product of these vectors is

(a) 20 (b) 23

(c) 26 (d) $5\sqrt{33}$.

15. Given : $\vec{A} = \hat{i} + 2\hat{j} - 3\hat{k}$. When a vector $\vec{B}$ is added to $\vec{A}$, we get a unit vector along X-axis. Then, $\vec{B}$ is

(a) $-2\hat{j} + 3\hat{k}$ (b) $-\hat{i} - 2\hat{j}$

(c) $-\hat{i} + 3\hat{k}$ (d) $2\hat{j} - 3\hat{k}$.

16. Given : $\vec{A} = 3\hat{i} - \hat{j} + 7\hat{k}$ and $\vec{B} = 5\hat{i} + \hat{j} + 9\hat{k}$. The direction cosine '$m$' of the vector $\vec{A} + \vec{B}$ is

(a) 0 (b) $\frac{3}{\sqrt{31}}$

(c) $\frac{9}{\sqrt{107}}$ (d) 5.

17. If $\vec{P}.\vec{Q} = 0$, then $|\vec{P}\times\vec{Q}|$ is

(a) $|\vec{P}|\,|\vec{Q}|$ (b) zero

(c) 1 (d) $\sqrt{PQ}$.

18. What are the direction cosines of $\hat{i}+\hat{j}+\hat{k}$?

(a) 1, 1, 1 (b) 2, 2, 2

(c) $\frac{1}{2}, \frac{1}{\sqrt{2}}, \frac{1}{2\sqrt{2}}$ (d) $\frac{1}{\sqrt{3}}, \frac{1}{\sqrt{3}}, \frac{1}{\sqrt{3}}$.

19. Two forces $\vec{F_1}$ and $\vec{F_2}$ are acting at right angles to each other. Then their resultant is

(a) $F_1 + F_2$ (b) $\sqrt{F_1^2 + F_2^2}$

(c) $\sqrt{F_1^2 - F_2^2}$ (d) $\frac{F_1+F_2}{2}$.

20. The resultant of two vectors of magnitudes 2A and $\sqrt{2}$ A acting at an angle θ is $\sqrt{10}$ A. The correct value of θ is

(a) 30° (b) 45°

(c) 60° (d) 90°.

21. Given : $\vec{A} = 2\hat{i} + 3\hat{j}$ and $\vec{B} = 5\hat{i} - 6\hat{j}$. The magnitude of $\vec{A}+\vec{B}$ is

(a) 4 units (b) 10 units

(c) $\sqrt{58}$ units (d) $\sqrt{61}$ units.

22. The x and y components of a force are 2 N and – 3 N. The force is

(a) $2\hat{i} - 3\hat{j}$ (b) $2\hat{i} + 3\hat{j}$

(c) $-2\hat{i} - 3\hat{j}$ (d) $3\hat{i} + 2\hat{j}$.

23. Given : $\vec{A} = 3\hat{i} + 2\hat{j} + \hat{k}$ and $\vec{B} = 6\hat{i} + 4\hat{j} + 2\hat{k}$. The two vectors $\vec{A}$ and $\vec{B}$

(a) are parallel (b) are perpendicular

(c) make an angle of 30° to each other

(d) make an angle of 60° to each other.

24. $|\vec{b}|$ can be written as

(a) $\sqrt{\vec{b}.\vec{b}}$ (b) $\vec{b}.\vec{b}$

(c) $(\vec{b}.\vec{b})^2$ (d) $(\vec{b}.\vec{b})^{3/2}$.

25. Given : $\vec{c} = \vec{a}\times\vec{b}$. The angle which $\vec{a}$ makes with $\vec{c}$ is

(a) 0° (b) 45°

(c) 90° (d) 180°.

26. The magnitudes of the two vectors $\vec{a}$ and $\vec{b}$ are a and b respectively. The vector product of $\vec{a}$ and $\vec{b}$ cannot be

(a) equal to zero (b) less than ab

(c) equal to ab (d) greater than ab.

27. Given : $\vec{P} = \vec{A} - \vec{B}$ and P = A + B. The angle between $\vec{A}$ and $\vec{B}$ is

(a) 0° (b) 90°

(c) 180° (d) 270°.

28. The angle between the z-axis and the vector $\hat{i}+\hat{j}+\sqrt{2}\hat{k}$ is

(a) 30° (b) 45°

(c) 60° (d) 90°.

29. The angle between $\hat{i}+\hat{j}+\hat{k}$ and $2\hat{i} + 2\hat{j} + 2\hat{k}$ is

(a) 0° (b) 30°

(c) 60° (d) 90°.

30. The dot product of $(\hat{i}+\hat{j}+\hat{k})$ and $(\hat{i}-\hat{j}+\hat{k})$ is

(a) 1 (b) 2

(c) 4 (d) – 1.

31. $(\vec{A}.\vec{B})^2 + |\vec{A}\times\vec{B}|^2 =$

(a) zero (b) A^2B^2

(c) AB (d) $\sqrt{AB}$.

32. If $\vec{P} = 2\hat{i} - 3\hat{j} + \hat{k}$ and $\vec{Q} = 3\hat{i} + 2\hat{j}$, then $\vec{P}.\vec{Q}$ is

(a) zero (b) 6

(c) 12 (d) 15.

33. Which of the following is neither a scalar nor a vector ?

(a) Torque (b) Force

(c) Angular momentum (d) Dielectric constant.

34. A man walks 40 m North, then 30 m East and then 40 m South. What is his displacement from the starting point ?

(a) 30 m East (b) 150 m West

(c) 40 m West (d) 150 m East.

35. A vector is represented by $\hat{i} + 3\hat{j} + 5\hat{k}$. Its length in X-Y plane is

(a) 1 (b) 3

(c) 5 (d) $\sqrt{10}$.

36. Given : $\vec{R} = \vec{A} + \vec{B}$ and R = A = B. The angle between $\vec{A}$ and $\vec{B}$ is

(a) 60° (b) 90°

(c) 120° (d) 180°.

37. If the angle between $\vec{A}$ and $\vec{B}$ is 270°, then $\vec{A}.\vec{B}$ is

(a) 0 (b) 1

(c) – 1 (d) $\frac{1}{3}$.

38. Given : $\vec{A}.\vec{B} = 0$ and $\vec{A}\times\vec{C} = 0$. The angle between $\vec{B}$ and $\vec{C}$ is

(a) 0° (b) 90°

(c) 180° (d) 270°.

39. Given : $\vec{P} = \vec{A} + \vec{B}$ and P = A + B. The angle between $\vec{A}$ and $\vec{B}$ is

(a) 0° (b) $\frac{\pi}{4}$

(c) $\frac{\pi}{2}$ (d) π.

40. Given : $\vec{P} \cdot \vec{Q} = 0$ and $\vec{P} \times \vec{Q} = \vec{0}$. Which of the following is correct ?

(a) $\vec{P} \perp \vec{Q}$. (b) $\vec{P} \parallel \vec{Q}$.

(c) $\vec{P}$ is inclined to $\vec{Q}$ at an angle of 45°.

(d) Statement is wrong.

41. Three forces $(2\hat{i} - 3\hat{j} + 4\hat{k})$, $(8\hat{i} - 7\hat{j} + 6\hat{k})$ and $m(\hat{i} - \hat{j} + \hat{k})$ keep a body in equilibrium. The value of m is

(a) 10 (b) – 10

(c) 20 (d) – 20.

42. What is the angle between $\vec{A} \times \vec{B}$ and $\vec{B} \times \vec{A}$?

(a) 0° (b) 180°

(c) 45° (d) 90°.

43. If $\vec{A} = \vec{B} + \vec{C}$ and $\vec{B} = \vec{C} + \vec{A}$, then the vector $\vec{C}$ is

(a) $= \vec{A} + \vec{B}$ (b) $= \vec{B} - \vec{A}$

(c) $= \vec{A} - \vec{B}$ (d) null vector.

44. Projection of $\vec{P}$ on $\vec{Q}$ is

(a) $\vec{P} \cdot \hat{Q}$ (b) $\hat{P} \cdot \vec{Q}$

(c) $\vec{P} \times \hat{Q}$ (d) $\vec{P} \times \vec{Q}$.

45. If $\vec{A} \cdot \vec{B} = AB$, then the angle between $\vec{A}$ and $\vec{B}$ is

(a) 0° (b) 45°

(c) 90° (d) 180°.

46. A boy walks 8 m towards east and 6 m towards north. The magnitude of displacement is

(a) 2 m (b) 10 m

(c) 14 m (d) $[(8)^2 + (6)^2]$ m.

47. A vector $\vec{P_1}$ is along the positive x-axis. If its vector product with another vector $\vec{P_2}$ is zero, then $\vec{P_2}$ could be

(a) $4\hat{j}$ (b) $-4\hat{i}$

(c) $(\hat{j} + \hat{k})$ (d) $-(\hat{i} + \hat{j})$.

48. The magnitude of $\vec{P} - \vec{Q}$ is P + Q. The angle between $\vec{P}$ and $\vec{Q}$ is

(a) $\frac{\pi}{2}$ radian (b) π radian

(c) $\frac{3\pi}{2}$ radian (d) $\frac{5\pi}{2}$ radian.

49. If $\vec{A} \times \vec{B} = 0$ and $\vec{B} \times \vec{C} = 0$, then the value of $\vec{A} \times \vec{C}$ is

(a) zero (b) AC sin θ$\hat{n}$

(c) AC cos θ (d) AB tan θ.

50. Given : $\vec{C} = \vec{A} + \vec{B}$. Also, the magnitudes of $\vec{A}$, $\vec{B}$ and $\vec{C}$ are 12, 5 and 13 units respectively. The angle between $\vec{A}$ and $\vec{B}$ is

(a) 0° (b) $\frac{\pi}{4}$

(c) $\frac{\pi}{2}$ (d) π.

51. If $\vec{P} + \vec{Q} = \vec{P} - \vec{Q}$ and θ is the angle between $\vec{P}$ and $\vec{Q}$, then

(a) θ = 0° (b) θ = 90°

(c) P = 0 (d) Q = 0.

52. The force vector which has components (– 2, 1, 3) is

(a) $2\hat{i} + \hat{j} + 3\hat{k}$ (b) $2\hat{i} - \hat{j} - 3\hat{k}$

(c) $-2\hat{i} + \hat{j} + 3\hat{k}$ (d) $\hat{i} - \hat{j} + 3\hat{k}$.

53. A hall has the dimensions 10 m × 12 m × 14 m. A fly starting at one corner ends up at a diametrically opposite corner. The magnitude of its displacement is nearly

(a) 16 m (b) 17 m

(c) 18 m (d) 21 m.

54. The value of $\hat{j} \times (\hat{i} \times \hat{k})$ is

(a) zero (b) 1

(c) 2 (d) $\sqrt{2}$.

55. If $\vec{A}$ is parallel to $\vec{B}$, then

(a) $\vec{A} \cdot (\vec{A} + \vec{B}) = A^2 + AB$ (b) $\vec{A} \times (\vec{A} + \vec{B}) = AB$

(c) $\vec{A} \cdot \vec{B} = 0$ (d) $\vec{A} \times (\vec{A} \times \vec{B}) = 9$.

56. The co-ordinates of the end points of a vector are (4, – 4, 0) and (– 2, – 2, 0). The magnitude of this vector is

(a) $2\sqrt{10}$ units (b) $3\sqrt{10}$ units

(c) $4\sqrt{10}$ units (d) $5\sqrt{10}$ units.

57. The resultant of two forces of magnitudes 8 N and 15 N is 17 N. The angle between the forces of 8 N and 15 N is

(a) 30° (b) 45°

(c) 60° (d) 90°.

58. The sum and difference of two perpendicular vectors of equal lengths are

(a) of equal lengths and have an acute angle between them

(b) of equal length and have an obtuse angle between them

(c) also perpendicular and of different lengths

(d) also perpendicular and of equal lengths.

59. Two vectors $\vec{a}$ and $\vec{b}$ are such that $|\vec{a}+\vec{b}| = |\vec{a}-\vec{b}|$. What is the angle between $\vec{a}$ and $\vec{b}$?

(a) 0° (b) 90°
(c) 60° (d) 180°.

60. Given : $\vec{A} = 2\hat{i} + 3\hat{j}$. The angle between $\vec{A}$ and y-axis is

(a) $\sin^{-1}\frac{2}{3}$ (b) $\cos^{-1}\frac{2}{3}$
(c) $\tan^{-1}\frac{2}{3}$ (d) $\tan^{-1}\frac{3}{2}$.

61. What is the angle between $(\vec{A}-\vec{B})$ and $(\vec{A}\times\vec{B})$?

(a) 0° (b) $\frac{\pi}{2}$ radian
(c) π radian (d) $\frac{3\pi}{2}$ radian.

QUESTION BANK

Level II

Choose the correct answer from the four alternatives given in each of the following questions :

1. A particle is moving on a circular path with constant speed v. What is the change in its velocity after it has described an angle of 60° ?

(a) $v\sqrt{2}$ (b) $v\sqrt{3}$
(c) v (d) $2v$.

2. A car with a vertical wind shield moves along in a rain storm at speed of 40 km h^{-1}. The rain drops fall vertically with a terminal speed of 20 m s^{-1}. The angle at which the rain drops strike the wind shield is

(a) $\tan^{-1}$ (2/3) (b) $\tan^{-1}$ (3/2)
(c) $\tan^{-1}$ (5/9) (d) $\tan^{-1}$ (9/5).

3. A force vector applied on a mass is represented as $\vec{F} = 6\hat{i} - 8\hat{j} + 10\hat{k}$ N and accelerates the mass at 1 m s^{-2}. The mass of the body is

(a) 10 kg (b) 20 kg
(c) $2\sqrt{10}$ kg (d) $10\sqrt{2}$ kg.

4. The resultant of two forces makes angles of 30° and 60° with them and has a magnitude of 40 N. The magnitudes of the two forces are

(a) 20 N, 20 N (b) 20 N, 28 N
(c) 20 N, $20\sqrt{3}$ N (d) 20 N, 60 N.

5. A force of $(2\hat{i} + 3\hat{j} + \hat{k})$ N and another force of $(\hat{i} + \hat{j} + \hat{k})$ N are acting on a body. The magnitude of total force acting on the body is

(a) $\sqrt{3}$ N (b) $\sqrt{14}$ N
(c) $\sqrt{17}$ N (d) $\sqrt{29}$ N.

6. The resultant $\vec{C}$ of $\vec{A}$ and $\vec{B}$ is perpendicular to $\vec{A}$. Also, $|\vec{A}| = |\vec{C}|$. The angle between $\vec{A}$ and $\vec{B}$ is

(a) $\frac{\pi}{4}$ radian (b) $\frac{3\pi}{4}$ radian
(c) $\frac{5\pi}{4}$ radian (d) $\frac{7\pi}{4}$ radian.

7. A boy moves on a cycle with a velocity of 4 km h^{-1}. The rain appears to fall on him with a velocity of 3 km h^{-1} vertically. The actual velocity of the rain is

(a) 7 km h^{-1} (b) 5 km h^{-1}
(c) 4/3 km h^{-1} (d) 3/4 km h^{-1}.

8. A boat which has a speed of 5 km h^{-1} in still water crosses a river of width 1 km along the shortest possible path in 15 minutes. The velocity of the river water is

(a) 1 km h^{-1} (b) 3 km h^{-1}
(c) 4 km h^{-1} (d) $\sqrt{41}$ km h^{-1}.

9. A particle has position vector $(3\hat{i} - \hat{j} + 2\hat{k})$ metre at time $t = 0$. It moves with constant velocity $(-\hat{i} + \hat{j} - 3\hat{k})$ m s^{-1}. The position vector (in m) of the particle after 3 second is

(a) $2\hat{j} - 7\hat{k}$ (b) $2\hat{i} - \hat{k}$
(c) $\hat{j}$ (d) $3\hat{k}$.

10. Rain is falling vertically downwards with a speed of 4 km h^{-1}. A girl moves on a straight road with a velocity of 3 km h^{-1}. The apparent velocity of rain with respect to the girl is

(a) 3 km h^{-1} (b) 4 km h^{-1}
(c) 5 km h^{-1} (d) 7 km h^{-1}.

11. The resultant of two vectors $\vec{A}$ and $\vec{B}$ is perpendicular to $\vec{A}$. The magnitude of the resultant is equal to half of the magnitude of $\vec{B}$. The angle between $\vec{A}$ and $\vec{B}$ is

(a) 0° (b) 60°
(c) 150° (d) 180°.

12. In an equilateral triangle ABC, AL, BM and CN are the medians. Which of the following would correctly represent the resultant of two forces represented by BC and BA ?

(a) AC (b) 2AL

(c) 2BM (d) 2CN.

13. To a man walking at the rate of 4 km h^{-1}, the rain appears to fall vertically. If the apparent velocity of rain is $4\sqrt{3}$ km h^{-1}, then inclination of the real velocity of the rain with the vertical will be

(a) 60° (b) 30°

(c) 45° (d) 90°.

14. The resultant of two vectors of magnitudes 3 units and 4 units is $\sqrt{37}$. The angle between the two vectors is

(a) 0° (b) 30°

(c) 60° (d) 90°.

15. A force of $4\hat{i} + 5\hat{j}$ newton displaces a body through $3\hat{i} + 6\hat{k}$ metre. The work done is

(a) 12 J (b) 30 J

(c) 42 J (d) 0 J.

16. If $|\vec{A} \times \vec{B}| = \vec{A}.\vec{B}$, then the angle between $\vec{A}$ and $\vec{B}$ is

(a) $\frac{\pi}{4}$ (b) $\frac{\pi}{2}$

(c) $\frac{3\pi}{4}$ (d) $\frac{3\pi}{2}$.

17. Given : $\vec{A} = 2\hat{i} - \hat{j} + 2\hat{k}$ and $\vec{B} = -\hat{i} - \hat{j} + \hat{k}$. The unit vector of $\vec{A} - \vec{B}$ is

(a) $\frac{3\hat{i} + \hat{k}}{\sqrt{10}}$ (b) $\frac{3\hat{i}}{\sqrt{10}}$

(c) $\frac{\hat{k}}{\sqrt{10}}$ (d) $\frac{-3\hat{i} - \hat{k}}{\sqrt{10}}$.

18. Given : $\vec{a}.\vec{b} = \frac{\sqrt{3}\,ab}{2}$. The angle between $\vec{a}$ and $\vec{b}$ is

(a) 0° (b) 30°

(c) 60° (d) 90°.

19. A force $\vec{F} = 3\hat{i} - 2\hat{j} + 4\hat{k}$ displaces a body from a point A(8, – 2, – 3) to the point B(– 2, 0, 6). The work done is

(a) 1 unit (b) 2 units

(c) 3 units (d) 4 units.

20. The angles which $\hat{i} + \hat{j} + \sqrt{2}\hat{k}$ makes with x-axis, y-axis and z-axis respectively are

(a) 60°, 60° and 45° (b) 60°, 45° and 45°

(c) 45°, 45° and 45° (d) 60°, 60° and 60°.

21. The angle between two vectors $-3\hat{i} + 6\hat{k}$ and $2\hat{i} + 3\hat{j} + \hat{k}$ is

(a) 45° (b) 60°

(c) 90° (d) 180°.

22. Two forces of $\vec{F_1}$ = 500 N due east and $\vec{F_2}$ = 250 N due north have their common initial point. $\vec{F_2} - \vec{F_1}$ is

(a) $250\sqrt{5}$ N, $\tan^{-1}(2)$ W of N (b) 250 N, $\tan^{-1}(2)$ W of N

(c) zero (d) 750 N $\tan^{-1}\left(\frac{3}{4}\right)$ N of W.

23. A river is flowing from west to east at a speed of 5 units. A man on the south bank of the river, capable of swimming at 10 units in still water, wants to swim across the river in shortest time. He should swim in a direction

(a) due north (b) 30° east of north

(c) 30° west of north (d) 60° east of north.

24. Ship A is travelling with a velocity of 5 km h^{-1} due east. A second ship is heading 30° east of north. What should be the speed of second ship if it is to remain always due north with respect to the first ship ?

(a) 10 km h^{-1} (b) 9 km h^{-1}

(c) 8 km h^{-1} (d) 7 km h^{-1}.

25. A man swims from a point A on one bank of a river of width 100 m. When he swims perpendicular to the water current, he reaches the other bank 50 m downstream. The angle to the bank at which he should swim, to reach the directly opposite point B on the other bank is

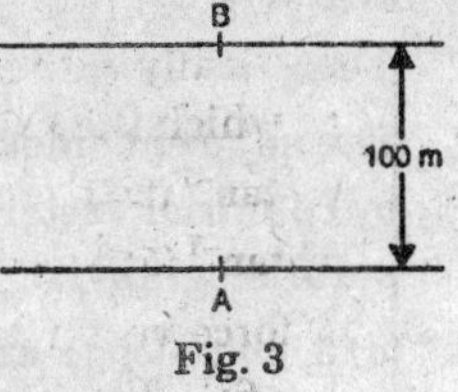

Fig. 3

(a) 10° upstream (b) 20° upstream

(c) 30° upstream (d) 60° upstream.

26. A vector of length m is turned through an angle β about its tail. The change in the position vector of its head is

(a) $2m \sin\frac{\beta}{2}$ (b) $2m \cos\frac{\beta}{2}$

(c) $2m \tan\frac{\beta}{2}$ (d) $2m \cot\frac{\beta}{2}$.

27. Two forces, each numerically equal to 5 N, are acting as shown in the Fig. 4. Then the resultant is

(a) 2.5 N (b) 5 N

(c) $5\sqrt{3}$ N (d) 10 N.

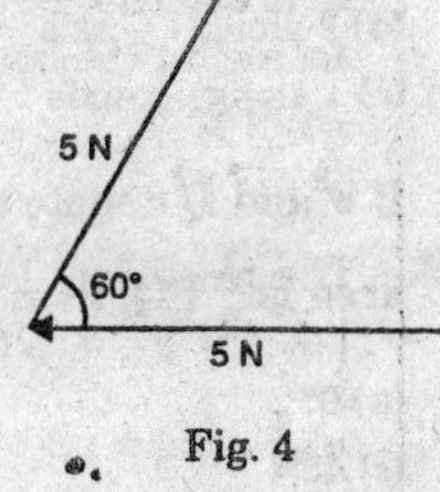

Fig. 4

28. The vector $5\hat{i} + 2\hat{j} - l\hat{k}$ is perpendicular to the vector $3\hat{i} + \hat{j} + 2\hat{k}$ for l =

(a) 1 (b) 4.7

(c) 6.3 (d) 8.5.

29. If $|\vec{A}+\vec{B}| = |\vec{A}| = |\vec{B}|$, then the angle between $\vec{A}$ and $\vec{B}$ is

(a) 0° (b) 60°
(c) 90° (d) 120°.

30. Given : $\vec{A} = 4\hat{i} + 6\hat{j}$ and $\vec{B} = 2\hat{i} + 3\hat{j}$. Which of the following is correct ?

(a) $\vec{A} \times \vec{B} = \vec{0}$ (b) $\vec{A}\cdot\vec{B} = 24$
(c) $\frac{|\vec{A}|}{|\vec{B}|} = \frac{1}{2}$ (d) $\vec{A}$ and $\vec{B}$ are anti-parallel.

31. If $3\hat{i} - 2\hat{j} + \hat{k}$ is perpendicular to $2\hat{i} + b\hat{j} + 6\hat{k}$, then the value of b is

(a) 2 (b) 4
(c) 5 (d) 6.

32. The resultant of the three vectors $\vec{OA}$, $\vec{OB}$ and $\vec{OC}$ shown in Fig. 5 is

(a) r
(b) $2r$
(c) $r(1 + \sqrt{2})$
(d) $r(\sqrt{2} - 1)$.

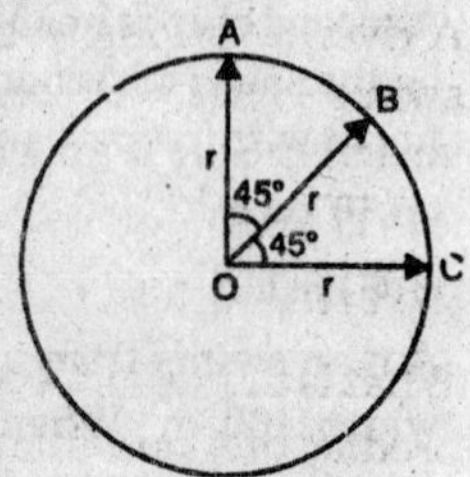

Fig. 5

33. The adjacent sides of a parallelogram are represented by co-initial vectors $2\hat{i} + 3\hat{j}$ and $\hat{i} + 4\hat{j}$. The area of the parallelogram is

(a) 5 units along z-axis (b) 5 units in x-y plane
(c) 3 units in x-z plane (d) 3 units in y-z plane.

34. If $\vec{A} + \vec{B} = \vec{C}$ and $A = \sqrt{3}$, $B = \sqrt{3}$ and $C = 3$, then the angle between $\vec{A}$ and $\vec{B}$ is

(a) 0° (b) 30°
(c) 60° (d) 90°.

35. Vector $\vec{A}$ is 2 cm long and is 60° above the x-axis in the first quadrant. Vector $\vec{B}$ is 2 cm long and is 60° below the x-axis in the fourth quadrant. The sum $\vec{A} + \vec{B}$ is a vector of magnitude

(a) 2 along + y-axis (b) 2 along + x-axis
(c) 1 along – x-axis (d) 2 along – x-axis.

36. If $\vec{P}$ and $\vec{Q}$ denote the sides of a parallelogram and its area is $\frac{1}{2}$ PQ, then the angle between $\vec{P}$ and $\vec{Q}$ is

(a) 0° (b) 30°
(c) 45° (d) 60°.

37. If $\vec{A}\cdot\vec{B} = 0$ and $\vec{A} \times \vec{B} = \vec{1}$, then $\vec{A}$ and $\vec{B}$ are

(a) perpendicular unit vectors
(b) parallel unit vectors
(c) parallel (d) perpendicular.

38. In the adjoining vector diagram, what is the angle between $\vec{A}$ and $\vec{B}$? (Given : $C = \frac{B}{2}$)

(a) 30° (b) 60°
(c) 120° (d) 150°.

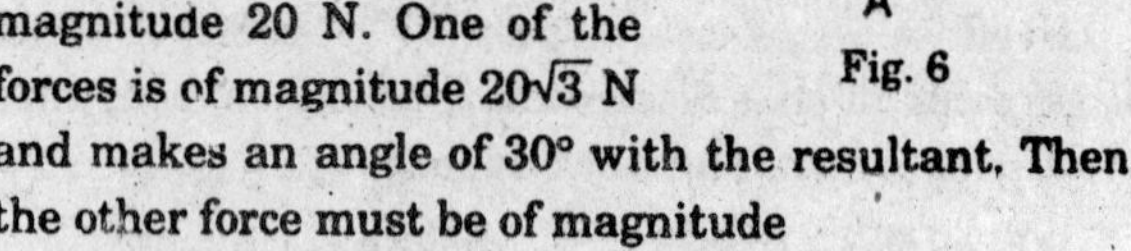

Fig. 6

39. The resultant of two forces has magnitude 20 N. One of the forces is of magnitude $20\sqrt{3}$ N and makes an angle of 30° with the resultant. Then, the other force must be of magnitude

(a) 10 N (b) $10\sqrt{3}$ N
(c) 20 N (d) $20\sqrt{3}$ N.

40. What is the angle between two vector forces of equal magnitude such that the resultant is one-third as much as either of the original forces ?

(a) $\cos^{-1}\left(-\frac{17}{18}\right)$ (b) $\cos^{-1}\left(\frac{1}{3}\right)$
(c) 45° (d) 120°.

41. Two forces, each of magnitude F, have a resultant of the same magnitude F. The angle between the two forces is

(a) 45° (b) 120°
(c) 150° (d) 180°.

42. What is the angle made by $3\hat{i} + 4\hat{j}$ with x-axis ?

(a) 0° (b) 180°
(c) $\tan^{-1}(3)$ (d) $\tan^{-1}\left(\frac{4}{3}\right)$.

43. Three vectors $\vec{A}, \vec{B}, \vec{C}$ satisfy the relation $\vec{A}\cdot\vec{B} = 0$ and $\vec{A}\cdot\vec{C} = 0$. The vector $\vec{A}$ is parallel to

(a) $\vec{B}$ (b) $\vec{C}$
(c) $\vec{B}\cdot\vec{C}$ (d) $\vec{B} \times \vec{C}$.

44. If the magnitude of the sum of the two vectors is equal to the difference of their magnitudes, then the angle between vectors is

(a) 0° (b) 45°
(c) 90° (d) 180°.

45. If $\vec{A} = \vec{B} + \vec{C}$, and the magnitudes of $\vec{A}, \vec{B}, \vec{C}$ are 5, 4 and 3 units, then angle between $\vec{A}$ and $\vec{C}$ is

(a) $\cos^{-1}\left(\frac{3}{5}\right)$ (b) $\cos^{-1}\left(\frac{4}{5}\right)$

(c) $\sin^{-1}\left(\frac{3}{4}\right)$ (d) $\frac{\pi}{2}$.

46. What is the angle between $\hat{i}+\hat{j}+\hat{k}$ and $\hat{i}$?

(a) 0° (b) 45°

(c) 60° (d) none of these.

47. What is the unit vector along $\hat{i}+\hat{j}$?

(a) $\frac{\hat{i}+\hat{j}}{\sqrt{2}}$ (b) $\sqrt{2}(\hat{i}+\hat{j})$

(c) $\hat{i}+\hat{j}$ (d) $\hat{k}$.

48. Given : $\vec{A} = A\cos\theta\,\hat{i} + A\sin\theta\,\hat{j}$. A vector $\vec{B}$ which is perpendicular to $\vec{A}$ is given by

(a) $B\cos\theta\hat{i} - B\sin\theta\hat{j}$ (b) $B\sin\theta\hat{i} - B\cos\theta\hat{j}$

(c) $B\cos\theta\hat{i} + B\sin\theta\hat{j}$ (d) $B\sin\theta\hat{i} + B\cos\theta\hat{j}$.

49. If the resultant of two vectors $\vec{A}$ and $\vec{B}$ is perpendicular to $\vec{A}$, then the angle between $\vec{A}$ and $\vec{B}$ is

(a) $\tan^{-1}\left(-\frac{A}{B}\right)$ (b) $\sin^{-1}\left(\frac{A}{B}\right)$

(c) $\tan^{-1}\left(\frac{A}{B}\right)$ (d) $\cos^{-1}\left(-\frac{A}{B}\right)$.

50. A region is specified by the potential function $V = 5x + 4y - 6z$. The potential gradient is

(a) $5\hat{i} + 4\hat{j} - 6\hat{k}$ (b) $5\hat{i} - 4\hat{j} - 6\hat{k}$

(c) $-5\hat{i} + 4\hat{j} + 6\hat{k}$ (d) $-5\hat{i} - 4\hat{j} + 7\hat{k}$.

51. One of the rectangular components of a velocity of 60 km h^{-1} is 30 km h^{-1}. The other rectangular component is

(a) 30 km h^{-1} (b) $30\sqrt{3}$ km h^{-1}

(c) $30\sqrt{2}$ km h^{-1} (d) zero.

52. The torque of a force $\vec{F} = -3\hat{i} + \hat{j} + 5\hat{k}$ acting at a point is $\vec{\tau}$. If the position vector of the point is $7\hat{i} + 3\hat{j} + \hat{k}$, then $\vec{\tau}$ is

(a) $7\hat{i} - 8\hat{j} + 9\hat{k}$ (b) $14\hat{i} - \hat{j} + 3\hat{k}$

(c) $2\hat{i} - 3\hat{j} + 8\hat{k}$ (d) $14\hat{i} - 38\hat{j} + 16\hat{k}$.

53. The square of the resultant of two equal co-initial vectors is three times the product of the magnitudes of the vectors. The angle between the vectors is

(a) 0° (b) 60°

(c) 90° (d) 120°.

54. Each components of a vector $\vec{A}$ making equal angles with x, y and z axes is

(a) $\frac{A}{\sqrt{3}}$ (b) $\sqrt{3}$ A

(c) $\sqrt{2}$ A (d) $\frac{A}{\sqrt{2}}$.

55. The following four forces act simultaneously on a particle at rest at the origin of the co-ordinate system.

$\vec{F}_1 = 2\hat{i} - 3\hat{j} - 2\hat{k}$, $\vec{F}_2 = 5\hat{i} + 8\hat{j} + 6\hat{k}$,

$\vec{F}_3 = -4\hat{i} - 5\hat{j} + 5\hat{k}$ and $\vec{F}_4 = -3\hat{i} + 4\hat{j} - 7\hat{k}$

The particle will move in

(a) XY plane (b) YZ plane

(c) ZX plane (d) space.

56. A vector $\vec{A}$ of magnitude $5\sqrt{3}$ units, another vector $\vec{B}$ of magnitude 10 units are inclined to each other at an angle of 30°. The magnitude of the vector product of the two vectors is

(a) 1 unit (b) $5\sqrt{3}$ units

(c) 75 units (d) $25\sqrt{3}$ units.

57. A river is flowing at the rate of 6 km h^{-1}. A swimmer swims across with a velocity of 9 km h^{-1}. The resultant velocity of the man will be

(a) $\sqrt{17}$ km h^{-1} (b) $\sqrt{117}$ km h^{-1}

(c) $\sqrt{340}$ km h^{-1} (d) $3\sqrt{40}$ km h^{-1}.

58. What is the scalar product of $\vec{A}$ and $(\vec{A}+\vec{B})$ if the vector product of $\vec{A}$ and $\vec{B}$ is zero ?

(a) $A^2 + AB$ (b) AB

(c) A^2 (d) zero.

59. Given : $\vec{r} = 4\hat{j}$ and $\vec{p} = 2\hat{i} + 3\hat{j} + \hat{k}$. The angular momentum is

(a) $4\hat{i} - 8\hat{k}$ (b) $8\hat{i} - 4\hat{k}$

(c) $8\hat{j}$ (d) $9\hat{k}$.

60. The maximum and minimum resultants of two forces are in the ratio 7 : 3. The ratio of the forces is

(a) 4 : 1 (b) 5 : 2

(c) $\sqrt{7} : \sqrt{3}$ (d) 49 : 9.

61. The vector which must be added to the sum of the two vectors $\hat{i} + 2\hat{j} - \hat{k}$ and $\hat{i} - 2\hat{j} + 2\hat{k}$ to get a resultant of unit vector along z-axis is

(a) $2\hat{i} + \hat{j}$ (b) $-2\hat{i}$

(c) $\hat{i} + \hat{j} + \hat{k}$ (d) $\hat{i} - \hat{j} - \hat{k}$.

62. Forces $(\hat{i} + \hat{j} - 2\hat{k})$, $(2\hat{i} - \hat{j} + \hat{k})$, $(3\hat{i} - \hat{j} + 4\hat{k})$, $(4\hat{i} - 2\hat{j} + 3\hat{k})$ act on a body simultaneously. The resultant force has a magnitude of

(a) $\sqrt{145}$ (b) 145

(c) 72.5 (d) $\sqrt{290}$.

63. A boat is moving with a velocity of $3\hat{i} + 4\hat{j}$ with respect to ground. The water in the river is moving with a velocity $-3\hat{i} - 4\hat{j}$ with respect to ground. The relative velocity of boat with respect to water is

(a) $6\hat{i} + 8\hat{j}$ (b) $-6\hat{i} - 8\hat{j}$

(c) $6\hat{i}$ (d) $8\hat{j} + \hat{k}$.

64. A force of $(10\hat{i} - 3\hat{j} + 6\hat{k})$ newton acts on a body of mass 100 g and displaces it from $(6\hat{i} + 5\hat{j} - 3\hat{k})$ metre to $(10\hat{i} - 2\hat{j} + 7\hat{k})$ metre. The work done is

(*a*) 21 J (*b*) 121 J
(*c*) 361 J (*d*) 1000 J.

65. The sum of the magnitudes of two vectors is 18. The magnitude of their resultant is 12. If the resultant is perpendicular to one of the vectors, then the magnitudes of the two vectors are

(*a*) 5 and 13 (*b*) 6 and 12
(*c*) 7 and 11 (*d*) 8 and 10.

66. A carrom board block is hit by a force $(5\hat{i} - \hat{j} + 3\hat{k})$. It gets displaced through $(2\hat{i} + \hat{j} - \hat{k})$. The kinetic energy received by the block is

(*a*) 6 J (*b*) 7 J
(*c*) 9 J (*d*) 14 J.

67. Two forces 8 N and 12 N act at 120°. The third force required to keep the body in equilibrium is

(*a*) 4 N (*b*) $4\sqrt{7}$ N
(*c*) 20 N (*d*) None of these.

68. A force F makes an angle 20° with another force F. The resultant of two forces is

(*a*) F cos 20° (*b*) 2F cos 20°
(*c*) 2F cos 10° (*d*) F cos 10°.

69. Given : $\vec{R} = \vec{P} + \vec{Q}$. Also, $\vec{P}$, $\vec{Q}$ and $\vec{R}$ have magnitudes 5, 12 and 13 units respectively. The angle between $\vec{Q}$ and $\vec{R}$ is

(*a*) $\cos^{-1}\left(\frac{12}{13}\right)$ (*b*) $\cos^{-1}\left(\frac{5}{13}\right)$
(*c*) $\cos^{-1}\left(\frac{7}{13}\right)$ (*d*) 90°.

70. If unit vectors $\vec{A}$ and $\vec{B}$ are inclined at an angle *q*, then $|\vec{A} - \vec{B}|$ is

(*a*) $2 \sin \frac{\theta}{2}$ (*b*) $2 \cos \frac{\theta}{2}$
(*c*) $2 \tan \frac{\theta}{2}$ (*d*) $\tan \theta$.

71. If the sum of two unit vectors is also a unit vector, then the magnitude of their difference is

(*a*) $\sqrt{2}$ (*b*) $\sqrt{3}$
(*c*) $\sqrt{5}$ (*d*) $\sqrt{7}$.

72. Which vector should be added to $2\hat{i} + 4\hat{j} - 3\hat{k}$ and $3\hat{i} - 5\hat{j} + 7\hat{k}$ to get a unit vector along y-axis ?

(*a*) $5\hat{i} + 9\hat{j} + \hat{k}$ (*b*) $\hat{i}$
(*c*) $-3\hat{k}$ (*d*) $-5\hat{i} + 2\hat{j} - 4\hat{k}$.

73. Forces 5 N, 12 N and 13 N are in equilibrium. If sin 23° = $\frac{5}{13}$, then the angle between the 5 N force and 13 N force is

(*a*) 67° (*b*) 90°
(*c*) 23° (*d*) 113°.

74. If $0.5\hat{i} + 0.8\hat{j} + c\hat{k}$ is a unit vector, then the value of *c* is

(*a*) $\sqrt{0.11}$ (*b*) $\sqrt{0.22}$
(*c*) $\sqrt{0.33}$ (*d*) $\sqrt{0.89}$.

75. The ratio of maximum and minimum magnitudes of the resultant of two vectors $\vec{a}$ and $\vec{b}$ is 3 : 1. Now, $|\vec{a}|$ =

(*a*) $|\vec{b}|$ (*b*) $2\,|\vec{b}|$
(*c*) $3\,|\vec{b}|$ (*d*) $4\,|\vec{b}|$.

76. Two forces, each equal to F, act as shown in Fig. 7. Their resultant is

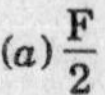

(*a*) $\frac{F}{2}$

(*b*) F

(*c*) $\sqrt{3}$ F

(*d*) $\sqrt{5}$ F.

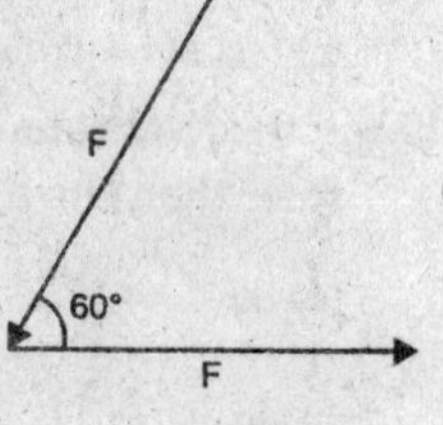

Fig. 7

77. A car is moving on a straight road due north with a uniform speed of 50 km h^{-1} when it turns left through 90°. If the speed remains unchanged after turning, the change in the velocity of the car in the turning process is

(*a*) zero
(*b*) $50\sqrt{2}$ km h^{-1} in S-W direction
(*c*) $50\sqrt{2}$ km h^{-1} in N-W direction
(*d*) 50 km h^{-1} due west.

78. Angle between two vectors $(\hat{i} + \hat{j})$ and $(\hat{i} - \hat{j})$ is

(*a*) 30° (*b*) 60°
(*c*) 45° (*d*) 90°.

79. The angle which the vector $\vec{A} = 2\hat{i} + 3\hat{j}$ makes with *y*-axis, where $\hat{i}$ and $\hat{j}$ are unit vectors along *x* and *y*-axes respectively is

(*a*) $\cos^{-1}$ (3/5) (*b*) $\cos^{-1}$ (2/3)
(*c*) $\tan^{-1}$ (2/3) (*d*) $\sin^{-1}$ (2/3).

80. If $\vec{P} = 4\hat{i} - 2\hat{j} + 6\hat{k}$ and $\vec{Q} = \hat{i} - 2\hat{j} - 3\hat{k}$, then the angle which $\vec{P} + \vec{Q}$ makes with *x*-axis is

(*a*) $\cos^{-1}\left(\frac{3}{\sqrt{50}}\right)$ (*b*) $\cos^{-1}\left(\frac{4}{\sqrt{50}}\right)$
(*c*) $\cos^{-1}\left(\frac{5}{\sqrt{50}}\right)$ (*d*) $\cos^{-1}\left(\frac{12}{\sqrt{50}}\right)$.

81. A boat is sent across a river with a velocity of 8 km h^{-1}. If the resultant velocity of the boat is 10 km h^{-1}, the river is flowing with a velocity

(a) 6 km h^{-1} (b) 8 km h^{-1}
(c) 10 km h^{-1} (d) 128 km h^{-1}.

82. A man is walking on a road with a velocity 3 km h^{-1}. Suddenly rain starts falling. The velocity of rain is 10 km h^{-1} in vertically downward direction. The relative velocity of the rain is

(a) $\sqrt{7}$ km h^{-1} (b) $\sqrt{13}$ km h^{-1}
(c) 13 km h^{-1} (d) $\sqrt{109}$ km h^{-1}.

83. Three forces $\vec{A} = (\hat{i}+\hat{j}+\hat{k})$, $\vec{B} = (2\hat{i}-\hat{j}+3\hat{k})$ and $\vec{C}$ acting on a body keep it in equilibrium. Then $\vec{C}$ is

(a) $\sqrt{3}$ (b) $\sqrt{14}$
(c) $-(3\hat{i}+4\hat{k})$ (d) $3\hat{i}+4\hat{j}+5\hat{k}$.

84. Given : $0.3\hat{i}+0.4\hat{j}+n\hat{k}$ is a unit vector. The value of n is

(a) $\sqrt{0.25}$ (b) $\sqrt{0.50}$
(c) $\sqrt{0.75}$ (d) $\sqrt{0.80}$.

85. A man walks 4 m towards East and then 3 m towards North and there he fixes a pole 12 m high. The distance between the starting point and tip of the pole in space is

(a) 7 m (b) 11 m
(c) 13 m (d) 19 m.

86. At what angle the forces of 2 N and $\sqrt{2}$ N act so that their combined effect is that of a single force of $\sqrt{10}$ N ?

(a) 0° (b) 30°
(c) 45° (d) 60°.

87. Given : $\vec{P} = 3\hat{i} - 4\hat{j}$. Which of the following is perpendicular to $\vec{P}$?

(a) $3\hat{i}$ (b) $4\hat{j}$
(c) $4\hat{i}+3\hat{j}$ (d) $4\hat{i}-3\hat{j}$.

88. The angle between $\hat{i}+\hat{j}$ and $\hat{j}+\hat{k}$ is

(a) 0° (b) 90°
(c) 45° (d) 60°.

89. The rectangular components of a vector are 2, 2. The corresponding rectangular components of another vector are 1, $\sqrt{3}$. The angle between the vectors is

(a) 15° (b) 30°
(c) 45° (d) 60°.

90. The linear velocity of a rotating body is given by

$$\vec{v} = \vec{\omega} \times \vec{r}$$

If $\vec{\omega} = \hat{i}-2\hat{j}+2\hat{k}$ and $\vec{r} = 4\hat{j}-3\hat{k}$, then the magnitude of $\vec{v}$ is

(a) $\sqrt{29}$ units (b) $\sqrt{31}$ units
(c) $\sqrt{37}$ units (d) $\sqrt{41}$ units.

91. Two point masses 1 and 2 move with uniform velocities $\vec{v}_1$ and $\vec{v}_2$ respectively. Their initial position vectors are $\vec{r}_1$ and $\vec{r}_2$ respectively. Which of the following should be satisfied for the collision of the point masses ?

(a) $\dfrac{\vec{r}_2-\vec{r}_1}{|\vec{r}_2+\vec{r}_1|} = \dfrac{\vec{v}_2+\vec{v}_1}{|\vec{v}_2-\vec{v}_1|}$ (b) $\dfrac{\vec{r}_1-\vec{r}_2}{|\vec{r}_1-\vec{r}_2|} = \dfrac{\vec{v}_2-\vec{v}_1}{|\vec{v}_2-\vec{v}_1|}$

(c) $\dfrac{\vec{r}_2-\vec{r}_1}{|\vec{r}_2+\vec{r}_1|} = \dfrac{\vec{v}_2-\vec{v}_1}{|\vec{v}_2+\vec{v}_1|}$ (d) $\dfrac{\vec{r}_2+\vec{r}_1}{|\vec{r}_1-\vec{r}_2|} = \dfrac{\vec{v}_2-\vec{v}_1}{|\vec{v}_2+\vec{v}_1|}$.

92. The adjacent sides of a parallelogram are represented by $\hat{i}+4\hat{j}$ and $7\hat{i}+3\hat{j}$. The area of the parallelogram is

(a) 10 units (b) 15 units
(c) 25 units (d) 30 units.

93. The sum of the magnitudes of two forces acting at a point is 16 N. The resultant of these forces is perpendicular to the smaller force and has a magnitude of 8 N. If the smaller force is of magnitude x, then the value of x is

(a) 2 N (b) 4 N
(c) 6 N (d) 7 N.

94. A force $\vec{F} = 2\hat{i}+2\hat{j}$ newton displaces a particle through $\vec{S} = 2\hat{i}+2\hat{k}$ metre in 16 second. The power developed by $\vec{F}$ is

(a) 0.25 J s^{-1} (b) 25 J s^{-1}
(c) 225 J s^{-1} (d) 450 J s^{-1}.

95. In going from one city to another, a car travels 75 km north, 60 km north-west and 20 km east. The magnitude of displacement between the two cities is (Take $1/\sqrt{2} = 0.7$)

(a) 170 km (b) 137 km
(c) 119 km (d) 140 km.

96. The area of the parallelogram whose diagonals are represented by the vectors $3\hat{i}+\hat{j}+2\hat{k}$ and $\hat{i}-3\hat{j}+4\hat{k}$ is

(a) 28 units (b) 8.66 units
(c) 21 units (d) 18 units.

97. Given : $\vec{A} = 2\hat{i}+p\hat{j}+q\hat{k}$ and $\vec{B} = 5\hat{i}+7\hat{j}+3\hat{k}$. If $\vec{A} \parallel \vec{B}$, then the values of p and q are respectively

(a) $\dfrac{14}{5}$ and $\dfrac{6}{5}$ (b) $\dfrac{14}{3}$ and $\dfrac{6}{5}$
(c) $\dfrac{6}{5}$ and $\dfrac{1}{3}$ (d) $\dfrac{3}{4}$ and $\dfrac{1}{4}$.

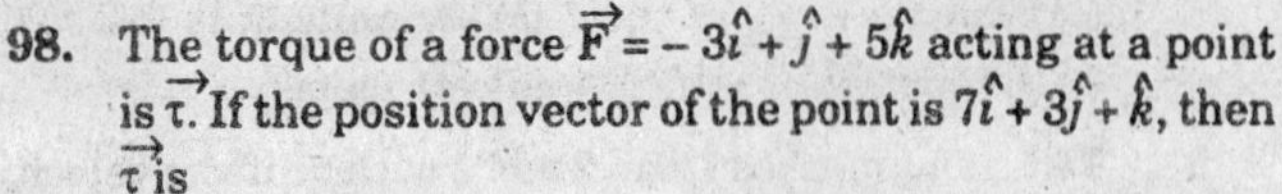

98. The torque of a force $\vec{F} = -3\hat{i} + \hat{j} + 5\hat{k}$ acting at a point is $\vec{\tau}$. If the position vector of the point is $7\hat{i} + 3\hat{j} + \hat{k}$, then $\vec{\tau}$ is

(a) $7\hat{i} - 8\hat{j} + 9\hat{k}$ (b) $14\hat{i} - \hat{j} + 3\hat{k}$
(c) $2\hat{i} - 3\hat{j} + 8\hat{k}$ (d) $14\hat{i} - 38\hat{j} + 16\hat{k}$.

99. If the resultant of two vectors having magnitudes of 6 and 3 is 9, then the dot product of two vectors is

(a) 6 (b) 3
(c) 9 (d) 18.

100. Two forces P and Q acting at a point are such that if P is reversed, the direction of the resultant is turned through 90°. Then

(a) P = Q (b) P = 2Q
(c) $P = \frac{Q}{2}$ (d) No relation between P and Q.

101. Two vectors $\vec{a}$ and $\vec{b}$ are at an angle of 60° with each other. Their resultant makes an angle of 45° with $\vec{a}$. If $|\vec{b}| = 2$ units, then $|\vec{a}|$ is

(a) $\sqrt{3}$ (b) $\sqrt{3} - 1$
(c) $\sqrt{3} + 1$ (d) $\frac{\sqrt{3}}{2}$.

102. A carrom board (4′ × 4′) has the queen at the centre. The queen hit by the striker moves to the front edge, rebounds and goes in the hole behind the striking line. The displacement of the queen from the centre to the hole is

(a) 4′ (b) $2\sqrt{2}$′
(c) $\frac{2}{3}\sqrt{10}$′ (d) $\frac{4}{3}\sqrt{10}$′.

103. In Q. 102, the displacement of the queen from the centre to the front edge is

(a) 4′ (b) $2\sqrt{2}$′
(c) $\frac{2}{3}\sqrt{10}$′ (d) $\frac{4}{3}\sqrt{10}$′.

104. In Q. 102, the displacement of the queen from the front edge to the hole is

(a) 4′ (b) $2\sqrt{2}$′
(c) $\frac{2}{3}\sqrt{10}$′ (d) $\frac{4}{3}\sqrt{10}$′.

105. Given : $\vec{a} + \vec{b} + \vec{c} = 0$. Out of the three vectors $\vec{a}$, $\vec{b}$ and $\vec{c}$, two are equal in magnitude. The magnitude of the third vector is $\sqrt{2}$ times that of either of the two having equal magnitude. The angles between the vectors are

(a) 90°, 135°, 135° (b) 30°, 60°, 90°
(c) 45°, 45°, 90° (d) 45°, 60°, 90°.

106. A person walks in the following pattern : 3 km north, then 2 km west and finally 5 km south. How far would a bird fly in a straight line from the same starting point to the same final point ?

(a) 2 km (b) $2\sqrt{2}$ km
(c) $3\sqrt{2}$ km (d) $4\sqrt{2}$ km.

107. The resultant of two forces P and Q is of magnitude P. If P be doubled, the resultant will be inclined to Q at an angle

(a) 0° (b) 30°
(c) 60° (d) 90°.

108. The area of a parallelogram formed from the vectors $\vec{A} = \hat{i} - 2\hat{j} + 3\hat{k}$ and $\vec{B} = 3\hat{i} - 2\hat{j} + \hat{k}$ as adjacent sides is

(a) $8\sqrt{3}$ units (b) 64 units
(c) 32 units (d) $4\sqrt{6}$ units.

109. The position vectors of the head and tail of radius vector are $2\hat{i} + \hat{j} + \hat{k}$ and $2\hat{i} - 3\hat{j} + \hat{k}$. The linear momentum is $2\hat{i} + 3\hat{j} + \hat{k}$. The angular momentum is

(a) $4\hat{i} - 8\hat{k}$ (b) $2\hat{i} + \hat{j} + \hat{k}$
(c) $2\hat{i} - 3\hat{j} + \hat{k}$ (d) $2\hat{i} + 3\hat{j} + \hat{k}$.

110. The resultant of three vectors 1, 2 and 3 units whose directions are those of the sides of an equilateral triangle is at an angle of

(a) 30° with the first vector
(b) 15° with the first vector
(c) 100° with the first vector
(d) 150° with the first vector.

111. Given : $\vec{r} = 2\hat{i} + \hat{j} + \hat{k}$ and $\vec{p} = 2\hat{i} - 3\hat{j} + \hat{k}$. Then, angular momentum $\vec{L}$ is

(a) $4\hat{i}$ (b) $8\hat{k}$
(c) $4\hat{i} + 8\hat{k}$ (d) $4\hat{i} - 8\hat{k}$.

112. Given : $\vec{A} = 3\hat{i} - 4\hat{j} - 2\hat{k}$ and $\vec{B} = 2\hat{i} + 4\hat{j} - 5\hat{k}$. The angle which $\vec{A} + \vec{B}$ makes with Y-axis is

(a) 0° (b) 45°
(c) 60° (d) 90°.

113. The area of the triangle formed by the adjacent sides with $\vec{A} = -3\hat{i} + 2\hat{j} - 4\hat{k}$ and $\vec{B} = -\hat{i} + 2\hat{j} + \hat{k}$ is

(a) $\frac{\sqrt{165}}{2}$ units (b) $\frac{\sqrt{137}}{2}$ units
(c) $\sqrt{165}$ units (d) $\sqrt{137}$ units.

114. A particle moves in the xy plane with only an x-component of acceleration of 2 m s^{-2}. The particle starts from the origin at $t = 0$ with an initial velocity having an x-component of 8 m s^{-1} and y-component of – 15 m s^{-1}. The total velocity vector at any time t is

(a) $[(8 + 2t)\hat{i} - 15\hat{j}]$ m s^{-1} (b) zero

(c) $2t\hat{i} + 15\hat{j}$ (d) directed along z-axis.

115. A force of $(10\hat{i} - 3\hat{j} + 6\hat{k})$ newton acts on a body of mass 100 g and displaces it from $(6\hat{i} + 5\hat{j} - 3\hat{k})$ metre to $(10\hat{i} - 2\hat{j} + 7\hat{k})$ metre. The work done is

(a) 21 J (b) 121 J

(c) 361 J (d) 1000 J.

116. The angle between two vectors $\vec{a}$ and $\vec{b}$ is q. The resultant of $\vec{a}$ and $\vec{b}$ is making an angle $\frac{\theta}{2}$ with $\vec{a}$. Then

(a) $a = b$ (b) $ab = 1$

(c) $a = \frac{b}{2}$ (d) $a = 2b$.

117. A region is specified by the potential function $2x - 3y + 4z$. Then, the potential gradient is

(a) $2\hat{i} + 3\hat{j} + 0\hat{k}$ (b) $2\hat{i} - 3\hat{j} + 4\hat{k}$

(c) $-3\hat{j}$ (d) $4\hat{k}$.

118. The resultant of two vectors $\vec{P}$ and $\vec{Q}$ is $\vec{R}$. If the magnitude of $\vec{Q}$ is doubled, the new resultant becomes perpendicular to $\vec{P}$, then the magnitude of $\vec{R}$ is

(a) $\frac{P^2 - Q}{2PQ}$ (b) $\frac{P + Q}{P - Q}$

(c) Q (d) $\frac{P}{Q}$.

119. What is the angle made by $\hat{j} + \hat{i} + 0\hat{k}$ with X-axis ?

(a) 0° (b) 30°

(c) 45° (d) 75°.

120. Two forces of magnitude F and $\sqrt{3}$ F act at right angles to each other. Their resultant makes an angle β with F. The value of β is

(a) 30° (b) 45°

(c) 60° (d) 135°.

121. Two vectors of equal magnitude have a resultant equal to one and half times the magnitude of either vector. The cosine of the angle between the two vectors is

(a) 0.125 (b) 0.250

(c) – 2.7 (d) 0.8.

122. The magnitude of the vector product of two vectors is 4. The magnitude of their scalar product is $4\sqrt{3}$. The angle between the two vectors is

(a) 30° (b) 45°

(c) 60° (d) 75°.

123. Given that $\vec{A} + \vec{B} + \vec{C} = 0$. Out of three vectors, two are equal in magnitude and the magnitude of the third vector is $\sqrt{2}$ times that of either of the two having equal magnitude. Then the angles between vectors are given by

(a) 45°, 45°, 90° (b) 90°, 135°, 135°

(c) 30°, 60°, 90° (d) 45°, 60°, 90°.

124. $(\vec{a} + \vec{b}) \times (\vec{a} - \vec{b})$ equals

(a) zero (b) $\vec{a} . \vec{b}$

(c) $-2(\vec{a} \times \vec{b})$ (d) $\vec{b}$.

125. A unit vector along incident ray of light is $\hat{i}$. The unit vector for the corresponding refracted ray of light is $\hat{r}$. $\hat{n}$ is a unit vector normal to the boundary of the medium and directed towards the incident medium. If m be the refractive index of the medium, then Snell's law (2nd) of refraction is

(a) $\hat{i} \times \hat{n} = \mu(\hat{n} \times \vec{r})$ (b) $\hat{i} . \hat{n} = \mu(\hat{r} . \hat{n})$

(c) $\hat{i} \times \hat{n} = \mu(\hat{r} \times \hat{n})$ (d) $\mu(\hat{i} \times \hat{n}) = \hat{r} \times \hat{n}$.

126. The simple sum of two co-initial vectors is 16 units. Their vector sum is 8 units. The resultant of the vectors is perpendicular to the smaller vector. The magnitudes of the two vectors are

(a) 2 units and 14 units (b) 4 units and 12 units

(c) 6 units and 10 units (d) 8 units and 8 units.

127. The components of a vector along x and y directions are $(n + 1)$ and 1 respectively. If the co-ordinate system is rotated by an angle θ = 60°, then the components change to n and 3. The value of n is

(a) 2 (b) cos 60°

(c) sin 60° (d) 3.5.

128. ABCDEF is a regular hexagon with point O as centre. The value of $\vec{AB} + \vec{AC} + \vec{AD} + \vec{AE} + \vec{AF}$ is

(a) $2\vec{AO}$ (b) $4\vec{AO}$

(c) $6\vec{AO}$ (d) 0.

ANSWERS

QUESTION BANK (Level I)

1. (b) 2. (a) 3. (c) 4. (b) 5. (b) 6. (c) 7. (b) 8. (b)
9. (b) 10. (c) 11. (d) 12. (c) 13. (b) 14. (a) 15. (a) 16. (a)

17. (a)	18. (d)	19. (b)	20. (b)	21. (c)	22. (a)	23. (a)	24. (a)
25. (c)	26. (d)	27. (c)	28. (b)	29. (a)	30. (a)	31. (b)	32. (a)
33. (d)	34. (a)	35. (d)	36. (c)	37. (a)	38. (b)	39. (a)	40. (d)
41. (b)	42. (b)	43. (d)	44. (a)	45. (a)	46. (b)	47. (b)	48. (b)
49. (a)	50. (c)	51. (d)	52. (c)	53. (d)	54. (a)	55. (a)	56. (a)
57. (d)	58. (d)	59. (b)	60. (c)	61. (b).			

QUESTION BANK (Level II)

1. (c)	2. (c)	3. (d)	4. (c)	5. (d)	6. (b)	7. (b)	8. (b)
9. (a)	10. (c)	11. (c)	12. (c)	13. (b)	14. (c)	15. (a)	16. (a)
17. (a)	18. (b)	19. (b)	20. (a)	21. (c)	22. (a)	23. (a)	24. (a)
25. (d)	26. (a)	27. (b)	28. (d)	29. (d)	30. (a)	31. (d)	32. (c)
33. (a)	34. (c)	35. (b)	36. (b)	37. (a)	38. (d)	39. (c)	40. (a)
41. (b)	42. (d)	43. (d)	44. (d)	45. (a)	46. (d)	47. (a)	48. (b)
49. (d)	50. (a)	51. (b)	52. (d)	53. (b)	54. (a)	55. (b)	56. (d)
57. (b)	58. (a)	59. (a)	60. (b)	61. (b)	62. (a)	63. (a)	64. (b)
65. (a)	66. (a)	67. (b)	68. (c)	69. (a)	70. (a)	71. (b)	72. (d)
73. (d)	74. (a)	75. (b)	76. (b)	77. (b)	78. (d)	79. (c)	80. (c)
81. (a)	82. (d)	83. (c)	84. (c)	85. (c)	86. (c)	87. (c)	88. (d)
89. (a)	90. (a)	91. (b)	92. (c)	93. (c)	94. (a)	95. (c)	96. (b)
97. (a)	98. (d)	99. (d)	100. (a)	101. (b)	102. (b)	103. (c)	104. (d)
105. (a)	106. (b)	107. (d)	108. (d)	109. (a)	110. (a)	111. (d)	112. (d)
113. (a)	114. (a)	115. (b)	116. (a)	117. (b)	118. (c)	119. (c)	120. (c)
121. (a)	122. (a)	123. (b)	124. (c)	125. (c)	126. (c)	127. (d)	128. (c).

HINTS/SOLUTIONS

QUESTION BANK (Level I)

1. (a) If a vector $\vec{A}$ makes angles α, β and γ with x-axis, y-axis and z-axis respectively, then

$$\frac{A_x}{A} = \cos\alpha = l\,; \quad \frac{A_y}{A} = \cos\beta = m; \quad \frac{A_z}{A} = \cos\gamma = n$$

$$l^2 + m^2 + n^2 = \frac{A_x^{\,2}}{A^2} + \frac{A_y^{\,2}}{A^2} + \frac{A_z^{\,2}}{A^2}$$

$$= \frac{A_x^{\,2} + A_y^{\,2} + A_z^{\,2}}{A^2} = \frac{A^2}{A^2} = 1$$

2. (a) $|\hat{A} \times \hat{B}| = (1)(1)\sin\theta = \sin\theta$.

3. (c) The condition $C^2 = A^2 + B^2$ is satisfied if $\vec{A}$ is perpendicular to $\vec{B}$.

4. (b) ☞ ***If any two vectors are parallel or equal, then the scalar triple product is zero.***

5. (b) Four vectors in different planes may give zero resultant.

6. (c) $\vec{A}\cdot(\vec{A}+\vec{B}) = \vec{A}\cdot\vec{A} + \vec{A}\cdot\vec{B} = A^2 + O = A^2$

7. (b) Required numerical value is $\sqrt{3^2 + 4^2 + 5^2}$, *i.e.*, $\sqrt{50}$ or $5\sqrt{2}$

8. (b) ⊗ in the Fig. 8 represents $\vec{A}$ directed vertically downwards. Hold your right-hand parallel to $\vec{B}$. Curl your fingers towards $\vec{A}$. The direction of erect thumb will give the direction of $\vec{B}\times\vec{A}$. The erect thumb will point towards west.

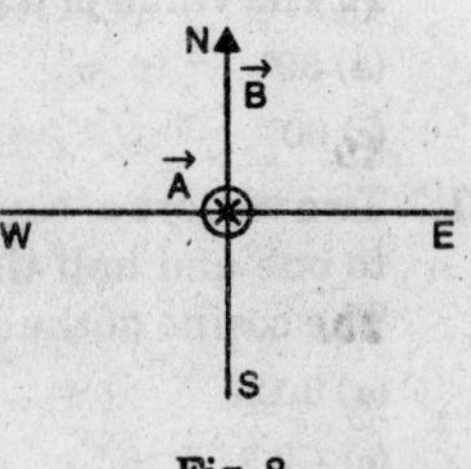

Fig. 8

9. (b) $R^2 = P^2 + P^2 + 2P^2\cos\theta$ or $R^2 = 2P^2 + 2P^2\cos\theta$

or $R^2 = 2P^2(1+\cos\theta)$ or $R^2 = 2P^2\left(1 + 2\cos^2\frac{\theta}{2} - 1\right)$

or $R^2 = 4P^2 \cos^2 \frac{\theta}{2}$ or $R = 2P \cos \frac{\theta}{2}$

10. (c) $F = \sqrt{\left(\frac{P}{2}\right)^2 + \left(\frac{P}{2}\right)^2} = \sqrt{2}\frac{P}{2} = \frac{P}{\sqrt{2}}$

11. (d) $AB \sin\theta = AB \cos\theta$ or $\tan\theta = 1$ or $\theta = 45°$

$\therefore$ $C^2 = A^2 + B^2 + 2AB\cos 45° = A^2 + B^2 + \sqrt{2}AB$

or $C = \sqrt{A^2 + B^2 + \sqrt{2}\,AB}$

12. (c) $AB\sin\theta = \sqrt{3}\,AB\cos\theta$, $\tan\theta = \sqrt{3}$, $\theta = 60° = \frac{\pi}{3}$ radian.

13. (b) $R_{max.} = (10 + 6)\text{ N} = 16\text{ N}$; $R_{min.} = (10 - 6)\text{ N} = 4\text{ N}$

Clearly, 8 N is possible.

> *Maximum possible resultant of $\vec{P}$ and $\vec{Q}$ is (P + Q).*
>
> *Minimum possible resultant of $\vec{P}$ (> $\vec{Q}$) and $\vec{Q}$ is (P – Q)*
>
> *The resultant of $\vec{P}$ and $\vec{Q}$ can vary from (P – Q) to (P + Q).*

14. (a) $\vec{P}.\vec{Q} = (3\hat{j} + 4\hat{k}).(2\hat{i} + 5\hat{k})$

$= 20(\hat{k}.\hat{k}) = 20$

15. (a) $\vec{B} + (\hat{i} + 2\hat{j} - 3\hat{k}) = \hat{i}$

or $\vec{B} = -2\hat{j} + 3\hat{k}$

16. (a) $\vec{A} + \vec{B} = 8\hat{i} + 0\hat{j} + 16\hat{k}$

$m = \frac{0}{|\vec{A} + \vec{B}|} = 0$

17. (a) $\vec{P}.\vec{Q} = 0 \Rightarrow \vec{P} \perp \vec{Q}$ or $\theta = 90°$

$|\vec{P} \times \vec{Q}| = PQ \sin 90° = PQ$ or $|\vec{P}|\,|\vec{Q}|$

18. (d) $\cos\alpha = \cos\beta = \cos\gamma = \frac{1}{\sqrt{1^2 + 1^2 + 1^2}} = \frac{1}{\sqrt{3}}$

19. (b) $F^2 = F_1{}^2 + F_2{}^2 + 2F_1F_2 \cos 90°$

or $F^2 = F_1{}^2 + F_2{}^2$

20. (b) $10A^2 = 4A^2 + 2A^2 + 2 \times 2A \times \sqrt{2}A \times \cos\theta$

or $4A^2 = 4\sqrt{2}A^2 \cos\theta$

or $\cos\theta = \frac{1}{\sqrt{2}} \Rightarrow \theta = 45°$

21. (c) $\vec{A} + \vec{B} = 7\hat{i} - 3\hat{j}$

$|\vec{A} + \vec{B}| = \sqrt{7^2 + 3^2} = \sqrt{49 + 9} = \sqrt{58}$

22. (a) $\vec{F} = F_x\hat{i} + F_y\hat{j}$ or $\vec{F} = 2\hat{i} - 3\hat{j}$

23. (a) $\vec{B} = 2\vec{A}$. So $\vec{B} \parallel \vec{A}$.

24. (a) $\sqrt{\vec{b}.\vec{b}} = \sqrt{b^2} = b = |\vec{b}|$

25. (c) The result follows from the definition of cross product.

26. (d) $|\vec{a} \times \vec{b}| = ab \sin\theta$

$\sin\theta$ cannot be greater than 1.

$\therefore$ $|\vec{a} \times \vec{b}|$ cannot be greater than ab.

27. (c) $|\vec{P}| = A + B$

$|\vec{A} - \vec{B}|^2 = (A + B)^2$

$A^2 + B^2 - 2AB\cos\theta = A^2 + B^2 + 2AB$

or $\cos\theta = -1$ or $\theta = 180°$

28. (b) $\cos\theta = \frac{(\hat{k}).(\hat{i} + \hat{j} + \sqrt{2}\,\hat{k})}{1\sqrt{1^2 + 1^2 + 2}}$

or $\cos\theta = \frac{\sqrt{2}}{2} = \frac{1}{\sqrt{2}}$ or $\theta = 45°$

29. (a) $\vec{A} = \hat{i} + \hat{j} + \hat{k}$; $\vec{B} = 2(\hat{i} + \hat{j} + \hat{k}) = 2\vec{A}$

Clearly, the direction of $\vec{B}$ is the same as that of $\vec{A}$

$\therefore$ $\theta = 0°$

30. (a) $(\hat{i} + \hat{j} + \hat{k}).(\hat{i} - \hat{j} + \hat{k}) = \hat{i}.\hat{i} - \hat{j}.\hat{j} + \hat{k}.\hat{k} = 1 - 1 + 1 = 1$

31. (b) $A^2B^2\cos^2\theta + A^2B^2\sin^2\theta = A^2B^2(\cos^2\theta + \sin^2\theta) = A^2B^2$

32. (a) $(2\hat{i} - 3\hat{j} + \hat{k}).(3\hat{i} + 2\hat{j}) = 6(\hat{i}.\hat{i}) - 6(\hat{j}.\hat{j}) = 0$

33. (d) Dielectric constant is neither a scalar nor a vector. It is a tensor. In a non-isotropic medium, it has different values in different directions.

34. (a) $S_1 = 40\hat{j} + 30\hat{i} - 40\hat{j}$

$= 30\hat{i}$.

So, displacement is 30 m east.

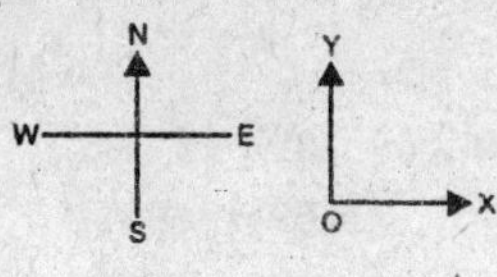

Fig. 9

35. (d) Length in XY plane $= \sqrt{1^2 + 3^2} = \sqrt{10}$

36. (c) $R^2 = R^2 + R^2 + 2R^2\cos\theta$ or $R^2 = 2R^2 + 2R^2\cos\theta$

$\frac{1}{2} = 1 + \cos\theta$ or $\cos\theta = -\frac{1}{2}$ or $\theta = 120°$

37. (a) $\vec{A}.\vec{B} = AB\cos 270°$

$= AB\cos(360° - 90°) = AB\cos 90° = 0.$

38. (b) $\vec{A}.\vec{B} = 0 \Rightarrow \vec{A} \perp \vec{B}$

$\vec{A} \times \vec{C} = 0 \Rightarrow \vec{A} \parallel \vec{C}$

$\therefore$ $\vec{B} \perp \vec{C}$ $\therefore$ $\theta = 90°$

39. (a) $|\vec{P}| = A + B \Rightarrow |\vec{P}|^2 = (A + B)^2$

$|\vec{A} + \vec{B}|^2 = (A + B)^2$

or $A^2 + B^2 + 2AB\cos\theta = A^2 + B^2 + 2AB$

or $\cos\theta = 1$ or $\theta = 0°$.

40. (d) $\vec{P}.\vec{Q} = 0 \Rightarrow \vec{P} \perp \vec{Q}$

$\vec{P} \times \vec{Q} = 0 \Rightarrow \vec{P} \parallel \vec{Q}$

These are contradictory results.

So, statement is wrong.

41. (*b*) The vector sum of given forces should be zero.

$\therefore\ 10\hat{i} + m\hat{i} - 10\hat{j} - m\hat{j} + 10\hat{k} + m\hat{k} = 0$

$(10 + m)\hat{i} - (10 + m)\hat{j} + (10 + m)\hat{k} = 0$

This equation will be satisfied if

$10 + m = 0$ or $m = -10$

42. (*b*) $\vec{A} \times \vec{B}$ and $\vec{B} \times \vec{A}$ are oppositely directed. So, angle is 180°.

43. (*d*) $\vec{A} = \vec{C} + \vec{A} + \vec{C}$ or $2\vec{C} = \vec{0}$ or $\vec{C} = \vec{0}$

So, $\vec{C}$ is a null vector.

44. (*a*) Projection of $\vec{P}$ on $\vec{Q}$ is P cos θ.

$$P\cos\theta = \frac{PQ\cos\theta}{Q} = \frac{\vec{P}.\vec{Q}}{Q} = \vec{P}.\hat{Q}$$

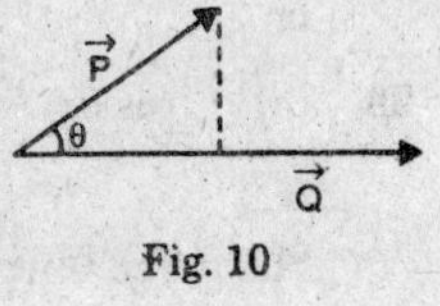

Fig. 10

45. (*a*) AB cos θ = AB or cos θ = 1 or θ = 0°

46. (*b*) $|8\hat{i} + 6\hat{j}| = \sqrt{8^2 + 6^2} = \sqrt{100}$ m = 10 m

47. (*b*) Note that vector product of parallel vectors is zero.

48. (*b*) $|\vec{P} - \vec{Q}|^2 = (P + Q)^2$

$P^2 + Q^2 - 2PQ\cos\theta = P^2 + Q^2 + 2PQ$

$2PQ + 2PQ\cos\theta = 0$ or $\cos\theta = -1$

or θ = 180° or π radian.

49. (*a*) $\vec{A} \times \vec{B} = 0 \Rightarrow \vec{A} \parallel \vec{B}$; $\vec{B} \times \vec{C} = 0 \Rightarrow \vec{B} \parallel \vec{C}$

$\therefore\ \vec{A} \parallel \vec{C}$ $\quad \therefore\ \vec{A} \times \vec{C} = 0$

50. (*c*) $13^2 = 12^2 + 5^2$

169 = 169

So, $\vec{A} \perp \vec{B}$. $\therefore\ \theta = \frac{\pi}{2}$ radian.

51. (*d*) $\vec{P} + \vec{Q} = \vec{P} - \vec{Q}$ or $2\vec{Q} = 0$ or $\vec{Q} = 0$

52. (*c*) $\vec{F} = F_x\hat{i} + F_y\hat{j} + F_z\hat{k}$

53. (*d*) $\vec{S} = 10\hat{i} + 12\hat{j} + 14\hat{k}$

or $|\vec{S}| = \sqrt{10^2 + 12^2 + 14^2}$ m

$= \sqrt{100 + 144 + 196}$ m $= \sqrt{440}$ m ≈ 21 m

54. (*a*) $\vec{A} \times (\vec{B} \times \vec{C}) = (\vec{A}.\vec{C})\vec{B} - (\vec{A}.\vec{B})\vec{C}$

$\hat{j} \times (\hat{i} \times \hat{k}) = (\hat{j}.\hat{k})\hat{i} - (\hat{j}.\hat{i})\hat{k} = 0\hat{i} - 0\hat{k} = 0$

55. (*a*) $\vec{A}.(\vec{A} + \vec{B}) = \vec{A}.\vec{A} + \vec{A}.\vec{B}$

$= A^2 + AB\cos 0° = A^2 + AB$

56. (*a*) $\vec{r} = (x_2 - x_1)\hat{i} + (y_2 - y_1)\hat{j} + (z_2 - z_1)\hat{k}$

$\vec{r} = [-2 - 4]\hat{i} + [-2 + 4]\hat{j} + [0 - 0]\hat{k} = -6\hat{i} + 2\hat{j}$

$|\vec{r}|$ or $r = \sqrt{6^2 + 2^2} = \sqrt{36 + 4} = \sqrt{40}$

$= \sqrt{4 \times 10} = 2\sqrt{10}$ units

57. (*d*) $17^2 = 8^2 + 15^2$ or $289 = 64 + 225 = 289$

Condition for perpendicularity is satisfied. So, the angle between 8 N and 15 N is 90°.

58. (*d*) Think of the diagonals of a square.

59. (*b*) $a^2 + b^2 + 2ab\cos\theta = a^2 + b^2 - 2ab\cos\theta$

or $4ab\cos\theta = 0$

But $4ab \neq 0$ $\quad \therefore\ \cos\theta = 0$ or $\theta = 90°$

Aliter

$(\vec{a} + \vec{b})$ and $(\vec{a} - \vec{b})$ are the diagonals of a parallelogram whose adjacent sides are $\vec{a}$ and $\vec{b}$.

Since $|\vec{a} + \vec{b}| = |\vec{a} - \vec{b}|$, therefore, the two diagonals of a parallelogram are equal. So, think of square. This leads to θ = 90°.

60. (*c*) $\tan\theta = \frac{2}{3}$

$\theta = \tan^{-1}\left(\frac{2}{3}\right)$

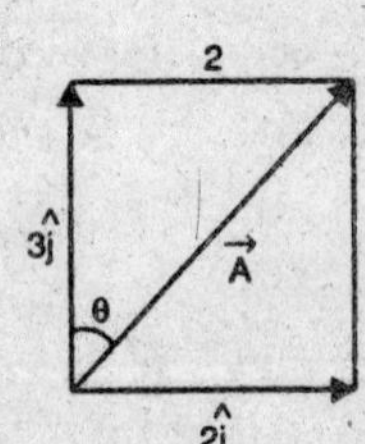

Fig. 11

61. (*b*) $\vec{A} \times \vec{B}$ is perpendicular to the plane of $\vec{A}$ and $\vec{B}$.

Also, $\vec{A} - \vec{B}$ is in the plane of $\vec{A}$ and $\vec{B}$. So, $\vec{A} \times \vec{B}$ is perpendicular to $\vec{A} - \vec{B}$.

$\therefore\ \theta = \frac{\pi}{2}$ radian

QUESTION BANK (Level II)

1. (*c*) $\Delta\vec{v} = \vec{v_2} - \vec{v_1} = \vec{v_2} + (-\vec{v_1})$

The angle between $\vec{v_1}$ and $\vec{v_2}$ is 120°.

$\Delta v^2 = v^2 + v^2 + 2v^2\cos 120°$

$= 2v^2\left[1 - \frac{1}{2}\right] = v^2$

or $\Delta v = v$

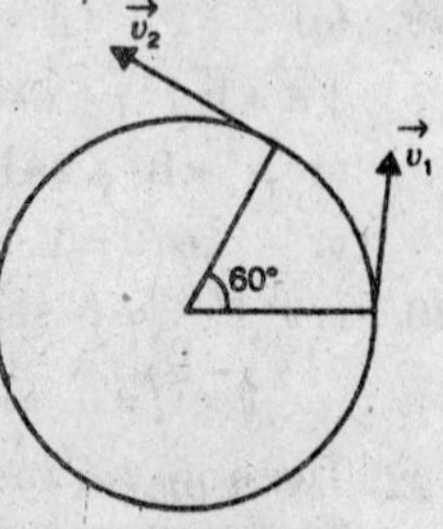

Fig. 12

2. (*c*) $\tan\beta = \frac{v_c}{v_r} = \frac{40 \times \frac{5}{18}}{20}$

$= \frac{5}{9}$

$\beta = \tan^{-1}\left(\frac{5}{9}\right)$

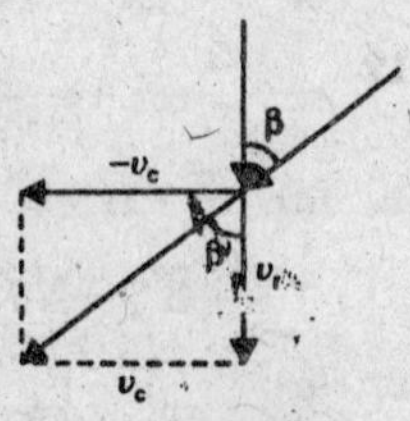

Fig. 13

3. (*d*) $F = \sqrt{6^2 + 8^2 + 10^2} = \sqrt{36 + 64 + 100} = 10\sqrt{2}$ N

$a = 1 \text{ m s}^{-2}, m = \frac{F}{a} = \frac{10\sqrt{2}}{1} \text{ kg} = 10\sqrt{2}$ kg

4. (*c*) $\tan 60° = \frac{Q \sin 90°}{P + Q \cos 90°}$ or $\sqrt{3} = \frac{Q}{P}$ or $Q = \sqrt{3}P$

Also, $P^2 + Q^2 = 1600$ or $P^2 + 3P^2 = 1600$ or $P^2 = 400$ or $P = 20$ N, Again, $Q = 20\sqrt{3}$ N

5. (*d*) $\vec{F} = 2\hat{i} + 3\hat{j} + \hat{k} + \hat{i} + \hat{j} + \hat{k} = 3\hat{i} + 4\hat{j} + 2\hat{k}$

$|\vec{F}|$ or $F = \sqrt{3^2 + 4^2 + 2^2} = \sqrt{9 + 16 + 4} = \sqrt{29}$ N

6. (*b*) $\tan \theta' = \frac{C}{A} = 1$

$\theta' = 45° = \frac{\pi}{4}$

$\theta = \pi - \frac{\pi}{4} = \frac{3\pi}{4}$

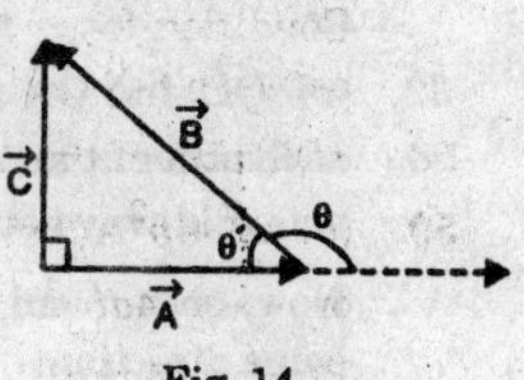

Fig. 14

7. (*b*) $v_r^2 = v_{rc}^2 + v_c^2 = 9 + 16 = 25$ or $v_r = 5$ km h^{-1}

8. (*b*) $v_r = \frac{1 \text{ km}}{\frac{1}{4} \text{ h}} = 4$ km h^{-1}

$v_b = 5$ km h^{-1}

$v_w = \sqrt{v_b^2 - v_r^2}$

$= \sqrt{25 - 16} = \sqrt{9} = 3$ km h^{-1}

Fig. 15

9. (*a*) $\vec{r_2} = \vec{r_1} + \vec{v}\,t$

or $\vec{r_2} = (3\hat{i} - \hat{j} + 2\hat{k}) + (-\hat{i} + \hat{j} - 3\hat{k})3 = 2\hat{j} - 7\hat{k}$

10. (*c*) $\vec{v_{rg}} = \vec{v_r} + (-\vec{v_g})$

$v_{rg} = \sqrt{v_r^2 + v_g^2}$

$= \sqrt{16 + 9}$ km h^{-1} = 5 km h^{-1}

11. (*c*) $\sin \theta' = \frac{R}{B}$

or $\sin \theta' = \frac{B/2}{B} = \frac{1}{2}$

or $\theta' = 30°$,

$\theta = 180° - 30°$

$= 150°$

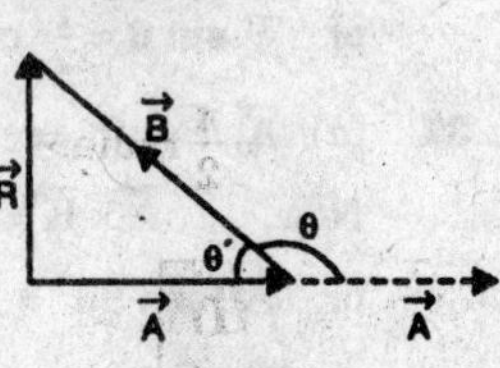

Fig. 16

12. (*c*) Let the forces represented by BC and BA be $\vec{F_1}$ and $\vec{F_2}$ respectively. As is clear from Fig. 17, the resultant $\vec{R}$ of $\vec{F_1}$ and $\vec{F_2}$ is represented by BM′ which is equal to 2 BM.

Fig. 17

13. (*b*) $\tan \beta = \frac{|-\vec{v_m}|}{|\vec{v_{rp}}|} = \frac{4}{4\sqrt{3}} = \frac{1}{\sqrt{3}}$ or $\beta = 30°$

14. (*c*) $37 = 9 + 16 + 2 \times 3 \times 4 \cos \theta$

$\frac{37 - 25}{24} = \cos \theta$ or $\cos \theta = \frac{12}{24} = \frac{1}{2}$ or $\theta = 60°$

15. (*a*) $W = \vec{F} . \vec{S} = (4\hat{i} + 5\hat{j}) . (3\hat{i} + 6\hat{k}) = 12$

16. (*a*) $AB \sin \theta = AB \cos \theta$ or $\tan \theta = 1$ or $\theta = \frac{\pi}{4}$

17. (*a*) $\vec{A} - \vec{B} = 3\hat{i} + \hat{k}$

$|\vec{A} - \vec{B}| = \sqrt{3^2 + 1^2} = \sqrt{10}$

Required unit vector is $\frac{3\hat{i} + \hat{k}}{\sqrt{10}}$ [Note that $\hat{A} = \frac{\vec{A}}{A}$]

18. (*b*) $ab \cos \theta = ab \times \frac{\sqrt{3}}{2}$

or $\cos \theta = \frac{\sqrt{3}}{2}$ or $\theta = 30°$

19. (*b*) $W = (3\hat{i} - 2\hat{j} + 4\hat{k}) . (-10\hat{i} + 2\hat{j} + 9\hat{k})$

$= -30 - 4 + 36 = 2$ units

20. (*a*) $\vec{A} = \hat{i} + \hat{j} + \sqrt{2}\hat{k}$

$A = \sqrt{1^2 + 1^2 + 2} = \sqrt{4} = 2$

$\cos \alpha = \frac{1}{2} \Rightarrow \alpha = 60°$; $\cos \beta = \frac{1}{2} \Rightarrow \beta = 60°$

$\cos \gamma = \frac{\sqrt{2}}{2} = \frac{1}{\sqrt{2}} \Rightarrow \gamma = 45°$

21. (*c*) $\cos \theta = \frac{\vec{A} . \vec{B}}{AB} = \frac{A_xB_x + A_yB_y + A_zB_z}{\sqrt{A_x^2 + A_y^2 + A_z^2}\sqrt{B_x^2 + B_y^2 + B_z^2}}$

$= \frac{(-3)(2) + (0)(3) + (6)(1)}{\sqrt{9 + 36} + \sqrt{4 + 9 + 1}} = 0$

$\therefore \quad \vec{A} \perp \vec{B}$

22. (*a*) $\vec{F_2} - \vec{F_1} = \vec{F_2} + (-\vec{F_1})$

= 250 N due north + 500 N due west

$\tan \theta = \frac{500}{250} = 2$

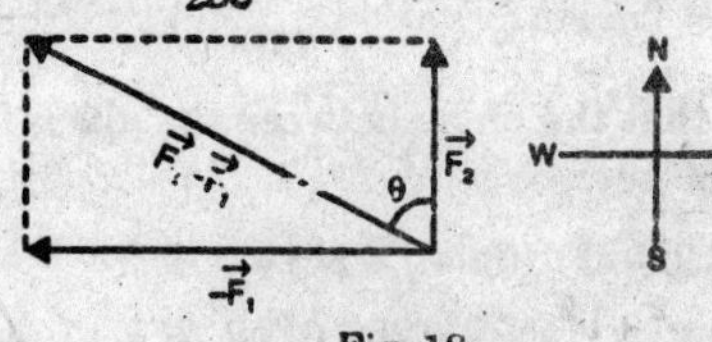

Fig. 18

$|\vec{F_2} - \vec{F_1}| = \sqrt{(500)^2 + (250)^2} = 250\sqrt{5}$ N

23. (*a*) $v \cos \theta = \frac{l}{t}$ or $t = \frac{l}{v \cos \theta}$

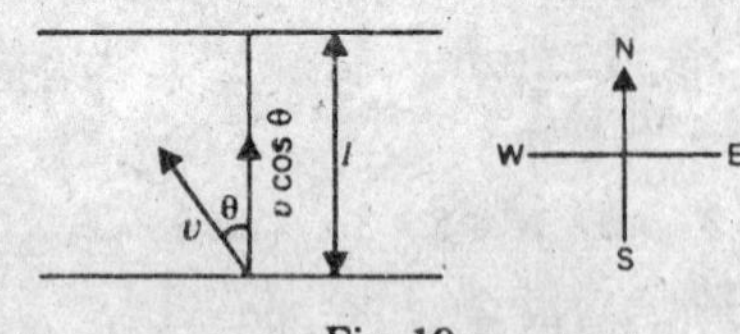

Fig. 19

When t is minimum, $\cos\theta$ is maximum, *i.e.*, $1 \Rightarrow \theta = 0°$. So, the man should swim due north.

24. (*a*) $\sin 30° = \dfrac{v_1}{v_2} = \dfrac{5}{v_2}$ or $\dfrac{1}{2} = \dfrac{5}{v_2}$ or $v_2 = 10 \text{ km h}^{-1}$

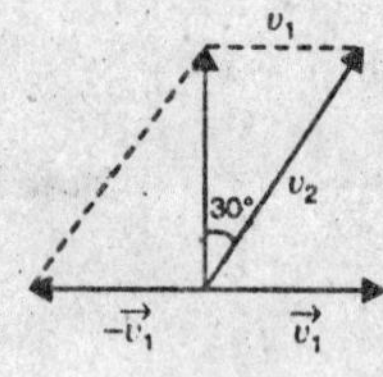

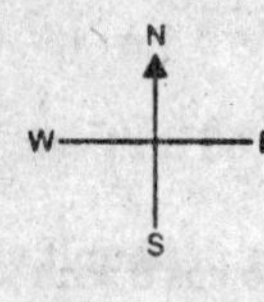

Fig. 20

25. (*d*) Refer to Fig. 21 (*a*), $\tan\theta = \dfrac{v_w}{v_m} = \dfrac{50}{100} = \dfrac{1}{2}$ or $v_m = 2v_w$

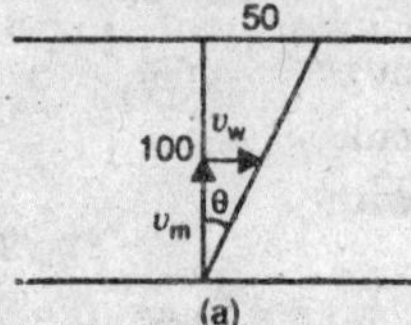

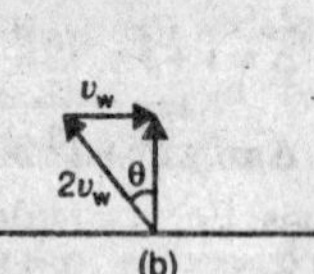

Fig. 21

Refer to Fig. 21 (*b*),

$$\sin\theta = \frac{v_w}{2v_w} = \frac{1}{2} \text{ or } \theta = 30°$$

So, it is 60° upstream.

26. (*a*) $\Delta m^2 = m^2 + m^2 - 2m^2 \cos\beta$

$$= 2m^2(1 - \cos\beta)$$

$$= 2m^2\left(1 - 1 + 2\sin^2\frac{\beta}{2}\right)$$

$$= 4m^2 \sin^2\frac{\beta}{2}$$

or $\Delta m = 2m \sin\dfrac{\beta}{2}$

Fig. 22

27. (*b*) Note that the angle between the given forces is 120° and not 60°.

28. (*d*) $(5\hat{i} + 2\hat{j} - l\hat{k}).(3\hat{i} + \hat{j} + 2\hat{k}) = 0$

or $15 + 2 - 2l = 0$ or $2l = 17$ or $l = 8.5$

29. (*d*) $A^2 = A^2 + A^2 + 2A^2\cos\theta$ or $2A^2\cos\theta = -A^2$

or $\cos\theta = -\dfrac{1}{2}$ or $\theta = 120°$

30. (*a*) $\vec{A} \times \vec{B} = (4\hat{i} + 6\hat{j}) \times (2\hat{i} + 3\hat{j})$

$$= 12(\hat{i} \times \hat{j}) + 12(\hat{j} \times \hat{i}) = 12(\hat{i} \times \hat{j}) - 12(\hat{i} \times \hat{j}) = 0$$

Again, $\vec{A}.\vec{B} = (4\hat{i} + 6\hat{j}).(2\hat{i} + 3\hat{j}) = 8 + 18 = 26$

Again, $\dfrac{|\vec{A}|}{|\vec{B}|} = \dfrac{\sqrt{16+36}}{\sqrt{4+9}} \neq \dfrac{1}{2}$

Again, $\vec{B} = \dfrac{1}{2}\vec{A}$

$\vec{A}$ and $\vec{B}$ are parallel and not anti-parallel.

31. (*d*) $(3\hat{i} - 2\hat{j} + \hat{k}).(2\hat{i} + b\hat{j} + 6\hat{k}) = 0$

or $6 - 2b + 6 = 0$ or $2b = 12$ or $b = 6$

32. (*c*) $\vec{OC}$ and $\vec{OA}$ are equal in magnitude and inclined to each other at an angle of 90°. So, their resultant is $\sqrt{2}r$. It acts mid-way between $\vec{OC}$ and $\vec{OA}$, *i.e.*, along OB.

Now, both r and $\sqrt{2}r$ are along the same line and in the same direction.

$\therefore$ Resultant $= r + \sqrt{2}r = r(1 + \sqrt{2})$

33. (*a*) $\vec{A} \times \vec{B} = (2\hat{i} + 3\hat{j}) \times (\hat{i} + 4\hat{j})$

$$= 8(\hat{i} \times \hat{j}) + 3(\hat{j} \times \hat{i}) = 8\hat{k} - 3\hat{k} = 5\hat{k}$$

34. (*c*) $\vec{A} + \vec{B} = \vec{C}$ (given)

So, it is given that $\vec{C}$ is the resultant of $\vec{A}$ and $\vec{B}$

$\therefore$ $C^2 = A^2 + B^2 + 2AB\cos\theta$

$$3^2 = 3 + 3 + 2 \times 3 \times \cos\theta$$

$$3 = 6\cos\theta \text{ or } \cos\theta = \frac{1}{2} \Rightarrow \theta = 60°$$

35. (*b*) The angle between two vectors of equal magnitude is 120°. So, resultant has the same magnitude as either of the given vectors. Moreover, it is mid-way between the two vectors, *i.e.*, it is along x-axis.

36. (*b*) $\frac{1}{2}PQ = |\vec{P} \times \vec{Q}| = PQ\sin\theta$

or $\sin\theta = \frac{1}{2}$ or $\theta = 30°$

37. (*a*) $\vec{A}.\vec{B} = 0 \Rightarrow \vec{A} \perp \vec{B}$

Now, $\vec{A} \times \vec{B} = \vec{1}$ or $AB\sin\theta = 1$

or $AB\sin 90° = 1$ or $AB = 1$ $\Rightarrow$ $A = 1$ and $B = 1$

So, $\vec{A}$ and $\vec{B}$ are perpendicular unit vectors.

38. (*d*) $\sin\theta' = \dfrac{C}{B}$

or $\sin\theta' = \dfrac{B/2}{B}$

or $\sin\theta' = \dfrac{1}{2}$ or $\theta' = 30°$

Fig. 23

Now, $\theta + \theta' = 180°$

or $\theta + 30° = 180°$ or $\theta = 150°$

39. (c) Let P be the unknown force

$P^2 = (20\sqrt{3})^2 + 20^2 - 2(20)(20\sqrt{3})\cos 30°$

or $P^2 = 1600 - 1200 = 400$

or $P = 20$ N

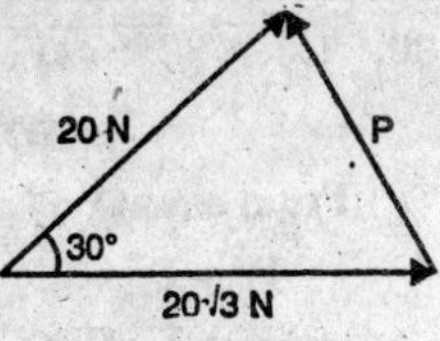

Fig. 24

40. (a) $\left(\frac{1}{3}\right)^2 = 1^2 + 1^2 + 2 \times 1 \times 1 \times \cos\theta$

$\frac{1}{9} = 2(1 + \cos\theta)$

or $1 + \cos\theta = \frac{1}{18}$

or $\cos\theta = \frac{1}{18} - 1 = -\frac{17}{18}$

or $\theta = \cos^{-1}\left(-\frac{17}{18}\right)$

Fig. 25

41. (b) $F^2 = F^2 + F^2 + 2F^2\cos\theta$

or $F^2 = 2F^2(1 + \cos\theta)$

or $1 + \cos\theta = \frac{1}{2}$

or $\cos\theta = -\frac{1}{2}$ or $\theta = 120°$ $\quad | \because \cos 120° = -\frac{1}{2}$

42. (d) $\tan\beta = \frac{4}{3}$ or $\beta = \tan^{-1}\left(\frac{4}{3}\right)$

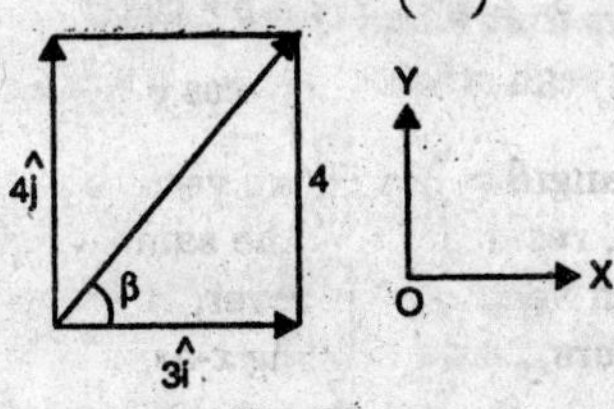

Fig. 26

43. (d) $\vec{A}\cdot\vec{B} = 0$ (given) $\Rightarrow \vec{A} \perp \vec{B}$

$\vec{A}\cdot\vec{C} = 0$ (given) $\Rightarrow \vec{A} \perp \vec{C}$

$\vec{A}$ is perpendicular to both $\vec{B}$ and $\vec{C}$.

We know, from the definition of cross product, that $\vec{B}\times\vec{C}$ is perpendicular to both $\vec{B}$ and $\vec{C}$.

So, $\vec{A}$ is parallel to $\vec{B}\times\vec{C}$.

44. (d) $\sqrt{P^2 + Q^2 + 2PQ\cos\theta} = P - Q$

or $P^2 + Q^2 + 2PQ\cos\theta = P^2 + Q^2 - 2PQ$

$2PQ(1 + \cos\theta) = 0$

But $2PQ \neq 0$

$\therefore$ $1 + \cos\theta = 0$ or $\cos\theta = -1$

or $\theta = 180°$

45. (a) $\cos\theta = \frac{C}{A} = \frac{3}{5}$

or $\theta = \cos^{-1}\left(\frac{3}{5}\right)$

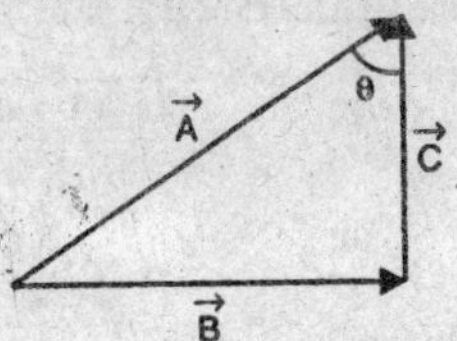

Fig. 27

46. (d) If θ is the required angle, then

$\cos\theta = \frac{(\hat{i}+\hat{j}+\hat{k})\cdot\hat{i}}{\sqrt{1^2+1^2+1^2}\sqrt{1^2}} = \frac{1}{\sqrt{3}}$ $\quad \because \cos\theta = \frac{\vec{A}\cdot\vec{B}}{AB}$

or $\theta = \cos^{-1}\left(\frac{1}{\sqrt{3}}\right)$

This is 54.7°

47. (a) $\vec{A} = A\hat{A}$ or $\hat{A} = \frac{\vec{A}}{A}$

$\therefore$ Required unit vector is $\frac{\hat{i}+\hat{j}}{|\hat{i}+\hat{j}|} = \frac{\hat{i}+\hat{j}}{\sqrt{2}}$

48. (b) Clearly, $\vec{B}$ should be either in second quadrant or fourth quadrant. In none of the given options, we have '$-\hat{i}$ term'. So, second quadrant is ruled out. Also, $\vec{B}$ should make an angle of 90° − θ with X-axis. So, $\vec{B}$ should be

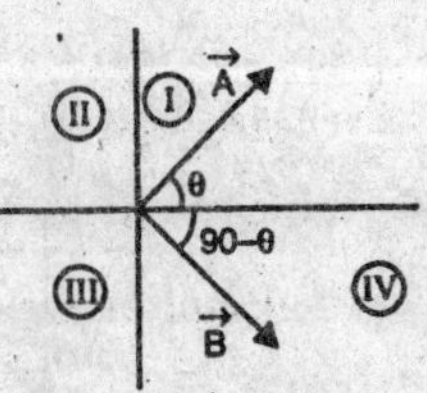

Fig. 28

$B\cos(90° - \theta)\hat{i} - B\sin(90° - \theta)\hat{j}$

or $B\sin\theta\hat{i} - B\cos\theta\hat{j}$

49. (d) If θ is the angle between $\vec{A}$ and $\vec{B}$, then

$\tan\beta = \frac{B\sin\theta}{A + B\cos\theta}$

or $\tan 90° = \frac{B\sin\theta}{A + B\cos\theta}$

or $\infty = \frac{B\sin\theta}{A + B\cos\theta}$

$\Rightarrow A + B\cos\theta = 0$

or $\cos\theta = -\frac{A}{B}$ or $\theta = \cos^{-1}\left(-\frac{A}{B}\right)$

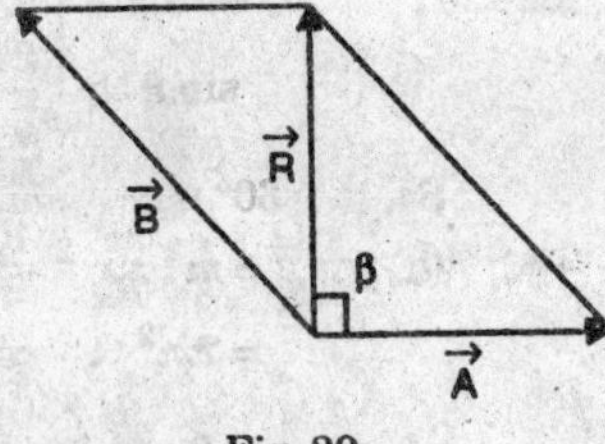

Fig. 29

50. (a) $\frac{\partial V}{\partial x}\hat{i} + \frac{\partial V}{\partial y}\hat{j} + \frac{\partial V}{\partial z}\hat{k} = 5\hat{i} + 4\hat{j} - 6\hat{k}$

51. (b) $60^2 = 30^2 + v^2$ or $v^2 = 90 \times 30$ or $v = 30\sqrt{3}$ kmh^{-1}

52. $\vec{\tau} = \vec{r} \times \vec{F}$

$$\vec{\tau} = \begin{vmatrix} \hat{i} & \hat{j} & \hat{k} \\ 7 & 3 & 1 \\ -3 & 1 & 5 \end{vmatrix}$$

$= \hat{i}(15 - 1) + \hat{j}(-3 - 35) + \hat{k}(7 + 9) = 14\hat{i} - 38\hat{j} + 16\hat{k}$

53. (*b*) $P^2 + P^2 + 2P^2 \cos\theta = 3P^2$

or $2P^2(1+\cos\theta) = 3P^2$ or $1+\cos\theta = \frac{3}{2}$

or $\cos\theta = \frac{3}{2} - 1 = \frac{1}{2}$ or $\theta = 60°$

54. (*a*) $\frac{A_x}{A} = \cos\alpha,\ \frac{A_y}{A} = \cos\beta,\ \frac{A_z}{A} = \cos\gamma$

But $\cos\alpha = \cos\beta = \cos\gamma$

$\therefore\ A_x = A_y = A_z$

Now, $\cos\alpha = \frac{A_x}{\sqrt{A_x^2 + A_x^2 + A_x^2}} = \frac{1}{\sqrt{3}}$

Again, $A_x = A\cos\alpha = \frac{A}{\sqrt{3}}$

Similarly, $A_y = A_z = \frac{A}{\sqrt{3}}$

55. (*b*) $\vec{F} = \vec{F_1} + \vec{F_2} + \vec{F_3} + \vec{F_4}$

or $\vec{F} = 4\hat{j} + 2\hat{k}$

Clearly, the particle shall move in YZ plane.

56. (*d*) $|\vec{A} \times \vec{B}| = (5\sqrt{3})(10)\sin 30° = 25\sqrt{3}$ units.

57. (*b*) Resultant velocity $= \sqrt{6^2 + 9^2} = \sqrt{117}$ units.

58. (*a*) $\vec{A} \times \vec{B} = 0 \Rightarrow \vec{A} \parallel \vec{B}$ or $\theta = 0°$

$\vec{A}.(\vec{A} + \vec{B}) = \vec{A}.\vec{A} + \vec{A}.\vec{B}$

$= A^2 + AB\cos 0° = A^2 + AB$

59. (*a*) $\vec{L} = \vec{r} \times \vec{p} = \begin{vmatrix} \hat{i} & \hat{j} & \hat{k} \\ 0 & 4 & 0 \\ 2 & 3 & 1 \end{vmatrix}$

$= \hat{i}[4-0] + \hat{j}[0-0] + \hat{k}[0-8] = 4\hat{i} - 8\hat{k}$

60. (*b*) $\frac{P+Q}{P-Q} = \frac{7}{3}, \frac{2P}{2Q} = \frac{10}{4}$ or $\frac{P}{Q} = \frac{5}{2}$

61. (*b*) $\vec{A} + (\hat{i} + 2\hat{j} - \hat{k}) + (\hat{i} - 2\hat{j} + 2\hat{k}) = \hat{k}$ or $\vec{A} = -2\hat{i}$

62. (*a*) $\vec{F} = 10\hat{i} - 3\hat{j} + 6\hat{k}$

$F = \sqrt{10^2 + 3^2 + 6^2} = \sqrt{100 + 9 + 36} = \sqrt{145}$

63. (*a*) $v_{bw} = v_b + (-v_w)$

$= (3\hat{i} + 4\hat{j}) + (3\hat{i} + 4\hat{j}) = 6\hat{i} + 8\hat{j}$

64. (*b*) $\vec{S} = (10\hat{i} - 2\hat{j} + 7\hat{k}) - (6\hat{i} + 5\hat{j} - 3\hat{k}) = 4\hat{i} - 7\hat{j} + 10\hat{k}$

$W = \vec{F}.\vec{S}$

$= (10\hat{i} - 3\hat{j} + 6\hat{k}).(4\hat{i} - 7\hat{j} + 10\hat{k})$

$= (40 + 21 + 60)\text{ J} = 121\text{ J}$

65. (*a*) $P + Q = 18$ ...(*i*)

$P^2 + Q^2 + 2PQ\cos\theta = 144$ ...(*ii*)

$\frac{Q\sin\theta}{P + Q\cos\theta} = \tan 90° = \infty$

$\Rightarrow P + Q\cos\theta = 0$ or $Q\cos\theta = -P$

From equation (*ii*),

$P^2 + Q^2 + 2P(-P) = 144$ or $Q^2 - P^2 = 144$

or $(Q-P)(Q+P) = 144$

or $(Q-P)\,18 = 144$ [From eqn. (*i*)]

or $Q - P = 8$ ...(*iii*)

Adding (*i*) and (*iii*),

$2Q = 26$ or $Q = 13$

Now, $P + 13 = 18$ [From eqn. (*i*)]

or $P = 5$

66. (*a*) Kinetic energy = work done

$= \vec{F}.\vec{S} = (5\hat{i} - \hat{j} + 3\hat{k}).(2\hat{i} + \hat{j} - \hat{k}) = 10 - 1 - 3 = 6$ units

67. (*b*) The required force should be equal in magnitude to the resultant R of 8 N and 12 N

$R^2 = 8^2 + 12^2 + 2 \times 8 \times 12 \times \cos 120°$

or $R^2 = 64 + 144 - 96 = 112$ or $R = \sqrt{112}\text{ N} = 4\sqrt{7}\text{ N}$

68. (*c*) $R^2 = F^2 + F^2 + 2F^2\cos 20° = 2F^2(1 + \cos 20°)$

or $R^2 = 2F^2(1 + 2\cos^2 10° - 1) = 4F^2\cos^2 10°$

or $R = 2F\cos 10°$

69. (*a*) $13^2 = 5^2 + 12^2$

$169 = 169$

This shows that $\vec{P}$ and $\vec{Q}$ are perpendicular to each other.

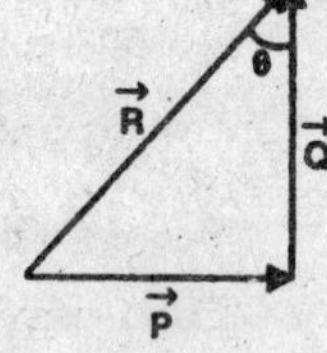

Fig. 30

$\cos\theta = \frac{Q}{R} = \frac{12}{13}$

or $\theta = \cos^{-1}\left(\frac{12}{13}\right)$

70. (*a*) $|\hat{A} - \hat{B}|^2 = (\hat{A} - \hat{B}).(\hat{A} - \hat{B})$

$= \hat{A}.\hat{A} - \hat{A}.\hat{B} - \hat{B}.\hat{A} + \hat{B}.\hat{B}$

$= 1 - \hat{A}.\hat{B} - \hat{A}.\hat{B} + 1 = 2 - 2\cos\theta$

$= 2[1 - \cos\theta] = 2\left[1 - 1 + 2\sin^2\frac{\theta}{2}\right] = 4\sin^2\frac{\theta}{2}$

or $|\hat{A} - \hat{B}| = 2\sin\frac{\theta}{2}$

71. (*b*) We know that when the magnitude of the sum (resultant) of two vectors is equal to the magnitude of either vector, then the angle between the two vectors is 120°.

Difference of unit vectors is

$\sqrt{1^2 + 1^2 - 2 \times 1 \times 1 \times \cos 120°}$

$= \sqrt{1 + 1 - 2\left(-\frac{1}{2}\right)} = \sqrt{3}$

72. (*d*) $\vec{A} + 2\hat{i} + 4\hat{j} - 3\hat{k} + 3\hat{i} - 5\hat{j} + 7\hat{k} = \hat{j}$

or $\vec{A} = -5\hat{i} + 2\hat{j} - 4\hat{k}$

73. (*d*) $\sin\theta = \frac{5}{13}$

or $\theta = 23°$

$\theta' = 90° - 23° = 67°$

$\theta'' = 180° - 67° = 113°$

Fig. 31

74. (*a*) $(0.5)^2 + (0.8)^2 + c^2 = 1$

$0.25 + 0.64 + c^2 = 1$

or $c^2 = 1 - 0.25 - 0.64 = 0.11$

or $c = \sqrt{0.11}$

75. (*b*) $\frac{a+b}{a-b} = \frac{3}{1}$ or $3a - 3b = a + b$

or $2a = 4b$ or $a = 2b$

76. (*b*) Note that the angle between two forces is 120° and not 60°.

$R^2 = F^2 + F^2 + 2F^2 \cos 120°$

or $R^2 = 2F^2 + 2F^2\left(-\frac{1}{2}\right) = F^2$

or $R = F$

77. (*b*) $\vec{v_1} = 50$ km h^{-1} due north

$\vec{v_2} = 50$ km h^{-1} due west

$\Delta v = \sqrt{{v_1}^2 + {v_2}^2} = 50\sqrt{2}$ km h^{-1}

Also, $\vec{\Delta v} = \vec{v_2} + (-\vec{v_1})$

This confirms S-W direction.

78. (*d*) $\cos\theta = \frac{(\hat{i}+\hat{j}).(\hat{i}-\hat{j})}{\sqrt{1^2+1^2}\sqrt{1^2+1^2}} = 0, \therefore \theta = 90°$

79. (*c*) $\tan\beta = \frac{2}{3}$ or $\beta = \tan^{-1}\left(\frac{2}{3}\right)$

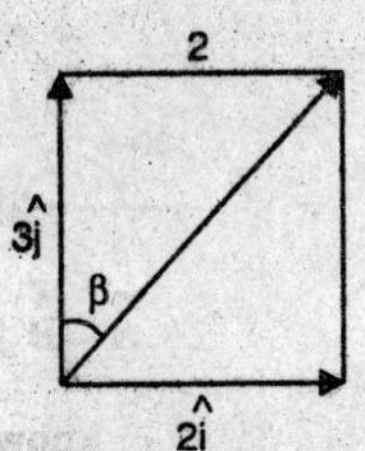

Fig. 32

80. (*c*) $\vec{P} + \vec{Q} = 5\hat{i} - 4\hat{j} + 3\hat{k}$

$\cos\alpha = \frac{5}{\sqrt{5^2+4^2+3^2}}$

$= \frac{5}{\sqrt{50}}$

or $\alpha = \cos^{-1}\left(\frac{5}{\sqrt{50}}\right)$

81. (*a*) $10^2 = v^2 + 8^2$

or $v^2 = 100 - 64 = 36$

or $v = 6$ km h^{-1}

Fig. 33

82. (*d*) $v_r = \sqrt{10^2 + 3^2}$

$= \sqrt{109}$ km h^{-1}

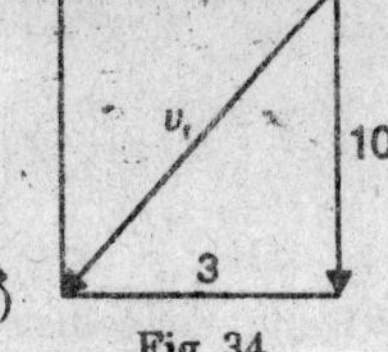

Fig. 34

83. (*c*) $\vec{A} + \vec{B} + \vec{C} = 0$ or $\vec{C} = -(\vec{A} + \vec{B})$

$= -(3\hat{i} + 4\hat{k})$

84. (*c*) $\sqrt{(0.3)^2 + (0.4)^2 + n^2} = 1$ or $0.09 + 0.16 + n^2 = 1$

or $n^2 = 1 - 0.25 = 0.75$ or $n = \sqrt{0.75}$

85. (*c*) $\vec{s} = 4\hat{i} + 3\hat{j} + 12\hat{k}$

$|\vec{s}| = s = \sqrt{4^2 + 3^2 + 12^2}$ m $= \sqrt{169}$ m $= 13$ m

86. (*c*) $10 = 4 + 2 + 4\sqrt{2}\cos\theta$ or $\cos\theta = \frac{4}{\sqrt{2}}$ or $\theta = 45°$

87. (*c*) $\vec{P}$ is in fourth quadrant.

$4\hat{i} + 3\hat{j}$ is in the first quadrant.

Clearly, $4\hat{i} + 3\hat{j}$ can be perpendicular to $\vec{P}$. For confirmation, let us check whether their dot product is zero.

$(3\hat{i} - 4\hat{j}).(4\hat{i} + 3\hat{j}) = 12 - 12 = 0$

This shows that $4\hat{i} + 3\hat{j}$ is perpendicular to $3\hat{i} - 4\hat{j}$

88. (*d*) $\cos\theta = \frac{\vec{A}.\vec{B}}{AB} = \frac{A_xB_x + A_yB_y + A_zB_z}{\sqrt{{A_x}^2 + {A_y}^2 + {A_z}^2}\sqrt{{B_x}^2 + {B_y}^2 + {B_z}^2}}$

$= \frac{(1)(0) + (1)(1) + (0)(1)}{\sqrt{1^2+1^2}\sqrt{1^2+1^2}} = \frac{1}{2} \Rightarrow \theta = 60°$

89. (*a*) $\vec{A} = 2\hat{i} + 2\hat{j}, \vec{B} = \hat{i} + \sqrt{3}\hat{j}$

$\cos\theta = \frac{\vec{A}.\vec{B}}{AB}$

or $\cos\theta = \frac{(2\hat{i}+2\hat{j}).(\hat{i}+\sqrt{3}\hat{j})}{\sqrt{2^2+2^2}\sqrt{1^2+3}} = \frac{2+2\sqrt{3}}{2\times2\sqrt{2}} = \frac{1+\sqrt{3}}{2\sqrt{2}}$

$= \frac{1+1.732}{2\times1.414} = \frac{2.732}{2.828} = 0.966$

$\theta = \cos^{-1}(0.966) = 14.97° \approx 15°$

Alternative method

$\tan\theta_1 = \frac{2}{2} = 1$ or $\theta_1 = 45°$

$\tan\theta_2 = \frac{\sqrt{3}}{1} = \sqrt{3}$ or $\theta_2 = 60°$

$\theta = \theta_2 - \theta_1 = 60° - 45° = 15°$

Note. $\tan\beta = \frac{A_y}{A_x}$

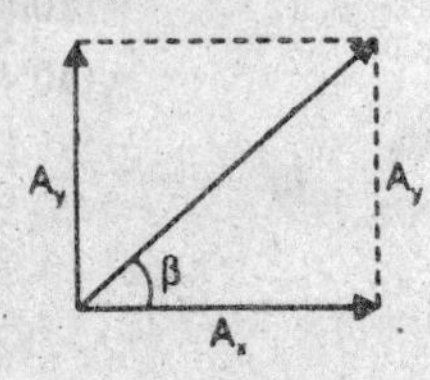

Fig. 35

90. (*a*) $\vec{v} = \begin{vmatrix} \hat{i} & \hat{j} & \hat{k} \\ 1 & -2 & 2 \\ 0 & 4 & -3 \end{vmatrix}$

$= \hat{i}[6-8] + \hat{j}[0+3] + \hat{k}[4-0] = -2\hat{i} + 3\hat{j} + 4\hat{k}$

$|\vec{v}| = v = \sqrt{4+9+16} = \sqrt{29}$ units

91. (*b*) For collision,

$\vec{r_1} + \vec{v_1}t = \vec{r_2} + \vec{v_2}t$

or $\vec{r_1} - \vec{r_2} = (\vec{v_2} - \vec{v_1})t$

Equating unit vectors, we get

$$\frac{\vec{r_1} - \vec{r_2}}{|\vec{r_1} - \vec{r_2}|} = \frac{\vec{v_2} - \vec{v_1}}{|\vec{v_2} - \vec{v_1}|}$$

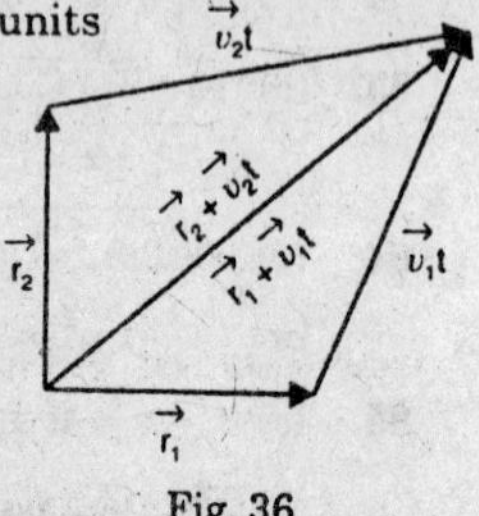

Fig. 36

92. (*c*) $\vec{A} = \hat{i} + 4\hat{j}$

$\vec{B} = 7\hat{i} + 3\hat{j}$

$\vec{A} \times \vec{B} = \begin{vmatrix} \hat{i} & \hat{j} & \hat{k} \\ 1 & 4 & 0 \\ 7 & 3 & 0 \end{vmatrix} = \hat{k}(3-28) = -25\hat{k}$

$|\vec{A} \times \vec{B}| = 25$

$\therefore$ Area of parallelogram = 25 units.

93. (*c*) $x + y = 16$

Also, $y^2 = 8^2 + x^2$

or $y^2 = 64 + (16-y)^2$

$[\because x = 16 - y]$

or $y^2 = 64 + 256 + y^2 - 32y$

or $32y = 320$ or $y = 10$ N

$\therefore$ $x + 10 = 16$ or $x = 6$ N

Fig. 37

94. (*a*) $P = \dfrac{\vec{F}.\vec{S}}{t}$

$= \dfrac{(2\hat{i} + 2\hat{j}).(2\hat{i} + 2\hat{k})}{16}$ J s^{-1} $= \dfrac{4}{16}$ J s^{-1} = 0.25 J s^{-1}

95. (*c*) $\vec{S} = 75\hat{j} + [60 \cos 45°\hat{j} - 60 \sin 45°\hat{i}] + 20\hat{i}$

or $\vec{S} = (20 - 60 \times 0.7)\hat{i} + (60 \times 0.7 + 75)\hat{j}$

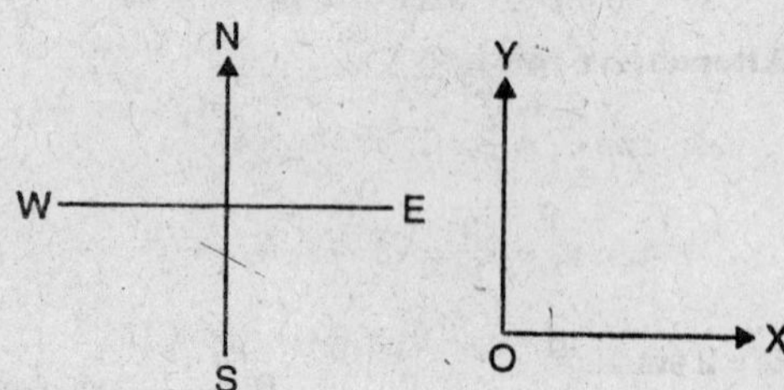

Fig. 38

or $\vec{S} = -22\hat{i} + 117\hat{j}$

$S = \sqrt{22^2 + 117^2} = \sqrt{484 + 13689}$

$= \sqrt{14173} = 119$ km

96. (*b*) $\vec{A} + \vec{B} = 3\hat{i} + \hat{j} + 2\hat{k}$

$\vec{A} - \vec{B} = \hat{i} - 3\hat{j} + 4\hat{k}$

Now, calculate $|\vec{A} \times \vec{B}|$.

97. (*a*) $\vec{A} \times \vec{B} = \begin{vmatrix} \hat{i} & \hat{j} & \hat{k} \\ 2 & p & q \\ 5 & 7 & 3 \end{vmatrix} = 0$

or $\hat{i}(3p - 7q) + \hat{j}(5q - 6) + \hat{k}(14 - 5p) = 0$

$\Rightarrow$ $3p = 7q,\ 5q - 6 = 0$ or $q = \dfrac{6}{5}$

$14 - 5p = 0$ or $5p = 14$ or $p = \dfrac{14}{5}$

98. (*d*) $\vec{\tau} = \vec{r} \times \vec{F} = \begin{vmatrix} \hat{i} & \hat{j} & \hat{k} \\ 7 & 3 & 1 \\ -3 & 1 & 5 \end{vmatrix}$

$= \hat{i}[15-1] + \hat{j}[-3-35] + \hat{k}[7+9] = 14\hat{i} - 38\hat{j} + 16\hat{k}$

99. (*d*) Clearly, R = P + Q

$\therefore$ $\theta = 0°$

$\vec{P}.\vec{Q} = PQ \cos 0° = 6 \times 3 = 18$

100. (*a*) $\tan \beta = \dfrac{Q \sin \theta}{P + Q \cos \theta}$ ...(*i*)

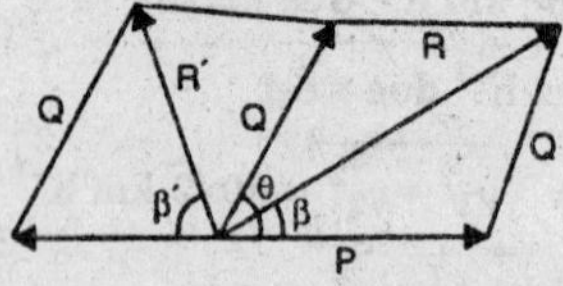

Fig. 39

$\tan \beta' = \dfrac{Q \sin (180 - \theta)}{P + Q \cos (180 - \theta)}$

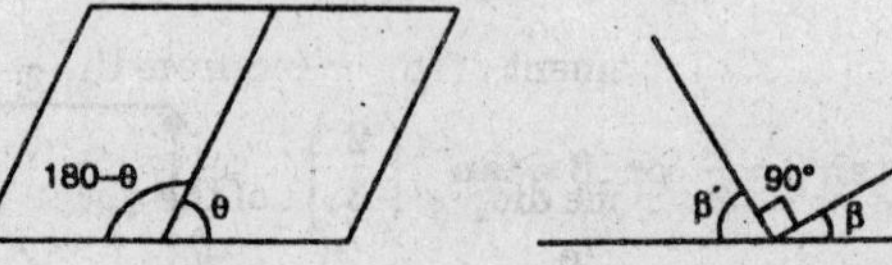

Fig. 40 Fig. 41

or $\tan \beta' = \dfrac{Q \sin \theta}{P - Q \cos \theta}$

But $\beta' + 90° + \beta = 180°$ or $\beta' = 90° - \beta$

$\tan \beta' = \tan (90° - \beta) = \cot \beta$

$\therefore$ $\cot \beta = \dfrac{Q \sin \theta}{P - Q \cos \theta}$ ...(*ii*)

Multiplying (*i*) and (*ii*),

$$\frac{Q^2 \sin^2 \theta}{P^2 - Q^2 \cos^2 \theta} = 1$$

or $Q^2 \sin^2 \theta = P^2 - Q^2 \cos^2 \theta$

or $Q^2(\sin^2 \theta + \cos^2 \theta) = P^2$

But $\sin^2\theta + \cos^2\theta = 1$

$\therefore \quad Q^2 = P^2$ or $Q = P$ | No need for negative sign.

101. (*b*) $\tan 45° = \dfrac{2\sin 60°}{a + 2\cos 60°} = \dfrac{\sqrt{3}}{a+1}$

or $\quad 1 = \dfrac{\sqrt{3}}{a+1}$ or $a + 1 = \sqrt{3}$

or $\quad a = \sqrt{3} - 1$

102-104. $CQ^2 = 2^2 + 2^2$ or $CQ = 2\sqrt{2}'$

This gives the displacement from the centre to the hole.

Again, let P be the point on the front edge from where the queen rebounds.

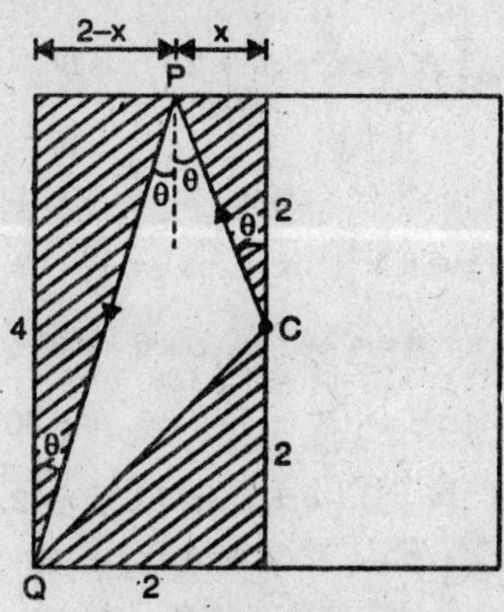

Fig. 42

Now, $\quad \tan\theta = \dfrac{x}{2} = \dfrac{2-x}{4}$ or $4x = 4 - 2x$

or $\quad 6x = 4$ or $x = \dfrac{2'}{3}$

$CP^2 = \left(\dfrac{2}{3}\right)^2 + 2^2$ or $CP^2 = \dfrac{4}{9} + 4 = \dfrac{40}{9}$ or $CP = \dfrac{2}{3}\sqrt{10}'$

This gives the displacement of the queen from the centre to the front edge.

Again, PQ represents the displacement of the queen from the front edge to the hole.

$PQ^2 = 4^2 + \left(2 - \dfrac{2}{3}\right)^2$ or $PQ^2 = 4^2 + \left(\dfrac{4}{3}\right)^2$ or $PQ = \dfrac{4}{3}\sqrt{10}'$

105. (*a*) Suppose $\quad |\vec{a}| = |\vec{b}| = a$

Then $\quad |\vec{c}| = \sqrt{2}a$

Given : $\quad \vec{a} + \vec{b} + \vec{c} = 0$ or $\vec{a} + \vec{b} = -\vec{c}$

or $\quad (\vec{a} + \vec{b}) \cdot (\vec{a} + \vec{b}) = (-\vec{c}) \cdot (-\vec{c})$

or $\quad a^2 + a^2 + 2a^2\cos\theta = 2a^2$ or $\cos\theta = 0$ or $\theta = 90°$

Again, $\quad \vec{b} + \vec{c} = -\vec{a}$

or $\quad (\vec{b} + \vec{c}) \cdot (\vec{b} + \vec{c}) = (-\vec{a}) \cdot (-\vec{a})$

or $\quad a^2 + 2a^2 + 2(a)(\sqrt{2}a)\cos\theta = a^2$

or $\quad \cos\theta = -\dfrac{2a^2}{2\sqrt{2}\,a^2} = -\dfrac{1}{\sqrt{2}}$ or $\theta = 135°$

Similarly, we can work out the third angle.

106. (*b*) $\quad \vec{r} = 3\hat{j} - 2\hat{i} - 5\hat{j}$

or $\quad \vec{r} = -2\hat{j} - 2\hat{i}$

$r = \sqrt{2^2 + 2^2}$

$= 2\sqrt{2}$ km

Fig. 43

107. (*d*) Resultant of P and Q is given to be P.

$\therefore \quad P^2 = P^2 + Q^2 + 2PQ\cos\theta$

or $\quad Q^2 + 2PQ\cos\theta = 0$ or $Q(Q + 2P\cos\theta) = 0$

But $\quad Q \neq 0 \quad \therefore \quad Q + 2P\cos\theta = 0$

If β is the angle which the resultant makes with $\vec{Q}$, then

$$\tan\beta = \frac{2P\sin\theta}{Q + 2P\cos\theta}$$

But $Q + 2P\cos\theta = 0$

$\therefore \quad \tan\beta = \infty$ or $\beta = 90°$

108. (*d*) $(\hat{i} - 2\hat{j} + 3\hat{k}) \times (3\hat{i} - 2\hat{j} + \hat{k})$

$= -2\hat{k} - \hat{j} - 6(-\hat{k}) - 2\hat{i} + 9\hat{j} - 6(-\hat{i}) = 4\hat{i} + 8\hat{j} + 4\hat{k}$

Modulus is $\sqrt{4^2 + 8^2 + 4^2} = \sqrt{32 + 64}$

$= \sqrt{96} = 4\sqrt{6}$ units

109. (*a*) $\quad \vec{\Delta r} = (2\hat{i} + \hat{j} + \hat{k}) - (2\hat{i} - 3\hat{j} + \hat{k}) = 4\hat{j}$

$\vec{L} = \vec{\Delta r} \times \vec{p} = 4\hat{j} \times (2\hat{i} + 3\hat{j} + \hat{k}) = -8\hat{k} + 4\hat{i} = 4\hat{i} - 8\hat{k}$

110. (*a*) $\quad R_x = 1 + 2\cos 120° + 3\cos 240°$

or $\quad R_x = 1 + 2\left(-\dfrac{1}{2}\right) + 3\left(-\dfrac{1}{2}\right) = -\dfrac{3}{2}$

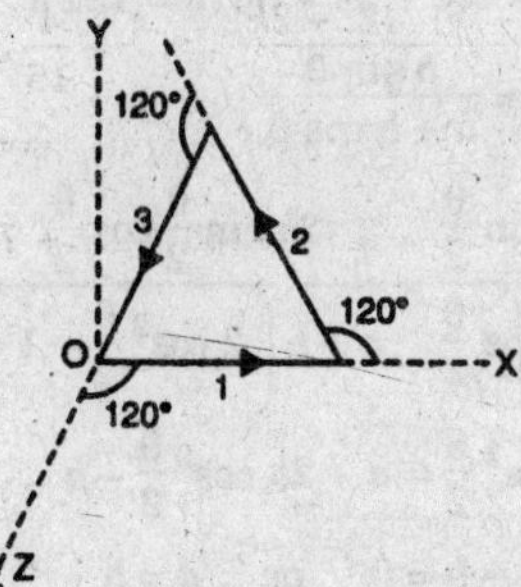

Fig. 44

$R_y = 2\sin 120°$

$+ 3\sin 240°$

$= 2 \times \dfrac{\sqrt{3}}{2} + 3 \times \left(-\dfrac{\sqrt{3}}{2}\right)$

$= -\dfrac{\sqrt{3}}{2}$

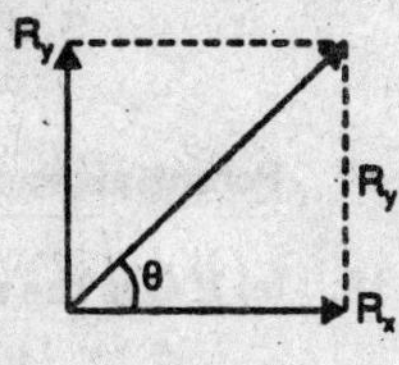

Fig. 45

$$\tan\theta = \frac{R_y}{R_x} = \frac{-\frac{\sqrt{3}}{2}}{-\frac{3}{2}} = \frac{\sqrt{3}}{3} = \frac{1}{\sqrt{3}}$$

$$\theta = \tan^{-1}\left(\frac{1}{\sqrt{3}}\right) = 30^\circ$$

111. (*d*) $\vec{L} = \vec{r}\times\vec{p} = \begin{vmatrix} \hat{i} & \hat{j} & \hat{k} \\ 2 & 1 & 1 \\ 2 & -3 & 1 \end{vmatrix}$

$= \hat{i}[1+3] + \hat{j}[2-2] + \hat{k}[-6-2] = 4\hat{i} - 8\hat{k}$

112. (*d*) $\vec{A}+\vec{B} = 5\hat{i} - 7\hat{k}$

$\vec{A}+\vec{B}$ is in the XZ plane. So, angle with Y-axis is 90°.

113. (*a*) $\vec{A}\times\vec{B} = \begin{vmatrix} \hat{i} & \hat{j} & \hat{k} \\ -3 & 2 & -4 \\ -1 & 2 & 1 \end{vmatrix}$

$= \hat{i}[2+8] + \hat{j}[4+3] + \hat{k}[-6+2] = 10\hat{i} + 7\hat{j} - 4\hat{k}$

$|\vec{A}\times\vec{B}| = \sqrt{10^2+7^2+4^2} = \sqrt{100+49+16} = \sqrt{165}$

Area of triangle $= \frac{1}{2}|\vec{A}\times\vec{B}| = \frac{\sqrt{165}}{2}$

114. (*a*) $u_x = 8 \text{ m s}^{-1}, a_x = 2 \text{ m s}^{-2}$

$\vec{v_x} = \vec{u_x} + \vec{a_x}t = 8\hat{i} + 2t\hat{i}$

$\vec{v_y} = -15\hat{j},\ \vec{V} = \vec{v_x} + \vec{v_y}$

or $\vec{V} = [(8+2t)\hat{i} - 15\hat{j}] \text{ m s}^{-1}$

115. (*b*) $\vec{S} = (10\hat{i} - 2\hat{j} + 7\hat{k}) - (6\hat{i} + 5\hat{j} - 3\hat{k}) = 4\hat{i} - 7\hat{j} + 10\hat{k}$

$W = \vec{F}.\vec{S}$

$= (10\hat{i} - 3\hat{j} + 6\hat{k}).(4\hat{i} - 7\hat{j} + 10\hat{k})$

$= (40 + 21 + 60)\text{ J} = 121\text{ J}$

116. (*a*) $\tan\frac{\theta}{2} = \frac{b\sin\theta}{a + b\cos\theta}$

$$\frac{\sin\frac{\theta}{2}}{\cos\frac{\theta}{2}} = \frac{2b\sin\frac{\theta}{2}\cos\frac{\theta}{2}}{a + b\left(2\cos^2\frac{\theta}{2} - 1\right)}$$

or $2b\cos^2\frac{\theta}{2} = a + 2b\cos^2\frac{\theta}{2} - b$

or $a - b = 0$ or $a = b$

117. (*b*) $V = 2x - 3y + 4z$

$$\frac{\partial V}{\partial x} = 2,\ \frac{\partial V}{\partial y} = -3,\ \frac{\partial V}{\partial z} = 4$$

Potential gradient $= \hat{i}\frac{\partial V}{\partial x} + \hat{j}\frac{\partial V}{\partial y} + \hat{k}\frac{\partial V}{\partial z} = 2\hat{i} - 3\hat{j} + 4\hat{k}$

118. (*c*) If Q is the angle between $\vec{P}$ and $\vec{Q}$, then

$$R^2 = P^2 + Q^2 + 2PQ\cos\theta$$

When Q is doubled,

$$\tan 90^\circ = \frac{2Q\sin\theta}{P + 2Q\cos\theta} \text{ or } \infty = \frac{2Q\sin\theta}{P + 2Q\cos\theta}$$

$\Rightarrow$ $P + 2Q\cos\theta = 0$

Again, $R^2 = P[P + 2Q\cos\theta] + Q^2$ or $R = Q$

119. (*c*) $\cos\theta = \frac{A_x}{A} = \frac{1}{\sqrt{1^2+1^2}} = \frac{1}{\sqrt{2}}$

$\therefore$ $\theta = 45^\circ$

120. (*c*) $\tan\beta = \frac{\sqrt{3}F\sin 90^\circ}{F + \sqrt{3}F\cos 90^\circ} = \frac{\sqrt{3}}{1}$

$\Rightarrow$ $\beta = 60^\circ$

121. (*a*) $\frac{9}{4}A^2 = A^2 + A^2 + 2A^2\cos\theta$

or $\frac{9}{4} = 2 + 2\cos\theta$ or $\frac{9}{4} = 2(1+\cos\theta)$

or $1 + \cos\theta = \frac{9}{8}$ or $\cos\theta = \frac{9}{8} - 1 = \frac{1}{8} = 0.125$

122. (*a*) $AB\sin\theta = 4$ and $AB\cos\theta = 4\sqrt{3}$

$\tan\theta = \frac{4}{4\sqrt{3}} = \frac{1}{\sqrt{3}}$ $\Rightarrow$ $\theta = 30^\circ$

123. (*b*) If $|\vec{A}| = |\vec{B}| = x$, then $|\vec{C}| = \sqrt{2}\,x$

Now, $\vec{A}+\vec{B} = -\vec{C}$

or $(\vec{A}+\vec{B}).(\vec{A}+\vec{B}) = (-\vec{C}).(-\vec{C})$

or $x^2 + x^2 + 2x^2\cos\theta = 2x^2$

or $\cos\theta = 0$ or $\theta = 90^\circ$

Now, $\vec{A}+\vec{C} = -\vec{B}$

or $(\vec{A}+\vec{C}).(\vec{A}+\vec{C}) = (-\vec{B}).(-\vec{B})$

or $\vec{A}.\vec{A} + \vec{C}.\vec{C} + 2\vec{A}.\vec{C} = B^2$

or $x^2 + 2x^2 + 2\sqrt{2}x^2\cos\theta = x^2$

or $\cos\theta = -\frac{2x^2}{2\sqrt{2}\,x^2} = -\frac{1}{\sqrt{2}}$

or $\cos(180^\circ - \theta) = \frac{1}{\sqrt{2}} = \cos 45^\circ$

$\therefore$ $\theta = 180^\circ - 45^\circ = 135^\circ$

Again, $\vec{B}+\vec{C} = -\vec{A}$

or $(\vec{B}+\vec{C}).(\vec{B}+\vec{C}) = (-\vec{A}).(-\vec{A})$

or $x^2 + 2x^2 + 2\sqrt{2}\,x^2\cos\theta = x^2$

or $\cos\theta = -\frac{2x^2}{2\sqrt{2}\,x^2} = -\frac{1}{2}$ or $\theta = 135^\circ$

124. (*c*) $(\vec{a}+\vec{b})\times(\vec{a}-\vec{b}) = \vec{a}\times\vec{a} - \vec{a}\times\vec{b} + \vec{b}\times\vec{a} - \vec{b}\times\vec{b}$

$= -\vec{a}\times\vec{b} - \vec{a}\times\vec{b} = -2(\vec{a}\times\vec{b})$

125. (*c*) You have to try all the options. Let us discuss the correct option only.

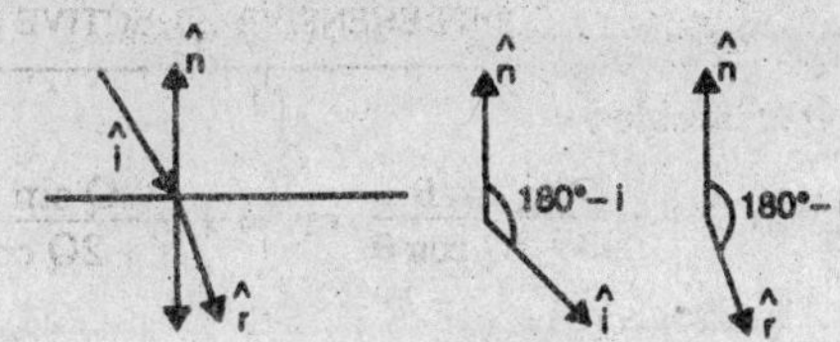

Fig. 46

$$\hat{i} \times \hat{n} = \mu(\hat{r} \times \hat{n})$$

$$(1)(1) \sin (180° - i) = \mu(1)(1) \sin (180° - r)$$

$$\sin i = \mu \sin r$$

126. (c) $P + Q = 16$ ...(i)

$P^2 + Q^2 + 2PQ \cos \theta = 64$ ...(ii)

$$\tan 90° = \frac{Q \sin \theta}{P + Q \cos \theta}$$

$$\infty = \frac{Q \sin \theta}{P + Q \cos \theta}$$

$\Rightarrow$ $P + Q \cos \theta = 0$ or $Q \cos \theta = -P$

From Eqn. (ii),

$P^2 + Q^2 + 2P(-P) = 64$ or $Q^2 - P^2 = 64$

or $(Q - P)(Q + P) = 64$

or $Q - P = \frac{64}{16} = 4$ ...(iii)

Adding Eqns. (i) and (iii), we get

$2Q = 20$ or $Q = 10$ units

From (i), $P + 10 = 16$ or $P = 6$ units

127. (d) The length of the vector is not changed by the rotation of the co-ordinate axes.

$\therefore$ $\sqrt{(n+1)^2 + 1^2} = \sqrt{n^2 + 3^2}$

or $n^2 + 2n + 2 = n^2 + 9$

or $2n = 7$ or $n = \frac{7}{2} = 3.5$

129. (c) $\vec{AB} + \vec{AF} = \vec{AO}$

$\vec{AC} = \vec{AB} + \vec{AO}$

$\vec{AD} = 2\vec{AO}$

$\vec{AE} = \vec{AO} + \vec{AF}$

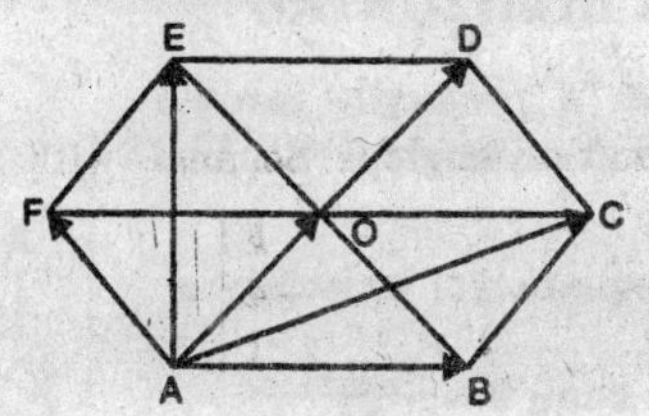

Fig. 47

Adding, $\vec{AB} + \vec{AC} + \vec{AD} + \vec{AE} + \vec{AF}$

$= 5\vec{AO} + \vec{AB} + \vec{AF}$

$= 5\vec{AO} + \vec{AO}$ $[\because \vec{AB} + \vec{AF} = \vec{AO}]$

$= 6\vec{AO}$

3.2 MOTION IN TWO AND THREE DIMENSIONS

Important Terms, Facts and Formulae

IMPORTANT POINTS TO REMEMBER FOR THE CASE OF A PROJECTILE THROWN AT AN ANGLE θ WITH THE HORIZONTAL

Consider a projectile thrown with velocity v at an angle θ with the horizontal.

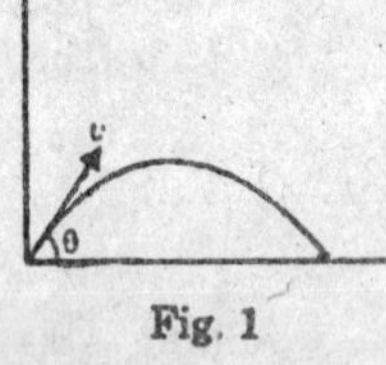

Fig. 1

1. The equation of trajectory is

$$y = x \tan\theta - \frac{gx^2}{2v^2 \cos^2\theta}$$

This is the equation of a parabola.

2. Resultant velocity at any time t is

$$\sqrt{v^2 + g^2t^2 - 2vgt \sin\theta}$$

3. If β is the angle which the resultant velocity makes with the horizontal, then

$$\tan\beta = \frac{v \sin\theta - gt}{v \cos\theta}$$

4. Time of ascent $= \dfrac{v \sin\theta}{g}$

5. Time of descent $= \dfrac{v \sin\theta}{g}$

6. Time of flight, $T = \dfrac{2v \sin\theta}{g}$. **It is largest for $\theta = 90°$.**

7. Maximum height, $h_{max.}$ **or** $H = \dfrac{v^2 \sin^2\theta}{2g}$

8. Maximum height, also known as vertical range, attains the maximum value for $\theta = 90°$.

9. Horizontal range, $R = \dfrac{v^2 \sin 2\theta}{g}$

10. The horizontal range is maximum for an angle of projection of 45°.

11. The maximum value of horizontal range is $\dfrac{v^2}{g}$

12. The horizontal range remains the same whether the projectile is thrown at an angle θ with the horizontal or at an angle of (90° – θ) with the horizontal.

13. The horizontal range remains the same whether the projectile is thrown at an angle θ with the horizontal or at an angle θ with the vertical.

A GENERAL FORM OF FORMULA FOR HORIZONTAL RANGE

$$R = \frac{v^2 \sin 2\theta}{g}$$

or $$R = \frac{2v^2 \sin\theta \cos\theta}{g}$$

or $$R = \frac{2(v \cos\theta)(v \sin\theta)}{g}$$

or $$\boxed{R = \frac{2v_{\parallel}v_{\perp}}{g}}$$

Fig. 2

IMPORTANT FACTUAL INFORMATION FOR THE CASE OF AN OBLIQUE PROJECTILE THROWN AT AN ANGLE θ WITH THE VERTICAL

Let us now consider the case of a projectile thrown with velocity v at an angle θ with the vertical.

Fig. 3

1. $x = v \sin\theta t$

2. $y = v \cos\theta t - \dfrac{1}{2}gt^2$

3. The equation of trajectory is

$$y = x \cot\theta - \frac{gx^2}{2v^2 \sin^2\theta}$$

4. Resultant velocity at any time t is

$$\sqrt{v^2 + g^2t^2 - 2v\, gt \cos\theta}$$

5. If β is the angle which the resultant velocity makes with the horizontal, then

$$\tan\beta = \frac{v \cos\theta - gt}{v \sin\theta}$$

6. Time of ascent $= \dfrac{v \cos \theta}{g}$

7. Time of descent $= \dfrac{v \cos \theta}{g}$

8. Time of flight $T = \dfrac{2v \cos \theta}{g}$. It is largest for $\theta = 0°$

9. Maximum height, $h_{max.}$ or $H = \dfrac{v^2 \cos^2 \theta}{2g}$

10. Horizontal range, $R = \dfrac{v^2 \sin 2\theta}{g}$

BASIC CONCEPTS

1. The horizontal motion of both horizontal and oblique projectiles is uniform motion. So, horizontal velocity remains constant.
2. In the case of horizontal projectile, the vertical velocity increases with time.
3. In the case of oblique projectile, the vertical velocity decreases with time up to the highest point of trajectory. Beyond the highest point, the vertical velocity begins to increase with time.
4. If acceleration due to gravity is not the same at all points of the trajectory of the projectile, then the trajectory cannot be a symmetric parabola.
5. If angle of projection with the horizontal is 90°, then the trajectory is a vertical straight line.
6. At the highest point of the trajectory of an oblique projectile, the velocity and acceleration are at right angles to each other.
7. The time taken by a projectile, thrown with velocity v at an angle θ with horizontal, to reach the maximum height is equal to the time taken by the body, projected vertically upwards with a velocity $v \sin \theta$, to reach the highest point.

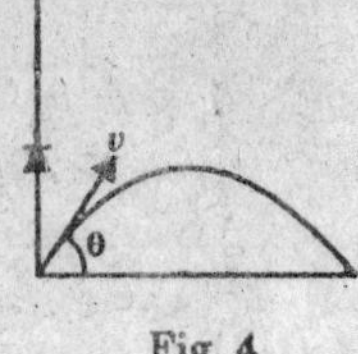

Fig. 4

8. When a body is projected so that its horizontal range R is maximum, then the maximum height attained by the projectile is $\dfrac{R}{4}$.
9. When a body is projected horizontally from the top of a tower, the time taken by it to reach the ground does not depend upon the velocity of projection. It depends upon the height of the tower.
10. The time of flight and maximum height of an oblique projectile depend upon the vertical component of the velocity of projection.
11. The trajectory of a projectile is a parabola only when the acceleration is constant. Also, the direction of acceleration should be different from the direction of velocity.
12. If velocity of projection is large, then acceleration due to gravity cannot be taken as constant in magnitude and direction. In this case, the path will be a conic section.
13. If velocity of projection is equal to orbital velocity, then the trajectory will be circular.
14. If the velocity of projection is greater than orbital velocity but less than escape velocity, then the trajectory will be elliptical.
15. If the velocity of projection is greater than escape velocity, then the trajectory is a hyperbola.

QUESTION BANK

Level I

Choose the correct answer from the four alternatives given in each of the following questions :

1. If R is the range for an angle of projection of 15° with the horizontal, then the other angle of projection for which the range is R, is

   (*a*) 75° (*b*) 60°
   (*c*) 45° (*d*) 30°.

2. An object is thrown along a direction inclined at an angle of 45° with the horizontal direction. The horizontal range of the particle is

   (*a*) four times the vertical height
   (*b*) thrice the vertical height
   (*c*) twice the vertical height
   (*d*) equal to vertical height.

3. A projectile thrown with a velocity v at an angle θ has a range R on the surface of earth. For same v and θ, its range on the surface of moon will be

   (*a*) 36 R (*b*) R/36
   (*c*) R/6 (*d*) 6R.

4. If air resistance is ignored, then the horizontal motion of the oblique projectile takes place at

   (*a*) uniform acceleration (*b*) variable acceleration
   (*c*) uniform retardation (*d*) uniform velocity.

5. A ball is projected with kinetic energy K at an angle of 45° to the horizontal. At the highest point during its flight, its kinetic energy will be

   (*a*) K (*b*) $K/\sqrt{2}$
   (*c*) K/2 (*d*) zero.

6. Two anti-aircraft guns fire two shells, one at an angle of 20° with the horizontal and the other at an angle of 70° with the horizontal. If muzzle velocity is same in both the cases, then

   (*a*) the first shell has a smaller horizontal range
   (*b*) the second shell has a smaller horizontal range
   (*c*) both the shells have the same vertical range
   (*d*) both the shells have the same horizontal range.

7. An aeroplane flying horizontally with a speed of 360 km h^{-1} releases a bomb at a height of 490 m from the ground. When will the bomb strike the ground ?

   (*a*) 8 s (*b*) 6 s
   (*c*) 7 s (*d*) 10 s.

8. In Q. 7, the bomb will strike the ground at a distance of

   (*a*) 50 m (*b*) 100 m
   (*c*) 1000 m (*d*) 267.9 m.

9. A hunter takes an aim at a monkey sitting on a tree and fires a bullet. Just when the bullet leaves the barrel of the gun, it so happens that the monkey begins to fall freely. The bullet will

   (*a*) go above the monkey
   (*b*) go below the monkey
   (*c*) may or may not hit the monkey. It will depend upon the velocity of the bullet
   (*d*) hit the monkey.

10. If the velocity of projection of a projectile is trebled, then its maximum range will be

    (*a*) quadrupled (*b*) nine times
    (*c*) six times (*d*) eight times.

11. The time of flight of a projectile is maximum when angle of projection is

    (*a*) 30° (*b*) 45°
    (*c*) 60° (*d*) 90°.

12. If air resistance and air buoyancy are to be ignored, then which of the following factors determines the time of flight of the projectile ?

    (*a*) initial velocity
    (*b*) horizontal component of initial velocity
    (*c*) vertical component of initial velocity
    (*d*) horizontal component of acceleration.

13. A body is thrown with a velocity of 10 m s^{-1} at an angle of 60° with the horizontal. Its velocity at the highest point is

    (*a*) 7 m s^{-1} (*b*) 9 m s^{-1}
    (c) 18.7 m s^{-1} (*d*) 5 m s^{-1}.

14. A ball is projected upwards. Its acceleration at the highest point is

    (*a*) infinite (*b*) zero
    (c) directed upwards (*d*) directed downwards.

15. Two projectiles are thrown with the same initial velocity at angles α and (90° – α) with the horizontal. The maximum heights attained by them are h_1 and h_2 respectively. Then $\frac{h_1}{h_2}$ is equal to

(a) $\sin^2 \alpha$ (b) $\cos^2 \alpha$
(c) $\tan^2 \alpha$ (d) 1.

16. The greatest height to which a man can throw a ball is h. What is the greatest distance to which he can throw the ball ?

(a) $h/4$ (b) $h/2$
(c) h (d) $2h$.

17. An oblique projectile has the smallest speed on its trajectory when

(a) it is at the highest point
(b) it is at the point of projection
(c) it hits the ground
(d) the horizontal distance travelled by it is one-eighth of the maximum height.

18. Two bodies are projected at angles θ and $(90 - \theta)$ with the horizontal at the same speed. The ratio of their maximum heights is

(a) 1 : 1 (b) $1 : \tan \theta$
(c) $\tan \theta : 1$ (d) $\tan^2 \theta : 1$.

19. A body is projected at an angle of 30° with the horizontal with momentum p. At its highest point, the momentum is

(a) p (b) $\frac{p}{2}$
(c) $\frac{\sqrt{3}}{2} p$ (d) $\frac{2}{\sqrt{3}} p$.

20. It was calculated that a shell when fired from a gun with a certain velocity and at an angle of elevation $\frac{5\pi}{36}$ rad should strike a given target. In actual practice, it was found that a hill just prevented the trajectory. At what angle of elevation should the gun be fired to hit the target ?

(a) $\frac{5\pi}{36}$ rad (b) $\frac{11\pi}{36}$ rad
(c) $\frac{7\pi}{36}$ rad (d) $\frac{13\pi}{36}$ rad.

21. Two balls are projected from the same point in directions at 30° and 60° with the horizontal. Both the balls attain the same height. The ratio of their velocities of projection is

(a) $\sqrt{3} : 2$ (b) $\sqrt{3} : 3$
(c) $\sqrt{3} : 5$ (d) $\sqrt{3} : 1$.

22. A glass marble projected horizontally from the top of a table falls at a distance x from the edge of the table. If h is the height of the table, then the velocity of projection is

(a) $h\sqrt{\frac{g}{2x}}$ (b) $x\sqrt{\frac{g}{2h}}$
(c) gxh (d) $gx + h$.

23. In the case of an oblique projectile, the velocity is perpendicular to acceleration

(a) once only (b) twice
(c) thrice (d) four times.

QUESTION BANK

Level II

Choose the correct answer from the four alternatives given in each of the following questions :

1. Two projectiles thrown from the same point at angles 60° and 30° with the horizontal attain the same height. The ratio of their initial velocities is

(a) 1 (b) 2
(c) $\sqrt{3}$ (d) $\frac{1}{\sqrt{3}}$.

2. A bomb is dropped on an enemy post by an aeroplane flying horizontally with a velocity of 60 km h^{-1} and at a height of 490 m. At the time of dropping the bomb, how far the aeroplane should be from the enemy post so that the bomb may directly hit the target ?

(a) $\frac{400}{3}$ m (b) $\frac{500}{3}$ m
(c) $\frac{1700}{3}$ m (d) 498 m.

3. A ball rolls off the top of a stairway horizontally with a velocity of 4.5 m s^{-1}. Each step is 0.2 m high and 0.3 m wide. If g is 10 m s^{-2}, then the ball will strike the nth step where n is equal to

(a) 9 (b) 10
(c) 11 (d) 12.

4. A body is projected at 30° angle with the horizontal with velocity 30 m s^{-1}. What is the angle with the horizontal after 1.5 s ? (Take g = 10 m s^{-2}).

(a) 0° (b) 30°
(c) 60° (d) 90°.

5. A body of mass 1 kg is thrown with a velocity of 10 m s^{-1} at an angle of 60° with the horizontal. Its momentum at the highest point is
(a) 2 kg m s^{-1} (b) 3 kg m s^{-1}
(c) 4 kg m s^{-1} (d) 5 kg m s^{-1}.

6. An aeroplane flying horizontally with a velocity of 216 km h^{-1} drops a food packet while flying at a height of 490 m. The total horizontal distance travelled by the packet is
(a) 600 m (b) 490 m
(c) 216 m (d) 490 × 216 m.

7. The x and y coordinates of a particle at any time t are given by $x = 2t + 4t^2$ and $y = 5t$, where x and y are in metre and t in second. The acceleration of the particle at $t = 5$ s is
(a) 40 m s^{-2} (b) 20 m s^{-2}
(c) 8 m s^{-2} (d) zero.

8. A body is projected at an angle θ to the horizontal with kinetic energy E. Neglecting air friction, the kinetic energy of the body at the topmost point of the path will be
(a) zero (b) E cos θ
(c) E $\cos^2 \theta$ (d) E.

9. A helicopter is flying horizontally at an altitude of 2 km with a speed of 100 m s^{-1}. A packet is dropped from it. The horizontal distance between the point where the packet is dropped and the point where it hits the ground is (g = 10 m s^{-2})
(a) 2 km (b) 0.2 km
(c) 20 km (d) 4 km.

10. A body is projected from the ground at an angle of 60° with the horizontal. The ratio of the maximum height attained by it to its horizontal range is
(a) $\sqrt{3} : 1$ (b) $\sqrt{3} : 2$
(c) $1 : \sqrt{3}$ (d) $\sqrt{3} : 4$.

11. A monkey can jump a maximum horizontal distance of 20 m. Then the velocity of the monkey is
(a) 10 m s^{-1} (b) 14 m s^{-1}
(c) 20 m s^{-1} (d) 24 m s^{-1}.

12. A bullet is fired horizontally with a velocity of 80 m s^{-1}. During the first second,
(a) it falls 9.8 m (b) it falls $\frac{80}{9.8}$ m
(c) it does not fall at all (d) it falls 4.9 m.

13. A cricket ball is hit with a velocity 25 m s^{-1}, 60° above the horizontal. How far above the ground, ball passes over a fielder 50 m from the bat (consider the ball is struck very close to the ground) Take $\sqrt{3}$ = 1.7 and g = 10 m s^{-2}.
(a) 6.8 m (b) 7 m
(c) 5 m (d) 10 m.

14. A projectile is projected with kinetic energy of 800 J. If it has the maximum possible horizontal range, then its kinetic energy at the highest point will be
(a) 800 J (b) 400 J
(c) 200 J (d) 100 J.

15. A stone is thrown with a velocity v making an angle θ with the horizontal. At some instant, its velocity V is perpendicular to the initial velocity v. Then V is
(a) v sin θ (b) v cos θ
(c) v tan θ (d) v cot θ.

16. A body is projected at an angle of 30° with the horizontal with kinetic energy E. The kinetic energy at the topmost point is
(a) E (b) $\frac{E}{2}$
(c) $\frac{E}{4}$ (d) $\frac{3E}{4}$.

17. A bomb is dropped from an aeroplane flying horizontally with a velocity 469 m s^{-1} at an altitude of 980 m. The bomb will hit the ground after a time
(a) 2 s (b) $\sqrt{2}$ s
(c) $5\sqrt{2}$ s (d) $10\sqrt{2}$ s.

18. A projectile is thrown at angle β with vertical. It reaches a maximum height H. The time taken to reach the highest point of its path is
(a) $\sqrt{\frac{H}{g}}$ (b) $\sqrt{\frac{2H}{g}}$
(c) $\sqrt{\frac{H}{2g}}$ (d) $\sqrt{\frac{2H}{g \cos \beta}}$

19. Two projectiles, one fired from earth with 5 m s^{-1} and the other fired from a planet with 3 m s^{-1} trace identical trajectories. If acceleration due to gravity on earth is 9.8 m s^{-2}, then the acceleration due to gravity on the planet is
(a) 1.5 m s^{-2} (b) 3.5 m s^{-2}
(c) 7.5 m s^{-2} (d) 9.5 m s^{-2}.

20. A plane flying horizontally at 98 m s^{-1} releases an object which reaches the ground in 10 second. The angle made by the velocity of the object with the horizontal at the time of hitting the ground is

(*a*) 30° (*b*) 45°
(*c*) 60° (*d*) 75°.

21. The equation of a projectile is $y = \sqrt{3}x - \frac{gx^2}{2}$

The angle of projection is given by

(*a*) $\tan\theta = \frac{1}{\sqrt{3}}$ (*b*) $\tan\theta = \sqrt{3}$

(*c*) $\frac{\pi}{2}$ (*d*) zero.

22. In Q. 21, the initial velocity is

(*a*) 4 m s^{-1} (*b*) 2 m s^{-1}
(*c*) 1 m s^{-1} (*d*) 5 m s^{-1}.

23. An arrow is shot into air. Its range is 200 m and its time of flight is 5 second. If g = 10 m s^{-2}, then the horizontal component of the velocity of arrow is

(*a*) 12.5 m s^{-1} (*b*) 25 m s^{-1}
(*c*) 31.25 m s^{-1} (*d*) 40 m s^{-1}.

24. In Q. 23, the maximum height attained by the arrow is

(*a*) 12.5 m (*b*) 25 m
(*c*) 31.25 m (*d*) 40 m.

25. A large number of bullets are fired in all directions with the same speed v. What is the maximum area on the ground on which these bullets will spread ?

(*a*) $\frac{\pi v^2}{g}$ (*b*) $\frac{\pi v^4}{g^2}$

(*c*) $\frac{\pi^2 v^4}{g^2}$ (*d*) $\frac{\pi^2 v^2}{g^2}$.

26. A projectile of mass m is thrown with a velocity v making an angle of 45° with the horizontal. The change in momentum from departure to arrival along vertical direction, is

(*a*) $2mv$ (*b*) $\sqrt{2}\, mv$

(*c*) mv (*d*) $\frac{mv}{\sqrt{2}}$.

27. A body is thrown horizontally with a velocity $\sqrt{2gh}$ from the top of a tower of height h. It strikes the level ground through the foot of the tower at a distance x from the tower. The value of x is

(*a*) h (*b*) $\frac{h}{2}$

(*c*) $2h$ (*d*) $\frac{2h}{3}$.

28. A particle having a mass of 0.5 kg is projected under gravity with a speed of 98 m s^{-1} at an angle of 60°. The magnitude of the change of momentum of the particle after 10 s is

(*a*) 0.5 N s (*b*) 49 N s
(*c*) 98 N s (*d*) 490 N s.

29. Two stones thrown at different angles have same initial velocity and same range. If H is the maximum height attained by one stone thrown at an angle of 30°, then the maximum height attained by the other stone is

(*a*) $\frac{H}{2}$ (*b*) H

(*c*) 2H (*d*) 3H.

30. The maximum range of a gun on a horizontal terrain is 4 km. The muzzle velocity of the shell is (Given : g = 10 m s^{-2}).

(*a*) 200 m s^{-1} (*b*) 300 m s^{-1}
(*c*) 400 m s^{-1} (*d*) Data is incomplete.

31. If the air resistance causes a vertical retardation of 10% of value of acceleration due to gravity, then the time of flight of an oblique projectile will be decreased by nearly

(*a*) 6% (*b*) 7%
(*c*) 8% (*d*) 9%.

32. A projectile shot into air at some angle with the horizontal has a range of 200 m. If the time of flight is 5 second, then the horizontal component of the velocity of the projectile at the highest point of trajectory is

(*a*) 40 m s^{-1} (*b*) 0 m s^{-1}
(*c*) 9.8 m s^{-1}
(*d*) equal to the velocity of projection of the projectile.

33. A projectile is thrown horizontally with a speed of 20 m s^{-1}. If g is 10 m s^{-2}, then the speed of the projectile after 5 second will be nearly

(*a*) 0.5 m s^{-1} (*b*) 5 m s^{-1}
(*c*) 54 m s^{-1} (*d*) 500 m s^{-1}.

34. If the angle of projection of a projectile is 30°, then how many times the horizontal range is larger than the maximum height ?

(*a*) 2 (*b*) 3
(*c*) $3\sqrt{4}$ (*d*) $4\sqrt{3}$.

35. The momentum mv of a projectile at the point of projection makes an angle of 45° with the vertical. If air is assumed to be absent, then the change in momentum, along vertical, from the point of projection to the point where the projectile returns to the plane of projection is given by

(*a*) $\sqrt{2}\, mv$ (*b*) $\sqrt{3}\, mv$
(*c*) $\sqrt{6}\, mv$ (*d*) zero.

36. The kinetic energy of a projectile at the highest point is half of the initial kinetic energy. What is the angle of projection with the horizontal ?

(*a*) 30° (*b*) 45°
(*c*) 60° (*d*) 90°.

37. A ball is projected from a certain point on the surface of a planet at a certain angle with the horizontal surface. The horizontal and vertical displacements x and y vary with time t in second as :

$$x = 10\sqrt{3}t \text{ and } y = 10t - t^2$$

The maximum height attained by the ball is

(*a*) 100 m (*b*) 75 m
(*c*) 50 m (*d*) 25 m.

38. A projectile A is thrown at an angle of 30° to the horizontal from point P. At the same time, another projectile B is thrown with velocity v^2 upwards from the point Q vertically below the highest point. For B to collide with A, $\frac{v_2}{v_1}$ should be

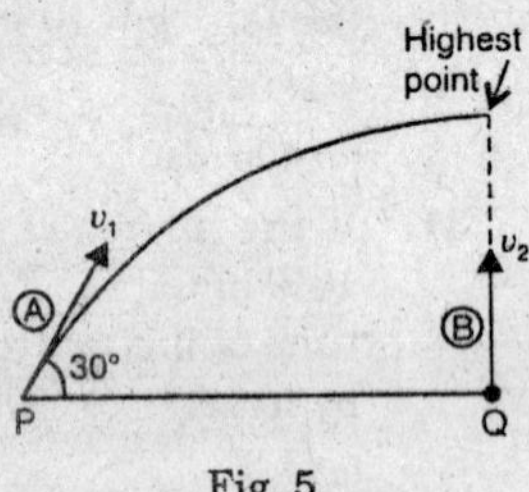

Fig. 5

(*a*) 1 (*b*) 2
(*c*) $\frac{1}{2}$ (*d*) 4.

39. For a projectile thrown into space with a speed v, the horizontal range is $\frac{\sqrt{3}v^2}{2g}$. The vertical range is $\frac{v^2}{8g}$. The angle which the projectile makes with the horizontal initially is

(*a*) 15° (*b*) 30°
(*c*) 45° (*d*) 60°.

40. If the height attained by a projectile is the greatest, then which of the following is the maximum ?

(*a*) Horizontal range
(*b*) Angle of projection with the vertical
(*c*) Force of gravity
(*d*) Time of flight.

41. The velocity of projection of an oblique projectile is $(6\hat{i} + 8\hat{j})$ m s^{-1}. The horizontal range of the projectile is

(*a*) 4.9 m (*b*) 9.6 m
(*c*) 19.6 m (*d*) 14 m.

42. A ball is projected with a velocity v so that its range on a horizontal plane is twice the greatest height attained. Then, its range is

(*a*) $\frac{5}{4}\frac{v^2}{g}$ (*b*) $\frac{v^2}{g}$
(*c*) $\frac{4v^2}{5g}$ (*d*) $\frac{2v^2}{g}$.

43. A body has an initial velocity of 3 m s^{-1} and has an acceleration of 1 m s^{-2} normal to the direction of the initial velocity. Then its velocity, 4 second after the start is

(*a*) 7 m s^{-1} along the direction of initial velocity
(*b*) 7 m s^{-1} along the normal to the direction of the initial velocity
(*c*) 7 m s^{-1} mid-way between the two directions
(*d*) 5 m s^{-1} at an angle of $\tan^{-1}\frac{4}{3}$ with the direction of the initial velocity.

44. A body is projected at an angle q to the horizontal with kinetic energy E_k. The potential energy at the highest point of the trajectory is

(*a*) E_k (*b*) $E_k \cos^2 \theta$
(*c*) $E_k \sin^2 \theta$ (*d*) $E_k \tan^2 \theta$.

45. A cricket ball is hit at 30° with the horizontal with kinetic energy E_k. What is the kinetic energy at the highest point ?

(*a*) $E_k/2$ (*b*) $3E_k/4$
(*c*) $E_k/4$ (*d*) zero.

46. An aeroplane moving horizontally at a speed of 200 m s^{-1} and at a height of 8×10^3 m is to drop a bomb on a target. At what horizontal distance from the target should the bomb be released ? (Take g = 10 m s^{-2}).

(*a*) 9124 m (*b*) 8714 m
(*c*) 8000 m (*d*) 7234 m.

47. The range of a projectile fired at an angle of 15° is 40 m. If it is fired with the same speed at an angle of 45°, its range will be

(*a*) 60 m (*b*) 70 m
(*c*) 80 m (*d*) 100 m.

48. A stone of mass m is thrown vertically upwards. Another stone of mass $2m$ is thrown at an angle θ with the vertical. Both of them stay in air for same period of time. The heights attained by the two stones are in the ratio of

(*a*) 1 : 1 (*b*) 2 : 1
(*c*) 1 : cos θ (*d*) cos θ : 1.

49. A particle can be projected with a given speed in two possible ways so as to make it pass through a point at a distance r from the point of projection. The product of the times taken to reach this point in the two possible ways is then proportional to

(a) r (b) $1/r$

(c) $1/r^2$ (d) $1/r^3$.

50. A stone is thrown with a velocity v at an angle θ with the horizontal. Its speed when it makes an angle β with the horizontal is

(a) $v \cos\theta$ (b) $\frac{v}{\cos\beta}$

(c) $v \cos\theta \cos\beta$ (d) $\frac{v\cos\theta}{\cos\beta}$.

51. A small ball suspended from a string is set into oscillation. When the ball passes through the lowest point of the motion, the string is cut. If the ball is then moving with the velocity 0.8 m s^{-1} at a height of 5 m above the ground, the horizontal distance travelled by the ball is (Given : g = 10 m s^{-2})

(a) 0.2 m (b) 0.4 m

(c) 0.6 m (d) 0.8 m.

52. Two projectiles A and B are thrown with velocities v and $\frac{v}{2}$ respectively. They have the same range. If B is thrown at an angle of 15° to the horizontal, A must have been thrown at an angle

(a) $\sin^{-1}\left(\frac{1}{16}\right)$ (b) $\sin^{-1}\left(\frac{1}{4}\right)$

(c) $2\sin^{-1}\left(\frac{1}{4}\right)$ (d) $\frac{1}{2}\sin^{-1}\left(\frac{1}{8}\right)$.

53. A ball is projected horizontally at 20 m s^{-1}. The approximate speed of the ball after 5 second is (Take g = 10 m s^{-2})

(a) 24 m s^{-1} (b) 34 m s^{-1}

(c) 44 m s^{-1} (d) 54 m s^{-1}.

54. A man can throw a stone 100 m away. The maximum height to which he can throw vertically is

(a) 200 m (b) 100 m

(c) 50 m (d) 25 m.

55. A ball is projected horizontally with a velocity of 4 m s^{-1}. The velocity of the ball after 0.7 s (g = 10 m s^{-2}) is nearly

(a) 11 m s^{-1} (b) 10 m s^{-1}

(c) 8 m s^{-1} (d) 3 m s^{-1}.

56. Two balls are thrown with the same speed but making complementary angles with the horizontal. If the angle of projection of one is 30° and its maximum height is h, then the maximum height of the other will be

(a) $\frac{h}{3}$ (b) $2h$

(c) $3h$ (d) $7h$.

57. When a projectile is moving at 60 m s^{-1} at the highest point of its trajectory, it explodes into two equal parts. One part moves vertically up with a velocity of 50 m s^{-1}. The magnitude of velocity of other part is

(a) 50 m s^{-1} (b) 60 m s^{-1}

(c) 120 m s^{-1} (d) 130 m s^{-1}.

58. A ball thrown by one player to another is in air for two second. The maximum height attained by the ball is (g = 10 m s^{-2})

(a) 2 m (b) 5 m

(c) 7 m (d) 9 m.

59. A javelin thrown at an angle of 45° to the horizontal has range of 32.4 m (g = 10 m s^{-2}).

(a) Its velocity at the highest point is zero.

(b) Its velocity at the highest point is nearly 13 m s^{-1}.

(c) 32.4 m is not the maximum horizontal range.

(d) For no other angle of projection, the maximum height will be greater.

60. The horizontal range of an oblique projectile is equal to the distance through which a projectile has to fall freely from rest to acquire a velocity equal to the velocity of projection in magnitude. The angle of projection is

(a) 75° (b) 60°

(c) 45° (d) 30°.

61. A particle starts from the origin at t = 0. It moves in a plane with a velocity given by : $\vec{v} = v_0\hat{i} + (a\omega\cos\omega t)\hat{j}$. The equation of trajectory of the particle is

(a) $y = a\sin\omega t$ (b) $y = a\cos\omega t$

(c) $y = a\sin\left(\frac{\omega x}{v_0}\right)$ (d) $y = a\cos\left(\frac{\omega x}{v_0}\right)$.

62. A projectile is thrown with velocity v making an angle q with the horizontal. It just crosses the tops of two poles, each of height h, after 1 second and 3 second respectively. The time of flight of the projectile is

(a) 1 s (b) 3 s

(c) 4 s (d) 7.8 s.

63. In Q. 62, maximum height attained by the projectile is

(a) g (b) $2g$

(c) $3g$ (d) $4g$.

64. In Q. 62, h is

(a) $\frac{g}{2}$ (b) g

(c) $\frac{3g}{2}$ (d) $2g$.

65. A projectile is thrown with velocity v at an angle θ with the horizontal. When the projectile is at a height equal to half of the maximum height, then the vertical component of the velocity of projectile is

(a) $v \sin\theta \times 3$ (b) $v \sin\theta + 3$

(c) $\frac{v \sin\theta}{\sqrt{2}}$ (d) $\frac{v \sin\theta}{\sqrt{3}}$.

66. In Q. 65, the velocity of the projectile when it is at a height equal to half of the maximum height is

(a) $v\sqrt{\cos^2\theta + \frac{\sin^2\theta}{2}}$ (c) $\sqrt{2}v\cos\theta$

(c) $\sqrt{2}v\sin\theta$ (d) $v\tan\theta\sec\theta$.

67. In Q. 65, what is the angle of projection with the vertical if the velocity at the highest point is $\sqrt{\frac{2}{5}}$ times the velocity when it is at a height equal to half of the maximum height ?

(a) 15° (b) 30°

(c) 45° (d) 60°.

68. The path of one projectile in motion as seen from another moving projectile is

(a) a straight line (b) a circle

(c) an ellipse (d) a parabola.

69. A ball is projected upwards from the top of a tower with a velocity of 50 m s^{-1} making an angle of 60° with the vertical. If the height of the tower is 70 m, then the ball will reach the ground in (Take g = 10 m s^{-2})

(a) 2 s (b) 3 s

(c) 5 s (d) 7 s.

70. A stone is projected from the ground with velocity 50 m s^{-1} and angle of 30°. It crosses a wall after 3 s. How far beyond the wall the stone will strike the ground ? (Take g = 10 m s^{-2})

(a) 80.5 m (b) 85.6 m

(c) 86.6 m (d) 75.2 m.

71. A plane surface is inclined making an angle θ with the horizontal. From the bottom of this inclined plane, a bullet is fired with velocity v. The maximum possible range of the bullet on the inclined plane is

(a) $\frac{v^2}{g}$ (b) $\frac{v^2}{g(1+\sin\theta)}$

(c) $\frac{v^2}{g(1-\sin\theta)}$ (d) $\frac{v^2}{g(1+\cos\theta)}$.

72. A particle is projected horizontally with a speed of 14 m s^{-1} from a point O at time $t = 0$. The displacement of the particle from O at $t = 2\frac{6}{7}$ s is

(a) 9.8 m (b) 40 m

(c) $40\sqrt{2}$ m (d) $40\sqrt{3}$ m.

73. A body is projected horizontally with speed 20 m s^{-1}. The approximate displacement of the body after 5 second is

(a) 80 m (b) 120 m

(c) 160 m (d) 320 m.

74. If the time of flight of a bullet over a horizontal range R is T, then the inclination of the direction of projection with the horizontal is

(a) $\tan^{-1}\left[\frac{gT^2}{2R}\right]$ (b) $\tan^{-1}\left[\frac{2R^2}{gT}\right]$

(c) $\tan^{-1}\left[\frac{2R}{g^2\pi}\right]$ (d) $\tan^{-1}\left[\frac{2R}{gT}\right]$.

75. A ball is projected horizontally with a speed v from the top of a plane inclined at an angle 45° with the horizontal. How far from the point of projection will the ball strike the plane ?

(a) $\frac{v^2}{g}$ (b) $\sqrt{2}\frac{v^2}{g}$

(c) $\frac{2v^2}{g}$ (d) $\sqrt{2}\left[\frac{2v^2}{g}\right]$.

76. A projectile is fired with a velocity v at an angle θ with the horizontal. The speed of the projectile when its direction of motion makes an angle β with the horizontal is

(a) $v\cos\theta$ (b) $v\cos\theta\cos\beta$

(c) $v\cos\theta\sec\beta$ (d) $v\cos\theta\tan\beta$.

77. A ball rolls off the top of a stairway with horizontal velocity v_0 m s^{-1}. If the steps are h metre high and w metre wide, the ball will hit the edge of nth step if

(a) $n = \frac{2hv_0}{gw^2}$ (b) $n = \frac{2hv_0^2}{gw}$

(c) $n = \frac{hv_0^2}{gw^2}$ (d) $n = \frac{2hv_0^2}{gw^2}$.

78. With a given velocity, a projectile can have the same range R for two angles of projection. If t_1 and t_2 be the times of flight in the two cases, then what is the product of the two times of flight ?

(a) $\propto R^2$ (b) $\propto R$

(c) $\propto \frac{1}{R}$ (d) $\propto \frac{1}{R^2}$.

79. A body of mass m thrown horizontally with velocity v, from the top of tower of height h touches the level ground at a distance of 250 m from the foot of the tower. A body of mass 2 m thrown horizontally with velocity

$\frac{v}{2}$, from the top of tower of height $4h$ will touch the level ground at a distance x from the foot of tower. The value of x is

(a) 250 m (b) 500 m

(c) 125 m (d) $250\sqrt{2}$ m.

80. A projectile is thrown in the upward direction making an angle of 60° with the horizontal with a velocity of 147 m s^{-1}. Then the time after which its inclination with horizontal is 45° is

(a) 15 s (b) 10.98 s

(c) 5.49 s (d) 2.745 s.

81. From the top of a tower of height 40 m, a ball is projected upwards with a speed of 20 m s^{-1} at an angle of elevation of 30°. Then the ratio of the total time taken by the ball to hit the ground to its time of flight is

(a) 2 : 1 (b) 3 : 1

(c) 3 : 2 (d) 4 : 1.

82. Two tall buildings are 40 m apart. With what speed must a ball be thrown horizontally from a window 145 m above the ground in one building, so that it will enter a window 22.5 m from the ground in the other ?

(a) 5 m s^{-1} (b) 8 m s^{-1}

(c) 10 m s^{-1} (d) 16 m s^{-1}.

83. On an inclined plane of inclination 30°, a ball is thrown at an angle of 60° with the horizontal from the foot of the incline with a velocity of $10\sqrt{3}$ m s^{-1}. If g = 10 m s^{-2}, then the ball will hit the inclined plane in

(a) 1 s (b) 2 s

(c) $2\sqrt{3}$ s (d) $4\sqrt{3}$ s.

84. The velocity at the maximum height of a projectile is half of its initial velocity u. Its range on the horizontal plane is

(a) $\frac{2u^2}{3g}$ (b) $\frac{3u^2}{g}$

(c) $\frac{\sqrt{3}u^2}{2g}$ (d) $\frac{u^2}{3g}$.

85. The horizontal range is four times the maximum height attained by a projectile. The angle of projection is

(a) 30° (b) 45°

(c) 60° (d) 90°.

86. If the range of a gun which fires a shell with muzzle speed v is R, then the angle of elevation of the gun is

(a) $\cos^{-1}\left[\frac{v^2}{gR}\right]$ (b) $\cos^{-1}\left(\frac{gR}{v^2}\right)$

(c) $\frac{1}{2}\left[\frac{v^2}{gR}\right]$ (d) $\frac{1}{2}\sin^{-1}\left(\frac{gR}{v^2}\right)$.

87. The initial velocity of a particle along x-axis is 3 m s^{-1}. It moves with a uniform acceleration of $(\hat{i}-\hat{j}+\hat{k})$ m s^{-2}. The displacement of the particle after 2 second is

(a) $2\hat{k}$ (b) $-2\hat{j}$

(c) $8\hat{i}-2\hat{j}$ (d) $8\hat{i}-2\hat{j}+2\hat{k}$.

88. A particle has an initial velocity of $3\hat{i}+4\hat{j}$. If the acceleration of this particle is $0.4\hat{i}+0.3\hat{j}$, then the velocity of the particle after 10 second is

(a) 7 units (b) 8.5 units

(c) $7\sqrt{2}$ units (d) 10 units.

89. A ball is projected up an incline of 30° with a velocity of 30 m s^{-1} at an angle of 30° with reference to the inclined plane from the bottom of the inclined plane. If g = 10 m s^{-2}, then the range on the inclined plane is

(a) 12 m (b) 60 m

(c) 120 m (d) 600 m.

90. The range of a projectile is 5 times its maximum height. Angle of projection is

(a) $\sin^{-1}\left(\frac{3}{5}\right)$ (b) $\tan^{-1}\left(\frac{4}{5}\right)$

(c) $\sin^{-1}\left(\frac{4}{5}\right)$ (d) $\cos^{-1}\left(\frac{3}{5}\right)$.

91. The velocity of projection of an oblique projectile is : $\vec{v}=3\hat{i}+2\hat{j}$ (in m s^{-1}). The speed of the projectile at the highest point of the trajectory is

(a) 3 m s^{-1} (b) 2 m s^{-1}

(c) 1 m s^{-1} (d) zero.

92. In Q. 91, if g = 12 m s^{-2}, then the horizontal range of the projectile is

(a) 1 m (b) 2 m

(c) 3 m (d) 12 m.

93. A 0.098 kg block slides down a frictionless track as shown. The vertical component of the velocity of block at A is

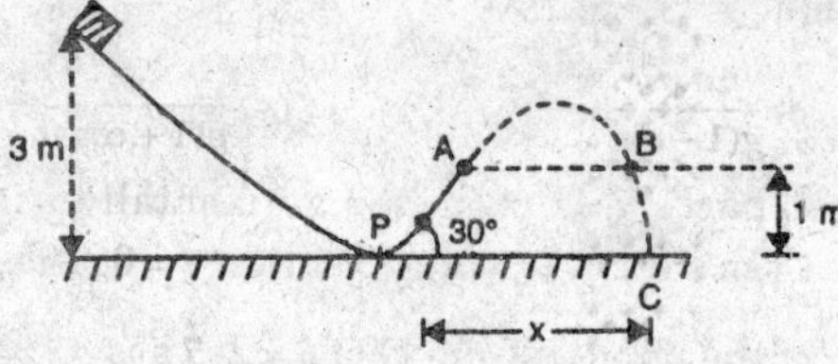

Fig. 6

(a) $\sqrt{g}$ (b) $2\sqrt{g}$
(c) $3\sqrt{g}$ (d) $4\sqrt{g}$.

94. In Q. 93, the time taken by the block to move from A to B is

(a) $\frac{1}{\sqrt{g}}$ (b) $\frac{2}{\sqrt{g}}$
(c) $\frac{3}{\sqrt{g}}$ (d) $\frac{4}{\sqrt{g}}$.

95. In Q. 93, the time taken by the block to move from A to C is

(a) $\sqrt{\frac{3}{g}}$ (b) $\sqrt{\frac{2}{g}}$
(c) $\frac{1}{\sqrt{g}}$ (d) $\frac{1+\sqrt{3}}{\sqrt{g}}$.

96. In Q. 93, the horizontal distance x travelled, by the block in moving from A to C is

(a) $(1+\sqrt{3})$ m (b) $(1-\sqrt{3})$ m
(c) $(\sqrt{3}+3)$ m (d) g metre.

97. When a projectile is projected at a certain angle with the horizontal, its horizontal range is R and time of flight is T_1. When the same projectile is thrown with the same speed at some other angle with the horizontal, its horizontal range is R and time of flight is T_2. The product of T_1 and T_2 is

(a) $\frac{R}{g}$ (b) $\frac{2R}{g}$
(c) $\frac{3R}{g}$ (d) $\frac{4R}{g}$.

98. A projectile is thrown with an initial velocity of $(v_x\hat{i}+v_y\hat{j})$ m s^{-1}. If the range of the projectile is double the maximum height, then v_y =

(a) v_x (b) $2v_x$
(c) $3v_x$ (d) $4v_x$.

ANSWERS

QUESTION BANK (Level I)

1. (a)	2. (a)	3. (d)	4. (d)	5. (c)	6. (d)	7. (d)	8. (c)
9. (d)	10. (b)	11. (d)	12. (c)	13. (d)	14. (d)	15. (c)	16. (d)
17. (a)	18. (d)	19. (c)	20. (d)	21. (d)	22. (b)	23. (a).	

QUESTION BANK (Level II)

1. (d)	2. (b)	3. (a)	4. (a)	5. (d)	6. (a)	7. (c)	8. (c)
9. (a)	10. (d)	11. (b)	12. (d)	13. (c)	14. (b)	15. (d)	16. (d)
17. (d)	18. (b)	19. (b)	20. (b)	21. (b)	22. (b)	23. (d)	24. (c)
25. (b)	26. (b)	27. (c)	28. (b)	29. (d)	30. (a)	31. (d)	32. (a)
33. (c)	34. (d)	35. (a)	36. (b)	37. (d)	38. (c)	39. (b)	40. (d)
41. (b)	42. (c)	43. (d)	44. (c)	45. (b)	46. (c)	47. (c)	48. (a)
49. (a)	50. (d)	51. (d)	52. (d)	53. (d)	54. (c)	55. (c)	56. (c)
57. (d)	58. (b)	59. (b)	60. (a)	61. (c)	62. (c)	63. (b)	64. (c)
65. (c)	66. (a)	67. (b)	68. (a)	69. (d)	70. (c)	71. (b)	72. (c)
73. (c)	74. (a)	75. (d)	76. (c)	77. (d)	78. (b)	79. (c)	80. (c)
81. (a)	82. (b)	83. (b)	84. (c)	85. (b)	86. (d)	87. (d)	88. (c)
89. (d)	90. (b)	91. (a)	92. (a)	93. (a)	94. (b)	95. (d)	96. (c)
97. (b)	98. (b).						

HINTS/SOLUTIONS

QUESTION BANK (Level I)

1. (*a*) $\theta = (90° - 15°) = 75°$.

2. (*a*) $R = \frac{v^2}{g}$, $h_{max.} = \frac{v^2 \sin^2 45°}{2g} = \frac{v^2}{4g}$

Clearly, $R = 4\, h_{max.}$

3. (*d*) $R \propto \frac{1}{g}$. g is reduced by a factor of 6 on the moon.

4. (*d*) The only force acting on a projectile is the force of gravity which acts vertically downwards. It has no horizontal component. So, horizontal motion of the oblique projectile takes place at uniform velocity.

5. (*c*) Kinetic energy at highest point $= K \cos^2 45° = \frac{K}{2}$

6. (*d*) The horizontal range is the same whether the projectile is thrown at an angle θ with the horizontal or at an angle (90 – θ) with the horizontal.

7. (*d*) $t = \sqrt{\frac{2 \times 490}{9.8}}\ s = \sqrt{\frac{2 \times 49 \times 100}{98}}\ s = \sqrt{100}\ s = 10\ s$

8. (*c*) Horizontal distance, $x = 360\ km\ h^{-1} \times 10\ s$

$$= \frac{360 \times 1000}{3600} \times 10\ m = 1000\ m$$

9. (*d*) As is clear from Fig. 7,

$$y = x \tan\theta - \frac{1}{2}gt^2 \quad ...(1)$$

(for the monkey)

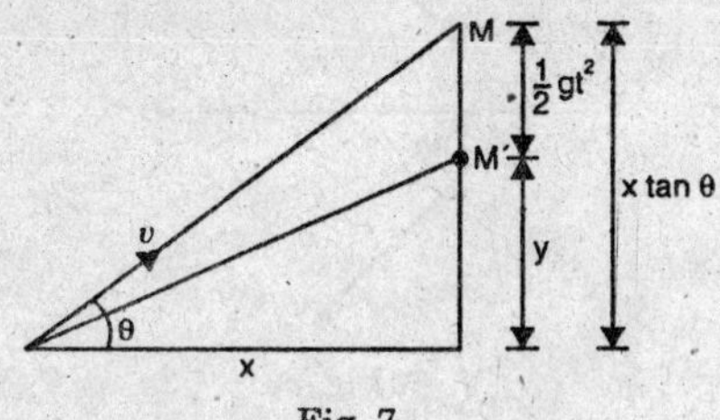

Fig. 7

For the bullet, $y = x \tan\theta - \frac{gx^2}{2v^2 \cos^2\theta}$

or $\quad y = x \tan\theta - \frac{1}{2}g\left[\frac{x}{v \cos\theta}\right]^2$

or $\quad y = x \tan\theta - \frac{1}{2}gt^2 \quad ...(2) \quad \left[\because\ v\cos\theta = \frac{x}{t}\right]$

The result follows from (1) and (2).

10. (*b*) Max. range $\propto v^2$.

11. (*d*) $T = \frac{2v \sin\theta}{g}$, For maximum T, sin θ should be maximum *i.e.* θ should be 90°.

12. (*c*) $T = \frac{2v \sin\theta}{g} = \frac{2v_{\perp}}{g}$

13. (*d*) $v \cos\theta = 10 \cos 60° = 5\ m\ s^{-1}$

14. (*d*) Acceleration due to gravity is always directed vertically downward.

15. (*c*) $h_1 = \frac{v^2 \sin^2\alpha}{2g}$, $h_2 = \frac{v^2 \cos^2\alpha}{2g}$, $\frac{h_1}{h_2} = \tan^2\alpha$

16. (*d*) $v = \sqrt{2gh}$, greatest distance $= \frac{v^2}{g} = \frac{2gh}{g} = 2h$

17. (*a*) Only the horizontal component of velocity exists at the highest point.

18. (*d*) $\quad h_{max.} = \frac{v^2 \sin^2\theta}{2g}$

$$h'_{max.} = \frac{v^2 \sin^2(90° - \theta)}{2g} \quad \text{or} \quad h'_{max.} = \frac{v^2 \cos^2\theta}{2g}$$

Dividing, $\frac{h_{max.}}{h'_{max.}} = \frac{\sin^2\theta}{\cos^2\theta} = \tan^2\theta$

19. (*c*) $p = mv$

$$p' = mv \cos 30° = \frac{\sqrt{3}mv}{2} = \frac{\sqrt{3}p}{2}$$

20. (*d*) Required angle $= \frac{\pi}{2} - \frac{5\pi}{36} = \frac{18\pi - 5\pi}{36} = \frac{13\pi}{36}$ radian

21. (*d*) $\quad \frac{v_1^2 \sin^2 30°}{2g} = \frac{v_2^2 \sin^2 60°}{2g}$

or $\quad \frac{v_1^2}{v_2^2} = \frac{\sin^2 60°}{\sin^2 30°} = \frac{3}{4} \times \frac{4}{1} = 3$

$\therefore \quad v_1 : v_2 :: \sqrt{3} : 1$

22. (*b*) $t = \sqrt{\frac{2h}{g}}$, $x = v\sqrt{\frac{2h}{g}}$ or $v = x\sqrt{\frac{g}{2h}}$

23. (*a*) At the highest point of trajectory, the velocity becomes horizontal. So, it is perpendicular to acceleration (which is directed vertically downwards).

QUESTION BANK (Level II)

1. (*d*) $h_{max.} = \frac{v^2 \sin^2 \theta}{2g}$

In the given problem, $h_{max.}$ is same in both the cases.

$\therefore \quad v_1^2 \sin^2 60° = v_2^2 \sin^2 30°$

or $\quad \frac{v_1}{v_2} = \frac{\sin 30°}{\sin 60°} = \frac{1}{2} \times \frac{2}{\sqrt{3}} = \frac{1}{\sqrt{3}}$

2. (*b*) $t = \sqrt{\frac{2h}{g}} = \sqrt{\frac{2 \times 490}{9.8}}$ s $= \sqrt{100}$ s $= 10$ s

$x = vt = \left(60 \times \frac{5}{18}\right)$ m s^{-1} $\times 10$ s $= \frac{500}{3}$ m

3. (*a*) $0.2n = \frac{1}{2} \times 10 \times t^2 = 5t^2$

$0.3n = 4.5t$

Dividing, $\quad \frac{2}{3} = \frac{50}{45} t$ or $t = 0.6$ s

Now, $\quad 0.3n = 4.5 \times 0.6 = 2.7$ or $n = \frac{2.7}{0.3} = 9$

4. (*a*) $\tan \beta = \frac{v \sin \theta - gt}{v \cos \theta} = \frac{30\left(\frac{1}{2}\right) - 10(1.5)}{30\left(\frac{\sqrt{3}}{2}\right)}$

or $\tan \beta = 0$ or $\beta = 0°$

5. (*d*) $p = mv \cos \theta$

$= 1 \times 10 \times \cos 60° = 10\left(\frac{1}{2}\right)$ kg m s^{-1} = 5 kg m s^{-1}

6. (*a*) $\quad t = \sqrt{\frac{2 \times 490}{9.8}}$ s = 10 s

$x = \left(216 \times \frac{5}{18}\right)$ m s^{-1} $\times$ 10 s

$= 60$ m s^{-1} $\times$ 10 s = 600 m

7. (*c*) $\quad v_x = \frac{d}{dt}(2t + 4t^2) = 2 + 8t$

$a_x = \frac{d}{dt}(2 + 8t) = 8$

$v_y = \frac{d}{dt}(5t) = 5, \; a_y = \frac{d}{dt}(v_y) = 0$

So, acceleration is along x-axis and its value is 8 m s^{-2}.

8. (*c*) $E' = \frac{1}{2} mv^2 \cos^2 \theta = E \cos^2 \theta$

9. (*a*) $t = \sqrt{\frac{2 \times 2000}{10}}$ s $= \sqrt{400}$ s = 20 s

$x = 100$ m s^{-1} $\times$ 20 s = 2000 m = 2 km

10. (*d*) $\frac{h_{max.}}{R} = \frac{v^2 \sin^2 \theta}{2g} \times \frac{g}{v^2 \sin 2\theta}$

$= \frac{\sin \theta}{4 \cos \theta} = \frac{\tan \theta}{4} = \frac{\tan 60°}{4} = \frac{\sqrt{3}}{4}$

> ☞ **CTM.** $\frac{h_{max.}}{R} = \frac{\tan \theta}{4}$

11. (*b*) $\frac{v}{g} = 20$ or $v^2 = 20\,g = 20 \times 9.8 = 196$, $v = 14$ m s^{-1}

12. (*d*) $S = 0 \times 1 + \frac{1}{2} \times 9.8 \times 1 \times 1 = 4.9$ m

13. (*c*) $\quad y = 50 \tan 60° - \frac{10 \times 50 \times 50}{2 \times 25 \times 25 \times \cos^2 60°}$

$= (50 \times 1.7 - 80)$ m $= (35 - 80)$ m = 5 m

14. (*b*) For maximum possible horizontal range, θ = 45°

Kinetic energy at the highest point

$= \frac{1}{2} m (v \cos \theta)^2 = \frac{1}{2} mv^2 \cos^2 45°$

$= \frac{1}{2}\left(\frac{1}{2} mv^2\right) = \frac{1}{2} \times 800 \text{ J} = 400 \text{ J}$

15. (*d*) Equating horizontal components, we get

$V \cos (90° - \theta) = v \cos \theta$

or $\quad V \sin \theta = v \cos \theta$ or $V = v \cot \theta$.

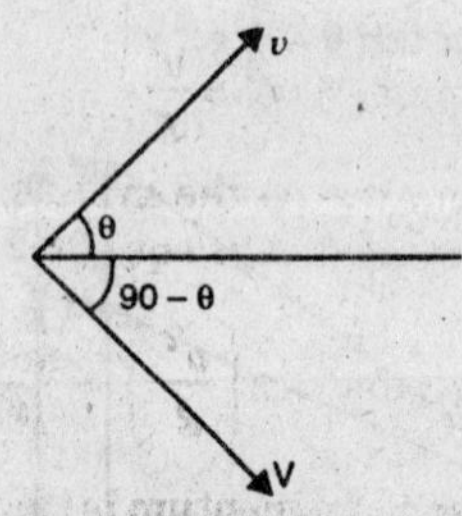

Fig. 8

16. (*d*) $\quad E_H = \frac{1}{2} m(v \cos 30°)^2 = \frac{3}{4}\left(\frac{1}{2} mv^2\right) = \frac{3E}{4}$

17. (*d*) $t = \sqrt{\frac{2h}{g}} = \sqrt{\frac{2 \times 980}{9.8}}$ s $= 10\sqrt{2}$ s

18. (*b*) $\quad H = \frac{v^2 \cos^2 \beta}{2g}$

or $v \cos \beta = \sqrt{2gH}$

$t = \frac{v \cos \beta}{g} = \frac{\sqrt{2gH}}{g}$ or $t = \sqrt{\frac{2H}{g}}$

19. (*b*) $y = x\tan\theta - \dfrac{gx^2}{2v^2\cos^2\theta}$

For the given problem, x, y and θ are constants.

$\therefore \quad \dfrac{g}{v^2} = \text{constant}$

$\dfrac{g'}{3\times3} = \dfrac{9.8}{5\times5}$ or $g' = 3.5\ \text{m s}^{-2}$

20. (*b*) $\tan\theta = \dfrac{v_y}{v_x} = \dfrac{0+9.8\times10}{98} = 1$ or $\theta = 45°$

21. (*b*) Comparing the given equation with

$y = x\tan\theta - \dfrac{gx^2}{2v^2\cos^2\theta}$, we get

$\tan\theta = \sqrt{3}$

22. (*b*) Clearly, $\theta = 60°$

Now, $\dfrac{gx^2}{2v^2\cos^2\theta} = \dfrac{gx^2}{2}$

or $v^2\cos^2\theta = 1$ or $v\cos\theta = 1$

or $v\cos 60° = 1$ or $\dfrac{v}{2} = 1$ or $v = 2\ \text{m s}^{-1}$

23. (*b*) $200 = \dfrac{v^2\sin 2\theta}{g}$; $5 = \dfrac{2v\sin\theta}{g}$

Dividing, $40 = v\cos\theta$

24. (*c*) $v\sin\theta = \dfrac{5g}{2} = \dfrac{5\times10}{2} = 25\ \text{m s}^{-1}$

$h_{max.} = \dfrac{v^2\sin^2\theta}{2g} = \dfrac{25\times25}{2\times10}\ \text{m} = 31.25\ \text{m}$

25. (*b*) Maximum range $= \dfrac{v^2}{g}$

Maximum area on the ground is equal to the area of a circle whose radius is equal to maximum range.

Area on ground $= \pi\left(\dfrac{v^2}{g}\right)^2 = \dfrac{\pi v^4}{g^2}$

26. (*b*) Change in momentum is the product of force and time.

$\Delta p = mg\times\dfrac{2v\sin\theta}{g}$

$= 2mv\sin\theta = 2mv\sin 45° = \dfrac{2mv}{\sqrt{2}} = \sqrt{2}\,mv$

27. (*c*) $t = \sqrt{\dfrac{2h}{g}}$

$x = \sqrt{2gh}\sqrt{\dfrac{2h}{g}} = 2h$

28. (*b*) Change in momentum

= weight × time

= 0.5 × 9.8 × 10 N s = 49 N s

29. (*d*) Since range is given to be the same therefore the other angle is (90° – 30°), *i.e.*, 60°.

$\text{H} = \dfrac{v^2\sin^2 30°}{2g} = \dfrac{1}{4}\left[\dfrac{v^2}{2g}\right]$

$\text{H}' = \dfrac{v^2\sin^2 60°}{2g} = \dfrac{3}{4}\left[\dfrac{v^2}{2g}\right]$

$\dfrac{\text{H}'}{\text{H}} = \dfrac{3}{4}\times\dfrac{4}{1} = 3$ or $\text{H}' = 3\text{H}$

30. (*a*) $\dfrac{v^2}{g} = 4000$ or $v^2 = 4000\times10$ or $v = 200\ \text{m s}^{-1}$

31. (*d*) $\text{T} = \dfrac{2v\sin\theta}{g}$

$\text{T}' = \dfrac{2v\sin\theta}{g+\dfrac{g}{10}}$ or $\dfrac{\text{T}}{\text{T}'} = \dfrac{11}{10}$

Percentage decrease in time of flight

$= \dfrac{\text{T}-\text{T}'}{\text{T}}\times100 = \dfrac{11-10}{11}\times100 \approx 9$

32. (*a*) $\dfrac{v^2\sin 2\theta}{g} = 200$, $\dfrac{2v\sin\theta}{g} = 5$

Dividing, $\dfrac{v^2\times2\sin\theta\cos\theta}{g}\times\dfrac{g}{2v\sin\theta} = \dfrac{200}{5} = 40$

or $v\cos\theta = 40\ \text{m s}^{-1}$

It may be noted here that the horizontal component of the velocity of projection remains the same during the flight of the projectile.

33. (*c*) Even after 5 second, the horizontal velocity v_x will be 20 m s^{-1}. The vertical velocity v_y is given by

$v_y = 0 + 10\times5 = 50\ \text{m s}^{-1}$

Now, $\text{V} = \sqrt{v_x^2+v_y^2} = \sqrt{20^2+50^2} \approx 54\ \text{m s}^{-1}$

34. (*d*) $\theta = 30°$

$\dfrac{\text{R}}{\text{H}} = \dfrac{v^2(2\sin\theta\cos\theta)}{g}\times\dfrac{2g}{v^2\sin^2\theta} = \dfrac{4\cos\theta}{\sin\theta}$

or $\text{R} = 4\cot 30°\times\text{H} = 4\sqrt{3}\ \text{H}$

35. (*a*) $\Delta p = mg\times\text{T} = mg\times\dfrac{2v\sin\theta}{g}$

$= 2mv\sin\theta = 2mv\sin 45° = \sqrt{2}\,mv$

36. (*b*) $(\text{KE})_\text{H} = \dfrac{1}{2}(\text{KE})_i$

$\dfrac{1}{2}mv^2\cos^2\theta = \dfrac{1}{2}\left(\dfrac{1}{2}mv^2\right) = \dfrac{1}{4}mv^2$

or $\cos^2\theta = \dfrac{1}{2}$ or $\cos\theta = \dfrac{1}{\sqrt{2}}$ or $\theta = 45°$

37. (*d*) $v_y = \frac{d}{dt}(y) = \frac{d}{dt}(10t) - \frac{d}{dt}(t^2) = 10 - 2t$

At maximum height, $v_y = 0$

$\therefore \quad 10 - 2t = 0$ or $2t = 10$ or $t = 5$ s

$\therefore \quad y = (10 \times 5 - 5 \times 5)$ m $= 25$ m

38. (*c*) Equating velocities along the vertical,

$$v_2 = v_1 \sin 30° \quad \text{or} \quad \frac{v_2}{v_1} = \frac{1}{2}$$

39. (*b*) $\frac{v^2 \sin 2\theta}{g} = \frac{\sqrt{3}v^2}{2g}$

or $\sin 2\theta = \frac{\sqrt{3}}{2}$ or $2\theta = 60°$ or $\theta = 30°$

Let us cross-check with the help of data for vertical range.

$$\frac{v^2 \sin^2 \theta}{2g} = \frac{v^2}{8g} \quad \text{or} \quad \sin^2 \theta = \frac{1}{4}$$

or $\sin \theta = \frac{1}{2}$ or $\theta = 30°$

40. (*d*) A projectile shall attain the greatest height when its angle θ with the horizontal is 90°. So, time of flight $\left(\frac{2v \sin \theta}{g}\right)$ is largest when $\theta = 90°$. Note that for $\theta = 90°$, $\sin \theta$ has maximum possible value 1.

41. (*b*) $\vec{v} = 6\hat{i} + 8\hat{j}$

Comparing with

$\vec{v} = v_x\hat{i} = v_y\hat{j}$, we get

$v_x = 6$ m s^{-1}

and $v_y = 8$ m s^{-1}

Also, $v^2 = v_x^2 + v_y^2$

$= 36 + 64 = 100$

or $v = 10$ m s^{-1}

Fig. 9

$$\sin \theta = \frac{8}{10} \text{ and } \cos \theta = \frac{6}{10}$$

$$R = \frac{v^2 \sin 2\theta}{g} = \frac{2v^2 \sin \theta \cos \theta}{g}$$

$$R = 2 \times 10 \times 10 \times \frac{8}{10} \times \frac{6}{10} \times \frac{1}{10} \text{ m} = 9.6 \text{ m}$$

42. (*c*) $\frac{v^2 \sin 2\theta}{g} = 2\frac{v^2 \sin^2 \theta}{2g}$

or $2 \cos \theta = \sin \theta$

or $\tan \theta = 2$, $\sin \theta = \frac{2}{\sqrt{5}}$

$$\cos \theta = \frac{1}{\sqrt{5}}$$

$$R = \frac{v^2}{g}\left[2 \times \frac{2}{\sqrt{5}} \times \frac{1}{\sqrt{5}}\right] = \frac{4v^2}{5g}$$

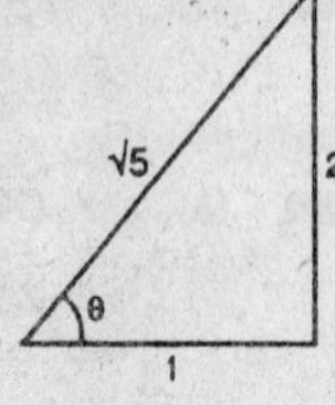

Fig. 10

43. (*d*) Let the initial velocity be along x-axis. Then acceleration is along y-axis.

After 4 second from start,

$v_x = 3$ m s^{-1},

$v_y = 0 + 1 \times 4 = 4$ m s^{-1}

$v = \sqrt{v_x^2 + v_y^2}$

$= \sqrt{9 + 16} = \sqrt{25}$ m s^{-1} $= 5$ m s^{-1}

Fig. 11

$$\tan \beta = \frac{4}{3} \quad \text{or} \quad \beta = \tan^{-1}\left(\frac{4}{3}\right)$$

44. (*c*) Let v be the velocity of projection and θ the angle of projection.

Kinetic energy at highest point is

$$\frac{1}{2}mv^2 \cos^2 \theta \quad \text{or} \quad E_k \cos^2 \theta$$

Potential energy at highest point

$$= E_k - E_k \cos^2 \theta = E_k(1 - \cos^2 \theta) = E_k \sin^2 \theta$$

45. (*b*) $E_k' = E_k \cos^2 30° = \frac{3E_k}{4}$

46. (*c*) $t = \sqrt{\frac{2 \times 8 \times 10^3}{10}}$ s $= 40$ s

$x = vt = 200 \times 40$ m $= 8000$ m

47. (*c*) $R = \frac{v^2 \sin 2\theta}{g}$ or $R \propto \sin 2\theta$

$$\frac{R'}{R} = \frac{\sin 90°}{\sin 30°} = 2 \quad \text{or} \quad R' = 2R$$

48. (*a*) Since time of flight is same, therefore, '$v \cos \theta$' is same.

Again, $h_{max} = \frac{v^2 \cos^2 \theta}{2g}$. Clearly, h_{max} will be same in both the cases. [Note that θ is angle with the vertical.]

49. (*a*) $T_1 = \frac{2v \sin \theta}{g}$ and $T_2 = \frac{2v \cos \theta}{g}$

$$T_1T_2 = \frac{2(v^2 \sin 2\theta)}{g \times g} \quad \text{or} \quad T_1T_2 \propto r$$

50. (*d*) $V \cos \beta = v \cos \theta$

$$V = \frac{v \cos \theta}{\cos \beta}$$

☞ *Note that the horizontal velocity remains unchanged.*

51. (*d*) $t = \sqrt{\frac{2 \times 5}{10}}$ s $= 1$ s

$x = 0.8$ m s^{-1} $\times 1$ s or $x = 0.8$ m

52. (*d*) $R = \frac{v^2 \sin 2\theta}{g}$

In the given problem, $v^2 \sin 2\theta$ = constant

$$v^2 \sin 2\theta = \left(\frac{v}{2}\right)^2 \sin 30° = \frac{v^2}{8}$$

or $\sin 2\theta = \frac{1}{8}$ or $2\theta = \sin^{-1}\left[\frac{1}{8}\right]$ or $\theta = \frac{1}{2}\sin^{-1}\left[\frac{1}{8}\right]$

53. (*d*) $v_y = 0 + 10 \times 5 = 50 \text{ m s}^{-1}$

$v_x = 20 \text{ m s}^{-1}$

$V^2 = v_x^2 + v_y^2 = 20^2 + 50^2 = 400 + 2500 = 2900$

or $V = 53.8 \text{ m s}^{-1} \approx 54 \text{ m s}^{-1}$

54. (c) $\frac{v^2}{g} = 100$ or $v^2 = 100\text{ g}$

$$h_{max} = \frac{v^2}{2g} = \frac{100\text{ g}}{2\text{ g}} = 50 \text{ m}$$

55. (c) $v_x = 4 \text{ m s}^{-1}$, $v_y = 0 + 0.7 \times 10 = 7 \text{ m s}^{-1}$

$v = \sqrt{4^2 + 7^2} \text{ m s}^{-1} = \sqrt{65} \text{ m s}^{-1}$

$\approx \sqrt{64} \text{ m s}^{-1} = 8 \text{ m s}^{-1}$

56. (c) $\frac{h'}{h} = \frac{\sin^2 60°}{\sin^2 30°} = \frac{3}{4} \times \frac{4}{1} = 3$ or $h' = 3h$

57. (*d*) In this problem, we have to keep in mind that the momentum has to be conserved both along horizontal and vertical directions.

Let v_x be the horizontal velocity of the other part.

Applying momentum conservation along X-axis, we get

$mv_x = (2m)60$

or $v_x = 120 \text{ m s}^{-1}$

Applying momentum conservation along Y-axis, we get

$mv_y = m(50)$ or $v_y = 50 \text{ m s}^{-1}$

Velocity of the other part, v

$= \sqrt{v_x^2 + v_y^2} = \sqrt{120^2 + 50^2} \text{ m s}^{-1}$

$= \sqrt{16900} \text{ m s}^{-1} = 130 \text{ m s}^{-1}$

58. (*b*) $\frac{2v \sin\theta}{g} = 2$ or $v \sin\theta = g = 10$

Max. height $= \frac{v^2 \sin^2\theta}{2g} = \frac{10 \times 10}{2 \times 10} \text{ m} = 5 \text{ m}$

59. (*b*) Maximum horizontal range $= \frac{v^2}{g}$

$\therefore \frac{v^2}{g} = 32.4$ or $v^2 = 324$ or $v = 18 \text{ m s}^{-1}$

Horizontal component

$= 18 \cos 45° = \frac{18}{\sqrt{2}} \text{ m s}^{-1} \approx 13 \text{ m s}^{-1}$

At the maximum height, the vertical velocity is zero. So, the velocity of projectile is equal to horizontal component.

60. (*a*) Using $v^2 - u^2 = 2aS$, we get

$$S = \frac{v^2}{2g}$$

Now, $\frac{v^2 \sin 2\theta}{g} = \frac{v^2}{2g}$ or $\sin 2\theta = \frac{1}{2}$

or $\sin 2\theta = \sin 30°$ or $\theta = 15°$

The other possible angle of projection is (90° – 15°), *i.e.*, 75°.

61. (c) $v_x = \frac{dx}{dt} = v_0$ or $dx = v_0\, dt$ or $x = v_0 t$ or $t = \frac{x}{v_0}$

Again, $v_y = a\omega \cos \omega t$

$\frac{dy}{dt} = a\omega \cos \omega t$ or $y = a\omega \frac{\sin \omega t}{\omega}$

or $y = a \sin \omega t = a \sin \omega \frac{x}{v_0}$

or $y = a \sin \frac{\omega x}{v_0}$

62. (c) $h = v \sin\theta t - \frac{1}{2}gt^2$

or $\frac{1}{2}gt^2 - v \sin\theta t + h = 0$

$t_1 + t_2 = -\frac{-v \sin\theta}{\frac{1}{2}g}$ or $t_1 + t_2 = \frac{2v \sin\theta}{g} = T$

or $T = (1 + 3)\text{ s} = 4 \text{ s}$

63. (*b*) $h_{max.} = \frac{v^2 \sin^2\theta}{2g}$

$= \frac{g^2T^2}{8g} = \frac{1}{8}gT^2 = \frac{1}{8}g \times 4 \times 4 = 2g$ $\left[\because v \sin\theta = \frac{gT}{2}\right]$

64. (c) $h = v \sin\theta t - \frac{1}{2}gt^2$

when $t = 1$ s, then

$h = v \sin\theta - \frac{1}{2}g$

or $h = \frac{g \times 4}{2} - \frac{g}{2} = 2g - \frac{g}{2}$ or $h = \frac{3g}{2}$

Aliter. $\frac{1}{2}gt^2 - v \sin\theta t + h = 0$

$t_1 t_2 = \frac{h}{\frac{1}{2}g}$ or $1 \times 3 = \frac{h}{g/2}$ or $h = \frac{3g}{2}$

65. (c) $v_y^2 - v^2 \sin^2\theta = -2g\left[\frac{1}{2}\frac{v^2 \sin^2\theta}{2g}\right]$

or $$v_y{}^2 = v^2 \sin^2\theta - \frac{v^2 \sin^2\theta}{2}$$

or $$v_y{}^2 = \frac{v^2 \sin^2\theta}{2} \text{ or } v_y = \frac{v \sin\theta}{\sqrt{2}}$$

66. (*a*) $$v'^2 = (v\cos\theta)^2 + \left(\frac{v\sin\theta}{\sqrt{2}}\right)^2$$

or $$v'^2 = v^2\cos^2\theta + \frac{v^2\sin^2\theta}{2}$$

or $$v'^2 = v^2\left(\cos^2\theta + \frac{\sin^2\theta}{2}\right)$$

or $$v' = v\sqrt{\cos^2\theta + \frac{\sin^2\theta}{2}}$$

67. (*b*) $$v^2\cos^2\theta = \frac{2}{5}v^2\left[\cos^2\theta + \frac{\sin^2\theta}{2}\right]$$

or $5\cos^2\theta = 2\cos^2\theta + \sin^2\theta$ or $3\cos^2\theta = \sin^2\theta$

or $\tan^2\theta = 3$ or $\tan\theta = \sqrt{3}$ or $\theta = 60°$

Angle of projection with vertical

$$= 90° - 60° = 30°.$$

68. (*a*) $$x_1 = v_1\cos\theta_1 t\,;\quad y_1 = v_1\sin\theta_1 t - \frac{1}{2}gt^2$$

$$x_2 = v_2\cos\theta_2 t\,;\quad y_2 = v_2\sin\theta_2 t - \frac{1}{2}gt^2$$

$$\frac{y_2 - y_1}{x_2 - x_1} = \frac{v_2\sin\theta_2 - v_1\sin\theta_1}{v_2\cos\theta_2 - v_1\cos\theta_1}$$

or $y_2 - y_1 = (\text{constant})(x_2 - x_1)$

This is the equation of a straight line.

Alternative Explanation

$$\vec{v_1} = v_1\cos\theta_1\,\hat{i} + (v_1\sin\theta_1 - gt)\hat{j}$$

and $$\vec{v_2} = v_2\cos\theta_2\,\hat{i} + (v_2\sin\theta_2 - gt)\hat{j}$$

Relative velocity, $\vec{v}_{12} = \vec{v_1} - \vec{v_2}$

$$= (v_1\cos\theta_1 - v_2\cos\theta_2)\hat{i} + (v_1\sin\theta_1 - v_2\sin\theta_2)\hat{j}$$

Clearly, the relative velocity is time-independent. In other words, relative velocity is constant. This leads us to the required result.

69. (*d*) 'u' $= -50\cos 60°$ m s^{-1} $= -25$ m s^{-1}

'a' = 10 m s^{-2}, t = ?, S = 70 m

Using $S = ut + \frac{1}{2}at^2$, we get

$$70 = -25t + 5t^2 \quad\text{or}\quad 5t^2 - 25t - 70 = 0$$

or $t^2 - 5t - 14 = 0$ or $t^2 - 7t + 2t - 14 = 0$

or $t(t-7) + 2(t-7) = 0$ or $(t-7)(t+2) = 0$

or $t = 7$ s [Rejecting –ve value of t]

70. (*c*) $$T = \frac{2\times 50\times\frac{1}{2}}{10}\text{ s} = 5\text{ s}$$

Horizontal distance travelled in last 2 second

$$= 50\times\cos 30°\times 2\text{ m}$$

$$= 100\times\frac{\sqrt{3}}{2}\text{ m} = 50\sqrt{3}\text{ m} = 86.6\text{ m}$$

71. (*b*) $v_x = v\cos(\alpha - \theta)\,;\quad v_y = v\sin(\alpha - \theta)$

$a_x = -g\sin\theta\,;\quad a_y = -g\cos\theta$

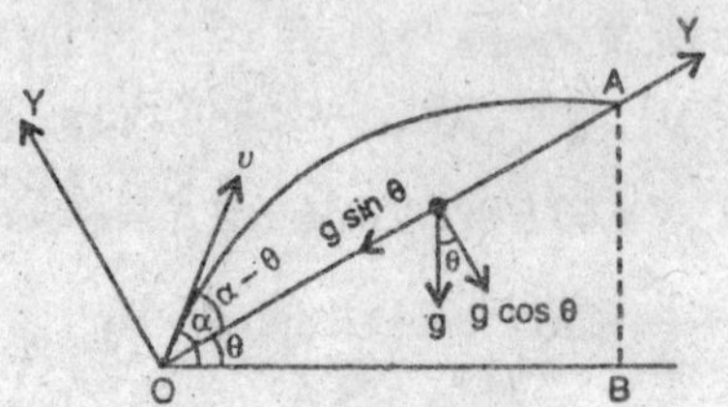

Fig. 12

If T is the time of flight, then

$$0 = v\sin(\alpha - \theta)\,.\,T - \frac{1}{2}g\cos\theta\,.\,T^2$$

or $$T = \frac{2v\sin(\alpha - \theta)}{g\cos\theta}$$

$$OB = v\cos\alpha\times T$$

Now, $\cos\theta = \frac{OB}{OA}$ or $OA = \frac{OB}{\cos\theta}$

or $$OA = \frac{v\cos\alpha\,.\,T}{\cos\theta}$$

or $$OA = v\cos\alpha\times\frac{2v\sin(\alpha - \theta)}{g\cos\theta}\times\frac{1}{\cos\theta}$$

or $$OA = \frac{v^2}{g\cos^2\theta}\,[2\sin(\alpha - \theta)\cos\alpha]$$

or $$OA = \frac{v^2}{g\cos^2\theta}\,[\sin(2\alpha - \theta) + \sin(-\theta)]$$

or $$OA = \frac{v^2}{g\cos^2\theta}\,[\sin(2\alpha - \theta) - \sin\theta]$$

Clearly, the range R(= OA) will be maximum when $\sin(2\alpha - \theta)$ is maximum, *i.e.*, 1. This would mean

$$2\alpha - \theta = \frac{\pi}{2} \quad\text{or}\quad \alpha = \frac{\theta}{2} + \frac{\pi}{4}$$

Maximum range up the inclined plane,

$$R_{max} = \frac{v^2}{g\cos^2\theta}(1 - \sin\theta) = \frac{v^2(1 - \sin\theta)}{g(1 - \sin^2\theta)}$$

$$= \frac{v^2(1 - \sin\theta)}{g(1 + \sin\theta)(1 - \sin\theta)} = \frac{v^2}{g(1 + \sin\theta)}$$

72. (c) $x = 14 \text{ m s}^{-1} \times \frac{20}{7} \text{ s} = 40 \text{ m}$

$$y = 0 + \frac{1}{2} \times 9.8 \times \frac{20}{7} \times \frac{20}{7} \text{ m}$$

or $$y = 4.9 \times \frac{400}{49} \text{ m} = 40 \text{ m}$$

$$r = \sqrt{x^2 + y^2} = 40\sqrt{2} \text{ m}$$

73. (c) $x = 20 \times 5 \text{ m} = 100 \text{ m}$

$$y = \frac{1}{2} \times 10 \times 5 \times 5 = 125 \text{ m}$$

$$r = \sqrt{100^2 + 125^2} \text{ m} \approx 160 \text{ m}$$

74. (a) $T = \frac{2v \sin\theta}{g} \Rightarrow T^2 = \frac{4v^2 \sin^2\theta}{g^2}$

$$R = \frac{v^2 (2 \sin\theta \cos\theta)}{g}$$

$$\frac{T^2}{R} = \frac{2}{g} \tan\theta$$

or $$\tan\theta = \frac{gT^2}{2R} \text{ or } \theta = \tan^{-1}\left[\frac{gT^2}{2R}\right]$$

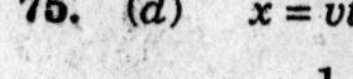

75. (d) $x = vt$

$$y = \frac{1}{2} gt^2$$

or $$y = \frac{gx^2}{2v^2}$$

$$\tan 45° = \frac{y}{x} \quad \text{or} \quad y = x$$

or $$x = \frac{gx^2}{2v^2}$$

or $$x = \frac{2v^2}{g}$$

Fig. 13

Also, $$y = \frac{2v^2}{g}$$

$$l^2 = x^2 + y^2$$

or $$l = \sqrt{2}\left[\frac{2v^2}{g}\right]$$

Fig. 14

76. (c) $V \cos\beta = v \cos\theta$

or $V = v \cos\theta \sec\beta$

77. (d) For horizontal motion,

$$nw = v_0 t \text{ or } t = \frac{nw}{v_0}$$

For vertical motion, $nh = \frac{1}{2} gt^2$

or $$\frac{1}{2} g \left(\frac{n^2 w^2}{v_0^2}\right) = nh \text{ or } n = \frac{2hv_0^2}{gw^2}$$

78. (b) $t_1 = \frac{2v \sin\theta}{g}, t_2 = \frac{2v \cos\theta}{g}$

$$t_1 t_2 \propto \frac{4v^2 \sin\theta \cos\theta}{g^2} \Rightarrow t_1 t_2 \propto \frac{2}{g}\left(\frac{v^2 \sin 2\theta}{g}\right)$$

$$t_1 t_2 \propto R$$

79. (a) $t = \sqrt{\frac{2h}{g}}$

$$x = vt = v\sqrt{\frac{2h}{g}}$$

$$x' = \frac{v}{2}\sqrt{\frac{2(4h)}{g}} = x = 250 \text{ m}$$

80. (c) $\tan\beta = \frac{v \sin\theta - gt}{v \cos\theta}$

$$\tan 45° = \frac{v \sin\theta - gt}{v \cos\theta}$$

or $$gt = v(\sin\theta - \cos\theta)$$

or $$t = \frac{v (\sin\theta - \cos\theta)}{g}$$

$$= \frac{147\left[\frac{\sqrt{3}}{2} - \frac{1}{2}\right]}{9.8} \text{ s} = \frac{147 \times 0.732}{2 \times 9.8} \text{ s} = 5.49 \text{ s}$$

81. (a) If t is the total time taken, then

$$40 = -20 \sin 30° \, t + \frac{1}{2} \times 10 \times t^2$$

or $40 = -10t + 5t^2$ or $5t^2 - 10t - 40 = 0$

or $t^2 - 2t - 8 = 0$ or $t^2 - 4t + 2t - 8 = 0$

or $t(t - 4) + 2(t - 4) = 0$ or $(t + 2)(t - 4) = 0$

$\Rightarrow$ $t = 4$ s [Negative time is not allowed]

$$T = \frac{2v \sin\theta}{g} = \frac{2 \times 20 \sin 30°}{10} \text{ s} = 2 \text{ s}$$

$$\therefore \quad \frac{t}{T} = \frac{4}{2} = \frac{2}{1}$$

82. (b) $h = 122.5$ m

Now, $$40 = v\sqrt{\frac{2 \times 122.5}{9.8}}$$

or $$40 = v \times 5 \text{ or } v = 8 \text{ m s}^{-1}$$

83. (b) $T = \frac{2 \times 10\sqrt{3} \sin(60° - 30°)}{10 \cos 30°} \text{ s}$

$$= 4 \sin 30° = 2 \text{ s}$$

84. (c) $u \cos\theta = \frac{1}{2} u$ or $\cos\theta = \frac{1}{2}$ or $\theta = 60°$

$$R = \frac{u^2 \sin 120°}{g} = \frac{u^2 \sin 60°}{g} = \frac{\sqrt{3}\, u^2}{2g}$$

85. (b) $\frac{v^2 (2 \sin\theta \cos\theta)}{g} = 4 \frac{v^2 \sin^2\theta}{2g}$

$$2 \sin\theta \cos\theta = 2 \sin^2\theta \text{ or } \tan\theta = 1 \text{ or } \theta = 45°$$

86. (*d*) $R = \frac{v^2 \sin 2\theta}{g}$ or $\sin 2\theta = \frac{gR}{v^2}$

or $2\theta = \sin^{-1}\left[\frac{gR}{v^2}\right]$ or $\theta = \frac{1}{2}\sin^{-1}\left[\frac{gR}{v^2}\right]$

87. (*d*) $\vec{S} = \vec{u}t + \frac{1}{2}\vec{a}t^2$

$= (3\hat{i})2 + \frac{1}{2}(\hat{i} - \hat{j} + \hat{k})4 = 8\hat{i} - 2\hat{j} + 2\hat{k}$

88. (*c*) $v_x = 3 + 0.4 \times 10 = 7$

$v_y = 4 + 0.3 \times 10 = 7$

$v = \sqrt{v_x^2 + v_y^2} = \sqrt{7^2 + 7^2} = 7\sqrt{2}$ units

89. (*b*) $R = \frac{2 \times 30 \times 30 \sin 30° \cos 60°}{10 \cos^2 30°}$

$= 180 \times \frac{1}{2} \times \frac{1}{2} \times \frac{2 \times 2}{3}$ m = 60 m

90. (*b*) $\tan\theta = \frac{4}{n} = \frac{4}{5}$ or $\theta = \tan^{-1}\left(\frac{4}{5}\right)$

91. (*a*) At the highest point, velocity is horizontal.

92. (*a*) $R = \frac{2v_1v_1}{g} = \frac{2 \times 3 \times 2}{12}$ m = 1 m

93. (*a*) $\frac{1}{2}mv^2 = mg(3 - 1) = 2mg$

or $v = \sqrt{4g} = 2\sqrt{g}$

Vertical component at A = $2\sqrt{g} \sin 30° = \sqrt{g}$

94. (*b*) $T = \frac{2v \sin\theta}{g} = \frac{2\sqrt{g}}{g} = \frac{2}{\sqrt{g}}$

95. (*d*) Using $S = ut + \frac{1}{2}at^2$, we get

$-1 = \sqrt{g}t - \frac{1}{2}gt^2$ or $\frac{1}{2}gt^2 - \sqrt{g}t - 1 = 0$

$t = \frac{\sqrt{g} \pm \sqrt{g + 4 \times \frac{1}{2}g}}{2 \times \frac{1}{2}g}$ or $t = \frac{1 \pm \sqrt{3}}{\sqrt{g}}$

Neglecting – ve time,

$t = \frac{1 + \sqrt{3}}{\sqrt{g}}$

96. (*c*) $x = 2\sqrt{g} \cos 30°\left[\frac{1 + \sqrt{3}}{\sqrt{g}}\right]$ or $x = (\sqrt{3} + 3)$ metre

97. (*b*) The two angles of projection are clearly θ and $(90 - \theta)$.

$T_1 = \frac{2v \sin\theta}{g}$ and $T_2 = \frac{2v \sin(90 - \theta)}{g}$

$T_1T_2 = \frac{2(v)^2 (2 \sin\theta \cos\theta)}{g \times g} = \frac{2R}{g}$

98. (*b*) $\tan\theta = \frac{4}{n} = \frac{4}{2} = 2$

Also, $\tan\theta = \frac{v_y}{v_x}$ or $2 = \frac{v_y}{v_x}$ or $v_y = 2v_x$

4 LAWS OF MOTION

Important Terms, Facts and Formulae

BASIC CONCEPTS AND IMPORTANT POINTS TO REMEMBER

1. Newton's first law of motion gives the definition of force.
2. Newton's second law of motion gives the measure of force.
3. Newton's third law of motion gives the properties of force.
4. The absolute units of force remain the same throughout the universe. On the other hand, gravitational units of force are not constant. This is because the gravitational units of force depend upon the value of g which is different at different places.
5. Both μ_k and μ_s represent the ratio of the magnitudes of two forces. So, they are dimensionless.
6. The actual values of μ_s and μ_k depend on the nature of both the surfaces in contact.
7. Both μ_s and μ_k can exceed unity, although commonly they are less than one.
8. When a body is in equilibrium, it does not mean that no force is acting on the body. It simply means that the resultant of all the forces acting on the body is zero.
9. The line of action of a force is the line along which the force acts.
10. The force exerted on the system by bodies outside the system is called external force.
11. If no external force acts on a system, then the system is called closed system or isolated system.
12. The forces of interaction between bodies constituting the system are called internal forces.
13. If two bodies of different masses have the same momentum, then the lighter body possesses greater velocity as compared to the heavier body.
14. If two bodies of equal masses are moving with different velocities, then the one moving with greater velocity will have a greater momentum.
15. Mass of a body is a measure of the opposition offered by the body to the change in velocity of the body. In other words, mass is a measure of the inertia of the body.
16. The area under the force-time graph gives the magnitude of the impulse of the given force in the given time.
17. The forces of action and reaction act along the line joining the centres of two bodies.
18. One of the two forces involved in the interaction between two bodies may be called 'action' force. The other force will be called the 'reaction' force. The forces of action and reaction constitute a mutual simultaneous interaction. It cannot be said that action is the cause of reaction or reaction is the effect of action.
19. Forces always exist in pairs. They never exist singly. A single isolated force is an impossibility.
20. The forces of action and reaction always act on different bodies. Therefore they cannot balance each other.
21. Two forces, equal in magnitude and opposite in direction, acting on the same body balance each other.
22. Whenever we talk of a single force, we are merely referring to one aspect of mutual interaction.
23. Whenever the applied force exceeds the limiting frictional force, the force of friction is equal to the force of limiting friction and not equal to the applied force.
24. Momentum is conserved in the following cases :

    (*a*) Firing of gun　　(*b*) Explosion of bomb.
25. Force of friction is a self-adjusting force.
26. Coefficient of static friction, $\mu_s = \frac{f_{ms}}{R}$
27. Coefficient of kinetic friction, $\mu_k = \frac{f_k}{R}$
28. Relation between momentum and kinetic energy is

    $$p = \sqrt{2mE_k}$$
29. According to Newton's second law of motion,

    $$\vec{F} \propto \frac{d\vec{p}}{dt} \text{ or } \vec{F} = k\frac{d\vec{p}}{dt}$$

Choose units in such a way that $k = 1$

$$\therefore \quad \vec{F} = \frac{d\vec{p}}{dt} \quad \text{or} \quad \vec{F} = \frac{d}{dt}(m\vec{v})$$

Case I. If mass is constant, then

$$\vec{F} = m\frac{d}{dt}(\vec{v}) \quad \text{or} \quad \vec{F} = m\vec{a}$$

Case II. If velocity is constant, then

$$\vec{F} = \vec{v}\frac{dm}{dt}$$

30. Refer to system in figure.

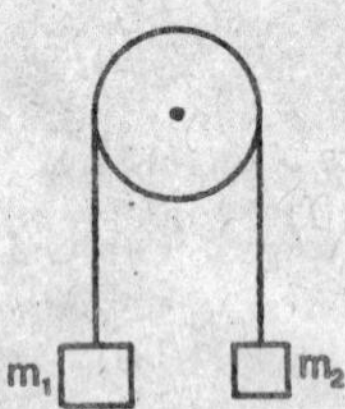

$$a = \frac{m_1 - m_2}{m_1 + m_2}g$$

$$T = \frac{2m_1m_2}{m_1 + m_2}g$$

31. Acceleration of a body sliding down an inclined plane of inclination θ is given by

$$a = g(\sin\theta - \mu_k \cos\theta)$$

32. Work done in dragging a body up a rough inclined plane of inclination θ is $mg(\sin\theta + \mu_k\cos\theta)$ S.

QUESTION BANK

Level I

Choose the correct answer from the four alternatives given in each of the following questions :

1. The coefficient of friction between two surfaces is 0.2. The angle of friction is

(*a*) $\sin^{-1}(0.2)$ (*b*) $\cos^{-1}(0.2)$

(*c*) $\tan^{-1}(0.1)$ (*d*) $\cot^{-1}(5)$.

2. Block A of mass 2 kg is placed over a block B of mass 8 kg. The combination is placed on a rough horizontal surface. If $g = 10\text{ m s}^{-2}$, coefficient of friction between B and floor = 0.5, coefficient of friction between A and B = 0.4 and a horizontal force of 10 N is applied on 8 kg block, then the force of friction between A and B is

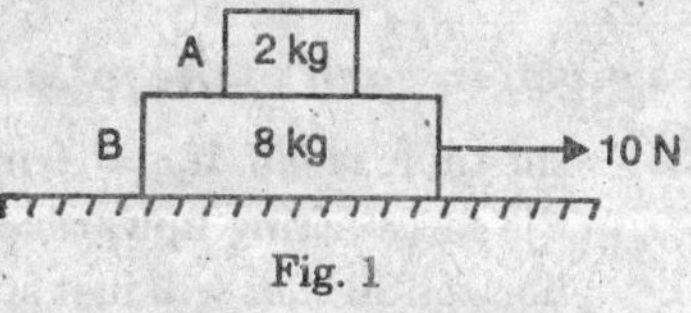

Fig. 1

(*a*) 100 N (*b*) 50 N

(*c*) 40 N (*d*) None of these.

3. A block of mass m is placed on a smooth inclined plane of inclination θ with the horizontal. The force exerted by the plane on the block has a magnitude

(*a*) mg (*b*) $\frac{mg}{\cos\theta}$

(*c*) $mg\cos\theta$ (*d*) $mg\tan\theta$.

4. A force $\vec{F} = 8\hat{i} - 6\hat{j} - 10\hat{k}$ newton produces an acceleration of 1 m s^{-2} in a body. The mass of the body is

(*a*) 10 kg (*b*) $10\sqrt{2}$ kg

(*c*) $10\sqrt{3}$ kg (*d*) 200 kg.

5. A block of mass 4 kg rests on an inclined plane. The inclination of the plane is gradually increased. It is found that when the inclination is 3 in 5, the block just begins to slide down the plane. The coefficient of friction between the block and the plane is

(*a*) 0.4 (*b*) 0.6

(*c*) 0.8 (*d*) 0.75.

6. A block of weight W is held against a vertical wall by applying a horizontal force of 75 N. The surface of the wall is rough. Now,

(*a*) 75 > W (*b*) 75 = W

(*c*) 75 < W (*d*) $75 \gtrless$ W.

7. A block of mass 2 kg is sliding with a constant velocity of 8 m s^{-1} on a frictionless horizontal surface. The force exerted on the horizontal surface is nearly

(*a*) 20 N (*b*) 10 N

(*c*) 40 N (*d*) 16 N.

8. A machine gun fires n bullets per second, each of mass m. If the speed of each bullet is u, then the force of recoil is

(*a*) $m\,n\,g$ (*b*) $m\,n\,v$

(*c*) $m\,n\,v\,g$ (*d*) $\frac{m\,n\,v}{g}$.

9. A force of 6 N acts on a mass of 1 kg which acquires a velocity 30 m s^{-1}. The time for which the force acts is

(*a*) 26 s (*b*) 6 s

(*c*) 5 s (*d*) 2 s.

10. A force of 5 newton acts on a body of weight 9.80 newton. What is the acceleration produced in m s^{-2} ?

(*a*) 0.51 (*b*) 1.96

(*c*) 5.00 (*d*) 49.00.

11. Gravel is dropped on to a conveyor belt at the rate of 0.5 kg s^{-1}. The extra force in newton required to keep the belt moving at 2 m s^{-1} is

(*a*) 1 (*b*) 2

(*c*) 4 (*d*) 5

(*e*) 10.

12. The engine of a car produces an acceleration of 6 m s^{-2} in the car. If this car pulls another car of the same mass, then the acceleration would be

(*a*) 6 m s^{-2} (*b*) 12 m s^{-2}

(*c*) 3 m s^{-2} (*d*) 1.5 m s^{-2}.

13. A shell of mass m moving with velocity v suddenly explodes into two pieces. One part of mass $\frac{m}{4}$ remains stationary. The velocity of the other part will be

(*a*) v (*b*) $2v$

(*c*) $\frac{3v}{4}$ (*d*) $\frac{4v}{3}$.

14. A body of mass 1 kg is acted upon by two forces, each of magnitude 1 N, making an angle of 60° with each other. The net acceleration of the body is

(*a*) 1 m s^{-2} (*b*) $\sqrt{3}\text{ m s}^{-2}$

(*c*) $\sqrt{5}\text{ m s}^{-2}$ (*d*) $\sqrt{7}\text{ m s}^{-2}$.

15. Two bodies of masses m_1 and m_2 are connected by a light, inextensible string which passes over a frictionless pulley. If the pulley is moving upward with uniform acceleration g, then the tension in the string is

(a) $\frac{4m_1m_2}{m_1+m_2}g$ (b) $\frac{m_1+m_2}{4m_1m_2}g$

(c) $\frac{m_1m_2}{m_1+m_2}g$ (d) $\frac{m_1-m_2}{m_1+m_2}g^2$.

16. An inclined plane makes an angle θ with the horizontal. A groove cut in the plane makes an angle α with OX. The acceleration of a body which is free to slide down the groove under the influence of gravity is

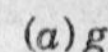

(a) g

(b) $g \sin \theta$

(c) $g \sin \theta \cos \alpha$

(d) $g \sin \theta \sin \alpha$.

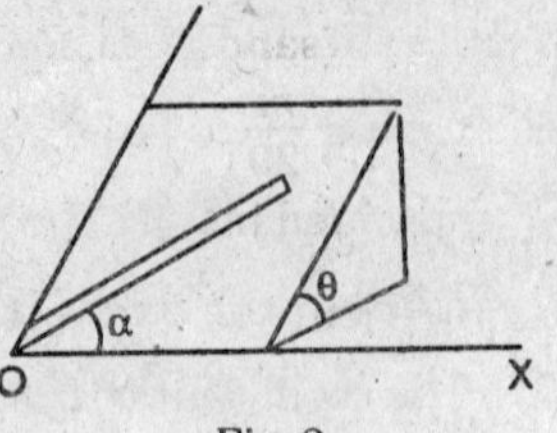

Fig. 2

QUESTION BANK

Level II

Choose the correct answer from the four alternatives given in each of the following questions :

1. Steady rain, giving 5 mm an hour, turns suddenly into a downpour giving 20 mm an hour and the speed of the rain drops falling vertically on to a flat roof simultaneously doubles. The pressure exerted by the falling rain on the roof is raised by a factor of

(a) 4 (b) 8

(c) 16 (d) 32.

2. The linear momentum of a particle varies with time t as

$$p = a + bt + ct^2$$

Which of the following statements is correct ?

(a) Force varies with time in a quadratic manner.

(b) Force is time-dependent.

(c) The velocity of the particle is proportional to time.

(d) The displacement of the particle is proportional to t.

3. A machine gun mounted on a 2000 kg car on a horizontal frictionless surface fires 10 bullets per second. If 10 g be the mass of each bullet and 500 m s^{-1} the velocity of each bullet, then the acceleration of the car will be

(a) $\frac{1}{10}$ m s^{-2} (b) $\frac{1}{20}$ m s^{-2}

(c) $\frac{1}{40}$ m s^{-2} (d) $\frac{1}{60}$ m s^{-2}.

4. In Q. 3, the average thrust due to ejected bullets on the system will be

(a) 550 N (b) 50 N

(c) 250 N (d) 250 dyne.

5. A block of mass 1 kg is at rest on a horizontal table. The coefficient of static friction between the block and the table is 0.5. If g = 10 m s^{-2}, then the magnitude of the force acting upwards at an angle of 60° from the horizontal that will just start the block moving is

(a) 5 N (b) 5.36 N

(c) 74.6 N (d) 10 N.

6. A block of mass 2 kg rests on a rough inclined plane making an angle of 30° with the horizontal. The coefficient of static friction between the block and the plane is 0.7. The frictional force on the block is

(a) 9.8 N (b) $0.7 \times 9.8 \times \sqrt{3}$ N

(c) $9.8 \times \sqrt{3}$ N (d) 0.7×9.8 N.

7. Starting from rest, a body slides down a 45° inclined plane in twice the time it takes to slide down the same distance in the absence of friction. The coefficient of friction between the body and the inclined plane is

(a) 0.33 (b) 0.75

(c) 0.25 (d) 0.80.

8. A heavy uniform chain lies on a horizontal table top. If the coefficient of friction between the chain and the table is 0.25, then the maximum percentage of the length of the chain that can hang over one edge of the table is

(a) 20% (b) 25%

(c) 35% (d) 15%.

9. A body of mass 2 kg is placed on a horizontal surface having kinetic friction 0.4 and static friction 0.5. If the force applied on the body is 2.5 N, then the frictional force acting on the body will be [g = 10 m s^{-2}]

(a) 8 N (b) 10 N

(c) 20 N (d) 2.5 N.

10. A bag of sand of mass m is suspended by a rope. A bullet of mass $\frac{m}{20}$ is fired at it with a velocity v and gets embedded into it. The velocity of the bag finally is

(a) $\frac{v}{20} \times 21$ (b) $\frac{20v}{21}$

(c) $\frac{v}{20}$ (d) $\frac{v}{21}$.

11. An elevator is moving up with an acceleration $\frac{g}{5}$. The apparent weight of a 60 kg man standing in the lift is [Take $g = 10$ m s^{-2}]

(a) 480 N (b) 720 N

(c) 600 N (d) 1000 N.

12. An iron chain lies on a rough horizontal table. It starts sliding when one-fourth of its length hangs over the edge of the table. The coefficient of static friction between the chain and the surface of the table is

(a) $\frac{1}{2}$ (b) $\frac{1}{3}$

(c) $\frac{1}{4}$ (d) $\frac{1}{5}$.

13. Fig. 3 shows a system of three masses being pulled with a force F. The masses are connected to each other by strings. The horizontal surface is frictionless. The tension T_1 in the first string is 16 N. The acceleration of the system is

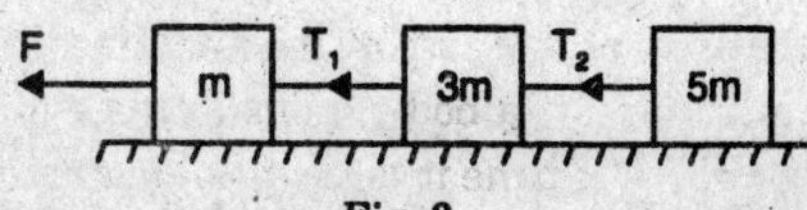

Fig. 3

(a) $\frac{1}{m}$ (b) $\frac{2}{m}$

(c) $\frac{3}{m}$ (d) $\frac{4}{m}$.

14. In Q. 13, the tension T_2 is

(a) 16 N (b) 8 N

(c) 10 N (d) 4 N.

15. In Q. 13, F is equal to

(a) 6 N (b) 8 N

(c) 12 N (d) 18 N.

16. A 40 N force pulls a system of three masses on a horizontal frictionless surface as shown in Fig. 4. Tension T_1 is

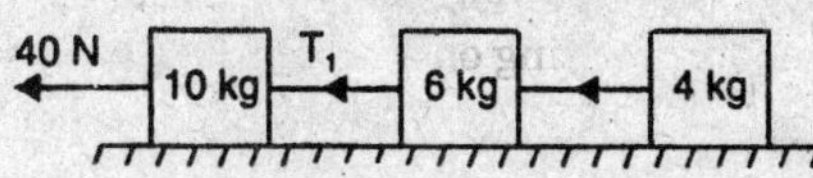

Fig. 4

(a) 40 N (b) 20 N

(c) 10 N (d) 5 N.

17. For the arrangement shown in the Fig. 5 the tension in the string is [Given : $\tan^{-1}(0.8) \approx 39°$].

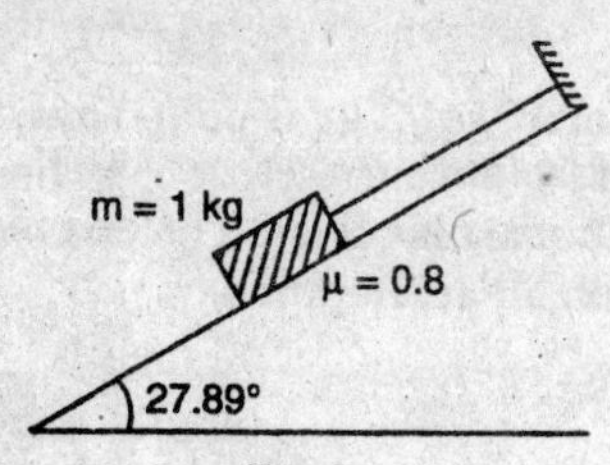

Fig. 5

(a) 6 N (b) 6.4 N

(c) 0.4 N (d) zero.

18. A box weighing 20 N is moved along a horizontal surface through a distance of 25 m. If the coefficient of friction is 0.4, the work done against friction is

(a) 200 J (b) 250 J

(c) 500 J (d) 100 J.

19. Fig. 6 shows a block of mass m_1 resting on a smooth surface. It is connected to a mass m_2 by a string passing over a massless and frictionless pulley $m_2 > m_1$. The acceleration of the hanging mass m_2 is

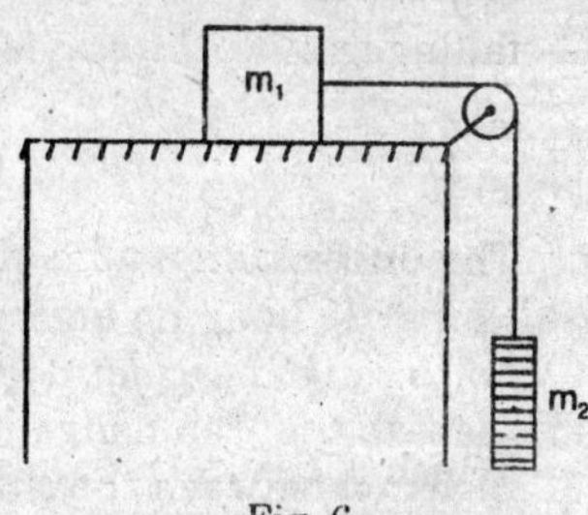

Fig. 6

(a) $\frac{m_1 - m_2}{m_1 + m_2} g$ (b) $\frac{m_2 - m_1}{m_2 + m_1} g$

(c) $\frac{m_1 g}{m_1 + m_2}$ (d) $\frac{m_2 g}{m_1 + m_2}$.

20. A disc of mass 10 g is kept floating horizontally in air by firing bullets, each of mass 5 g, with the same velocity at the same rate of 10 bullets per second. The bullets rebound with the same speed in opposite direction. The velocity of each bullet at the time of impact is

(a) 196 cm s^{-1} (b) 98 cm s^{-1}

(c) 49 cm s^{-1} (d) 392 cm s^{-1}.

21. A vehicle of mass 120 kg is moving with a uniform velocity of 108 km h^{-1}. The force required to stop the vehicle in 10 second is

(a) 90 N (b) 180 N

(c) 360 N (d) 720 N.

22. A 1000 kg lift is supported by a cable that can support 2000 kg. The shortest distance in which the lift can be stopped when it is descending with a speed of 2.5 m s^{-1} is [Take g = 10 m s^{-2}]

(a) 1 m (b) 2 m

(c) $\frac{5}{32}$ m (d) $\frac{5}{16}$ m.

23. A toy cart of mass $\sqrt{3}$ kg is pulled by a force of 20 N at an angle of 30° with the frictionless horizontal surface on which the cart is placed. The cart shall move on the surface with an acceleration

(a) 0 m s^{-2} (b) 1 m s^{-2}

(c) 10 m s^{-2} (d) $10\sqrt{3}$ m s^{-2}.

24. A balloon has 5 g of air. A small hole is pierced into it. The air escapes at a uniform rate with a velocity of 4 cm s^{-1}. If the balloon shrinks completely in 2.5 second, then the average force acting on the balloon is

(a) 2 dyne (b) 50 dyne

(c) 8 dyne (d) 8 N.

25. A U^{238} nucleus decays by emitting an alpha particle of speed v m s^{-1}. The recoil speed of the residual nucleus is

(a) $\frac{v}{4}$ m s^{-1} (b) $-\frac{4}{238}v$ m s^{-1}

(c) $-\frac{4}{234}v$ m s^{-1} (d) $\frac{4}{238}v$ m s^{-1}.

26. A partly hanging uniform chain of length L is resting on a rough horizontal table. l is the maximum possible length that can hang in equilibrium. The coefficient of friction between the chain and table is

(a) $\frac{l}{L-l}$ (b) $\frac{L}{l}$

(c) $\frac{l}{L}$ (d) $\frac{lL}{L+l}$.

27. A system is accelerated upward with an acceleration 5 g. The apparent weight of a body of mass m in the system is

(a) mg N (b) $5mg$ N

(c) $6mg$ N (d) $8mg$ N.

28. A body of mass 32 kg is suspended by a spring balance from the roof of a vertically operating lift and going downward from rest. At the instant the lift has covered 20 m and 50 m, the spring balance showed 30 kg and 36 kg respectively. Then the velocity of the lift is

(a) decreasing at 20 m and increasing at 50 m

(b) increasing at 20 m and decreasing at 50 m

(c) continuously decreasing at a steady rate throughout the journey

(d) constantly increasing at constant rate throughout the journey.

29. A rocket standing vertically on a launch pad has to start moving up with practically zero velocity. If the initial mass of the rocket is 5×10^5 kg, then the rate of burning of the fuel should be [Take g = 10 m s^{-2} and velocity of exhaust gases = 10 km s^{-1}]

(a) 10 kg s^{-1} (b) 50 kg s^{-1}

(c) 500 kg s^{-1} (d) 5000 kg s^{-1}.

30. A block of mass M is suspended from the ceiling of a room through a metal chain of mass m. The ceiling pulls the chain upwards with force

(a) mg (b) Mg

(c) $(m + M)g$ (d) zero.

31. In Q. 30, the tension at the mid-point of the metal chain is

(a) mg (b) $\frac{mg}{2}$

(c) $(M + m)g$ (d) $\left[M + \frac{m}{2}\right]g$.

32. An object at rest in space suddenly explodes into three parts of same mass. The momentum of the two parts are $2p\hat{i}$ and $p\hat{j}$. The momentum of the third part

(a) will have a magnitude $p\sqrt{3}$

(b) will have a magnitude $p\sqrt{5}$

(c) will have a magnitude p

(d) will have a magnitude $2p$.

33. An empty plastic box of mass 5 kg is observed to accelerate up at the rate of $\frac{g}{6}$ when placed deep inside water. What mass of sand should be put inside the box so that it may accelerate down at the rate of $\frac{g}{6}$?

(a) 1 kg (b) 1.5 kg

(c) 2 kg (d) 2.5 kg.

34. The acceleration of the system shown in Fig. 7 is

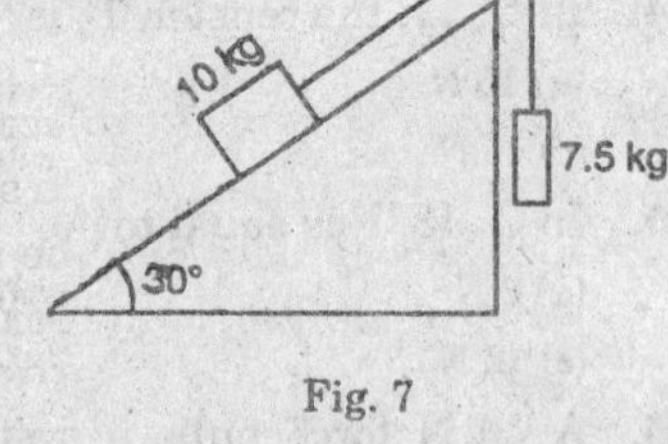

Fig. 7

(a) $\frac{3.5}{17.5}$ g

(b) $\frac{7.5}{17.5}$ g

(c) $\frac{14.5}{17.5}$ g

(d) $\frac{g}{7}$.

35. In Q. 34, if 7.5 kg is released from rest at t = 0, then the distance it will travel in 2 second is

(a) $\frac{g}{7}$ (b) $\frac{2g}{7}$

(c) $\frac{3g}{7}$ (d) $\frac{4g}{7}$.

36. A ball of mass 0.2 kg is thrown normally against a wall with a speed of 20 m s^{-1}. It rebounds normally with a speed of 15 m s^{-1}. The impulse of the force by the ball on the wall is

(a) 0.35 N s (b) 0.7 N s
(c) 7 N s (d) 70 N s.

37. A body of mass 2 kg is at rest on a horizontal table. The coefficient of friction between the body and the table is 0.3. A force of 5 N is applied on the body. The acceleration of the body is

(a) 0 m s^{-2} (b) 2.5 m s^{-2}
(c) 5 m s^{-2} (d) 7.5 m s^{-2}.

38. In Q. 37, the force of friction is

(a) 5 N (b) 5.88 N
(c) 6 N (d) 20 N.

39. A truck moving at $\frac{250}{9}$ m s^{-1} carries a steel girder which rests on its wooden floor. The minimum time in which the truck can come to a stop without the girder moving forward is : Given : $\mu_s = 0.5$.

(a) 3 s (b) 4 s
(c) 5 s (d) 5.7 s.

40. A block slides down an inclined plane of slope angle θ with constant velocity. It is then projected up the same plane with an initial speed v_0. How far up the incline will it move before coming to rest ?

(a) $\frac{v_0^2}{4g \sin \theta}$ (b) $\frac{4g \sin \theta}{v_0^2}$

(c) $g \sin \theta$ (d) $\frac{g}{\sin \theta}$.

41. A body released from the top of a smooth inclined plane reaches the bottom of the plane in 4 second. The time taken by the body to cover the first half of the inclined plane is

(a) 2 s (b) $2\sqrt{2}$ s
(c) $3\sqrt{2}$ s (d) 5 s.

42. A box of mass m kg is placed on the rear side of an open truck accelerating at 4 m s^{-2}. The coefficient of friction between the box and the surface below it is 0.4. The net acceleration of the box with respect to the truck is zero. The value of m is [Given : g = 10 m s^{-2}].

(a) 4 kg (b) 8 kg
(c) 9.78 kg (d) It could be any value.

43. A 2 kg body moving with a velocity of 10 m s^{-1} due east collides and sticks with a 2 kg body moving with a velocity of 10 m s^{-1} due north. The compound body moves north-east with velocity

(a) $2\sqrt{2}$ m s^{-1} (b) $3\sqrt{2}$ m s^{-1}
(c) $4\sqrt{2}$ m s^{-1} (d) $5\sqrt{2}$ m s^{-1}.

44. In a rocket, fuel is ejected with a velocity of 6 km s^{-1}. If the fuel burns at the rate of 1 kg s^{-1}, then the force exerted by the exhaust gases on the rocket is

(a) 6×10^3 N (b) 6×10^2 N
(c) 60 N (d) 6 N.

45. Two masses A and B of 10 kg and 5 kg respectively, are connected with a string passing over a frictionless pulley fixed at the corner of a table as shown in Fig. 8. The coefficient of friction of A with the table is 0.2. The minimum mass of C that may be placed on A to prevent it from moving is

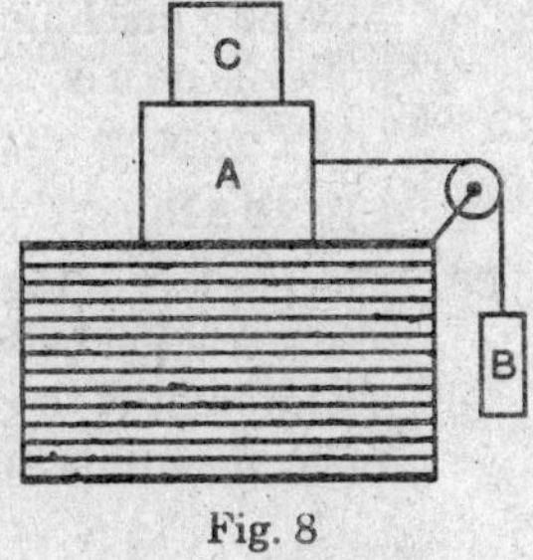

Fig. 8

(a) 15 kg (b) 5 kg
(c) 10 kg (d) 0 kg.

46. A triangular block of mass M with angles 30°, 60° and 90° rests with its 30°—90° side on a horizontal table. A cubical block of mass m rests on the 60°—30° side. The acceleration which M must have relative to the table to keep m stationary relative to the triangular block assuming frictionless contact is

(a) g (b) $\frac{g}{\sqrt{2}}$

(c) $\frac{g}{\sqrt{3}}$ (d) $\frac{g}{\sqrt{5}}$.

47. Fig. 9 shows the thrust-time graph of a rocket. If the mass of the rocket is 1 ton, then the velocity of the rocket after 1 minute, starting from rest, is

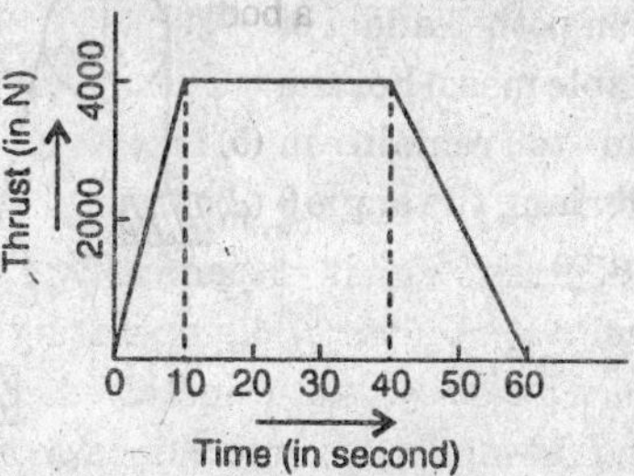

Fig. 9

(a) 18 m s^{-1} (b) 180 m s^{-1}
(c) 1800 m s^{-1} (d) None of these.

48. Three solids of masses m_1, m_2 and m_3 are connected with weightless strings in succession and are placed on a frictionless table. If the mass m_3 is dragged with a force T, the tension in the string between m_2 and m_3 is

(a) $\frac{m_2}{m_1 + m_2 + m_3}$ T (b) $\frac{m_3}{m_1 + m_2 + m_3}$ T

(c) $\frac{m_1 + m_2}{m_1 + m_2 + m_3}$ T (d) $\frac{m_2 + m_3}{m_1 + m_2 + m_3}$ T.

49. Two bodies of masses 6 kg and 3 kg are tied to the ends of a string which passes over a fixed pulley of the Atwood's machine. The total downward thrust on the pulley is nearly

(a) 5×9.8 N (b) 6×9.8 N

(c) 7×9.8 N (d) 8×9.8 N.

50. The block A in Fig. 10 weighs 100 N. The coefficient of static friction between the block and the table is 0.25. The weight of the block B is maximum for the system to be in equilibrium. The value of T_1 is

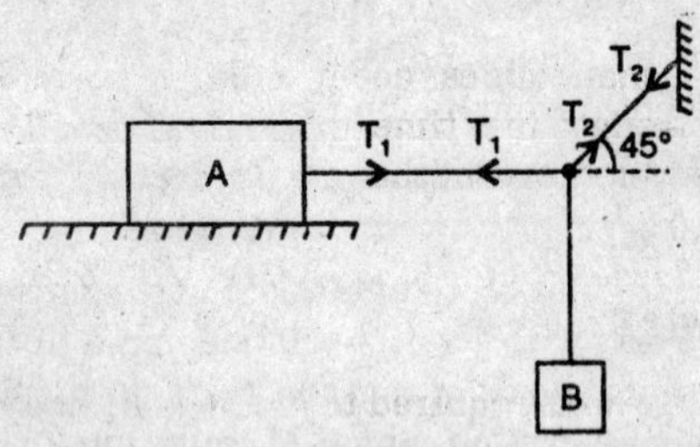

Fig. 10

(a) 0.25 N (b) 25 N

(c) 100 N (d) 100.25 N.

51. In Q. 50, the weight of the block B is

(a) 0.25 N (b) 25 N

(c) 100 N (d) 100.25 N.

52. The pulleys and the strings shown in Fig. 11 are smooth and of negligible mass. For the system to remain in equilibrium, the angle θ should be

(a) 0°

(b) 30°

(c) 45°

(d) 60°.

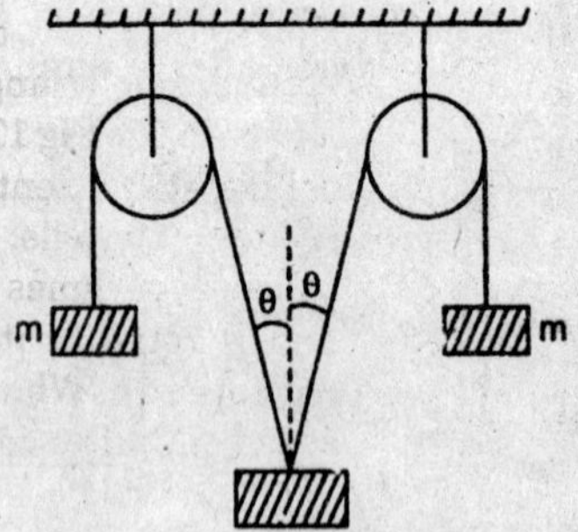

Fig. 11

53. A body of mass 8 kg lies on a rough horizontal table. It is observed that a certain horizontal force gives the body an acceleration of 4 m s^{-2}. When this force is doubled, the acceleration of the body is 16 m s^{-2}. The coefficient of friction is [Given : g = 10 m s^{-2}]

(a) 0.2 (b) 0.3

(c) 0.4 (d) 0.8.

54. Refer to the system shown in Fig. 12. The ratio of tensions T_1 and T_2 is

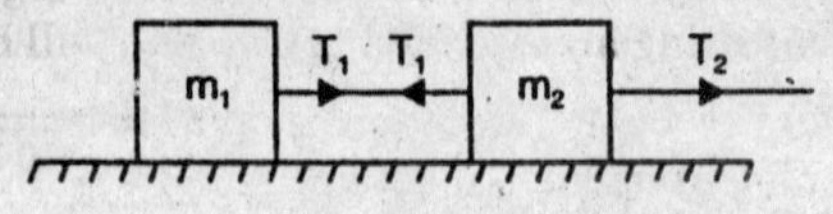

Fig. 12

(a) $\frac{m_1}{m_1 + m_2}$ (b) $\frac{m_2}{m_1 + m_2}$

(c) $\frac{m_1}{m_2}$ (d) $\frac{m_2}{m_1}$.

55. A block of mass 3 kg is in contact with a block of mass 2 kg. Both the blocks rest on a frictionless floor. A horizontal force of 10 N is applied to push the block of mass 3 kg. The force with which the first block pushes the second block is

(a) 0 N (b) 4 N

(c) 6 N (d) 10 N.

56. A bird of weight 5 N sits at the mid-point of a clothesline. The line is so distorted that it makes 30° angles with the horizontal at each end as shown in Fig. 13. The tension in the clothesline is

(a) 5 N

(b) 5 cos 30° N

(c) 5 sin 30° N

(d) 5.98 N.

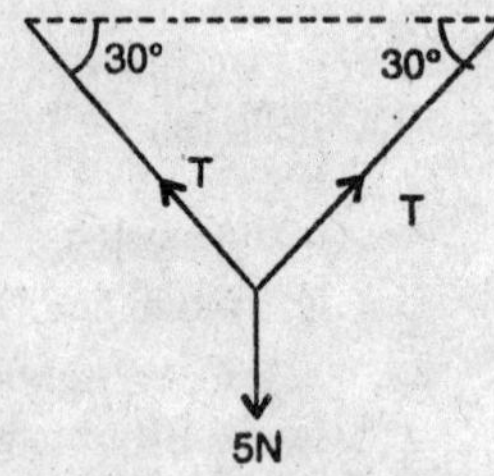

Fig. 13

57. A 100 kg block is suspended with the help of three strings A, B and C. The tension in the string C is

(a) 50 g N

(b) 100 g N

(c) 20 g N

(d) 100$\sqrt{2}$ g N.

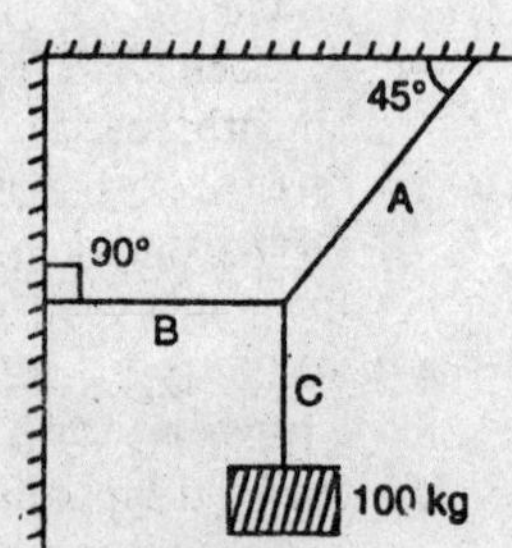

Fig. 14

58. In Q. 57, the tension in the string B is

(a) 50 g N (b) 100 g N

(c) gN (d) 100$\sqrt{2}$ g N.

59. In Q. 57, the tension in the string A is

(a) 50g N (b) 100g N

(c) g N (d) 100$\sqrt{2}$ g N.

60. A 40 kg slab rests on a frictionless floor. A 10 kg block rests on top of the slab. The static coefficient of friction between the block and the slab is 0.60 while the kinetic coefficient of friction is 0.40. The 10 kg block is acted upon by a horizontal force of 100 N. If $g = 9.8 \text{ m s}^{-2}$, the resulting acceleration of the slab will be

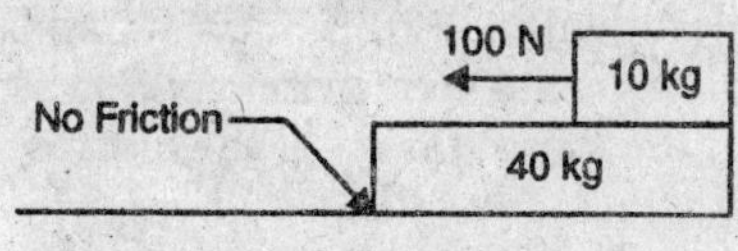

Fig. 15

(*a*) 1.47 m s^{-2} (*b*) 1.69 m s^{-2}
(*c*) 9.8 m s^{-2} (*d*) 0.98 m s^{-2}.

61. A rocket is fired upward from the earth's surface such that it creates an acceleration of 19.6 m s^{-2}. If after 5 s, its engine is switched off, the maximum height of the rocket from earth's surface would be

(*a*) 245 m (*b*) 490 m
(*c*) 980 m (*d*) 735 m.

62. A body of mass 6 kg moves in a straight line according to the equation

$$x = t^3 - 75t,$$

where x denotes the distance in metre and t the time in second. The force on the body at $t = 4$ second is

(*a*) 64 newton (*b*) 72 newton
(*c*) 144 newton (*d*) 36 newton.

63. A 500 kg horse pulls a cart of mass 1500 kg along a horizontal rough road with an acceleration of 1 m s^{-2}, $\mu_s = 0.2$. The forward force on the horse is [Take $g = 10 \text{ m s}^{-2}$]

(*a*) zero (*b*) 2000 N
(*c*) 4000 N (*d*) 6000 N.

64. A block of mass m kg is pulled along a horizontal smooth surface by a horizontal rope of mass 1 kg. A force of 15 N is applied at the free end of the rope. The force exerted by the block on the rope is 13.5 N. The mass m of the block is

(*a*) 1.5 kg (*b*) 3 kg
(*c*) 6 kg (*d*) 9 kg.

65. Two blocks of masses 1 kg and 2 kg rest on a smooth horizontal table. When the 2 kg block is pulled by a certain force F, the tension T in the string is 1.5 N. The value of F is

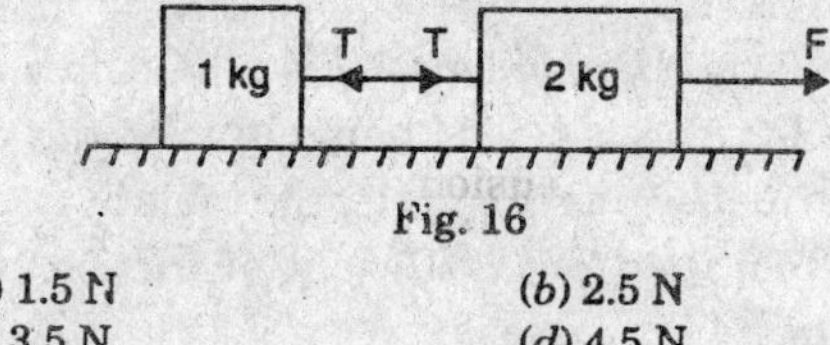

Fig. 16

(*a*) 1.5 N (*b*) 2.5 N
(*c*) 3.5 N (*d*) 4.5 N.

66. A box of mass 10 kg slides down a rough inclined plane of inclination 30° with uniform speed. The coefficient of kinetic friction in this case is

(*a*) 0.577 (*b*) 0.87
(*c*) 0.58 (*d*) 0.33.

67. A frictionless inclined plane of length l having inclination θ is placed inside a lift which is accelerating downward with on acceleration $a(< g)$. If a block is allowed to move, down the inclined plane, from rest, then the time taken by the block to slide from top of the inclined plane to the bottom of the inclined plane is

(*a*) $\sqrt{\dfrac{2l}{g}}$ (*b*) $\sqrt{\dfrac{2l}{g-a}}$

(*c*) $\sqrt{\dfrac{2l}{g+a}}$ (*d*) $\sqrt{\dfrac{2l}{(g-a)\sin\theta}}$.

68. A man slides down a light rope whose breaking strength is η times his weight ($\eta < 1$). The maximum acceleration of the man so that the rope just breaks is

(*a*) $g(1-\eta)$ (*b*) $g(1+\eta)$
(*c*) $g\,\eta$ (*d*) $\dfrac{g}{\eta}$.

69. The force required to just move a body up the inclined plane is double the force required to just prevent the body from sliding down the plane. The coefficient of friction is μ. The inclination θ of the plane is

(*a*) $\tan^{-1}\mu$ (*b*) $\tan^{-1}(\mu/2)$
(*c*) $\tan^{-1}2\mu$ (*d*) $\tan^{-1}3\mu$.

70. Three particles A, B and C of equal mass, move with equal speed v along the medians of an equilateral triangle as shown in the Fig 17. They collide at the centroid G of the triangle. After collision, A comes to rest and B retraces its path with speed v. What is the speed of C after collision ?

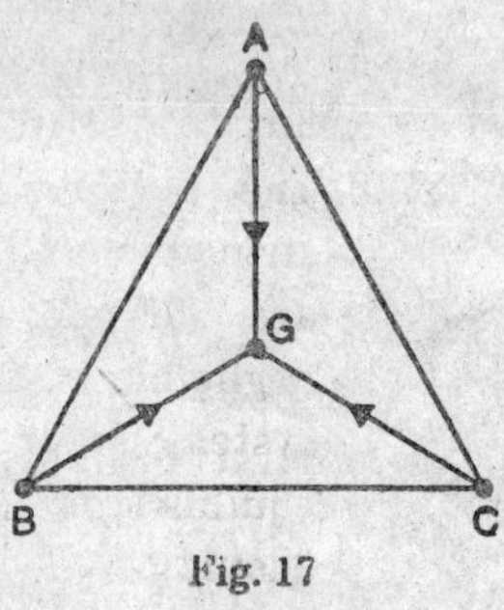

Fig. 17

(*a*) 0 (*b*) $\dfrac{v}{2}$
(*c*) v (*d*) $2v$.

71. In which direction will the particle C in Q. 70, move after the collision ?

(*a*) the same as that of B (*b*) opposite to that of B
(*c*) perpendicular to that of B
(*d*) at an angle of 120° to the direction of B.

72. A car starts from rest to cover a distance x. The coefficient of friction between the road and tyres is μ. The minimum time in which the car can cover distance x is proportional to

(a) μ (b) $\frac{1}{\sqrt{\mu}}$

(c) $\sqrt{\mu}$ (d) $\frac{1}{\mu}$.

73. A stationary bomb explodes into three pieces. One piece of 2 kg mass moves with a velocity of 8 m s^{-1} at right angles to the other piece of mass 1 kg moving with a velocity of 12 m s^{-1}. If the mass of the third piece is 0.5 kg, then the velocity of the third piece is

(a) 10 m s^{-1} (b) 20 m s^{-1}

(c) 30 m s^{-1} (d) 40 m s^{-1}.

74. An open carriage in a goods train is moving with a uniform velocity of 10 m s^{-1}. If the rain adds water with zero velocity at the rate of 5 kg s^{-1}, then the additional force applied by the engine to maintain the same velocity of the train is

(a) 0.5 N (b) 2.0 N

(c) 50 N (d) 25 N.

75. A block of metal weighing 2 kg is resting on a frictionless plane. It is struck by a jet releasing water at a rate of 1 kg s^{-1} and at a speed of 5 m s^{-1}. The initial acceleration of the block will be

(a) 2.5 m s^{-2} (b) 5.0 m s^{-2}

(c) 10 m s^{-2} (d) 10^3 km h^{-2}.

76. A wooden box of mass 8 kg slides down an inclined plane of inclination 30° to the horizontal with a constant acceleration of 0.4 m s^{-2}. What is the force of friction between the box and inclined plane? [g = 10 m s^{-2}]

(a) 36.8 N (b) 76.8 N

(c) 65.6 N (d) 97.8 N.

77. A block of mass M placed on a horizontal surface is connected to a nylon rope of mass M. A force of 1 N is applied to the free end of the rope in the horizontal direction. The tension at the mid-point of the rope is

(a) 0.25 N (b) 0.50 N

(c) 0.75 N (d) 1 N.

78. A body of mass 40 kg resting on a rough horizontal surface is subjected to a force P which is just enough to start the motion of the body. If $\mu_s = 0.5$, $\mu_k = 0.4$, $g = 10$ m s^{-2} and the force P is continuously applied on the body, then the acceleration of the body is

(a) zero (b) 1 m s^{-2}

(c) 2 m s^{-2} (d) 2.4 m s^{-2}.

79. A car is moving along a straight horizontal road with a velocity of 72 km h^{-1}. If $\mu_s = 0.5$, then the shortest distance in which the car can be stopped is [Take g = 10 m s^{-2}.]

(a) 40 m (b) 80 m

(c) 100 m (d) 120 m.

80. A rocket with a lift-off mass 10^5 kg is blasted upward with an initial acceleration of 5 m s^{-2}. If g = 10 m s^{-2}, then the initial thrust of the blast is

(a) 1.5 × 10^2 N (b) 1.5 × 10^3 N

(c) 1.5 × 10^5 N (d) 1.5 × 10^6 N.

81. A block of mass $\sqrt{3}$ kg rests on a horizontal frictionless XY plane. What would be the acceleration of the block if it is subjected to two forces as shown in Fig. 18?

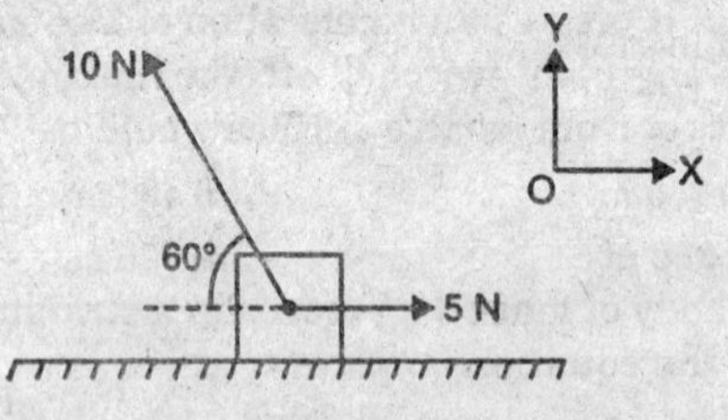

Fig. 18

(a) 2.5 m s^{-2} along Y-axis (b) 5 m s^{-2} along Y-axis

(c) 10 m s^{-2} along X-axis (d) 15 m s^{-2} along Y-axis.

82. A box weighing 20 kg is being pulled along a wooden floor with a force of 500 N. If $\mu_k = 0.5$, the acceleration of the block is [g = 10 m s^{-2}]

(a) 5 m s^{-2} (b) 25 m s^{-2}

(c) 20 m s^{-2} (d) 12.5 m s^{-2}.

83. A long platform of infinite mass is moving horizontally with a velocity of 6 m s^{-1}. A 10 g small block is gently lowered on it. The coefficient of friction between the block and the platform is 0.3. The distance moved by the block before coming to rest with respect to the platform is [Take g = 10 m s^{-2}]

(a) 3 m (b) 6 m

(c) 12 m (d) 18 m.

84. Two masses 3 kg and 5 kg are suspended from the ends of an unstretchable massless cord passing over a frictionless pulley. When the masses are released, the pressure on the pulley is

(a) 2 kg f (b) 7.5 kg f

(c) 8 kg f (d) 15 kg f.

85. A uniform chain of length l is placed on a rough table with length $l/n (n > 1)$ hanging over the edge. If the chain just begins to slide off the table by itself from this position, then the coefficient of friction between the chain and the table is

(a) $\frac{1}{n}$ (b) $\frac{1}{n-1}$

(c) $\frac{1}{n+1}$ (d) $\frac{n-1}{n+1}$.

86. A 25 g bullet is fired with a velocity of 400 m s^{-1} into a sand-bag of mass 4.975 kg suspended by a rope. The bullet gets embedded into the bag. The loss of kinetic energy is

(a) 1990 J (b) 2110 J

(c) 19 J (d) 8 J.

87. A block of mass m on a rough horizontal surface is acted upon by two forces as shown in Fig. 19. For equilibrium of block, the coefficient of friction between block and surface is

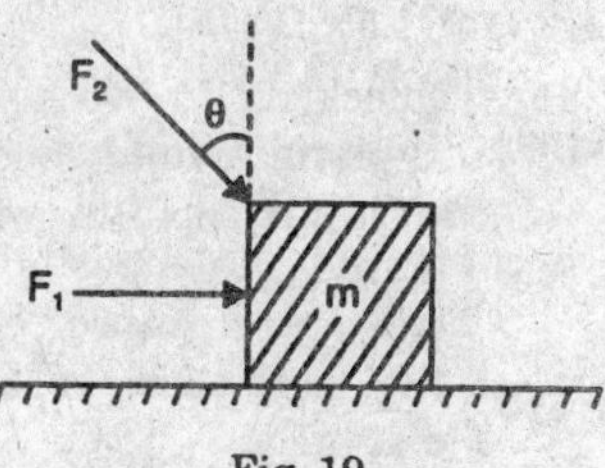

Fig. 19

(a) $\frac{F_1 + F_2 \sin\theta}{mg + F_2 \cos\theta}$ (b) $\frac{F_1 \cos\theta + F_2}{mg - F_2 \sin\theta}$

(c) $\frac{F_1 + F_2 \cos\theta}{mg + F_2 \sin\theta}$ (d) $\frac{F_1 \sin\theta - F_2}{mg - F_2 \cos\theta}$.

88. Refer to the system shown in Fig. 20. Assume that the inclined plane is smooth. Which of the following is correct ?

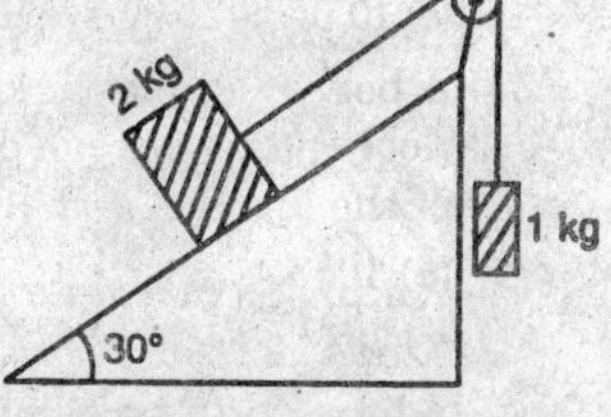

Fig. 20

(a) 1 kg body moves downward with acceleration g.

(b) 2 kg body moves down the inclined plane with acceleration $g \sin 30°$.

(c) The system is at rest.

(d) All of the above.

89. A vertical board having a rough surface is accelerated horizontally with acceleration a such that a 5 kg block pressing against it does not fall. If μ be the coefficient of friction between the block and the board, then which of the following is the correct condition so that the block does not fall ?

(a) $\mu \le \frac{g}{a}$ (b) $\mu \ge \frac{g}{a}$

(c) $\mu \ge g$ (d) $\mu \le g$.

90. A 0.5 kg ball thrown up in air with velocity 30 m s^{-1} reaches the highest point in 2.5 second. If g = 10 ms^{-2}, then the average air resistance encountered by the ball during upward motion is

(a) zero (b) 0.5 N

(c) 1 N (d) 1.5 N.

91. An open knife edge of mass M is dropped from a height h on a wooden floor. If the blade penetrates distance S into the wood, the average resistance offered by the wood to the blade is

(a) Mg (b) $Mg\left(1 - \frac{h}{S}\right)$

(c) $Mg\left(1 + \frac{h}{S}\right)$ (d) Mgh.

92. A weight W rests an a rough horizontal plane. If the angle of friction be θ, the least force that will move the body along the plane will be

(a) W cos θ (b) W cot θ

(c) W tan θ (d) W sin θ.

93. A stone is accelerated upwards by a cord whose breaking strength is three times the weight of the stone. The maximum acceleration with which the stone can be moved up without breaking the cord is

(a) g (b) $2g$

(c) $3g$ (d) $4g$.

94. A block of weight W is suspended from the mid-point of a rope whose ends are at the same horizontal level. The force required to straighten the rope is

(a) W (b) 2W

(c) $\frac{W}{2}$ (d) infinitely large.

95. A block weighing 10 N is resting on a rough inclined plane of inclination 30°). $\mu_s = \frac{\sqrt{3}}{2}$. The force of friction on the block is [Take : g = 10 m s^{-2}]

(a) 3 N (b) 5 N

(c) 7.5 N (d) 10 N.

96. A bob of mass 0.450 kg hangs from the massless string of a long simple pendulum. A bullet of mass 0.05 kg is fired vertically from below into the bob. The bullet gets embedded into the bob and the combination rises vertically through a height of 1.8 m. If g = 10 m s^{-2}, then the velocity of the bullet is

(a) 6 m s^{-1} (b) 60 m s^{-1}

(c) 600 m s^{-1} (d) 6000 m s^{-1}.

97. A bullet fired from a gun experiences a force of $600 - 2 \times 10^5\, t$ in the barrel of the gun. Here t is in second. The force becomes zero as soon as the bullet leaves the barrel of the gun. The time for which the bullet was in the barrel of the gun is

(a) 2×10^5 s (b) 10^5 s

(c) 3×10^5 s (d) 3×10^{-3} s.

98. In Q. 97, the average impulse imparted to the bullet is [Assume that force is given in newton]

(*a*) 0.6 Ns (*b*) 0.7 Ns

(*c*) 0.9 Ns (*d*) 0.98 Ns.

99. A horizontal jet of water of cross-sectional area 10 cm^2 hits a vertical wall with a velocity of 10 m s^{-1}. After having hit the wall, water moves parallel to the wall. The force exerted by water on the wall is

(*a*) 10^2 N (*b*) 10^3 N

(*c*) 10^4 N (*d*) 10^5 N.

100. A block of mass m_1 kg is resting on a rough horizontal plane, coefficient of kinetic friction between block and surface is μ. If m_1 is connected to another mass m_2 with the help of string and pulley as shown in the diagram, then the common acceleration when released from rest will be

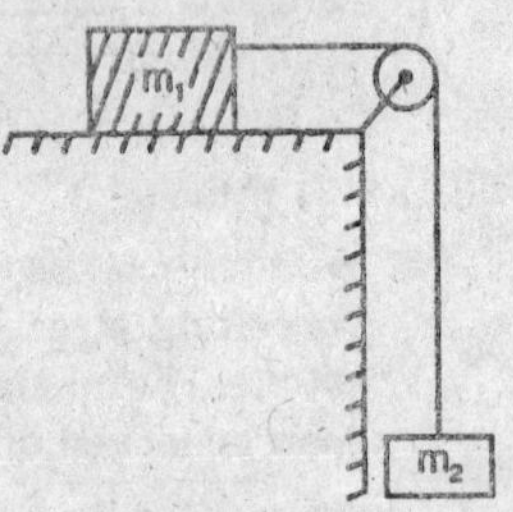

Fig. 21

(*a*) $\frac{m_2 g}{m_1 + m_2}$ (*b*) $\left[\frac{m_2 - m_1}{m_1 + \mu m_2}\right]g$

(*c*) $\left[\frac{\mu m_2 + m_1}{m_1 + m_2}\right]g$ (*d*) $\left[\frac{m_2 - \mu m_1}{m_1 + m_2}\right]g$.

101. A block is lying on an inclined plane which makes an angle of 60° with the horizontal. If coefficient of friction between block and plane is 0.25 and g = 10 m s^{-2}, then the acceleration of block when it moves along the plane will be [Take $\sqrt{3}$ = 1.7]

(*a*) 2.50 m s^{-2} (*b*) 5.00 m s^{-2}

(*c*) 7.25 m s^{-2} (*d*) 8.66 m s^{-2}.

102. A trailer of mass 1000 kg is towed by means of a rope attached to a car moving at a steady speed along a level road. The tension in the rope is 400 N. The car starts to accelerate steadily. If the tension in the rope is now 1650 N, with what acceleration is the trailer moving ?

(*a*) 1.75 m s^{-2} (*b*) 0.75 m s^{-2}

(*c*) 2.5 m s^{-2} (*d*) 1.25 m s^{-2}.

103. A stuntman weighing 80 kg drops himself gently into an open car weighing 720 kg and moving with a velocity of 80 km h^{-1}. The velocity of the car after the event is

(*a*) 36 km h^{-1} (*b*) 72 km h^{-1}

(*c*) 80 km h^{-1} (*d*) None of these.

104. An object of mass 5 kg is acted upon by a force that varies with the position of the object as shown in Fig. 22. If the object starts out from rest at the point x = 0, then its speed at x = 25 m is

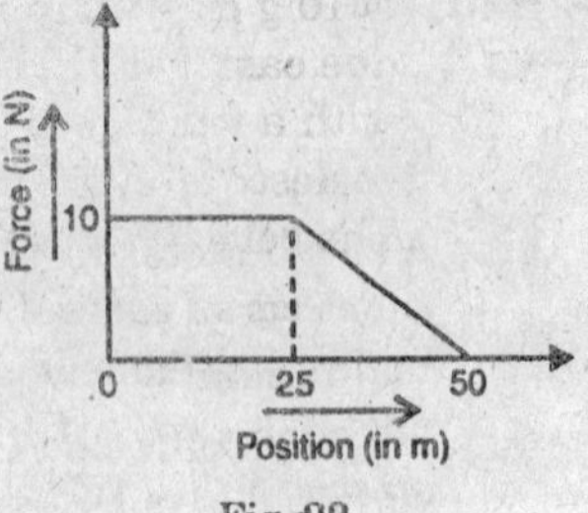

Fig. 22

(*a*) 5 m s^{-1} (*b*) 10 m s^{-1}

(*c*) 20 m s^{-1} (*d*) 28 m s^{-1}.

105. Two trolleys of masses m and 3 m are connected by a spring. When they are compressed and released, they move off in opposite directions and come to rest after covering distances S_1 and S_2 respectively. Assuming the coefficient of friction to be uniform, ratio of distances $S_1 : S_2$ is

(*a*) 1 : 9 (*b*) 1 : 3

(*c*) 3 : 1 (*d*) 9 : 1.

106. A ball of mass 0.05 kg travelling at 4 m s^{-1} hits a wall and rebounds without any change in its speed. If the ball remained in contact with the wall for 0.01 s, then the force exerted by the ball on the wall is

(*a*) 0.05 N (*b*) 0.01 N

(*c*) 50 N (*d*) 40 N.

107. Two blocks of masses 3 m and 2 m are in contact on a smooth table. A force P is first applied horizontally on block of mass 3 m and then on mass 2 m. The contact forces between the two blocks in the two cases are in the ratio

Fig. 23

(*a*) 1 : 2 (*b*) 2 : 3

(*c*) 3 : 2 (*d*) 5 : 3.

108. A lineman of mass 50 kg is holding a vertical pole. The coefficient of friction between his hands and the pole is 0.5. If he is to climb the pole, the minimum force with which he should press the pole with his hands is [Take g = 10 m s^{-2}]

(*a*) 500 N (*b*) 250 N

(*c*) 750 N (*d*) 1000 N.

109. A block can slide on a smooth inclined plane of inclination θ kept on the floor of a lift. When the lift is descending with retardation a, the acceleration of the block relative to the incline is

(*a*) $(g + a)$ sin θ (*b*) $g - a$

(*c*) g sin θ (*d*) $(g - a)$ sin θ.

110. A 10 g clay lump moving with a velocity of 10 cm s^{-1} due east collides head on with 10 g clay lump moving with a velocity of 20 cm s^{-1} due west. The two lumps coalesce after the collision. No external force has acted. The velocity of the compound lump is

(a) 5 cm s^{-1} due west (b) 5 cm s^{-1} due east

(c) 15 cm s^{-1} due west (d) 15 cm s^{-1} due east.

111. A 5000 kg rocket is set for vertical launching with an upward acceleration of 20 m s^{-2}. If the exhaust speed of gases is to be kept at 800 m s^{-1}, then the amount of gases ejected per second to supply the required thrust should be [Take g = 10 m s^{-2}]

(a) 187.5 kg s^{-1} (b) 62.5 kg s^{-1}

(c) 125 kg s^{-1} (d) 1 kg s^{-1}.

112. A 6000 kg rocket is set for vertical firing. If the exhaust speed is 1000 m s^{-1}, the gas ejected per second to supply the thrust needed to overcome the weight of the rocket is [Take g = 10 m s^{-2}]

(a) 6 kg s^{-1} (b) 60 kg s^{-1}

(c) 600 kg s^{-1} (d) 6000 kg s^{-1}.

113. A 1 kg object suspended by a thread is (i) lifted up with an acceleration of $\frac{g}{2}$ (ii) lowered with an acceleration of $\frac{g}{2}$. The ratio of the tensions in the two cases is

(a) 1 : 2 (b) 1 : 3

(c) 2 : 1 (d) 3 : 1.

114. A rope of length L is pulled by a constant force F. What is the tension in the rope at a distance x from the end where the force is applied ?

(a) $\frac{FL}{x}$ (b) $\frac{F(L-x)}{L}$

(c) $\frac{FL}{L-x}$ (d) $\frac{Fx}{L-x}$.

115. A rocket of mass 5000 kg is to be projected vertically upwards. The gases are exhausted vertically downwards with velocity 1000 m s^{-1} with respect to the rocket. What is the minimum rate of burning the fuel so as to just lift the rocket upwards against gravitational attraction ?

(a) 49 kg s^{-1} (b) 147 kg s^{-1}

(c) 98 kg s^{-1} (d) 196 kg s^{-1}.

116. In Q. 115, if the rocket is to be launched with acceleration $2g$, what is the minimum rate of burning the fuel ?

(a) 49 kg s^{-1} (b) 98 kg s^{-1}

(c) 147 kg s^{-1} (d) 196 kg s^{-1}.

117. The ratio of the weight of a man in a stationary lift and in a lift accelerating downwards with a uniform acceleration 'a' is 3 : 2. The acceleration of the lift is

(a) $g/3$ (b) $g/2$

(c) g (d) $2g$.

118. In the arrangement shown in Fig. 24, if the surface is smooth, the acceleration of the block m_2 will be

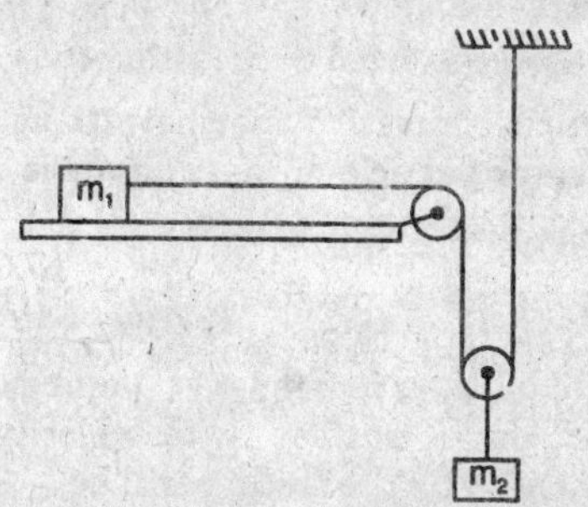

Fig. 24

(a) $\frac{m_2 g}{4m_1 + m_2}$ (b) $\frac{2m_2 g}{4m_1 + m_2}$

(c) $\frac{2m_2 g}{m_1 + 4m_2}$ (d) $\frac{2m_1 g}{m_1 + m_2}$.

119. In the Fig. 25, a smooth pulley of negligible weight is suspended by a spring balance. Weights of 1 kg and 5 kg are attached to the opposite ends of a string passing over the pulley and move with acceleration because of gravity. During their motion, the spring balance reads a weight of

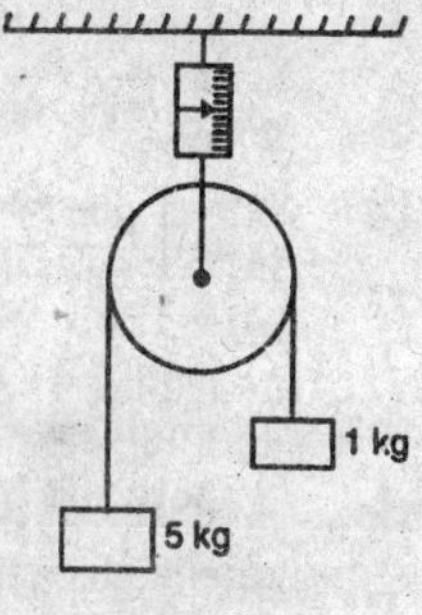

Fig. 25

(a) 6 kg

(b) less than 6 kg

(c) more than 6 kg

(d) may be more or less than 6 kg.

120. A ball hangs from a string inside a rail road car moving along a straight track. The string is observed to be inclined towards the rear of the car making a constant small angle with the vertical. It shows that the car is

(a) moving with a uniform acceleration

(b) moving with a uniform velocity

(c) moving with a uniform retardation

(d) moving with an acceleration which is increasing uniformly.

121. Three blocks of masses m_1, m_2 and m_3 are connected by massless strings as shown on a frictionless table in Fig. 26. They are pulled with a force T_3 = 40 N. If m_1 = 10 kg, m_2 = 6 kg and m_3 = 4 kg, the tension T_2 will be

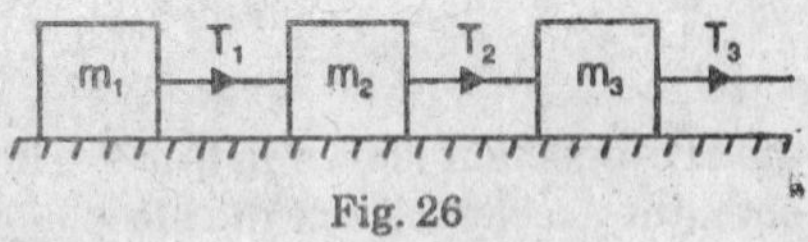

Fig. 26

(*a*) 20 N (*b*) 40 N
(*c*) 10 N (*d*) 32 N.

122. A particle of mass '*m*' moving with velocity *v* strikes a stationary particle of mass '2*m*' and sticks to it. The speed of the system will be

(*a*) *v*/2 (*b*) 2*v*
(*c*) *v*/3 (*d*) 3*v*.

123. *n* small balls each of mass '*m*' impinge elastically each second on a surface with velocity *u*. The force experienced by the surface will be

(*a*) *mnu* (*b*) 2 *mnu*
(*c*) 4 *mnu* (*d*) $\frac{1}{2}$ *mnu*.

124. A block starts to move on a rough horizontal surface having coefficient of friction μ = 0.5. The block stops after travelling a distance of 9.8 m. The initial velocity of the block was

(*a*) 4.9 m s^{-1} (*b*) 9.8 m s^{-1}
(*c*) 19.6 m s^{-1} (*d*) 39.2 m s^{-1}.

125. A particle moves so that its acceleration is always twice its velocity. If its initial velocity is 0.1 m s^{-1}, its velocity after it has gone 0.1 m is

(*a*) 0.3 m s^{-1} (*b*) 0.7 m s^{-1}
(*c*) 1.2 m s^{-1} (*d*) 3.6 m s^{-1}.

126. A block A slides down a smooth inclined plane when released from the top in time *t*. Another block falls freely from the same point and strikes the ground in time *t*/2. The value of angle θ is

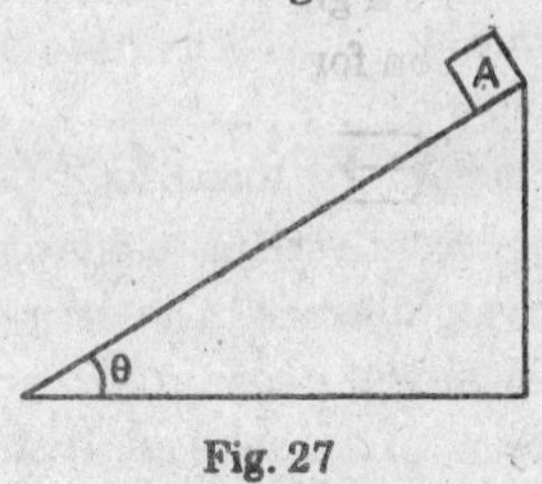

Fig. 27

(*a*) 15° (*b*) 30°
(*c*) 45° (*d*) 60°.

127. An object is resting at the bottom of two strings which are inclined at an angle of 120° with each other. Each string can withstand a tension of 20 N. The maximum weight of the object that can be supported without breaking the string is

(*a*) 5 N (*b*) 10 N
(*c*) 20 N (*d*) 40 N.

128. Refer to the system shown in Fig. 28. The acceleration of the masses is

(*a*) $\frac{g}{3}$ (*b*) $\frac{g}{6}$
(*c*) $\frac{g}{9}$ (*d*) $\frac{g}{12}$.

129. The tension T_1 in the string between B and C is

(*a*) $\frac{2g}{9}$ (*b*) $\frac{4g}{9}$
(*c*) $\frac{6g}{9}$ (*d*) $\frac{10g}{9}$.

Fig. 28

130. The tension T_2 in the string between A and B is

(*a*) $\frac{10g}{9}$ (*b*) $\frac{20g}{9}$
(*c*) $\frac{30g}{9}$ (*d*) $\frac{40g}{9}$.

131. A body of mass *m* has its position *x* at a time *t*, expressed by the equation :

$$x = 3t^{3/2} + 2t - \frac{1}{2}$$

The instantaneous force F on the body is proportional to

(*a*) $t^{3/2}$ (*b*) t
(*c*) $t^{-1/2}$ (*d*) t^0.

132. A bomb explodes in air when it has a horizontal speed of 100 km h^{-1}. It breaks into two pieces A, B of mass ratio 1 : 2. If A goes vertically at speed 400 km h^{-1}, the speed of B will be

(*a*) 200 km h^{-1} (*b*) 250 km h^{-1}
(*c*) 300 km h^{-1} (*d*) 500 km h^{-1}.

133. On a smooth plane surface (Fig. 29), two blocks A and B are accelerated up by applying a force 15 N on A. If mass of B is twice that of A, the force on B is

Fig. 29

(*a*) 30 N (*b*) 15 N
(*c*) 10 N (*d*) 5 N.

134. A boy stands on a weighing machine inside a lift. When the lift is going down with acceleration *g*/4, the machine shows a reading 30 kg. When the lift goes upwards at acceleration *g*/4, the reading would be

(a) 18 kgf (b) 37.5 kgf

(c) 50 kgf (d) 67.5 kgf.

135. A block of wood is kept on the floor of a stationary elevator. The elevator begins to descend with an acceleration of 12 m s^{-2}. If g = 10 m s^{-2}, then the displacement of the block during the first 0.2 second after the start is

(a) 0.02 m (b) 0.1 m

(c) 0.2 m (d) 0.4 m.

136. Two weights w_1 and w_2 are connected by a light thread which passes over a light smooth pulley. If the pulley is raised upwards with an acceleration equal to g, then the tension in the thread will be

(a) $\frac{2w_1w_2}{w_1+w_2}$ (b) $\frac{w_1w_2}{w_1+w_2}$

(c) $\frac{4w_1w_2}{w_1+w_2}$ (d) $\frac{4w_1w_2}{w_1-w_2}$.

137. A dynamometer D is attached to two blocks of masses 6 kg and 4 kg. Forces of 20 N and 10 N are applied on the blocks as shown in Fig. 30. The dynamometer reads

Fig. 30

(a) 10 N (b) 20 N

(c) 6 N (d) 14 N.

138. Two masses 5 kg and 3 kg are suspended from the ends of an inextensible light string passing over a frictionless pulley. When the masses are released, the force exerted by the string on the pulley is

(a) $5g$ (b) $8g$

(c) $7.5g$ (d) $15g$.

139. A block has weight W and it is held against a vertical wall by applying a horizontal force F. The minimum value of F is

(a) less than W (b) equal to W

(c) greater than W (d) not settled by the data.

140. A 10 kg stone is suspended with a rope of breaking strength 30 kg-wt. The minimum time in which the stone can be raised through a height 10 m starting from rest is (Take g = 10 N/kg).

(a) 0.5 s (b) 1.0 s

(c) $\sqrt{2/3}$ s (d) 2 s.

141. A block Q of mass M is placed on a horizontal frictionless surface FF, and a body P of mass m is released on its frictionless slope. As P slides by length L on this slope of inclination θ, the block Q would slide by distance

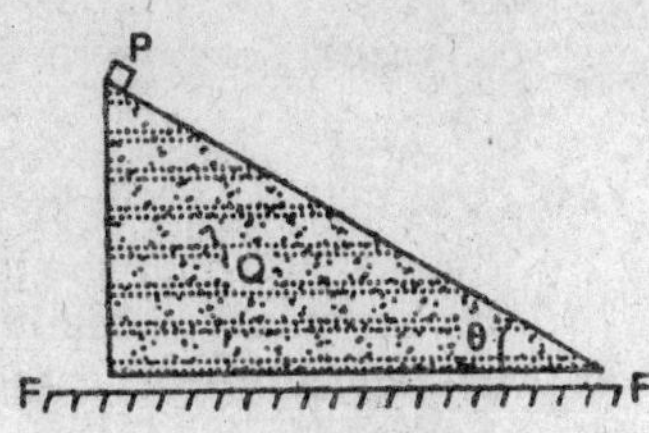

Fig. 31

(a) (m/M)L cos θ (b) (mL cos θ)/(m + M)

(c) mL/(m + M) (d) None of these.

142. A trolley T (mass 5 kg) on a horizontal smooth surface is pulled by a load L (2 kg) through a uniform rope ABC of length 2 m and mass 1 kg. As the load falls from BC = 0 to BC = 2 m, its acceleration (in m s^{-2}) changes from (Take g = 10 m s^{-2})

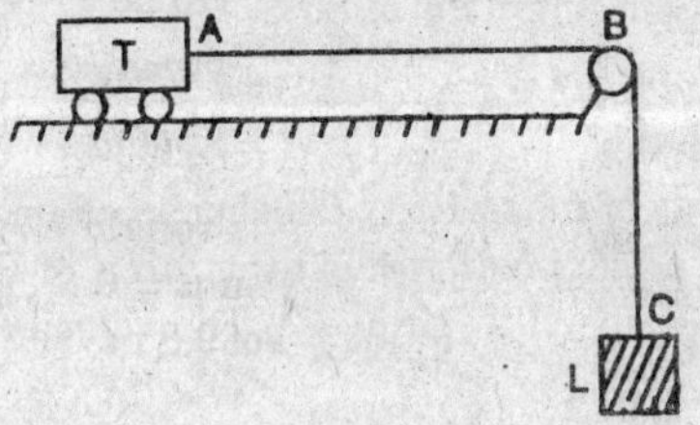

Fig. 32

(a) 20/6 to 20/5 (b) 20/8 to 30/8

(c) 20/5 to 30/6 (d) None of these.

143. A block moving up an inclined plane of inclination 30° with a velocity of 5 m s^{-1} stops after 0.5 second. If g = 10 m s^{-2}, then the approximate value of coefficient of friction is

(a) 0.2 (b) 0.4

(c) 0.6 (d) 0.8.

144. In Fig. 33 shows a girder supporting a load of 10,000 N. The compression force F in newton in the girder is

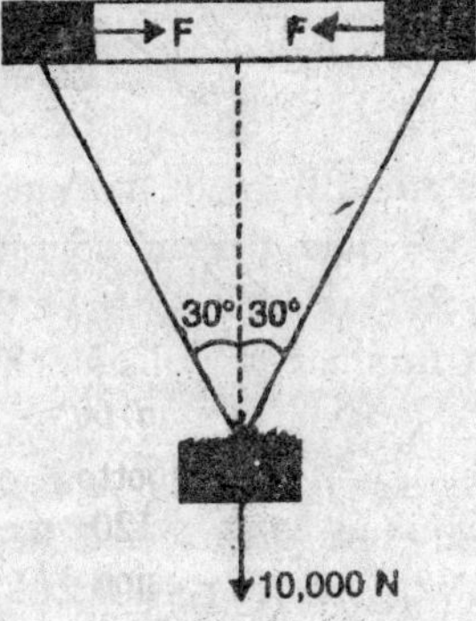

Fig. 33

(a) $\frac{5000}{\sqrt{3}}$ (b) $\frac{10000}{\sqrt{3}}$

(c) 5000 (d) $5000\sqrt{3}$

(e) 10000.

145. A ball of mass 0.4 kg thrown up in air with velocity 30 m s^{-1} reaches the highest point in 2.5 second. The air resistance encountered by the ball during upward motion is

(a) 0.88 N (b) 8800 N

(c) 300 dyne (d) 300 N.

146. A body of mass 2 kg has an initial velocity of 3 m s^{-1} along OE and it is subjected to a force of 4 newton in OF direction perpendicular to OE. The distance of the body from O after 4 s will be

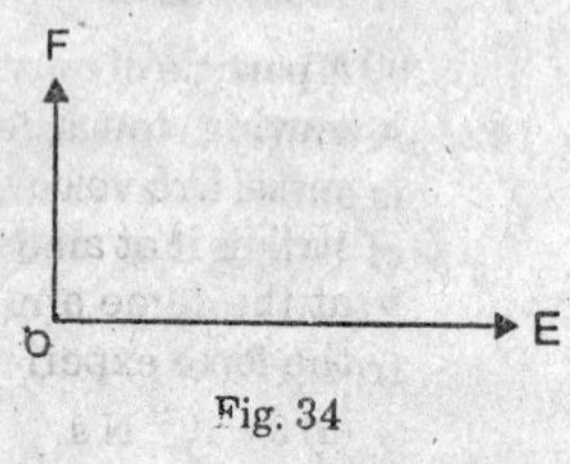

Fig. 34

(a) 12 m (b) 20 m

(c) 28 m (d) 48 m.

147. A body of mass 10 kg is being acted upon by a force $3t^2$ and an opposing constant force of 32 newton. The initial speed is 10 m s^{-1}. The velocity of body after 5 second is

(a) 14.5 m s^{-1} (b) 6.5 m s^{-1}

(c) 3.5 m s^{-1} (d) 4.5 m s^{-1}.

148. A hammer of mass M strikes a nail of mass m with velocity of v m s^{-1} and drives it 'a' metre into fixed block of wood. The average resistance of the wood to the penetration of the nail is

(a) $\frac{M}{M+m} \cdot \frac{v^2}{2a}$ (b) $\frac{M^2}{M+m^2} \cdot \frac{v^2}{2a}$

(c) $\frac{M+m}{M} \cdot \frac{v^2}{2a}$ (d) $\frac{M^2}{M+m} \cdot \frac{v^2}{2a}$.

149. A football player kicks a 0.5 kg ball and imparts it a velocity of 10 m s^{-1}. The contact between foot and ball is only for one-fiftieth of a second. The kicking force is

(a) 250 N (b) 125 N

(c) 0 N (d) 3.78 N.

150. A machine gun fires a bullet of mass 40 g with velocity 1200 m s^{-1}. The person holding it can apply a maximum force of 144 newton on the gun. What is the maximum number of bullets that can be fired per second ?

(a) 3 (b) 4

(c) 5 (d) 7.

151. A block of mass M is pulled along a horizontal frictionless surface by a rope of mass m. Force P is applied at one end of the rope. The force which the rope exerts on the block is

(a) $\frac{P}{M(m+M)}$ (b) $\frac{P}{M-m}$

(c) $\frac{Pm}{M-m}$ (d) $\frac{PM}{m+M}$.

152. Husband and wife stand facing each other and pull on a light rope stretched between them. How far each has moved when they meet ?

Given : Husband's weight = 60 kg wt.

Wife's weight = 40 kg wt.

Initial separation = 5 m.

(a) Wife moves 3 m while husband moves 2 m

(b) Wife moves 2 m while husband moves 3 m

(c) Both move 2.5 m towards each other

(d) Wife does not move while the husband moves 5 m.

153. An astronaut of weight M g is in a rocket accelerating upward with an acceleration of 4 g. The apparent weight of the astronaut will be

(a) 5 Mg (b) 4 Mg

(c) Mg (d) zero.

154. Fig. 35 shows x and y are the two strings. The tension in the string x is

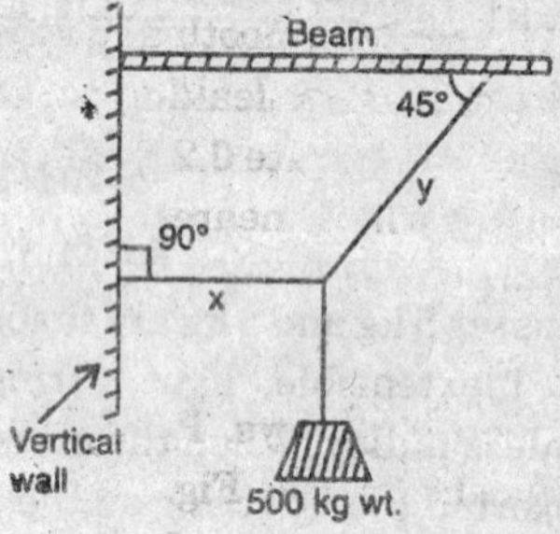

Fig. 35

(a) zero (b) 500 cos 45° kg wt.

(c) 500 sin 45° kg wt. (d) 500 kg wt.

155. A smooth inclined plane of length L having inclination θ with the horizontal is inside a lift which is moving down with retardation a. The time taken by a body to slide down the inclined plane, from rest, will be

(a) $\sqrt{\frac{2L}{(g+a)\sin\theta}}$ (b) $\sqrt{\frac{2L}{(g-a)\sin\theta}}$

(c) $\sqrt{\frac{2L}{g\sin\theta}}$ (d) $\sqrt{\frac{2L}{a\sin\theta}}$.

156. A plate of mass M is placed on a horizontal frictionless surface (see Fig. 36), and a body of mass m is placed on this plate. The coefficient of dynamic friction between this body and the plate is μ. If a force 2μ *mg* is applied to the body of mass *m* along the horizontal, the acceleration of the plate will be

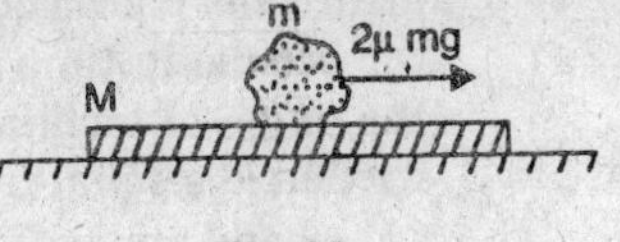

Fig. 36

(*a*) $\frac{\mu m}{M} g$ (*b*) $\frac{\mu m}{(M+m)} g$

(*c*) $\frac{2\mu m}{M} g$ (*d*) $\frac{2\mu m}{(M+m)} g$.

157. A two-car train, with data as given in Fig. 37, is pulled by a child on a horizontal frictionless surface. The strings are massless. The force exerted by car A on the string S is

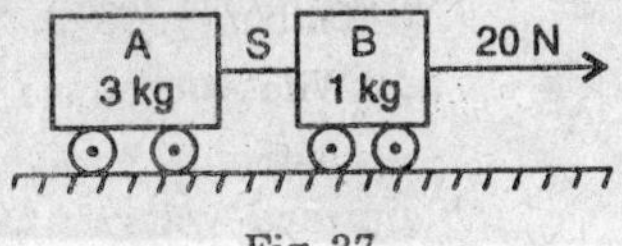

Fig. 37

(*a*) 10 N (*b*) 20 N

(*c*) zero (*d*) 15 N.

158. A trolley of mass 20 kg carries 5 kg grain and moves on a horizontal, smooth and straight track at 20 ms^{-1}. If the grain starts leaking out of a hole at the bottom at time $t = 0$, at rate 0.2 kg s^{-1}, the speed of the trolley at $t = 10$ s will be nearest to

(*a*) 20 ms^{-1} (*b*) 25 ms^{-1}

(*c*) 15 ms^{-1} (*d*) 13.3 ms^{-1}.

159. A system of pulleys, P_1 and P_2 is shown in Fig. 38. Neglecting weight of the pulleys, the pull needed at A to lift a load W will be

(*a*) W/2 (*b*) W/3

(*c*) a little above W/2

(*d*) a little below W/3.

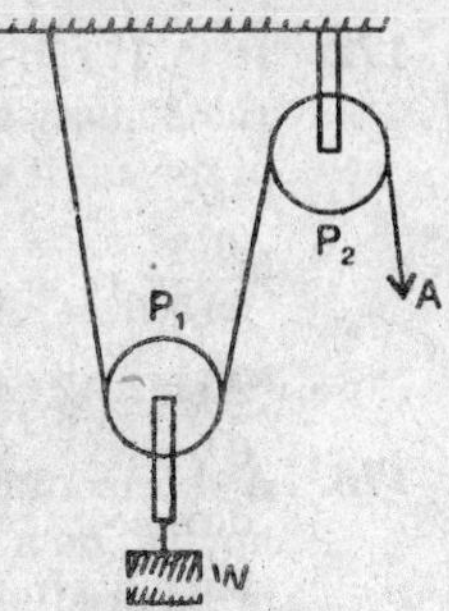

Fig. 38

160. A father (60 kg) and his daughter (20 kg) are both at rest on a frictionless ice pond. The father lifts a 1 kg ball and throws it to his daughter with horizontal speed 5 ms^{-1}; the daughter catches it. The speeds of father and daughter are (in ms^{-1}) is

(*a*) 1/12, 5/21 (*b*) 5/59, 1/4

(*c*) 5/61, 1/4 (*d*) 5/59, 5/21.

161. In the Atwood machine (data as in Fig. 39), P is a massless pulley, and springs S_1 and S_2 are also massless. If the blocks are let go, the readings in S_1 and S_2 will respectively be

(*a*) 1/3 kg, 2/3 kg

(*b*) 4/3 kg, 4/3 kg

(*c*) 2/3 kg, 4/3 kg

(*d*) 2/3 kg, 2/3 kg.

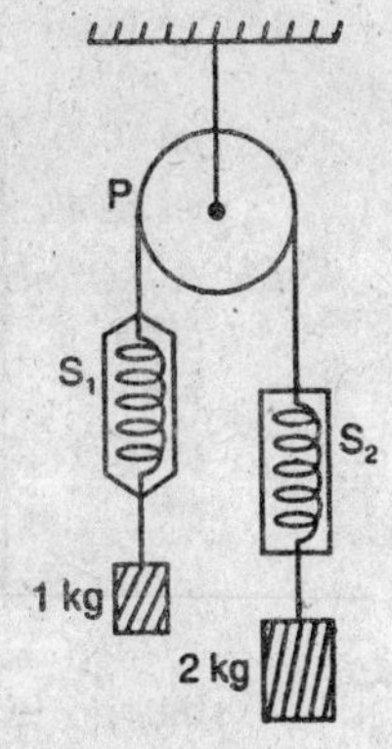

Fig. 39

162. A particle of mass $m = 9 \times 10^{-31}$ kg moving towards the wall of a vessel at a velocity of $v = 600$ m s^{-1} strikes it at an angle of 60° to the normal and rebounds at the same angle at the same speed. The impulse of the force experienced by the wall during the impact is

(*a*) 3×10^{21} N s (*b*) 9×10^{-28} N s

(*c*) 5.4×10^{-28} N s (*d*) 5.4×10^{-27} N s.

163. In the system shown in Fig. 40, the pulley is smooth and massless, the string has a total mass 5 g, and the two suspended blocks have masses 25 g and 15 g. The system is released from state $l = 0$ and is studied upto stage $l' = 0$. During the process, the acceleration of block A will (with g_E for acceleration due to gravity) be

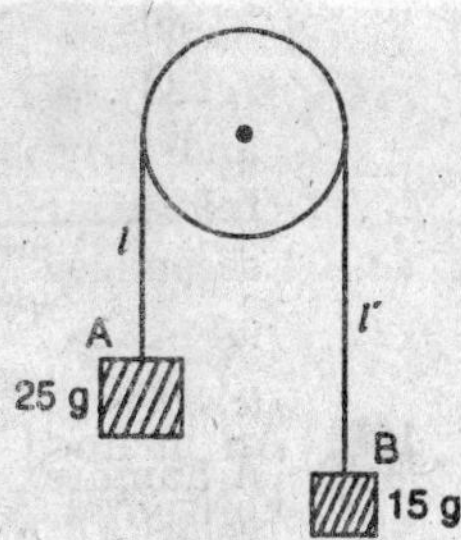

Fig. 40

(*a*) constant at $g_E/9$ (*b*) constant at $g_E/4$

(*c*) increasing by factor of 3 (*d*) increasing by factor of 2.

164. Fig. 41 shows a man pulling a rope with such a force that the plank is held in position. The weights of the ropes, plank and pulleys are to be neglected. If $g = 10$ m s^{-2} and the weight of the man is 80 kgf, then the force with which the man pulls the rope is

Fig. 41

(*a*) 100 N (*b*) 200 N

(*c*) 400 N (*d*) 800 N.

165. The diagram shows a cord supporting a picture. Which of the graphs shown in Fig. 43 correctly represents the relationship between the tension T in the cord and the angle θ ?

Fig. 42

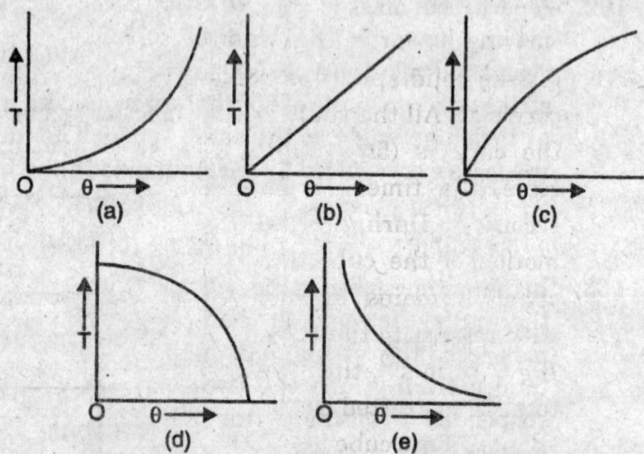

Fig. 43

166. An inclined plane is inclined at angle θ with the horizontal. A body of weight one newton rests on it, μ is the coefficient of friction. The minimum force in SI units that has to be applied parallel to the inclined plane to make the body just move up the inclined plane is

(*a*) sin θ (*b*) μ cos θ

(*c*) μ cos θ – sin θ (*d*) μ cos θ + sin θ.

167. A jet of water with a cross sectional area '*a*' is striking against a wall at an angle θ to the horizontal and rebounds elastically. If the velocity of water jet is *v* and the density is ρ, the normal force acting on the wall is

(*a*) $2av^2$ ρ cos θ (*b*) *av* ρ cos θ

(*c*) 2*av* ρ cos θ (*d*) *av* cos θ.

168. A toy train consists of three identical compartments *x*, *y* and *z*. It is pulled by a constant force F from the side *z* horizontally. Assume friction to be absent. T_2 is the tension in the string connecting *x* and *y*. T_1 is the tension in the string connecting *y* and *z*. Then $\frac{T_2}{T_1}$ =

(*a*) 1 : 2 (*b*) 2 : 1

(*c*) 1 : 1 (*d*) 1 : 3.

169. A body placed on the surface of a smooth inclined plane reaches the bottom of the plane in time *t*. When it is identically placed on a rough inclined plane, it reaches the bottom in time *nt* where *n* > 1. In both the cases, the angle of inclination θ is same. The coefficient of friction μ is equal to

(*a*) $\sin\theta\left[1-\frac{1}{n^2}\right]$ (*b*) $\cos\theta\left[1-\frac{1}{n^2}\right]$

(*c*) $\tan\theta\left[1-\frac{1}{n^2}\right]$ (*d*) tan θ.

170. A block rests on a rough horizontal surface. μ is the coefficient of friction between the block and the surface. A force *mg* pulls the block. This force acts at an angle θ with the vertical side of the cube. The mass of the block is *m*. The block can be pulled along the surface if $\cot\frac{\theta}{2}$

(*a*) > μ (*b*) = μ

(*c*) < μ (*d*) $\frac{\mu}{6}$.

171. An open-topped rail road car of mass M kg has an initial velocity *v* along a straight horizontal smooth track. The velocity *v* remains constant until, at time *t* = 0, it suddenly starts to rain. The rain drops fall vertically with a velocity v_y and add a mass μ kg s^{-1} of water to the car. The velocity of the car *t* second after the rain starts, assuming that it has not filled up with water is

(*a*) $v_0+\frac{\mu t v_y}{M+\mu t}$ (*b*) $\frac{Mv+\mu t v}{M+\mu t}$

(*c*) $\frac{Mv}{M+\mu t}$ (*d*) $v+\frac{\mu v}{M}$.

172. In Q. 171, if the car holds upto V cm^3 of water and ρ is the density of water in kg m^{-3}, then the final velocity of the car if the rain continues long enough is

(*a*) $2v_0$ (*b*) $\frac{Mv}{M+V\rho}$

(*c*) $\frac{Mv}{M+v_y\rho}+u$ (*d*) $v_0+\frac{v\rho}{m+v_y\rho}$.

173. A horizontal force F is applied on a block of mass *m* placed on a rough inclined plane of inclination θ. The normal reaction N is

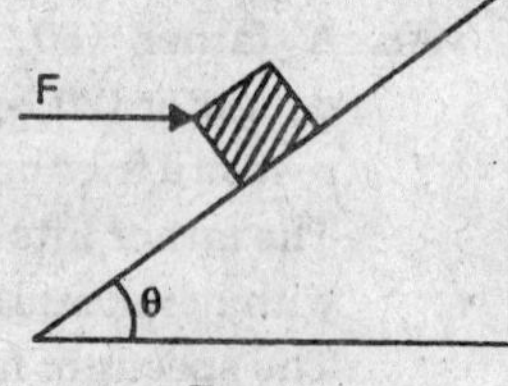

Fig. 44

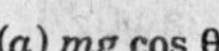

(*a*) *mg* cos θ

(*b*) *mg* sin θ

(*c*) *mg* cos θ – F cos θ

(*d*) *mg* cos θ + F sin θ.

174. In Q. 173, if μ_s is the coefficient of static friction, then the force of friction is

(a) $\mu_s\, mg \cos \theta$ (b) $\mu_s\, mg \sin \theta$

(c) $\mu_s\, (mg \cos \theta - F \cos \theta)$ (d) $\mu_s\, (mg \cos \theta + F \sin \theta)$.

175. A 30 kg block rests on a rough horizontal surface. A force of 200 N is applied on the body. The block acquires a speed of 4 m s^{-1}, starting from rest, in 2 second. The value of coefficient of friction is [Take g = 10 m s^{-2}]

(a) $10/\sqrt{3}$ (b) $\sqrt{3}/10$

(c) 0.47 (d) 0.185.

176. Refer to Fig. 45. All surfaces are frictionless. The ratio of T_1 and T_2 is

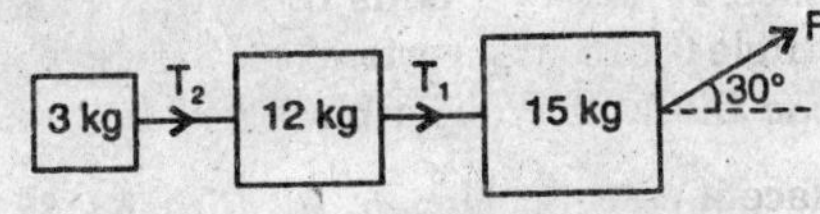

Fig. 45

(a) 1 : 5 (b) 5 : 1

(c) 4 : 1 (d) 1 : 4.

177. A man weighing 60 kg is standing on a trolley weighing 240 kg. The trolley is resting on frictionless horizontal rails. If the man starts walking on the trolley with a velocity of 1 m s^{-1}, then after 4 second, his displacement relative to the ground is

(a) 6 m (b) 4.8 m

(c) 3.2 m (d) 2.4 m.

178. A weight W hangs from a rope that is tied to two other ropes that are fastened to the ceiling as shown in Fig. 46. The upper ropes make angles θ and ϕ with the horizontal. Now, the values of T_1 and T_2 are

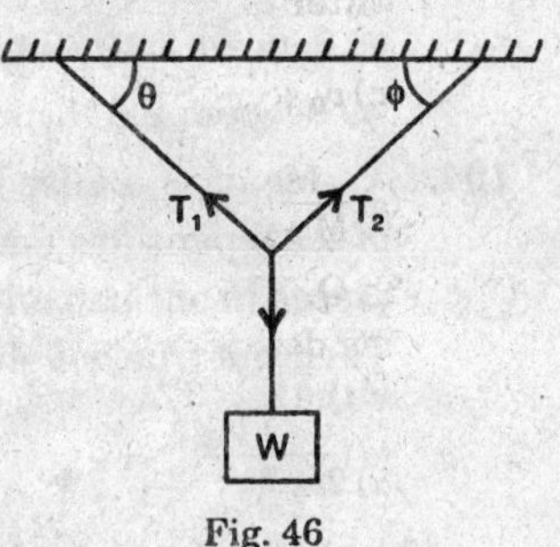

Fig. 46

(a) $\dfrac{W \sin \phi}{\sin (\theta + \phi)}, \dfrac{W \sin \theta}{\sin (\theta + \phi)}$ (b) $\dfrac{W \sin \phi}{\cos (\theta + \phi)}, \dfrac{W \sin \theta}{\cos (\theta + \phi)}$

(c) $\dfrac{W \cos \phi}{\sin (\theta + \phi)}, \dfrac{W \cos \theta}{\sin (\theta + \phi)}$ (d) $\dfrac{W \cos \phi}{\tan (\theta + \phi)}, \dfrac{W \cos \theta}{\tan (\theta + \phi)}$.

179. For the system shown in Fig. 47, the pulleys are light and frictionless. The tension in the string will be

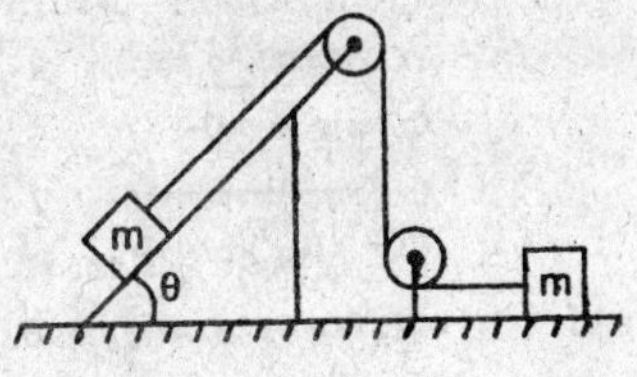

Fig. 47

(a) $\dfrac{2}{3}\, mg \sin \theta$ (b) $\dfrac{3}{2}\, mg \sin \theta$

(c) $\dfrac{1}{2}\, mg \sin \theta$ (d) $2\, mg \sin \theta$.

180. A grenade having mass of 10 kg flying horizontally with a velocity of 10 m s^{-1} explodes into two fragments. The larger fragment has a velocity 25 m s^{-1} in the direction of the initial velocity of the grenade. The smaller fragment has a velocity of 12.5 m s^{-1} in the opposite direction. The masses of the fragments are

(a) 2 kg, 8 kg (b) 3 kg, 7 kg

(c) 4 kg, 6 kg (d) 4.5 kg, 5.5 kg.

181. A 2 kg solid sphere is resting inside a cube at the corner C. All the surfaces are smooth. The velocity of the cube is $(5t\hat{i} + 2t\hat{j})$. Here, the time t is in second. During the motion of the cube, the sphere remains at rest with respect to the cube. If g = 10 m s^{-2}, then the total force exerted by the sphere on the cube is

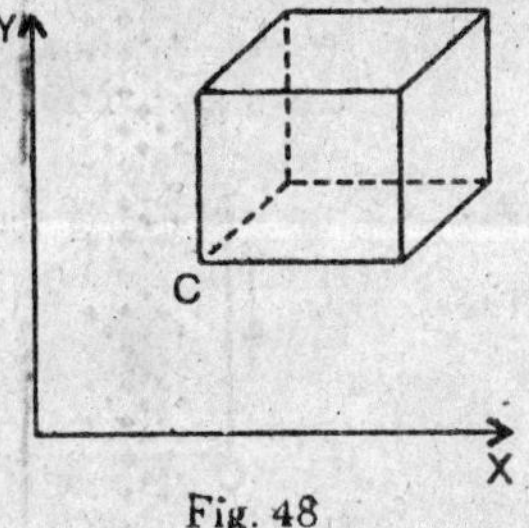

Fig. 48

(a) 23 N (b) 24 N

(c) 25 N (d) 26 N.

182. Fig. 49 represents a painter in a crate which hangs along the side of a building. When the painter of mass 100 kg pulls the rope, the force exerted by him on the floor of the crate is 450 N. If the crate weighs 25 kg, then the acceleration in the rope is [Given g = 10 m s^{-2}].

(a) 1 m s^{-2} (b) 2 m s^{-2}

(c) 3 m s^{-2} (d) 4 m s^{-2}.

183. In Q. 182, the tension in the rope is

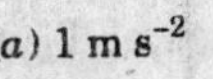

(a) 250 N (b) 200 N

(c) 550 N (d) 750 N.

Fig. 49

(Fig. 49 labels: T, 450 N, T, 250 N, 1000 N)

184. The rear side of a truck is open and a box of mass 20 kg is placed on the truck 4 m away from the open end. If μ = 0.15 and g = 10 m s^{-2} and the truck starts from rest with an acceleration of 2 m s^{-2} on a straight road, then the box will fall off the truck when it is at a distance of x metre from the starting point. The value of x is

(a) 4 (b) 8

(c) 16 (d) 32.

185. A block of mass 5 kg, resting on a horizontal surface, is connected by a cord, passing over a light frictionless pulley to a hanging block of mass 5 kg. The coefficient of kinetic friction between the block and the surface is 0.5. Tension in the cord is [Take g = 9.8 m s^{-2}]

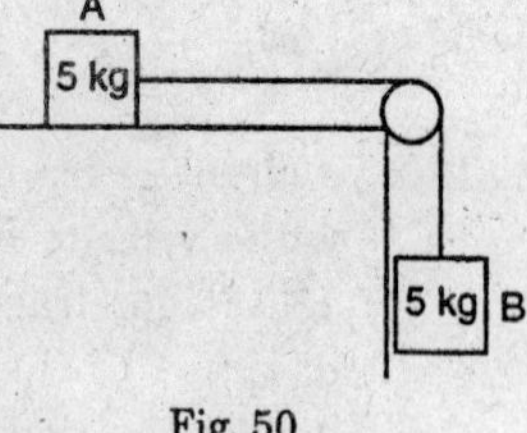

Fig. 50

(a) 49 N (b) 36 N
(c) 36.75 N (d) 2.45 N.

186. In Q. 185, the acceleration of the block is

(a) 4.9 m s^{-2} (b) 2.45 m s^{-2}
(c) Zero (d) 9.8 m s^{-2}.

187. What acceleration must the cart in Fig. 51 have in order that the block A will not fall ? (μ is coefficient of friction between cart and block)

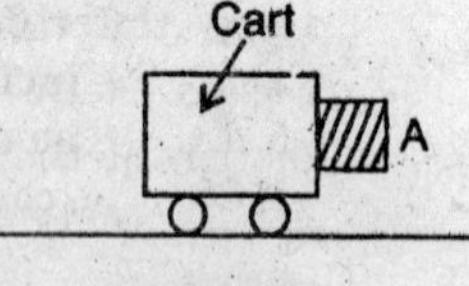

Fig. 51

(a) μg (b) $\frac{g}{\mu}$
(c) $\frac{\mu}{g}$ (d) $\mu + g$.

188. Coefficient of kinetic friction between 3 kg and 2 kg blocks is 0.3. The horizontal table surface is smooth. The acceleration of masses is

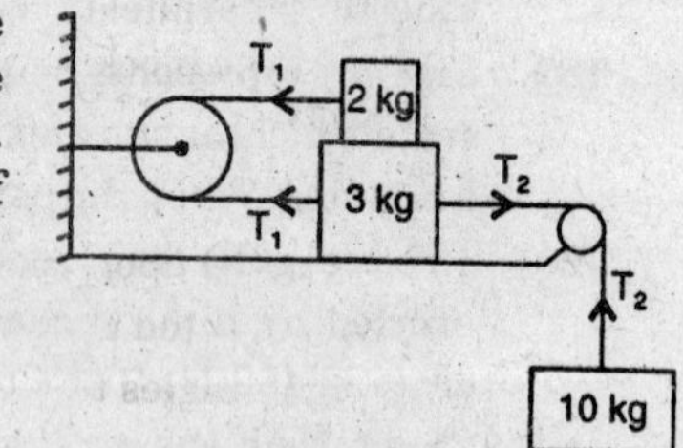

Fig. 52

(a) 5.75 m s^{-2}
(b) 2.25 m s^{-2}
(c) 1.75 m s^{-2}
(d) 0.75 m s^{-2}.

189. In Q. 188, the approximate values of T_1 and T_2 are respectively

(a) 17 N and 41 N (b) 41 N and 17 N
(c) 21 N and 29 N (d) 35 N and 17 N.

190. A metal plate weighing 200 g is balanced in mid-air by throwing 40 balls per second vertically upwards from below. The collision of balls with the plate is elastic. What is the speed with which the balls strike the plate ? [Given : mass of each ball = 200 g]

(a) 5.7 cm s^{-1} (b) 12.25 cm s^{-1}
(c) 9.8 cm s^{-1} (d) 14 cm s^{-1}.

191. A block is projected up an inclined plane of inclination 30° with velocity v. If μ is the coefficient of friction, then the block moves up a distance of

(a) $\frac{v^2}{g(1+\sqrt{3}\mu)}$ (b) $\frac{v^2}{g}$
(c) $\frac{v^2}{1+\sqrt{3}\mu}$ (d) $\frac{v^2}{\mu}$.

192. The respective accelerations of A and B in Fig. 53 are

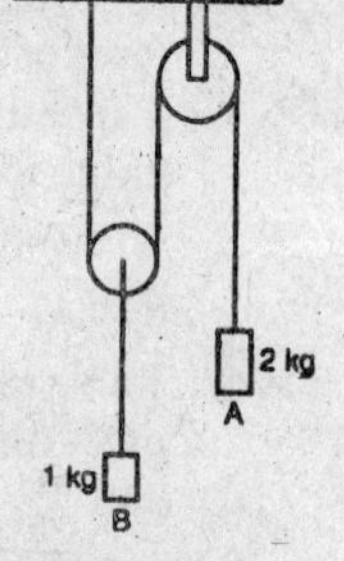

Fig. 53

(a) $\frac{2g}{3}$ downwards and $\frac{g}{3}$ upwards

(b) $\frac{g}{3}$ downwards and $\frac{g}{3}$ upwards

(c) $\frac{2g}{7}$ downwards and $\frac{g}{7}$ upwards

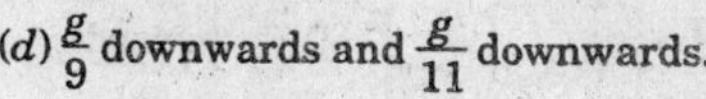
(d) $\frac{g}{9}$ downwards and $\frac{g}{11}$ downwards.

193. The monkey B shown in Fig. 54 is holding on to the tail of the monkey A which is climbing up a rope. The masses of the monkeys A and B are 5 kg and 2 kg respectively. If A can tolerate a tension of 30 N in its tail, what force should it apply on the rope in order to carry the monkey B with it ? [Take g = 10 m s^{-2}].

Fig. 54

(a) Between 70 N and 105 N
(b) Between 50 N and 69 N
(c) Between 30 N and 50 N
(d) Between 106 N and 116 N.

194. The arms of a false balance are equal but the weights of scale pans are not equal. An object weighs w_1 when placed in one pan and w_2 when placed in the other pan. Then the true weight of the object is

(a) $\frac{w_1 + w_2}{2}$ (b) $\sqrt{w_1 w_2}$
(c) $\sqrt{w_1^2 + w_2^2}$ (d) $\sqrt{\frac{w_1^2 + w_2^2}{2}}$.

195. A false balance is such that the beam remains horizontal when the pans are empty. An object weighs w_1 when placed in one pan and w_2 when placed in the other pan. The true weight w of the object is

(a) $\frac{w_1 + w_2}{2}$ (b) $\sqrt{w_1 w_2}$
(c) $\sqrt{w_1^2 + w_2^2}$ (d) $\sqrt{\frac{w_1^2 + w_2^2}{2}}$.

196. The acceleration of the 500 g block in Fig. 55 is

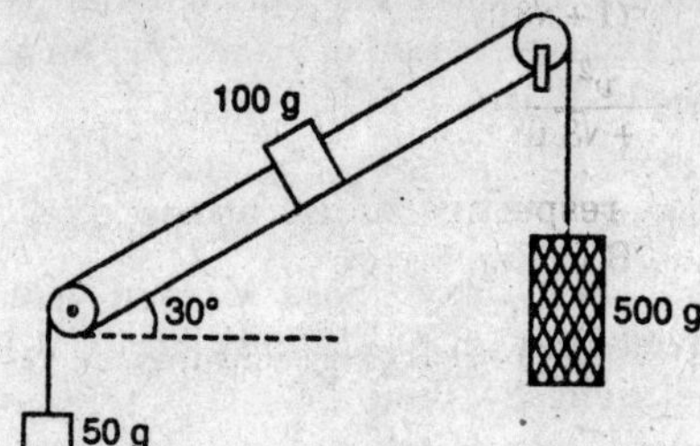

Fig. 55

(a) $\frac{6g}{13}$ downwards (b) $\frac{7g}{13}$ downwards

(c) $\frac{8g}{13}$ downwards (d) $\frac{9g}{13}$ upwards.

197. A passenger is travelling in a train moving at 72 km h^{-1}. His suitcase is kept on the berth. The driver of the train applies brakes such that the speed of the train decreases at a constant rate of 36 km h^{-1} in 5 second. What should be the minimum coefficient of friction between the suitcase and the berth if the suitcase is not to slide during retardation of the train ?

(a) $\frac{10}{49}$ (b) $\frac{10}{98}$

(c) $\frac{28}{49}$ (d) $\frac{30}{98}$.

198. A particle is at rest at $x = a$. A force $\vec{F} = -\frac{b}{x^2}\hat{i}$, directed towards the origin, begins to act on the particle. The particle starts its motion, towards the origin, along x-axis. The velocity of the particle, when it reaches a distance x from the origin, is

(a) $\sqrt{\frac{2b}{m}\left(\frac{a-x}{ax}\right)}$ (b) $\frac{m}{2b}\sqrt{\frac{a+x}{ax}}$

(c) zero (d) infinity.

199. A horizontal force F is applied to a frictionless pulley of mass m_2. The horizontal surface is smooth. The acceleration of the block of mass m_1 is n times the acceleration of the pulley. The value of n is

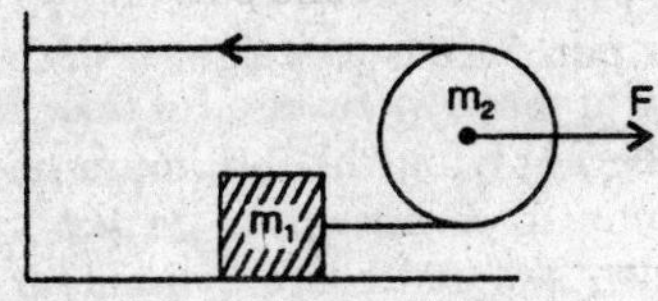

Fig. 56

(a) 1 (b) 2

(c) 3 (d) 4.

200. In Q. 198, the acceleration of the block is

(a) $\frac{F}{m_2 + 4m_1}$ (b) $\frac{2F}{m_2 + 4m_1}$

(c) $\frac{F}{4m_2 + m_1}$ (d) $\frac{2F}{4m_2 + m_1}$.

201. An isolated particle of mass m is moving in a horizontal plane (XY plane), along x-axis, at a certain height above the ground. It suddenly explodes into two fragments of masses $\frac{m}{4}$ and $\frac{3m}{4}$. An instant later, the smaller fragment is at $y = +15$ cm. The larger fragment at this instant is at $y =$

(a) – 5 cm (b) + 5 cm

(c) + 20 cm (d) – 20 cm.

202. A block projected from the bottom of an inclined plane stops, after 0.5 s, after travelling a certain distance along the inclined plane. The velocity of projection is 5 m s^{-1}. The angle of inclination of the inclined plane is 30°. The coefficient of friction is [Take g = 10 m s^{-2}]

(a) $\frac{1}{\sqrt{2}}$ (b) $\frac{1}{\sqrt{3}}$

(c) $\frac{1}{\sqrt{5}}$ (d) $\frac{1}{\sqrt{7}}$.

203. A 4 kg block is placed on a rough floor. The coefficient of static friction is 0.8. If a force of 19 N is applied on the block parallel to the floor, then the force of friction between the block and floor is

(a) 32 N (b) 18 N

(c) 19 N (d) 9.8 N.

204. A body acted upon by a force F has an acceleration 'A'. When it is acted upon by two forces each of magnitude F at right angles to each other, its acceleration will be

(a) 2A (b) A

(c) A$\sqrt{2}$ (d) $\sqrt{2}$.

205. As shown in the Fig. 57, six steel balls of identical size are lined up along a straight groove (frictionless). Two

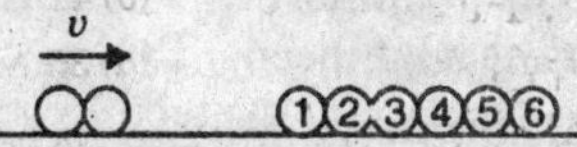

Fig. 57

similar balls moving with speed v along the groove collide with this row on the extreme left hand end. Then

(a) one ball from the right end will move on with speed $2v$, all the remaining balls will be at rest

(b) two balls from the extreme right end will move on with speed v each and the remaining balls will be at rest (*c*) all the balls will start moving to the right with speed $v/8$ each

(*d*) all the six balls originally at rest will move on with speed $v/6$ each and the two incident balls will come to rest.

206. A spacecraft of mass M moving with velocity v in free space explodes and breaks into two pieces. After the explosion, a mass m of the space craft is left stationary. The velocity of the other part is

(*a*) $\frac{mv}{M-m}$ (*b*) $\frac{Mv}{M-m}$

(*c*) $\frac{M+m}{M}$ (*d*) $\frac{Mv}{M}$.

207. A one kg mass suspended by a thread is first lifted up with an acceleration $\frac{g}{2}$ and then lowered downward with an acceleration $\frac{g}{2}$. The ratio of the tensions in the first and second case is

(*a*) 1 : 2 (*b*) 2 : 1

(*c*) 1 : 3 (*d*) 3 : 1.

208. A ball of mass 400 g is dropped from a height of 5 m. A boy on the ground hits the ball vertically upwards with a bat with an average force of 100 newton so that it attains a vertical height of 20 m. The time for which the ball remains in contact with the bat is [g = 10 m s^{-2}]

(*a*) 0.12 s (*b*) 0.08 s

(*c*) 0.04 s (*d*) 12 s.

209. A person standing on the floor of a lift drops a coin. The coin touches the floor after t_1 second if the lift is at rest and after t_2 second if the lift is moving with a uniform velocity v upwards. Then

(*a*) $t_1 = t_2$ (*b*) $t_1 > t_2$

(*c*) $t_1 < t_2$ (*d*) $t_1 \gtrless t_2$.

210. In Fig. 58 a block of weight 10 N is shown resting on a horizontal surface. The coefficient of static friction between the block and surface is μ_s = 0.4. A force of 3.5 N will keep the block in uniform motion, once it has been set in motion. A horizontal force of 3 N is applied to the block. The block will then

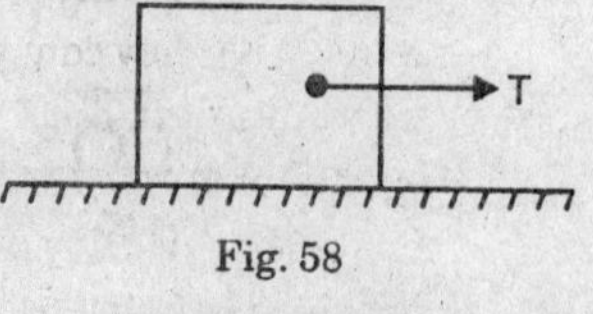

Fig. 58

(*a*) move over the surface with a constant velocity.

(*b*) move having accelerated motion over the surface.

(*c*) not move.

(*d*) first move with a constant velocity for sometime and then will have accelerated motion.

211. A wooden block of mass 1 kg is attached to the hook of a spring balance. The spring balance is then raised with an acceleration of 9.8 m s^{-2}. The apparent weight of the body is

(*a*) 1 kg wt (*b*) 2 kg wt

(*c*) 3 kg wt (*d*) 4 kg wt.

212. If a lighter body (mass M_1 and velocity V_1) and a heavier body (mass M_2 and velocity V_2) have the same kinetic energy, then

(*a*) $M_2V_2 < M_1V_1$ (*b*) $M_2V_2 = M_1V_1$

(*c*) $M_2V_1 = M_1V_2$ (*d*) $M_2V_2 > M_1V_1$.

213. A boy of mass 40 kg stands in a frame of mass 360 kg. He pulls on a light rope which passes over a pulley. The other end of the rope is firmly attached to the frame. What force should be exerted by the boy on the rope for the system to be in equilibrium ?

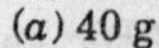

(*a*) 40 g (*b*) 400 g

(*c*) 200 g (*d*) 100 g.

Fig. 59

214. Refer to Fig. 60. When the block of mass 1.4 m is released,

(*a*) the system will remain in equilibrium

(*b*) the block of mass 1.4 m begin to move down with acceleration $g/4$

Fig. 60

(*c*) the block of mass 1.4 m will begin to move with acceleration $g/3$

(*d*) the block of mass 1.4 m will undergo periodic motion.

215. A uniform rope of mass M, length L is fixed at its upper end vertically from a rigid support. Then the tension in the rope at the distance l from the rigid support is

(*a*) Mg (*b*) $\frac{M}{L}(L-l)g$

(*c*) $\frac{l}{L}Mg$ (*d*) $\frac{M}{l}Lg$.

216. A body of mass m rests on a floor. It is just made to move by applying the minimum necessary force and the same force continues to act afterwards. If the coefficients of static and dynamic friction are 0.4 and 0.3 respectively, then the acceleration of the body is

(*a*) 2.94 m s^{-2} (*b*) 0.98 m s^{-2}

(*c*) 3.98 m s^{-2} (*d*) zero m s^{-2}.

217. A bullet of mass 25 g moving with a velocity of 2 m s^{-1} is stopped within 5 cm of the target. The average resistance offered by the target is

(a) 1 N (b) 2 N
(c) 4 N (d) 8 N.

218. A particle moves in the *x-y* plane under the action of a force $\vec{F}$ such that the value of its linear momentum $\vec{p}$ at any time *t* is

$$p_x = 2 \cos t,\ p_y = 2 \sin t$$

The angle *q* between $\vec{F}$ and $\vec{p}$ at any given time *t* will be

(a) 90° (b) 0°
(c) 180° (d) 30°.

219. A man of mass 100 kg enters a lift and selects the desired floor. The lift now accelerates upwards at 0.2 m s^{-2} for 2 second and then moves with constant velocity. As the lift approaches his floor, it decelerates at the same rate as it previously accelerated. The lift cables can safely withstand a tension of 2×10^4 N and the lift itself has mass of 500 kg. How many men, each of mass 100 kg, could it safely carry at one time ?

(a) 22 (b) 15
(c) 18 (d) 12.

220. A body slides down the length PQ of a smooth incline at angle θ from the horizontal (see Fig. 61). P may be varied along the vertical line RS, so that length PQ and angle θ vary, but base length QR is kept fixed. As θ is varied from near 0° to near 90°, the time of passage from P to Q

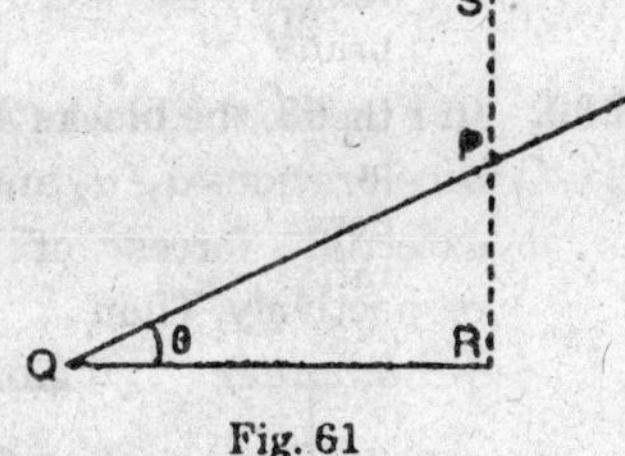

Fig. 61

(a) decreases continuously
(b) increase continuously
(c) first decreases, later increases
(d) first increases, later decreases.

221. An iron nail is dropped from a height y_1 from the level of a sand bed. It penetrates through a distance y_2 in the sand before coming to rest. If *m* is the mass of the nail, then the average force exerted by the sand on the nail is

(a) $mg\left(\frac{y_1}{y_2}+1\right)$ (b) $mg\, y_1$
(c) $mg\frac{y_1}{y_2}$ (d) $mg\left(\frac{y_2}{y_1}-1\right)$.

222. A uniform rope is moving with constant acceleration on a smooth horizontal surface. The ratio of the tension in the rope at its mid-point to the applied force is

(a) 1 : 2 (b) 2 : 1
(c) 1 : 1 (d) 1 : ∞.

223. Fig. 62 shows a block of mass M resting on a horizontal surface. The acceleration with which a boy of mass *m* should climb the rope (neglect mass) so as to lift the block should be greater than

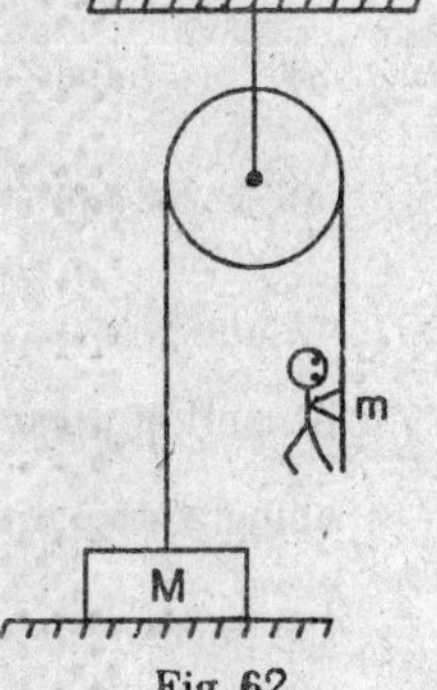

Fig. 62

(a) $\frac{M}{m}g$ (b) $\left(\frac{M}{m}-1\right)g$
(c) $\left(\frac{M}{m}+1\right)g$ (d) $\frac{mg}{M}$.

224. A block of mass *m*, starting from rest from the top of a smooth inclined plane of inclination θ, reaches A at time t_1 and B at time t_2. The change in momentum in the time interval $(t_2 - t_1)$ is

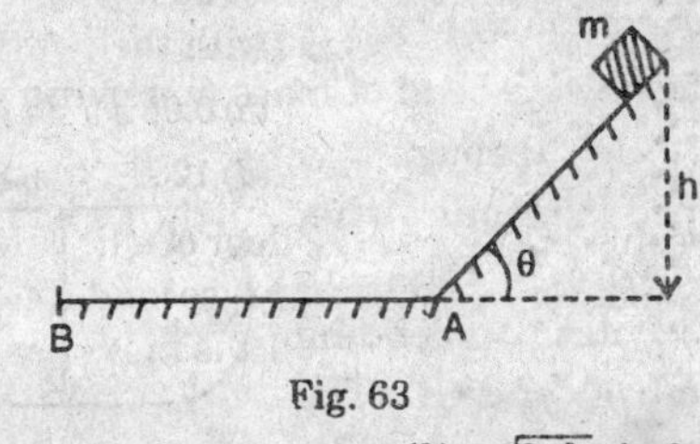

Fig. 63

(a) zero (b) $m\sqrt{2gh}\sin\theta$
(c) $m\sqrt{2gh}\cos\theta$ (d) $2m\sqrt{2gh}\sin\frac{\theta}{2}$.

225. Two boats set on parallel courses move under their own momentum through the stagnant water of a lake towards each other and with the same speed of 6 m s^{-1}. As soon as they come sufficiently close, a load is shifted from the first boat on to the second. After that, the second boat continues to move in the original direction but with a velocity 4 m s^{-1}. Find the mass of the second boat if the first boat weighs 500 kgf when empty and the weight of the load is 60 kgf. (Neglect the resistance of water)

(a) 1500 kg (b) 400 kg
(c) 300 kg (d) 350 kg.

226. An object of mass M is moved on a frictionless fixed inclined plane surface as shown in the Fig. 64. The thread is assumed to be massless and the pulley frictionless. [Take g = 10 m s^{-2}]. The respective accelerations of M and m are

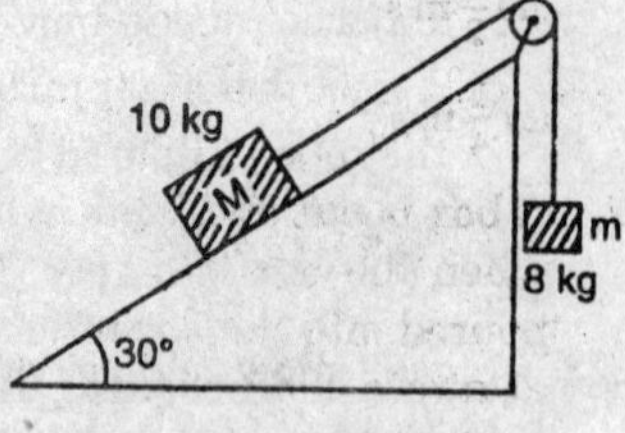

Fig. 64

(a) $\frac{5}{3}$ m s^{-2} and $\frac{5}{3}$ m s^{-}

(b) $\frac{5}{3}$ m s^{-2} and $\frac{2}{5}$ m s^{-2}

(c) $\frac{1}{5}$ m s^{-2} and $\frac{9}{7}$ m s^{-2} (d) 98 m s^{-2} and 9.8 m s^{-2}.

227. A shell of mass 100 g moving with speed of 50 m s^{-1} along a straight line inclined at an angle $\tan^{-1}\left(\frac{4}{3}\right)$ to the horizontal bursts into two pieces. The piece of mass 60 g starts moving horizontally with a speed of 50 m s^{-1}. The other piece moves with a speed of

(a) 50 m s^{-1} opposite to the first piece

(b) 100 m s^{-1} in vertically downward direction

(c) 50 m s^{-1} in the same direction as the first piece

(d) zero : (remains at rest).

228. Fig. 65 shows a bead of mass m moving with uniform speed v through a U-shaped smooth wire. The wire has a semi-circular bending between A and B. The average force exerted by the bead on the part AB of the wire is

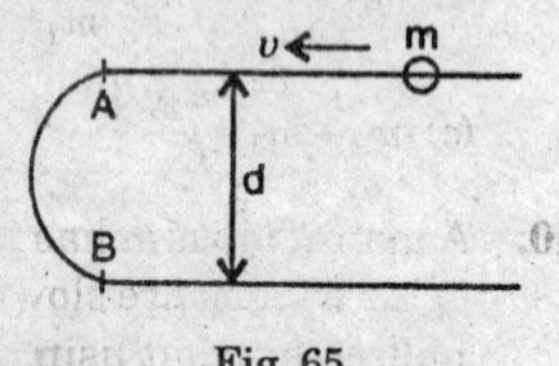

Fig. 65

(a) $2mv$ (b) $\frac{\pi d}{2}$

(c) $\frac{\pi d}{2v}$ (d) $\frac{4mv^2}{\pi d}$.

229. A pendulum is hanging from the ceiling of a cage. If the cage moves up with constant acceleration a, its tension is T_1. If it moves down with the same acceleration a then the tension is T_2. If the cage moves horizontally with the same acceleration a, then the tension is T. Now, $2T_2$ =

(a) $T_1^2 + T_2^2$ (b) $T_1^2 - T_2^2$

(c) $2T_1^2 + T_2^2$ (d) $T_1^2 - 2T_2^2$.

230. Two blocks of masses 7 kg and 3 kg are connected by a spring of stiffness 1000 N m^{-1} and placed on a smooth horizontal surface. They are acted by horizontal forces of 72 N and 32 N in opposite directions as shown in the Fig. 66. When accelerations of the blocks are equal and constant, the extension of the spring is

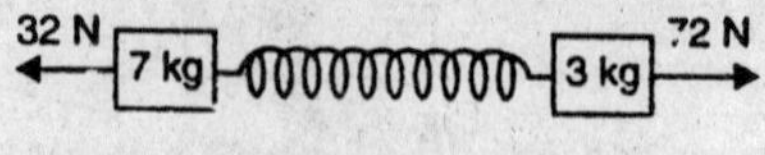

Fig. 66

(a) 4 cm (b) 5 cm

(c) 6 cm (d) 8 cm.

231. A body of mass m starting from rest slides down a smooth inclined plane of gradient θ, fixed in a lift moving upwards with an acceleration a_0 as shown in Fig. 67. If the base of the inclined plane has length L, the time taken by the body to slide from top to bottom of the inclined plane, will be

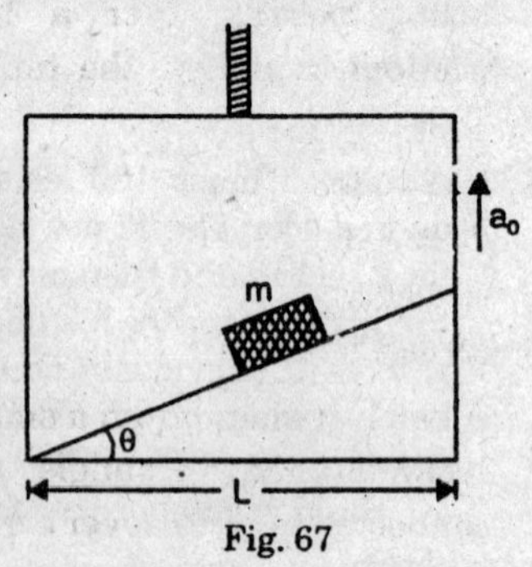

Fig. 67

(a) $\frac{L}{(g + a_0)\sin 2\theta}$ (b) $\left[\frac{4L}{(g + a_0)\sin 2\theta}\right]^{1/2}$

(c) $\frac{(g + a_0)\sin 2\theta}{2L}$ (d) $\left[\frac{2L}{(g + a_0)\sin\theta}\right]^{-2}$.

232. In Fig. 68, the blocks A, B and C of mass m each have accelerations a_1, a_2 and a_3 respectively. F_1 and F_2 are external forces of magnitudes $2mg$ and mg respectively. Then

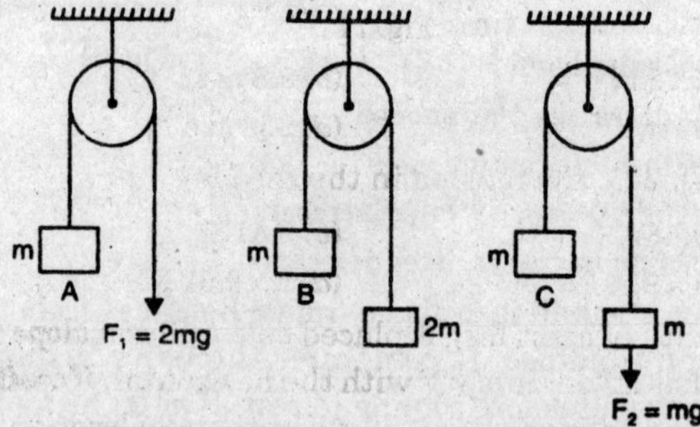

Fig. 68

(a) $a_1 = a_2 = a_3$ (b) $a_1 > a_3 > a_2$

(c) $a_1 = a_2, a_2 > a_3$ (d) $a_1 > a_2, a_2 = a_3$.

233. In the arrangement shown in Fig. 69, the ends P and Q of an unstretchable string move downwards with uniform speed v. Pulleys A and B are fixed. Mass m moves upwards with a speed of

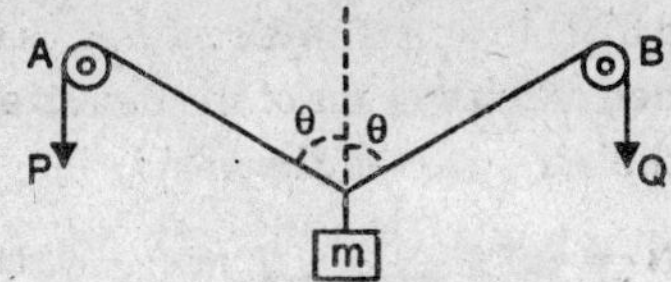

Fig. 69

(a) $2v \cos\theta$ (b) $v \cos\theta$

(c) $\frac{2v}{\cos\theta}$ (d) $\frac{v}{\cos\theta}$.

234. With reference to the Fig. 70 showing a light inextensible string passing over a fixed frictionless pulley, the tension T_3 is

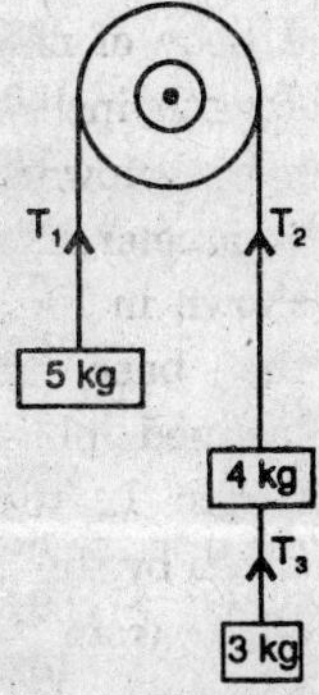

Fig. 70

(a) 24.5 N

(b) 29.4 N

(c) 18.6 N

(d) 68.6 N.

235. A block of mass m_1 on a smooth inclined plane of angle α is connected by a cord over a small frictionless pulley to a second block of mass m_2 hanging vertically. The acceleration of either body is [Given : $m_1 = 3$ kg, $m_2 = 2$ kg and $\alpha = 30°$]

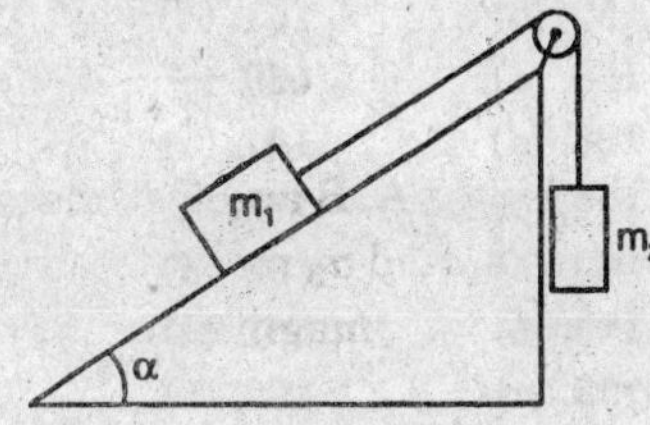

Fig. 71

(a) 0.98 m s^{-2} (b) 9.8 m s^{-2}

(c) 98 m s^{-2} (d) 980 m s^{-2}.

236. In Q. 235, the tension in the cord is

(a) 17.64 N (b) 9.81 N

(c) 3×9.81 N (d) 2×9.81 N.

237. A brick of mass 2 kg is placed on a uniform slope which makes an angle of 20° with the horizontal. If coefficient of kinetic friction is 0.25, then the acceleration of the brick is nearly [Take sin 20° = 0.3, cos 20° = 0.9 and $g = 10$ m s^{-2}]

(a) $\frac{1}{2}$ m s^{-2} (b) $\frac{3}{4}$ m s^{-2}

(c) $\frac{1}{4}$ m s^{-2} (d) 1 m s^{-2}.

238. A box is put on a scale which is adjusted to read zero, when the box is empty. A stream of pebbles is then poured into the base from a height h above its bottom at a rate of n pebbles/s. Each pebble has a mass m. If the pebbles collide with the box such that they immediately come to rest after collision, then the scale reading at time t after the pebbles begin to fill the box is

(a) mnt (b) $mn[\sqrt{2gh} + gt]$

(c) $mngt$ (d) zero.

239. A frictionless cart carries two other frictionless carts as shown in Fig. 72 connected by a string over a pulley. The horizontal force F that must be applied so that m_1 and m_2 do not move relative to m_3 is

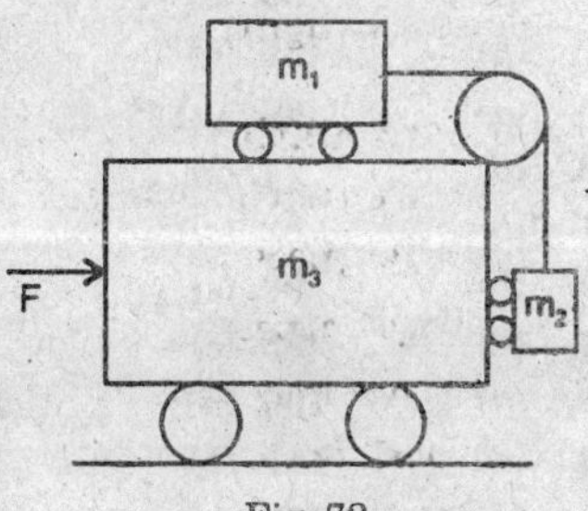

Fig. 72

(a) $(m_1 + m_2 + m_3)\frac{m_2 g}{m_1}$ (b) $(m_1 + m_2)\frac{m_2 g}{m_1}$

(c) $(m_2 + m_3)\frac{m_1 g}{m_2}$ (d) $(m_1 + m_3)\frac{m_1 g}{m_2}$.

240. A man of mass m has fallen into a ditch of width d. Two of his friends are slowly pulling him out using a light rope and two fixed pulleys as shown in Fig. 73. Both the friends exert forces of equal magnitude F. When the man is at a depth h, the value of F is

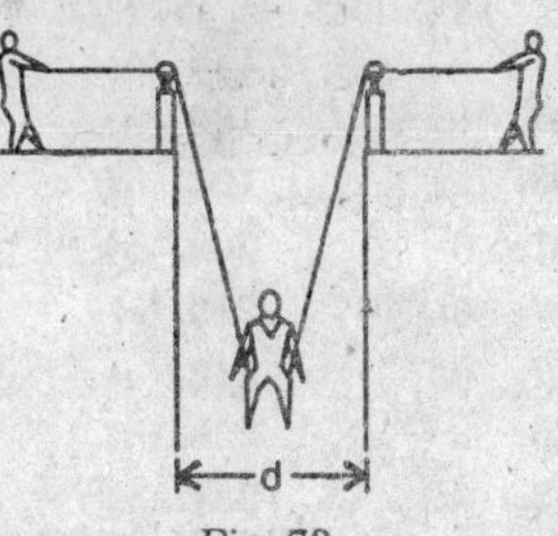

Fig. 73

(a) $\frac{mg}{4h}\sqrt{d^2 + 4h^2}$ (b) hmg

(c) dmg (d) infinite.

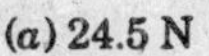

ANSWERS

QUESTION BANK (Level I)

1. (*d*)	**2.** (*d*)	**3.** (*c*)	**4.** (*b*)	**5.** (*d*)	**6.** (*a*)	**7.** (*a*)	**8.** (*b*)
9. (*c*)	**10.** (*c*)	**11.** (*a*)	**12.** (*c*)	**13.** (*d*)	**14.** (*b*)	**15.** (*a*)	**16.** (*d*)

QUESTION BANK (Level II)

1. (*b*)	**2.** (*b*)	**3.** (*c*)	**4.** (*b*)	**5.** (*b*)	**6.** (*a*)	**7.** (*b*)	**8.** (*a*)
9. (*d*)	**10.** (*d*)	**11.** (*b*)	**12.** (*b*)	**13.** (*b*)	**14.** (*c*)	**15.** (*d*)	**16.** (*b*)
17. (*d*)	**18.** (*a*)	**19.** (*d*)	**20.** (*b*)	**21.** (*c*)	**22.** (*d*)	**23.** (*c*)	**24.** (*c*)
25. (*c*)	**26.** (*a*)	**27.** (*c*)	**28.** (*b*)	**29.** (*c*)	**30.** (*c*)	**31.** (*d*)	**32.** (*b*)
33. (*c*)	**34.** (*d*)	**35.** (*b*)	**36.** (*c*)	**37.** (*a*)	**38.** (*a*)	**39.** (*d*)	**40.** (*a*)
41. (*b*)	**42.** (*d*)	**43.** (*d*)	**44.** (*a*)	**45.** (*a*)	**46.** (*c*)	**47.** (*b*)	**48.** (*c*)
49. (*d*)	**50.** (*b*)	**51.** (*b*)	**52.** (*c*)	**53.** (*d*)	**54.** (*a*)	**55.** (*b*)	**56.** (*a*)
57. (*b*)	**58.** (*b*)	**59.** (*d*)	**60.** (*d*)	**61.** (*d*)	**62.** (*c*)	**63.** (*d*)	**64.** (*d*)
65. (*d*)	**66.** (*a*)	**67.** (*d*)	**68.** (*a*)	**69.** (*d*)	**70.** (*c*)	**71.** (*b*)	**72.** (*b*)
73. (*d*)	**74.** (*c*)	**75.** (*a*)	**76.** (*a*)	**77.** (*c*)	**78.** (*b*)	**79.** (*a*)	**80.** (*d*)
81. (*b*)	**82.** (*c*)	**83.** (*b*)	**84.** (*b*)	**85.** (*b*)	**86.** (*a*)	**87.** (*a*)	**88.** (*c*)
89. (*b*)	**90.** (*c*)	**91.** (*c*)	**92.** (*c*)	**93.** (*b*)	**94.** (*d*)	**95.** (*b*)	**96.** (*b*)
97. (*d*)	**98.** (*c*)	**99.** (*a*)	**100.** (*d*)	**101.** (*c*)	**102.** (*d*)	**103.** (*b*)	**104.** (*b*)
105. (*d*)	**106.** (*d*)	**107.** (*b*)	**108.** (*d*)	**109.** (*a*)	**110.** (*a*)	**111.** (*a*)	**112.** (*b*)
113. (*d*)	**114.** (*b*)	**115.** (*a*)	**116.** (*c*)	**117.** (*a*)	**118.** (*a*)	**119.** (*b*)	**120.** (*a*)
121. (*d*)	**122.** (*c*)	**123.** (*b*)	**124.** (*b*)	**125.** (*a*)	**126.** (*b*)	**127.** (*c*)	**128.** (*c*)
129. (*d*)	**130.** (*d*)	**131.** (*c*)	**132.** (*b*)	**133.** (*c*)	**134.** (*c*)	**135.** (*c*)	**136.** (*c*)
137. (*d*)	**138.** (*c*)	**139.** (*d*)	**140.** (*b*)	**141.** (*b*)	**142.** (*b*)	**143.** (*c*)	**144.** (*a*)
145. (*a*)	**146.** (*b*)	**147.** (*b*)	**148.** (*d*)	**149.** (*a*)	**150.** (*a*)	**151.** (*d*)	**152.** (*a*)
153. (*a*)	**154.** (*d*)	**155.** (*a*)	**156.** (*a*)	**157.** (*d*)	**158.** (*a*)	**159.** (*c*)	**160.** (*a*)
161. (*b*)	**162.** (*c*)	**163.** (*c*)	**164.** (*b*)	**165.** (*e*)	**166.** (*d*)	**167.** (*a*)	**168.** (*a*)
169. (*c*)	**170.** (*a*)	**171.** (*c*)	**172.** (*b*)	**173.** (*d*)	**174.** (*d*)	**175.** (*c*)	**176.** (*b*)
177. (*c*)	**178.** (*c*)	**179.** (*c*)	**180.** (*c*)	**181.** (*d*)	**182.** (*b*)	**183.** (*d*)	**184.** (*c*)
185. (*c*)	**186.** (*b*)	**187.** (*b*)	**188.** (*a*)	**189.** (*a*)	**190.** (*b*)	**191.** (*a*)	**192.** (*a*)
193. (*a*)	**194.** (*a*)	**195.** (*b*)	**196.** (*c*)	**197.** (*a*)	**198.** (*a*)	**199.** (*b*)	**200.** (*b*)
201. (*a*)	**202.** (*b*)	**203.** (*c*)	**204.** (*c*)	**205.** (*b*)	**206.** (*b*)	**207.** (*d*)	**208.** (*a*)
209. (*a*)	**210.** (*c*)	**211.** (*b*)	**212.** (*d*)	**213.** (*c*)	**214.** (*d*)	**215.** (*b*)	**216.** (*b*)
217. (*c*)	**218.** (*a*)	**219.** (*c*)	**220.** (*d*)	**221.** (*a*)	**222.** (*a*)	**223.** (*b*)	**224.** (*d*)
225. (*c*)	**226.** (*a*)	**227.** (*b*)	**228.** (*d*)	**229.** (*a*)	**230.** (*c*)	**231.** (*b*)	**232.** (*b*)
233. (*d*)	**234.** (*a*)	**235.** (*a*)	**236.** (*a*)	**237.** (*d*)	**238.** (*b*)	**239.** (*a*)	**240.** (*a*).

HINTS/SOLUTIONS

QUESTION BANK (Level I)

1. (*d*) $\tan\theta = 0.2 = \frac{1}{5}$

or $\cot\theta = 5$ or $\theta = \cot^{-1}(5)$

2. (*d*) $f = \mu R = \mu mg$ | m is mass of the combination

$f = 0.5 \times 10 \times 10 \text{ N} = 50 \text{ N}$

So, a force of 10 N is unable to start the motion of the system. There is no relative motion between A and B.

3. (*c*) Component of weight mg in a direction perpendicular to the inclined plane is $mg \cos \theta$. According to Newton's third law of motion, the force exerted by the plane on the block is $mg \cos \theta$.

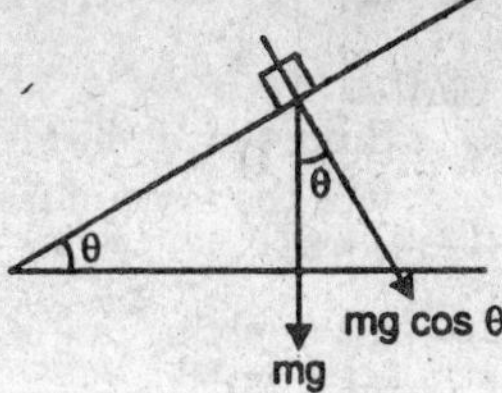

Fig. 74

4. (*b*) $m = \dfrac{\sqrt{8^2 + (-6)^2 + (-10)^2}}{1}$ kg $= \sqrt{200}$ kg $= 10\sqrt{2}$ kg

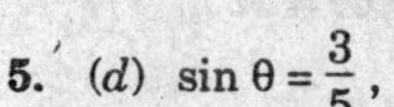

5. (*d*) $\sin \theta = \dfrac{3}{5}$,

Clearly, base of the triangle is 4 units

$$\tan \theta = \frac{3}{4}$$

$$\mu = \tan \theta = \frac{3}{4} = 0.75$$

Fig. 75

6. (*a*) $f \geq W$

$\mu R \geq W \Rightarrow \mu \times 75 \geq W$

or $75 \geq \dfrac{W}{\mu}$

$\Rightarrow \quad 75 = \dfrac{W}{\mu}$

or $75 > \dfrac{W}{\mu}$

But μ is generally less than 1

$\therefore \quad 75 > W$

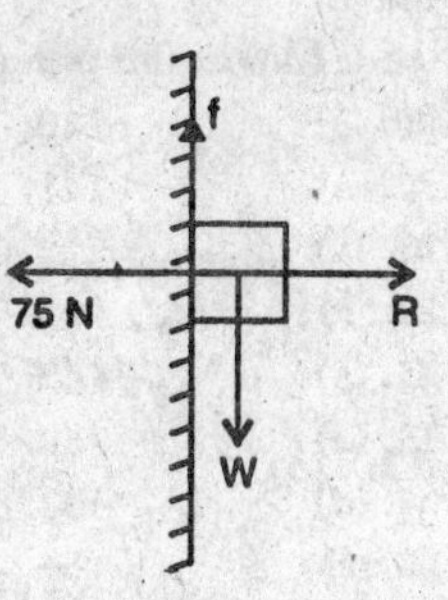

Fig. 76

7. (*a*) Force on horizontal surface = Weight of block = $2g$

$$g \approx 10 \text{ m s}^{-2}$$

$\therefore$ Weight of block is nearly 20 N

8. (*b*) Change of momentum of one bullet is mv.

Time for 1 bullet $= \dfrac{1}{n}$

$\therefore$ force = time rate of change of momentum $= \dfrac{mv}{1/n} = mnv$

9. (*c*) $u = 0, v = 30 \text{ m s}^{-1}, t = ?, a = 6 \text{ m s}^{-2}$

$$30 = 6t \text{ or } t = 5 \text{ s}$$

10. (*c*) $mg = 9.80$ N or $m = \dfrac{9.80}{9.8}$ kg $= 1$ kg

Acceleration $= \dfrac{5 \text{ N}}{1 \text{ kg}} = 5 \text{ m s}^{-2}$

11. (*a*) $F = v\dfrac{dm}{dt}$

or $F = 2 \times 0.5 \text{ kg m s}^{-2}$ or N or $F = 1$ N

12. (*c*) Force applied by engine = 6 m

When two cars are pulled,

$$(m + m)a = 6 \text{ m}$$

or $2ma = 6$ m or $a = 3 \text{ m s}^{-2}$

13. (*d*) Using law of conservation of momentum,

$$mv = \frac{m}{4} \times 0 + \frac{3m}{4} V \text{ or } V = \frac{4v}{3}$$

14. (*b*) $F = \sqrt{1^2 + 1^2 + 2 \times 1 \times 1 \times \cos 60°} = \sqrt{3}$ N

$$a = \frac{\sqrt{3} \text{ N}}{1 \text{ kg}} = \sqrt{3} \text{ m s}^{-2}$$

15. (*a*) When the system accelerates upwards, the effective value of acceleration due to gravity is given by

$$g' = g + a = g + g = 2g$$

Now, $T = \dfrac{2(m_1)(m_2)}{m_1 + m_2}(2g)$ or $T = \dfrac{4m_1 m_2}{m_1 + m_2} g$

16. (*d*) Acceleration along the line of maximum slope

$= g \sin \theta$

Component of this acceleration along AO

$= g \sin \theta \cos (90° - \alpha)$

$= g \sin \theta \sin \alpha$

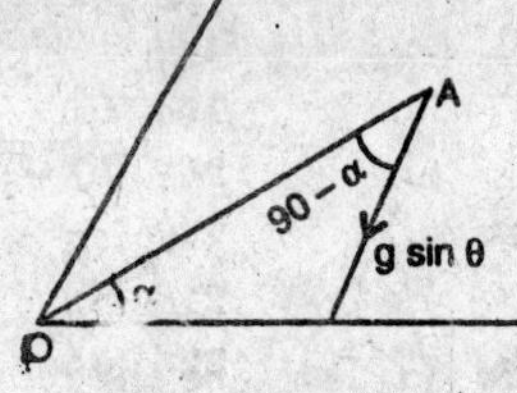

Fig. 77

QUESTION BANK (Level II)

1. (*b*) We know that $F = v\dfrac{dm}{dt}$

$\dfrac{dm}{dt}$ is increased by a factor of 4 and v is doubled.

So, F is increased by a factor of 8.

2. (*b*) $F = \dfrac{d}{dt}(p)$

$F = \dfrac{d}{dt}(a + bt + ct^2)$ or $F = b + 2ct$

Clearly, the force is time-dependent.

3. (*c*) Acceleration of the car $= \dfrac{\text{Force on the car}}{\text{Mass of the car}}$

$$= \frac{mnv}{M} = \frac{0.01 \times 10 \times 500}{2000} \text{ m s}^{-2} = \frac{5}{200} \text{ m s}^{-2} = \frac{1}{40} \text{ m s}^{-2}$$

4. (*b*) Average thrust = mnv

$$= 0.01 \times 10 \times 500 \text{ N} = 50 \text{ N}$$

5. (*b*)

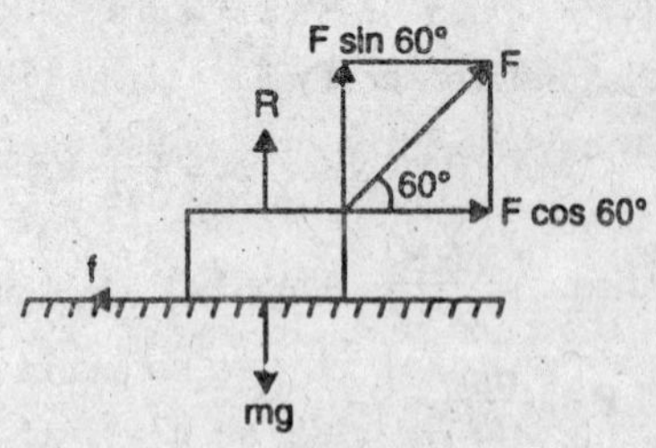

Fig. 78

$$R + F \sin 60° = mg \quad \text{or} \quad R = mg - \frac{\sqrt{3}F}{2}$$

$$F \cos 60° = f = \mu R$$

or $$\frac{F}{2} = 0.5\left[1 \times 10 - \frac{\sqrt{3}F}{2}\right]$$

or $$F + \frac{\sqrt{3}F}{2} = 10 \quad \text{or} \quad F = \frac{20}{2+\sqrt{3}}$$

or $$F = \frac{20}{3.732} \text{ N} = 5.36 \text{ N}$$

6. (*a*) Since the block is at rest, therefore force of friction

$$= mg \sin \theta = 2 \times 9.8 \times \sin 30° = 9.8 \text{ N}$$

7. (*b*) $$S = ut + \frac{1}{2}at^2$$

$$l = \frac{1}{2}at^2 \quad \text{or} \quad \left(\frac{t_1}{t_2}\right)^2 = \frac{a_2}{a_1}$$

or $$\left(\frac{2t_2}{t_2}\right)^2 = \frac{g \sin \theta}{g(\sin \theta - \mu \cos \theta)}$$

or $$4 = \frac{1}{1-\mu}$$ [Note that sin 45° = cos 45°]

or $$1 - \mu = \frac{1}{4} \quad \text{or} \quad \mu = 1 - \frac{1}{4} = \frac{3}{4} = 0.75$$

8. (*a*) Let L be the total length of the chain. Let l be the maximum length that can hang vertically. Let M be the mass of the chain. Weight of hanging length = force of limiting friction

$$\frac{M}{L} lg = \mu \frac{M}{L}(L - l)g$$

or $$l = \mu(L - l) \quad \text{or} \quad l(1 + \mu) = \mu L$$

or $$\frac{l}{L} = \frac{\mu}{1+\mu} = \frac{0.25}{1.25} = \frac{1}{5}$$

Maximum percentage of the length of the chain that can hang vertically = $\frac{1}{5} \times 100 = 20$

9. (*d*) $f_k = 0.4 \times 2 \times 10 \text{ N} = 8 \text{ N}$

$f_{ms} = 0.5 \times 2 \times 10 \text{ N} = 10 \text{ N}$

Since the applied force is less than force of friction, therefore the force of friction is equal to the applied force *i.e.* 2.5 N.

10. (*d*) Applying law of conservation of momentum,

$$\frac{m}{20} v = \left(m + \frac{m}{20}\right)V \quad \text{or} \quad V = \frac{v}{20} \times \frac{20}{21} = \frac{v}{21}$$

11. (*b*) Apparent weight = $mg + ma$

$$= 60\left(g + \frac{g}{5}\right) = \frac{6g}{5} \times 60 = 72 \text{ g}$$

$$= 72 \times 10 \text{ N} = 720 \text{ N}$$

12. (*b*) Weight of hanging portion = $\frac{1}{4}mg$

Weight of chain on table $= \frac{3}{4}mg$

Normal reaction, R $= \frac{3}{4}mg$

Maximum possible force of friction,

$$f_{ms} = \mu_s R = \frac{3}{4}\mu_s mg$$

Equating f_{ms} with weight of hanging portion,

$$\frac{3}{4}\mu_s mg = \frac{1}{4}mg \quad \text{or} \quad \mu_s = \frac{1}{3}$$

13. (*b*)
$T_2 = 5ma$ ...(*i*)
$T_1 - T_2 = 3ma$ ...(*ii*)
$F - T_1 = ma$ ...(*iii*)

Adding (*i*) and (*ii*),

$$T_1 = 8ma$$

But $T_1 = 16$ N

$\therefore$ $16 = 8ma$

or $$a = \frac{16}{8m} = \frac{2}{m}$$

Fig. 79

14. (*c*) From equation (*i*), $T_2 = 5\,m\left(\frac{2}{m}\right) = 10$ N

15. (*d*) From (*iii*), $F - 16 = m\left(\frac{2}{m}\right) = 2$

or $F = 18$ N

16. (*b*) $$a = \frac{40\text{N}}{(10+6+4)\text{ kg}}$$

$$= \frac{40 \text{ N}}{20 \text{ kg}} = 2 \text{ m s}^{-2}$$

Fig. 80

$$40 - T_1 = 10 \times 2$$

or $T_1 = (40 - 20) \text{ N} = 20 \text{ N}$

17. (*d*) If α represents angle of repose, then

$$\tan\alpha = 0.8 \text{ or } \alpha = \tan^{-1}(0.8) = 39°$$

The given angle of inclination is less than the angle of repose. So, the 1 kg block has no tendency to move. [Note that $mg\sin\theta$ is exactly balanced by the force of friction. So, T = 0.]

18. (*a*) Work done against friction = $\mu\, mg \times S$

$$= 0.4 \times 20 \times 25 \text{ J} = 200 \text{ J} \quad [\because mg = 20 \text{ N}]$$

19. (*d*) $T = m_1 a$

Also, $m_2 g - T = m_2 a$

or $m_2 g - m_1 a = m_2 a$

or $m_2 g = (m_1 + m_2)a$

or $a = \dfrac{m_2 g}{m_1 + m_2}$

Aliter. Weight $m_2 g$ is the accelerating force. Total mass to be accelerated $= m_1 + m_2$

$\therefore$ acceleration $= \dfrac{m_2 g}{m_1 + m_2}$

Fig. 81

20. (*b*) $2mnv = Mg$

or $v = \dfrac{Mg}{2mn} = \dfrac{10 \times 980}{2 \times 5 \times 10} \text{ cm s}^{-1}$

$= \dfrac{9800}{100} \text{ cm s}^{-1} = 98 \text{ cm s}^{-1}$

21. (*c*) $v = 108 \times \dfrac{5}{18} \text{ m s}^{-1} = 30 \text{ m s}^{-1}$

Momentum $= 120 \times 30 \text{ kg m s}^{-1}$ (initial)

Final momentum $= 0$

Force $= \dfrac{\text{Change in momentum}}{\text{time}}$

$= \dfrac{120 \times 30}{10} \text{ N} = 360 \text{ N}$

22. (*d*) $T = mg + ma$

$2000g = 1000g + 1000a$

or $a = g$

Direction is upward.

Now, $0^2 - 2.5^2 = -2 \times 10 \times S$

or $S = \dfrac{2.5 \times 2.5}{20} \text{ m} = \dfrac{625}{100 \times 20} = \dfrac{25}{80} \text{ m} = \dfrac{5}{16} \text{ m}$

23. (*c*) Horizontal component of force

$= 20\cos 30° \text{ N} = 20 \times \dfrac{\sqrt{3}}{2} \text{ N} = 10\sqrt{3} \text{ N}$

Acceleration along the surface $= \dfrac{10\sqrt{3}\text{ N}}{\sqrt{3}\text{ kg}} = 10 \text{ m s}^{-2}$

24. (*c*) $F = v\dfrac{dm}{dt} = \dfrac{4 \times 5}{2.5} \text{ dyne} = 8 \text{ dyne}$

25. (*c*) Using law of conservation of momentum,

$$234V + 4v = 0 \text{ or } V = -\frac{4}{234}v$$

26. (*a*) If μ is the mass/length, then

Weight of hanging length $= \mu\, lg$

Weight of chain on table $= \mu(L - l)g$

$R = \mu(L - l)g.$

$f = \mu_s R = \mu_s \mu(L - l)g$

Equating, $\mu_s\, \mu(L - l)g = \mu lg$ or $\mu_s = \dfrac{l}{L - l}$

27. (*c*) Apparent weight $= m(g + a)$

$= m(g + 5g) = 6mg$

28. (*b*) $30 \text{ kg f} < 32 \text{ kg f} \Rightarrow mg' < mg$

So, it is a case of downward acceleration *i.e.* the velocity is increasing.

Again, $36 \text{ kg f} > 32 \text{ kg f} \Rightarrow mg' > mg$

So, it is a case of downward retardation (*i.e.* upward acceleration). In other words, the velocity is decreasing.

29. (*c*) $v_r \dfrac{dm}{dt} = m_0 g$ or $\dfrac{dm}{dt} = \dfrac{m_0 g}{v_r}$

or $\dfrac{dm}{dt} = \dfrac{5 \times 10^5 \times 10}{10 \times 10^3} \text{ kg s}^{-1} = 500 \text{ kg s}^{-1}$

30. (*c*) Total weight of system $= (m + M)g$

The system pulls the ceiling downward with force $(m + M)g$. Using Newton's third law of motion, the ceiling shall pull the system upward with force $(m + M)g$.

31. (*d*) Tension at mid-point

= Weight of half the chain below the mid-point + Weight of block

$= \dfrac{mg}{2} + Mg = \left[\dfrac{m}{2} + M\right]g$

32. (*b*) Combined momentum $= 2p\hat{i} + p\hat{j}$

Magnitude of combined momentum

$= \sqrt{(2p)^2 + p^2} = \sqrt{5p^2} = \sqrt{5}p$

This must be equal to the momentum of the third part.

33. (*c*) $f - 5g = \dfrac{5g}{6}$ ...(*i*)

$(5 + m)g - f = (5 + m)\dfrac{g}{6}$ ...(*ii*)

Adding (*i*) and (*ii*), $mg = \dfrac{g}{6}(10 + m)$

or $6m = 10 + m$ or $5m = 10$ kg or $m = 2$ kg

34. (*d*) Refer to the free-body diagrams.

$T - 10g\sin 30° = 10a$ or $T - 5g = 10a$

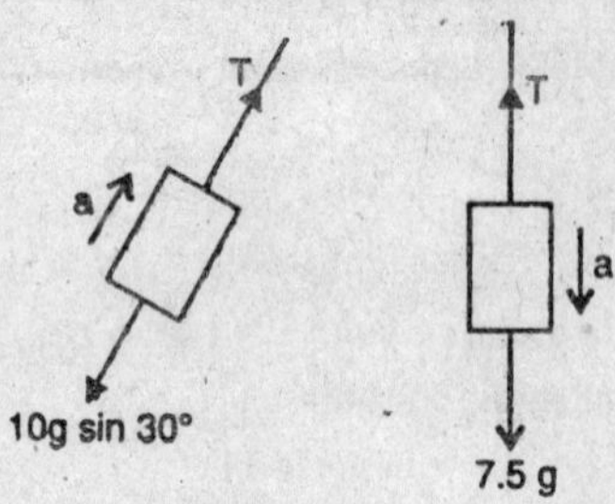

Fig. 82

Again, $7.5g - T = 7.5a$

Adding, $2.5g = 17.5a$

or $a = \frac{25g}{175} = \frac{g}{7}$

35. (*b*) $S = \frac{1}{2}at^2 = \frac{1}{2} \times \frac{g}{7} \times 2 \times 2 = \frac{2g}{7}$

36. (*c*) $|\vec{J}| = |\Delta\vec{p}| = 0.2(20 + 15)$ N s = 7 N s

37. (*a*) $f_{ms} = 0.3 \times 2 \times 10 = 6$ N

Applied force $< f_{ms}$

So, body would not move.

38. (*a*) We know that friction is a self-adjusting force. Since the applied force is less than limiting friction, therefore the force of friction is equal to the applied force *i.e.* 5 N.

39. (*d*) $u = \frac{250}{9}$ m s^{-1}, $v = 0$

$a = -\mu g = -0.5 \times 9.8$ m s^{-2}

$= -4.9$ m s^{-2}, $t = ?$

Using $t = \frac{v - u}{a}$,

$t = \frac{250}{9} \times \frac{1}{4.9}$ s = 5.7 s

40. (*a*) Clearly, θ can be treated as angle of repose. $\mu = \tan\theta$

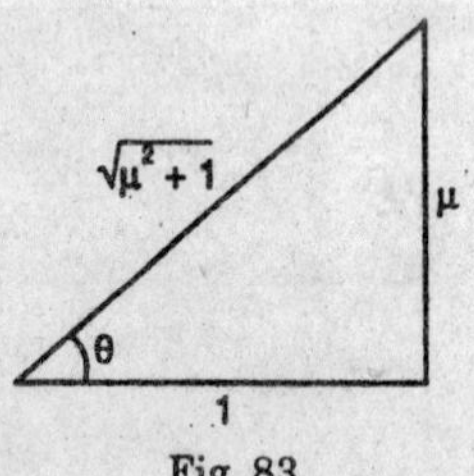

Fig. 83

Also, $\sin\theta = \frac{\mu}{\sqrt{\mu^2 + 1}}$

and $\cos\theta = \frac{1}{\sqrt{\mu^2 + 1}}$

Now, $0^2 - v_0^2 = -2[g\sin\theta + \mu g\cos\theta]$ S

or $S = \frac{v_0^2}{2g\left[\frac{\mu}{\sqrt{\mu^2+1}} + \frac{\mu}{\sqrt{\mu^2+1}}\right]}$

or $S = \frac{v_0^2}{4g\frac{\mu}{\sqrt{\mu^2+1}}}$ or $S = \frac{v_0^2}{4g\sin\theta}$

41. (*b*) $l = \frac{1}{2}at^2$ $[\because u = 0]$

or $t \propto \sqrt{l},\ t' \propto \sqrt{\frac{l}{2}}$

Dividing, $\frac{t'}{t} = \frac{1}{\sqrt{2}}$ or $t' = \frac{4}{\sqrt{2}}$ s $= 2\sqrt{2}$ s

42. (*d*) Pseudo force on the block = $m \times 4$ newton (backward)

Force of friction $= 0.4 \times m \times 10$ newton (forward)

Equating, $m \times 4 = 0.4 \times m \times 10$ or $4m = 4m$

Clearly, the equation holds good for all values of m.

43. (*d*) Applying conservation of momentum,

$4v = [20^2 + 20^2]^{1/2} = \sqrt{2} \times 20$

or $v = \frac{\sqrt{2} \times 20}{4}$ m s^{-1} = $5\sqrt{2}$ m s^{-1}

44. (*a*) Force = $v_r \frac{dm}{dt} = 6 \times 10^3 \times 1$ N $= 6 \times 10^3$ N

45. (*a*) Since the given system is in equilibrium therefore tension in the string is $5g$.

When we consider the combination of A and C, then

$T = \mu R$

or $T = 0.2[10 + \text{mass of C}]g$

or $5g = 0.2[10 + \text{mass of C}]g$

or mass of C = 15 kg

Fig. 84

46. (*c*) $ma\cos 30° = mg\sin 30°$

or $a \times \frac{\sqrt{3}}{2} = \frac{g}{2}$ or $a = \frac{g}{\sqrt{3}}$

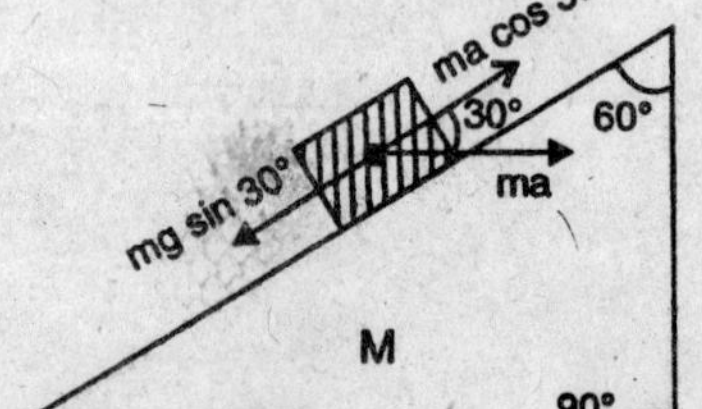

Fig. 85

47. (*b*) Change in momentum = impulse

= Area between thrust – time graph and time axis

$= \frac{1}{2}[60 + 30] \times 4000$ kg m s^{-1} $= 1.8 \times 10^5$ kg m s^{-1}

Also, change in momentum = $10^3 v$

Equating, $10^3 v = 1.8 \times 10^5$

or $v = 1.8 \times 10^2$ m s^{-1} = 180 m s^{-1}

48. (c) Common acceleration, $a = \dfrac{T}{m_1 + m_2 + m_3}$

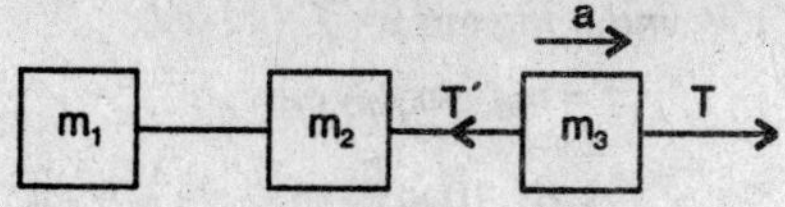

Fig. 86

$$T - T' = m_3 a \quad \text{or} \quad T' = T - m_3 a$$

or $$T' = T - m_3 \frac{T}{m_1 + m_2 + m_3}$$

or $$T' = \frac{m_1 + m_2}{m_1 + m_2 + m_3} T$$

49. (d) $$T = \frac{2 \times 6 \times 3}{6 + 3} g \text{ newton} = 4 \times 9.8 \text{ newton}$$

Total downward thrust on the pulley = 2T

$$= 2 \times 4 \times 9.8 \text{ newton} = 8 \times 9.8 \text{ newton}$$

50. (b) $T_1 = 0.25 \times 100$ N

or $T_1 = 25$ N

51. (b) $W_B = T_2 \sin 45°$

$25 = T_2 \cos 45°$

Dividing, $\dfrac{W_B}{25} = 1$ or $W_B = 25$ N

52. (c) $2mg \cos \theta = \sqrt{2}\, mg$

or $\cos \theta = \dfrac{1}{\sqrt{2}}$

or $\theta = 45°$

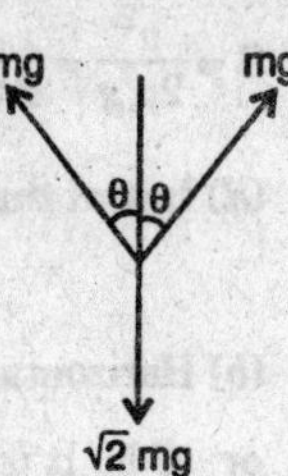

Fig. 87

53. (d) $F - f = 8 \times 4 = 32$ ...(i)

$2F - f = 128$ ...(ii)

Multiplying (i) by (ii), we get

$2F - 2f = 64$ ...(iii)

(ii) – (iii) gives

$f = 128 - 64 = 64$ N

$\mu mg = 64$

$\mu \times 8 \times 10 = 64$

or $\mu = \dfrac{64}{80} = \dfrac{8}{10} = 0.8$

54. (a) $T_2 - T_1 = m_2 a$

$T_1 = m_1 a$

Dividing,

$$\frac{T_2 - T_1}{T_1} = \frac{m_2}{m_1}$$

or $$\frac{T_2}{T_1} = \frac{m_2}{m_1} + 1 = \frac{m_2 + m_1}{m_1}$$

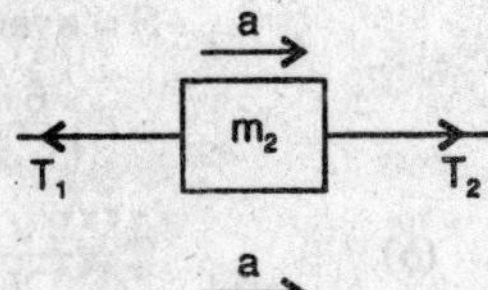

a m1 T1

Fig. 88

or $$\frac{T_1}{T_2} = \frac{m_1}{m_1 + m_2}$$

55. (b) $10 = 5a$

or $a = 2 \text{ m s}^{-2}$

$f = 2 \text{ kg} \times 2 \text{ m s}^{-2}$

$= 4$ N

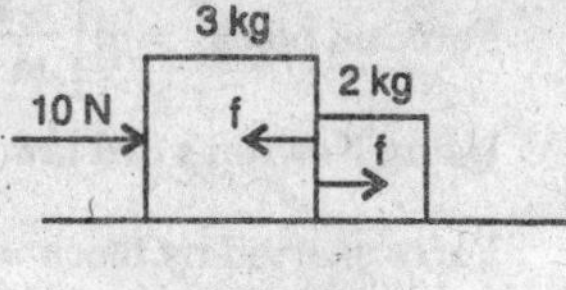

Fig. 89

56. (a) Horizontal components (each equal to T cos 30°) cancel out.

For vertical equilibrium, $5 = 2T \sin 30° = T$

57. (b) $T_C = 100$ g

58-59. $T_A \sin 45°$

$= T_C = 100$ g

or $T_A = \dfrac{100 \text{ g}}{\sin 45°}$

$= 100\sqrt{2}$ g

$T_B = T_A \cos 45°$

$= 100\sqrt{2} \times \dfrac{1}{\sqrt{2}}$ g

$= 100$ g

Tc 100 g $T_A \sin 45°$ T_B $T_A \cos 45°$ Tc

Fig. 90

60. (d) $f_{ms} = 0.6 \times 10 \times 9.8$ N

$= 58.8$ N

Since the applied force is greater than f_{ms} therefore the block will be in motion. So, we should consider f_k.

$$f_k = 0.4 \times 10 \times 9.8 \text{ N} \quad \text{or} \quad f_k = 4 \times 9.8 \text{ N}$$

This would cause acceleration of 40 kg block.

Acceleration $= \dfrac{4 \times 9.8 \text{ N}}{40 \text{ kg}} = 0.98 \text{ m s}^{-2}$

61. (d) $h_1 = \dfrac{1}{2} \times 19.6 \times 5 \times 5 \text{ m} = 245$ m

$v_1 = 19.6 \times 5 \text{ m s}^{-1}$

Again $h_2 = \dfrac{19.6 \times 5 \times 19.6 \times 5}{2 \times 9.8}$ m

$= 25 \times 19.6$ m $= 490$ m

Total height, $h = h_1 + h_2$

$= (245 + 490)$ m $= 735$ m

Note. In this problem of rocket propulsion, gravitational field is assumed to be uniform.

62. (c) $v = 3t^2 - 75$; $a = 6t$.

$F = ma = 36t$. At $t = 4$ s, $F = 144$ N

63. (d) Forward force on the horse

$= \mu mg + ma = m[\mu g + a]$

$= 2000[0.2 \times 10 + 1]$ N $= 6000$ N

64. (*d*) Acceleration $= \frac{15}{m+1}$

Force on block $= m\left[\frac{15}{m+1}\right]$

Using Newton's 3rd law of motion,

Force exerted by block $= \frac{15m}{m+1}$

or $13.5 = \frac{15m}{m+1}$ or $(15 - 13.5)m = 13.5$

or $m = \frac{13.5}{1.5}$ kg = 9 kg

65. (*d*) $T = 1 \times a$ or $1.5 = a$

$F = (1 + 22)$ kg $\times 1.5$ m s^{-2} = 4.5 N

66. (*a*) $\mu = \tan 30° = \frac{1}{\sqrt{3}} = \frac{\sqrt{3}}{3} = \frac{1.732}{3} = 0.577$

67. (*d*) Effective value of acceleration due to gravity in the lift $= g - a$.

Acceleration down the inclined plane $= (g - a) \sin \theta$

Using, $S = ut + \frac{1}{2}at^2$, we get

$l = \frac{1}{2}(g - a) \sin \theta t^2$, we get

$t = \sqrt{\frac{2l}{(g-a)\sin\theta}}$

68. (*a*) $mg - \eta mg = ma$

$a = g(1 - \eta)$

69. (*d*) $mg \sin\theta + \mu\, mg \cos\theta$

$= 2(mg \sin\theta - \mu\, m\, g \cos\theta)$

$mg \sin\theta = 3\mu\, m\, g \cos\theta$

$\tan\theta = 3\mu$ or $\theta = \tan^{-1}(3\mu)$

70-71. In order to conserve momentum, C should move with speed υ in a direction opposite to that of B.

72. (*b*) $S = \frac{1}{2}\mu g t^2$ or $t \propto \frac{i}{\sqrt{\mu}}$

73. (*d*) Net momentum of two pieces moving at right angles

$= \sqrt{16^2 + 12^2} = \sqrt{256 + 144} = \sqrt{400} = 20$ kg m s^{-1}

Momentum of third piece $= 0.5v$

Using law of conservation of momentum,

$0.5v = 20$ or $v = \frac{20}{0.5}$ m s^{-1} = 40 m s^{-1}

74. (*c*) $F = v\frac{dm}{dt} = 10 \times 5$ N = 50 N

75. (*a*) $F = v\frac{dm}{dt} = 5$ m s$^{-1} \times 1$ kg s^{-1} = 5 N

Initial acceleration $= \frac{5\text{ N}}{2\text{ kg}} = 2.5$ m s^{-2}

76. (*a*) $ma = mg \sin\theta - f$

or $f = mg \sin\theta - ma$

$= 8\left[10 \times \frac{1}{2} - 0.4\right]$N $= 8 \times 4.6$ N = 36.8 N

77. (*c*) Common acceleration of block and rope $= \frac{1}{2M}$

Tension at mid-point

= [Mass of Block + Mass of half the rope] a

$= \left[M + \frac{M}{2}\right]a = \frac{3Ma}{2} = \frac{3M}{2} \times \frac{1}{2M} = \frac{3}{4}$ N = 0.75 N

78. (*b*) $P = f_{ms} = \mu_s mg$

When the body starts moving with acceleration a, then

$P - f_k = ma$

$\mu_s mg - \mu_k mg = ma$ or $a = (\mu_s - \mu_k)g$

or $a = (0.5 - 0.4)10 = 0.1 \times 10$ m s^{-2} = 1 m s^{-2}

79. (*a*) $v = 72$ km h^{-1} $= 72 \times \frac{5}{18}$ m s^{-1} = 20 m s^{-1}

$S = \frac{v^2}{2\mu g} = \frac{20 \times 20}{2 \times 0.5 \times 10}$ m = 40 m

80. (*d*) Initial thrust $= mg + ma = m(g + a)$

$= 10^5(10 + 5)$ N $= 15 \times 10^5$ N $= 1.5 \times 10^6$ N.

81. (*b*) Horizontal component of 10 N is 10 cos 60° *i.e.* $10 \times \frac{1}{2}$ N or 5 N. It is balanced by 5 N force along *x*-axis.

Y-component of 10 N force is 10 sin 60°.

i.e. $10 \times \frac{\sqrt{3}}{2}$ N or $5\sqrt{3}$ N

$a_y = \frac{5\sqrt{3}\text{ N}}{\sqrt{3}\text{ kg}} = 5$ m s^{-2}

82. (*c*) $a = \frac{500 - f}{20} = \frac{500 - 0.5 \times 20 \times 10}{20}$ m s^{-2}

$= \frac{500 - 100}{20}$ m s^{-2} $= \frac{400}{20}$ m s^{-2} = 20 m s^{-2}

83. (*b*) $t = \frac{v - u}{\mu g} = \frac{6}{0.3 \times 10} = 2$ s

S = average velocity × time

$S = \left(\frac{6 + 0}{2}\right)2$ m = 6 m

84. (*b*) $T = \frac{2 \times 3 \times 5}{3 + 5}$ kgf $= \frac{15}{4}$ kgf

Pressure on the pulley $= 2T = 2 \times \frac{15}{4}$ kgf = 7.5 kgf

☞ **An Important Result**

Pressure on the pulley = 2 × Tension in the string passing over the pulley.

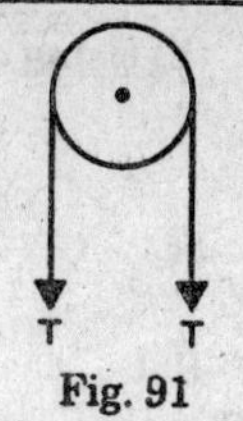

Fig. 91

85. (*b*) $\left(\frac{M}{l}\right)\frac{l}{n}g = \mu\left[\frac{M}{l}\left(l - \frac{l}{n}\right)\right]g$ or $\frac{l}{n} = \mu l\left(1 - \frac{1}{n}\right)$

or $\frac{1}{n} = \frac{\mu(n-1)}{n}$ or $\mu = \frac{1}{n-1}$

86. (*a*) Loss of kinetic energy $= \frac{1}{2}mv^2 - \frac{1}{2}(m + M)V^2$

$$= \frac{1}{2}mv^2 - \frac{1}{2}(m + M)\left(\frac{mv}{m + M}\right)^2$$

$$= \frac{1}{2}mv^2\left[1 - \frac{m}{m + M}\right] = \frac{1}{2}\left[\frac{mM}{m + M}\right]v^2$$

Substituting values, we get the result.

87. (*a*) $R = mg + F_2 \cos\theta$, $f = \mu R$

$f = \mu[mg + F_2 \cos\theta]$

Also,

$f = F_1 + F_2 \sin\theta$

Equating,

$\mu\,[mg + F_2 \cos\theta]$

$= F_1 + F_2 \sin\theta$

or $\mu = \frac{F_1 + F_2 \sin\theta}{mg + F_2 \cos\theta}$

Fig. 92

88. (*c*) Component of weight of 2 kg body along the inclined plane $= 2g \sin 30° = g$

Weight of 1 kg body $= 1 \times g = g$

Clearly, there can be no motion.

89. (*b*) In order that the block may not fall,

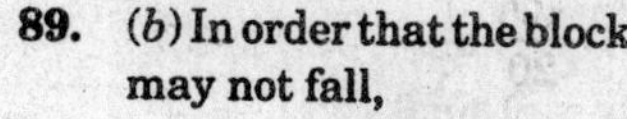

$f \geq mg$

$\mu R \geq mg$

or $\mu ma \geq mg$

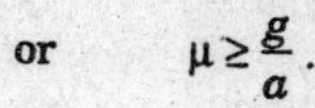

or $\mu \geq \frac{g}{a}$.

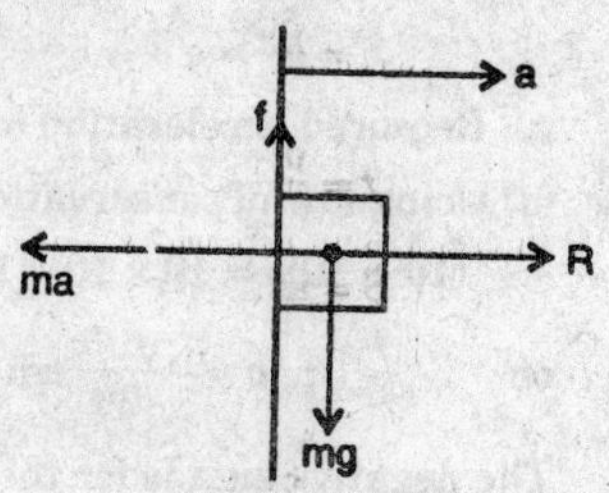

Fig. 93

90. (*c*) $v = u + at$

$0 = 30 + 2.5a$ or $a = -\frac{30}{2.5}\text{ m s}^{-2}$

or $-a = 12\text{ m s}^{-2}$

This is total retardation (due to air resistance and gravity).

Retardation due to air resistance

$= (12 - 10)\text{ m s}^{-2} = 2\text{ m s}^{-2}$

Air resistance $= 0.5\text{ kg} \times 2\text{ m s}^{-2}$

$= 1\text{ kg m s}^{-2} = 1\text{ N}$

91. (*c*) Velocity acquired in falling through height $h = \sqrt{2gh}$

Now, $u = \sqrt{2gh}$, $v = 0$, 'S' = S, a = ?

$0^2 - 2gh = 2aS$ or $a = -\frac{gh}{S}$

This is total retardation due to gravity and air resistance. If a_r is the retardation due to air resistance, then

$a_r - g = -a$ or $a_r = -a + g$ or $a_r = \frac{gh}{S} + g$

Force due to air resistance $= Mg\left(1 + \frac{h}{S}\right)$

92. (*c*) $F = \mu mg$

But $\mu = \tan\theta$

$\therefore$ $F = mg \tan\theta$ or $F = W \tan\theta$

93. (*b*) $3mg = mg + ma$

or $2mg = ma$ or $a = 2g$

94. (*d*) $2T \sin\theta = W$ or $T = \frac{W}{2 \sin\theta}$

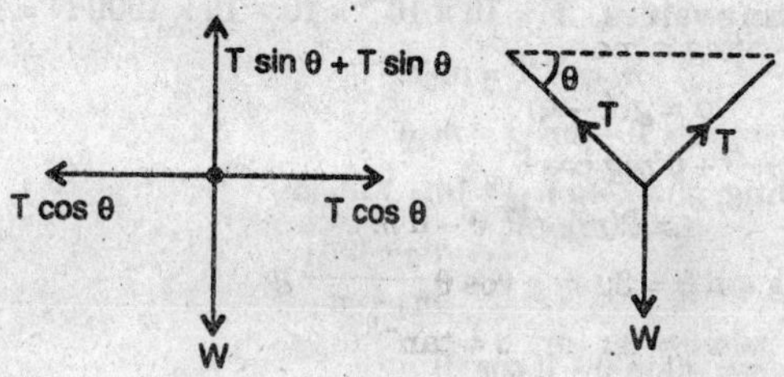

Fig. 94

For the straight rope, $\theta = 0°$, $\sin\theta = 0$

$\therefore$ $T = \infty$

Clearly, infinite force is required.

95. (*b*) $f_{ms} = \mu_s mg \cos\theta = \frac{\sqrt{3}}{2} \times 10 \times \frac{\sqrt{3}}{2}\text{ N} = 7.5\text{ N}$

Force pulling the block down the inclined plane

$= mg \sin\theta = 10 \sin 30°\text{ N} = 5\text{ N}$

The force on the block is less than f_{ms}. So, force of friction is 5 N.

Note that if the applied force is less than f_{ms}, then the force of friction is exactly equal to the applied force.

96. (*b*) $0.05v = (0.450 + 0.05)\,V$

or $V = \frac{0.05}{0.50}v = \frac{v}{10}$

Using $v^2 - u^2 = 2aS$, we get

$$0^2 - \left(\frac{v}{10}\right)^2 = -2 \times 10 \times 1.8$$

or $$\left(\frac{v}{10}\right)^2 = 36 \text{ or } v = 60 \text{ m s}^{-1}$$

97. (*d*) Since the force becomes zero as soon as the bullet leaves the barrel of the gun therefore t can be obtained by putting F = 0 in the equation

$$F = 600 - 2 \times 10^5 t$$

$\therefore$ $$2 \times 10^5 t = 600 \text{ or } t = 3 \times 10^{-3} \text{ s}$$

98. (*c*) $$J = \int_0^t Ft = \int_0^{3\times10^{-3}} (600 - 2 \times 10^5 t)\, dt$$

$$= \left| 600t - 2 \times 10^5 \frac{t^2}{2} \right|_0^{3\times10^{-3}}$$

$$= 600 \times 3 \times 10^{-3} - 10^5 \times 3 \times 10^{-3} \times 3 \times 10^{-3}$$

$$= 3 \times 10^{-3} [600 - 300] = 0.9 \text{ N s}$$

99. (*a*) Volume of water/s = av

Mass of water/s = $av\rho$

Momentum of water/s = $av^2\rho$

$$F = av^2\rho$$

Putting values, $F = 10 \times 10^{-4} \times 10 \times 10 \times 1000 \text{ N} = 10^2 \text{ N}$

100. (*d*) $$m_2 g - T = m_2 a$$

$$T - \mu m_1 g = m_1 a$$

Adding, $g(m_2 - \mu m_1) = (m_1 + m_2)a$

or $$a = \frac{m_2 - \mu m_1}{m_1 + m_2} g$$

101. (*c*) $a = g(\sin\theta - \mu \cos\theta)$

$$a = 10[\sin 60° - 0.25 \cos 60°] = 10\left[\frac{\sqrt{3}}{2} - \frac{0.25}{2}\right]$$

$$a = 5[1.7 - 0.25]$$

$$a = 5 \times 1.45 \text{ m s}^{-2} = 7.25 \text{ m s}^{-2}$$

102. (*d*) Clearly, the force of friction is 400 N.

Acceleration force = (1650 – 400) N = 1250 N

Acceleration $= \frac{1250}{1000} \text{ m s}^{-2} = 1.25 \text{ m s}^{-2}$

103. (*b*) Applying conservation of momentum,

$$(720 + 80)v = 720 \times 80$$

or $$v = \frac{720 \times 80}{800} \text{ km h}^{-1} \text{ or } v = 72 \text{ km h}^{-1}$$

104. (*b*) $$a = \frac{10 \text{ N}}{5 \text{ kg}} = 2 \text{ m s}^{-2}$$

Now, $v^2 - 0^2 = 2 \times 2 \times 25 = 100$ or $v = 10 \text{ m s}^{-1}$

105. (*d*) Using conservation of momentum,

$$mv_1 = 3mv_2$$

or $$\frac{v_1}{v_2} = \frac{3}{1} \text{ or } \frac{v_1^2}{v_2^2} = \frac{9}{1}$$

But $v^2 \propto S$

$\therefore$ $$\frac{S_1}{S_2} = \frac{9}{1}$$

106. (*d*) $|\Delta \vec{p}| = 2 \times 0.05 \times 4 = 0.4 \text{ kg m s}^{-1}$

$$Ft = 0.4 \text{ or } F = \frac{0.40}{0.01} \text{ N} = 40 \text{ N}$$

107. (*b*) If f_1 is the contact force in the first case, then

$$f_1 = (2m)(\text{acceleration})$$

$$= 2m \times \frac{P}{5m} = \frac{2P}{5}$$

If f_2 is the contact force in the second case, then

$$f_2 = (3m)(\text{acceleration})$$

$$= 3m \times \frac{P}{5m} = \frac{3P}{5}$$

Required ratio is $\frac{2P}{5} : \frac{3P}{5}$ or 2 : 3

Note that the total acceleration of the system is the same as the acceleration of mass $2m$ or mass $3m$.

108. (*d*) $f = mg$

$\mu R = mg$

$\mu F = mg$

$$F = \frac{mg}{\mu}$$

$$= \frac{50 \times 10}{0.5} \text{ N}$$

$= 1000$ N

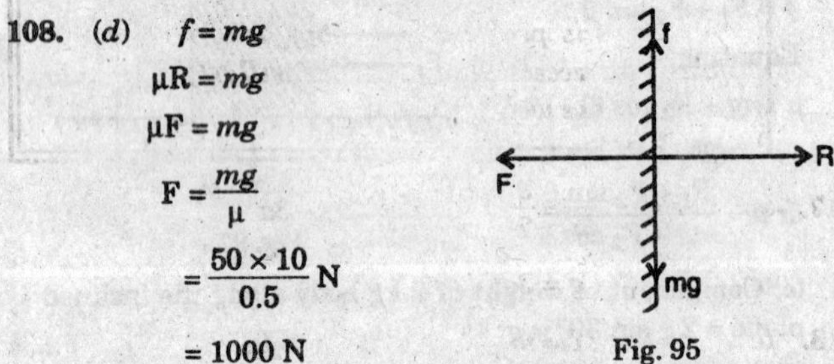

Fig. 95

109. (*a*) Downward retardation may be regarded as equivalent to upward acceleration.

$\therefore$ $$g' = g + a$$

$\therefore$ Required acceleration = $(g + a) \sin\theta$

110. (*a*) Using law of conservation of momentum,

$$(10 + 10)v = 10 \times 10 - 10 \times 20 = -100$$

or $$v = -\frac{100}{20} \text{ cm s}^{-1} = -5 \text{ cm s}^{-1}$$

The negative sign is for the west direction.

111. (*a*) $v_r \frac{dm}{dt} = mg + ma = m[g + a]$

or $$\frac{dm}{dt} = \frac{5000 \times 30}{800} \text{ kg s}^{-1} = 187.5 \text{ kg s}^{-1}$$

112. (*b*) $v_r \dfrac{dm}{dt} = mg$ or $\dfrac{dm}{dt} = \dfrac{mg}{v_r}$

$$= \frac{6000 \times 10}{1000} \text{ kg s}^{-1} = 60 \text{ kg s}^{-1}$$

113. (*d*) $\dfrac{T_1}{T_2} = \dfrac{m\left(g + \dfrac{g}{2}\right)}{m\left(g - \dfrac{g}{2}\right)} = \dfrac{3g/2}{g/2} = \dfrac{3}{1}$

114. (*b*) If M be the mass of the rope, then acceleration $= \dfrac{F}{M}$

$$T = \frac{M}{L}(L - x)\frac{F}{M} = \frac{F(L - x)}{L}$$

Fig. 96

115. (*a*) $\dfrac{dm}{dt} = \dfrac{mg}{v_r} = \dfrac{5000 \times 9.8}{1000} \text{ kg s}^{-1}$

$$= 5 \times 9.8 \text{ kg s}^{-1} = 49 \text{ kg s}^{-1}$$

116. (*c*) $v_r \dfrac{dm}{dt} = 3mg$

$$\frac{dm}{dt} = 3\left[\frac{mg}{v_r}\right] = 3 \times 49 \text{ kg s}^{-1} = 147 \text{ kg s}^{-1}$$

> ☞ *In this problem, note that '2mg' will be required for acceleration and 'mg' will be required to overcome the weight.*

117. (*a*) $\dfrac{mg}{mg - ma} = \dfrac{3}{2}$ or $2g = 3g - 3a$

or $g = 3a$ or $a = \dfrac{g}{3}$

118. (*a*) $m_2g - 2T = m_2a$ or $2T = m_2(g - a)$ ...(*i*)

Again, $T = m_1(2a)$ or $2T = 4m_1a$ ...(*ii*)

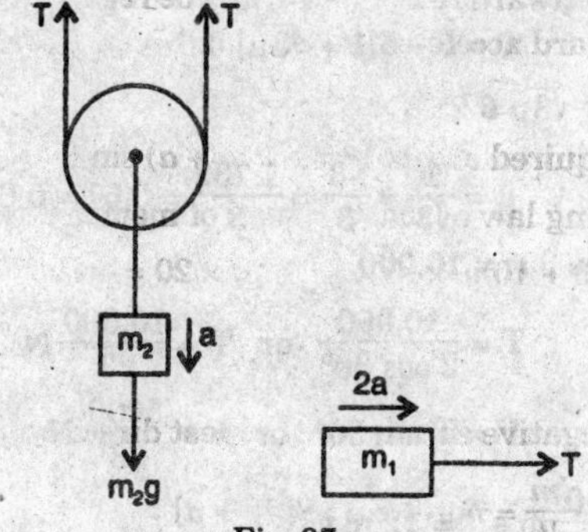

Fig. 97

Equating (*i*) and (*ii*),

$$m_2g - m_2a = 4m_1a \text{ or } (4m_1 + m_2)a = m_2g$$

or $$a = \frac{m_2g}{4m_1 + m_2}$$

119. (*b*) Spring balance reading in terms of kgf

$$= \frac{4m_1m_2}{m_1 + m_2} = \frac{4 \times 5 \times 1}{6} = \frac{10}{3}$$

This is less than 6 kgf.

121. (*d*) Common acceleration, $a = \dfrac{40}{10 + 6 + 4} \text{ m s}^{-2} = 2 \text{ m s}^{-2}$

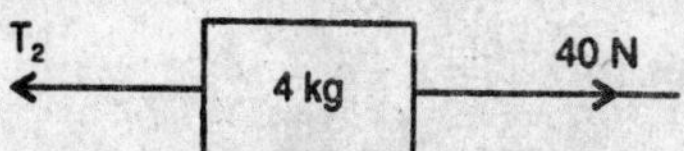

Fig. 98

Now, $40 - T_2 = 4 \times 2$ or $T_2 = (40 - 8)$ N $= 32$ N

124. (*b*) $v^2 = 2\mu gS$ or $v^2 = 2 \times 0.5 \times 9.8 \times 9.8$

or $v^2 = 9.8 \times 9.8$ or $v = 9.8 \text{ m s}^{-1}$

125. (*a*) $a = 2v \Rightarrow \dfrac{dv}{dt} = 2v$

$$\frac{dv}{ds} \cdot \frac{ds}{dt} = 2v$$

or $v\dfrac{dv}{ds} = 2v$ or $dv = 2ds$

$$\int_{0.1}^{v} dv = 2 \left| s \right|_0^{0.1} = 0.2$$

$$v - 0.1 = 0.2 \text{ or } v = 0.3 \text{ m s}^{-1}$$

126. (*b*) $t = \sqrt{\dfrac{2l}{g \sin\theta}}$

$$\frac{t}{2} = \sqrt{\frac{2h}{g}}$$

$$4 = \frac{2l}{g \sin\theta} \times \frac{g}{2h}$$

$$4 = \frac{1}{\sin\theta\left(\dfrac{h}{l}\right)} = \frac{1}{\sin^2\theta}$$

or $\sin\theta = \dfrac{1}{2}$ or $\theta = 30°$

Fig. 99

127. (*c*) If W is the maximum weight, then

$$W = 2T\cos 60° \text{ or } W = T = 20 \text{ N}$$

128. (*c*) $5g - T_2 = 5a$ ...(*i*)

$T_2 - T_1 - 3g = 3a$ ...(*ii*)

$T_1 - g = a$ ...(*iii*)

Adding (*i*) and (*iii*),

$$-T_2 + T_1 + 4g = 6a$$

Adding this to (*ii*), we get

$$g = 9a \text{ or } a = \frac{g}{9}$$

Fig. 100

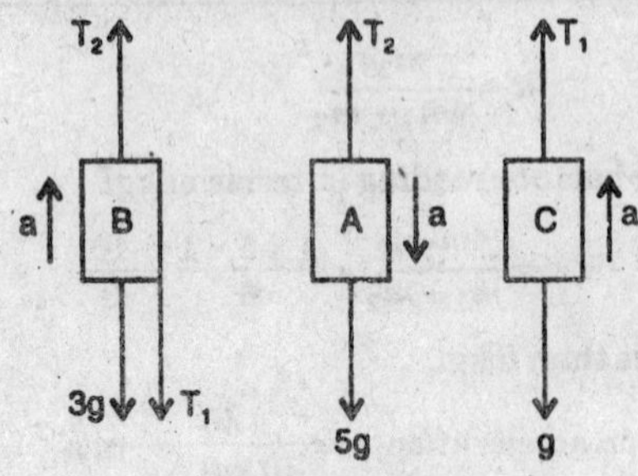

Fig. 101

129. (*d*) From eqn. (*iii*), $T_1 = g + a$

or $$T_1 = g + \frac{g}{9} = \frac{10g}{9}$$

130. (*d*) From eqn. (*i*), $T_2 = 5g - 5a = 5g - \frac{5g}{9} = \frac{40g}{9}$

131. (*c*) $$x = 3t^{3/2} + 2t - \frac{1}{2}$$

$$v = 3 \times \frac{3}{2} \times t^{1/2} + 2 \quad \text{or} \quad v = \frac{9}{2} t^{1/2} + 2$$

Acceleration $= \frac{9}{2} \times \frac{1}{2} t^{1/2-1} = \frac{9}{4} t^{-1/2}$

or Acceleration $\propto t^{-1/2}$.

132. (*b*) Note that the initial vertical momentum is zero.

Now, $3m \times 100 = 2m \times v_x$ or $v_x = 150$ km h^{-1}

Again, $m \times 400 = 2mv_y$ or $v_y = 200$ km h^{-1}

Now, $V = \sqrt{200^2 + 150^2} = 250$ km h^{-1}

133. (*c*) Common acceleration of A and B $= \frac{15}{3} = 5$ units

Force on B = 5 × 2 N = 10 N

134. (*c*) The reading would change from $\left(g - \frac{g}{4}\right)$ to $\left(g + \frac{g}{4}\right)$ *i.e.* from $\frac{3g}{4}$ to $\frac{5g}{4}$. The factor is $\frac{5}{3}$

∴ New reading $= \frac{5}{3} \times 30$ kgf = 50 kgf

135. (*c*) Displacement of lift $= \frac{1}{2} \times 12 \times 0.2 \times 0.2$ m = 0.24 m

Displacement of block $= \frac{1}{2} \times 2 \times 0.2 \times 0.2$ m = 0.04 m

Required displacement = (0.24 – 0.04) m = 0.2 m

> ☞ *Displacements of lift and block are opposite.*

136. (*c*) $$T = \frac{2m_1m_2}{m_1 + m_2}(2g)$$

or $$T = \frac{2\frac{w_1}{g}\frac{w_2}{g}}{\frac{w_1}{g} + \frac{w_2}{g}}(2g)$$

or $$T = \frac{2w_1w_2}{g^2} \times \frac{g}{w_1 + w_2}(2g) \quad \text{or} \quad T = \frac{4w_1w_2}{w_1 + w_2}$$

137. (*d*) $20 - T = 6a$

$T - 10 = 4a$

Adding, $10 = 10a$ or $a = 1$ m s^{-2}

Now, $T - 10 = 4 \times 1 = 4$ or $T = 14$ N

138. (*c*) Required force = 2T

$$= 2\frac{2m_1m_2}{m_1 + m_2} g = \frac{4 \times 5 \times 3}{8} g = 7.5 g$$

139. (*d*) Frictional force = μF = W in the limiting case.

Since μ can have any value (like 0.01 or 5.0) therefore none of *a*, *b*, *c* holds.

140. (*b*) $30g = 10g + 10a$

or $a = 2g = 20$ m s^{-2}

Using $S = \frac{1}{2}at^2$,

we get $t = \sqrt{\frac{2S}{a}} = \sqrt{\frac{2 \times 10}{20}}$ s = 1 s

141. (*b*) Consider (M + *m*) moving to the left by *x* and *m* moving to the right by L cos θ.

142. (*b*) Initially, the weight of load L is the force on the system of mass 8 kg.

Acceleration $= \frac{2 \times 10}{8} = \frac{20}{8}$ units

Towards the end, force = (2 + 1) × 10 N = 30 N

So, acceleration now is $\frac{30}{8}$ units

143. (*c*) Retardation = 10[sin 30° + μ cos 30°]

$$= 10\left[\frac{1}{2} + \frac{\sqrt{3}\mu}{2}\right] = 5[1 + \sqrt{3}\mu]$$

Now using $v = u + at$, we get

$$0 = 5 - 5[1 + \sqrt{3}\mu]\, 0.5$$

or $1 + \sqrt{3}\mu = 2$

or $$\mu = \frac{1}{\sqrt{3}} = \frac{\sqrt{3}}{3} = \frac{1.732}{3} = 0.57 \approx 0.6$$

144. (*a*) 2T cos 30° = 10,000

or $$T = \frac{10,000}{2 \cos 30^\circ} \text{ or } T = \frac{10,000}{\sqrt{3}} \text{ N}$$

Also, F = T sin 30° or $F = \frac{5000}{\sqrt{3}}$ N

145. (*a*) $a = \frac{30}{2.5}$ m s^{-2} = 12 m s^{-2}

Retardation due to air = (12 – 9.8) m s^{-2} = 2.2 m s^{-2}

Air resistance = 0.4 kg × 2.2 m s^{-2} = 0.88 N

146. (*b*) Acceleration along OF $= \frac{4}{2}$ m s^{-2} = 2 m s^{-2}

Distance along OF $= \frac{1}{2} \times 2 \times 4 \times 4$ m = 16 m

Distance along OE $= 3 \times 4$ m = 12 m

Distance from O $= \sqrt{16^2 + 12^2} = \sqrt{256 + 144}$

$= \sqrt{400}$ m = 20 m

147. (*b*) $F = 3t^2 - 32$

$$m\frac{dv}{dt} = 3t^2 - 32 \quad \text{or} \quad m\,dv = (3t^2 - 32)\,dt$$

$$\int_{10}^{v} m\,dv = \int_0^5 (3t^2 - 32)\,dt$$

$$10(v - 10) = \left| t^3 - 32t \right|_0^5$$

or $10v - 100 = 125 - 160$ or $10v = 100 - 35 = 65$

or $v = 6.5$ m s^{-1}

148. (*d*) $(M + m)V = Mv$ or $V = \frac{Mv}{M + m}$

Using $v^2 - u^2 = 2AS$, we get

$$0^2 - \frac{M^2v^2}{(M+m)^2} = -2Aa$$

or $$A = \frac{M^2v^2}{(M+m)^2\, 2a}$$

Force of Resistance = (M + *m*)A

$$= (M+m)\left(\frac{M^2v^2}{(M+m)^2\, 2a}\right)$$

$$= \frac{M^2v^2}{(M+m)\, 2a} = \frac{M^2}{M+m}\frac{v^2}{2a}$$

149. (*a*) Here, it is assumed that the ball was originally at rest.

∴ Change in momentum

$= 0.5 \times 10$ kg m s^{-1} = 5 kg m s^{-1}

$5 = F \times \frac{1}{50}$ or F = 250 N

150. (*a*) $m = 40$ g, $v = 1200$ m s^{-1}, F = 144 N

Let *n* be the maximum number of bullets that can be fired per second.

Rate of change of momentum $= nmv = n \times \frac{40}{1000} \times 1200$

This is equal to F, according to Newton's second law of motion.

∴ $n \times \frac{40}{1000} \times 1200 = 144$ or $n = \frac{12}{4} = 3$

151. (*d*) Acceleration $= \frac{P}{M + m}$

Force on the block = Mass of block × Acceleration

$= \frac{MP}{M + m}$

152. (*a*) Let F be the force experienced by each.

acceleration of husband $= \frac{F}{60}$ m s^{-2}

acceleration of wife $= \frac{F}{40}$ m s^{-2}

We know that $S = ut + \frac{1}{2}at^2$

Initially, both are at rest, ∴ $u = 0$ ∴ $S = \frac{1}{2}at^2$

In the given problem, *t* is constant ∴ $S \propto a$

∴ Distances moved by husband and wife will be in the ratio of their respective accelerations.

i.e., $\frac{F}{60} : \frac{F}{40}$ or 4 : 6 or 2 : 3.

Aliter. Force exerted on each other is same. But acceleration of each is different.

∴ Acceleration $\propto \frac{1}{\text{Mass}}$

Also, Distance ∝ Acceleration

∴ Distance $\propto \frac{1}{\text{Mass}}$

Mass ratio is 60 : 40 or 3 : 2

Distance ratio is 2 : 3

153. (*a*) $Mg' = Mg + M(4g)$ or $Mg' = 5Mg$

154. (*d*) $T_y \sin 45° = 500$

or $T_y = 500\sqrt{2}$ kg wt

Again, $T_x = T_y \cos 45°$

or $T_x = 500\sqrt{2} \times \frac{1}{\sqrt{2}}$ kgf

or $T_x = 500$ kgf

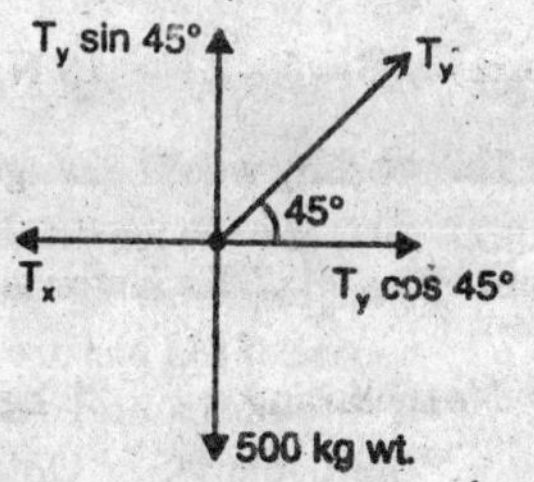

Fig. 102

155. (*a*) Downward retardation means upward acceleration.

$g' = g + a$

Now, $t = \sqrt{\frac{2L}{g' \sin\theta}}$ or $t = \sqrt{\frac{2L}{(g + a)\sin\theta}}$

156. (*a*) Reaction on *m* is *mg*. Maximum friction force by M on *m* is μ*mg*. So, the force on M is μ*mg* forward.

Acceleration $= \frac{\mu mg}{M}$

157. (*d*) Acceleration of system $= \frac{20\text{ N}}{4\text{ kg}} = 5$ m s^{-2}

Force exerted by car A

$$= 3 \text{ kg} \times 5 \text{ m s}^{-2} = 15 \text{ kg m s}^{-2} = 15 \text{ N}$$

158. (*a*) Velocity of released grain relative to trolley is zero. So, no force is experienced by trolley. Thus, its velocity remains unchanged.

159. (*c*) $\frac{W}{2}$ if string elements supporting W were vertical. Here sloping. So, (*c*) is the correct choice.

160. (*a*) For the father, after throwing the ball, $m = 60$ kg.

So, $v = \frac{5}{60} \text{ m s}^{-1} = \frac{1}{12} \text{ m s}^{-1}$

For the daughter, after catching,

$m = 21$ kg

So, $v = \frac{5}{21} \text{ m s}^{-1}$

161. (*b*) Tension uniform, accelerations equal

$$\frac{2 \text{ kg} - T}{2} = \frac{T - 1 \text{ kg}}{1}$$

So, $T = \frac{4}{3}$ kg

162. (*c*) Impulse = Change in momentum

$= 2mv \cos \theta$

$= 2 \times 9 \times 10^{-31} \times 600 \times \cos 60° \text{ N s}$

$= 5.4 \times 10^{-28} \text{ N s}$

163. (*c*) Initial net force = 25 – (15 + 5)*g* wt = 5*g* wt

Final is (25 + 5) – 15 = 15*g* wt.

So, it is 3-fold.

164. (*b*) Clearly, $4F = 80 \times 10$

or $F = 200$ N

165. (*e*) Draw the forces acting (T in each cord and *mg* through the centre of the picture). Resolve T into horizontal and vertical components. Obtain an equation for T by equating the vertical forces. Then consider the value of T when θ = 0°, only one graph satisfies the condition.

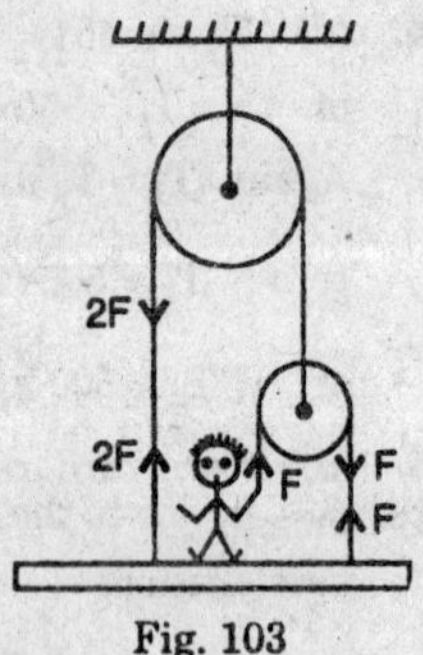

Fig. 103

166. (*d*) Minimum force required to be applied parallel to the inclined plane $= mg(\sin \theta + \mu \cos \theta)$

$= \sin \theta + \mu \cos \theta$ [∵ $mg = 1$]

167. (*a*) Volume of water/s = *a*

Mass of water/s = $av\rho$

Momentum of water/s = $av^2\rho$

Initial horizontal component = $av^2\rho \cos \theta$

Final vertical component = $-av^2\rho \cos \theta$

| Change in momentum/s | = $2av^2\rho \cos \theta$

168. (*a*) $T_2 = ma$ ...(*i*)

$T_1 - T_2 = ma$

Fig. 104

Adding, $T_1 = 2ma$ ...(*ii*)

Dividing (*i*) by (*ii*), $\frac{T_2}{T_1} = \frac{1}{2}$

169. (*c*) $t = \sqrt{\frac{2l}{g \sin \theta}}$

$nt = \sqrt{\frac{2l}{g \sin \theta - \mu g \cos \theta}}$

$\frac{1}{n^2} = \frac{g \sin \theta - \mu g \cos \theta}{g \sin \theta}$ or $\frac{1}{n^2} = 1 - \mu \cot \theta$

or $\mu \cot \theta = 1 - \frac{1}{n^2}$ or $\mu = \tan \theta \left[1 - \frac{1}{n^2}\right]$

170. (*a*) $R + mg \cos \theta = mg$

or $R = mg - mg \cos \theta$ or $R = mg(1 - \cos \theta)$

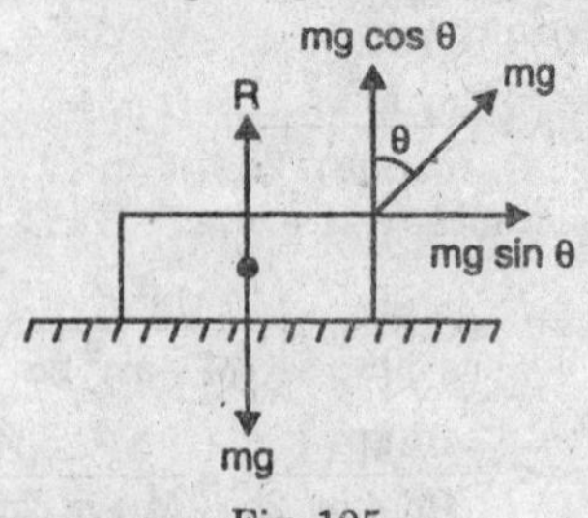

Fig. 105

For pulling the block, $mg \sin \theta > \mu R$

or $mg \sin \theta > \mu\, mg(1 - \cos \theta)$ or $\sin \theta > \mu(1 - \cos \theta)$

or $2 \sin \frac{\theta}{2} \cos \frac{\theta}{2} > \mu \left(1 - 1 + 2 \sin^2 \frac{\theta}{2}\right)$

or $\cos \frac{\theta}{2} > \mu \sin \frac{\theta}{2}$ or $\cot \frac{\theta}{2} > \mu$

171. (*c*) $Mv = [M + \mu t]\, v'$ or $v' = \frac{Mv}{M + \mu t}$

172. (*b*) $v' = \frac{Mv}{M + [\text{Total mass of water}]}$

$v' = \frac{Mv}{M + V\rho}$

> ☞ *The vertical velocity v_y of the rain has no effect on horizontal momenta.*

173. (*d*) As is clear from Fig. 106

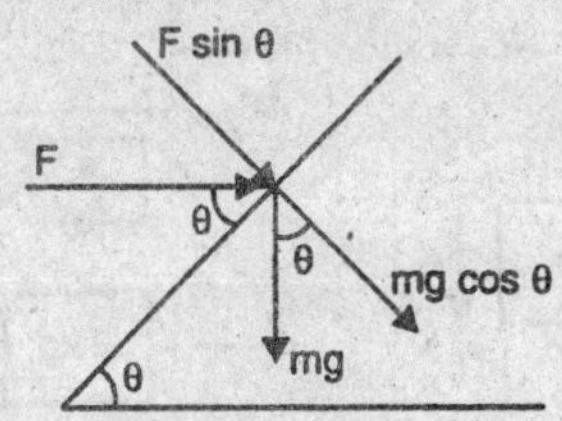

Fig. 106

$$R = mg \cos\theta + F \sin\theta$$

174. (*d*) $f = \mu_s R$

or $f = \mu_s(mg \cos\theta + F \sin\theta)$

175. (*c*) $200 - f = 30 \times 2$ or $f = (200 - 60)N = 140\ N$

Again, $140 = \mu \times 30 \times 10$

or $\mu = \dfrac{140}{300} = 0.47$

176. (*b*) $T_1 = (12 + 3)a$; $T_2 = 3a$

$$\frac{T_1}{T_2} = \frac{15a}{3a} = \frac{5}{1}$$

177. (*c*) The trolley shall move backwards to conserve momentum. The backward momentum would be shared by both the trolley and man.

Applying conservation of momentum,

$$60 \times 1 = (240 + 60)v \text{ or } 60 = 300v$$

or $v = \dfrac{60}{300} = \dfrac{1}{5}\ m\ s^{-1} = 0.2\ m\ s^{-1}$

Speed of man w.r.t. ground

$$= (1 - 0.2)\ m\ s^{-1} = 0.8\ m\ s^{-1}$$

Displacement of man $= 0.8 \times 4\ m = 3.2\ m$

178. (*c*) $W = T_1 \sin\theta + T_2 \sin\phi$

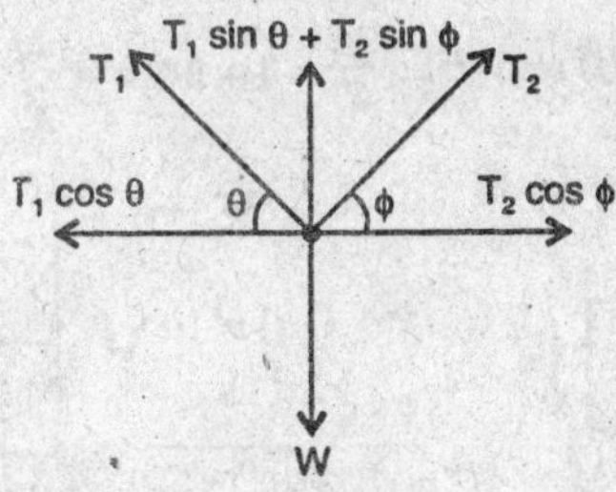

Fig. 107

But $T_2 \cos\phi = T_1 \cos\theta$ or $T_2 = \dfrac{T_1 \cos\theta}{\cos\phi}$

$\therefore$ $W = T_1 \sin\theta + \dfrac{T_1 \cos\theta}{\cos\phi} \sin\phi$

or $W \cos\phi = T_1 \sin\theta \cos\phi + T_1 \cos\theta \sin\phi$

or $T_1[\sin\theta \cos\phi + \cos\theta \sin\phi] = W \cos\phi$

or $T_1 \sin(\theta + \phi) = W \cos\phi$

or $T_1 = \dfrac{W \cos\phi}{\sin(\theta + \phi)}$

Similarly, $T_2 = \dfrac{W \cos\theta}{\sin(\theta + \phi)}$

179. (*c*) $mg \sin\theta - T = ma$ ...(*i*)

$T = ma$ ...(*ii*)

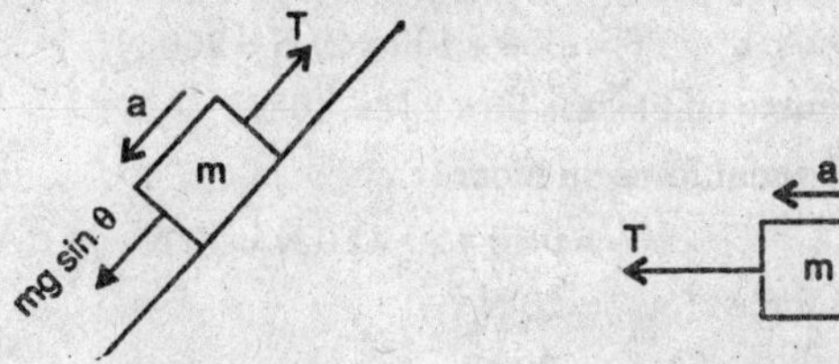

Fig. 108 (*a*) Fig. 108 (*b*)

From (*i*) and (*ii*),

$$mg \sin\theta - T = T$$

or $2T = mg \sin\theta$ or $T = \dfrac{1}{2} mg \sin\theta$

180. (*c*) Applying conservation of momentum,

$10 \times 10 = m \times 25 - (10 - m)12.5$ | mass of heavier fragment is taken as *m*.

$8 = 2m - 10 + m$

or $3m = 18$ or $m = 6$ kg

Mass of other fragment is (10 – 6) kg *i.e.* 4 kg.

181. (*d*) $a_x = \dfrac{d}{dt}(5t) = 5\ m\ s^{-2}$

$N_x = 2\ kg \times 5\ m\ s^{-2} = 10\ N$

$a_y = \dfrac{d}{dt}(2t) = 2\ m\ s^{-2}$

$N_y = m(a_y + g) = 2(2 + 10)\ N = 24\ N$

$N = \sqrt{10^2 + 24^2} = \sqrt{100 + 576}\ N$

$= \sqrt{676}\ N = 26\ N$

182-183. Refer to the free-body diagram of painter.

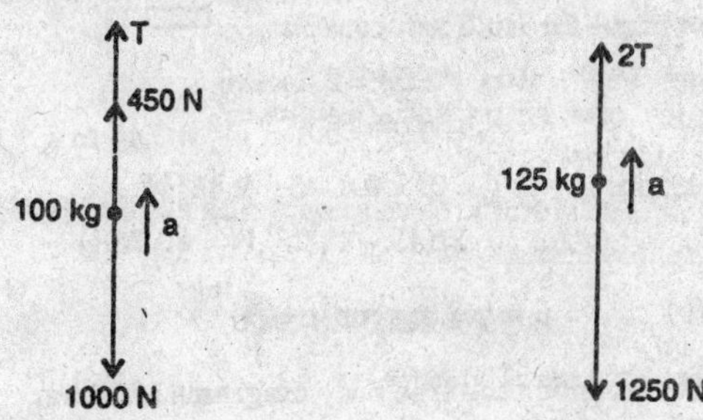

Fig. 109 (*a*) Fig. 109 (*b*)

$$T + 450 - 1000 = 100a$$

or $T - 550 = 100a$ ...(*i*)

Refer to the free body diagram of the system.

$$2T - 1250 = 125a \quad ...(ii)$$

From (*ii*) and (*i*),

$$2[550 + 100a] - 1250 = 125a$$

or $75a = 1250 - 1100 = 150$

or $a = \frac{150}{75} \text{ m s}^{-2} = 2 \text{ m s}^{-2}$

From (*i*), $T = (550 + 100 \times 2) \text{ N} = 750 \text{ N}$

184. (*c*) Force on block $= 20 \times 2 \text{ N} = 40 \text{ N}$

Frictional force on block

$= 0.15 \times 20 \times 10 \text{ N} = 30 \text{ N}$

Net force $= 10 \text{ N}$

Acceleration $= \frac{10 \text{ N}}{20 \text{ kg}} = \frac{1}{2} \text{ m s}^{-2}$

Now, $4 = 0 \times t + \frac{1}{2} \times \frac{1}{2} t^2$ or $t = 4 \text{ s}$

Let us calculate the distance travelled by the truck.

$$x = 0 \times 4 + \frac{1}{2} \times 2 \times 4 \times 4 = 16 \text{ m}$$

185-186. Refer to the free body diagram of block B

$5g - T = 5a$ or $T = 5g - 5a$

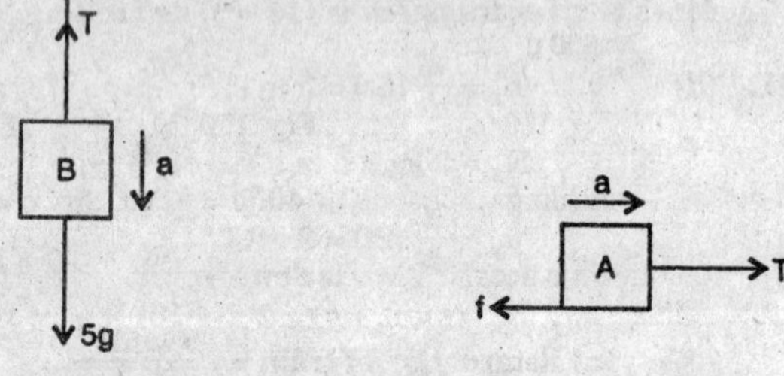

Fig. 110 (*a*) Fig. 110 (*b*)

Refer to the free body diagram of block A

$T - f = 5a$

or $5g - 5a - 0.5 \times 5 \times g = 5a$

or $10a = 2.5g = 2.5 \times 9.8$

or $a = 0.25 \times 9.8 \text{ m s}^{-2} = 2.45 \text{ m s}^{-2}$

Again, $T = (5 \times 9.8 - 5 \times 2.45) \text{ N}$

$= (49 - 12.25) \text{ N} = 36.75 \text{ N}$

187. (*b*) $\mu ma = mg$ or $a = \frac{g}{\mu}$

188-189. Consider the free body diagrams of 10 kg, 3 kg and 2 kg.

$$10g - T_2 = 10a \quad ...(i)$$

$$T_2 - T_1 - 0.6g = 3a \quad ...(ii)$$

$$T_1 - 0.6g = 2a \quad ...(iii)$$

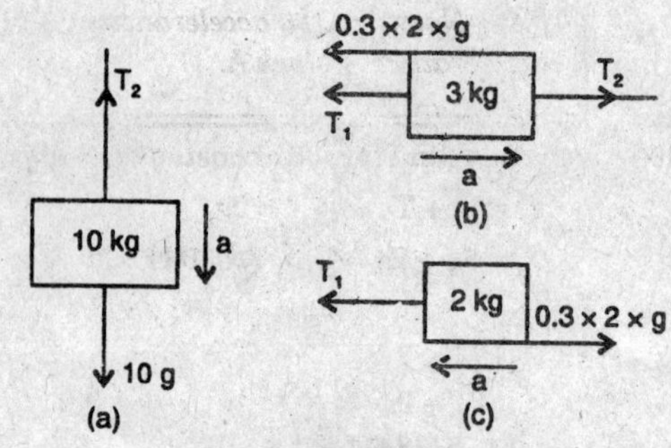

Fig. 111

Adding, $15a = 8.8g$

or $a = \frac{8.8 \times 9.8}{15} \text{ m s}^{-2} = 5.75 \text{ m s}^{-2}$

From eqn. (*iii*), $T_1 = 0.6 \times 9.8 + 2 \times 5.75$

$= 5.88 + 11.5 \approx 17 \text{ N}$

From eqn. (*i*),

$T_2 = 10g - 10a = (98 - 57.5) \text{ N}$

$= 40.5 \text{ N} \approx 41 \text{ N}$

190. (*b*) $2mnv = Mg$

or $v = \frac{Mg}{2mn}$

$= \frac{200 \times 980}{2 \times 200 \times 40} = \frac{980}{80} \text{ cm s}^{-1} = 12.25 \text{ cm s}^{-1}$

191. (*a*) Retardation, $-a = g \sin\theta + \mu g \cos\theta$

$$0^2 - v^2 = -2(g \sin\theta + \mu g \cos\theta) S$$

or $S = \frac{v^2}{2g(\sin 30° + \mu \cos 30°)} = \frac{v^2}{g(1 + \sqrt{3}\mu)}$

192. (*a*) $2g - T = 2a$

or $4g - 2T = 4a \quad ...(i)$

Again, $2T - g = \frac{a}{2} \quad ...(ii)$

Adding (*i*) and (*ii*), $\frac{9a}{2} = 3g$ or $a = \frac{2g}{3}$

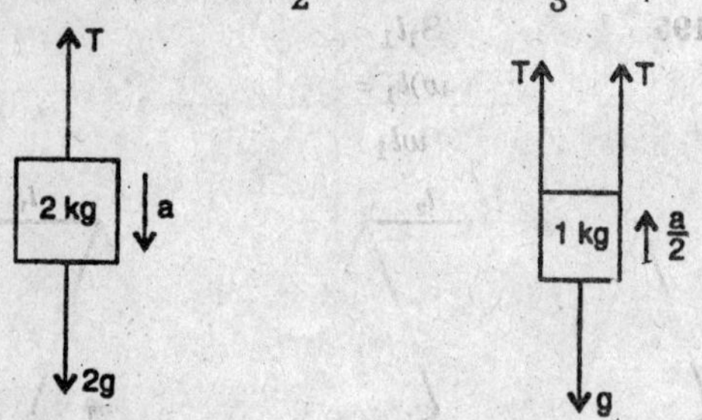

Fig. 112 (*a*) Fig. 112 (*b*)

It is downwards.

Acceleration of 1 kg block is $\frac{1}{2} \times \frac{2g}{3}$ *i.e.* $\frac{g}{3}$. It is upwards.

> ☞ *Note that the acceleration of block B is half of acceleration of block A.*

193. (*a*) If A is climbing with constant velocity, then

$T' = 5g + T$ and $T = 2g$

$T' = 5g + 2g = 7g = 7 \times 10\text{ N} = 70\text{ N}$

Suppose A is climbing with acceleration '*a*' such that $T = 30\text{ N}$

$T - 2g = 2a$

$30 - 2 \times 10 = 2a$

or $a = 5\text{ m s}^{-2}$

Again, $T' - T - 5g = 5a$ or $T' = T + 5g + 5a$

or $T' = (30 + 50 + 25)\text{ N} = 105\text{ N}$

Fig. 113

194. (*a*) Let S_1 and S_2 be the weights of left pan and right pan respectively. When the object is in left pan,

$(S_1 + w)l = (S_2 + w_1)l$

or $S_1 + w = S_2 + w_1$ ...(*i*)

When the object is in the right pan,

$(S_1 + w_2)l = (S_2 + w)l$

or $S_1 + w_2 = S_2 + w$ ...(*ii*)

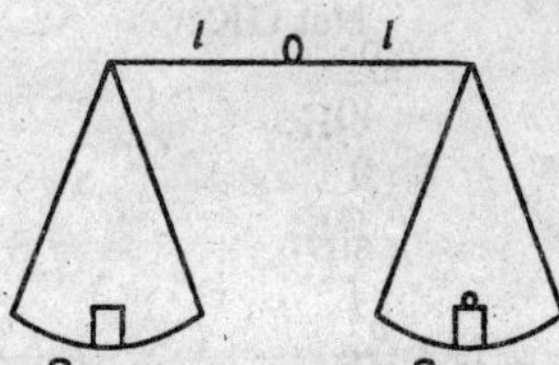

Fig. 114 (*a*)

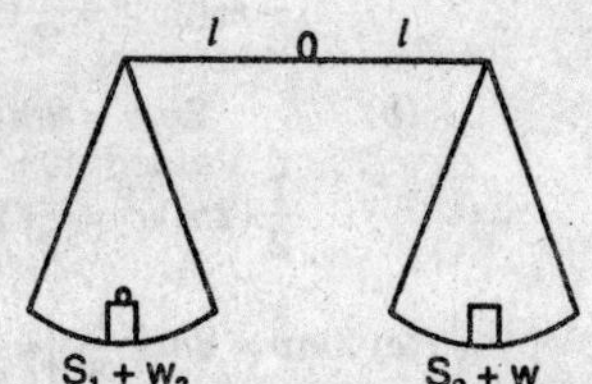

Fig. 114 (*b*)

(*i*) – (*ii*) gives

$w - w_2 = w_1 - w$ or $2w = w_1 + w_2$

or $w = \dfrac{w_1 + w_2}{2}$

195. (*b*) $S_1 l_1 = S_2 l_2$ ...(*i*)

$(S_1 + w)l_1 = (S_2 + w_1)l_2$

or $w l_1 = w_1 l_2$ ...(*ii*) [Using (*i*)]

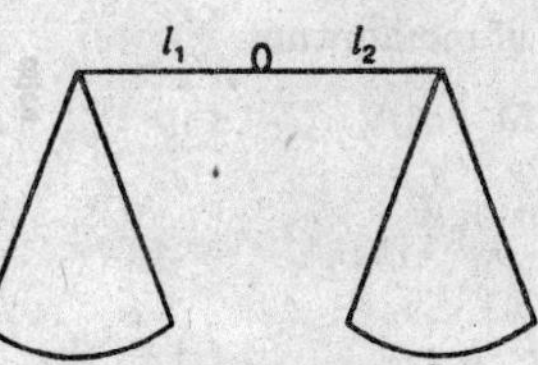

Fig. 115 (*a*)

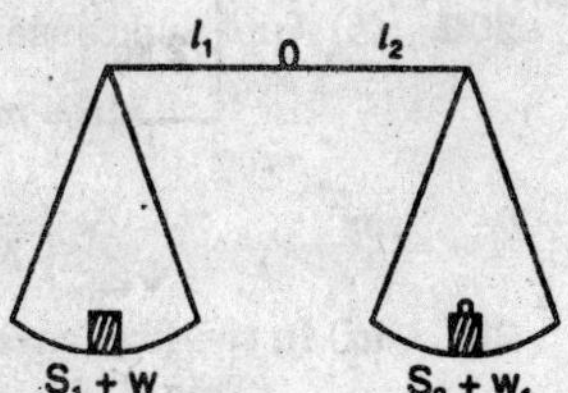

Fig. 115 (*b*)

Again, $(S_1 + w_2)l_1 = (S_2 + w)l_2$

or $w_2 l_1 = w l_2$ ...(*iii*) [Using (*i*)]

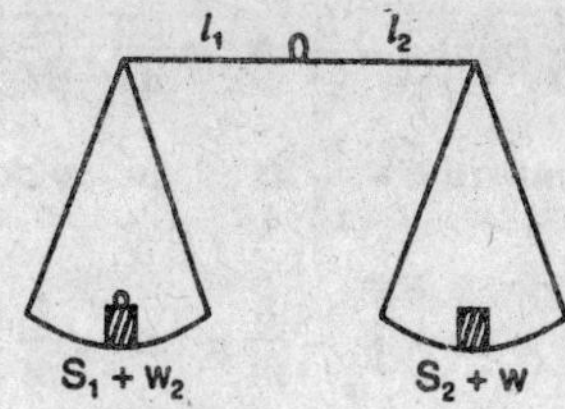

Fig. 115 (*c*)

Dividing (*ii*) by (*iii*),

$\dfrac{w}{w_2} = \dfrac{w_1}{w}$ or $w^2 = w_1 w_2$ or $w = \sqrt{w_1 w_2}$

196. (*c*) $500g - T = 500a$ ...(*i*)

$T - 100g \sin 30° - T' = 100a$

or $T - T' - 50g = 100a$ ...(*ii*)

Again, $T' - 50g = 50a$ ...(*iii*)

From (*ii*) and (*iii*),

$T - 100g = 150a$ ...(*iv*)

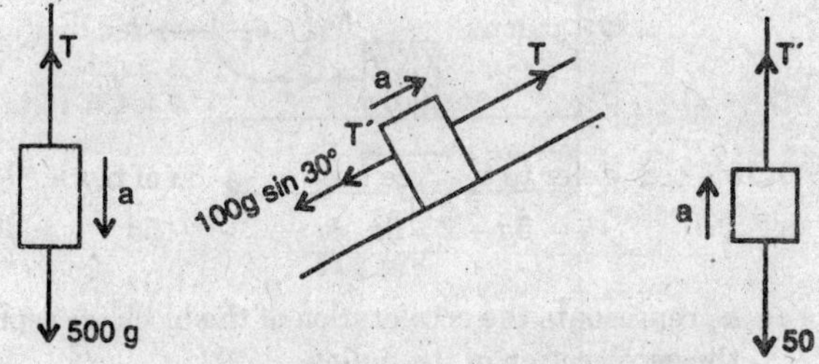

Fig. 116 (*a*) Fig. 116 (*b*) Fig. 116 (*c*)

Adding (*i*) and (*iv*), $400g = 650a$ or $a = \dfrac{400g}{650} = \dfrac{8g}{13}$

This acceleration is downwards.

197. (*a*) Retardation of train $= \dfrac{36\text{ kmh}^{-1}}{5\text{ s}}$

$= \dfrac{36 \times \frac{5}{18}\text{ m s}^{-1}}{5\text{ s}} = 2\text{ m s}^{-2}$

It acts in the backward direction.

Fictitious force on suitcase = $2m$ newton, where m is the mass of suitcase.

It acts in the forward direction.

Due to this force, the suitcase has a tendency to slide forward. If suitcase is not to slide, then

$2m$ = force 'f' of friction

or $2m = \mu mg$ or $\mu = \dfrac{2}{g} = \dfrac{2}{9.8} = \dfrac{20}{98} = \dfrac{10}{49}$

198. (*a*) $F = -\dfrac{b}{x^2},\ \dfrac{d}{dt}(p) = -\dfrac{b}{x^2}$

or $\frac{d}{dt}(mv) = -\frac{b}{x^2}$ or $m\frac{dv}{dx}\cdot\frac{dx}{dt} = -\frac{b}{x^2}$

or $mv\,dv = -\frac{b}{x^2}dx$ or $v\,dv = -\frac{b}{mx^2}dx$

$$\int_0^v v\,dv = \int_a^x -\frac{b}{m}x^{-2}\,dx = \frac{b}{m}\left|\frac{1}{x}\right|_a^x$$

or $\frac{v^2}{2} = \frac{b}{m}\left[\frac{1}{x} - \frac{1}{a}\right]$

or $v = \sqrt{\frac{2b}{m}\left(\frac{a-x}{xa}\right)}$

199-200. Length of string, $L = 2y - x$

Differentiating twice,

$$2\frac{d^2y}{dt^2} - \frac{d^2x}{dt^2} = 0 \text{ or } 2a_2 = a_1$$

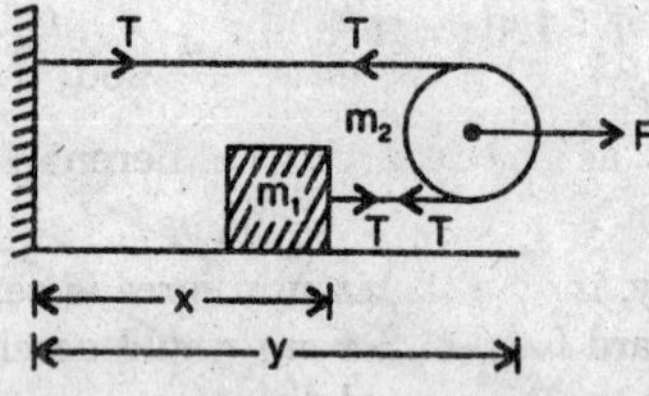

Fig. 117

a_1 represents the acceleration of the block, a_2 represents the acceleration of the pulley.

So, the acceleration of the block is *two times* the acceleration of the pulley.

Again, $F - 2T = m_2 a_2$

or $F - 2(m_1a_1) = m_2\left[\frac{a_1}{2}\right]$

or $F = 2m_1a_1 + \frac{m_2a_1}{2}$

or $2F = a_1[4m_1 + m_2]$ or $a_1 = \frac{2F}{4m_1 + m_2}$

201. (*a*) Applying law of conservation of momentum,

$$\frac{m}{4}\times v_s + \frac{3m}{4}\times v_l = 0$$

or $\frac{v_s}{v_l} = -3$ or $v_s = -3v_l$

or $15t = -3yt$ or $y = -5$ cm

202. (*b*) Using $v = u + at$, we get

$$0 = 5 + 0.5a \text{ or } a = -\frac{5}{0.5}\text{ m s}^{-2}$$

or $a = -10$ m s^{-2}

Also, $a = g(\sin 30° + \mu\cos 30°)$

$$10 = 10\left[\frac{1}{2} + \frac{\sqrt{3}\mu}{2}\right]$$

or $1 + \sqrt{3}\mu = 2$ or $\mu = \frac{1}{\sqrt{3}}$

203. (*c*) $f_{ms} = \mu_s mg = 0.8 \times 4 \times 9.8\text{ N} = 31.36\text{ N}$

Applied force, $F < f_{ms}$

So, the force of friction is equal to the applied force *i.e.* 19 N.

> ☞ *In this type of questions, the step I should always be to calculate f_{ms}. If the applied force is less than f_{ms}, then the force of friction is equal to the applied force.*

204. (*c*) $m = \frac{F}{A}$

Now, $mA' = \sqrt{2}F$

or $\frac{F}{A}A' = \sqrt{2}F$ or $A' = \sqrt{2}A$.

205. (*b*)

(*a*) $2mv = m(2v)$ OK

$\frac{1}{2}m(2v)^2 = \frac{1}{2}(2m)v^2$ Not OK

(*b*) $2mv = 2mv$ OK

$\frac{1}{2}(2m)v^2 = \frac{1}{2}(2m)v^2$ OK

(*c*) $2mv = 8m\left(\frac{v}{8}\right) = mv$ Not OK

(*d*) $2mv = (6m)\frac{v}{6}$ Not OK.

> ☞ **Author's line of argument in the above solution**
>
> *First check on the basis of conservation of momentum and then conserve energy.*

206. (*b*) Applying conservation of momentum,

$$Mv = m \times 0 + (M - m)V$$

or $V = \frac{M}{M-m}v$

207. (*d*) In the first case,

$$T_1 - mg = \frac{mg}{2}$$

or $T_1 = \frac{3mg}{2}$ ...(*i*)

In the second case,

$$mg - T_2 = \frac{mg}{2}$$

or $$T_2 = mg - \frac{mg}{2} = \frac{mg}{2} \quad ...(ii)$$

Dividing (*i*) by (*ii*),

$$\frac{T_1}{T_2} = \frac{3mg}{2} \times \frac{2}{mg} = \frac{3}{1}$$

208. (*a*) $$v_i = \sqrt{2 \times 10 \times 5} \text{ m s}^{-1} = 10 \text{ m s}^{-1}$$

$$p_i = mv_i = 0.4 \text{ kg} \times 10 \text{ m s}^{-1}$$

or $$p_i = 4 \text{ kg m s}^{-1}$$

It is directed downwards.

$$v_f = \sqrt{2 \times 10 \times 20} \text{ m s}^{-1} = 20 \text{ m s}^{-1}$$

$$p_f = 0.4 \text{ kg} \times 20 \text{ m s}^{-1} = 8 \text{ kg m s}^{-1}$$

It is directed vertically upwards.

Now, impulse = change of momentum

$$-100 \times \Delta t = -8 - 4$$

[***Note that both force and velocity are directed upward.***]

[Giving negative sign to the vertically upward direction.]

Now, $$\Delta t = \frac{12}{100} \text{ s} = 0.12 \text{ s}$$

A common mistake. Avoid it.

| *Change of momentum* | $= 8 - 4 = 4$ *kg m* s^{-1}.

209. (*a*) In the first case, $$h = \frac{1}{2} g t_1^2 \quad ...(i)$$

In the second case,

$$h - vt_2 = -vt_2 + \frac{1}{2} g t_2^2$$

or $$h = \frac{1}{2} g t_2^2 \quad ...(ii)$$

Comparing (*i*) and (*ii*), $t_1 = t_2$

210. (*c*) $$f_{ms} = 0.4 \times 10 \text{ N} = 4 \text{ N}$$

The applied force is less than f_{ms}. So, the block would not move.

211. (*b*) $$R' = m(g + a) = m(g + g)$$

$$= 2mg \text{ newton} = 2m \text{ kg wt}$$

$$= 2 \text{ kg wt} \quad [\because m = 1]$$

212. (*d*) $$p = \sqrt{2mE_k}$$

Here, E_k is constant

$\therefore$ $$p \propto \sqrt{m}$$

So, in the given problem, the heavier body shall have more momentum.

213. (*c*) If T is the tension in the rope, then the force exerted by the boy on the rope is equal to the force exerted by the rope on the boy. Let R be the normal reaction between the boy and the frame.

Fig. 118 (*a*) Fig. 118 (*b*)

Refer to the free body diagram of boy

$$T + R = 40g \quad ...(i)$$

Refer to the free body diagram of frame.

$$T = R + 360g \quad ...(ii)$$

Adding, $$2T = 400g \text{ or } T = 200g$$

214. (*d*) Let us consider three different situations in this question.

Initially, the net horizontal force is zero. The vertically downward force of 1.4 *mg* would accelerate the central block in the downward direction.

A stage will be reached when 2T cos θ = 1.4 *mg*.

Due to inertia of motion, the block of mass 1.4 *m* will overhoot a possible equilibrium situation.

Then, 2T cos θ′ would be greater than 1.4 *mg*. So, a net vertical force would accelerate the central block upwards.

215. (*b*) Tension at the given point = Weight of portion of rope between the given point and the free end

$$= \frac{M}{L}(L - l)g$$

216. (*b*) Clearly, the applied force is μ_s *mg i.e.* 0.4 *mg*. When the body is in motion, the force of friction is $\mu_k mg$ *i.e.* 0.3 *mg*. So, accelerating force is (0.4 *mg* – 0.3 *mg*) *i.e.* 0.1 *mg*

$$\text{Acceleration} = \frac{0.1\,mg}{m} = 0.1\,g$$

$$= 0.1 \times 9.8 \text{ m s}^{-2} = 0.98 \text{ m s}^{-2}$$

217. (*c*) $$u = 2 \text{ m s}^{-1}, v = 0, S = \frac{5}{100} \text{ m}$$

$$0^2 - 2^2 = 2 \times a \times \frac{5}{100}$$

$$a = -40 \text{ m s}^{-2}$$

$$\text{Resistance force, } F = \frac{25}{1000} \times 40 \text{ N} = 1 \text{ N}$$

218. (c) $p = \sqrt{p_x^2 + p_y^2}$

$= \sqrt{4\cos^2 t + 4\sin^2 t} = \sqrt{4(\cos^2 t + \sin^2 t)} = 2$

Clearly, momentum is time-independent.

So, force must be perpendicular to momentum.

219. (b) If n is the required number of men, then total mass to be considered is $(n \times 100 + 500)$ kg. Maximum tension will be there only in the case of upward acceleration.

We know that $T - mg = ma$

$\therefore \; T - (n \times 100 + 500)g = (n \times 100 + 500)a$

or $T = (n \times 100 + 500)(g + a)$

Substituting values,

$2 \times 10^4 = (n \times 100 + 500)(9.8 + 0.2)$

or $2000 = 100n + 500$

or $n = \dfrac{1500}{100} = 15$

220. (c) If b is base length, then slope length l is $\dfrac{b}{\cos\theta}$

Also, acceleration along the inclined plane $a = g\sin\theta$

Now, $l = \dfrac{1}{2}at^2$

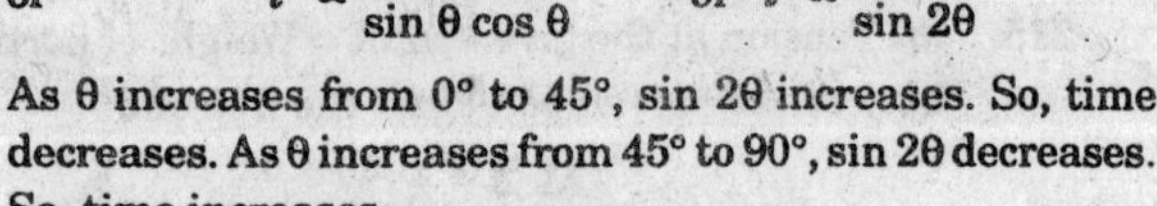

or $t^2 \propto \dfrac{l}{a}$

Fig. 119

or $t^2 \propto \dfrac{b}{\cos\theta} \times \dfrac{1}{g\sin\theta}$

or $t^2 \propto \dfrac{1}{\sin\theta\cos\theta}$ or $t^2 \propto \dfrac{1}{\sin 2\theta}$

As θ increases from 0° to 45°, $\sin 2\theta$ increases. So, time decreases. As θ increases from 45° to 90°, $\sin 2\theta$ decreases. So, time increases.

221. (a) $0^2 - v_0^2 = -2ay_2$ or $v_0^2 = 2ay_2$

or $2gy_1 = 2ay_2$ or $a = \dfrac{gy_1}{y_2}$

$R = m[g + a] = mg\left[1 + \dfrac{y_1}{y_2}\right]$

222. (a) $F = ma$

$F - T = \dfrac{m}{2}a$

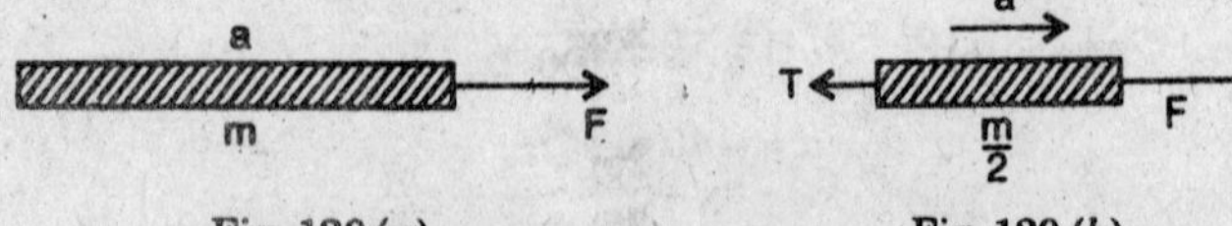

Fig. 120 (a) Fig. 120 (b)

Dividing, $\dfrac{F-T}{F} = \dfrac{\frac{m}{2}a}{ma} = \dfrac{1}{2}$

$2F - 2T = F$ or $F = 2T$ or $\dfrac{T}{F} = \dfrac{1}{2}$

> ☞ *An important note. If the rope in this question is massless, then $\dfrac{T}{F} = 1$.*

223. (b) Refer to the free body diagram of the boy

$T - mg = ma$

or $T = mg + ma$

or $T = m(g + a)$

Fig. 121 (a) Fig. 121 (b)

The condition for lifting block of mass M is

$T > Mg$ or $m(g + a) > Mg$

or $ma > (M - m)g$ or $a > \left(\dfrac{M}{m} - 1\right)g$

224. (d) Suppose the block reaches A with velocity v. Resolving momentum mv at A into horizontal and vertical components, we get $mv\cos\theta$ and $mv\sin\theta$.

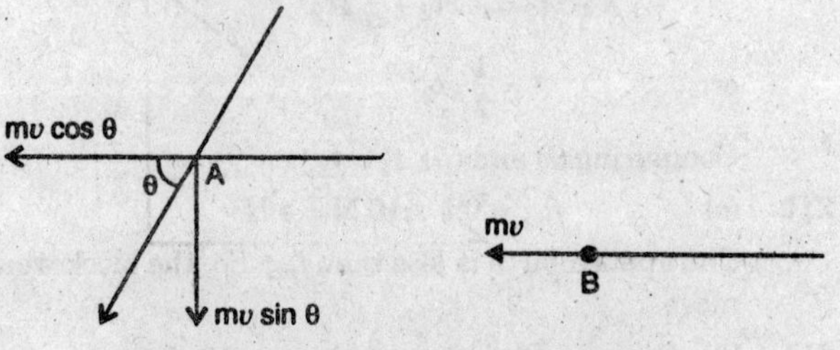

Fig. 122 (a) Fig. 122 (b)

At B, momentum mv is along horizontal.

$\Delta p_x = mv - mv\cos\theta$

$\Delta p_y = 0 - mv\sin\theta$

$\Delta p = \sqrt{[mv(1-\cos\theta)]^2 + [-mv\sin\theta]^2}$

or $\Delta p = mv\sqrt{1 + \cos^2\theta - 2\cos\theta + \sin^2\theta}$

or $\Delta p = \sqrt{2}\,mv\sqrt{1 - \left(1 - 2\sin^2\dfrac{\theta}{2}\right)} = 2mv\sin\dfrac{\theta}{2}$

But $\quad v = \sqrt{2gh}$

$\therefore \quad \Delta p = 2m\sqrt{2gh}\,\sin\frac{\theta}{2}$

225. (*c*) If m is the mass of the second boat,

Initial momentum $= -60 \times 6 + m \times 6$

Final momentum $= (m + 60)4$

Applying conservation of momentum,

$$(m + 60)4 = -360 + 6m$$

or $\quad 2m = 360 + 240 = 600$

or $\quad m = 300$ kg

226. (*a*) Refer to the free body diagram of 8 kg

$$8g - T = 8a \qquad \text{...}(i)$$

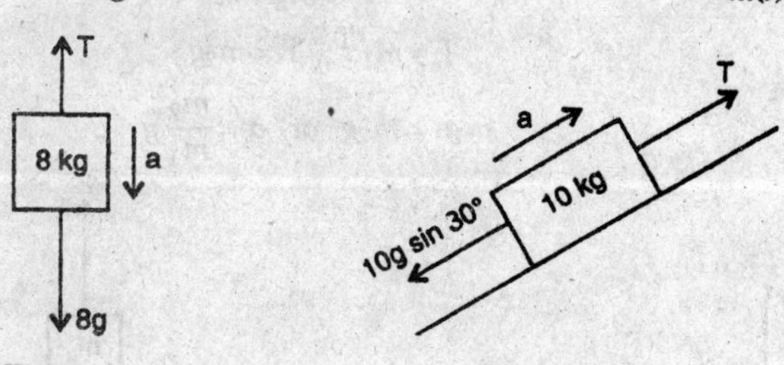

Fig. 123 (*a*) Fig. 123 (*b*)

Refer to the free body diagram of 10 kg

$$T - 10g \sin 30° = 10a$$

or $\quad T - 5g = 10a \qquad \text{...}(ii)$

Adding (*i*) and (*ii*),

$$3g = 18a \quad \text{or} \quad a = \frac{3g}{18} = \frac{g}{6} = \frac{10}{6} = \frac{5}{3}\ \text{m s}^{-2}$$

227. (*b*) Initial momentum $= \frac{100}{1000} \times 50$ kg m s^{-1} $= 5$ kg m s^{-1}

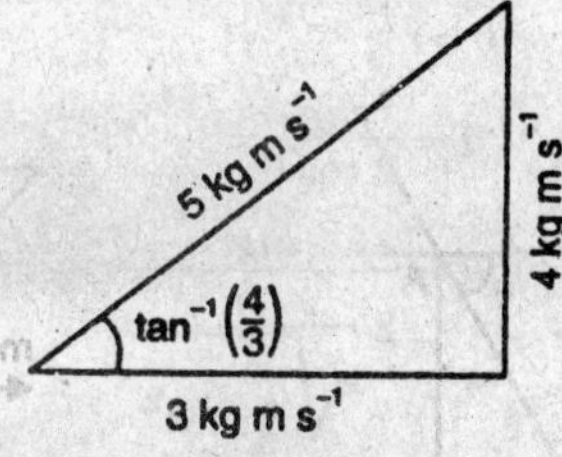

Fig. 124

Horizontal momentum = 3 kg m s^{-1}

Vertical momentum = 4 kg m s^{-1}

$$\text{Velocity} = \frac{4\ \text{kg m s}^{-1}}{0.04\ \text{kg}} = 100\ \text{m s}^{-1}$$

To conserve momentum, it should be directed vertically downward.

228. (*d*) | Change of momentum | $= 2mv$

$$\text{Time taken from A to B} = \frac{\pi d/2}{v} = \frac{\pi d}{2v}$$

$$\text{Average force} = \frac{2mv}{\pi d/2v} = \frac{4mv^2}{\pi d}$$

229. (*d*) $T_1 = m(g + a)$ or $T_1^2 = m^2(g + a)^2$

$T_2 = m(g - a)$ or $T_2^2 = m^2(g - a)^2$

Again, $T \cos\theta = ma$

and $\quad T \sin\theta = mg$

Squaring and adding,

$$T^2 = m^2(g^2 + a^2)$$

Now, $T_1^2 + T_2^2$

$= m^2[(g + a)^2 + (g - a)^2]$

or $T_1^2 + T_2^2 = 2m^2(g^2 + a^2)$

or $\quad T_1^2 + T_2^2 = 2T^2$

Fig. 125

230. (*c*) $\quad 72 - T = 3a \qquad \text{...}(i)$

$T - 32 = 7a \qquad \text{...}(ii)$

Fig. 126 (*a*)

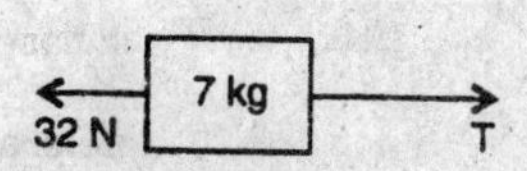

Fig. 126 (*b*)

Adding (*i*) and (*ii*), we get

$$40 = 10a \quad \text{or} \quad a = 4\ \text{m s}^{-2}$$

From eqn. (*ii*),

$$T - 32 = 7 \times 4 = 28$$

or $\quad T = 60$ N

$$\xi = \frac{60\ \text{N}}{1000\ \text{N m}^{-1}} = \frac{60}{1000} \times 100\ \text{cm} = 6\ \text{cm}$$

231. (*b*) Acceleration along the inclined plane, a

$$= (g + a_0)\sin\theta$$

$$\text{Slope length} = \frac{L}{\cos\theta}$$

Using $S = ut + \frac{1}{2}at^2$, we get

$$\frac{L}{\cos\theta} = 0 \times t + \frac{1}{2}(g + a_0)\sin\theta t^2$$

or $\quad t = \sqrt{\frac{2L}{(g + a_0)\sin\theta\cos\theta}}$

or $\quad t = \sqrt{\frac{4L}{(g + a_0)\sin 2\theta}}$

232. (*b*) $\quad a_1 = \frac{2mg}{m} = 2g$

$$a_2 = \frac{2m - m}{2m + m}g = \frac{g}{3}$$

$$a_3 = \frac{2mg}{2m} = g$$

Clearly, $a_1 > a_3 > a_2$

233. (*d*) $l^2 = x^2 + y^2$

Differentiating,

$$2l\frac{dl}{dt} = 2y\frac{dy}{dt}$$

$$\left[\because \frac{d}{dt}(x^2) = 0\right]$$

or $lv = yV$

or $V = \frac{lv}{y} = \frac{v}{y/l} = \frac{v}{\cos\theta}$

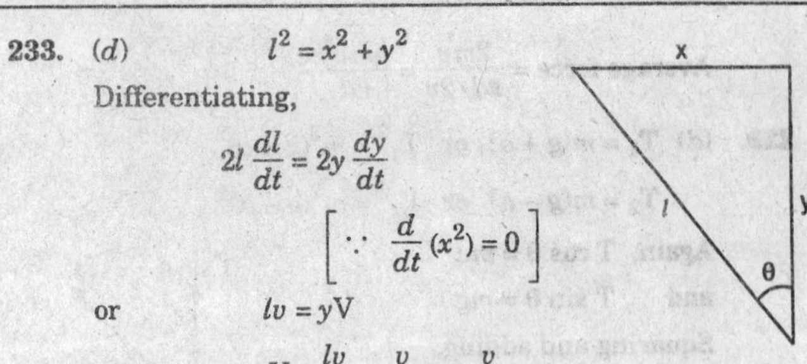

Fig. 127

234. (*a*) $a = \frac{7-5}{7+5}g = \frac{2g}{12} = \frac{g}{6}$

Let us consider motion of 3 kg mass

$3g - T = 3a$

or $T = 3g - 3\left(\frac{g}{6}\right)$

or $T = 3g - \frac{g}{2}$

or $T = \frac{5g}{2} = \frac{5}{2} \times 9.8 \text{ N} = 24.5 \text{ N}$

Fig. 128

235. (*a*) $2g - T = 2a$ ...(*i*)

$T - 3g \sin 30° = 3a$

or $T - \frac{3g}{2} = 3a$ ...(*ii*)

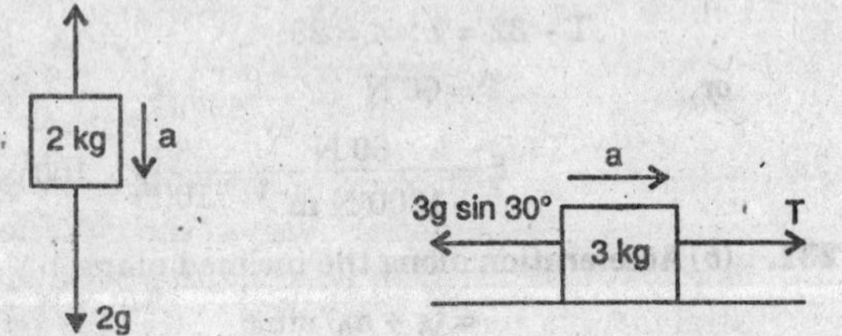

Fig. 129 (*a*) Fig. 129 (*b*)

Adding (*i*) and (*ii*), we get

$$2g - \frac{3g}{2} = 5a \quad \text{or} \quad 5a = \frac{g}{2} \quad \text{or} \quad a = \frac{g}{10} \text{ m s}^{-2}$$

or $a = \frac{9.8}{10} \text{ m s}^{-2} = 0.98 \text{ m s}^{-2}$

236. (*a*) From eqn. (*i*),

$2 \times 9.8 - T = 2 \times 0.98$

or $T = 2 \times 9.8 - 2 \times 0.98$

$= (19.6 - 1.96) \text{ N} = 17.64 \text{ N}$

237. (*d*) $a = g(\sin\theta - \mu\cos\theta)$

$$a = 10\left[\sin 20° - \frac{1}{4}\cos 20°\right]$$

$$a = 10\left[0.3 - \frac{0.9}{4}\right]$$

$= 10[0.3 - 0.2] = 10 \times 0.1 = 1 \text{ m s}^{-2}$

238. (*b*) **Scale reading at time *t* will be due to time rate of change of momentum and weight of pebbles reaching the box in time *t***

Scale reading is $v\frac{dm}{dt} + (mnt)g$

i.e. $\sqrt{2gh} \times mn + mngt$

i.e. $mn[\sqrt{2gh} + gt]$

239. (*a*) $a = \frac{F}{m_1 + m_2 + m_3}$

$T = m_1 a$; $T = m_2 g$

$\therefore$ $m_1 a = m_2 g$ or $a = \frac{m_2}{m_1}g$

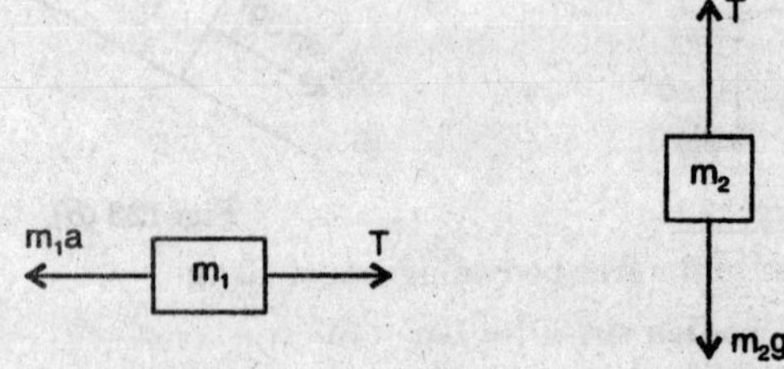

Fig. 130 (*a*) Fig. 130 (*b*)

$\therefore$ $F = \frac{m_2}{m_1}g(m_1 + m_2 + m_3)$

or $F = (m_1 + m_2 + m_3)\frac{m_2 g}{m_1}$

240. $2T\cos\theta = mg$

$T = \frac{mg}{2\cos\theta}$

or $T = \frac{mg}{2\frac{h}{\sqrt{\frac{d^2}{4} + h^2}}}$

$T = \frac{mg}{\frac{4h}{\sqrt{d^2 + 4h^2}}}$

$= \frac{mg\sqrt{d^2 + 4h^2}}{4h}$

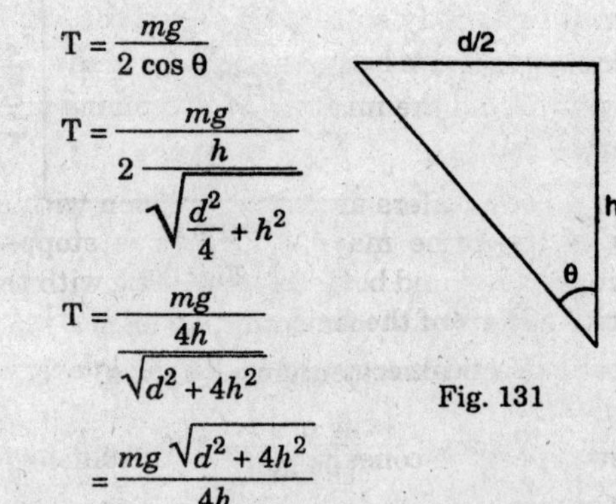

Fig. 131

5 WORK, ENERGY AND POWER

Important Terms, Facts and Formulae

IMPORTANT POINTS TO REMEMBER

1. In one-dimensional elastic collision, the relative velocity after impact is equal in magnitude and opposite in direction to the relative velocity before impact. [This is Newton's experimental law of impact]
2. Loss of energy in an inelastic collision
$$= \frac{1}{2}\frac{m_1 m_2}{m_1 + m_2}(v_{1i} - v_{2i})^2 (1 - e)^2$$
3. In the case of perfectly inelastic collision, the value of coefficient of restitution is zero. So, the loss of kinetic energy in a perfectly inelastic collision is
$$\frac{1}{2}\frac{m_1 m_2}{m_1 + m_2}(v_{1i} - v_{2i})^2$$
4. When two bodies of equal masses suffer one-dimensional elastic collision, they interchange their velocities.
5. When a massive body suffers an elastic collision with a stationary light body, there is practically no change in the velocity of the massive body but the light body acquires a velocity which is nearly double the initial velocity of massive body.
6. When a light body suffers an elastic collision with a stationary massive body, the velocity of the light body is reversed and the massive body remains practically at rest.
7. When a body suffers an elastic collision with another body of the same mass at rest, it is stopped dead whereas the second body starts moving with the same velocity as that of the first.
8. Momentum and energy are conserved in every type of collision.
9. Kinetic energy is conserved in elastic collision.
10. Work done by centripetal force is always zero. So, kinetic energy and speed remain constant.

BASIC CONCEPTS

1. Kinetic energy depends more on velocity than on mass. [Note that $E_k \propto v^2$ and $E_k \propto m$]
2. A body cannot have momentum without kinetic energy. If $E_k = 0$, then $p = \sqrt{2mE_k}$ = zero.
3. The value of coefficient of restitution depends on the nature of material of the colliding bodies.
4. For an inelastic collision, $0 < e < 1$.
5. The velocity v may be acquired by the body in any manner. The kinetic energy of a group of particles or bodies is the sum of the kinetic energies of the individual particles. Consider a system consisting of n particles of masses $m_1, m_2, \ldots\ldots, m_n$. Let $\vec{v_1}, \vec{v_2}, \ldots\ldots, \vec{v_n}$ be their respective velocities. Then, the total kinetic energy E_k of the system is given by
$$E_k = \frac{1}{2}m_1 v_1^2 + \frac{1}{2}m_2 v_2^2 + \ldots\ldots + \frac{1}{2}m_n v_n^2$$
6. When a particle moves with constant speed in a circle, there is no change in the kinetic energy of the particle. So, according to work-energy principle, the work done by centripetal force is zero.
7. If the kinetic energy of the body decreases, then the work done on the body is negative. In this case, force and displacement are opposite directed *i.e.*, the force opposes the motion of the body.
8. Just as the area under the force-displacement graph gives the work done, similarly the area under the force-displacement graph gives the kinetic energy of the particle.
9. All the central forces are conservative.
10. Frictional force is a non-conservative force.
11. The work done in moving a body against a conservative force is independent of the path followed.
12. The work done in moving a body against a conservative force along a closed path is always zero.
13. Gravitational force is a conservative force.

14. Consider an inclined plane of height h. Two different paths 1 and 2 are shown in Fig. 1. Work done in moving a body from A to B along any of the two paths is mgh.

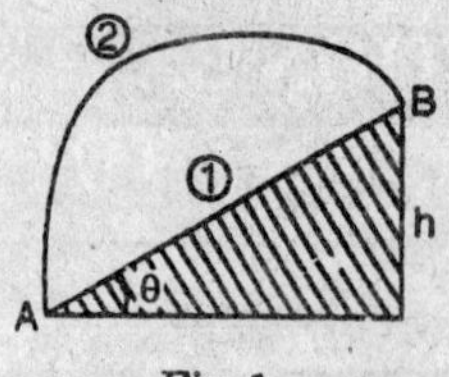

Fig. 1

15. For conservative force, the sum of kinetic and potential energies remains constant throughout the motion. It does not depend upon time. This is known as law of conservation of mechanical energy.

16. Total energy is always conserved. But the mechanical energy is conserved only when force is conservative.

17. If you change the frame of reference, the force does not change. But displacement may change. So, work done by a force is different in different frames.

18. A body cannot have momentum without having kinetic energy and vice-versa.

19. Internal energy is the sum of kinetic and potential energies of the molecules/atoms of a system.

20. The ratio of load and effort is called mechanical advantage.

QUESTION BANK

Level I

Choose the correct answer from the four alternatives given in each of the following questions :

1. A body of mass m moving with velocity v makes a head on elastic collision with another body of mass $2m$ which is initially at rest. The loss of kinetic energy of the colliding body (mass m) is

(a) 1/2 of its initial kinetic energy
(b) 1/9 of its initial kinetic energy
(c) 8/9 of its initial kinetic energy
(d) 1/4 of its initial kinetic energy.

2. In the non-relativistic regime, if the momentum, is increased by 100%, the percentage increase in kinetic energy is

(a) 100 (b) 200
(c) 300 (d) 400.

3. The power developed by a 50 kg boy when moving with a velocity of 2 m s^{-1} upwards in a direction making an angle of 60° with the vertical is

(a) 980 W (b) 490 W
(c) 490 $\sqrt{3}$ W (d) 245 W.

4. A 4 kg mass is moving on a frictionless horizontal table with a velocity of 20 m s^{-1}. It strikes an ideal spring and comes to rest. If the spring constant of the spring is 100 newton per metre, then the compression of the spring is

(a) 4 mm (b) 4 cm
(c) 4 m (d) 0.4 m.

5. A 5 kg body falls to the ground from a height of 30 m. If all the mechanical energy is converted into heat, then the heat produced will be

(a) 350 cal (b) 150 cal
(c) 60 cal (d) 6 cal.

6. A bullet fired into a target loses half its velocity after penetrating 25 cm. How much further will it penetrate before coming to rest ?

(a) $\sqrt{25}$ cm (b) 25 cm
(c) 8.3 cm (d) 75 cm.

7. An engine develops 10 kW of power. How much time will it take to lift a mass of 200 kg to a height of 40 m ? [Given : $g = 10$ m s^{-2}]

(a) 4 s (b) 5 s
(c) 8 s (d) 10 s.

8. A body of mass m_1 moving with uniform velocity of 40 m s^{-1} collides with another body of mass m_2 at rest and then the two together begin to move with uniform velocity of 30 m s^{-1}. The ratio of their masses $\frac{m_1}{m_2}$ is

(a) 0.75 (b) 1.33
(c) 3.0 (d) 4.0.

9. Two springs have force constants K_1 and K_2. Both are stretched till their elastic energies are equal. If the stretching forces are F_1 and F_2, then $F_1 : F_2$ is

(a) $K_1 : K_2$ (b) $K_2 : K_1$
(c) $\sqrt{K_1} : \sqrt{K_2}$ (d) $K_1^2 : K_2^2$.

10. Two springs have force constants K_1 and K_2. These are extended through the same distance x. If their elastic energies are E_1 and E_2, then $\frac{E_1}{E_2}$ is equal to

(a) $K_1 : K_2$ (b) $K_2 : K_1$
(c) $\sqrt{K_1} : \sqrt{K_2}$ (d) $K_1^2 : K_2^2$.

11. A body moves a distance of 5 m along a straight line under the action of a force of 10 N. If the work done is 25 J, then the angle which the force makes with the direction of motion of the body is

(a) 0° (b) 30°
(c) 60° (d) 90°.

12. An electric motor creates a tension of 9000 N in a hoisting cable and reels it in at the rate of 2 m s^{-1}. The power of the electric motor is

(a) 18 kW (b) 15 kW
(c) 81 W (d) 225 W.

13. A block of mass 10 kg slides down a rough slope which is inclined at 45° to the horizontal. The coefficient of sliding friction is 0.30. When the block has slid 5 m, the work done on the block by the force of friction is nearly

(a) 115 J (b) $-75\sqrt{2}$ J
(c) 321.4 J (d) -321.4 J.

14. Two solid rubber balls, A and B having masses 200 g and 400 g respectively are moving in opposite directions with velocity of A equal to 0.3 m s^{-1}. After collision the two balls come to rest when the velocity of B is

(a) 1.5 m s^{-1} (b) 0.15 m s^{-1}
(c) -0.15 m s^{-1} (d) none of these.

15. A bullet when fired at a target with a velocity of 100 m s^{-1}, penetrates one metre into it. If the bullet is fired with the same velocity at a similar target with a thickness 0.5 metre, then it will emerge from it with a velocity of

(*a*) $50\sqrt{2}$ m s^{-1} (*b*) $\frac{50}{\sqrt{2}}$ m s^{-1}

(*c*) 50 m s^{-1} (*d*) 10 m s^{-1}.

16. A body dropped from a height 1 m on to a floor rises to a height of 25 cm after the first rebound. The coefficient of restitution is

(*a*) $\frac{3}{4}$ (*b*) $\frac{1}{4}$

(*c*) $\frac{1}{2}$ (*d*) $\frac{2}{3}$.

17. A body of mass 2 kg is thrown up vertically with kinetic energy of 490 J. The height at which the kinetic energy of the body becomes half of its original value is

(*a*) 50 m (*b*) 12.25 m

(*c*) 25 m (*d*) 10 m.

18. A ball is allowed to fall from a height of 10 m. If there is 40% loss of energy due to impact, then after one impact, the ball will go upto

(*a*) 10 m (*b*) 8 m

(*c*) 4 m (*d*) 6 m.

19. A sub-atomic particle travelling with velocity v and kinetic energy E collides elastically head-on with a stationary nucleus of mass number A. The fraction of the total energy retained by the sub-atomic particle is

(*a*) $\frac{(1-A)^2}{(1+A)^2}$ (*b*) $\frac{4A}{(1+A)^2}$

(*c*) $\left(\frac{A+1}{A}\right)^2$ (*d*) $\left(\frac{A}{A-1}\right)^2$.

20. A uniform chain of length l and mass m is resting on a smooth table. One-third of its length is hanging vertically down over the edge of the table. The amount of work required to pull the hanging part back to the table is

(*a*) $\frac{mgl}{3}$ (*b*) $\frac{mgl}{6}$

(*c*) $\frac{mgl}{9}$ (*d*) $\frac{mgl}{18}$.

21. A uniform force of 4 N acts on a body of mass 40 kg for a distance of 2.0 m. The kinetic energy acquired by the body is

(*a*) $4 \times 2 \times 2$ J (*b*) $4 \times 4 \times 2 \times 10^8$ erg

(*c*) 4×2 J (*d*) $4 \times 4 \times 2$ erg.

22. Two bodies of masses 1 kg and 4 kg are moving with equal kinetic energies. The ratio of their linear momenta is

(*a*) 1 : 4 (*b*) 1 : 2

(*c*) 4 : 1 (*d*) 2 : 1.

23. Two balls at the same temperature collide. What is conserved ?

(*a*) momentum (*b*) kinetic energy

(*c*) velocity (*d*) temperature.

24. The momentum of a body increases by 20%. The percentage increase in its kinetic energy is

(*a*) 20 (*b*) 44

(*c*) 66 (*d*) 88.

25. A long spring when stretched by x cm has a potential energy V. On increasing the stretching to 'nx', the potential energy stored in the spring will be

(*a*) V/n (*b*) nV

(*c*) n^2V (*d*) V/n^2.

26. If the kinetic energy of a body increases by 0.1%, the percent increase of its momentum will be

(*a*) 0.05% (*b*) 0.1%

(*c*) 1.0% (*d*) 10%.

27. The linear momentum p of a body moving in one dimension varies with time according to the equation $p = a + bt^2$, where a and b are positive constants. The net force acting on the body is

(*a*) proportional to t^2 (*b*) a constant

(*c*) proportional to t (*d*) inversely proportional to t.

28. A particle is displaced from a position $2\hat{i} - \hat{j} + \hat{k}$ to another position $3\hat{i} + 2\hat{j} - 2\hat{k}$ under the action of a force $2\hat{i} + \hat{j} - \hat{k}$. The work done by the force (in arbitrary units) is

(*a*) 8 (*b*) 10

(*c*) 12 (*d*) 36.

29. A body of mass 2 kg is thrown vertically upwards with K.E. 245 J. The acceleration due to gravity is 9.8 m s^{-2}. The K.E. of the body will become half at a height of

(*a*) 25 m (*b*) 12.5 m

(*c*) 6.25 m (*d*) 5 m.

30. A mass of 10 g moving with a velocity of 100 cm s^{-1} strikes a pendulum bob of mass 10 g. The two masses stick together. The maximum height reached by the system now is (g = 10 m s^{-2})

(*a*) Zero (*b*) 5 cm

(*c*) 2.5 cm (*d*) 1.25 cm.

31. A car is moving with a speed of 100 km h^{-1}. If the mass of the car is 950 kg, then its kinetic energy is

(*a*) 0.367 M J (*b*) 3.67 J
(*c*) 3.67 M J (*d*) 367 J.

32. A body is dropped from a height of 20 m and rebounds to a height of 10 m. The loss of energy is
(*a*) 10% (*b*) 45%
(*c*) 50% (*d*) 75%.

33. Tripling the velocity of your scooter multiplies the distance required for stopping it by
(*a*) 3 (*b*) 6
(*c*) 9 (*d*) 12.

34. For the same value of kinetic energy, the momentum shall be maximum for
(*a*) a proton (*b*) an electron
(*c*) a deutron (*d*) an α-particle.

35. At a certain instant, a 0.3 kg body has a velocity of $(4\hat{i} + 3\hat{j})$ m s^{-1}. The kinetic energy of the body is
(*a*) 3.75 J (*b*) 4.75 J
(*c*) 9 J (*d*) 0 J.

36. A motor boat is travelling with a speed of 3.0 m s^{-1}. If the force on it due to water flow is 500 N, the power of the boat is
(*a*) 150 kW (*b*) 15 kW
(*c*) 1.5 kW (*d*) 150 W.

37. Which one of the following is 'watt' ?
(*a*) kg m s^{-2} (*b*) N m s^{-2}
(*c*) kg m^2 s^{-3} (*d*) N m s^{-1}.

38. A 50 g bullet moving with a speed of 10 m s^{-1} strikes a stationary body of mass 950 g and enters it. The percentage loss of kinetic energy of the bullet is
(*a*) 100 (*b*) 95
(*c*) 75 (*d*) 60.

39. A 4 kg ball moving with a velocity of 10 m s^{-1} makes a head- on elastic collision with another ball of mass 12 kg at rest. The velocity of the 12 kg ball after collision is
(*a*) 4 m s^{-1} (*b*) 5 m s^{-1}
(*c*) 6 m s^{-1} (*d*) 3 m s^{-1}.

40. In Q. 39, the velocity of the 4 kg mass after the collision is
(*a*) + 5 m s^{-1} (*b*) – 5 m s^{-1}
(*c*) 4 m s^{-1} (*d*) – 4 m s^{-1}.

41. If the momentum of a body is numerically equal to its kinetic energy, then the speed of the body is
(*a*) 1 m s^{-1} (*b*) 2 m s^{-1}
(*c*) 4 m s^{-1} (*d*) 8 m s^{-1}.

42. Two ball bearings, each of mass *m*, move in opposite directions with equal speeds *v*. They collide head-on elastically. Now, they would
(*a*) stick (*b*) come to rest
(*c*) move at right angles to each other
(*d*) rebound from each other with speed *v*.

43. A block of mass *m* slides down along the surface of the bowl (radius R) from the rim to the bottom. The velocity of the block at the bottom will be
(*a*) $\sqrt{gR}$ (*b*) $\sqrt{2Rg}$
(*c*) $2\sqrt{\pi Rg}$ (*d*) $\sqrt{\pi Rg}$.

QUESTION BANK

Level II

Choose the correct answer from the four alternatives given in each of the following questions :

1. A particle of mass M is moving in a horizontal circle of radius R under a centripetal force given by $(-k/R^2)$, where *k* is a constant. Then
(*a*) the kinetic energy of the particle is *k*/12R
(*b*) the total energy of the particle is (– *k*/2R)
(*c*) the kinetic energy of the particle is (– *k*/R)
(*d*) the potential energy of the particle is *k*/2R.

2. A pendulum bob has a speed of 3 m s^{-1} at its lowest position. The pendulum is 0.5 m long. The speed of the bob, when the length makes an angle of 60° to the vertical, will be (G = 10 m s^{-2})
(*a*) 3 m s^{-1} (*b*) 1/3 m s^{-1}
(*c*) 1/2 m s^{-1} (*d*) 2 m s^{-1}.

3. A moving body with a mass m_1 strikes a stationary body of mass m_2. The masses m_1 and m_2 should be in the ratio m_1/m_2 so as to decrease the velocity of the first body 1.5 times assuming a perfectly elastic impact. Then the ratio m_1/m_2 is
(*a*) 5 (*b*) 1/5
(*c*) 1/25 (*d*) 25.

4. In a certain situation, $\vec{F}$ and $\vec{S}$ are not equal to zero but the work done is zero. From this, we conclude that

(a) $\vec{F}$ and $\vec{S}$ are in same direction

(b) $\vec{F}$ and $\vec{S}$ are in opposite direction

(c) $\vec{F}$ and $\vec{S}$ are at right angles

(d) $\vec{F} > \vec{S}$.

5. In a children's park, there is a slide which has a total length of 10 m and a height of 8.0 m. A vertical ladder is provided to reach the top. A boy weighing 200 N climbs up the ladder to the top of the slide and slides down to the ground. The average friction offered by the slide is three-tenth of his weight. The work done by the slide on the boy as he comes down is

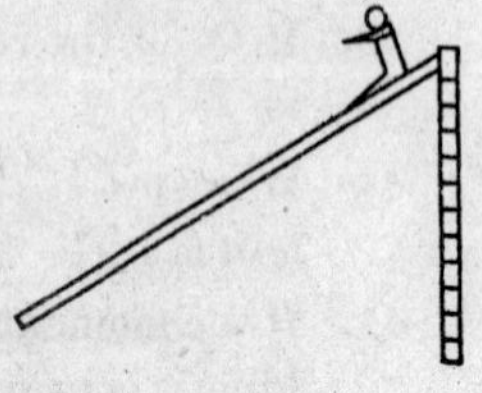

Fig. 2

(a) 0 J (b) + 600 J

(c) – 600 J (d) + 1600 J.

6. A body of mass m is suspended from a massless spring of natural length l. It stretches the spring through a vertical distance y. The potential energy of the stretched spring is

(a) $mg(l + y)$ (b) $\frac{1}{2}mg(l + y)$

(c) $\frac{1}{2}mgy$ (d) mgy.

7. A bomb of mass M at rest explodes into two fragments of masses m_1 and m_2. The total energy released in the explosion is E. If E_1 and E_2 represent the energies carried by masses m_1 and m_2 respectively, then which of the following is correct ?

(a) $E_1 = \frac{m_2}{M}E$ (b) $E_1 = \frac{m_1}{m_2}E$

(c) $E_1 = \frac{m_1}{M}E$ (d) $E_1 = \frac{m_2}{m_1}E$.

8. The F-ξ graph of a spring is shown in Fig. 3. If the same experiment is performed with half the length of the spring, then the F – ξ graph will make an angle θ with ξ-axis such that

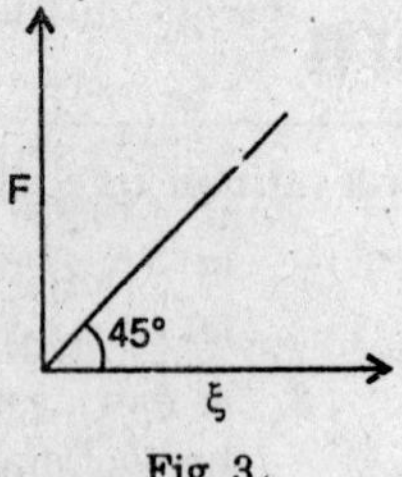

Fig. 3.

(a) θ = 45° (b) θ < 45°

(c) θ = 0° (d) θ > 45°.

9. A glass marble dropped from a certain height above the horizontal surface reaches the surface in time t and then continues to bounce up and down. The time in which the marble finally comes to rest is

(a) $e^n t$ (b) $e^2 t$

(c) $t\left[\frac{1+e}{1-e}\right]$ (d) $t\left[\frac{1-e}{1+e}\right]$.

10. A weight suspended from the free end of a vertically hanging spring produces an extension of 3 cm. The spring is cut into two parts so that the length of the longer part is $\frac{2}{3}$ of the original length. If the same weight is now suspended from the longer part of the spring, the extension produced will be

(a) 0.1 cm (b) 0.5 cm

(c) 1 cm (d) 2 cm.

11. Ten litre of water per second is lifted from a well through 10 m and delivered with a velocity of 10 m s^{-1}. If g = 10 m s^{-2}, then the power of the motor is

(a) 1 kW (b) 1.5 kW

(c) 2 kW (d) 2.5 kW.

12. The energy required to accelerate a car from rest to 10 m s^{-1} is E. What energy will be required to accelerate the car from 10 m s^{-1} to 20 m s^{-1} ?

(a) E (b) 3E

(c) 5E (d) 7E.

13. In a molecule, the potential energy between two atoms is given by : $U(x) = \frac{a}{x^{12}} - \frac{b}{x^6}$, where a and b are positive constants and x is the distance between atoms. For stable equilibrium of atom, the value of x is

(a) zero (b) $\left[\frac{a}{2b}\right]^{1/6}$

(c) $\left[\frac{2a}{b}\right]^{1/6}$ (d) $\left[\frac{11a}{5b}\right]^{1/6}$.

14. If the force acting on a body is inversely proportional to its velocity, then the kinetic energy acquired by the body in time t is proportional to

(a) t^0 (b) t^1

(c) t^2 (d) t^4.

15. A position-dependent force $F = 3x^2 - 2x + 7$ acts on a body of mass 7 kg and displaces it from $x = 0$ m to $x = 5$ m. The work done on the body is x' joule. If both F and x are measured in SI units, the value of x' is

(a) 135 (b) 235

(c) 335 (d) 935.

16. A nucleus ruptures into two nuclear parts which have their velocity ratio equal to 2 : 1. The ratio of their respective nuclear sizes (nuclear radii) is

(a) 1 : 2 (b) 1 : $\sqrt{2}$

(c) 1 : $2^{1/3}$ (d) 1 : 8.

17. A 10 m long iron chain of linear mass density 0.8 kg m^{-1} is hanging freely from a rigid support. If g = 10 m s^{-2}, then the power required to lift the chain upto the point of support in 10 second is

(a) 10 W (b) 20 W

(c) 30 W (d) 40 W.

18. A pump motor is used to deliver water at a certain rate from a given pipe. To obtain thrice as much water from the same pipe in the same time, power of the motor has to be increased

(a) 3 times (b) 9 times

(c) 27 times (d) 81 times.

19. A ball is thrown from a height of h metre with an initial downward velocity v_0. It hits the ground, loses half of its kinetic energy and bounces back to the same height. The value of v_0 is

(a) $\sqrt{2gh}$ (b) $\sqrt{gh}$

(c) $\sqrt{3gh}$ (d) $\sqrt{2.5gh}$.

20. An engine of power 7500 W makes a train move on a horizontal surface with constant velocity of 20 m s^{-1}. The force involved in the problem is

(a) 375 N (b) 400 N

(c) 500 N (d) 600 N.

21. A gas expands from 5 litre to 105 litre at a constant pressure 100 N m^{-2}. The work done is

(a) 1 joule (b) 4 joule

(c) 8 joule (d) 10 joule.

22. A force of 5 N moves the particle through a distance of 10 m. If 25 joule of work is performed, then the angle between the force and the direction of motion is

(a) 0° (b) 90°

(c) 30° (d) 60°.

23. A 5 kg brick of dimensions 20 cm × 10 cm × 8 cm is lying on the largest base. It is now made to stand with length vertical. If g = 10 m s^{-2}, then the amount of work done is

(a) 3 J (b) 5 J

(c) 7 J (d) 9 J.

24. The displacement x of a body of mass 1 kg on horizontal smooth surface as a function of time t is given by : $x = \frac{t^3}{3}$. The work done by the external agent for the first one second is

(a) 0.25 J (b) 0.5 J

(c) 0.75 J (d) 1 J.

25. A ball of mass m approaches a moving wall of infinite mass with speed v along the normal to the wall. The speed of the wall is u towards the ball. The speed of the ball after an elastic collision with the wall is

(a) $v - 2u$ away from the wall

(b) $2u + v$ away from the wall

(c) $u - v$ away from the wall

(d) $u + v$ away from the wall.

26. An automobile weighing 1200 kg climbs up a hill that rises 1 in 20. Neglecting frictional effects, the minimum power developed by the engine is 9000 W. If g = 10 m s^{-2}, then the velocity of the automobile is

(a) 36 km h^{-1} (b) 54 km h^{-1}

(c) 72 km h^{-1} (d) 90 km h^{-1}.

27. A 500 kg car, moving with a velocity of 36 km h^{-1} on a straight road unidirectionally, doubles its velocity in one minute. The power delivered by the engine for doubling the velocity is

(a) 750 W (b) 1050 W

(c) 1150 W (d) 1250 W.

28. A 1 kg body is projected vertically upward with a speed of 200 m s^{-1}. It rises to a height of 1500 m. If gravitational field intensity is uniform and equal to 10 N kg^{-1}, then the energy used up in overcoming air resistance is

(a) 2 kJ (b) 4 kJ

(c) 5 kJ (d) 50 kJ.

29. A moving neutron collides with a stationary α-particle. The fraction of the kinetic energy lost by the neutron is

(a) $\frac{16}{25}$ (b) $\frac{9}{25}$

(c) $\frac{3}{5}$ (d) $\frac{2}{5}$.

30. At a given time, the momentum of a body of mass 5 kg is 10 kg m s^{-1}. Now, a force of 0.2 N acts on the body in the direction of motion for 10 second. The increase in the kinetic energy is

(a) 2.2 J (b) 3.3 J

(c) 4.4 J (d) 5.5 J.

31. A ship weighing 0.3 × 10^8 kg wt is pulled by a force of 0.5 × 10^5 N through a distance of 3 m. If the ship were originally at rest and water-resistance is negligibly small, then the ship will acquire a speed of

(a) 0.1 m s^{-1} (b) 1 m s^{-1}

(c) 1.5 m s^{-1} (d) 12 m s^{-1}.

32. Under the action of a force, a 2 kg body moves such that its position x as a function of time is given by $x = \frac{t^3}{3}$ where x is in metre and t in second. The work done by force in first two second is

(a) 1.6 J (b) 16 J

(c) 160 J (d) 1600 J.

33. If clouds are 1 km above the earth and rain fell sufficient to cover 1 km^2 at sea level, 1 cm deep, then the work done in raising water to the clouds is

(a) 98×10^{16} erg (b) 98 erg
(c) 10^6 erg (d) 69.73 erg.

34. A man, by working a hand pump fixed to a well, pumps out 10 cubic metre water in 1 second. If the water in the well is 10 m below the ground level, then the work done by the man is (g = 10 m s^{-2})

(a) 10^3 J (b) 10^4 J
(c) 10^5 J (d) 10^6 J.

35. A fat person weighing 75 kg falls on a concrete floor from 2 m. If whole of the mechanical energy is converted into heat energy, then heat produced in SI units will be nearly

(a) 1500 joule (b) 1200 joule
(c) 1500 k cal (d) 1200 k cal.

36. A one kilowatt motor is used to pump water from a well 10 m deep. The quantity of water pumped out per second is nearly

(a) 1 kg (b) 10 kg
(c) 100 kg (d) 1000 kg.

37. A 20 newton stone falls accidentally from a height of 4 m on to a spring of stiffness constant 1960 N m^{-1}. The maximum distance to which the spring shall suffer compression is

(a) $\frac{2}{7}$ m (b) $\frac{3}{7}$ m
(c) $\frac{4}{6}$ m (d) 0.3295 m.

38. The power of a water jet flowing through an orifice of radius r with velocity v is

(a) zero (b) $500\,\pi r^2 v^2$
(c) $500\,\pi r^2 v^3$ (d) $\pi r^4 v$.

39. Under the action of a constant force, a particle is experiencing a constant acceleration. The power is

(a) zero (b) positive
(c) negative
(d) increasing uniformly with time.

40. The velocity of the bob of an oscillating simple pendulum of length 1 m is 2 m s^{-1} at the lowest position. When the bob is at the extreme position, the angle made by the string of the pendulum with the vertical is [g = 10 m s^{-2}]

(a) $\cos^{-1}$ (0.4) (b) $\cos^{-1}$ (0.6)
(c) $\cos^{-1}$ (0.8) (d) 30°.

41. A horse pulls a wagon with a force of 360 N at an angle of 60° E with the horizontal at a speed of 10 km h^{-1}. The power of the horse is

(a) 1000 W (b) 2000 W
(c) 500 W (d) 750 W.

42. A car manufacturer claims that his car can be accelerated from rest to a velocity of 10 m s^{-1} in 5 second. If the total mass of the car and its occupants is 1000 kg, then the average horse power developed by the engine is

(a) $\frac{10^3}{746}$ (b) $\frac{10^4}{746}$
(c) $\frac{10^5}{746}$ (d) 8.

43. A ball is dropped from a height h on the ground. If the coefficient of restitution is e, the height to which the ball goes up after it rebounds for the nth time is

(a) he^{2n} (b) he^n
(c) $\frac{e^{2n}}{h}$ (d) $\frac{h}{e^{2n}}$.

44. A water pump driven by petrol raises water at a rate of 0.5 m^3 min^{-1} from a depth of 30 m. If the pump is 70% efficient, what power is developed by the engine ?

(a) 1750 W (b) 2450 W
(c) 3500 W (d) 7000 W.

45. A 1 kg block moving with a velocity of 4 m s^{-1} collides with a stationary 2 kg block. The lighter block comes to rest after the collision. The loss of kinetic energy of the system is

(a) 1 J (b) 2 J
(c) 3 J (d) 4 J.

46. An iron chain of length l and mass m is placed on a frictionless table with $\frac{1}{n}$ th part hanging over the edge. The work done in pulling the hanging chain to the table is

(a) mgl (b) $\frac{mgl}{2}$
(c) $\frac{mgl}{2n^2}$ (d) $\frac{mgl}{2n^4}$.

47. A body falls on the ground from a height of 10 metre and rebounds to a height of 2.5 m. The ratio of the velocity of the body before collision to the velocity of the body after collision is

(a) 2 : 1 (b) 1 : 2
(c) 4 : 1 (d) 3 : 1.

48. In Q. 47, the percentage loss of kinetic energy of the body during its collision with the ground is

(a) 25% (b) 50%
(c) 75% (d) 99%.

49. The mechanical energy required to accelerate a body from 10 m s^{-1} to 20 m s^{-1} compared with that required to accelerate from 0 to 10 m s^{-1} in the same interval of time covering same distance is

(a) two times (b) three times
(c) four times (d) five times.

50. Sand falls vertically at the rate of 2 kg s^{-1} on to a conveyor belt moving horizontally with a velocity of 0.2 m s^{-1}. The extra power required to keep the belt moving is

(a) 0.08 W (b) 0.04 W
(c) 4 W (d) 1 W.

51. A body of mass 5 kg collides elastically with a stationary body of mass 2.5 kg. After the collision, the 2.5 kg body begins to move with a kinetic energy of 8 J. Assuming the collision to be one-dimensional, the kinetic energy of the 5 kg body before collision is

(a) 3 J (b) 6 J
(c) 9 J (d) 11 J.

52. A lift pump works at 200 V, 10 A. It pumps water at 4°C to an average height of 15 m to fill a tank of volume 3 m × 2 m × 1 m. If g = 10 m s^{-2} and the efficiency of the pump is 75%, then the time required to fill the tank is

(a) 10 minute (b) 20 minute
(c) 24 minute (d) 28 minute.

53. What power must a sprinter, weighing 80 kg, develop from the start if he has to impart a velocity of 10 m s^{-1} to his body in 4 second ?

(a) 1 kW (b) 2 kW
(c) 3 kW (d) 4 kW.

54. A 1 kg block is attached (and held at rest with outside support) to the free end of a vertically hanging spring of force constant 10 N cm^{-1}. When the block is released, what maximum extension does it cause when it comes to rest instantaneously ? [g = 10 m s^{-2}]

(a) 1 cm (b) 2 cm
(c) 3 cm (d) 4 cm.

55. A block weighing 10 N travels down a smooth curved track AB joined to a rough horizontal surface (Fig. 4). The rough surface has a friction coefficient of 0.20 with the block. If the block starts slipping on the track from a point 1.0 m above the horizontal surface, then it would move a distance S on the rough surface. The value of S is [g = 10 m s^{-2}]

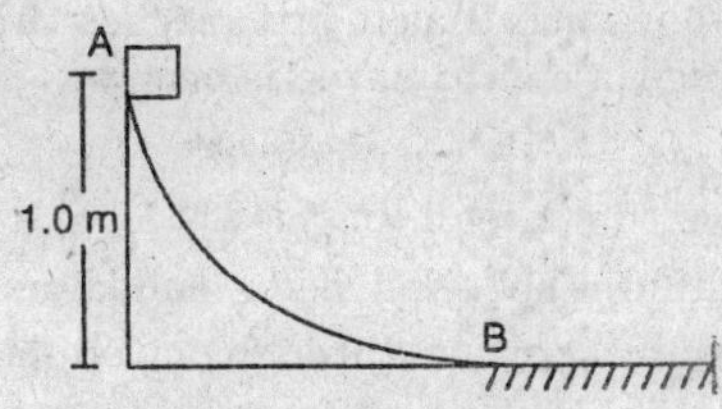

Fig. 4

(a) m (b) $2m$
(c) $3m$ (d) $5m$.

56. A coolie 1.5 m tall raises a load of 80 kg in 2 second from the ground to his head and then walks a distance of 40 m in another 2 second. The power developed by the coolie is [g = 10 m s^{-2}]

(a) 0.2 kW (b) 0.4 kW
(c) 0.6 kW (d) 0.8 kW.

57. A ball moving with a velocity of 10 m s^{-1} impinges on a vertical wall at an angle of 45° with the normal to the wall. After the collision, the ball moves on the other side of the normal at an angle of 45° with the normal. The coefficient of restitution is 0.5. The velocity of the ball after the collision will be

(a) 2 m s^{-1} (b) 3 m s^{-1}
(c) 5 m s^{-1} (d) 50 m s^{-1}.

58. A uniform chain of length L and mass M overhangs a horizontal table with its two third part on the table. The friction coefficient between the table and the chain is μ. The work done by the friction during the period the chain slips off the table is

(a) $-\frac{1}{4}\mu MgL$ (b) $-\frac{2}{9}\mu MgL$
(c) $-\frac{4}{9}\mu MgL$ (d) $-\frac{6}{7}\mu MgL$.

59. Two springs A and B(k_A = $2k_B$) are stretched by applying forces of equal magnitudes at the four ends. If the energy stored in A is E, that in B is

(a) E/2 (b) 2E
(c) E (d) E/4.

60. Two carts on horizontal straight rails are pushed apart by an explosion of a gun powder placed between the carts. Suppose the coefficients of friction between the carts and the rails are identical. If the 200 kg cart travels a distance of 36 m and stops, then the distance covered by the cart weighing 300 kg is

(a) 32 m (b) 24 m
(c) 16 m (d) 12 m.

61. A man throws a piece of stone to a height of 12 m where it reaches with a speed of 12 m s^{-1}. If he throws the same stone such that it just reaches this height, the percentage of energy saved is nearly

(*a*) 19% (*b*) 38%
(*c*) 57% (*d*) 76%.

62. A pendulum bob has been pulled aside from its equilibrium position through an angle α and then released. If l is the length of the pendulum, then the bob will reach the equilibrium position with speed of

(*a*) $\sqrt{2gl}$ (*b*) $\sqrt{2gl \cos \alpha}$
(*c*) $\sqrt{2gl(1-\cos\alpha)}$ (*d*) $\sqrt{2gl(1-\sin\alpha)}$.

63. The potential energy of a certain spring when stretched through a distance S is 10 joule. The amount of work in joule that must be done on this spring to stretch it through an additional distance S will be

(*a*) 30 (*b*) 40
(*c*) 10 (*d*) 20.

64. The potential energy of a particle in a certain field has the form, $U = \frac{a}{r^2} - \frac{b}{r}$ where a and b are positive constants, r is the distance from the centre of the field. The distance of the particle in the stable equilibrium position is

(*a*) a/b (*b*) $-a/b$
(*c*) $2a/b$ (*d*) $-2a/b$.

65. A 30 g bullet initially travelling horizontally at 50 m s^{-1} penetrates 10 cm into a wooden block. The average force it exerts is

(*a*) 4000 N (*b*) 375 N
(*c*) 2500 N (*d*) 4500 N.

66. The F-x graph of a particle of mass 100 g is shown in Fig. 5. If the particle begins to move from rest at $x = 0$, its velocity at $x = 12$ m is

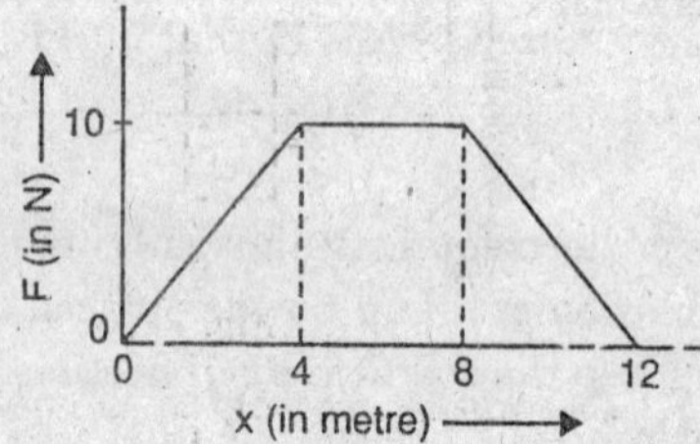

Fig. 5

(*a*) 10 m s^{-1} (*b*) 20 m s^{-1}
(*c*) 30 m s^{-1} (*d*) 40 m s^{-1}.

67. A 5 kg stone of relative density 3 is resting at the bed of a lake. It is lifted through a height of 5 metre in the lake. If $g = 10$ m s^{-2}, then the work done is

(*a*) $\frac{500}{3}$ J (*b*) $\frac{350}{3}$ J
(*c*) $\frac{750}{3}$ J (*d*) zero.

68. A block of mass m is suspended by a light thread from an elevator. The elevator is accelerating upward with uniform acceleration a. The work done during t second by the tension in the thread is

(*a*) zero (*b*) $\frac{m}{2}(g-a)at^2$
(*c*) $\frac{m}{2}(g+a)at^2$ (*d*) $\frac{m}{2}g\,at^2$.

69. It is observed that the force required to tow a boat at constant velocity is proportional to the velocity. It takes 16 hp to tow a boat at a velocity of 2 km h^{-1}. The horse power required to tow this boat at a velocity of 3 km h^{-1} is

(*a*) 9 (*b*) 18
(*c*) 36 (*d*) 72.

70. The velocity of a 2 kg body is changed from $(4\hat{i} + 3\hat{j})$ m s^{-1} to $6\hat{k}$ m s^{-1}. The work done on the body is

(*a*) 9 J (*b*) 11 J
(*c*) 1 J (*d*) 0 J.

71. A ball is allowed to fall from a height of 10 m. There is 40% loss of energy every time it hits the ground. After second impact with the ground, the ball will rise upto

(*a*) 10 m (*b*) 6 m
(*c*) 3.6 m (*d*) 2.4 m.

72. A force of 1 N acts on a body of mass 0.5 kg initially at rest. The ratio of the works done by the force in the first, second and third second is

(*a*) 1 : 2 : 3 (*b*) 1 : 3 : 4
(*c*) 1 : 3 : 5 (*d*) 1 : 5 : 9.

73. A body of density r and volume V is lifted through height h in a liquid of density σ (< ρ). The increase in potential energy of the body is

(*a*) $V(\rho - \sigma)gh$ (*b*) $V\rho gh$
(*c*) $V\sigma gh$ (*d*) zero.

74. A plate of mass m, length b and breadth a is initially lying on a horizontal floor with length parallel to the floor and breadth perpendicular to the floor. The work done to erect it on its breadth is

(*a*) $mg\left[\frac{b}{2}\right]$ (*b*) $mg\left[a + \frac{b}{2}\right]$
(*c*) $mg\left[\frac{b-a}{2}\right]$ (*d*) $mg\left[\frac{b+a}{2}\right]$.

75. A machine which is 75% efficient uses 12 J of energy in lifting up a 1 kg mass through a certain distance.

The mass is then allowed to fall through that distance. The velocity of the ball at the end of its fall is

(a) $\sqrt{24}$ m s^{-1} (b) $\sqrt{32}$ m s^{-1}
(c) $\sqrt{18}$ m s^{-1} (d) 3 m s^{-1}.

76. A metallic wire of length L metre extends by l metre when stretched by suspending a weight mg to it. The mechanical energy stored in the wire is

(a) $mg(L + l)$ (b) mgl
(c) $\frac{mgl}{2}$ (d) $\frac{mg(L+l)}{2}$.

77. A body of mass 2 kg slides down a curved track which is quadrant of a circle of radius 1 m. All the surfaces are frictionless. If the body starts from rest, its speed at the bottom of the track is

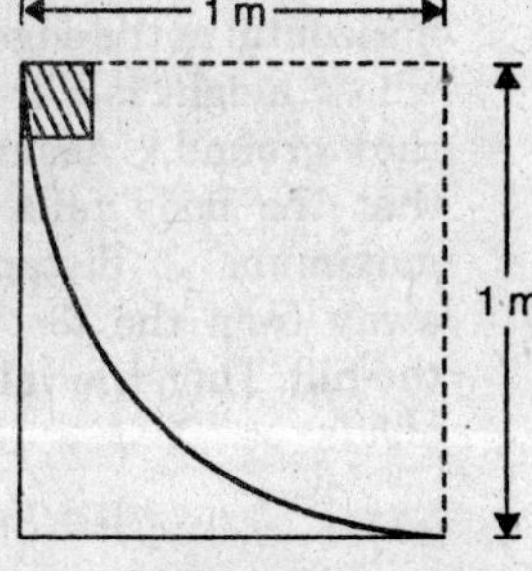

Fig. 6

(a) 4.43 m s^{-1} (b) 2 m s^{-1}
(c) 0.5 m s^{-1} (d) 19.6 m s^{-1}.

78. A sphere of mass 2 kg is moving on a frictionless horizontal table with velocity v. It strikes against a spring (force constant = 1 N m^{-1}) and compresses it by 4 m. The velocity of the sphere is

(a) 4 m s^{-1} (b) $2\sqrt{2}$ m s^{-1}
(c) 2 m s^{-1} (d) $\sqrt{2}$ m s^{-1}.

79. A dam is situated at a height of 550 m above sea level and supplies water to a power house which is at a height of 50 m above sea level. 2000 kg of water passes through the turbines per second. What would be the maximum electrical power output of the power house if the whole system were 80% efficient ?

(a) 8 MW (b) 10 MW
(c) 12.5 MW (d) 16 MW.

80. A car weighing 500 kg working against a resistance of 500 N, accelerates from rest to 20 m s^{-1} in 100 m. The work done by the engine of the car is [g = 10 m s^{-2}]

(a) 1.0×10^5 J (b) 1.5×10^5 J
(c) 1.05×10^5 J (d) None of these.

81. A bullet weighing 0.01 kg moving with a velocity of 1080 km h^{-1} strikes a 5 kg block of ice and drops dead. If the ice block was resting on a frictionless horizontal surface before the collision, then the velocity of the ice block after the collision is

(a) 0.2 m s^{-1} (b) 0.4 m s^{-1}
(c) 0.6 m s^{-1} (d) 0.8 m s^{-1}.

82. A bullet when fired at a target has its velocity decreased to 50% after penetrating 30 cm into it. Then the additional thickness it will penetrate before resting is

(a) 10 cm (b) 30 cm
(c) 40 cm (d) 60 cm.

83. A tennis ball dropped from a height of 2 m rebounds only 1.5 m after hitting the ground. What friction of its energy is lost in the impact ?

(a) $\frac{1}{4}$ (b) $\frac{1}{2}$
(c) $\frac{1}{3}$ (d) $\frac{1}{8}$.

84. A ball is thrown vertically upwards with a velocity of 10 m s^{-1}. It returns to the ground with a velocity of 9 m s^{-1}. If g = 9.8 m s^{-2}, then the maximum height attained by the ball is [Take g = 10 m s^{-2}]

(a) 5.1 m (b) 14.1 m
(c) 4.525 m (d) 15.0 m.

85. A 3 kg body is dropped from the top of a tower of height 135 m. If g = 10 m s^{-2}, then the kinetic energy of the body after 3 second will be

(a) 950 J (b) 10 J
(c) 1150 J (d) 1350 J.

86. A bullet moving with a velocity of $50\sqrt{2}$ m s^{-1} is fired into a target which it penetrates to the extent of d metre. If this bullet is fired into a target of $\frac{d}{2}$ metre thickness with the same velocity, it will come out of this target with a velocity of [Assume that resistance to motion is similar and uniform in both the cases]

(a) $40\sqrt{2}$ m s^{-1} (b) $25\sqrt{2}$ m s^{-1}
(c) 50 m s^{-1} (d) $5.0\sqrt{3}$ m s^{-1}.

87. Fig. 7 shows F-x graph of a particle. The work done is

(a) 0 J
(b) 10 J
(c) 20 J
(d) 30 J.

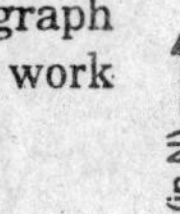
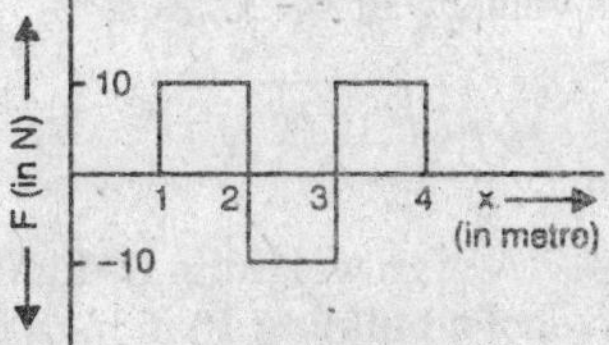

Fig. 7

88. A car weighing 1400 kg is moving at a speed of 54 km h^{-1} up a hill when the motor stops. If it is just able to reach the destination which is at a height of 10 m above the point, then the work done against friction (negative of the work done by the friction) is [Take g = 10 m s^{-2}]

(a) 10 kJ (b) 15 kJ
(c) 17.5 kJ (d) 25 kJ.

89. A body is moving with a velocity of 2 m s^{-1} at a height of 5 m from the ground. What will be its speed at a height of 2 m from the ground ? [Take g = 10 m s^{-2}]

(a) 2 m s^{-1} (b) 5.6 m s^{-1}
(c) 8 m s^{-1} (d) 9.2 m s^{-1}.

90. A metre scale of mass 0.1 kg pivoted at one end is turned so that it is inclined at 30° with the horizontal in the vertical plane. The work done is [Take g = 10 m s^{-2}]

(a) 0.05 joule (b) 0.025 joule
(c) 0.25 J (d) 0.5 J.

91. What percentage of kinetic energy of a moving particle is transferred to a stationary particle when it suffers one-dimensional elastic collision with the stationary particle ? [Given : mass of stationary particle = mass of moving particle.]

(a) 25% (b) 50%
(c) 75% (d) 100%.

92. In Q. 91, what percentage will be transferred if the mass of the stationary particle is three times the mass of moving particle ?

(a) 25% (b) 50%
(c) 75% (d) 100%.

93. A block of mass m is attached to two unstretched springs of spring constants k_1 and k_2 as shown in (Fig. 8). The block is displaced towards right through a distance x and is released. The speed of the block as it passes through the mean position shown is

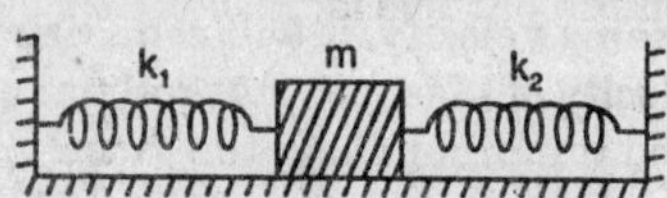

Fig. 8

(a) $\sqrt{\frac{k_1}{m}}\,x$ (b) $\sqrt{\frac{k_2}{m}}\,x$
(c) $\sqrt{\frac{k_1+k_2}{m}}\,x$ (d) $\sqrt{\frac{k_1-k_2}{m}}\,x$.

94. A man weighing 60 kg lifts a body of 15 kg to the top of a building 10 m high in 3 minutes. His efficiency is

(a) 10% (b) 20%
(c) 30% (d) 40%.

95. What is the work done in raising a stone of mass 4 kg and specific gravity 2 lying at the bed of a lake through a height of 5 m ? (g = 10 m s^{-2})

(a) 10 J (b) 50 J
(c) 100 J (d) 200 J.

96. A motor with an efficiency of 90% drives a pump whose efficiency is 60%. Then, the efficiency of the entire installation is

(a) 30% (b) 75%
(c) 54% (d) 45%.

97. A block of mass m moving at a speed v compresses a spring through a distance x before its speed is halved. The spring constant of the spring is

(a) $mv^2/4x^2$ (b) $3mv^2/4x^2$
(c) $5mv^2/4x^2$ (d) $7mv^2/4x^2$.

98. A small body of mass m slides down from the top A of a smooth hill of height H with initial velocity zero. The hill becomes horizontal at the edge B whose height is h from the ground. Assume that the body falls at maximum distance away from the foot of the hill. Then the value of h is

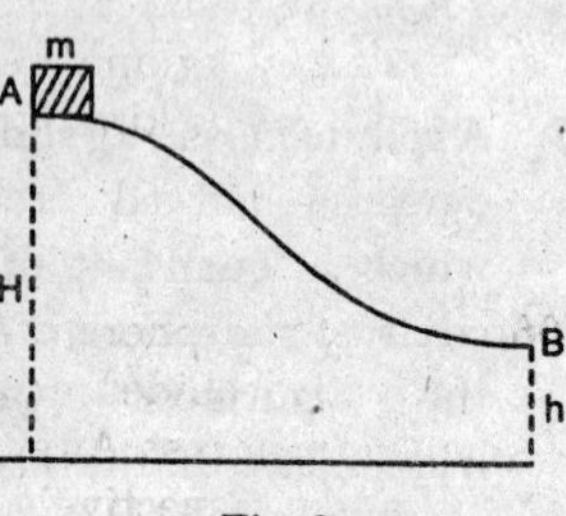

Fig. 9

(a) H (b) $\frac{H}{2}$
(c) $\frac{H}{3}$ (d) $\frac{H}{4}$.

99. In Q. 98, the maximum range is

(a) H (b) $\frac{H}{2}$
(c) $\frac{H}{3}$ (d) $\frac{H}{4}$.

100. A rope ladder with a length l carrying a man with a mass m at its end is attached to the basket of balloon with a mass M. The entire system is in equilibrium in the air. As the man climbs up the ladder into the balloon, the balloon descends by a height h. Then the potential energy of the man

(a) increases by $mg\,(l - h)$ (b) increases by mgl
(c) increases by mgh (d) increases by $mg(2l - h)$.

101. In Q. 100, the potential energy of the balloon

(a) decreases by mgh (b) increases by mgh
(c) increases by $mg\,(l + h)$ (d) increases by mgl.

102. In Q. 100, the work done by the man is

(a) mgl (b) mgh
(c) $\frac{1}{2}\,mgl$ (d) $mg(l + h)$.

103. The power of a water pump is 2 kW. If g = 10 m s^{-2}, then the amount of water it can raise in one minute to a height of 10 m is

(a) 2000 litre (b) 1000 litre
(c) 100 litre (d) 1200 litre.

104. An unloaded bus can be stopped by applying brakes on straight road after covering a distance x. Suppose the passengers add 50% of its weight as the load and the braking force remains unchanged. How long will the bus go after the application of the brakes ? The velocity of the bus is same in both the cases.

(*a*) x (*b*) $1.5x$

(*c*) $2x$ (*d*) $2.5x$.

105. A bomb of 12 kg mass explodes into two pieces of masses 4 kg and 8 kg. The velocity of 8 kg mass is 6 m s^{-1}. The kinetic energy of the other mass is

(*a*) 48 J (*b*) 32 J

(*c*) 24 J (*d*) 288 J.

106. A smooth sphere of mass M moving with velocity u directly collides elastically with another sphere of mass m at rest. After collision, their final velocities are V and v respectively. The value of v is

(*a*) $\frac{2uM}{m}$ (*b*) $\frac{2um}{M}$

(*c*) $\frac{2u}{1+\frac{m}{M}}$ (*d*) $\frac{2u}{1+\frac{M}{m}}$.

107. A frame of mass 200 g when suspended from a massless spring extends it by 10 cm. A lump of clay of mass 200 g is dropped from rest on to the frame from a height of 30 cm as shown in Fig. 10. As a result, the pan moves downward through a maximum distance of

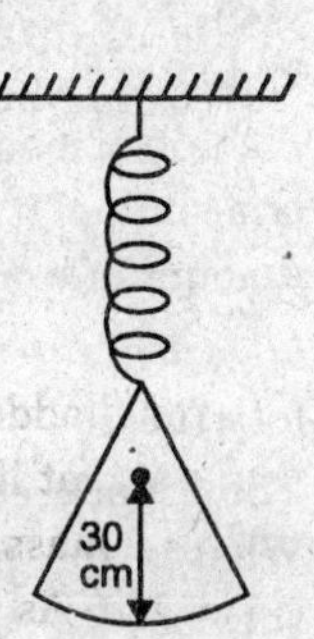

Fig. 10

(*a*) 10 cm (*b*) 20 cm

(*c*) 30 cm (*d*) 40 cm.

108. A running man has half the kinetic energy of that of a boy of half his mass. The man speeds up by 1 m s^{-1} and then has the same K.E. as that of boy. The original speeds of man and boy are, in m s^{-1}

(*a*) $(\sqrt{2}+1), (\sqrt{2}-1)$ (*b*) $(\sqrt{2}+1), 2(\sqrt{2}+1)$

(*c*) $\sqrt{2}, \sqrt{2}$ (*d*) $(\sqrt{2}+1), 2(\sqrt{2}-1)$.

109. The displacement x of a particle moving in one dimension under the action of constant force is related to the time by the equation : $t = \sqrt{x} + 3$ where x is in metre and t in second. What is the acceleration of the particle ?

(*a*) 1 m s^{-2} (*b*) 2 m s^{-2}

(*c*) 3 m s^{-2} (*d*) 4 m s^{-2}.

110. In Q. 109, the work done by the force in first 6 second is

(*a*) 2 m joule (*b*) 6 m joule

(*c*) 9 m joule (*d*) 18 m joule.

111. A human heart pumps 75 cm^3 of blood per beat. The blood pressure is 100 mm of mercury and beat frequency is 72 per minute. The density of mercury is 13.6 g cm^{-3}. The power of the heart is [g = 10 m s^{-2}]

(*a*) 1.2 W (*b*) 12 W

(*c*) 0.12 W (*d*) 0.012 W.

112. An installation consisting of an electric motor driving a water pump lifts 70 litres of water per second to a height of 5 metre. The motor consumes a power of 7 kW. The efficiency of the installation is [g = 10 m s^{-2}]

(*a*) 25% (*b*) 50%

(*c*) 75% (*d*) 98%.

113. A light spring of length l and spring constant k is placed vertically on a table. A small ball of mass m falls on it. The height h from the surface of the table at which the ball will have the maximum velocity is

(*a*) $\frac{l}{k}$ (*b*) $l - \frac{mg}{k}$

(*c*) $l + \frac{k}{mg}$ (*d*) $l - kmg$.

114. The kinetic energy of particle moving along a circle of radius R depends upon the distance covered S and is given by K = aS where a is a constant. Then the force acting on the particle is

(*a*) $\frac{aS}{R}$ (*b*) $\frac{2(aS)^2}{R}$

(*c*) $\frac{aS^2}{R^2}$ (*d*) $\frac{2aS}{R}$.

115. Water is drawn from a well in a 5 kg drum of capacity 55 litre by two ropes connected to the top of the drum. The linear mass density of each rope is 0.5 kg m^{-1}. The work done in lifting water to the ground from the surface of water in the well 20 m below is [g = 10 m s^{-2}]

(*a*) 1.4×10^4 J (*b*) 1.5×10^4 J

(*c*) $9.8 \times 10 \times 6$ J (*d*) 18 J.

116. Two billiard balls of the same size and mass are in contact on a billiard table. A third ball of the same size and mass strikes them symmetrically and remains at rest after the impact. The coefficient of restitution between the balls is

(*a*) $\frac{2}{3}$ (*b*) $\frac{2}{4.1}$

(*c*) $\frac{1}{7}$ (*d*) $\frac{1}{8}$.

117. Under the action of a force, a 3 kg body moves such that $x = \frac{t^2}{2}$, where position x is in metre and t is in second. The work done by the force in first 3 second is

(*a*) 13.5 J (*b*) 27 J
(*c*) 81 J (*d*) 109 J.

118. A technician is experimenting to construct laboratory model of a rifle. He observes that if the length of the barrel of the rifle is cut by 1 cm, the velocity of the bullet is reduced from $\sqrt{5}$ m s^{-1} to 2 m s^{-1}. If the bullet weighs 50 g, then the force acting on the bullet is

(*a*) 1 N (*b*) 1.5 N
(*c*) 2.5 N (*d*) 3.5 N.

119. Consider an oblique elastic collision between a moving ball and a stationary ball of the same mass. Both the balls move with the same speed after the collision. After the collision, the angle between the directions of motion of two balls is

(*a*) 0° (*b*) 30°
(*c*) 60° (*d*) 90°.

120. A ramp is constructed in parabolic shape such that the height y of any point on its surface is given in terms of the point's horizontal distance x from the bottom of the ramp by $y = x^2/2L$. A block of granite is to be set on the ramp ; the coefficient of static friction is 0.80. What is the maximum x coordinate x_M at which the block can be placed on the ramp and remain at rest, if L = 10 m ?

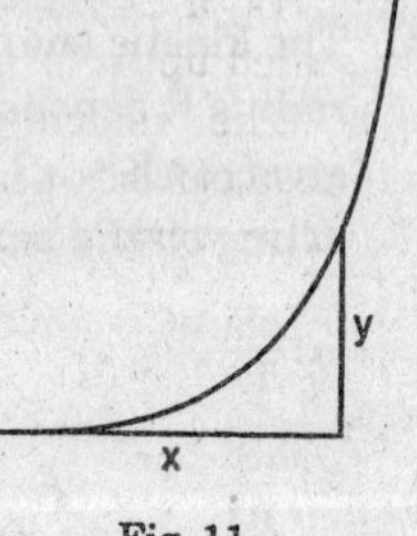

Fig. 11

(*a*) 8 m (*b*) 8.4 m
(*c*) 9 m (*d*) 9.4 m.

121. A stone is dropped from the top of a tall tower. The ratio of the kinetic energy of the stone at the end of three second to the increase in the kinetic energy of the stone during the next three second is

(*a*) 1 : 1 (*b*) 1 : 2
(*c*) 1 : 3 (*d*) 1 : 9.

122. A heavy steel ball of mass greater than 1 kg moving with a speed of 2 m s^{-1}, collides head-on with a stationary ping-pong ball of mass less than 0.1 g. The collision is elastic. After the collision, the ping-pong ball moves approximately with speed

(*a*) 2 m s^{-1} (*b*) 4 m s^{-1}
(*c*) 2×10^4 m s^{-1} (*d*) 2×10^3 m s^{-1}.

123. Consider the situation shown in Fig. 12. Initially the spring is unstretched when the system is released from rest. Assuming no friction in the pulley, the maximum elongation of the spring is

(*a*) $\frac{mg}{k}$ (*b*) $\frac{2mg}{k}$
(*c*) $\frac{3mg}{k}$ (*d*) $\frac{4mg}{k}$.

Fig. 12

124. The linear momentum of a body is increased by 50%. The kinetic energy will be increased by

(*a*) 25% (*b*) 50%
(*c*) 100% (*d*) 125%.

125. Two masses of 1 g and 9 g are moving with equal kinetic energies. The ratio of the magnitudes of their respective linear momenta is

(*a*) 3 : 1 (*b*) 1 : 3
(*c*) 1 : 9 (*d*) 9 : 1.

126. A chain of mass m and length l is placed on a table with one-sixth of it hanging freely from the table edge. The amount of work done to pull the chain on the table is

(*a*) $mgl/72$ (*b*) $mgl/36$
(*c*) $mgl/6$ (*d*) $mgl/4$.

127. The kinetic energy of a body is increased by 300%. What is the percentage increase in the momentum of the body ?

(*a*) 50% (*b*) 100%
(*c*) 150% (*d*) 200%.

128. A particle of mass 4 m which is at rest explodes into three fragments. Two of the fragments, each of mass m, are observed to move with a speed v each in mutually perpendicular directions. The total energy released in the process of explosion is

(*a*) mv^2 (*b*) $2mv^2$
(*c*) $\frac{1}{2}mv^2$ (*d*) $\frac{3}{2}mv^2$.

129. Two masses of 1 g and 4 g are moving with equal kinetic energies. The ratio of the magnitudes of their linear momenta is

(*a*) 4 : 1 (*b*) $\sqrt{2}$: 1
(*c*) 1 : 2 (*d*) 1 : 16.

130. A sphere of mass m moving horizontally with velocity v_0 collides against a pendulum bob of mass m. If the two masses stick together after the collision, then the maximum height attained is (Refer Fig. 13)

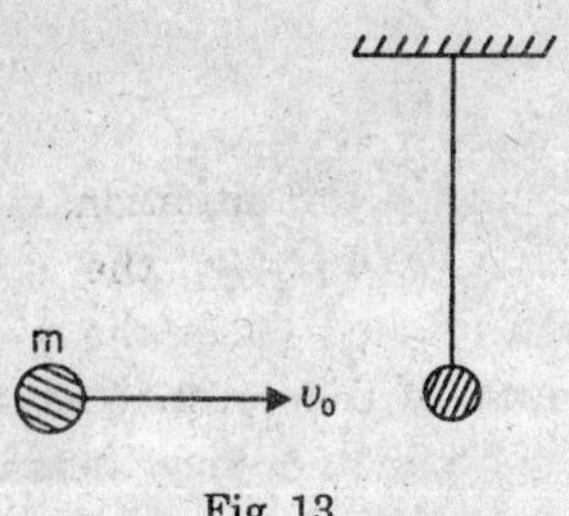

Fig. 13

(a) $\frac{v_0^2}{2g}$ (b) $\frac{v_0^2}{4g}$

(c) $\frac{v_0^2}{6g}$ (d) $\frac{v_0^2}{8g}$.

131. A body accelerates uniformly from rest to a velocity of 1 m s^{-1} in 15 second. The kinetic energy of the body will be $\frac{2}{9}$ J when 't' is equal to [Take mass of body as 1 kg]

(a) 4 s (b) 8 s

(c) 10 s (d) 12 s.

132. A rectangular plank of mass m_1 and height a is kept on a horizontal surface. Another rectangular plank of mass m_2 and height b is placed over the first plank. The gravitational potential energy of the system is

(a) $(m_1 + m_2)(a + b)$ (b) $\left[\frac{m_1 m_2}{2} a + m_2 \frac{b}{2}\right] g$

(c) $\left[\left(\frac{m_1}{2} + m_2\right) a + m_2 \frac{b}{2}\right] g$ (d) $\left[\left(\frac{m_1}{2} + m_2\right) a + m_1 \frac{b}{2}\right] g$.

133. A ball falling freely from a height of 4.9 m hits a horizontal surface. If $e = \frac{3}{4}$, then the ball will hit the surface second time after

(a) 0.5 s (b) 1.5 s

(c) 3.5 s (d) 3.4 s.

134. A 1 kg ball moving at 12 m s^{-1} collides head-on with a 2 kg ball moving in opposite direction with a speed of 24 m s^{-1}. If coefficient of restitution is $\frac{2}{3}$, then

(a) the respective velocities of the two balls after collision are -28 m s^{-1} and -4 m s^{-1}

(b) the respective velocities of the two balls after collision are -8 m s^{-1} and 9 m s^{-1}

(c) the energy lost in the collision is 1312 J

(d) the energy gained in the collision is 1312 J.

135. A man whose mass is 55 kg jumps vertically into air from sitting position in which his centre of mass is 40 cm above the ground. When his feet are just about to leave the ground, his centre of mass is 90 cm above the ground and finally rises to 120 cm when he is at the top of the jump. The average upward force exerted by the ground on him is [g = 10 m s^{-2}]

(a) 100 N (b) 110 N

(c) 220 N (d) 330 N.

136. In Q. 135, the maximum speed attained by the man is

(a) 15 m s^{-1} (b) 30 m s^{-1}

(c) 18 m s^{-1} (d) 2.45 m s^{-1}.

137. A wire of length L suspended vertically from a rigid support is made to suffer extension l in its length by applying a force F. The work done is

(a) $\frac{Fl}{2}$ (b) Fl

(c) $2Fl$ (d) FL.

138. A vertical, massless and elastic spring is suspended from a rigid support. A sphere is slowly attached to the free end of the spring. The sphere is slowly lowered to its equilibrium position so that the spring stretches through 1 cm. If the sphere were allowed to fall freely, then the maximum stretching of the spring would have been

(a) 2 cm (b) 3 cm

(c) 4 cm (d) 5 cm.

139. In Fig. 14, a carriage P is pulled up from A to B. The relevant coefficient of friction is 0.40. The work done will be

(a) 10 kJ
(b) 23 kJ
(c) 25 kJ
(d) 28 kJ.

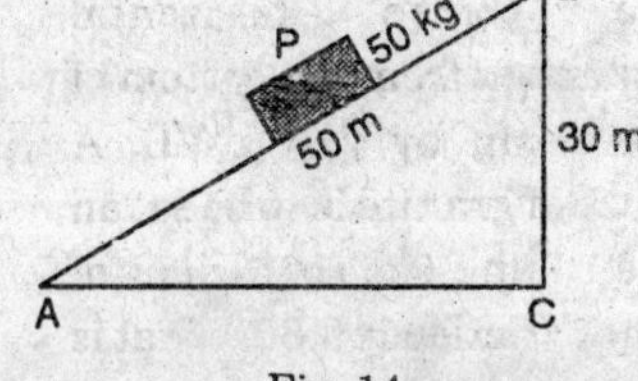

Fig. 14

For Qs. 140, 141 and 142, refer Fig. 15 which shows a ball of weight W hanging at the end of a light and inextensible string of length L. A force F, which is always kept horizontal, is applied to push the ball sideways, and its value is very slowly raised from 0 to F corresponding to rise h of the ball (angle θ). Such a process is called quasistatic process. The force is increased so slowly that the ball may be treated to be in equilibrium under the forces T, W and F all the time, where T is the tension in the string. Now answer the following questions.

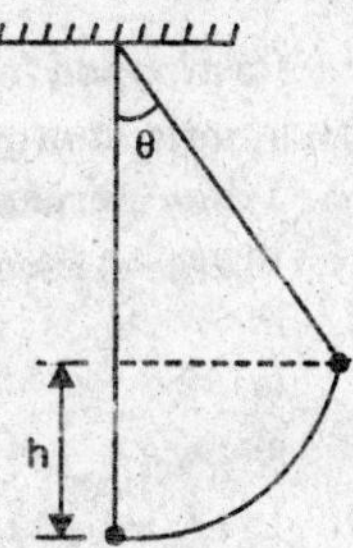

Fig. 15

140. At any stage, the magnitude of F is equal to

(*a*) W (*b*) W sin θ

(*c*) W cos θ (*d*) W tan θ.

141. The work done by the tension T in the above process is

(*a*) zero (*b*) T(L – L cos θ)

(*c*) – T*h* (*d*) – TL sin θ.

142. The work done by the varying force in changing the angular displacement from 0 to θ is

(*a*) W*h* (*b*) F′L sin θ

(*c*) F′*h* (*d*) $\frac{1}{2}$ F′L sin θ.

143. A sphere of mass *m* moving horizontally with velocity v_0 collides against a pendulum bob of mass *m*. The masses collide elastically along the line of initial motion. The maximum height reached by the pendulum bob is

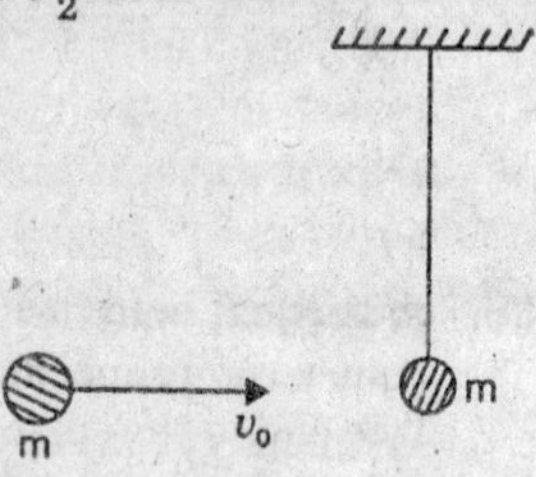

Fig. 16

(*a*) $\frac{v_0^2}{2g}$ (*b*) $\frac{v_0^2}{4g}$

(*c*) $\frac{v_0^2}{6g}$ (*d*) $\frac{v_0^2}{8g}$.

144. A bullet moving with a velocity of 800 m s^{-1} passes through two plates of widths d_1 and d_2. In passing through each of them, it loses a velocity of 200 m s^{-1}. If both the plates offer same resistance to the motion of bullet, then the ratio d_1/d_2 is

(*a*) $\frac{7}{5}$ (*b*) $\frac{5}{7}$

(*c*) $\frac{4}{3}$ (*d*) $\frac{3}{4}$.

145. If force acting on a body is proportional to the distance *x* covered by the body, then the work done by the force is proportional to

(*a*) x (*b*) x^2

(*c*) x^3 (*d*) $x^{3/2}$.

146. A mass 2 *m* is tied to one end of a light rod of length *l*. What horizontal velocity should be imparted to the lower end so that it may just take up the horizontal position ?

(*a*) $\sqrt{2gl}$ (*b*) $\sqrt{gl}$

(*c*) $2\sqrt{gl}$ (*d*) $4\sqrt{gl}$.

147. In Q. 146, if the mass is now rearranged with *m* at the lower end and the other *m* at the middle, then the velocity required to be imparted to mass *m* at the lower end so that it may just take up the horizontal position is

(*a*) $\sqrt{2gl}$ (*b*) $\sqrt{12gl}$

(*c*) $\sqrt{\frac{12}{5}gl}$ (*d*) $\sqrt{gl}$.

148. A body of 2 kg mass makes an elastic collision with another body at rest. The velocity of the 2 kg mass is reduced to one-fourth of the original velocity. However, direction remains unchanged. What is the mass of the body struck ?

(*a*) 1 kg (*b*) 1.2 kg

(*c*) 1.4 kg (*d*) 2 kg.

149. A fast moving neutron suffers one-dimensional elastic collision with a nucleus $_7N^{14}$. What approximate percentage of energy is lost by the neutron in the collision ?

(*a*) 5% (*b*) 10%

(*c*) 25% (*d*) 0%.

150. An 8 kg block accelerates uniformly from rest to a velocity of 4 m s^{-1} in 40 second. The instantaneous power at the end of 8 second is

(*a*) 0.64 W (*b*) 0.32 W

(*c*) 0.16 W (*d*) 0.08 W.

151. One end of a spring of natural length *h* and spring constant *k* is fixed at the ground and the other is fitted with a smooth ring of mass *m* which is allowed to slide on a horizontal rod fixed at a height *h*. Initially, the spring makes an angle of 37° with the vertical when the system is released from rest. The speed of the ring when the spring becomes vertical is

$\left[\text{Take } \cos 37° = \frac{1}{1.25}\right]$

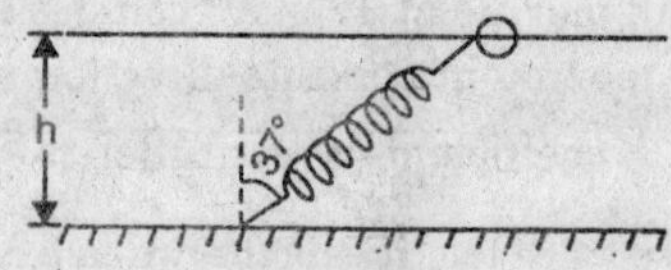

Fig. 17

(*a*) $\sqrt{\frac{k}{m}}$ (*b*) $\frac{h}{2}\sqrt{\frac{k}{m}}$

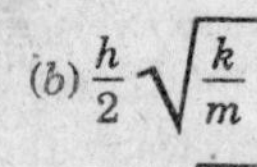

(*c*) $\frac{h}{3}\sqrt{\frac{k}{m}}$ (*d*) $\frac{h}{4}\sqrt{\frac{k}{m}}$.

ANSWERS

QUESTION BANK (Level I)

1. (*c*)	**2.** (*c*)	**3.** (*b*)	**4.** (*c*)	**5.** (*a*)	**6.** (*c*)	**7.** (*c*)	**8.** (*c*)
9. (*c*)	**10.** (*a*)	**11.** (*c*)	**12.** (*a*)	**13.** (*b*)	**14.** (*c*)	**15.** (*a*)	**16.** (*c*)
17. (*b*)	**18.** (*d*)	**19.** (*a*)	**20.** (*d*)	**21.** (*c*)	**22.** (*b*)	**23.** (*a*)	**24.** (*b*)
25. (*c*)	**26.** (*a*)	**27.** (*c*)	**28.** (*a*)	**29.** (*c*)	**30.** (*d*)	**31.** (*a*)	**32.** (*c*)
33. (*c*)	**34.** (*d*)	**35.** (*a*)	**36.** (*c*)	**37.** (*d*)	**38.** (*b*)	**39.** (*b*)	**40.** (*b*)
41. (*b*)	**42.** (*d*)	**43.** (*b*).					

QUESTION BANK (Level II)

1. (*b*)	**2.** (*d*)	**3.** (*a*)	**4.** (*c*)	**5.** (*c*)	**6.** (*c*)	**7.** (*a*)	**8.** (*d*)
9. (*c*)	**10.** (*d*)	**11.** (*b*)	**12.** (*b*)	**13.** (*c*)	**14.** (*b*)	**15.** (*a*)	**16.** (*c*)
17. (*d*)	**18.** (*c*)	**19.** (*a*)	**20.** (*a*)	**21.** (*d*)	**22.** (*d*)	**23.** (*a*)	**24.** (*b*)
25. (*b*)	**26.** (*b*)	**27.** (*d*)	**28.** (*c*)	**29.** (*a*)	**30.** (*c*)	**31.** (*a*)	**32.** (*b*)
33. (*a*)	**34.** (*d*)	**35.** (*a*)	**36.** (*b*)	**37.** (*a*)	**38.** (*c*)	**39.** (*d*)	**40.** (*c*)
41. (*c*)	**42.** (*b*)	**43.** (*a*)	**44.** (*c*)	**45.** (*d*)	**46.** (*c*)	**47.** (*a*)	**48.** (*c*)
49. (*b*)	**50.** (*b*)	**51.** (*c*)	**52.** (*a*)	**53.** (*a*)	**54.** (*b*)	**55.** (*d*)	**56.** (*c*)
57. (*c*)	**58.** (*b*)	**59.** (*b*)	**60.** (*c*)	**61.** (*b*)	**62.** (*c*)	**63.** (*a*)	**64.** (*c*)
65. (*b*)	**66.** (*d*)	**67.** (*a*)	**68.** (*c*)	**69.** (*c*)	**70.** (*b*)	**71.** (*c*)	**72.** (*c*)
73. (*a*)	**74.** (*c*)	**75.** (*c*)	**76.** (*c*)	**77.** (*a*)	**78.** (*b*)	**79.** (*a*)	**80.** (*b*)
81. (*c*)	**82.** (*a*)	**83.** (*a*)	**84.** (*c*)	**85.** (*d*)	**86.** (*c*)	**87.** (*b*)	**88.** (*c*)
89. (*c*)	**90.** (*c*)	**91.** (*d*)	**92.** (*c*)	**93.** (*c*)	**94.** (*b*)	**95.** (*c*)	**96.** (*c*)
97. (*b*)	**98.** (*b*)	**99.** (*a*)	**100.** (*a*)	**101.** (*b*)	**102.** (*a*)	**103.** (*a*)	**104.** (*b*)
105. (*d*)	**106.** (*c*)	**107.** (*c*)	**108.** (*b*)	**109.** (*b*)	**110.** (*d*)	**111.** (*a*)	**112.** (*b*)
113. (*b*)	**114.** (*d*)	**115.** (*a*)	**116.** (*a*)	**117.** (*a*)	**118.** (*c*)	**119.** (*d*)	**120.** (*a*)
121. (*c*)	**122.** (*b*)	**123.** (*b*)	**124.** (*d*)	**125.** (*b*)	**126.** (*a*)	**127.** (*b*)	**128.** (*d*)
129. (*c*)	**130.** (*d*)	**131.** (*c*)	**132.** (*c*)	**133.** (*b*)	**134.** (*a*)	**135.** (*d*)	**136.** (*d*)
137. (*a*)	**138.** (*a*)	**139.** (*b*)	**140.** (*d*)	**141.** (*a*)	**142.** (*a*)	**143.** (*a*)	**144.** (*a*)
145. (*b*)	**146.** (*a*)	**147.** (*c*)	**148.** (*b*)	**149.** (*c*)	**150.** (*a*)	**151.** (*d*).	

HINTS/SOLUTIONS

QUESTION BANK (Level I)

1. (*c*) Fractional loss of kinetic energy of the colliding body

$$= \text{Fraction of energy transferred} = \frac{4 \times 2}{(1+2)^2} = \frac{8}{9}$$

2. (*c*) $E = \frac{p^2}{2m}$ or $E \propto p^2$

or $\frac{E_1}{E_2} = \left(\frac{p_1}{p_2}\right)^2 = \left(\frac{p_1}{2p_1}\right)^2 = \frac{1}{4}$ or $E_2 = 4E_1$

So, increase is 300%.

3. (*b*) $P = mg \times 2 \times \cos 60°$

$$P = 50 \times 9.8 \times 2 \times \frac{1}{2}\text{ W} = 490\text{ W}$$

4. (*c*) $\frac{1}{2}k\xi^2 = \frac{1}{2}mv^2$ or $\xi = \sqrt{\frac{m}{k}}\,v$

or $\xi = \sqrt{\frac{4}{100}} \times 20 = \frac{20}{5}\text{ m} = 4\text{ m}$

5. (*a*) Work done = mgh

$$= 5 \times 9.8 \times 30\text{ J} = 5 \times 98 \times 3\text{ J} = 1470\text{ J}$$

Heat produced $= \frac{1470}{4.2}$ cal = 350 cal

6. (*c*) In the first case, $\frac{1}{2}mv^2 - \frac{1}{2}m\left(\frac{v}{2}\right)^2 = f \times 25$

or $\frac{3}{8}mv^2 = 25f$ ...(*i*)

In the second case, $\frac{1}{2}m\left(\frac{v}{2}\right)^2 - 0 = f \times S$

or $\frac{1}{8}mv^2 = f \times S$ ...(*ii*)

Dividing (*ii*) by (*i*),

$$\frac{S}{25} = \frac{1}{8} \times \frac{8}{3} \text{ or } S = \frac{25}{3} \text{ cm} = 8.3 \text{ cm.}$$

7. (*c*) $P = \frac{mgh}{t}$ or $t = \frac{mgh}{P}$

or $t = \frac{200 \times 10 \times 40}{10 \times 1000} \text{ s} = 8 \text{ s}$

8. (*c*) Applying conservation of momentum,

$m_1 \times 40 = (m_1 + m_2)\, 30$

or $40m_1 = 30m_1 + 30m_2$

or $10m_1 = 30m_2$ or $\frac{m_1}{m_2} = \frac{30}{10}$ or $\frac{m_1}{m_2} = 3$

9. (*c*) $E = \frac{F^2}{2K}$

$\therefore$ $F^2 \propto K$ or $F \propto \sqrt{K}$

$\therefore$ $\frac{F_1}{F_2} = \frac{\sqrt{K_1}}{\sqrt{K_2}}$

10. (*a*) $E = \frac{1}{2}Kx^2$

$\therefore$ $E \propto K$

$\therefore$ $\frac{E_1}{E_2} = \frac{K_1}{K_2}$

11. (*c*) $W = FS \cos\theta$

or $\cos\theta = \frac{W}{FS} = \frac{25}{10 \times 5} = \frac{1}{2}$ or $\theta = 60°$

12. (*a*) $P = Fv$

$= 9000 \text{ N} \times 2 \text{ m s}^{-1} = 18000 \text{ J s}^{-1}$

$= 18000 \text{ W} = 18 \text{ kW}$

13. (*b*) $f = \mu mg \cos\theta$

or $f = 0.30 \times 10 \times 10 \times \cos 45°$

or $f = \frac{30}{\sqrt{2}} \text{ N}$

$W = f \times S$

$= \frac{30}{\sqrt{2}} \times 5 \text{ J} = \frac{150}{\sqrt{2}} \times \frac{\sqrt{2}}{\sqrt{2}} \text{ J} = 75\sqrt{2} \text{ J}$

This is negative work because f and S are oppositely directed.

14. (*c*) $200 \times 0.3 + 400 \times v = 0$

or $v = -\frac{200 \times 0.3}{400} \text{ m s}^{-1} = -0.15 \text{ m s}^{-1}$

15. (*a*) $\frac{1}{2}m \times 100 \times 100 = f \times 1$ or $f = 5000m$

Again, $\frac{1}{2}mv^2 - \frac{1}{2}mv'^2 = f \times 0.5$

or $m[v^2 - v'^2] = f = 5000m$

or $v'^2 = v^2 - 5000$

or $v'^2 = 10000 - 5000$

or $v'^2 = 5000$

or $v'^2 = 2 \times 2500$

or $v' = 50\sqrt{2} \text{ m s}^{-1}$

16. (*c*) $h = 1 \text{ m},\ h_1 = \frac{1}{4} \text{ m}$

$h_1 = e^2 h$

$\frac{1}{4} = e^2$ or $e = \frac{1}{2}$

17. (*b*) Potential energy at the required height

$= \frac{490 \text{ J}}{2} = 245 \text{ J}$

Again, $245 = 2 \times 10 \times h$ or $h = \frac{245}{20} \text{ m} = 12.25 \text{ m}$

18. (*d*) Clearly, 60% energy is retained after impact.

So, $h' = \frac{60}{100} \times 10 = 6 \text{ m}$

19. (*a*) Fraction of total energy retained is $\left(\frac{1-x}{1+x}\right)^2$

Here x = A.

20. (*d*) Hanging mass $= \frac{m}{3}$

Height through which the centre of gravity is to be raised $= \frac{l}{6}$

Work done $= \left[\frac{m}{3}\right] g \left[\frac{l}{6}\right] = \frac{mgl}{18}$

21. (*c*) Work done = force × distance

$= 4 \text{ N} \times 2 \text{ m} = 8 \text{ J}$

This work is stored as kinetic energy.

22. (*d*) $\frac{m_1}{m_2} = \frac{1}{4}$; $\frac{\frac{1}{2}m_1 v_1^2}{\frac{1}{2}m_2 v_2^2} = 1$

On dividing, $\frac{v_1^2}{v_2^2} = \frac{4}{1}$ or $\frac{v_1}{v_2} = 2$

$\frac{m_1 v_1}{m_2 v_2} = \frac{2}{4} = \frac{1}{2}$

23. (*a*) In any type of collision, momentum is conserved.

24. (*b*) $\frac{E_k'}{E_k} = \left(1 + \frac{1}{5}\right)^2 = \frac{36}{25}$

$\left(\frac{E_k'}{E_k} - 1\right) \times 100 = \left(\frac{36}{25} - 1\right) \times 100 = 44$

25. (*c*) $V = \frac{1}{2} k\xi^2$; $V' = \frac{1}{2} k(n\xi)^2$

$\frac{V'}{V} = n^2$ or $V' = n^2 V$

26. (*a*) $E_k = \frac{p^2}{2m}$ or $p = \sqrt{2mE_k}$

Percentage increase in momentum $= \frac{1}{2} \times 0.1\% = 0.05\%$.

27. (*c*) $p = a + bt^2$

$F = \frac{d}{dt}(p) = \frac{d}{dt}[a + bt^2]$

or $F = b \times 2t$ or $F = 2bt$

$\therefore$ $F \propto t$

28. (*a*) $W = [2\hat{i} + \hat{j} - \hat{k}] \cdot [(3i + 2\hat{j} - 2\hat{k}) - (2\hat{i} - \hat{j} + \hat{k})]$

or $W = [2\hat{i} + \hat{j} - \hat{k}] \cdot [\hat{i} + 3\hat{j} - 3\hat{k}]$

or $W = (2 + 3 + 3)$ J = 8 J

29. (*c*) $\frac{1}{2} \times 245 = 2 \times 9.8 \times h$ or $h = \frac{245}{4 \times 9.8}$ m = 6.25 m

30. (*d*) Applying conservation of momentum,

$10 \times 100 = (10 + 10)v$ or $v = \frac{1000}{20}$ cm s^{-1}

or $v = 50$ cm s^{-1}

Now, $(2m)gh = \frac{1}{2}(2m)v^2$

or $h = \frac{v^2}{2g} = \frac{50 \times 50}{2 \times 1000} = 1.25$ cm

31. (*a*) Kinetic energy $= \frac{1}{2} \times 950 \times \left(100 \times \frac{5}{18}\right)^2$ J

$= 0.3665 \times 10^6$ J = 0.367 MJ

32. (*c*) $h' = \frac{1}{2} h$

$mgh' = \frac{1}{2} mgh$

So, there is 50% loss of energy.

33. (*c*) $\frac{1}{2} mv^2 - 0 = f \times S$ or $v^2 \propto S$

When v is increased by a factor of 3, S is increased by a factor of 9.

34. (*d*) $p = \sqrt{2mE_k}$ or $p \propto \sqrt{m}$

Out of the given particles, α-particle has maximum mass.

35. (*a*) $v = \sqrt{4^2 + 3^2} = \sqrt{16 + 9} = \sqrt{25} = 5$ m s^{-1}

Kinetic energy $= \frac{1}{2} \times 0.3 \times 5 \times 5$ J = 3.75 J

36. (*c*) $P = 500 \times 3 = 1500$ W = 1.5 kW

37. (*d*) 1 W = 1 J s^{-1} = 1 N × 1 m × 1 s^{-1} = 1 N m s^{-1}

38. (*b*) Applying law of conservation of momentum,

$\frac{50}{1000} \times 10 = \frac{50 + 950}{1000} v$ or $v = \frac{1}{2}$ m s^{-1}

39. (*b*) $v_{2f} = \frac{2 \times 4}{4 + 12} \times 10$ m s^{-1} $= \frac{80}{16}$ m s^{-1} = 5 m s^{-1}

40. (*b*) $v_{1f} = \frac{4 - 12}{4 + 12} \times 10$ m s^{-1} $= -\frac{80}{16}$ m s^{-1} = – 5 m s^{-1}

41. (*b*) $mv = \frac{1}{2} mv^2$ or $v = 2$ m s^{-1}

42. (*d*) Note that both momentum and kinetic energy are to be conserved.

43. (*b*) $\frac{1}{2} mv^2 = mgR$ or $v = \sqrt{2gR}$

QUESTION BANK (Level II)

1. (*b*) $\frac{mv^2}{R} = \frac{k}{R^2}$ or $mv^2 = \frac{k}{R}$

Kinetic energy $= \frac{1}{2} mv^2 = \frac{k}{2R}$

Potential energy $= -\int_{\infty}^{R} F\, dR$

Lower limit is taken as ∞ because potential energy is assumed to be zero at infinity.

$= -\int_{\infty}^{R} -\frac{k}{R^2} dR = k \left| \frac{R^{-1}}{-1} \right|_{\infty}^{R} = -\frac{k}{R}$

Total energy $= \frac{k}{2R} - \frac{k}{R} = -\frac{k}{2R}$

2. (*d*) $\frac{1}{2} mv^2 + mgl(1 - \cos\theta) = \frac{1}{2} mv'^2$

or $v^2 + 2gl(1 - \cos\theta) = v'^2$

or $v^2 + 2 \times 10 \times 0.5 \left(1 - \frac{1}{2}\right) = 9$

or $v^2 = 9 - 5 = 4$

or $v = 2$ m s^{-1}

3. (*a*) $m_1 u = m_1 \frac{2u}{3} + m_2 v$

or $\frac{1}{3} m_1 u = m_2 v$ ...(*i*)

Using Newton's experimental law of impact,

$$v - \frac{2u}{3} = u \text{ or } v = \frac{5}{3}u$$

From eqn. (1), $\frac{1}{3} m_1 u = \frac{5}{3} m_2 u$ or $\frac{m_1}{m_2} = 5$

4. (*c*) $W = F\,S \cos 90° = \text{zero}$

5. (*c*) $f = \frac{3}{10} mg$

$$W = -fS \text{ or } W = -\frac{3}{10} mgS$$

or $W = -\frac{3}{10} \times 200 \times 10 \text{ J} = -600 \text{ J}$

6. (*c*) $ky = mg$ or $k = \frac{mg}{y}$

$$U = \frac{1}{2}\left(\frac{mg}{y}\right)y^2 = \frac{1}{2} mgy$$

7. (*a*) $E = \frac{1}{2} m_1 v_1^2 + \frac{1}{2} m_2 v_2^2$

But $m_2 v_2 = m_1 v_1$ or $v_2 = \frac{m_1 v_1}{m_2}$

$\therefore$ $E = \frac{1}{2} m_1 v_1^2 + \frac{1}{2} m_2 \left[\frac{m_1 v_1}{m_2}\right]^2$

or $E = \frac{1}{2} m_1 v_1^2 \left[1 + \frac{m_1}{m_2}\right]$

or $E = E_1 \frac{m_2 + m_1}{m_2} = \frac{M}{m_2} E_1$ or $E_1 = \frac{m_2}{M} E$

8. (*d*) When spring is halved, the force constant will be doubled. So, slope would increase. $\therefore$ $\theta > 45°$.

9. (*c*) Time taken by body to fall through height $h = \sqrt{\frac{2h}{g}}$

After first rebound, time taken to rise to height h_1 and coming back

$$= 2\sqrt{\frac{2h_1}{g}} = 2\sqrt{\frac{2e^2 h}{g}} = 2e\sqrt{\frac{2h}{g}}$$

Time taken in rising to height h_2 and coming back

$$= 2e^2\sqrt{\frac{2h}{g}}$$

Total time taken by the body in coming to rest

$$= \sqrt{\frac{2h}{g}} + 2e\sqrt{\frac{2h}{g}} + 2e^2\sqrt{\frac{2h}{g}} + \ldots\ldots$$

$$= \sqrt{\frac{2h}{g}} + 2e\sqrt{\frac{2h}{g}}\,[1 + e + e^2 + \ldots\ldots]$$

$$= \sqrt{\frac{2h}{g}} + 2e\sqrt{\frac{2h}{g}} \times \frac{1}{1-e}$$

$$= \sqrt{\frac{2h}{g}}\left[1 + \frac{2e}{1-e}\right]$$

$$= \sqrt{\frac{2h}{g}}\left[\frac{1-e+2e}{1-e}\right] = \sqrt{\frac{2h}{g}}\left[\frac{1+e}{1-e}\right]$$

10. (*d*) $3k = mg$ or $k = \frac{mg}{3}$

Force constant is inversely proportional to length.

So, $k' = \frac{3}{2}k = \frac{3}{2} \times \frac{mg}{3} = \frac{mg}{2}$

If ξ is the required extension, then

$$k'\xi = mg \text{ or } \xi = \frac{mg}{k'} = \frac{mg}{mg} \times 2 = 2 \text{ cm}$$

11. (*b*) In this case, $P = \frac{mgh + \frac{1}{2}mv^2}{t}$

or $P = \frac{m}{t}\left[gh + \frac{v^2}{2}\right]$

or $P = \frac{10}{1}\left[10 \times 10 + \frac{10 \times 10}{2}\right] \text{W}$

$$= 10[100 + 50] = 1500 \text{ W}$$

$$= \frac{1500}{1000} \text{ kW} = 1.5 \text{ kW}$$

12. (*b*) $E' = \frac{1}{2} m\,[20^2 - 10^2] = \frac{1}{2} m \times 30 \times 10$

$$E = \frac{1}{2} m \times 10 \times 10$$

$$\frac{E'}{E} = \frac{\frac{1}{2} m \times 30 \times 10}{\frac{1}{2} m \times 10 \times 10} = 3 \text{ or } E' = 3E$$

13. (*c*) For stable equilibrium of atom, $\frac{dU}{dx} = 0$

$$a(-12)x^{-12-1} - b(-6)\,x^{-7} = 0$$

or $\frac{-12a}{x^{13}} + \frac{6b}{x^7} = 0$

or $\frac{-12a}{x^6} = -6b$

or $x^6 = \frac{12a}{6b} = \frac{2a}{b}$

or $x = \left[\frac{2a}{b}\right]^{1/6}$

14. (*b*) Kinetic energy acquired by the body is equal to the work done. Work done, $W = E_k$

$$= FS\cos\theta$$

$$F \propto \frac{1}{v}$$

$$\therefore \quad E_k \propto \frac{S}{v} \text{ or } E_k \propto \frac{S}{S/t} \text{ or } E_k \propto t$$

15. (*a*) This is the case of work done by a variable force

$$W = \int_0^5 (3x^2 - 2x + 7)\,dx$$

$$W = \left| x^3 - x^2 + 7x \right|_0^5 \text{ joule}$$

or $W = (5\times5\times5 - 5\times5 + 7\times5)$ joule

or $W = (125 - 25 + 35)$ joule $= 135$ joule

16. (*c*) Using conservation of momentum,

$$m_1v_1 = m_2v_2 \text{ or } \frac{m_1}{m_2} = \frac{v_2}{v_1}$$

or $$\frac{r_1^3}{r_2^3} = \frac{1}{2} \text{ or } \frac{r_1}{r_2} = \frac{1}{2^{1/3}}$$

17. (*d*) $m = 10 \times 0.8$ kg $= 8$ kg, $h = 5$ m

$$P = \frac{mgh}{t} = \frac{8\times10\times5}{10} W = 40 W$$

☞ *In this type of questions, focus should be on centre of gravity.*

18. (*c*) Power $= \dfrac{\frac{1}{2}mv^2}{t}$

Keeping t same, both m and v have to be increased by a factor of 3. So, the required factor is $3\times3\times3$ *i.e.* 27.

19. (*a*) Total energy at the time of throwing the ball

$$= mgh + \frac{1}{2}mv_0^2$$

Energy after collision with ground

$$= \frac{1}{2}(mgh + \frac{1}{2}mv_0^2).$$

The ball again rises to height h.

$$\therefore \quad \frac{1}{2}(mgh + \frac{1}{2}mv_0^2) = mgh$$

or $$mgh + \frac{1}{2}mv_0^2 = 2mgh$$

or $$\frac{1}{2}mv_0^2 = mgh \text{ or } v_0 = \sqrt{2gh}$$

20. (*a*) Power $= 7500$ W $= 7500$ J s^{-1}, Velocity, $v = 20$ m s^{-1}

$$P = Fv \text{ or } F = \frac{P}{v} = \frac{7500 \text{ J s}^{-1}}{20 \text{ m s}^{-1}} = 375 \text{ N}$$

20. (*a*) Change in volume, $dV = (105 - 5)$ litre

$= 100$ litre $= 10^{-1}$ m^3

Pressure, $P = 10^2$ N m^{-2}

Work done $= PdV = 10^2 \times 10^{-1}$ joule $= 10$ joule

21. (*d*) S = 10 m, F = 5 N, W = 25 J, θ = ?

$$\cos\theta = \frac{W}{FS} = \frac{25}{5\times10} = \frac{1}{2} \quad \therefore \theta = 60°$$

23. (*a*) Initial height of C.G. = 4 cm

Final height of C.G. = 10 cm

Increase in height = 6 cm = 0.06 m

Work done = 5 × 10 × 0.06 J = 3 J

24. (*b*) $$v = \frac{d}{dt}(x) = t^2$$

$$K.E. = \frac{1}{2}mt^4 = \frac{1}{2}\times1\times t^4 = \frac{t^4}{2}$$

In one second, gain of K.E. is $\frac{1}{2}$ J or 0.5 J. This is equal to work done.

25. (*b*) Assume that the wall moves to the right. Let it act as a reference for all the velocities. Velocity of the wall relative to the ball before collision

$$= u - (-v) = u + v$$

Let v' be the velocity of the ball to the right after collision. Velocity of wall relative to the ball after collision $= u - v'$.

Using Newton's law of collision,

$$u - v' = -(u + v) \text{ or } v' = 2u + v$$

26. (*b*) Minimum force is $mg\sin\theta$. So, minimum power is given by

$$P = mg\sin\theta\, v \text{ or } v = \frac{P}{mg\sin\theta}$$

or $$v = \frac{9000\times20}{1200\times10\times1} \text{ m s}^{-1} = 15 \text{ m s}^{-1}$$

$$= 15 \times \frac{18}{5} = 54 \text{ km h}^{-1}$$

27. (*d*) $u = 10$ m s^{-1}, $v = 20$ m s^{-1}

Work done = Increase in kinetic energy

$$= \frac{1}{2}\times500\,[20^2 - 10^2] = \frac{500\times30\times10}{2}$$

Power $$= \frac{500\times30\times10}{2\times60} W = 1250 W$$

28. (*c*) Energy required to overcome air resistance

= Kinetic energy – Potential energy

$$= \frac{1}{2}\times1\times200\times200 - 1\times10\times1500$$

$$= 20 \text{ kJ} - 15 \text{ kJ} = 5 \text{ kJ}$$

29. (*a*) $$\frac{4x}{(1+x)^2} = \frac{4\times4}{(1+4)^2} = \frac{16}{25}$$

Note that mass of α-particle is four times the mass of neutron. [Only an approximate idea.]

30. (*c*) Initial velocity, $u = \frac{10}{5}$ m s^{-1} = 2 m s^{-1}

Acceleration, $a = \frac{0.2}{5}$ m s^{-2} = 0.04 m s^{-2}

Using $S = ut + \frac{1}{2}at^2$, we get

$$S = 2 \times 10 + \frac{1}{2} \times 0.04 \times 10 \times 10 = 22 \text{ m}$$

Work done = 0.2 N × 22 m = 4.4 J

31. (*a*) $m = 0.3 \times 10^8$ kg, F = 0.5×10^5 N, S = 3 m, v = ?

Work done = F × S

This work becomes the kinetic energy of the ship.

$\therefore \ \frac{1}{2}mv^2 = F \times S$

or $v^2 = \frac{2FS}{m} = \frac{2 \times 0.5 \times 10^5 \times 3}{0.3 \times 10^8}$ or $v = 0.1$ m s^{-1}.

32. (*b*) Work done is equal to energy $\frac{1}{2}mv^2$

Also, $v = \frac{d}{dt}\left(\frac{t^3}{3}\right) = t^2$

$\therefore$ Work done $= \frac{1}{2}mt^4 = \frac{1}{2} \times 2 \times (2)^4 = 16$ joule

33. (*a*) h = 1 km = 10^5 cm, d = 1 cm

A = 1 km^2 = $(10^5 \text{ cm})^2 = 10^{10}$ cm^2

mass, m = volume × density

$= 10^{10} \times 1 \times 1 = 10^{10}$ g

Work done, W = $10^{10} \times 980 \times 10^5$ erg = 98×10^{16} erg

34. (*d*) Mass to be lifted = 10×10^3 kg

[∵ density of water = 10^3 kg m^{-3}]

Height, h = 10 m

Work done = $10^4 \times 10 \times 10$ J = 10^6 J

35. (*a*) Weight, $mg = 75 \times 10$ N = 750 N

Height, h = 2 m

Work, W = mgh = 750 N × 2 m = 1500 J

Heat energy = 1500 J

Note. SI unit of heat is joule and not kcal.

36. (*b*) $P = \frac{mgh}{t}$ or $\frac{m}{t} = \frac{P}{gh}$

or $\frac{m}{t} = \frac{1000}{10 \times 10}$ kg = 10 kg

37. (*a*) mg = 20 N, h = 4 m, P.E. = 80 J

$80 \text{ J} = \frac{1}{2}kx^2$

or $x^2 = \frac{80 \times 2}{1960} = \frac{16}{196}$ or $x = \frac{4}{14}$ m or $x = \frac{2}{7}$ m

38. (*c*) Volume/s = $av = \pi r^2 v$

Mass/s = $\pi r^2 v \times 1000$ SI units

Power of water jet

$$= \frac{\frac{1}{2}mv^2}{t} = \frac{1}{2} \times \pi r^2 v \times 1000 \times v^2 = 500\pi r^2 v^3$$

39. (*d*) P = Fv

$\frac{dP}{dt} = F\frac{dv}{dt}$ (∵ F is constant in the given problem.)

40. (*c*) $mgh = \frac{1}{2}mv^2$

or $h = \frac{v^2}{2g} = \frac{2 \times 2}{2 \times 10} = 0.2$ m

$\cos\theta = \frac{0.8}{1}$

or $\theta = \cos^{-1}(0.8)$

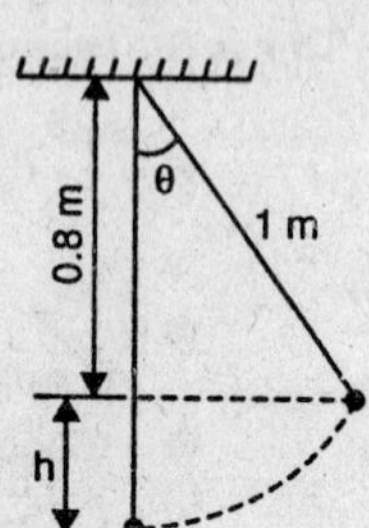

Fig. 18

41. (*c*) P = Fv cos θ

P = 360 N × 10 × $\frac{5}{18}$ × cos 60°

P = 360 × 10 × $\frac{5}{18} \times \frac{1}{2}$ W

or P = 500 W

42. (*b*) $a = \frac{10 - 0}{5}$ m s^{-2} = 2 m s^{-2};

F = ma or F = 1000 × 2 N = 2000 N

Average velocity $= \frac{0 + 10}{2}$ m s^{-1} = 5 m s^{-1}

Average power = 2000 × 5 W = 10^4 W

Required horse power is $\frac{10^4}{746}$

43. (*a*) $e = \frac{\text{relative velocity of separation}}{\text{relative velocity of approach}}$

$$e = \frac{\sqrt{2gh'}}{\sqrt{2gh}} = \sqrt{\frac{h'}{h}}$$

or $h' = e^2 h$

For second rebound

$e = \sqrt{\frac{h''}{h'}}$ or $h'' = e^2 h'$

or $h'' = e^4 h'$

Fig. 19

The result can be generalised for n rebounds.

44. (*c*) $\frac{dm}{dt} = 500$ kg min^{-1}

$= \frac{500}{60}$ kg s^{-1} = $\frac{50}{6}$ kg s^{-1} = $\frac{25}{3}$ kg s^{-1}

$\frac{70}{100} \times P = \frac{mgh}{t}$

or $$P = \frac{100\, mgh}{70\, t}$$

$$= 100 \times \frac{25}{3} \times 9.8 \times 30 \times \frac{1}{70}\text{ W} = 3500\text{ W}$$

45. (*d*) Applying conservation of momentum,

$$1 \times 4 = 2 \times v \quad \text{or} \quad v = 2\text{ m s}^{-1}$$

Loss of kinetic energy

$$= \frac{1}{2} \times 1 \times 4 \times 4 - \frac{1}{2} \times 2 \times 2 \times 2$$

$$= (8 - 4)\text{ J} = 4\text{ J}$$

46. (*c*) Mass of hanging portion $= \frac{m}{l} \times \frac{l}{n} = \frac{m}{n}$

Height through which the centre of gravity of the hanging length is to be moved $= \frac{l}{2n}$

$\therefore$ Work done $= \frac{m}{n} \times g \times \frac{l}{2n} = \frac{mgl}{2n^2}$

47. (*a*) If v_1 be the velocity of the body just before collision, then $v_1^2 = 2g \times 10 = 20g$

If v_2 be the velocity of the body after collision, then

$$0^2 = v_2^2 - 2g \times 2.5 \quad \text{or} \quad v_2^2 = 5g$$

Then, $\frac{v_1^2}{v_2^2} = \frac{20\,g}{5\,g} = 4$ or $\frac{v_1}{v_2} = 2$

48. (*c*) Percentage loss

$$= \frac{\text{K.E. before collision} - \text{K.E. after collision}}{\text{K.E. before collision}} \times 100$$

$$= \frac{\frac{1}{2} m \times 20\,g - \frac{1}{2} m \times 5g}{\frac{1}{2} m \times 20g} \times 100 = \frac{15}{20} \times 100 = 75$$

49. (*b*) Increase in K.E. in first case

$$= \frac{1}{2} m \times 20^2 - \frac{1}{2} m \times 10^2$$

$$= \frac{1}{2} m\,(400 - 100) = \frac{1}{2} m \times 300$$

Increase in K.E. in second case

$$= \frac{1}{2} m(100 - 0) = \frac{1}{2} m \times 100$$

50. (*b*) $F = v\frac{dm}{dt}$, $P = Fv = v^2\frac{dm}{dt}$

$$P = 0.2 \times 0.2 \times 2\text{ W} = 0.08\text{ W}$$

51. (*c*) Fraction of kinetic energy transferred

$$= \frac{4x}{(1+x)^2} = \frac{4 \times \frac{1}{2}}{\left(1 + \frac{1}{2}\right)^2} = \frac{8}{9}$$

$$\frac{\text{K.E. of 2.5 kg after collision}}{\text{K.E. of 5 kg before collision}} = \frac{8}{9}$$

Clearly K.E. of 5 kg before collision $= \frac{9}{8} \times 8\text{ J} = 9\text{ J}$

52. (*a*) $V = 6\text{ m}^3$, $m = 6000$ kg, $P = 2000$ W

Now, $\frac{75}{100} \times 2000 = \frac{6000 \times 10 \times 15}{t \times 60}$

$$\frac{3}{4} = \frac{3 \times 10}{t \times 4} \quad \text{or} \quad t = 10\text{ minute}$$

53. (*a*) $P = \frac{mv^2}{2t} = \frac{80 \times 10 \times 10}{2 \times 4}\text{ W} = 1000\text{ W}$

54. (*b*) Equating gain of elastic potential energy with loss of gravitational potential energy, we get

$$\frac{1}{2} k\xi^2 = mg\xi$$

or $$\xi = \frac{2mg}{k} = \frac{2 \times 1 \times 10}{1000}\text{ m}$$

$$= \frac{1}{50}\text{ m} = \frac{1}{50} \times 100\text{ cm} = 2\text{ cm}$$

55. (*d*) K.E. at B = P.E. at A

$$= mgh = 1 \times 10 \times 1\text{ J} = 10\text{ J}$$

Now, $10 = \mu mgS$

or $$S = \frac{10}{\mu\, mg} = \frac{10}{\frac{1}{5} \times 1 \times 10} = 5\text{ m}$$

56. (*c*) $$P = \frac{mgh}{t} = \frac{80 \times 10 \times 1.5}{2}\text{ W}$$

$$= 600\text{ W} = 0.6\text{ kW}$$

57. (*c*) $$e = \frac{\text{Normal component of relative velocity after collision}}{\text{Normal component of relative velocity before collision}}$$

$$0.5 = \frac{v \cos 45°}{10 \cos 45°} = \frac{v}{10}$$

or $$v = 0.5 \times 10\text{ m s}^{-1} = 5\text{ m s}^{-1}$$

58. (*b*) $$dW = -\mu\left[\frac{M}{L}\right] gl\, dl$$

$$W = \int_0^{\frac{2L}{3}} -\frac{\mu Mg}{L}\, l\, dl$$

or $$W = -\frac{\mu Mg}{L}\left|\frac{l^2}{2}\right|_0^{\frac{2L}{3}}$$

or $$W = -\frac{\mu Mg}{2L}\left[\frac{4L^2}{9} - 0\right]$$

or $$W = -\frac{2}{9}\mu MgL$$

59. (*b*) $E = \frac{1}{2} k\xi^2 = \frac{F^2}{2k}$

$$\frac{E_A}{E_B} = \frac{k_B}{k_A} = \frac{1}{2} \text{ or } E_B = 2E_A \text{ or } E_B = 2E$$

60. (*c*) Applying conservation of momentum,

$$300v' = 200v \text{ or } v' = \frac{2}{3}v$$

Again, $\frac{1}{2}mv^2 = \mu mgS$ or $v^2 = 2\mu gS$ or $v^2 \propto S$

Now $\frac{S'}{S} = \frac{v'^2}{v^2}$ or $S' = \frac{v'^2}{v^2} \times S = \frac{4}{9} \times 36 \text{ m} = 16 \text{ m}$

61. (*b*) Percentage of energy saved is

$$\frac{\frac{1}{2}mv^2 \times 100}{\frac{1}{2}mv^2 + mgh} \text{ or } \frac{v^2 \times 100}{v^2 + 2gh}$$

i.e. $\frac{12 \times 12 \times 100}{12 \times 12 + 2 \times 10 \times 12}$ or $\frac{14400}{144 + 240}$

or $\frac{14400}{384}$ or 37.5

62. (*c*) $\frac{1}{2}mv^2 = mg(l - l\cos\alpha)$

or $v^2 = 2gl(1 - \cos\alpha)$

or $v = \sqrt{2gl(1 - \cos\alpha)}$

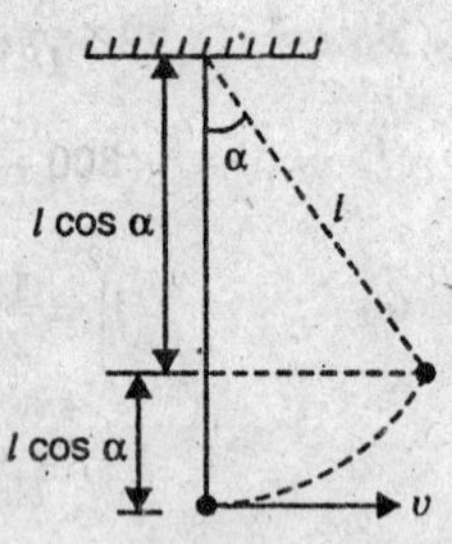

Fig. 20

63. (*a*) $10 = \frac{1}{2}kS^2$

Now, $W = \frac{1}{2}k[(2S)^2 - S^2]$

or $W = \frac{1}{2}k \times 3S^2 = 3 \times \frac{1}{2}kS^2$

$= 3 \times 10 \text{ J} = 30 \text{ J}$

> *For additional extension S,*
>
> $W = \frac{1}{2}k[(3S)^2 - (2S)^2]$
>
> *or* $W = \frac{1}{2}k \times 5S^2$
>
> *or* $W = 5 \times \frac{1}{2}kS^2 = 50 \text{ J}$
>
> ***Note that the ratio of work done for every additional distance S is 1 : 3 : 5 : ....***

64. (*c*) For equilibrium position, $\frac{dU}{dr} = 0$

$$\frac{d}{dr}\left(\frac{a}{r^2} - \frac{b}{r}\right) = 0$$

or $a\frac{d}{dr}(r^{-2}) - b\frac{d}{dr}(r^{-1}) = 0$

or $a(-2)r^{-3} - b(-1)r^{-1-1} = 0$

or $\frac{-2a}{r^3} + \frac{b}{r^2} = 0$

or $-\frac{2a}{r} = -b$ or $r = \frac{2a}{b}$

65. (*b*) $\frac{1}{2}mv^2 = f \times S$

or $f = \frac{mv^2}{2S}$

$= \frac{30}{1000} \times 50 \times 50 \times \frac{1 \times 100}{2 \times 10} \text{ N} = 375 \text{ N}$

66. (*d*) Work done = Area under F-*x* graph

$= \frac{1}{2}[12 + 4] \times 10 \text{ J} = 80 \text{ J}$

Also, work done = Increase in kinetic energy

$\therefore$ $\frac{1}{2} \times 0.1 \times v^2 = 80$

or $v^2 = 1600$ or $v = 40 \text{ m s}^{-1}$

67. (*a*) Relative density $= \frac{\text{Weight in air}}{\text{Loss of weight in water}}$

$\therefore$ Loss of weight in water $= \frac{5 \times 10}{3}$ N

Weight in water $= \left(50 - \frac{50}{3}\right)\text{N} = \frac{100}{3}$ N

Work done $= \frac{100}{3} \text{ N} \times 5 m = \frac{500}{3}$ J

68. (*c*) $S = \frac{1}{2}at^2$

Effective weight $= m(g + a)$

Work done $= m(g + a) \times \frac{1}{2}at^2$

69. (*c*) $P = Fv$ and $F \propto v$

$\therefore$ $P \propto v^2$; $\frac{P_2}{P_1} = \frac{v_2^2}{v_1^2}$

or $\frac{P_2}{16} = \frac{3 \times 3}{2 \times 2} = \frac{9}{4}$

or $P_2 = 16 \times \frac{9}{4} \text{ hp} = 36 \text{ hp}$

70. (*b*) $v_1 = \sqrt{4^2 + 3^2} = \sqrt{25} = 5 \text{ m s}^{-1}$

$v_2 = 6 \text{ m s}^{-1}$

Work done = Increase in kinetic energy

$= \frac{1}{2} \times 2\,[6^2 - 5^2]$ J

$= (36 - 25)$ J $= 11$ J

71. (*c*) $h_1 = \frac{60}{100} \times 10$ m $= 6$ m

$h_2 = \frac{60}{100} \times 6$ m $= 3.6$ m.

72. (*c*) $v = u + at = 0 + \frac{1}{0.5}t = 2t$

Kinetic energy $= \frac{1}{2}mv^2 = \frac{1}{2} \times 0.5 \times 4t^2 = t^2$

Ratio of work done in first second, first two second and first three second is 1 : 4 : 9.

$\therefore$ Ratio of work done in the first, second and third second is 1 : 3 : 5.

73. (*a*) Weight of body = $V\rho g$; Upthrust = $V\sigma g$

Effective weight $= V(\rho - \sigma)g$

Work done $= V(\rho - \sigma)gh$

74. (c) Initial height of C.G. $= \frac{a}{2}$

Final height of C.G. $= \frac{b}{2}$

Work done $= mg\left[\frac{b}{2} - \frac{a}{2}\right] = mg\left(\frac{b-a}{2}\right)$

75. (c) Useful work $= \frac{75}{100} \times 12$ J $= 9$ J

Now, $\frac{1}{2} \times 1 \times v^2 = 9$ or $v = \sqrt{18}$ m s^{-1}

76. (c) Effective height through which CG moves, $h = \frac{l}{2}$

Work done $= mgh = \frac{mgl}{2}$

77. (*a*) Equating gain of kinetic energy with loss of gravitational potential energy, we get

$$\frac{1}{2}mv^2 = mgh$$

or $v = \sqrt{2gh}$

or $v = \sqrt{2 \times 9.8 \times 1}$ m s^{-1}

$= \sqrt{19.6}$ m s^{-1} $= 4.43$ m s^{-1}

78. (*b*) $\frac{1}{2}mv^2 = \frac{1}{2}k\xi^2$

or $v = \sqrt{\frac{k}{m}}\,\xi = \sqrt{\frac{1}{2}} \times 4$

$= \sqrt{8}$ m s^{-1} $= 2\sqrt{2}$ m s^{-1}

> ☞ *Note that the kinetic energy gets transformed into elastic potential energy.*

79. (*a*) $h = 500$ m, $\frac{dm}{dt} = 2000$ kg s^{-1}

Power output $= \frac{80}{100} \times \frac{dm}{dt}gh$

$= \frac{4}{5} \times 2000 \times 10 \times 500$ W

$= 8 \times 10^6$ W $= 8$ MW

80. (*b*) $20^2 - 0^2 = 2 \times a \times 100$

or $a = \frac{400}{200} = 2$ m s^{-2}

Accelerating force = 500 kg × 2 m s^{-2} = 1000 N

Force required to overcome resistance to motion

= 500 N

Total required force, F = 1500 N

Work done = 1500 N × 100 m

$= 1.5 \times 10^5$ J

81. (*c*) $v = 1080 \times \frac{5}{18}$ m s^{-1} $= 300$ m s^{-1}

Now, $0.01 \times 300 + 5 \times 0 = 5v$ or $v = \frac{3}{5}$ m s^{-1} $= 0.6$ m s^{-1}

82. (*a*) $\frac{1}{2}m\left(\frac{v}{2}\right)^2 - \frac{1}{2}mv^2 = f \times 30$

or $f = -\frac{1}{30} \times \frac{3}{4}\left[\frac{1}{2}mv^2\right]$ or $f = -\frac{mv^2}{80}$

Again, $0^2 - \frac{1}{2}m\left(\frac{v}{2}\right)^2 = -\frac{mv^2}{80} \times S$ or $S = 10$ cm

> ☞ **An interesting solution**
>
> **Step I.** *When v becomes* $\frac{v}{2}$, E_k *becomes* $\frac{E_k}{4}$.
>
> **Step II.** *In 30 cm, loss of kinetic energy is* $\frac{E_k}{4}$.
>
> **Step III.** *For losing* $\frac{E_k}{4}$, *required distance is 3 times less i.e.* $\frac{30}{3}$ *cm or 10 cm.*

83. (*a*) $E_p = mgh$ or $E_p \propto h$

Fraction of energy lost $= \frac{E_p - E_p'}{E_p'}$

$= \frac{h - h'}{h} = \frac{2 - 1.5}{2} = \frac{0.5}{2} = \frac{1}{4}$

84. (c) Decrease in energy $= \frac{1}{2} m \times 10 \times 10 - \frac{1}{2} m \times 9 \times 9$

$= 50m - 40.5m = 9.5m$

This is total decrease of energy.

Decrease in energy in upward journey

$= \frac{1}{2} \times 9.5 \text{ m} = 4.75 \text{ m}$

Energy of ball at highest point

$= 50 \text{ m} - 4.75 \text{ m} = 45.25 \text{ m}$

If h is the maximum height attained, then

$mgh = 45.25 \text{ m}$

or $h = \frac{45.25}{10} = 4.525 \text{ m}$

85. (d) $h = 0 \times 3 + \frac{1}{2} \times 10 \times 3 \times 3 = 45 \text{ m}$

Loss of P.E. = $3 \times 10 \times 45$ J = 1350 J

This will be the kinetic energy at 't' = 3 s

86. (c) $\frac{1}{2} mv^2 = fd$

Also, $\frac{1}{2} mv^2 = f\frac{d}{2} + \frac{1}{2} mv'^2$

or $\frac{1}{2} mv^2 = \frac{1}{4} mv^2 + \frac{1}{2} mv'^2$

or $\frac{1}{2} mv'^2 = \frac{1}{4} mv^2$

or $v' = \frac{v}{\sqrt{2}} = \frac{50\sqrt{2}}{\sqrt{2}} \text{ m s}^{-1} = 50 \text{ m s}^{-1}$

87. (b) Let us calculate the area under F-x graph with due regard for positive and negative signs.

Area = + 10 × 1 − 10 × 1 + 10 × 1 = 10 units

So, work done is 10 J

88. (c) $1400 \times 10 \times 10 + W = \frac{1}{2} \times 1400 \times 15 \times 15$

or $W = 700 \times 15 \times 15 - 1400 \times 10 \times 10$

or $W = 700(225 - 200)$ J

or $W = 700 \times 25$ J = 17.5 kJ

89. (c) Using energy conservation,

$\frac{1}{2} mv^2 = \frac{1}{2} m \times 2 \times 2 + mg \times 3$

or $\frac{v^2}{2} = 2 + 30 = 32$ or $v = \sqrt{64} \text{ m s}^{-1} = 8 \text{ m s}^{-1}$

90. (c) $\theta = 60°$

$W = \frac{mgl}{2}(1 - \cos\theta)$

$W = 0.1 \times 10 \times \frac{1}{2}(1 - \cos 60°)$

or $W = 0.25$ J

☞ *In this type of questions, consider the height through which C.G. is raised.*

91. (d) $\frac{4x}{(1+x)^2} = \frac{4 \times 1}{(1+1)^2} = \frac{4}{4} = 1$

So, there is 100% transfer of energy.

92. (c) $\frac{4x}{(1+x)^2} = \frac{4 \times 3}{(1+3)^2} = \frac{12}{16} = \frac{3}{4}$

Percentage transferred is $\frac{3}{4} \times 100$ *i.e.* 75

93. (c) $\frac{1}{2} k_1 x^2 + \frac{1}{2} k_2 x^2 = \frac{1}{2} mv^2$

or $v^2 = \frac{k_1 + k_2}{m} x^2$ or $v = \sqrt{\frac{k_1 + k_2}{m}}\, x$

94. (b) Efficiency $= \frac{15gh}{75gh} \times 100 = \frac{1}{5} \times 100 = 20\%$

95. (c) Specific gravity $= \frac{\text{Weight in air}}{\text{Loss of weight in water}}$

or Loss of weight in water $= \frac{\text{Weight in air}}{\text{Specific gravity}}$

$= \frac{4 \text{ kg wt}}{2} = 2 \text{ kg wt}$

Work done = 2 × 10 × 5 J = 100 J

96. (c) If input to the motor is 100 units, then output is 90 units.

Input to pump is 90 units.

Output of pump is $\frac{60}{100} \times 90$ units

or 54 units.

So, efficiency of the entire installation is 54%.

97. (b) $\frac{1}{2} kx^2 = \frac{1}{2} mv^2 - \frac{1}{2} m\left(\frac{v}{2}\right)^2$

or $\frac{1}{2} kx^2 = \frac{1}{2} mv^2 - \frac{1}{4}\left(\frac{1}{2} mv^2\right)$

or $\frac{1}{2} kx^2 = \frac{3}{4} \times \frac{1}{2} mv^2 = \frac{3mv^2}{8}$

or $k = \frac{3mv^2}{4x^2}$

98-99. $v = \sqrt{2g(H - h)}$

$t = \sqrt{\frac{2h}{g}}$

$$x = vt$$

$$x = \sqrt{2g(H-h)} \times \sqrt{\frac{2h}{g}}$$

$$x = 2\sqrt{h(H-h)}$$

For maximum x,

$$h = H - h \text{ or } 2h = H \text{ or } h = \frac{H}{2}$$

Maximum value of x is

$$2\sqrt{\frac{H}{2}\left(H - \frac{H}{2}\right)} \text{ or } 2\frac{H}{2} \text{ or } H$$

100. (*a*) Effective height through which man moves up = $l - h$

Gain in potential energy = $mg(l - h)$.

101-102. Work done by man = mgl

Gain in potential energy of man = $mgl - mgh$

If ΔE be the change in potential energy of balloon, then

$$mgl = mgl - mgh + \Delta E$$

or $\Delta E = mgh$

So, the potential energy increases by mgh.

103. (*d*) $P = \frac{mgh}{t}$

$$m = \frac{Pt}{gh} = \frac{2000 \times 60}{10 \times 10} \text{ kg} = 1200 \text{ kg}$$

104. (*b*) For the unloaded bus, $\frac{1}{2}mv^2 = f \times S$

For the loaded bus, $\frac{1}{2}\left[m + \frac{m}{2}\right]v^2 = f \times S'$

Dividing, $\frac{S'}{S} = \frac{3}{2}$ or $S' = 1.5$ S

> **Time-saving solution**
>
> $\frac{1}{2}mv^2 = f \times S$
>
> *In the given problem, $m \propto S$*
>
> *If m becomes $\frac{3m}{2}$, then S becomes $\frac{3S}{2}$ i.e. 1.5 S.*

105. (*d*) Using conservation of momentum,

$4v = 8 \times 6 = 48$ or $v = 12 \text{ m s}^{-1}$

Kinctic energy of the other mass is

$\frac{1}{2} \times 4 \times 12 \times 12$ J or 288 J

106. (*c*) Using conservation of momentum,

$$Mu = MV + mv \quad \text{...}(i)$$

Using Newton's law of impact,

$$u = v - V$$

or $Mu = Mv - MV$ ...(*ii*)

Adding (*i*) and (*ii*),

$$2Mv = (M + m)v$$

or $v = \frac{2uM}{M+m}$ or $v = \frac{2u}{1 + \frac{m}{M}}$

107. (*c*) The collision of clay with pan is inelastic. So, energy will not be conserved. But momentum will be conserved.

Applying law of conservation of momentum,

$$400V = 200\sqrt{2gh} \text{ or } V = \frac{\sqrt{2gh}}{2}$$

Kinetic energy of putty and pan

$$= \frac{1}{2} \times 400 \times \frac{2gh}{4} = 100gh$$

$$= 100 \times 30 \times g = 3000g$$

Loss of gravitational potential energy = $400gx$

Gain of elastic potential energy

$$= \frac{1}{2}\frac{200g}{10}[(10 + x)^2 - 10^2]$$

Equating energies,

$$10g(20 + x)(x) = 3000g + 400gx$$

or $20x + x^2 = 300 + 40x$

or $x^2 - 20x - 300 = 0$

or $x^2 - 30x + 10x - 300 = 0$

or $x(x - 30) + 10(x - 30) = 0$

or $(x - 30)(x + 10) = 0$

or $x = 30$ or -10

Rejecting – ve value, $x = 30$ cm

108. (*b*) $\frac{1}{2}mV^2 = \frac{1}{2} \times \frac{1}{2}\frac{m}{2}v^2$

or $V^2 = \frac{v^2}{4}$ or $V = \frac{v}{2}$ ...(*i*)

Again, $\frac{1}{2}m(V + 1)^2 = \frac{1}{2}\frac{m}{2}v^2$

or $(V + 1)^2 = \frac{v^2}{2}$

or $V + 1 = \frac{v}{\sqrt{2}}$ ...(*ii*)

(*ii*) – (*i*) gives $1 = \frac{v}{\sqrt{2}} - \frac{v}{2}$

$$2\sqrt{2} = 2v - \sqrt{2}v$$

or $v = \frac{2\sqrt{2}}{2 - \sqrt{2}} \times \frac{2 + \sqrt{2}}{2 + \sqrt{2}}$ or $v = \sqrt{2}(2 + \sqrt{2})$

or $v = 2\sqrt{2} + 2$

or $v = 2(\sqrt{2} + 1)$

Again, $V = \frac{v}{2} = \frac{2(\sqrt{2}+1)}{2} = \sqrt{2} + 1$

109. (*b*) $x = (t-3)^2$

$v = \frac{dx}{dt} = 2(t-3)$

acceleration $= \frac{d}{dt}(v) = \frac{d}{dt}[2(t-3)] = 2 \text{ m s}^{-2}$

110. (*d*) Force, $F = 2m$

Work done $= 2m \times (6-3)^2 = 18$ m joule.

111. (*a*) $V = 75 \text{ cm}^3 = 75 \times 10^{-6} \text{ m}^3$

$P = 100$ mm of Hg

$= 10$ cm of Hg $= 0.1$ m of Hg

$= 0.1 \times 13600 \times 10 \text{ N m}^{-2}$

$= 13600 \text{ N m}^{-2}$

Power $= \frac{PV}{60/72}$

$= \frac{13600 \times 75 \times 10^{-6} \times 72}{60} \text{ W} = 1.2 \text{ W}$

112. (*b*) $\eta = \frac{\text{Useful power}}{\text{Input power}} \times 100$

$= \frac{70 \times 10 \times 5}{1} \times \frac{1}{7000} \times 100 = 50\%$

113. (*b*) Let the spring be compressed through Δx

Net downward force $= mg - k\Delta x$

Equation of motion of ball is $ma = mg - k\Delta x$

So long as a is positive, the velocity of the ball would go on increasing.

When $a = 0$, the velocity would become maximum and constant. Suppose acceleration becomes zero when the spring is compressed through Δl.

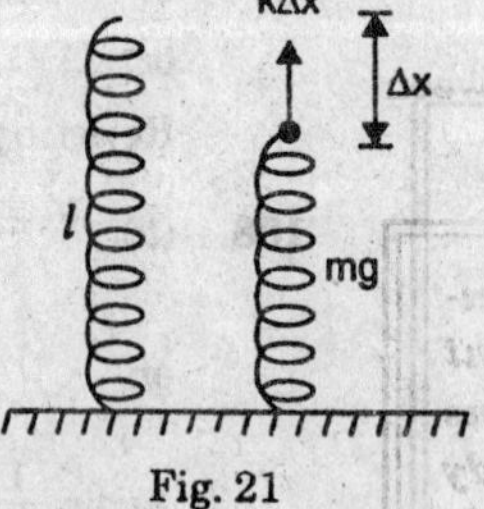

Fig. 21

$\therefore \quad 0 = mg - k\Delta l$

or $k\Delta l = mg$ or $\Delta l = \frac{mg}{k}$

Height above the table $= l - \Delta l = l - \frac{mg}{k}$

114. (*d*) Force on a particle moving on a circular path

$= \text{Centripetal force} = \frac{mv^2}{R}$

$= \left(\frac{1}{2}mv^2\right)\frac{2}{R} = \frac{2K}{R} = \frac{2aS}{R}$

115. (*a*) Work done in lifting water and drum

$= 60 \times 10 \times 20 \text{ J} = 12000 \text{ J}$

Total mass of ropes $= 40 \times 0.5 \text{ kg} = 20 \text{ kg}$

Work done in the case of ropes $= 20 \times 10 \times 10 = 2000$ J

Total work done $= 14000$ J

116. (*a*) $mu = 2mv \cos\theta$

But $\theta = 30°$

$\therefore \; u = 2v \cos 30° = \sqrt{3}v$

Using one ball for calculation,

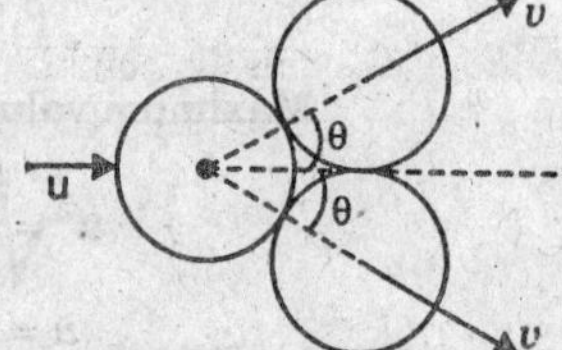

Fig. 22

$e = \frac{v - 0}{u \cos 30° - 0}$

$= \frac{v \times 2}{\sqrt{3}v \times \sqrt{3}} = \frac{2}{3}$

$\sin\theta = \frac{r}{2r}$

or $\sin\theta = \frac{1}{2}$

or $\theta = 30°$

Fig. 23

117. (*a*) $x = \frac{t^2}{2}$, $v = \frac{dx}{dt} = \frac{1}{2} \times 2t = t$

Kinetic energy at time $t = \frac{1}{2}mt^2$

Initial kinetic energy $= 0$

Final kinetic energy $= \frac{1}{2} \times 3 \times 3 \times 3 \text{ J} = 13.5 \text{ J}$

Work done = Increase in kinetic energy = 13.5 J

118. (*c*) $5 - 0 = 2a \times l \times 10^{-2}$ ...(*i*)

$4 - 0 = 2a(l-1) \times 10^{-2}$

Dividing, $\frac{5}{4} = \frac{l}{l-1}$ or $l = 5$ cm

From equation (*i*), $5 = 2a \times 5 \times 10^{-2}$

or $a = \frac{100}{2} \text{ m s}^{-2} = 50 \text{ m s}^{-2}$

$F = \frac{50}{1000} \times 50 \text{ N} = 2.5 \text{ N}$

119. (*d*) Applying conservation of momentum along Y-axis, we get

$mv \sin\theta_1 - mv \sin\theta_2 = 0$

or $\sin\theta_1 = \sin\theta_2$ or $\theta_1 = \theta_2$

Again, applying conservation of momentum along x-axis, we get

$mu = mv\cos\theta + mv\cos\theta$

or $u = 2v\cos\theta$ or $\cos\theta = \frac{u}{2v}$

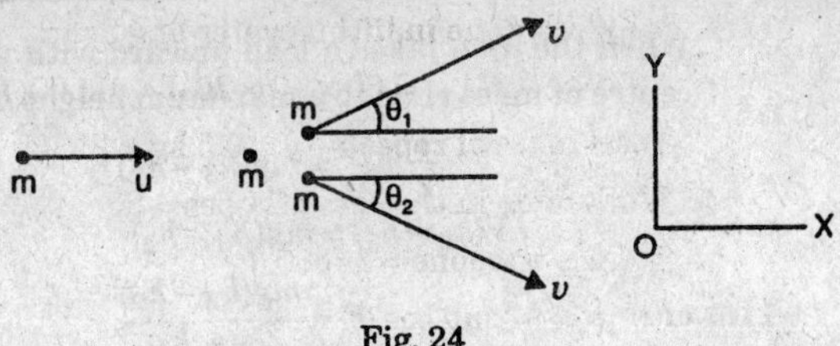

Fig. 24

Since collision is elastic therefore kinetic energy is conserved.

$$\therefore \quad \frac{1}{2}mu^2 = \frac{1}{2}mv^2 + \frac{1}{2}mv^2 \quad \text{or} \quad u^2 = 2v^2$$

$$\text{or} \quad u = \sqrt{2}v$$

$$\text{or} \quad \cos\theta = \frac{\sqrt{2}v}{2v} = \frac{1}{\sqrt{2}} \quad \text{or} \quad \theta = 45^\circ$$

So, the required angle is $2 \times 45^\circ$ or 90°

120. (*a*) For equilibrium at position P,

$$mg\sin\theta = \mu\, mg\cos\theta$$

$$\mu = \tan\theta$$

$$= \text{slope} = \frac{dy}{dx}$$

$$= \frac{d}{dx}\left(\frac{x^2}{2L}\right) = \frac{x}{L}$$

Fig. 25

$$\therefore \quad \frac{x}{L} = 0.8 \quad \text{or} \quad x = 0.8\,L = 0.8 \times 10 \text{ m} = 8 \text{ m}$$

121. (*c*) $E = \frac{1}{2}mg^2t^2$

$$\frac{E_1}{E_2} = \frac{\frac{1}{2}mg^2 \times 3^2}{\frac{1}{2}mg^2\,[6^2 - 3^2]} = \frac{9}{9 \times 3} = \frac{1}{3}$$

122. (*b*) It is clearly the case of elastic collision between massive body and a stationary light body.

> ☞ ***When a massive body suffers one-dimensional elastic collision with a stationary light body, the velocity of the massive body remains practically unchanged but the light body begins to move with a velocity which is double the velocity of the massive body.***

123. (*b*) Equating gain of elastic energy with loss of gravitational potential energy, we get

$$\frac{1}{2}k\xi^2 = mg\xi \quad \text{or} \quad \xi = \frac{2mg}{k}$$

124. (*d*) $E_k = \frac{p^2}{2m}$

p is increased by a factor of $\frac{3}{2}$. So, E_k is increased by a factor of $\frac{9}{4}$

Percentage increase of kinetic energy

$$= \frac{\frac{9}{4} - 1}{1} \times 100 = \frac{5}{4} \times 100 = 125$$

125. (*b*) $p = \sqrt{2mE_k}$

E_k is constant

$$\therefore \quad p \propto \sqrt{m}$$

$$\therefore \quad \frac{p_1}{p_2} = \sqrt{\frac{m_1}{m_2}} \quad \text{or} \quad \frac{p_1}{p_2} = \sqrt{\frac{1}{9}} = \frac{1}{3}$$

126. (*a*) Effective height through which C.G. is to be moved

$$= \frac{l}{12}$$

Effective mass to be lifted $= \frac{m}{6}$

$$\text{Work done} = \left(\frac{m}{6}\right)g\left(\frac{l}{12}\right) = \frac{mgl}{72}$$

127. (*b*) $p = \sqrt{2mE_k}$

E_k is increased by a factor of 4. p becomes double. So, percentage increase in momentum is 100%.

128. (*d*) Using conservation of momentum,

$$2mV = \sqrt{m^2v^2 + m^2v^2}$$

$$\text{or} \quad 2mV = \sqrt{2m^2v^2}$$

$$\text{or} \quad V = \frac{\sqrt{2}mv}{2m} = \frac{v}{\sqrt{2}}$$

Energy released in explosion

$$= 2 \times \frac{1}{2}mv^2 + \frac{1}{2} \times 2m \times \frac{v^2}{2} = \frac{3}{2}mv^2$$

129. (*c*) $p = \sqrt{2mE_k}$

$$\text{or} \quad p \propto \sqrt{m} \qquad [\because E_k \text{ is given to be constant.}]$$

$$\therefore \quad \frac{p_1}{p_2} = \sqrt{\frac{m_1}{m_2}} = \sqrt{\frac{1}{4}} = \frac{1}{2}$$

130. (*d*) Applying conservation of momentum,

$$m \times v_0 + m \times 0 = (2m)v \quad \text{or} \quad v = \frac{v_0}{2}$$

$$\text{Now,} \quad (2m)gh = \frac{1}{2}(2m)\left(\frac{v_0}{2}\right)^2 \quad \text{or} \quad h = \frac{v_0^2}{8g}$$

131. (*c*) $a = \frac{1-0}{15} \text{ m s}^{-2} = \frac{1}{15} \text{ m s}^{-2}$

Also, $\frac{1}{2} \times 1 \times v^2 = \frac{2}{9}$ or $v = \frac{2}{3} \text{ m s}^{-1}$

Now, $\frac{2}{3} = 0 + \frac{1}{15} \times t$ or $t = 10$ s

132. (*c*) Height of C.G. of mass $m_1 = \frac{a}{2}$

Height of C.G. of mass $m_2 = a + \frac{b}{2}$

$\therefore$ Gravitational potential energy of system

$$= m_1 g \frac{a}{2} + m_2 g \left(a + \frac{b}{2} \right)$$

$$= \left[\frac{m_1}{2} + m_2 \right] ga + m_2 g \frac{b}{a}$$

$$= \left[\left(\frac{m_1}{2} + m_2 \right) a + m_2 \frac{b}{a} \right] g$$

133. (*b*) Velocity on hitting the surface

$$= \sqrt{2 \times 9.8 \times 4.9} \text{ m s}^{-1} = 9.8 \text{ m s}^{-1}$$

Velocity after first bounce, $v = \frac{3}{4} \times 9.8$ m s^{-1}

Time taken from first bounce to the second bounce

$$= \frac{2v}{g} = 2 \times \frac{3}{4} \times 9.8 \times \frac{1}{9.8} \text{ s} = \frac{3}{2} \text{ s} = 1.5 \text{ s}$$

134. (*a*) Applying conservation of momentum,

$$1 \times 12 - 2 \times 24 = 1 \times v_1 + 2 \times v_2$$

or $v_1 + 2v_2 = -36$ ...(*i*)

Now, $\frac{v_2 - v_1}{12 - (-24)} = \frac{2}{3}$ or $v_2 - v_1 = 24$ ...(*ii*)

Adding (*i*) and (*ii*),

$3v_2 = -12$ or $v_2 = -4$ m s^{-1}

From equation (*ii*),

$-4 - v_1 = 24$ or $v_1 = -28$ m s^{-1}

135-136. These questions illustrate a motion where work is done essentially by internal forces, muscles etc. The internal energy of the man is converted into kinetic energy.

In order to give initial acceleration to the man, the upward force N must be greater than the weight mg of the man.

$$N - mg = ma = F \text{ (say)}$$

where a is the initial acceleration imparted to the man when his centre of mass rises from height h_1 to height h_2.

The force N acts at the point of contact between the feet of man and the ground. During the first phase, when the man has not yet left the ground, the point of contact does not move. So, no work is done by the force of reaction.

Before the man leaves the ground, the net work done is $F(h_2 - h_1)$. This work is done by muscles etc. of the body. The internal forces convert internal energy into kinetic energy $\frac{1}{2} mv^2$. $\therefore$ $F(h_2 - h_1) = \frac{1}{2} mv^2$

When the man takes a leap upward with velocity v, the centre of mass rises to a maximum height h_3.

$\therefore$ $\frac{1}{2} mv^2 = mg(h_3 - h_2)$

$\therefore$ $F(h_2 - h_1) = mg(h_3 - h_2)$

or $F = \frac{mg(h_3 - h_2)}{h_2 - h_1}$

Also, $v = \sqrt{2g(h_3 - h_2)}$

Substituting values,

$$F = \frac{55 \times 10(120 - 90)}{90 - 40} = \frac{55 \times 10 \times 30}{50} \text{ N} = 330 \text{ N}$$

Again, $v = \sqrt{\frac{2 \times 10 \times 30}{100}}$ m s^{-1} or $v = 2.45$ m s^{-1}

137. (*a*) $dW = F dl$

$W = \int_0^l F\, dl$ | $Y = \frac{FL}{dl}$

or $W = \int_0^l \frac{Yal}{L} dl$ | or $F = \frac{Yal}{L}$

or $W = \frac{Ya}{L} \int_0^l l\, dl$ or $W = \frac{Ya}{L} \left(\frac{l^2}{2} \right)$

or $W = \frac{1}{2} \frac{Yal}{L} l = \frac{1}{2} Fl$

138. (*a*) In the first case, $mg = kl$

In the second case, loss of P.E. = work done in stretching the spring through l'.

$$mgl' = \frac{1}{2} kl'^2 \text{ or } l' = \frac{2mg}{k} = \frac{2kl}{k} = 2l.$$

139. (*b*) Work done against gravity

$$= 50 \times 10 \times 30 \text{ J} = 15 \text{ kJ}$$

Work done against friction

$$= 0.4 \times 50 \times 10 \times \frac{4}{5} \times 50 \text{ J} = 8 \text{ kJ}$$

Total work done $= 15 \text{ kJ} + 8 \text{ kJ} = 23$ kJ

140. (*d*) $T \sin \theta = F$ and $T \cos \theta = W$

Dividing, $\frac{F}{W} = \tan \theta$ or $F = W \tan \theta$

141. (*a*) The instantaneous displacement all the time is normal to T. So, no work is done by or against T.

142. (*a*) Since F is always kept along the horizontal and its magnitude so adjusted that no kinetic energy is generated therefore the work is equal to increase of potential energy, which is Wh.

143. (*a*) Collision is elastic. So, the bob will acquire a velocity v_0 after the collision.

$\therefore \quad mgh = \frac{1}{2} m v_0^2$ or $h = \frac{v_0^2}{2g}$

144. (*a*) Loss of kinetic energy is equal to negative work done by resistance force *f*.

$$\frac{\frac{1}{2} m\,[800^2 - 600^2]}{\frac{1}{2} m\,[600^2 - 400^2]} = \frac{f d_1}{f d_2}$$

or $\quad \frac{64-36}{36-16} = \frac{d_1}{d_2}$ or $\frac{d_1}{d_2} = \frac{28}{20} = \frac{7}{5}$

145. (*b*) $F \propto x$ or $F = kx$

$$W = \int_0^x F\,dx = \int_0^x kx\,dx = \frac{kx^2}{2} \text{ or } W \propto x^2$$

146. (*a*) Loss of kinetic energy $= \frac{1}{2} \times 2m \times v^2 = mv^2$

Gain of gravitational potential energy = $(2m)gl$

Equating, $mv^2 = 2mgl$ or $v = \sqrt{2gl}$

147. (*c*) Using $v = r\omega$, we find that the upper mass m has velocity $\frac{v}{2}$

Using energy conservation,

$$\frac{1}{2} mv^2 + \frac{1}{2} m \left(\frac{v}{2}\right)^2 = mgl + \frac{mgl}{2}$$

or $\quad v^2 + \frac{v^2}{4} = 2gl + gl = 3gl$

or $\quad 5v^2 = 12gl$

or $\quad v = \sqrt{\frac{12gl}{5}}$

Fig. 26

148. (*b*) Applying conservation of momentum,

$$2v + m \times 0 = 2\left(\frac{v}{4}\right) + mV$$

or $\quad mV = 2v - \frac{v}{2} = \frac{3v}{2} \quad \quad ...(i)$

Since collision is elastic therefore kinetic energy is conserved

$$\frac{1}{2} \times 2 \times v^2 + \frac{1}{2} m \times 0 \times 0 = \frac{1}{2} \times 2\left(\frac{v}{4}\right)^2 + \frac{1}{2} mV^2$$

or $\quad v^2 = \frac{v^2}{16} + \frac{1}{2} mV^2 \quad \quad ...(ii)$

Relative velocity before impact $= v - 0 = v$

Relative velocity after impact $= \frac{v}{4} - V$

Now, $v = -\left(\frac{v}{4} - V\right)$ or $v + \frac{v}{4} = V$ or $V = \frac{5v}{4}$

From eqn. (*ii*),

$$v^2 = \frac{v^2}{16} + \frac{1}{2} m \frac{25v^2}{16}$$

or $\quad 1 = \frac{1}{16} + \frac{25m}{32}$ or $\frac{15}{16} = \frac{25m}{32}$

or $\quad m = \frac{30}{25} = \frac{6}{5} = 1.2$ kg

Let us cross-check with the help of equation (*i*)

$$\frac{6}{5} \times \frac{5v}{4} = \frac{3}{2} v \text{ or } \frac{3v}{2} = \frac{3v}{2}$$

Clearly, equation is satisfied.

149. (*c*) $\frac{4x}{(1+x)^2} \times 100 = \frac{4 \times 14}{(1+14)^2} \times 100 = \frac{5600}{225} = 24.9 \approx 25$

150. (*a*) $a = \frac{4-0}{40}$ m s^{-2} $= \frac{1}{10}$ m s^{-2}

Velocity at the end of 8 second $= 0 + \frac{1}{10} \times 8 = 0.8$ m s^{-1}

Required power $= (ma)v = 8 \times \frac{1}{10} \times 0.8$ W $= 0.64$ W

151. (*d*) $\cos 37° = \frac{h}{l}$ or $l = \frac{h}{\cos 37°}$

or $\quad l = 1.25h$

Extension $= l - h$

$= 1.25h - h = 0.25h$

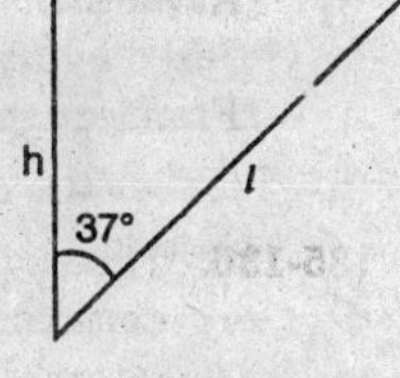

Fig. 27

Equating kinetic energy with potential energy,

$$\frac{1}{2} mv^2 = \frac{1}{2} k (0.25h)^2$$

or $\quad v^2 = \frac{k}{m} \times \frac{h^2}{16}$ or $v = \frac{h}{4}\sqrt{\frac{k}{m}}$

6 ROTATIONAL MOTION AND MOMENT OF INERTIA

Important Terms, Facts and Formulae

BASIC CONCEPTS

1. The position of the centre of mass depends upon the shape of the body. For a given shape, it depends upon the distribution of mass in the body. It is closer to the heavier part of the body.
2. If a body is symmetrical and has uniform distribution of mass, then the centre of mass coincides with the centre of symmetry or geometrical centre.
3. In a uniform gravitational field, the centre of mass and the centre of gravity coincide.
4. In translatory motion, the position of centre of mass changes. In purely rotational motion, the position of centre of mass remains unchanged.
5. The position of centre of mass does not depend upon the choice of co-ordinate system. As an example, the centre of mass of a solid sphere is at the geometrical centre, whatever may be the choice of co-ordinate system.
6. There may or may not be any mass at the centre of mass. As an illustration, the centre of mass of a hollow sphere is at its centre where there is no mass.
7. The moment of inertia of a rigid body is the opposition offered by the body to change in its state of rotational motion (or rest).
8. Farther the distribution of mass from the axis, more is the moment of inertia.
9. Rotational analogue of mass is moment of inertia.
10. Rotational analogues of linear displacement, linear velocity, linear acceleration and force are angular displacement, angular velocity, angular acceleration and torque respectively.
11. The centre of mass of two-particle system lies in between them on the line joining the two particles.
12. For applying the laws of mechanics, the whole mass of the body is supposed to be concentrated at the centre of mass.
13. In the absence of the external force, the velocity of the centre of mass remains constant.
14. In non-uniform circular motion, both radial and tangential accelerations are present.
15. In uniform circular motion, the directions of velocity, acceleration and tension change.
16. When a horizontal rod is rotated about a vertical axis passing through one of its ends, tension increases from the free end to the pivoted end.

IMPORTANT POINTS TO REMEMBER

1. Consider a solid cylinder of length l and radius r. If η is the modulus of rigidity of the material, then the torque required to twist, the cylinder through an angle θ is : $\tau = C\theta$, where C is twisting couple per unit twist of the cylinder. C is also called torsional rigidity of the material.

$$C = \frac{\pi\eta r^4}{2l}$$

2. For a particle in non-uniform circular motion.

$$\vec{a} = \vec{\alpha} \times \vec{r} + \vec{\omega} \times \vec{v}$$

where letters have usual meanings.

3. $\vec{\alpha} \times \vec{r}$ is the tangential component of $\vec{a}$.

$$\therefore \quad \vec{a_t} = \vec{\alpha} \times \vec{r}$$

4. $\vec{\omega} \times \vec{v}$ is the radial component of $\vec{a}$. It is also known as centripetal component of $\vec{a}$.

$$\vec{a_c} = \vec{\omega} \times \vec{v}$$

5. In the case of uniform circular motion,

$$a_t = 0 \text{ and } a_c \neq 0$$

6. When dealing with a system of discrete particles, the co-ordinates of the centre of mass are given by

$$X_{cm} = \frac{1}{M} \Sigma m_i x_i$$

$$Y_{cm} = \frac{1}{M} \Sigma m_i y_i$$

$$Z_{cm} = \frac{1}{M} \Sigma m_i z_i$$

In general, $\vec{R}_{cm} = \frac{1}{M} \Sigma m_i \vec{r_i}$

7. In the case of continuous distribution of mass, the centre of mass is given by

$$X_{cm} = \frac{1}{M}\int x\,dm$$

$$Y_{cm} = \frac{1}{M}\int y\,dm$$

$$Z_{cm} = \frac{1}{M}\int z\,dm$$

$$\vec{R}_{cm} = \frac{1}{M}\int \vec{r}\,dm$$

8. The velocity and acceleration of the centre of mass of a system of particles are given by

$$\vec{V}_{cm} = \frac{1}{M}\Sigma m_i\vec{v_i} \quad \text{and} \quad \vec{a}_{cm} = \frac{1}{M}\Sigma m_i\vec{a_i}$$

9. Moment of inertia, $I = \Sigma m_i r_i^{\,2}$

10. In the case of continuous distribution of mass, moment of inertia, $I = \int r^2\,dm$.

QUESTION BANK

Level I

Choose the correct answer from the four alternatives given in each of the following questions :

Circular Motion

1. A particle moves in a circle of radius 0.25 m at two revolutions per second. The acceleration of the particle in metre per second^2 is

(*a*) π^2 (*b*) $8\pi^2$
(*c*) $4\pi^2$ (*d*) $2\pi^2$.

2. The maximum frictional force between the tyres of a car and the road is 0.5 *mg*. The car negotiates a curve of radius 10 metre. The velocity is

(*a*) 10 m s^{-1} (*b*) 7 m s^{-1}
(*c*) 4.9 m s^{-1} (*d*) 14.2 m s^{-1}.

3. A stone of mass 0.5 kg tied with a string of length 1 metre is moving in a circular path with a speed of 4 m s^{-1}. The tension acting on the string in newton is

(*a*) 2 (*b*) 8
(*c*) 0.2 (*d*) 0.8.

4. What is the maximum speed of a car taking a turn of radius 30 m on a level road if the coefficient of friction between the tyres and the road is 0.4 ? (g = 9.8 m s^{-2})

(*a*) 5.42 m s^{-1} (*b*) 6.27 m s^{-1}
(*c*) 10.84 m s^{-1} (*d*) 21.68 m s^{-1}.

5. A body slides without friction from a height H along the track shown in Fig. 1 so that it loops the loop in the vertical plane. If the radius of the circular loop is R, what should be the minimum value of H in terms of R so that the frictionless body is just able to loop the loop ?

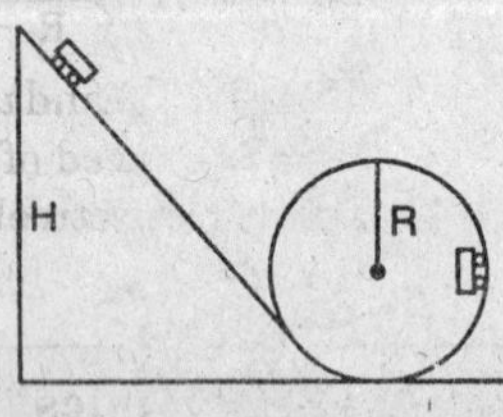

Fig. 1

(*a*) H = 2 R (*b*) $H = \frac{5R}{2}$
(*c*) H = 5 R (*d*) H = 4 R.

6. When the road is dry and the coefficient of friction is μ, the maximum speed of a car in a circular path is 10 m s^{-1}. If the road becomes wet and $\mu' = \frac{\mu}{2}$, what is the maximum speed permitted ?

(*a*) 5 m s^{-1} (*b*) 10 m s^{-1}
(*c*) $10\sqrt{2}$ m s^{-1} (*d*) $5\sqrt{2}$ m s^{-1}.

7. Assume that the earth goes round the sun in a circular orbit with a constant speed of 30 km s^{-1}.
Which of the following is correct ?

(*a*) The average velocity of the earth from 1st Jan., 90 to 30th June, 90 is zero.
(*b*) The average acceleration during the above period is 60 km s^{-2}.
(*c*) The average speed from 1st Jan., 90 to 31st Dec., 90 is zero.
(*d*) The instantaneous acceleration of the earth points towards the sun.

8. The length of the second hand of a watch is 1 cm. The velocity vector of the tip of the second hand in cm per second is

(*a*) 2π (*b*) $\frac{2\pi}{60}$
(*c*) $\frac{2\pi}{12 \times 60}$ (*d*) $\frac{2\pi}{24 \times 60}$.

9. A mass of 2 kg is whirled in a horizontal circle by means of a string at an initial speed of 5 r.p.m. Keeping the radius constant, the tension in the string is doubled. The new speed is nearly

(*a*) 14 r.p.m. (*b*) 10 r.p.m.
(*c*) 20 r.p.m. (*d*) 7 r.p.m.

10. A fly wheel rotates about a fixed axis and slows down from 300 r.p.m. to 100 r.p.m. in 2 minutes. Then its angular retardation in radian/minute^2 is

(*a*) $\frac{100}{\pi}$ (*b*) 100
(*c*) 100π (*d*) 200π.

11. Three identical cars A, B and C are moving at the same speed on three bridges. The car A goes on a plane bridge, B on a bridge convex upward and C goes on a bridge concave upward. Let F_A, F_B and F_C be the normal forces exerted by the cars on the bridges when they are at the middle of bridges. Then

(*a*) F_A is maximum of the three forces
(*b*) F_B is maximum of the three forces
(*c*) F_C is maximum of the three forces
(*d*) $F_A = F_B = F_C$.

12. A particle is projected so as to just move along a vertical circle. The ratio of the tensions in the string when the particle is at the lowest and highest point on the circle is

(*a*) 1 (*b*) finite but large

(*c*) zero (*d*) infinite.

13. A stone tied to the end of 20 cm long string is whirled in a horizontal circle. If the centripetal acceleration is 9.8 m s^{-2}, its angular speed in radian/s is

(*a*) 22/7 (*b*) 7

(*c*) 14 (*d*) 20.

14. A car moves at a constant speed on a road as shown in Fig. 2. The normal force by the road on the car is N_A and N_B when it is at the points A and B respectively.

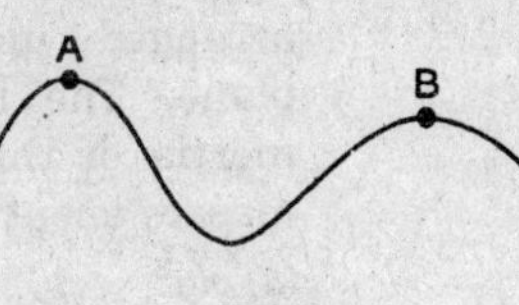

Fig. 2

(*a*) $N_A = N_B$ (*b*) $N_A > N_B$

(*c*) $N_A < N_B$

(*d*) insufficient information to decide the relation of N_A and N_B.

15. A simple pendulum is vibrating with an angular amplitude of 90° as shown in Fig. 3. For what value of θ is the acceleration directed

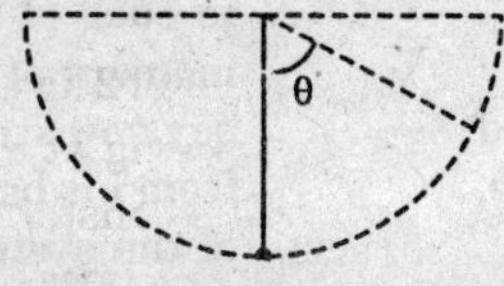

Fig. 3

(*i*) vertically upwards

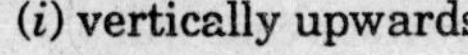

(*ii*) horizontally

(*iii*) vertically downwards.

(*a*) 0°, $\cos^{-1}\frac{1}{\sqrt{3}}$, 90° (*b*) $\cos^{-1}\frac{1}{\sqrt{3}}$, 0°, 90°

(*c*) 90°, $\cos^{-1}\frac{1}{\sqrt{3}}$, 0° (*d*) $\cos^{-1}\frac{1}{\sqrt{3}}$, 90°, 0°.

16. A flywheel gains a speed of 540 r.p.m. in 6 second. Its angular acceleration is

(*a*) 3π rad s^{-2} (*b*) 6π rad s^{-2}

(*c*) 9π rad s^{-2} (*d*) 12π rad s^{-2}.

17. A coin placed on a rotating turntable just slips if it is placed at a distance of 4 cm from the centre. If the angular velocity of the turntable is doubled, it will just slip at a distance of

(*a*) 1 cm (*b*) 2 cm

(*c*) 4 cm (*d*) 8 cm.

18. A curved road of radius *r* is perfectly banked for a speed *v*. When a car of weight W goes round the curve with speed *v* (friction neglected), then the normal reaction of the road is N. Then

(*a*) The centripetal force is the resultant of W and N

(*b*) Normal reaction is W

(*c*) The centripetal force is zero

(*d*) The centripetal force is W.

19. I revolve a stone of mass *m* tied to the end of a string in a vertical circle of radius R. The net forces at the lowest and highest points of the circle directed vertically downwards are

	Lowest point	*Highest point*
(*a*)	$mg - T_1$	$mg + T_2$
(*b*)	$mg + T_1$	$mg - T_2$
(*c*)	$mg + T_1 - \frac{mv_1^2}{R}$	$mg - T_2 + \frac{mv_2^2}{R}$
(*d*)	$mg - T_1 - \frac{m_1v_1^2}{R}$	$mg + T + \frac{mv_2^2}{R}$

Here T_1, T_2 (and v_1, v_2) denote the tension in the string (and the speed of the stone) at the lowest and highest points respectively.

Centre of Mass and Rotational Mechanics

20. Ring of mass 6 kg and radius 40 cm is revolving at the rate of 300 revolutions per minute. Its moment of inertia will be

(*a*) 0.92 kg m^2 (*b*) 0.96 kg m^2

(*c*) 2.4 kg m^2 (*d*) 2.98 kg m^2.

21. In Q. 20, the kinetic energy of rotation of the ring is

(*a*) $48\pi^2$ joule (*b*) 48 joule

(*c*) 48π joule (*d*) $\frac{48}{\pi}$ joule.

22. Two discs have same mass and thickness. Their materials are of densities ρ_1 and ρ_2. The ratio of their moment of inertia about central axis will be

(*a*) $\rho_1 : \rho_2$ (*b*) $\rho_1\rho_2 : 1$

(*c*) $1 : \rho_1\rho_2$ (*d*) $\rho_2 : \rho_1$.

23. Let I_A and I_B be moments of inertia of a body about two axes A and B respectively. The axis A passes through the centre of mass of the body but B does not

(*a*) $I_A < I_B$ (*b*) If $I_A < I_B$, the axes are parallel

(*c*) If the axes are parallel, $I_A < I_B$

(*d*) If the axes are not parallel, then $I_A \geq I_B$.

24. Fig. 4 shows a rectangular wire ABCD. About which axis, the moment of inertia of the wire will be minimum ? (Given : BC = 2AB)

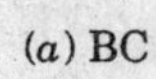

(*a*) BC (*b*) BD

(*c*) HF (*d*) EG.

Fig. 4

25. A disc of radius R is rotating with angular velocity ω about an axis. The moment of inertia of the disc about that axis is I. If a heavy point of mass *m* is gently attached to the rim of the disc, then the angular velocity of the system will be

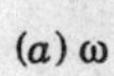

(*a*) ω (*b*) $\frac{I\omega}{mR^2}$

(*c*) $\frac{I\omega}{R^2}$ (*d*) $\frac{I\omega}{I + mR^2}$.

26. The moment of inertia of a 500 g cylinder of radius 10 cm about its natural axis is

(*a*) 0.5×10^{-3} kg m^2 (*b*) 1.5×10^{-3} kg m^2

(*c*) 2×10^{-3} kg m^2 (*d*) 2.5×10^{-3} kg m^2.

27. The angular momentum changes from 2 units to 6 units in 4 s. The torque is

(*a*) 1 unit (*b*) $\frac{1}{2}$ unit

(*c*) $\frac{3}{2}$ unit (*d*) 4 unit.

28. A ring of mass 10 kg and diameter 0.4 m is rotated about its axis. If it makes 2100 revolutions per minute, then its angular momentum will be

(*a*) 44 kg m^2 s^{-1} (*b*) 88 kg m^2 s^{-1}

(*c*) 4.4 kg m^2 s^{-1} (*d*) 0.4 kg m^2 s^{-1}.

29. If the Earth shrinks to half of the present radius, without any change in mass, then the duration of day and night becomes

(*a*) 24 hours (*b*) 12 hours

(*c*) 6 hours (*d*) 3 hours.

30. How is rotational kinetic energy of a rigid body related to angular momentum ?

(*a*) $\frac{L^2}{I} = E_k$ (*b*) $L = \sqrt{2IE_k}$

(*c*) $L = E_k$ (*d*) $L = \frac{E_k}{2}$.

31. A metre stick is held vertically with one end on the floor and is then allowed to fall. The speed of the other end when it hits the floor assuming that the end at floor does not slip (g = 9.8 m s^{-2})

(*a*) 3.2 m s^{-1} (*b*) 5.4 m s^{-1}

(*c*) 7.6 m s^{-1} (*d*) 9.2 m s^{-1}.

32. A wheel whose moment of inertia is 12 kg m^2 has an initial angular velocity of 40 rad/s. A constant torque of 20 Nm acts on the wheel. The time in which the wheel is accelerated to 100 rad/s is

(*a*) 72 second (*b*) 16 second

(*c*) 8 second (*d*) 36 second.

33. A constant couple of 400 Nm turns a wheel of moment of inertia 100 kgm^2 about an axis through its centre. The angular velocity gained in 4 second is

(*a*) 12 rad s^{-1} (*b*) 16 rad s^{-1}

(*c*) 20 rad s^{-1} (*d*) 24 rad s^{-1}.

34. Four point masses are arranged in the X-Y plane. The moment of inertia of this array of masses about Y-axis is

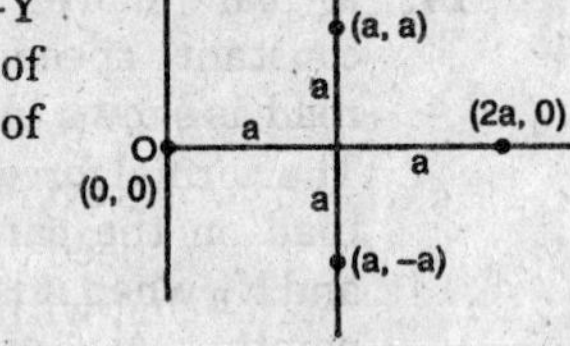

Fig. 5

(*a*) ma^2

(*b*) $2ma^2$

(*c*) $4ma^2$

(*d*) $6ma^2$.

35. A boy stands over the centre of a horizontal platform which is rotating freely with a speed of 2 revolutions/s about a vertical axis through the centre of the platform and straight up through the boy. He holds 2 kg masses in each of his hands close to his body. The combined moment of inertia of the system is 1 kg × metre^2. The boy now stretches his arms so as to hold the masses far from his body. In this situation, the moment of inertia of the system increases to 2 kg × metre^2. The kinetic energy of the system in the latter case as compared with that in the previous case will

(*a*) Remain unchanged (*b*) Decrease

(*c*) Increase (*d*) Remain uncertain.

36. The masses of two uniform discs are in the ratio 1 : 2 and their diameters in the ratio 2 : 1. The ratio of their moments of inertia about the axes passing through their respective centres and perpendicular to their planes is

(*a*) 1 : 1 (*b*) 1 : 2

(*c*) 2 : 1 (*d*) 1 : 4.

37. The root mean square distance of 5 identical particles of masses, 1 g each, and distributed along OX is 6 cm (from O). The total moment of inertia about OY is

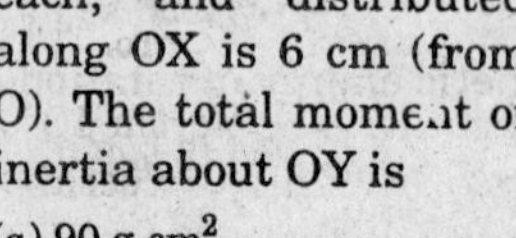

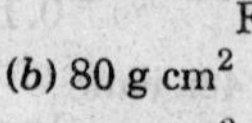

Fig. 6

(*a*) 90 g cm^2 (*b*) 80 g cm^2

(*c*) 180 g cm^2 (*d*) 36 g cm^2.

38. Which of the following is conserved when the torque acting on a system is zero ?

(*a*) K.E. (*b*) Angular momentum
(*c*) Angular K.E. (*d*) Linear momentum.

39. Two solid spheres are made of the same material. The ratio of their diameters is 1 : 2. The ratio of their moments of inertia about their respective diameters is

(*a*) 1 : 1 (*b*) 1 : 4
(*c*) 1 : 16 (*d*) 1 : 32.

40. The radius of gyration of a disc about its central axis is

(*a*) $\frac{R}{\sqrt{2}}$ (*b*) $\sqrt{2}R$
(*c*) $\frac{R}{2}$ (*d*) 2R.

41. A disc rolls over a horizontal floor without slipping with a linear speed of 5 cm s^{-1}. The linear speed of a particle on its rim, with respect to the floor, when it is in its highest position, is

(*a*) 10 cm s^{-1} (*b*) 5 cm s^{-1}
(*c*) 2.5 cm s^{-1} (*d*) 0.

42. A mass *m* is moving with a constant velocity parallel to the *x*-axis. Its angular momentum w.r.t. the origin

(*a*) remains constant (*b*) goes on increasing
(*c*) goes on decreasing (*d*) is zero.

43. A tangential force F acts at the rim of a ring of radius R and causes the ring to turn through an angle θ. The work done by the force will be

(*a*) $\frac{FR}{\theta}$ (*b*) FR θ
(*c*) $FR - \frac{1}{\theta}$ (*d*) FR – θ.

44. Radius of gyration of a body about an axis at a distance of 6 cm from its centre of mass is 10 cm. The radius of gyration about a parallel axis through its centre of mass is

(*a*) 2 cm (*b*) 4 cm
(*c*) 8 cm (*d*) 12 cm.

45. A star of mass M_1 (twice the solar mass M_s) and radius R_1 rotates about its axis with an angular speed of ω_1. What is the angular speed of the star when it collapses (due to inward gravitational forces) to a radius of R_2 ?

(*a*) $M_1 \div M_s$ (*b*) $M_s \div M_1$
(*c*) $R_1^2\omega_1 \div R_2^2$ (*d*) zero.

46. A spacecraft of rotational inertia 10^5 kg m^2 is having a spin of 0.1 rad s^{-1}. The spacecraft is free from all external fields. Inside the spacecraft, there is a rigidly mounted flywheel of moment of inertia 1000 kg m^2. The spin motion of the spacecraft can be stopped by changing the rotational rate of the flywheel by

(*a*) 1 rad s^{-1} (*b*) 10 rad s^{-1}
(*c*) 100 rad s^{-1} (*d*) 1000 rad s^{-1}.

47. A fan of moment of inertia 0.6 kg m^2 is turned upto a working speed of 0.5 r.p.s. The angular momentum of the fan is

(*a*) 0.6π kg m^2 s^{-1} (*b*) 6 kg m^2 s^{-1}
(*c*) 3 kg m^2 s^{-1} (*d*) $\frac{\pi}{6}$ kg m^2 s^{-1}.

48. A rigid body of moment of inertia I has an angular acceleration α. If the power supplied to the body is P, its instantaneous angular velocity is

(*a*) PIα (*b*) P/Iα
(*c*) PI/α (*d*) Pα/I.

49. Three children are sitting on a see-saw in such a way that it balances. A 20 kg and a 30 kg boy are on opposite sides at a distance of 2 m from the pivot. If the third boy jumps off, thereby destroying balance, then the initial angular acceleration of the board is : (Neglect weight of board)

(*a*) 0.098 rad s^{-2} (*b*) 0.98 rad s^{-2}
(*c*) 9.8 rad s^{-2} (*d*) 98 rad s^{-2}.

50. A triangular plate of uniform thickness and density is made to rotate about an axis perpendicular to the plane of the paper (*a*) passing through A, (*b*) passing through B, by the application of the same force F, at C (mid-point of AB) as shown. Now,

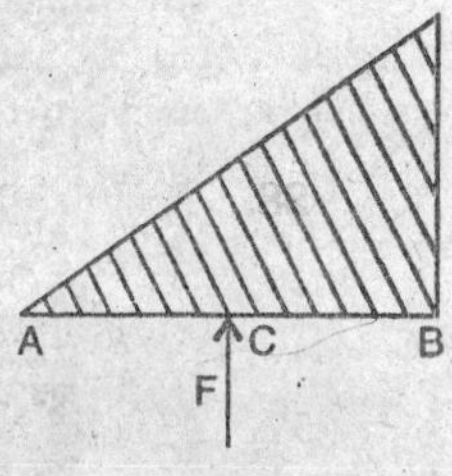

Fig. 7

(*a*) angular acceleration in both the cases is the same
(*b*) angular acceleration for case (*a*) is larger
(*c*) angular acceleration for case (*b*) is larger
(*d*) there would be no angular acceleration for case (*a*).

51. A circular disc rotates in a vertical plane about a fixed horizontal axis which passes through a point X on the circumference of the disc. When the centre of the disc moves with speed *v*, the speed of the opposite end of the diameter through X is

(*a*) 2*v* (*b*) $\sqrt{2}v$
(*c*) *v* (*d*) $\frac{v}{2}$.

52. The moment of inertia of a thin uniform rod about an axis passing through its centre and perpendicular to its length is I_0. What is the M.I. about an axis through one end and perpendicular to the rod ?

(a) $1/2\ I_0$ (b) $3I_0$

(c) $5I_0$ (d) $4I_C$.

53. A hollow phere of mass m starting from rest rolls down, without slipping, on an inclined plane of inclination θ. What is the total energy of the sphere after 10^{-5} s if velocity after 10^{-5} s is v ?

(a) $\frac{5}{6}mv^2$ (b) $\frac{2}{3}mv^2$

(c) $10^{-5}\ mv^2$ (d) $mv^2/10^{-5}$.

54. A particle of mass m and radius of gyration K is rotating with an angular acceleration α. The torque acting on the particle is

(a) $\frac{1}{2}mK^2\alpha$ (b) $mK^2\alpha$

(c) mK^2/α (d) $\frac{1}{4}mK^2\alpha^2$.

55. A thin hollow cylinder open at both ends slides without rotating and then rolls without slipping with the same speed. The ratio of the kinetic energies in the two cases is

(a) 1 : 1 (b) 1 : 2

(c) 2 : 1 (d) 1 : 4.

56. Two circular discs A and B are of equal masses and thicknesses but made of metals with densities d_A and d_B ($d_A > d_B$). If their moments of inertia about an axis passing through centres and normal to the circular faces be I_A and I_B, then

(a) $I_A = I_B$ (b) $I_A > I_B$

(c) $I_A < I_B$ (d) $I_A \gtrless I_B$.

57. A solid sphere rolls without slipping down a 30° inclined plane. If $g = 10$ m s^{-2}, then the acceleration of the rolling sphere is

(a) 5 m s^{-2} (b) $\frac{7}{25}$ m s^{-2}

(c) $\frac{25}{7}$ m s^{-2} (d) $\frac{15}{7}$ m s^{-2}.

58. O is the centre of an equilateral triangle ABC. F_1, F_2 and F_3 are the three forces acting along the sides AB, BC and AC respectively. What should be the value of F_3 so that the total torque about O is zero ?

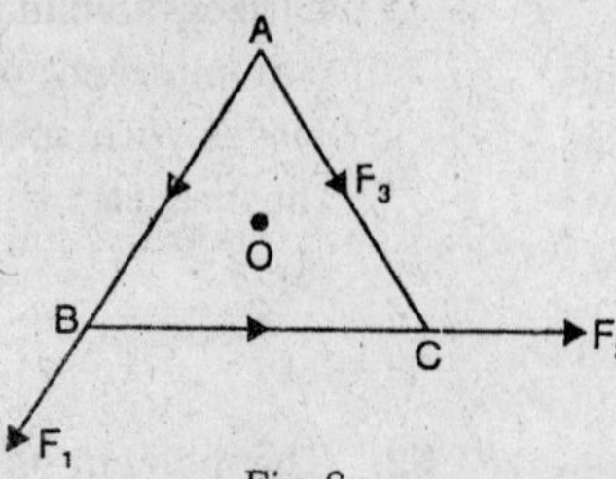

Fig. 8

(a) $2(F_1 + F_2)$ (b) $\frac{F_1 + F_2}{2}$

(c) $F_1 - F_2$ (d) $F_1 + F_2$.

59. A wheel of radius R has an axle of radius $\frac{R}{5}$. A force F is applied tangentially to the wheel. To keep the system in a state of "rotational" rest, a force F′ is applied tangentially to the axle. The value of F′ is

(a) F (b) 3F

(c) 5F (d) 7F.

60. The angular momentum of a 10 g particle moving with velocity $5\hat{i}$ m s^{-1} and having position vector $(10\hat{i} + 6\hat{j})$ metre about the origin is

(a) $-0.1\hat{k}$ Js (b) $-0.2\hat{k}$ Js

(c) $-0.3\hat{k}$ Js (d) $-0.4\hat{k}$ Js.

61. A body having moment of inertia about its axis of rotation equal to 3 kgm^2 is rotating with angular velocity equal to 3 rad s^{-1}. The kinetic energy of this rotating body is the same as that of a body of mass 27 kg moving with a velocity of

(a) 1 m s^{-1} (b) 1.5 m s^{-1}

(c) 2 m s^{-1} (d) 2.5 m s^{-1}.

62. A body rolls without slipping. The radius of gyration of the body about an axis passing through its centre of mass is K. The radius of the body is R. The ratio of rotational kinetic energy to translational kinetic energy is

(a) $\frac{K^2}{R^2}$ (b) $\frac{R^2}{K^2 + R^2}$

(c) $\frac{K^2}{K^2 + R^2}$ (d) $K^2 + R^2$.

63. A solid sphere of mass 1 kg rolls on a table with linear speed 1 m s^{-1}. Its total kinetic energy is

(a) 1 J (b) 0.5 J

(c) 0.7 J (d) 1.4 J.

64. The total kinetic energy of a rolling sphere is

(a) $\frac{1}{2}Mv^2$ (b) $\frac{1}{5}Mv^2$

(c) $\frac{7}{10}Mv^2$ (d) $\frac{7}{5}Mv^2$.

QUESTION BANK

Level II

Choose the correct answer from the four alternatives given in each of the following questions :

Circular Motion

1. A small ring P is threaded on a smooth wire bent in the form of a circle of radius a and centre O. The wire is rotating with constant angular speed ω about a vertical diameter XY, while the ring remains at rest relative to the wire at a distance $\frac{1}{2}a$ from XY.

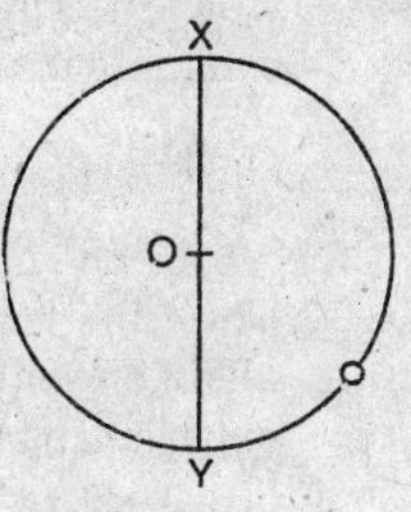

Fig. 9

(*a*) $\omega^2 = \frac{2g}{a}$ (*b*) $\omega^2 = \frac{g}{2a}$

(c) $\omega^2 = \frac{2g}{a\sqrt{3}}$ (*d*) $\omega^2 = \frac{g\sqrt{3}}{2a}$.

2. AB is an inclined plane at the foot of which there is a semi-circular hoop BCD of radius a. A body starts from rest at A and slides without friction. What should be the minimum height of A above the level of B so that the body may move along BCD without falling off?

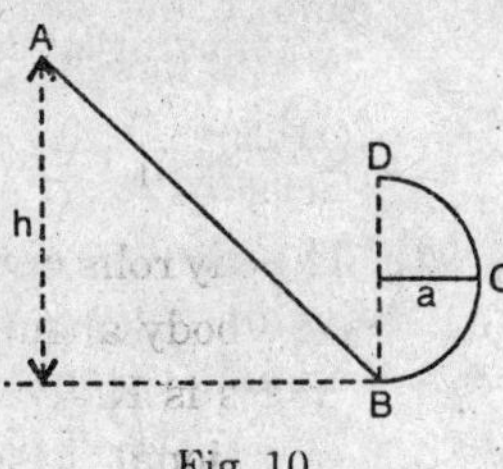

Fig. 10

(*a*) $\frac{3a}{2}$ (*b*) $\frac{5a}{2}$

(*c*) $\frac{3a}{4}$ (*d*) $\frac{5a}{4}$.

3. A person with his hands in his pockets is skating on ice at the rate of 10 m s^{-1} and describes a circle of 50 m radius. What is his inclination to the vertical ? (Given : g = 10 m s^{-2})

(*a*) $\tan^{-1}\left(\frac{1}{2}\right)$ (*b*) $\tan^{-1}\left(\frac{1}{5}\right)$

(c) $\tan^{-1}\left(\frac{3}{5}\right)$ (*d*) $\tan^{-1}\left(\frac{1}{10}\right)$.

4. When the string of a conical pendulum makes an angle of 45° with the vertical, its time period is T_1. When the string makes an angle of 60° with the vertical, its time period is T_2. Then, $\frac{T_1^2}{T_2^2}$ is

(*a*) $\sqrt{2}$ (*b*) $\sqrt{3}$

(*c*) $\sqrt{5}$ (*d*) $\sqrt{7}$.

5. The roadway bridge over a canal is in the form of an arc of a circle of radius 39.2 m. What is the maximum speed with which a car can move without leaving the ground at the highest point ? (Given : g = 9.8 m s^{-2})

(*a*) 9.8 m s^{-1} (*b*) 19.6 m s^{-1}

(*c*) 39.2 m s^{-1} (*d*) None of these.

6. A stone of mass 1 kg tied to a light inextensible string of length L = $\frac{10}{3}$ m is whirling in a circular path of radius L in vertical plane. If the ratio of the maximum tension in the string to the minimum tension in the string is 4 and if g is taken to be 10 m s^{-2}, the speed of the stone at the highest point of the circle is

(*a*) 10 m s^{-1} (*b*) $5\sqrt{2}$ m s^{-1}

(*c*) $10\sqrt{3}$ m s^{-1} (*d*) 20 m s^{-1}.

7. A particle moves in a circle of radius 4 cm clockwise at constant speed 2 cm s^{-1}. If $\hat{x}$ and $\hat{y}$ are unit acceleration vectors along $\hat{x}$ and $\hat{y}$ respectively (in cm s^{-2}), the acceleration of the particle at the instant half way between P and Q is given by

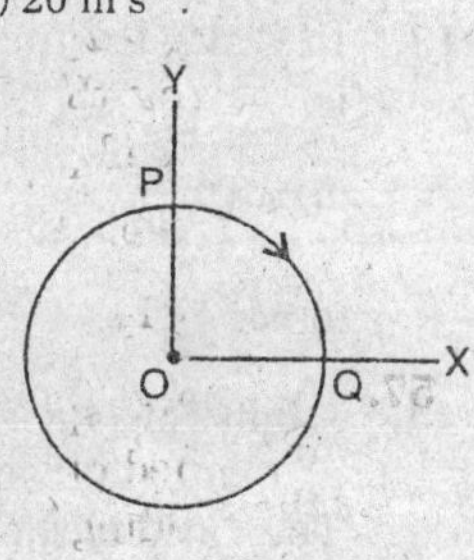

Fig. 11

(*a*) $-4(\hat{x}+\hat{y})$ (*b*) $4(\hat{x}+\hat{y})$

(*c*) $-(\hat{x}+\hat{y})/\sqrt{2}$ (*d*) $(\hat{x}-\hat{y})/4$.

8. A 4 kg stone tied at the end of a string 1 metre long is whirled in a vertical circle. At the instant when the string makes an angle q with the vertical, the speed of the stone is 4 m s^{-1} and the tension in the thread is 103.2 newton. Then θ is

(*a*) 0° (*b*) 30°

(*c*) 60° (*d*) 90°.

9. A car of mass 1000 kg moves on a circular track of radius 20 m. If the coefficient of friction is 0.64, what is the maximum velocity with which the car can be moved ?

(a) 112 m s^{-1} (b) 11.2 m s^{-1}

(c) $\frac{0.64 \times 20}{1000}$ m s^{-1} (d) $\frac{1000}{0.64 \times 20}$ m s^{-1}.

10. When the bob of a simple pendulum passes through the position of equilibrium, the string is subjected to a tension equal to twice the weight of the bob. Through what maximum angle from the vertical can the pendulum be deflected ?

(a) 30° (b) 45°

(c) 60° (d) 75°.

11. Two wheels, each with four spokes, are mounted one behind the other and aligned such that an observer sees only four spokes. At t = 0 minute, one wheel begins to spin at 5 revolutions/minute and the other at 4 revolutions/minute. Upto t = 1 minute, how many times the observer would see only four spokes ?

(a) 1 (b) 2

(c) 4 (d) 8.

12. A heavy particle hanging from a string of length l is projected horizontally with speed $\sqrt{gl}$. The speed of the particle at the point where the tension in the string equals the weight of the particle is

(a) $\sqrt{2gl}$ (b) $\sqrt{3gl}$

(c) $\sqrt{gl/2}$ (d) $\sqrt{gl/3}$.

13. In Q. 12, the inclination of the string with the vertical is

(a) $\cos^{-1}\frac{2}{3}$ (b) $\cos^{-1}\frac{3}{2}$

(c) $\cos^{-1}\frac{1}{2}$ (d) $\cos^{-1} 2$.

14. A mass of 2 kg is tied to a string 5 m long and is rotated at a constant speed of 30 m s^{-1} in a vertical circle. The ratio of the tensions at the top and the bottom is (g = 10 m s^{-2})

(a) 1 : 2 (b) 13 : 15

(c) 16 : 17 (d) 17 : 19.

15. A body crosses the topmost point of a vertical circle with critical speed. What will be its acceleration when the string is horizontal ?

(a) g (b) $2g$

(c) $3g$ (d) $6g$.

16. A particle of mass 0.1 kg is suspended from the end of a weightless string of length 1 m and is allowed to swing in a vertical plane. The speed of the mass is 2 m s^{-1} when the string makes an angle of 60° with the vertical. (g = 10 m s^{-2}). The tension in the string at θ = 60° with the vertical is

(a) 0.1 N (b) 0.4 N

(c) 0.6 N (d) 0.9 N.

17. In Q. 16, the speed of the particle when it is in the lowest position is

(a) $\sqrt{3}$ m s^{-1} (b) $\sqrt{7}$ m s^{-1}

(c) $\sqrt{91}$ m s^{-1} (d) $\sqrt{14}$ m s^{-1}

(e) $\sqrt{27.235}$ cm s^{-1}.

18. A particle is rotated in a vertical circle by connecting it to a string of length l and keeping the other end of the string fixed. The minimum speed of the particle when the string is horizontal for which the particle will complete the circle is

(a) $\sqrt{gl}$ (b) $\sqrt{2gl}$

(c) $\sqrt{3gl}$ (d) $\sqrt{5gl}$.

19. A tube one metre long is filled with liquid of mass 1 kg. The tube is closed at both the ends and is revolved about one end in a horizontal plane at 2 rev/s. The force experienced by the lid at the other end is

(a) $4\pi^2$ N (b) $8\pi^2$ N

(c) $16\pi^2$ N (d) 9.8 N.

20. A 1200 kg automobile rounds a level curve of radius 200 m, on an unbanked road, with a velocity of 72 km h^{-1}. The minimum coefficient of friction between the tyres and road in order that the automobile may not skid is (g = 10 m s^{-2})

(a) 0.3 (b) 0.2

(c) 0.6 (d) 0.5.

21. The moment of inertia of a uniform circular disc about its diameter is 200 g cm^2. Its moment of inertia about an axis passing through its centre and perpendicular to its circular face is

(a) 100 g cm^2 (b) 200 g cm^2

(c) 400 g cm^2 (d) 1000 cm^2.

22. A small mass of 10 g lies in a hemispherical bowl of radius 0.5 m at a height of 0.2 m from the bottom of the bowl. The mass will be in equilibrium if the bowl rotates at an angular speed of (g = 10 m s^{-2})

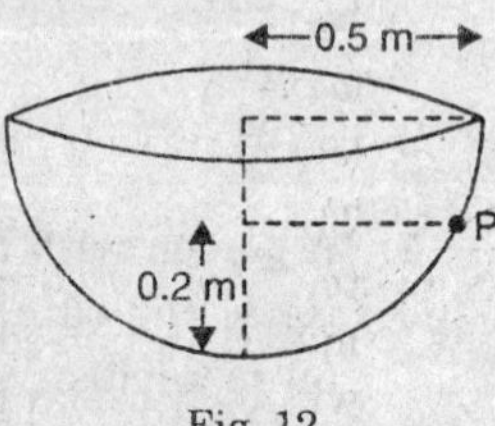

Fig. 12

(a) $\frac{10}{\sqrt{3}}$ rad/s (b) $10\sqrt{3}$ rad/s

(c) 10 rad/s (d) $\sqrt{20}$ rad/s.

23. A nail is located at a certain distance vertically below the point of suspension of a simple pendulum. The pendulum is released from a position where the string makes an angle of 60° with the vertical. The distance of the nail from the point of suspension such that the bob will just perform revolutions with the nail as the centre is (Given : length of pendulum = 1 m.)

(a) 0.2 m (b) 0.4 m

(c) 0.6 m (d) 0.8 m.

24. When a bucket containing water is rotated fast in a vertical circle of radius R, the water in the bucket doesn't spill provided

(a) The bucket is whirled with a maximum speed of $\sqrt{2gR}$

(b) The bucket is whirled around with a minimum speed of $\sqrt{(1/2)gR}$

(c) The bucket is having a r.p.m. of $\sqrt{900g/(\pi^2 R)}$

(d) The bucket is having a r.p.m. of $\sqrt{3600g/(\pi^2 R)}$.

25. A chain of 125 links is 1.25 m long and has a mass of 2 kg. With the ends fastened together, it is set rotating at 3000 rev/min. The centripetal force on each link is

(a) 3.14 N (b) 314 N

(c) $\frac{1}{3.14}$ N (d) $\frac{1}{314}$ N.

26. A smooth wire is bent into a vertical circle of radius a. A bead P can slide smoothly on the wire. The circle is rotated about diameter as axis with a speed ω. The bead P is at rest with respect to the circular ring in the position shown. Then ω^2 is equal to

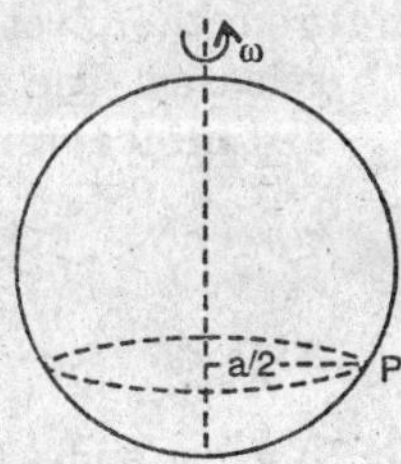

Fig. 13

(a) $2g/a$ (b) $2g/(a\sqrt{3})$

(c) $g\sqrt{3}/a$ (d) $2a/(g\sqrt{3})$.

27. A motor cyclist rides on a cylindrical wall of a velodrome (hollow metal cylinder kept with the axis vertical) of radius r rotating about its axis at the rate f rev/s. The least coefficient of friction necessary so that he does not step down is

(a) $g/(4\pi^2 f^2 r)$ (b) $(4\pi^2 f^2 r)/g$

(c) $(4\pi^2 f^2)/gr$ (d) $g/(4\pi^2 r^2)$.

28. The breaking tension of a string is 10 N. A particle tied to it is rotated along a circle of radius of 0.5 m. The maximum speed with which particle can be rotated without breaking the string is (The mass of particle is 100 g)

(a) $\sqrt{500}$ m s^{-1} (b) $\sqrt{50}$ m s^{-1}

(c) $\sqrt{5}$ m s^{-1} (d) $\sqrt{0.5}$ m s^{-1}.

29. A particle originally at rest at the highest point of a smooth vertical circle is slightly displaced. It will leave the circle at a vertical distance h below the highest point, such that

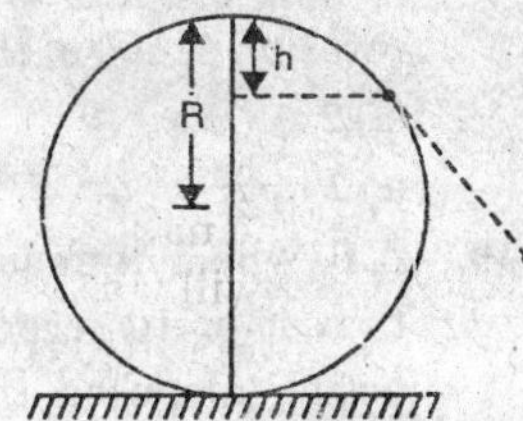

Fig. 14

(a) $h = R$ (b) $h = 2R$

(c) $h = R/2$ (d) $h = R/3$.

30. A particle of mass 0.1 kg is rotated at the end of a string in a vertical circle of radius 1.0 m at a constant speed of 5 m s^{-1}. The tension in newton in the string at the highest point of its path is ($g = 10$ m s^{-2})

(a) 0.5 (b) 1.0

(c) 1.5 (d) 3.5

(e) 15.

31. A racing car of mass 1000 kg moves round a banked track at a constant speed of 108 km h^{-1}. Assuming the total reaction at the wheels is normal to the track and the horizontal radius of the track is 100 m, the angle of inclination of the track to the horizontal is ($g = 10$ m s^{-2})

(a) 12° (b) 27°

(c) 42° (d) 65°.

32. A coin placed on a gramophone record rotating at 33 r.p.m. flies off the record, if it is placed at a distance of more than 16 cm from the axis of rotation. If the record is revolving at 66 r.p.m., the coin will fly off if it is placed at a distance not less than

(a) 1 cm (b) 2 cm

(c) 3 cm (d) 4 cm.

33. A body of mass 2 kg revolves around a circular path of radius 5 m. If it makes 10 revolutions per second, its centripetal acceleration is

(a) $3000\pi^2$ m s^{-2} (b) $2000\pi^2$ m s^{-2}

(c) $4000\pi^2$ m s^{-2} (d) $5000\pi^2$ m s^{-2}.

34. A motor car is travelling at 30 m s^{-1} on a circular road of radius 500 m. It is increasing its speed at the rate of 2 m s^{-2}. The acceleration is

(a) 1.8 m s^{-2} (b) 2 m s^{-2}

(c) 3.8 m s^{-2} (d) 2.7 m s^{-2}.

35. A vehicle is moving with a velocity v on a curved road of width b and radius of curvature R. For counteracting the centrifugal force on the vehicle, the difference in elevations required in between the outer and inner edges of the road is

(a) $\frac{v^2}{Rg}$ (b) $\frac{v^2 b}{Rg}$

(c) gR (d) bv^2.

36. A railway carriage has its centre of gravity at a height of 1 m above the rails which are 1 m apart. The maximum speed with which it would go round the unbanked track of 250 m radius is

(a) 10 m s^{-1} (b) 15 m s^{-1}

(c) 25 m s^{-1} (d) 35 m s^{-1}.

37. The string of a pendulum is horizontal. The mass of the bob attached to it is m. Now the string is released. The tension in the string in the lowest position is

(a) mg (b) $2mg$
(c) $3mg$ (d) $4mg$.

38. A 1 kg stone at the end of 1 m long string is whirled in a vertical circle at constant speed of 4 m s^{-1}. The tension in the string is 6 N when the stone is (g = 10 m s^{-2})

(a) at the top of the circle (b) at the bottom of the circle
(c) half way down (d) None of these.

Centre of Mass and Rotational Mechanics

39. The moment of inertia of a body about a given axis is 1.2 kg m^2. Initially, the body is at rest. In order to produce a rotational kinetic energy of 1500 J, an angular acceleration of 25 rad s^{-2} must be applied about the axis for a duration of

(a) 4 s (b) 2 s
(c) 8 s (d) 10 s.

40. A flywheel rotating about a fixed axis has a kinetic energy of 360 J, when its angular speed is 30 rad s^{-1}. The moment of inertia of the wheel about the axis of rotation is

(a) 0.6 kg m^2 (b) 0.15 kg m^2
(c) 0.8 kg m^2 (d) 0.75 kg m^2.

41. A thin uniform circular disc of mass M and radius R is rotating in a horizontal plane about an axis passing through its centre and perpendicular to its plane with an angular velocity ω. Another disc of the same dimensions but of mass M/4 is placed gently on the first disc coaxially. The angular velocity of the system now becomes

(a) $\frac{\omega}{2}$ (b) $\frac{\omega}{4}$
(c) $\frac{3\omega}{4}$ (d) $\frac{4\omega}{5}$.

42. A disc-like reel with massless thread unrolls itself while falling vertically downwards. The acceleration of its fall is

(a) zero (b) g
(c) $\frac{2}{3}g$ (d) $\frac{g}{2}$.

43. Four spheres, each of mass m and radius r, are placed with their centres on the four corners of a square of side b. The moment of inertia of the system about any side of the square is

(a) $\frac{8}{5}mr^2 + mb^2$ (b) $\frac{8}{5}mr^2 + 2mb^2$
(c) $\frac{8}{5}mr^2 + 4mb^2$ (d) None of these.

44. A flywheel of mass 25 kg has a radius of 0.2 m. What force should be applied tangentially to the rim of the flywheel so that it acquires an angular acceleration of 2 rad s^{-2} ?

(a) 1 N (b) 3 N
(c) 5 N (d) 8 N.

45. A nearly massless rod is pivoted at one end so that it can swing freely as a pendulum. Two masses $2m$ and m are attached to it at distances b and $3b$ respectively from the pivot. The rod is held horizontal and then released. The angular acceleration of the rod at the instant it is released is

(a) bg (b) $\frac{b}{g}$
(c) $gb + 5$ (d) $\frac{5g}{11b}$.

46. If a disc slides from top to bottom of an inclined plane, it takes time t_1. If it rolls, it takes time t_2. Now, $\frac{t_2^2}{t_1^2}$ is

(a) $\frac{1}{2}$ (b) $\frac{2}{3}$
(c) $\frac{3}{2}$ (d) $\frac{2}{5}$.

47. We have two spheres, one of which is hollow and the other solid. They have identical masses and moment of inertia about their respective diameters. The ratio of their radii is

(a) 5 : 7 (b) 3 : 5
(c) $\sqrt{3} : \sqrt{5}$ (d) $\sqrt{3} : \sqrt{7}$.

48. Two fly wheels A and B are mounted side by side with frictionless bearings on a common shaft. Their moments of inertia about the shaft are 5.0 kg m^2 and 20.0 kg m^2 respectively. Wheel A is made to rotate at 10 revolution per second. Wheel B, initially stationary, is now coupled to A with the help of a clutch. The rotation speed of the wheels will become

(a) $2\sqrt{5}$ r.p.s. (b) 0.5 r.p.s.
(c) 2 r.p.s. (d) None of these.

49. A flywheel of moment of inertia 0.5 kg m^2 is brought to rest in 10 second by a constant couple of 10 N m acting alone on the flywheel. What was the initial angular velocity of the flywheel in radian s^{-1} ?

(*a*) 0.5 (*b*) 2
(*c*) 60 (*d*) 100
(*e*) 200.

50. The density of a rod continuously increases from A to B. It is easier to set it into rotation by

(*a*) clamping the rod at A and applying a force F at B, perpendicular to the rod
(*b*) clamping the rod at B and applying a force F at A, perpendicular to the rod
(*c*) clamping the rod at mid-point of AB and applying a force F at A, perpendicular to the rod
(*d*) clamping the rod at mid-point of AB and applying a force F at B, perpendicular to the rod.

51. Two identical masses are connected to a horizontal thin massless rod as shown in Fig. 15. When their distance from the pivot is x, a torque produces an angular acceleration α_1. If the masses are now repositioned so that they are at distance $2x$ each from the pivot, the same torque will produce an angular acceleration α_2 such that

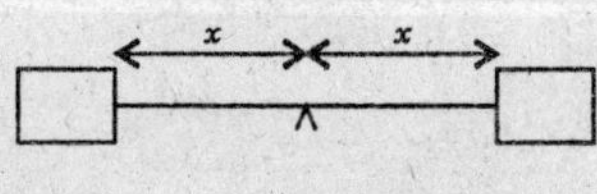

Fig. 15

(*a*) $\alpha_2 = 4\alpha_1$ (*b*) $\alpha_2 = \alpha_1$
(*c*) $\alpha_2 = \dfrac{\alpha_1}{2}$ (*d*) $\alpha_2 = \dfrac{\alpha_1}{4}$.

52. If the angular momentum of a body increases by 50%, its kinetic energy of rotation increases by

(*a*) 50% (*b*) 125%
(*c*) 200% (*d*) 100%.

53. A ring rolls down, starting from rest, down an inclined plane of length l and inclination q. The velocity of the centre of mass of the ring at the mid-point of the inclined plane is

(*a*) $\sqrt{gl \sin \theta}$ (*b*) $\sqrt{\dfrac{gl \sin \theta}{2}}$
(*c*) $\sqrt{2gl \sin \theta}$ (*d*) $\sqrt{\dfrac{4}{7} gl \sin \theta}$.

54. If the density of material of a square plate and a circular plate shown in Fig. 16 is same, the centre of mass of the composite system will be

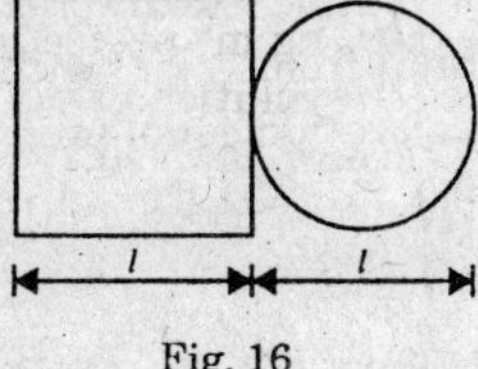

Fig. 16

(*a*) inside the square plate (*b*) inside the circular plate
(*c*) at the point of contact (*d*) outside the system.

55. A wheel is rolling straight on ground without slipping. If the axis of the wheel has speed v, the instantaneous velocity of a point P on the rim, defined by angle θ, relative to the ground will be

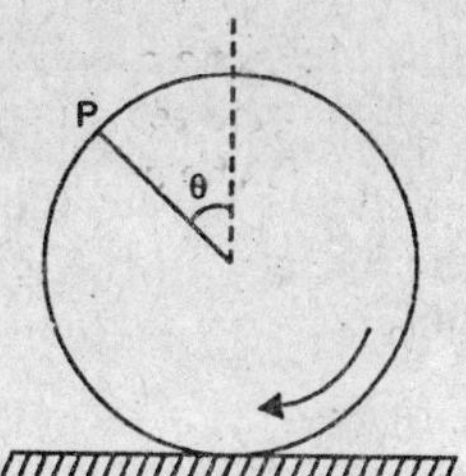

Fig. 17

(*a*) $v \cos\left(\dfrac{1}{2}\theta\right)$
(*b*) $2v \cos\left(\dfrac{1}{2}\theta\right)$
(*c*) $v(1 + \sin \theta)$ (*d*) $v(1 + \cos \theta)$.

56. Certain neutron starts (extremely dense stars) are believed to be rotating at about 1 rev/s. If such a star has a radius of 20 km, the acceleration of an object on the equator of the star will be

(*a*) 20×10^3 m s^{-2} (*b*) 120×10^3 m s^{-2}
(*c*) 8×10^5 m s^{-2} (*d*) 4×10^3 m s^{-2}.

57. The ratio of the time taken by a uniform solid sphere and a disc of the same mass and diameter to roll down from rest the same distance on an inclined plane is

(*a*) $\sqrt{\dfrac{15}{14}}$ (*b*) $\sqrt{\dfrac{14}{15}}$
(*c*) $\sqrt{\dfrac{11}{12}}$ (*d*) $\sqrt{\dfrac{12}{11}}$.

58. A solid flywheel has a moment of inertia of 0.1 kg m^2 about its axis of rotation. It is set in motion by applying a tangential force of 2 kg with a rope wound round its circumference. If the radius of the wheel is 10 cm, then the angular acceleration produced is

(*a*) 9.8 rad s^{-2} (*b*) 19.6 rad s^{-2}
(*c*) 3×9.8 rad s^{-2} (*d*) zero.

59. A child stands at the centre of a turn table with his two arms outstretched. The turntable is set rotating with an angular speed of 40 rev/min. Now, the child folds his hands back and thereby reduces his moment of inertia to $\dfrac{2}{5}$ times the initial value. The new kinetic energy of rotation is x times the initial kinetic energy of rotation. The value of x is

(*a*) 2.5 (*b*) 5
(*c*) 1 (*d*) 6.756.

60. If I_1 is the moment of inertia of a thin rod about an axis perpendicular to its length and passing through its centre of mass, and I_2 is the moment of inertia (about central axis) of the ring formed by bending the rod, then

(a) $I_1 : I_2 = 1 : 1$ (b) $I_1 : I_2 = \pi^2 : 3$
(c) $I_1 : I_2 = \pi : 4$ (d) $I_1 : I_2 = 3 : 5$.

61. A rod of length 1 m and mass 1/2 kg rotates at angular speed 6 rad s^{-1} about one of its ends. The kinetic energy of the rod is

(a) 1 joule (b) 2 joule
(c) 3 joule (d) 4 joule.

62. A certain bicycle can go up a gentle incline with constant speed when the frictional force of ground pushing the rear wheel is $F_2 = 4$ N. With what force F_1 must the chain pull on the sprocket wheel if $R_1 = 5$ cm and $R_2 = 30$ cm ?

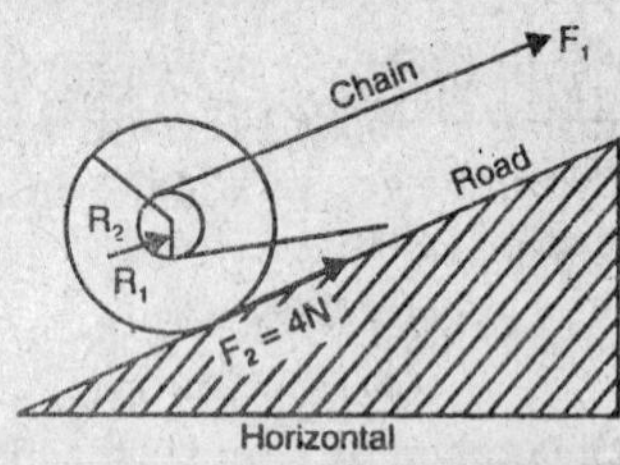

Fig. 18

(a) 4 N (b) 24 N
(c) 140 N (d) $\frac{35}{4}$ N.

63. If r denotes the distance between the sun and the earth, then the angular momentum of the earth around the sun is

(a) $\propto \sqrt{r}$ (b) $= mr$
(c) m^2r (d) $\propto r^{3/2}$.

64. A wheel of moment of inertia 30 kg m^2 is rotating at 10 rotations per minute. The work done in increasing its speed to 5 times its initial value will be

(a) $10\pi^2$ joule (b) $20\pi^2$ joule
(c) $30\pi^2$ joule (d) $40\pi^2$ joule.

65. A uniform rod of length l is free to rotate in a vertical plane about a fixed horizontal axis through. The rod begins rotating from rest from its unstable equilibrium position. When it has turned through an angle θ, its angular velocity ω is given as

Fig. 19

(a) $\sqrt{\frac{6g}{l}} \sin\theta$ (b) $\sqrt{\frac{6g}{l}} \sin\frac{\theta}{2}$

(c) $\sqrt{\frac{6g}{l}} \cos\frac{\theta}{2}$ (d) $\sqrt{\frac{6g}{l}} \cos\theta$.

66. A sphere of mass 2000 g and radius 5 cm is rotating at the rate of 300 r.p.m. Then the torque required to stop it in 2π revolutions, is

(a) 1.6×10^2 dyne cm (b) 1.6×10^3 dyne cm
(c) 2.5×10^4 dyne cm (d) 2.5×10^5 dyne cm.

67. A horizontal disc rotating freely about a vertical axis makes 90 revolutions per minute. A small piece of wax of mass m g falls vertically on the disc and sticks to it at a distance r cm from the axis. If the number of revolutions per minute reduce to 60, then the moment of inertia of the disc is

(a) mr^2 (b) $\frac{3}{2}mr^2$
(c) $2mr^2$ (d) $3mr^2$.

68. Three identical rings, each of mass M and radius R are arranged as shown in Fig. 20. The moment of inertia of the arrangement about YY′ is

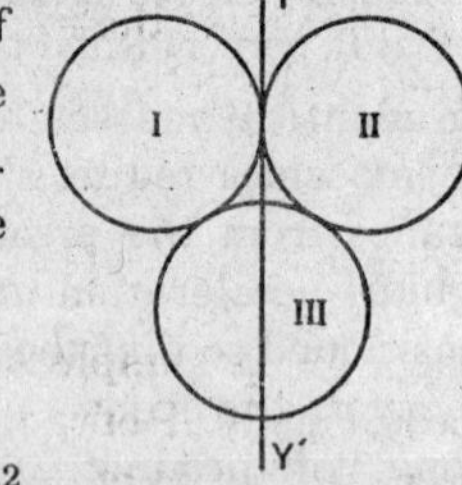

Fig. 20

(a) $\frac{1}{2}MR^2$ (b) MR^2
(c) $\frac{5}{2}MR^2$ (d) $\frac{7}{2}MR^2$.

69. An automobile engine develops 10^5 W when rotating at a speed of 1800 rev/min. The torque delivered by the engine is

(a) $\frac{10^2}{6\pi}$ Nm (b) $\frac{10^4}{6\pi}$ Nm
(c) $\frac{10^6}{6\pi}$ (d) $\frac{10^8}{6\pi}$.

70. A solid cylinder of mass 20 kg rotates about its axis with angular velocity 100 s^{-1}. The radius of the cylinder is 0.25 m. The moment of inertia of the cylinder is

(a) 1.25 kg m^2 (b) 0.625 kg m^2
(c) 0.314 kg m^2 (d) 2.5 kg m^2.

71. In Q. 70, the magnitude of the angular momentum of the cylinder about its axis of rotation is

(a) 62.5 Js (b) 625 Js
(c) 0.625 Js (d) Zero.

72. An inclined plane makes an angle of 30° with the horizontal. A solid sphere rolling down on this inclined plane from rest without slipping, has a linear acceleration equal to

(a) $\frac{5g}{7}$ (b) $\frac{5g}{12}$
(c) $\frac{5g}{14}$ (d) $\frac{5g}{17}$.

73. A ring of mass 1 kg and radius 1 m is moving with a velocity of 2 m s^{-1} by rolling on a frictionless horizontal plane. The total kinetic energy of the ring is

(a) 1 J (b) 2 J
(c) 4 J (d) 8 J.

74. A disc and a ring of the same mass are rolling. Their kinetic energies are equal. The ratio of their velocities is

(a) $1 : \sqrt{3}$ (b) $2 : \sqrt{3}$
(c) $3 : \sqrt{3}$ (d) $4 : \sqrt{3}$.

75. The moment of inertia of a thin uniform rod of length L and mass M about an axis passing through a point at a distance of $\frac{L}{3}$ from one of its ends and perpendicular to the rod is

(a) $\frac{ML^2}{3}$ (b) $\frac{ML^2}{6}$
(c) $\frac{ML^2}{9}$ (d) $\frac{ML^2}{12}$.

76. From a given sample of uniform wire, two circular loops are made ; P of radius r, and O of radius nr. If the moment of inertia of O about its axis is 4 times that of P about its axis, (assuming wire diameter much smaller than either radius), the value of n is

(a) $4^{1/3}$ (b) $4^{1/2}$
(c) $4^{1/4}$ (d) $4^{2/3}$.

77. A loop and a disc have the same mass and roll without slipping with the same linear velocity v. If the total kinetic energy of the loop is 8 J, the kinetic energy of the disc must be

(a) 8 J (b) 6 J
(c) 16 J (d) 4 J.

78. A sphere is rolling without slipping on a horizontal plane. The ratio of its rotational kinetic energy and translational kinetic energy is

(a) 2 : 3 (b) 2 : 5
(c) 2 : 7 (d) 2 : 9.

79. Two bodies of masses 4 kg and 2 kg are tied to the ends of a string which passes over a light frictionless pulley. The masses are initially at rest and released. If $g = 9$ m s^{-2}, then the acceleration of the centre of mass is

(a) 0 m s^{-2} (b) 1 m s^{-2}
(c) 2 m s^{-2} (d) 3 m s^{-2}.

80. A spherical ball rolls on a table without slipping. Then the fraction of its total energy associated with rotation is

(a) $\frac{2}{5}$ (b) $\frac{2}{7}$
(c) $\frac{3}{5}$ (d) $\frac{3}{7}$.

81. Two rigid bodies A and B rotate with angular momenta L_A and L_B respectively. The moments of inertia of A and B about the axes of rotation are I_A and I_B respectively. If $I_A = \frac{I_B}{4}$ and $L_A = 5L_B$, then the ratio of rotational kinetic energy K_A of A to the rotational kinetic energy K_B of B is given by

(a) $\frac{K_A}{K_B} = \frac{25}{4}$ (b) $\frac{K_A}{K_B} = \frac{5}{4}$
(c) $\frac{K_A}{K_B} = \frac{1}{4}$ (d) $\frac{K_A}{K_B} = 100$.

82. A solid cylinder of radius r at rest at the top of an inclined plane is released and rolls down the incline without slipping. If the height of the incline is h, then the linear speed at the bottom of the inclined plane is

(a) $\sqrt{2gh}$ (b) $\sqrt{gh}$
(c) $\sqrt{\frac{4gh}{3}}$ (d) $\sqrt{\frac{hrg}{4}}$.

83. A uniform circular disc of mass M is rotating freely with uniform angular speed ω in a horizontal plane about a fixed vertical axis passing through its centre. If a particle of mass M is gently attached to the rim of the disc, the angular speed of the disc changes to

(a) $\frac{\omega}{2}$ (b) $\frac{\omega}{3}$
(c) $\frac{\omega}{4}$ (d) $\frac{\omega}{5}$.

84. A flywheel at rest is to reach an angular velocity of 36 rad s^{-1} in 6 second, with a constant angular acceleration. The total angle turned through during this interval is

(a) 216 rad (b) 144 rad
(c) 108 rad (d) 72 rad.

85. A particle of mass m is moving in a plane along a circle of radius r. Its angular momentum about the axis of rotation is L. What is the centripetal force acting on the particle ?

(a) $\frac{L^2}{mr}$ (b) $\frac{L^2 m}{r}$
(c) $\frac{L^2}{mr^3}$ (d) $\frac{L^2}{mr^2}$.

86. A body rolls down an inclined plane. If its kinetic energy of rotational motion is 40% of its kinetic energy of translation, then the body is

(a) ring (b) solid sphere
(c) solid disc (d) cylinder.

87. A small hole is made in a disc of mass M and radius R at a distance $\frac{R}{4}$ from centre. The disc is supported on a horizontal peg through this hole. The moment of inertia of the disc about horizontal peg is

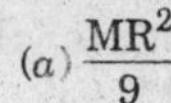

(a) $\frac{MR^2}{9}$ (b) $\frac{5}{16}MR^2$

(c) $\frac{9}{16}MR^2$ (d) $\frac{5}{4}MR^2$.

88. Two bodies with moment of inertia I_1 and I_2 ($I_1 > I_2$) have equal angular momenta. If E_1 and E_2 are the rotational kinetic energies, then

(a) $E_1 = E_2$ (b) $E_1 > E_2$

(c) $E_1 < E_2$ (d) $E_1 \geq E_2$.

89. A disc of radius 10 cm can rotate about an axis passing through its centre and perpendicular to its plane. A force of 10 N is applied along the tangent in the plane of the disc. If the moment of inertia of the disc about its axis is 5 kg m^2, the increase in the angular velocity of the disc in 10 s is

(a) 5 rad s^{-1} (b) 20 rad s^{-1}

(c) 2 rad s^{-1} (d) 50 rad s^{-1}.

90. In Q. 89, the change in the angular momentum of the disc is

(a) 20 kg m^2 s^{-1} (b) 250 kg m^2 s^{-1}

(c) 10 kg m^2 s^{-1} (d) 100 kg m^2 s^{-1}.

91. A weight W is attached to a string wrapped round a solid cylinder of weight W mounted on a frictionless horizontal axle. The weight W is released from rest. The velocity of the weight at any time t is (Take R to be the radius of the cylinder)

(a) αR (b) $\alpha \frac{1}{R}$

(c) $\alpha \frac{1}{R^2}$ (d) independent of R.

92. Which of the following correctly represents the relation between log L and log E_k ?

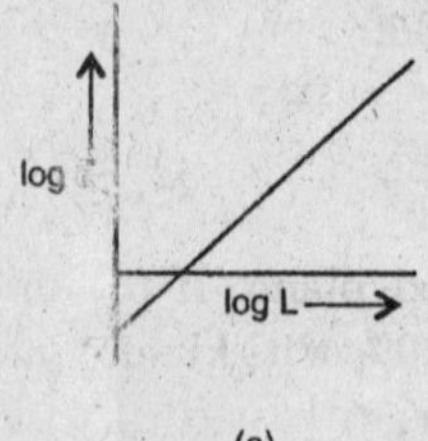

log E_k log L (b)

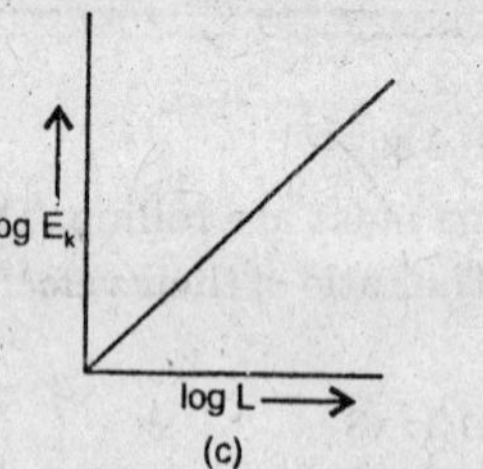

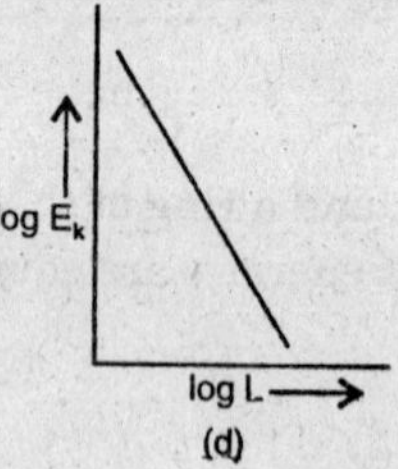

Fig. 21

93. The centre of a wheel rolling on a plane surface moves with a speed v_0. A particle on the rim of the wheel at the same level as the centre will be moving at speed

(a) zero (b) v_0

(c) $\sqrt{2}v_0$ (d) $2v_0$.

94. The moment of inertia of a thin spherical shell of mass M and radius, r, about a diameter is $\frac{2}{3}Mr^2$. Its radius of gyration K, about tangent will be

(a) $\frac{2}{3}r$ (b) $\sqrt{\frac{2}{3}}\,r$

(c) $\sqrt{\frac{5}{3}}\,r$ (d) $\sqrt{\frac{3}{5}}\,r$.

95. A thin uniform circular ring is rolling down an inclined plane of inclination 30° without slipping. Its linear acceleration along the inclined plane is

(a) g (b) $\frac{g}{2}$

(c) $\frac{g}{4}$ (d) $\frac{2g}{3}$.

96. Two identical discs are placed one over the other and both are rotating about a common axis through their common centre and perpendicular to their plane. The angular velocity is ω. If the upper disc is suddenly removed, the angular velocity of the other disc will be

(a) 2ω (b) ω

(c) 3ω (d) 2.5ω.

97. A thin metal hoop of radius 0.25 m and mass 2 kg starts from rest and rolls on an inclined plane. When the hoop reaches the foot of the inclined plane, its linear velocity is 2 m s^{-1}. The rotational kinetic energy at the foot of the inclined plane is

(a) 1 J (b) 2 J

(c) 3 J (d) 4 J.

98. A wheel of radius 40 cm rests against a step of height 20 cm as shown in Fig. 22. If the mass of the wheel is 2 kg, then the minimum horizontal force, applied perpendicular to the axle, required to make the wheel climb the step is

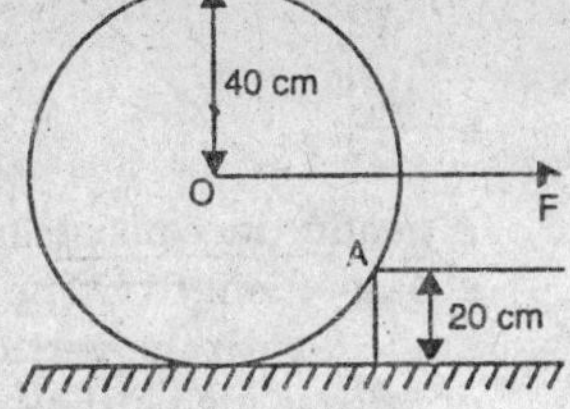

Fig. 22

(a) $19.6 \times \sqrt{3}$ N
(b) 19.6 N
(c) zero
(d) 20 N.

99. A thin circular ring of mass M and radius r is rotating about its axis with constant angular velocity ω. Two objects, each of mass m, are attached gently on the opposite ends of a diameter of the ring. The wheel now rotates with an angular velocity

(a) $\frac{M}{M+2m}\omega$
(b) $\frac{M}{M-2m}\omega$
(c) $\frac{M}{m}\omega$
(d) $\frac{m}{M}\omega$.

100. A circular disc of radius R has uniform thickness. A circular hole of diameter equal to the radius of the disc has been cut out from the disc as shown in (Fig. 23). The centre of gravity of the remaining portion of the disc lies on the diameter of the disc at a distance x to the left of the centre of the original disc. The value of x is

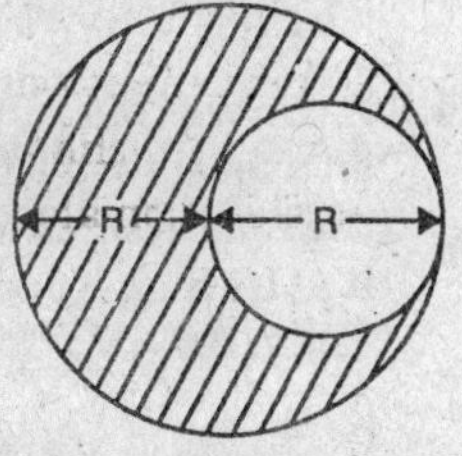

Fig. 23

(a) R
(b) $\frac{R}{2}$
(c) $\frac{R}{4}$
(d) $\frac{R}{6}$.

101. A rigid spherical body is spinning around an axis without any external torque. Due to change in temperature, the volume increases by 1% The angular speed will approximately

(a) increase by 1%
(b) decrease by 1%
(c) decrease by 0.67%
(d) decrease by 0.33%.

102. Which of the following represents graphically the relation between log K and log I where K and I represent the radius of gyration and moment of inertia respectively ?

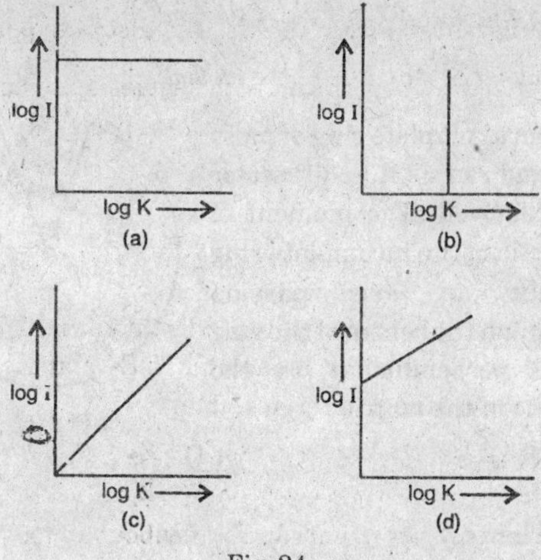

Fig. 24

103. A copper ring of mass M and radius R is suspended from a horizontal nail in a vertical wall. The moment of inertia of the ring about the nail is

(a) MR^2
(b) $2MR^2$
(c) $4MR^2$
(d) $8MR^2$.

104. Five masses are placed in a plane as shown in (Fig. 25). The coordinates of the centre of mass are nearest to

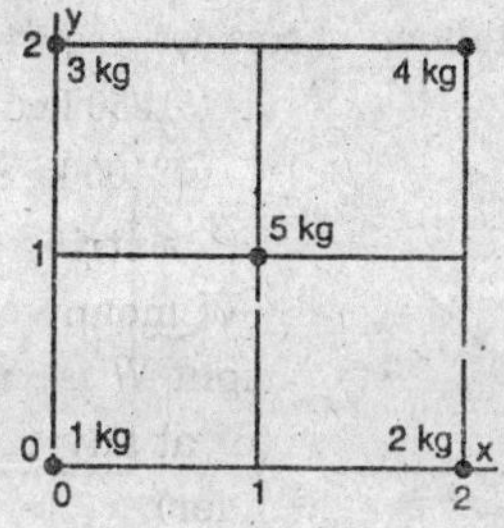

Fig. 25

(a) 1.2, 1.4
(b) 1.3, 1.1
(c) 1.1, 1.3
(d) 1.0, 1.0.

105. Fig. 26 shows a small wheel fixed coaxially on a bigger one of double the radius. The system rotates about the common axis. The strings supporting A and B do not slip on the wheels. If x and y be the distances travelled by A and B in the same time interval, then

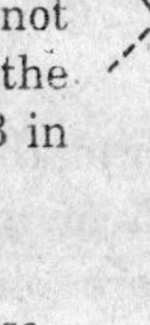

Fig. 26

(a) $x = 2y$
(b) $x = y$
(c) $y = 2x$
(d) None of these.

106. Three particles, each of mass m, are placed at the corners of an equilateral triangle of side l. The moment of inertia of the system about an axis coinciding with one side of the triangle is

(a) $\frac{2}{3}ml^2$ (b) $\frac{3}{4}ml^2$

(c) ml^2 (d) $3ml^2$.

107. From a complete ring of mass M and radius R, a 30° sector is removed. The moment of inertia of the incomplete ring about an axis passing through the centre of the ring and perpendicular to the plane of the ring is

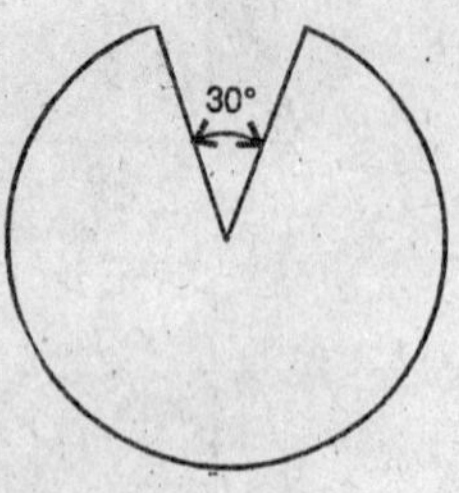

Fig. 27

(a) $\frac{9}{12}MR^2$ (b) $\frac{11}{12}MR^2$

(c) $\frac{11.3}{12}MR^2$ (d) MR^2.

108. The moment of inertia of a uniform circular disc about a diameter is I. Its moment of inertia about an axis perpendicular to its plane and passing through a point on its rim will be

(a) 2I (b) 4I

(c) 6I (d) 8I.

109. If L and E_{kr} represent the angular momentum and the rotational kinetic energy respectively of a body, then which of the following correctly represents the relation between L and $\sqrt{E_{kr}}$?

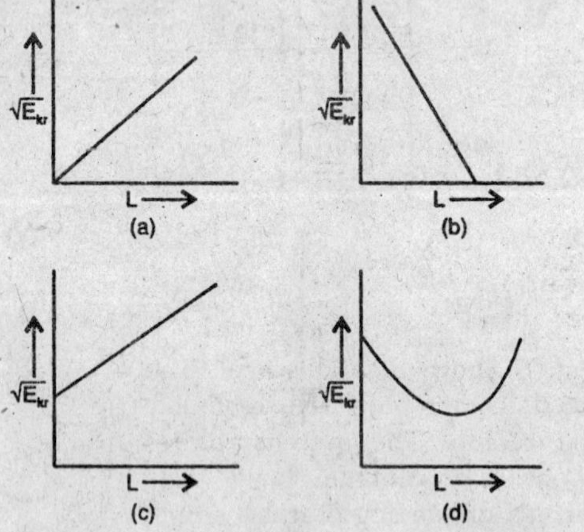

Fig. 28

110. Two uniform thin identical rods, each of mass M and length L are joined so as to form a cross as shown in (Fig. 29). The moment of inertia of the system about the bisector line AB is

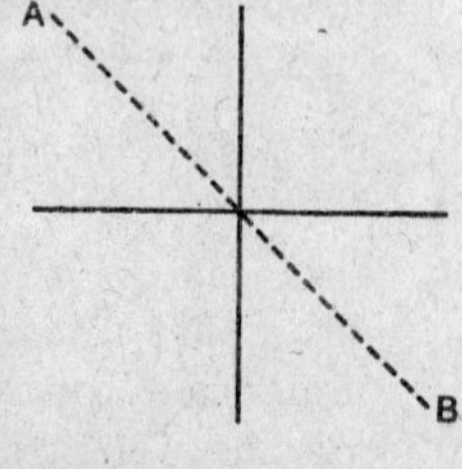

Fig. 29

(a) $\frac{ML^2}{12}$ (b) $\frac{ML^2}{6}$

(c) $\frac{ML^2}{3}$ (d) $\frac{2ML^2}{3}$.

111. A pendulum consists of a small mass m at the end of a string of length l. The pendulum is pulled aside to an angle θ with the vertical and released. At the instant of release, using the suspension point as axis, what is the torque on the pendulum ?

(a) mg (b) mgl

(c) $mgl \cos \theta$ (d) $mgl \sin \theta$.

112. In Q. 111, the angular acceleration at the instant of release is

(a) g (b) $g \sin \theta$

(c) $g \sin \theta/l$ (d) $g \cos \theta/l$.

113. Three thin uniform rods, each of mass m and length l lie along the x, y, z axes with one end of each at the origin. The moment of inertia about the z-axis for the three-rod system is

(a) $\frac{2}{3}ml^2$ (b) $\frac{1}{3}ml^2$

(c) $\frac{1}{2}ml^2$ (d) ml^2.

114. A car tyre of radius g metre is rolling without slipping on a level road with speed v. When it is at a point A on the road, a small stone stuck in its tread separates from it at its highest point and falls at a point B on the road. Then, AB =

(a) v (b) $2v$

(c) $3v$ (d) $4v$.

115. The velocity of a homogeneous solid sphere after rolling down an inclined plane of vertical height h, from rest without sliding is

(a) $\sqrt{gh}$ (b) $\sqrt{\frac{4gh}{3}}$

(c) $\sqrt{\frac{7gh}{3}}$ (d) $\sqrt{\frac{10gh}{7}}$

116. Two discs, each having moment of inertia 5 kg m^2 about its central axis, rotating with speeds 10 rad s^{-1} and 20 rad s^{-1}, are brought in contact face to face with their axes of rotation coincided. The loss of kinetic energy in the process is

(a) 2 J (b) 5 J

(c) 125 J (d) 0 J.

117. A cubical block of side L rests on a rough horizontal surface with coefficient of friction μ. A horizontal

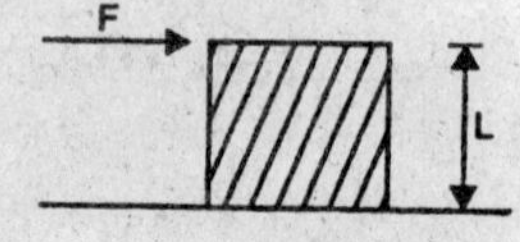

Fig. 30

force F is applied on the block as shown in (Fig. 30). If the coefficient of friction is sufficiently high so that the block does not slide before toppling, the minimum force required to topple the block is

(a) infinitesimal (b) $\frac{mg}{4}$

(c) $\frac{mg}{2}$ (d) $mg(1-\mu)$.

118. A hollow sphere of mass 200 g starting from rest rolls down, without slipping, an inclined plane which rises by 2 in every 5. What is the acceleration of the sphere? ($g = 10\text{ m s}^{-2}$.)

(a) 5 m s^{-2} (b) 12 m s^{-2}

(c) $\frac{12}{5}$ m s^{-2} (d) $\frac{12}{7}$ m s^{-2}.

119. In Q. 118, what is the velocity acquired by the sphere after 5 second?

(a) 5 m s^{-1} (b) 12 m s^{-1}

(c) $\frac{12}{5}$ m s^{-1} (d) $\frac{12}{7}$ m s^{-1}.

120. A body is rolling without slipping on a horizontal plane. The rotational energy of the body is 40% of the total kinetic energy. Identify the body.

(a) Ring (b) Hollow cylinder

(c) Solid cylinder (d) Hollow sphere.

121. The moment of inertia of a uniform semi-circular wire of mass M and radius R about a line perpendicular to the plane of the wire and passing through the centre is

(a) MR^2 (b) $2MR^2$

(c) $\frac{1}{2}MR^2$ (d) $4MR^2$.

122. A person supports a book between finger and thumb as shown in (Fig. 31) (the point of grip is assumed to be at the corner of the book). If the book has a weight of W, then the person is producing a torque, on the book, of

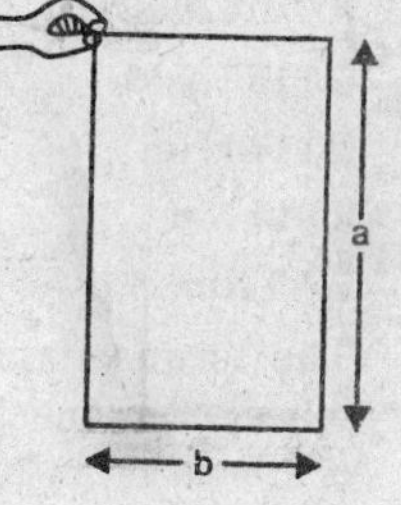

Fig. 31

(a) W $\frac{a}{2}$ anticlockwise

(b) W $\frac{b}{2}$ anticlockwise

(c) Wa anticlockwise (d) Wa clockwise.

123. A string of negligible thickness is wrapped several times around a cylinder kept on a rough horizontal surface. A man standing at a distance l from the cylinder holds one end of the string and pulls the cylinder towards him (Fig. 32). There is no slipping anywhere. The length of the string passed through the hand of the man while the cylinder reaches his hands is

Fig. 32

(a) l (b) $2l$

(c) $3l$ (d) $4l$.

124. A wire of uniform cross-section is bent in the shape shown in (Fig. 33). The co-ordinates of the centre of mass of each side are shown in (Fig. 33). The co-ordinates of the centre of mass of the system are

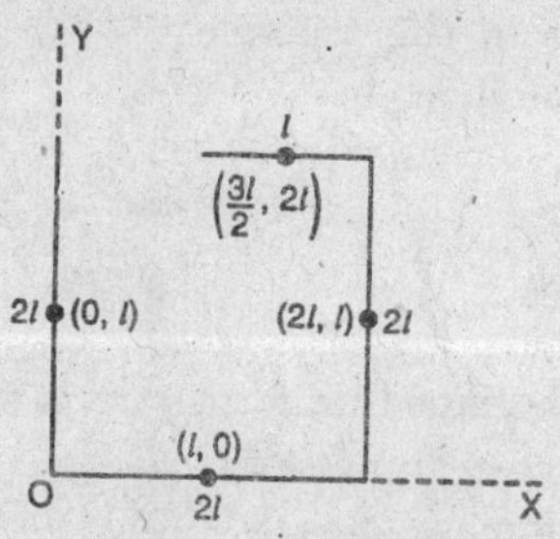

Fig. 33

(a) $\left(\frac{15l}{14}, \frac{6l}{7}\right)$ (b) $\left(\frac{15l}{14}, l\right)$

(c) $\left(l, \frac{l}{2}\right)$ (d) (l, l).

125. A sphere of mass 2 kg and radius 5 cm rolls without slipping along a horizontal plane. The velocity of its centre of mass is 10 cm s^{-1}. The kinetic energy of the sphere is

(a) 0.07 J (b) 0.014 J

(c) 0.025 J (d) 1.25 J.

126. The rectangular blocks A, B, C and D are piled one above the other in such a way that each block projects a little beyond the block below it. If the length of each block is l, then the maximum possible projection for A is

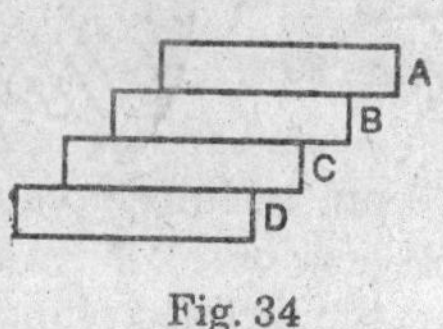

Fig. 34

(a) $\frac{l}{2}$ (b) $\frac{l}{4}$

(c) $\frac{l}{6}$ (d) $\frac{l}{8}$.

127. In Q. 126, the maximum possible projection for B is

(a) $\frac{l}{2}$ (b) $\frac{l}{4}$

(c) $\frac{l}{6}$ (d) $\frac{l}{8}$.

128. In Q. 126, the maximum possible projection for C is

(a) $\frac{l}{2}$ (b) $\frac{l}{4}$

(c) $\frac{l}{6}$ (d) $\frac{l}{8}$.

129. Three point masses m_1, m_2 and m_3 are located at the vertices of an equilateral triangle of side a. The moment of inertia of the system about an axis along the altitude of the triangle passing through m_1 is

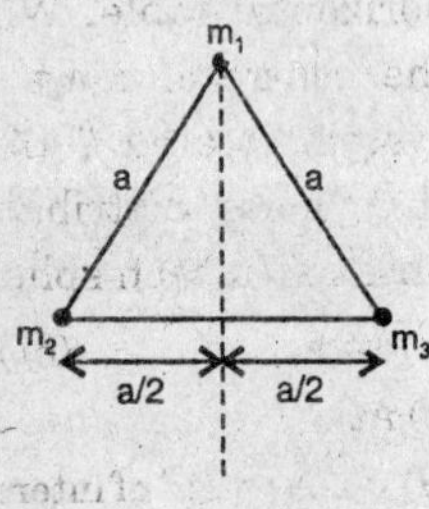

Fig. 35

(a) $\frac{a^2}{4}(m_2 + m_3)$ (b) $m_1 a^2$

(c) $(m_2 + m_3)a^2$ (d) $(m_1 + m_2 + m_3)a^2$.

130. Given : A U-shaped uniform wire of sides $2l$, l and l. The x and y co-ordinates of the centre of mass of each side are shown in (Fig. 36). The x and y co-ordinates of the centre of mass of wire are respectively

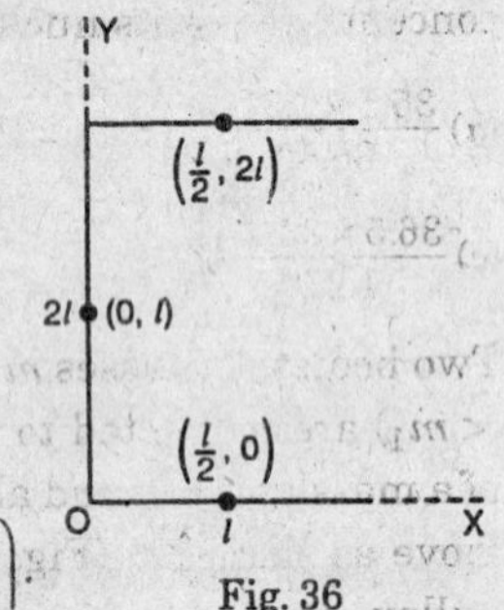

Fig. 36

(a) $\left(\frac{l}{4}, \frac{l}{4}\right)$ (b) (l, l)

(c) $\left(l, \frac{l}{4}\right)$ (d) $\left(\frac{l}{4}, l\right)$

131. The angular velocity of a body is $\vec{\omega} = 2\hat{i} + 3\hat{j} + 4\hat{k}$. A torque $\vec{\tau} = \hat{i} + 2\hat{j} + 3\hat{k}$ acts on it. The rotational power is (consider SI units only)

(a) 10 W (b) 20 W

(c) 30 W (d) 40 W.

132. A solid sphere of mass M rolls without slipping down, an inclined plane making an angle θ with the horizontal. Then the acceleration of the sphere down the inclined plane is

(a) g (b) $g \sin \theta$

(c) $\frac{5}{7} g \sin \theta$ (d) $\frac{2}{7} g \sin \theta$.

133. In Q. 132, the frictional force between the sphere of the inclined plane is

(a) $\frac{2}{7} mg \sin \theta$ (b) $\frac{5}{7} mg \sin \theta$

(c) $\mu\, mg \cos \theta$ (d) $mg \sin \theta$.

134. What is the moment of inertia of a wheel of mass m and radius R having two spokes each of mass m ? (The axis of rotation is perpendicular to the plane and passing through centre)

(a) $\frac{1}{2} mR^2$ (b) mR^2

(c) $\frac{2}{3} mR^2$ (d) $\frac{5}{3} mR^2$.

135. When a sphere rolls without slipping the ratio of kinetic energy of translation to its total kinetic energy is

(a) 1 : 7 (b) 1 : 2

(c) 1 : 1 (d) 5 : 7.

136. Calculate the force F that is applied horizontally at the axle of the wheel which is necessary to raise the wheel over the obstacle of height 4 m. Radius of wheel is 1 m and mass = 10 kg. F is (Take : $g = 10$ m s^{-2})

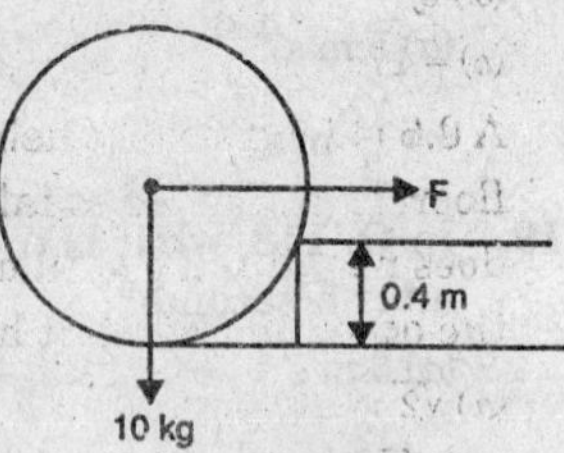

Fig. 37

(a) 100 N (b) 66 N

(c) 167 N (d) 133.3 N.

137. A turn table, initially at rest, is set in motion with a uniform angular acceleration α. Its angular velocity in rad s^{-1} at the end of one complete revolution is

(a) $\sqrt{\pi\alpha}$ (b) $2\sqrt{\pi\alpha}$

(c) $3\sqrt{\pi\alpha}$ (d) $4\sqrt{\pi\alpha}$.

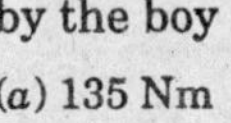

138. A boy pushes a merry-go-round of radius 3 m with a force 50 N as shown in (Fig. 38). If cos 20° = 0.9, then the torque applied by the boy is

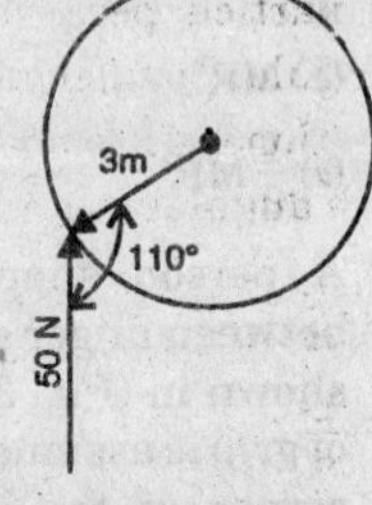

Fig. 38

(a) 135 Nm

(b) 140 Nm

(c) 40 Nm

(d) 0 Nm.

139. The torque required to stop a wheel of moment of inertia 5×10^{-3} kg m^2 from a speed of 20 rad s^{-1} in 10 s is

(a) 1 Nm (b) 0.1 Nm

(c) 0.01 Nm (d) 0.001 Nm.

140. The moment of inertia of a uniform disc about an axis, perpendicular to the disc, at the centre is $\frac{1}{2}$ MR^2 (M = mass, R = radius of disc). If the disc is rolling on its edge without slipping on a straight line path, the ratio of rotational kinetic energy to translational one is

(a) 1 (b) $\frac{1}{2}$

(c) $\frac{1}{4}$ (d) $\frac{1}{8}$.

141. The diameter of a flywheel increases by 1%. The percentage increase in moment of inertia about its axis passing through the centre will be

(a) 2% (b) 2.5%

(c) 1.5% (d) 3%.

142. The torque of force $(3\hat{i} + 2\hat{j} + 3\hat{k})$ newton acting at a point is : (Given : position vector of point is $[2\hat{i} - 3\hat{j} + 4\hat{k}]$ metre)

(a) $6\hat{j}$ (b) $13\hat{k}$

(c) $-17\hat{i}$ (d) $-17\hat{i} + 6\hat{j} + 13\hat{k}$.

143. A 0.6 m long stick is held vertically with one end on the floor. It is allowed to fall such that the end on the floor does not slip. If $g = 10$ m s^{-2}, then what is the speed of the other end when it hits the floor ?

(a) $\sqrt{2}$ m s^{-1} (b) $2\sqrt{2}$ m s^{-1}

(c) $3\sqrt{2}$ m s^{-1} (d) $4\sqrt{2}$ m s^{-1}.

144. A wheel of radius r and mass m stands in front of a step of height h. The least horizontal force which should be applied to the axle of the wheel to allow it to raise on to the step is

(a) $\frac{mgh\,(2r-h)}{r-h}$ (b) $mgh(r-h)$

(c) $\frac{mg\sqrt{h(2r-h)}}{r-h}$ (d) None of these.

145. A particle performs uniform circular motion with an angular momentum L. If the frequency of particle's motion is doubled and its kinetic energy is halved, then the angular momentum becomes

(a) $\frac{L}{2}$ (b) $\frac{L}{4}$

(c) $\frac{L}{6}$ (d) $\frac{L}{8}$.

146. A thin uniform wire of length l and mass m is bent in the form of a semi-circle. Its moment of inertia about an axis (Y′OY) passing through the free ends is

(a) $\frac{1}{2}ml^2$ (b) $\frac{ml^2}{2\pi^2}$

(c) $\frac{ml^2}{3\pi^2}$ (d) $\frac{ml^2}{4\pi^2}$.

147. The radius of gyration of a circular plate of radius R, about an axis perpendicular to its plane and bisecting a radius is

(a) R (b) $\frac{R}{2}$

(c) $\frac{\sqrt{3}\,R}{2}$ (d) $\frac{\sqrt{5}\,R}{2}$.

148. Three identical spheres, each of radius R, are placed touching each other on a horizontal table. Where is the centre of mass of the system located ? (Assume that mass distribution is uniform in each sphere)

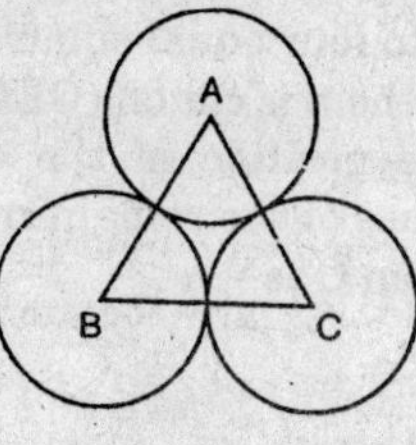

Fig. 39

(a) At A (b) At B

(c) At C

(d) At the point of intersection of the medians of the triangle.

149. In the HCl molecule, the separation between the nuclei of the two atoms is 1.27 Å. What is the approximate location of the centre of mass of the molecule ? (Given : chlorine atom is nearly 35.5 times as massive as a hydrogen atom and nearly all the mass of an atom is concentrated in its nucleus.)

(a) $\frac{35.5 \times 1.27}{36.5}$ Å (b) $\frac{36.5 \times 1.27}{35.5}$ Å

(c) $\frac{36.5 \times 35.5}{1.27}$ Å (d) $\frac{1.27}{36.5 \times 35.5}$ Å.

150. Two bodies of masses m_1 and m_2 ($< m_1$) are connected to the ends of a massless cord and allowed to move as shown in (Fig. 40). The pulley is both massless and frictionless. The acceleration of the centre of mass is

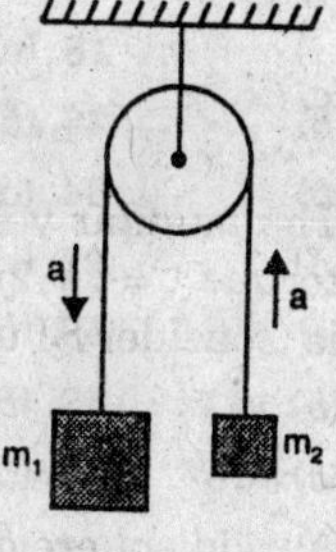

Fig. 40

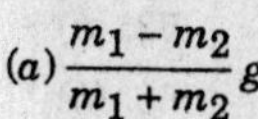

(a) $\frac{m_1 - m_2}{m_1 + m_2}g$ (b) $\frac{m_1 - m_2}{(m_1 + m_2)^2}g$

(c) $\frac{(m_1 - m_2)^2}{m_1 + m_2}g$ (d) $\frac{(m_1 - m_2)^2}{(m_1 + m_2)^2}\vec{g}$.

151. The co-ordinates of the centre of mass of a system of three particles of masses 1 μg, 2 μg and 3 μg are (2, 2, 2). Where should a fourth particle of mass 4 mg be positioned so that the centre of mass of the four particle system is at the origin of the three-dimensional rectangular co-ordinate system ?

(a) At (1, 1, 1) (b) At (2, 2, 2)

(c) At (3, 3, 3) (d) At (– 3, – 3, – 3).

152. A square of side 4 cm and of uniform thickness is divided into four equal squares. If one of them is cut off (OECF), then the position of the centre of mass of the remaining portion from O is

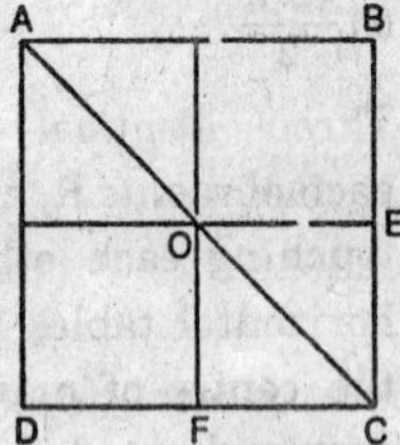

Fig. 41

(*a*) $\frac{1}{\sqrt{3}}$ cm (*b*) $\frac{2}{\sqrt{3}}$ cm

(*c*) $\frac{5}{\sqrt{3}}$ cm (*d*) $\frac{7}{\sqrt{3}}$ cm

(*e*) $\frac{\sqrt{2}}{3}$ cm.

ANSWERS

QUESTION BANK (Level I)

1. (*c*) **2.** (*b*) **3.** (*b*) **4.** (*c*) **5.** (*b*) **6.** (*d*) **7.** (*d*) **8.** (*b*)
9. (*d*) **10.** (*d*) **11.** (*c*) **12.** (*d*) **13.** (*b*) **14.** (*c*) **15.** (*a*) **16.** (*a*)
17. (*a*) **18.** (*a*) **19.** (*a*) **20.** (*b*) **21.** (*a*) **22.** (*d*) **23.** (*c*) **24.** (*d*)
25. (*d*) **26.** (*d*) **27.** (*a*) **28.** (*b*) **29.** (*c*) **30.** (*b*) **31.** (*b*) **32.** (*d*)
33. (*b*) **34.** (*d*) **35.** (*b*) **36.** (*c*) **37.** (*c*) **38.** (*b*) **39.** (*d*) **40.** (*a*)
41. (*a*) **42.** (*a*) **43.** (*b*) **44.** (*c*) **45.** (*c*) **46.** (*b*) **47.** (*a*) **48.** (*b*)
49. (*b*) **50.** (*c*) **51.** (*a*) **52.** (*d*) **53.** (*a*) **54.** (*b*) **55.** (*b*) **56.** (*c*)
57. (*c*) **58.** (*d*) **59.** (*c*) **60.** (*c*) **61.** (*a*) **62.** (*a*) **63.** (*c*) **64.** (*c*).

QUESTION BANK (Level II)

1. (*c*) **2.** (*b*) **3.** (*b*) **4.** (*a*) **5.** (*b*) **6.** (*a*) **7.** (*c*) **8.** (*a*)
9. (*b*) **10.** (*c*) **11.** (*c*) **12.** (*d*) **13.** (*a*) **14.** (*d*) **15.** (*c*) **16.** (*d*)
17. (*d*) **18.** (*c*) **19.** (*b*) **20.** (*b*) **21.** (*c*) **22.** (*a*) **23.** (*d*) **24.** (*c*)
25. (*b*) **26.** (*b*) **27.** (*a*) **28.** (*b*) **29.** (*d*) **30.** (*c*) **31.** (*c*) **32.** (*d*)
33. (*b*) **34.** (*d*) **35.** (*b*) **36.** (*d*) **37.** (*c*) **38.** (*a*). **39.** (*b*) **40.** (*c*)
41. (*d*) **42.** (*c*) **43.** (*b*) **44.** (*c*) **45.** (*d*) **46.** (*c*) **47.** (*c*) **48.** (*c*)
49. (*e*) **50.** (*b*) **51.** (*d*) **52.** (*b*) **53.** (*b*) **54.** (*a*) **55.** (*b*) **56.** (*c*)
57. (*b*) **58.** (*b*) **59.** (*a*) **60.** (*b*) **61.** (*c*) **62.** (*b*) **63.** (*a*) **64.** (*d*)
65. (*b*) **66.** (*d*) **67.** (*c*) **68.** (*d*) **69.** (*b*) **70.** (*b*) **71.** (*a*) **72.** (*c*)
73. (*c*) **74.** (*b*) **75.** (*c*) **76.** (*a*) **77.** (*b*) **78.** (*b*) **79.** (*b*) **80.** (*b*)
81. (*d*) **82.** (*c*) **83.** (*b*) **84.** (*c*) **85.** (*c*) **86.** (*b*) **87.** (*c*) **88.** (*c*)
89. (*c*) **90.** (*c*) **91.** (*d*) **92.** (*a*) **93.** (*c*) **94.** (*c*) **95.** (*c*) **96.** (*a*)
97. (*d*) **98.** (*a*) **99.** (*a*) **100.** (*d*) **101.** (*c*) **102.** (*d*) **103.** (*b*) **104.** (*c*)
105. (*c*) **106.** (*b*) **107.** (*b*) **108.** (*c*) **109.** (*a*) **110.** (*a*) **111.** (*d*) **112.** (*c*)
113. (*a*) **114.** (*d*) **115.** (*d*) **116.** (*c*) **117.** (*c*) **118.** (*c*) **119.** (*b*) **120.** (*d*)
121. (*a*) **122.** (*b*) **123.** (*b*) **124.** (*a*) **125.** (*b*) **126.** (*a*) **127.** (*b*) **128.** (*c*)
129. (*a*) **130.** (*d*) **131.** (*b*) **132.** (*c*) **133.** (*a*) **134.** (*d*) **135.** (*d*) **136.** (*d*)
137. (*b*) **138.** (*a*) **139.** (*c*) **140.** (*b*) **141.** (*a*) **142.** (*d*) **143.** (*c*) **144.** (*c*)
145. (*b*) **146.** (*b*) **147.** (*c*) **148.** (*d*) **149.** (*a*) **150.** (*d*) **151.** (*d*) **152.** (*e*).

HINTS/SOLUTIONS

QUESTION BANK (Level I)

1. (c) $a_c = \omega^2 r = 4\pi^2 v^2 r$

$= 4\pi^2 \times 2 \times 2 \times 0.25 = 4\pi^2 \text{ m s}^{-2}$

2. (b) $f_{ms} = \mu mg = 0.5mg, \mu = 0.5$

$v = \sqrt{\mu gr} = \sqrt{0.5 \times 9.8 \times 10} = \sqrt{49} = 7 \text{ m s}^{-1}$

3. (b) $T = \frac{mv^2}{r} = \frac{0.5 \times 4 \times 4}{1} \text{ N} = 8 \text{ N}$

4. (c) $v = \sqrt{0.4 \times 9.8 \times 30} \text{ m s}^{-1}$

$= \sqrt{117.6} \text{ m s}^{-1} = 10.84 \text{ m s}^{-1}$.

5. (b) $mgH = \frac{1}{2}mv^2$ or $H = \frac{v^2}{2g} = \frac{5gR}{2g} = \frac{5R}{2}$

6. (d) $v = \sqrt{\mu gr}$

When μ becomes $\frac{\mu}{2}$, v becomes $\frac{v}{\sqrt{2}}$ *i.e.* $\frac{10}{\sqrt{2}}$ or $\frac{10\sqrt{2}}{2}$

or $5\sqrt{2} \text{ m s}^{-1}$.

7. (d) Since the speed is given to be constant therefore there is no tangential acceleration.

8. (b) $\omega = \frac{2\pi}{60} \text{ rad s}^{-1}$

$v = 1 \times \omega = \frac{2\pi}{60} \text{ cm s}^{-1}$

9. (d) When tension is doubled, angular velocity becomes $\sqrt{2}$ times.

$\therefore$ New angular velocity $= \sqrt{2} \times 5$ rpm

$= 1.4 \times 5$ rpm $= 7$ rpm.

10. (d) $\omega_1 = 2\pi \times 300$ rad/min

$\omega_2 = 2\pi \times 100$ rad/min

$\alpha = \frac{\omega_1 - \omega_2}{2} = \frac{2\pi \times 300 - 2\pi \times 100}{2}$

$= 2\pi \times 100 \text{ rad/min}^2 = 200\pi \text{ rad/min}^2$.

11. (c) On a plane bridge, $F_A = mg$

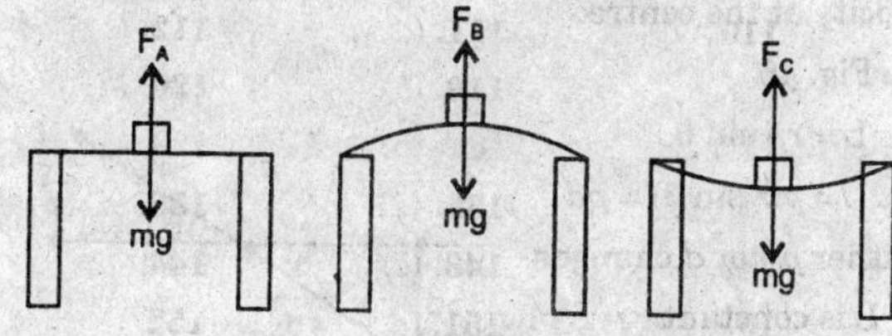

Fig. 42

On a convex bridge, $mg - F_B = \frac{mv^2}{R} \Rightarrow F_B = mg - \frac{mv^2}{R}$

On a concave bridge, $F_C - mg = \frac{mv^2}{R}$ or $F_C = mg + \frac{mv^2}{R}$

Clearly, F_C is maximum of the three forces.

12. (d) As per the conditions in the problem, tension at the highest point is zero.

$\therefore \quad \frac{T_L}{T_H} = \infty$

13. (b) $a_c = \omega^2 r$

or $\omega = \sqrt{\frac{a_c}{r}} = \sqrt{\frac{9.8 \times 100}{20}} = \sqrt{\frac{980}{20}} = \sqrt{49} = 7 \text{ rad s}^{-1}$

14. (c) For a convex surface in general,

$mg - N = \frac{mv^2}{r}$

or $\quad N = mg - \frac{mv^2}{r}$

At A, r less, $\frac{mv^2}{r}$ more, N less.

$\therefore \quad N_A < N_B$

Fig. 43

15. (a) When $\theta = 0°$, the net force $T - mg$ is directed vertically upwards. So, acceleration is directed vertically upwards. This alone decides that (a) is the correct choice.

16. (a) $v = \frac{540}{60} = 9, \omega = 2\pi v = 18\pi$

Angular acceleration $= \frac{\text{Gain in angular velocity}}{\text{time}}$

$= \frac{18\pi}{6} = 3\pi \text{ rad s}^{-2}$

17. (a) $f = mr\omega^2$, $r\omega^2$ = constant, ω doubled, ω^2 becomes four times, r is reduced by a factor of 4. So, the new radius is 1 cm.

18. (a) The angle between N and W is $180° - \theta$. Also,

W = N cos θ. Also, N sin θ provides the necessary centripetal force.

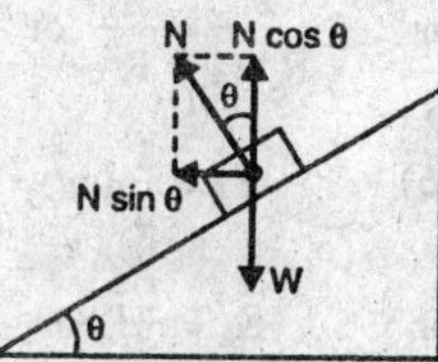

Fig. 44

Resultant of N and W

$= \sqrt{N^2 + W^2 + 2NW \cos(180° - \theta)}$

$= \sqrt{N^2 + N^2 \cos^2 \theta - 2N^2 \cos^2 \theta}$

$= \sqrt{N^2 (1 - \cos^2 \theta)} = N \sin \theta$

19. (*a*) $mg + T_2 = \frac{mv_2^2}{R}$, $T_1 - mg = \frac{mv_1^2}{R}$

Never confuse the actual force on a body (tension, force of gravity etc.) with the effects they produce : ma, $\frac{mv^2}{R}$ *etc.*

20. (*b*) $I = MR^2 = 6 \times \frac{40}{100} \times \frac{40}{100}$ kg m^2 = 0.96 kg m^2

21. (*a*) Rotational kinetic energy $= \frac{1}{2} I\omega^2$

$$= \frac{1}{2} \times 0.96 \times 4\pi^2 \times \frac{300}{60} \times \frac{300}{60} \text{ joule}$$
$$= 0.48 \times 100\pi^2 \text{ joule} = 48\pi^2 \text{ joule}$$

22. (*d*) $M = \pi R^2 t\rho$, M and t being same,

$R^2\rho$ = constant or $R^2 \propto \frac{1}{\rho}$

Again, $I = \frac{1}{2} MR^2$ or $I \propto R^2$ or $I \propto \frac{1}{\rho}$

$\therefore \frac{I_1}{I_2} = \frac{\rho_2}{\rho_1}$

23. (*c*) $I_B = I_A + Md^2$

24. (*d*) 'Distribution of mass is closest to axis' in the case of axis EG.

25. (*d*) Applying conservation of angular momentum, we get

$$I\omega = [I + mR^2]\omega' \text{ or } \omega' = \frac{I\omega}{I + mR^2}$$

26. (*d*) $I = \frac{1}{2} \times \frac{500}{1000} \times \frac{10}{100} \times \frac{10}{100}$ kg m^2

$= 2.5 \times 10^{-3}$ kg m^2

27. (*a*) $\tau = \frac{6-2}{4} = \frac{4}{4} = 1$ unit

28. (*b*) $L = I\omega = MR^2\omega$

$= 10 \times 0.2 \times 0.2 \times 2 \times \frac{22}{7} \times \frac{2100}{60}$ kg m^2 s^{-1}

$= 88$ kg m^2 s^{-1}

29. (*c*) $I\omega$ = constant,

$\frac{2}{5} MR^2 \times \frac{2\pi}{T}$ = constant or $\frac{T}{R^2}$ = constant

or $\frac{T'}{R^2/4} = \frac{T}{R^2}$ or $\frac{4T'}{R^2} = \frac{24}{R^2}$ or $T' = 6h$

30. (*b*) $E_k = \frac{1}{2} I\omega^2 = \frac{I^2\omega^2}{2I}$

or $E_k = \frac{L^2}{2I}$ or $L = \sqrt{2IE_k}$

31. (*b*) $\frac{1}{2} I\omega^2 = \frac{1}{2} mgl$ or $\frac{ml^2}{3} \times \frac{v^2}{l^2} = mgl$

But $l = 1$ m

$\therefore$ $v = \sqrt{3g}$ m s^{-1} = $\sqrt{3 \times 9.8}$ m s^{-1}

$= \sqrt{29.4}$ m s^{-1} = 5.4 m s^{-1}

32. (*d*) $\tau = \frac{I(\omega_2 - \omega_1)}{t}$ or $t = \frac{I(\omega_2 - \omega_1)}{\tau}$

or $t = \frac{12(100-40)}{20}$ s $= \frac{12 \times 60}{20}$ s = 36 s

33. (*b*) $\tau = I\alpha$ or $\alpha = \frac{\tau}{I}$ or $\alpha = \frac{400}{100} = 4$ rad s^{-2}

$\omega = \alpha t = 4 \times 4 = 16$ rad s^{-1}

34. (*d*) $I = m \times 0^2 + m \times a^2 + m \times a^2 + m(2a)^2$ or $I = 6ma^2$

35. (*b*) I is doubled. ω is halved.

Rotational K.E. $= \frac{1}{2} I\omega^2$

Clearly, rotational K.E. is decreased.

36. (*c*) $\frac{I_1}{I_2} = \frac{\frac{1}{2} \times 1 \times 2 \times 2}{\frac{1}{2} \times 2 \times 1 \times 1} = \frac{2}{1}$

37. (*c*) K = 6 cm, M = 5 g

$I = MK^2 = 5 \times 6 \times 6$ g cm^2 = 180 g cm^2

38. (*b*) $\tau = \frac{dL}{dt}$

If $\tau = 0$, then $\frac{dL}{dt} = 0$ or L = constant

39. (*d*) $I = \frac{2}{5} MR^2$

$I = \frac{2}{5} \times \frac{4}{3} \pi R^3 \rho R^2$ or $I \propto R^5$

$\frac{I_1}{I_2} = \frac{1}{2^5} = \frac{1}{32}$

40. (*a*) $MK^2 = \frac{1}{2} MR^2$ or $K = \frac{R}{\sqrt{2}}$

41. (*a*) $v = r\omega$ ω is constant

$\therefore$ $v \propto r$

The particle at the highest position is at double the distance as compared to the distance of the centre from the point of contact. So velocity is twice the velocity at the centre.

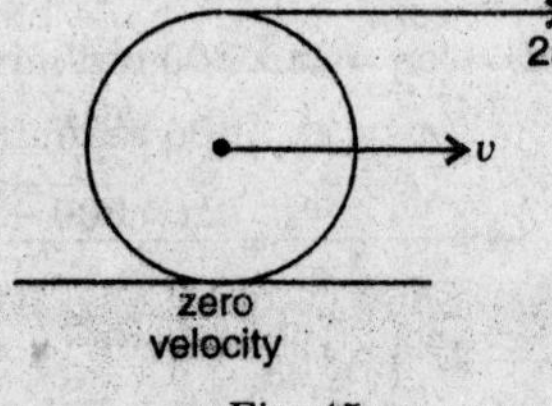

Fig. 45

42. See Fig. 46

$L = rp \sin\theta$

or $L = p(r \sin\theta) = pd$

Neither p nor d changes.

So, L is constant.

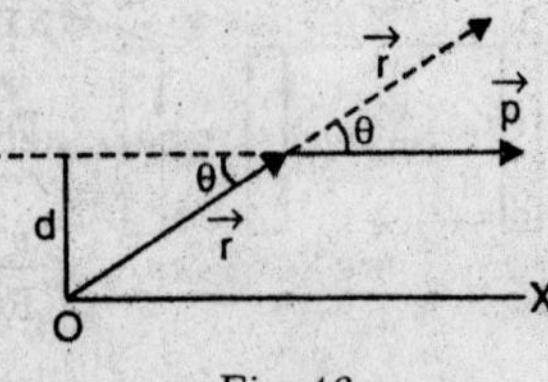

Fig. 46

43. (*b*) $\tau = FR$

$W = \tau\theta = FR\theta$

44. (*c*) $I = I_g + Md^2$

or $MK^2 = MK_g^2 + Md^2$ or $K^2 = K_g^2 + d^2$

$10^2 = K_g^2 + 6^2$ or $K_g^2 = 64$ or $K_g = 8$ cm

45. (*c*) Applying conservation of angular momentum,

$$\frac{2}{5}M_1R_2^2\omega_2 = \frac{2}{5}M_1R_1^2\omega_1 \text{ or } \omega_2 = \frac{R_1^2\omega_1}{R_2^2}$$

46. (*b*) No external torque is involved. Using conservation of angular momentum,

$I_f\omega_f = I_s\omega_s$

or $\omega_f = \frac{I_s\omega_s}{I_f} = \frac{10^5 \times 0.1}{1000}$ rad s^{-1} = 10 rad s^{-1}

47. (*a*) $L = I\omega$

$L = 0.6 \times 2\pi \times 0.5$ kg m^2 s^{-1}

or $L = 0.6\pi$ kg m^2 s^{-1}

48. (*b*) $P = \tau\omega = I\alpha\omega$ or $\omega = \frac{P}{I\alpha}$

49. Net torque, $\tau = 30g \times 2 - 20g \times 2 = 20g$

$I = 30 \times 2^2 + 20 \times 2^2 = 120 + 80$

$= 200$ kg m^2

$\tau = I\alpha,\ \alpha = \frac{\tau}{I} = \frac{20g}{200} = \frac{g}{10}$

$= \frac{9.8}{10} = 0.98$ rad s^{-2}

50. (*c*) $\tau = Fd$

τ is same in magnitude for both cases.

Also, $\tau = I\alpha$. So, $I\alpha$ is constant.

For case (*b*), I is small. So, α is large.

51. (*a*) $v = r\omega$; ω same, $v \propto r$.

52. (*d*) $I = \frac{ML^2}{3} = 4 \times \frac{ML^2}{12} = 4I_0$

53. (*a*) Energy $= \frac{1}{2}mv^2 + \frac{1}{2}I\omega^2$

$$= \frac{1}{2}mv^2 + \frac{1}{2}\left(\frac{2}{3}mr^2\right)\frac{v^2}{r^2} = \frac{1}{2}mv^2 + \frac{1}{3}mv^2 = \frac{5mv^2}{6}$$

54. (*b*) Use $\tau = I\alpha$ and $I = mK^2$

55. (*b*) In the first case, kinetic energy is $\frac{1}{2}mv^2$

In the second case, kinetic energy is $\frac{1}{2}mv^2 + \frac{1}{2}I\omega^2$

or $\frac{1}{2}mv^2 + \frac{1}{2} \times mR^2 \times \frac{v^2}{R^2}$ or mv^2

So, ratio is $\frac{\frac{1}{2}mv^2}{mv^2}$ or $\frac{1}{2} : 1$ or $1 : 2$

56. (*c*) $M = \pi R^2 td,\ R^2 \propto \frac{1}{d}$

$$\frac{I_A}{I_B} = \frac{\frac{1}{2}MR_A^2}{\frac{1}{2}MR_B^2} = \frac{d_B}{d_A}$$

But $d_B < d_A \ \therefore\ I_A < I_B$

57. (*c*) $a = \frac{10 \sin 30°}{1 + \frac{2}{5}} = \frac{5}{7} \times 5 = \frac{25}{7}$ m s^{-2}

58. (*d*) The torques F_1d and F_2d of F_1 and F_2 respectively are counter-clockwise. The torque F_3d is clockwise. Applying condition for rotational equilibrium,

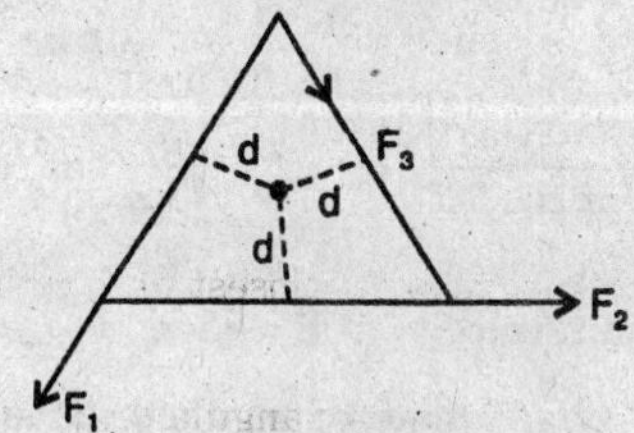

Fig. 47

$F_1d + F_2d - F_3d = 0$ or $F_3 = F_1 + F_2$

59. (*c*) $F' \times \frac{R}{5} = FR$ or $F' = 5F$

60. (*c*) $\vec{L} = \vec{r} \times \vec{p} = m(\vec{r} \times \vec{v})$

$$= \frac{10}{1000}[(10\hat{i} + 6\hat{j}) \times 5\hat{i}] = -0.3\hat{k} \text{ Js}$$

61. (*a*) $\frac{1}{2} \times 27 \times v^2 = \frac{1}{2} \times 3 \times 3 \times 3$ or $v = 1$ m s^{-1}

62. (*a*) $$\frac{\text{Rotational K.E.}}{\text{Translational K.E.}} = \frac{\frac{1}{2}I\omega^2}{\frac{1}{2}mv^2} = \frac{mK^2\frac{v^2}{R^2}}{mv^2} = \frac{K^2}{R^2}$$

63-64. $E = \frac{1}{2}Mv^2 + \frac{1}{2}I\omega^2$

$$= \frac{1}{2}Mv^2 + \frac{1}{2} \times \frac{2}{5}MR^2 \times \frac{v^2}{R^2} = \frac{1}{2}Mv^2 + \frac{1}{5}Mv^2 = \frac{7}{10}Mv^2$$

$= \frac{7}{10} \times 1 \times 1 \times 1$ J = 0.7 J

QUESTION BANK (Level II)

1. (c) $N \cos \theta = mg$

or $ma\omega^2 \cos \theta = mg$ or $\omega^2 = \dfrac{g}{a \cos \theta}$

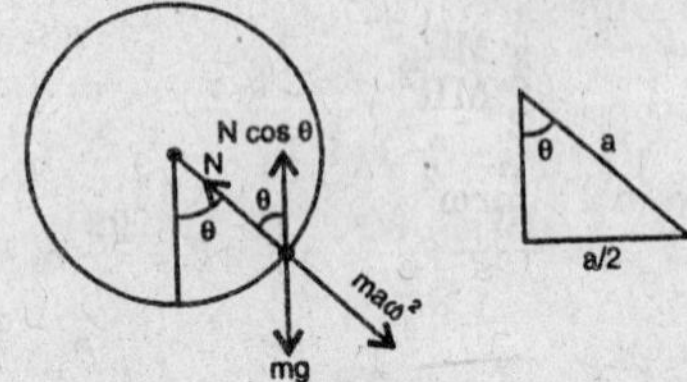

Fig. 48

Now, $\sin \theta = \dfrac{a}{2} \times \dfrac{1}{a}$ or $\sin \theta = \dfrac{1}{2}$ or $\theta = 30°$

$\therefore$ $\omega^2 = \dfrac{g}{a \cos 30°}$ or $\omega^2 = \dfrac{g}{\frac{\sqrt{3}\, a}{2}}$ or $\omega^2 = \dfrac{2g}{\sqrt{3}\, a}$

2. (b) Minimum velocity at B $= \sqrt{5ga}$

Now, $mgh = \dfrac{1}{2} m \times 5ga$ or $h = \dfrac{5a}{2}$

3. (b) $\tan \theta = \dfrac{v^2}{rg} = \dfrac{10 \times 10}{50 \times 10} = \dfrac{1}{5}$

or $\theta = \tan^{-1}\left[\dfrac{1}{5}\right]$

4. (a) $T \propto \sqrt{\cos \theta}$ or $T^2 \propto \cos \theta$

$$\frac{T_1^2}{T_2^2} = \frac{\cos 45°}{\cos 60°} = \frac{1}{\sqrt{2}} \times \frac{2}{1} = \sqrt{2}$$

5. (b) $v = \sqrt{9.8 \times 39.2}$ m s^{-1} = 19.6 m s^{-1}

6. (a) $\dfrac{T_{max.}}{T_{min.}} = \dfrac{\dfrac{mv_L^2}{L} + mg}{\dfrac{mv_H^2}{L} - mg} = 4$

or $\dfrac{v_L^2}{L} + g = \dfrac{4v_H^2}{L} - 4g$ or $\dfrac{v_L^2 + 2g(2L)}{L} = \dfrac{4v_H^2}{L} - 5g$

or $\dfrac{v_H^2}{L} + 9g = \dfrac{4v_H^2}{L}$ or $\dfrac{3v_H^2}{L} = 9g$ or $v_H = \sqrt{3gL}$

$$= \sqrt{\frac{3 \times 10 \times 10}{3}} \text{ m s}^{-1} = 10 \text{ m s}^{-1}$$

7. (c) The acceleration vector is in III quadrant. So, choice limits to (a) and (c). Acceleration magnitude $= \dfrac{v^2}{R} = 1$. Component $\dfrac{1}{\sqrt{2}}$ each. So, choice is (c).

8. (a) $T - mg \cos \theta = \dfrac{mv^2}{r}$,

$$103.2 - 4 \times 9.8 \cos \theta = \frac{4 \times 4 \times 4}{1}$$

$103.2 - 39.2 \cos \theta = 64$ or $39.2 \cos \theta = 103.2 - 64$

or $\cos \theta = \dfrac{39.2}{39.2} = 1$ $\therefore$ $\theta = 0°$

9. (b) $v = \sqrt{0.64 \times 20 \times 9.8}$ m s^{-1}

$= \sqrt{125.44}$ m s^{-1} = 11.2 m s^{-1}

10. (c) $\dfrac{1}{2} mv^2 = mgl(1 - \cos \theta)$

or $v^2 = 2gl(1 - \cos \theta)$

Now, $T - mg = \dfrac{mv^2}{l}$

$$2mg - mg = \frac{m}{l}[2gl(1 - \cos \theta)]$$

or $mg = \dfrac{m}{l}[2gl(1 - \cos \theta)]$

or $\dfrac{1}{2} = 1 - \cos \theta$

or $\cos \theta = 1 - \dfrac{1}{2} = \dfrac{1}{2}$ or $\theta = 60°$

11. (c) The spokes shall coincide when the faster wheel completes one- quarter rotation more than that of the slower wheel.

$$5t - 4t = \frac{1}{4} \text{ or } t = \frac{1}{4} \text{ minute}$$

So, the spokes coincide 4 times a minute.

12-13. $mg - mg \cos \theta = \dfrac{mv^2}{l}$

or $\dfrac{v^2}{l} = g(1 - \cos \theta)$

or $v^2 = gl(1 - \cos \theta)$ ...(1)

Fig. 49

Applying conservation of energy between L and P,

$$\frac{1}{2} mgl = \frac{1}{2} mv^2 + mgl(1 - \cos \theta)$$

$$gl = v^2 + 2gl(1 - \cos \theta)$$

or $v^2 = gl - 2gl(1 - \cos \theta)$ ...(2)

From (1) and (2),

$$gl(1 - \cos \theta) = gl[1 - 2 + 2 \cos \theta]$$

or $1 - \cos \theta = 2 \cos \theta - 1$ or $3 \cos \theta = 2$

or $\cos\theta = \frac{2}{3}$ or $\theta = \cos^{-1}\left(\frac{2}{3}\right)$

From equation (1),

$$v^2 = gl\left(1-\frac{2}{3}\right) \text{ or } v^2 = \frac{1}{3}gl \text{ or } v = \sqrt{\frac{gl}{3}}$$

14. (*d*) $mg = 2 \times 10 \text{ N} = 20 \text{ N}$

$$\frac{mv^2}{r} = \frac{2\times 30\times 30}{5} \text{ N} = 360 \text{ N}$$

$$\frac{T_t}{T_b} = \frac{\frac{mv^2}{r} - mg}{\frac{mv^2}{r} + mg} = \frac{360-20}{360+20} = \frac{340}{380} = \frac{17}{19}$$

15. (*c*) $v^2 - gr = 2gr$

or $v^2 = 3gr$

or $\frac{v^2}{r} = 3g$

So, the centripetal acceleration is $3g$.

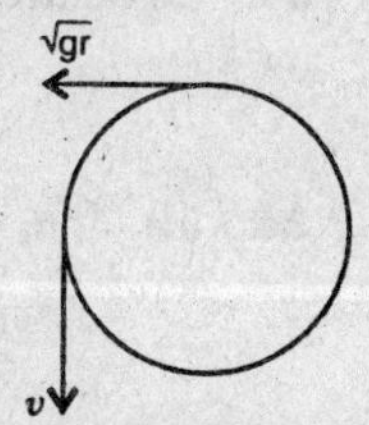

Fig. 50

16. (*d*) $T - mg\cos 60° = \frac{mv^2}{r}$

or $T = mg\cos 60° + \frac{mv^2}{r}$

or $T = \frac{mg}{2} + \frac{mv^2}{r}$

or $T = \frac{0.1\times 10}{2} + \frac{0.1\times 2\times 2}{1}$

or $T = (0.5 + 0.4) \text{ N} = 0.9 \text{ N}$

17. (*d*) $2^2 - u^2 = -2\times 10 \times [l(1-\cos 60°)]$

or $4 - u^2 = -20 \times \frac{l}{2} = -20 \times \frac{1}{2}$

or $u^2 = 4 + 10 = 14$

or $u = \sqrt{14} \text{ m s}^{-1}$

18. (*c*) $v_L = \sqrt{5gL}$ (known fact)

$v_M = ?$

$v_M^2 - 5gl = -2gl$

or $v_M^2 = 3gl$ or $v_M = \sqrt{3gl}$

19. (*b*) $F = mr\omega^2 = 1 \times \frac{1}{2} \times 4\pi^2 \times 2 \times 2 \text{ N} = 8\pi^2\text{N}$

20. (*b*) Since the road is unbanked therefore the centripetal force is provided by the frictional force.

$\therefore \quad f = \frac{mv^2}{r}$

But $f_{ms} \geq f$

$\therefore \quad \mu mg \geq \frac{mv^2}{r}$ or $\mu \geq \frac{v^2}{gr}$

$\therefore \quad \mu_{min} = \frac{v^2}{gr} = \frac{20\times 20}{10\times 200} = 0.2$

21. (*c*) $\frac{MR^2}{4} = 200$ or $MR^2 = 800$

Now, required moment of inertia

$$= \frac{1}{2}MR^2 = \frac{1}{2}\times 800 = 400 \text{ g cm}^2$$

22. (*a*) $N\sin\theta = mr\omega^2$

$N\cos\theta = mg$

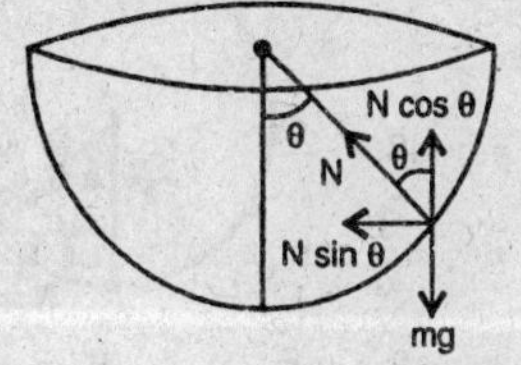

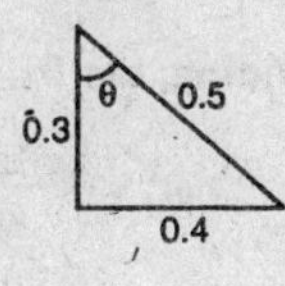

Fig. 51

$$\tan\theta = \frac{r\omega^2}{g}$$

$$\omega = \sqrt{\frac{g\tan\theta}{r}} \text{ or } \omega = \sqrt{\frac{10\times 0.4}{0.3\times 0.4}}$$

or $\omega = \sqrt{\frac{100}{3}} \text{ rad s}^{-1} = \frac{10}{\sqrt{3}} \text{ rad s}^{-1}$

23. (*d*) Loss of gravitational potential energy

= Gain of kinetic energy

$$mgl(1-\cos\theta) = \frac{1}{2}mv^2$$

$$mgl(1-\cos 60°) = \frac{1}{2}mv^2 \text{ or } v = \sqrt{gl}$$

Now, $\sqrt{gl} = \sqrt{5gh}$

or $h = \frac{l}{5}$ or $h = \frac{1 \text{ m}}{5} = 0.2 \text{ m}$

Fig. 52

Distance of nail from the point of suspension

$= (1 - 0.2) \text{ m} = 0.8 \text{ m}$

24. (*c*) At the highest point, $mg = mR\omega^2$ or $\omega = \sqrt{\frac{g}{R}} = 2\pi\nu$

$$\therefore \quad \nu = \frac{1}{2\pi}\sqrt{\frac{g}{R}} = \sqrt{\frac{g}{4\pi^2R}}$$

$$\text{r.p.m.} = 60\nu = 60\sqrt{\frac{g}{4\pi^2R}} = \sqrt{\frac{900\,g}{\pi^2R}}$$

25. (*b*) Mass of each link, $m = \frac{2}{125}$ kg

radius, $r = \frac{1.25}{2\pi}$ m

angular velocity, $\omega = 2\pi \times \frac{3000}{60}$ rad s^{-1} = 100π rad s^{-1}

$$F = mr\omega^2$$

or $$F = \frac{2}{125} \times \frac{1.25}{2\pi} \times (100\pi)^2 \text{ N}$$

$$= \frac{1}{100\pi} \times 100\pi \times 100\pi \text{ N}$$

$$= 100\pi \text{ N} = 100 \times 3.14 \text{ N} = 314 \text{ N}$$

26. (*b*) $N \sin\theta = m\left(\frac{a}{2}\right)\omega^2$

$N \cos\theta = mg$

Dividing, $\tan\theta = \frac{a\omega^2}{2g}$

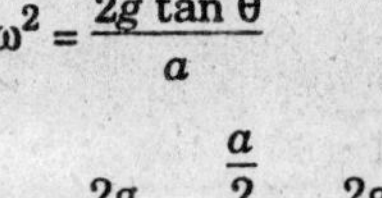

$$\omega^2 = \frac{2g \tan\theta}{a}$$

$$= \frac{2g}{a} \times \frac{\frac{a}{2}}{\frac{\sqrt{3}\,a}{2}} \quad \frac{2g}{\sqrt{3}\,a}$$

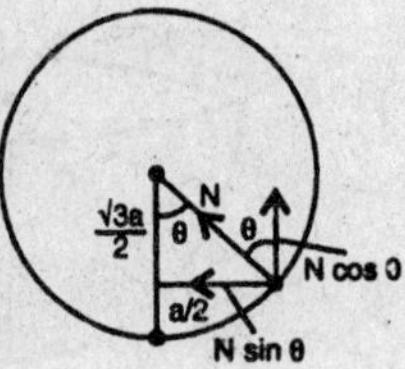

Fig. 53

27. (*a*) $mg = \mu N$ or $mg = \mu m r\omega^2$

or $g = \mu r \times (2\pi f)^2$ or $\mu = \frac{g}{4\pi^2 f^2 r}$

28. (*b*) $F = \frac{mv^2}{r}, v = \sqrt{\frac{rF}{m}}$

$$v = \sqrt{\frac{0.5 \times 10 \times 1000}{100}} \text{ m s}^{-1} = \sqrt{50} \text{ m s}^{-1}$$

29. (*d*) $mg \cos\theta = \frac{mv^2}{R}$

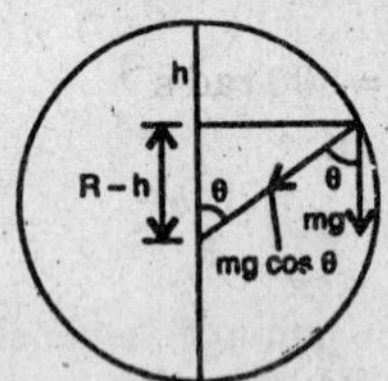

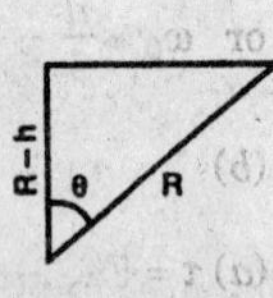

Fig. 54

or $g \cos\theta = \frac{v^2}{R} = \frac{2gh}{R}$

or $\cos\theta = \frac{2h}{R}$ or $\frac{R-h}{R} = \frac{2h}{R}$

or $3h = R$ or $h = \frac{R}{3}$

30. (*c*) $T = \frac{mv^2}{r} - mg = \frac{0.1 \times 5 \times 5}{1} - 0.1 \times 10 = 1.5$ N

31. (*c*) $v = 108$ km h^{-1} = $108 \times \frac{5}{18}$ m s^{-1} = 30 m s^{-1}

$\tan\theta = \frac{30 \times 30}{100 \times 10} = 0.9, \theta = \tan^{-1}(0.9) = 42°$

32. (*d*) $\mu mg = mr\omega^2$

μ and g are constants.

$\therefore$ $r\omega^2$ = constant or $4\pi^2 r\varpi^2$ = constant

or rv^2 = constant

or $r'v'^2 = rv^2$ or $r' = \frac{rv^2}{v'^2} = \frac{rv^2}{(2v)^2}$

or $r' = \frac{r}{4} = \frac{16 \text{ cm}}{4} = 4$ cm

33. (*b*) Centripetal acceleration

$= r\omega^2 = r(2\pi v)^2 = 4\pi^2 rv^2$

$= 4\pi^2 \times 5 \times 10 \times 10$ m s^{-2} = $2000\pi^2$ m s^{-2}

34. (*d*) $a_t = 2$ m s^{-2}

$a_r = \frac{30 \times 30}{500} = \frac{9}{5} = 1.8$ m s^{-2}

$a = \sqrt{a_t^2 + a_r^2} = \sqrt{2^2 + (1.8)^2}$ m s^{-2}

$= \sqrt{4 + 3.24}$ m s^{-2} = $\sqrt{7.24}$ m s^{-2} = 2.7 m s^{-2}

35. (*b*) $\tan\theta = \frac{v^2}{Rg}$

$\tan\theta \approx \sin\theta$

$\therefore$ $\sin\theta = \frac{v^2}{Rg}$

or $\frac{h}{b} = \frac{v^2}{Rg}$ or $h = \frac{v^2 b}{Rg}$

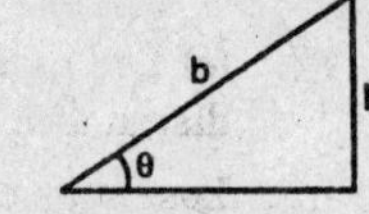

Fig. 55

36. (*d*) $\frac{\frac{mv^2}{R}}{\frac{b}{2}} = \frac{mg}{h}$

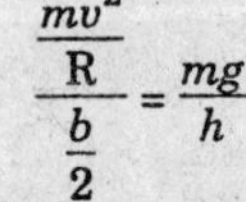

$\frac{2mv^2}{bR} = \frac{mg}{h}$

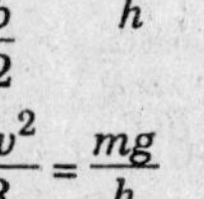

or $\frac{2v^2}{bR} = \frac{g}{h}$

or $v^2 = \frac{bgR}{2h}$ or $v = \sqrt{\frac{bgR}{2h}}$

or $v = \sqrt{\frac{1 \times 9.8 \times 250}{2 \times 1}}$ m s^{-1}

$= \sqrt{1225}$ m s^{-1} = 35 m s^{-1}

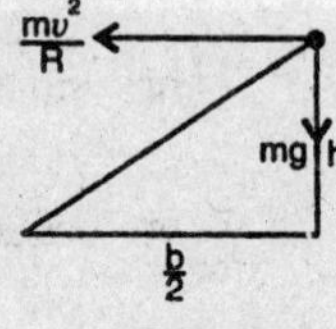

Fig. 56

37. (*c*) $T - mg = \frac{mv^2}{r}$

or $T = mg + \frac{m}{r} \times 2gr$

or $T = mg + 2mg$ or $T = 3mg$

38. (*a*) $F_c = \frac{1 \times 4 \times 4}{1} N = 16 N$

$mg = 1 \times 10 N = 10 N$

At the top,

$T + mg = F_c$

$T = F_c - mg = (16 - 10) N = 6 N$

39. (*b*) $\omega_0 = 0, \omega = \alpha t = 25t$

Now, $1500 = \frac{1}{2} \times 1.2 \times 625t^2$

or $t^2 = \frac{3000 \times 10}{12 \times 625} s = 4$ or $t = 2 s$

40. (*c*) Rotational kinetic energy $= \frac{1}{2} I\omega^2$

$360 = \frac{1}{2} I \times 30 \times 30$

or $I = \frac{360 \times 2}{30 \times 30} kg m^2 = 0.8 kg m^2$

41. (*d*) Applying conservation of angular momentum, we get

$\left(\frac{1}{2} MR^2\right)\omega = \frac{1}{2}\left(M + \frac{M}{4}\right)R^2\omega'$ or $\omega' = \frac{4\omega}{5}$

42. (*c*) Let us consider the instant the reel has descended distance *h*.

Loss of gravitational potential energy

= Translational kinetic energy

+ Rotational kinetic energy

$\therefore \quad mgh = \frac{1}{2} mv^2 + \frac{1}{2} I\omega^2$

or $mgh = \frac{1}{2} mv^2 + \frac{1}{2}\left(\frac{1}{2} mr^2\right)\frac{v^2}{r^2}$

or $mgh = \frac{1}{2} mv^2 + \frac{1}{4} mv^2$

or $gh = \frac{3}{4} v^2$ or $v^2 = \frac{4}{3} gh$

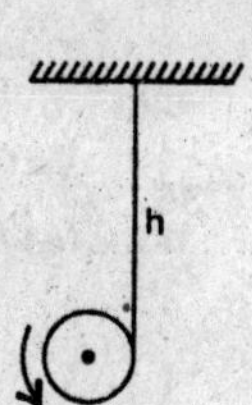

Fig. 57

Comparing with $v^2 = 2ah$, we get

$2a = \frac{4g}{3}$ or $a = \frac{2g}{3}$

43. (*b*) Combined moment of inertia of I and II

$= 2 \times \frac{2}{5} mr^2 = \frac{4}{5} mr^2$

Combined moment of inertia of III and IV

$= 2\left[\frac{2}{5} mr^2 + mb^2\right]$

$= \frac{4}{5} mr^2 + 2mb^2$

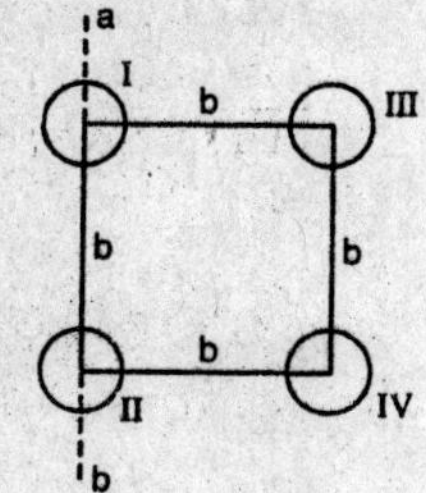

Fig. 58

Combined moment of inertia of the system

$= \frac{4}{5} mr^2 + \frac{4}{5} mr^2 + 2mb^2 = \frac{8}{5} mr^2 + 2mb^2.$

44. (*c*) $\tau = FR = I\alpha$ or $F = \frac{I\alpha}{R}$ or $F = \frac{\frac{1}{2} MR^2\alpha}{R}$

or $F = \frac{1}{2} MR\alpha = \frac{1}{2} \times 25 \times 0.2 \times 2 N = 5 N$

45. (*d*) Torque due to gravity, τ

$= 2mg \times b + mg \times 3b = 5mgb$

Moment of inertia,

$I = 2m \times b^2 + m \times 9b^2 = 11mb^2$

$\tau = I\alpha$ or $\alpha = \frac{\tau}{I}$

or $\alpha = \frac{5mgb}{11 mb^2} = \frac{5g}{11b}$

46. (*c*) $t_1 = \sqrt{\frac{2l}{g \sin \theta}}$; $t_2 = \sqrt{\frac{3l}{g \sin \theta}}$

$\frac{t_2^2}{t_1^2} = \frac{3}{2}$

47. (*c*) $\frac{2}{3} MR_h^2 = \frac{2}{5} MR_s^2$ or $\frac{R_h^2}{R_s^2} = \frac{3}{5}$ or $\frac{R_h}{R_s} = \sqrt{\frac{3}{5}}$

48. (*c*) Using conservation of angular momentum,

$5 \times 10 + 20 \times 0 = (5 + 20)\omega$

or $\omega = \frac{50}{25}$ r.p.s. = 2 r.p.s.

49. (*e*) $\tau = I\alpha, \tau = \frac{I\omega_0}{t}, t = \frac{I\omega_0}{\tau}$

or $\omega_0 = \frac{\tau t}{I} = \frac{10 \times 10}{0.5} = 200 \text{ rad s}^{-1}$

50. (*b*) $\tau = I\alpha$ or $\alpha = \frac{\tau}{I}$

(*a*) $\tau = Fd$, I large, α minimum

(*b*) $\tau = Fd$, I minimum, α maximum

(*c*) $\tau = \frac{Fd}{2}$

(*d*) $\tau = \frac{Fd}{2}$

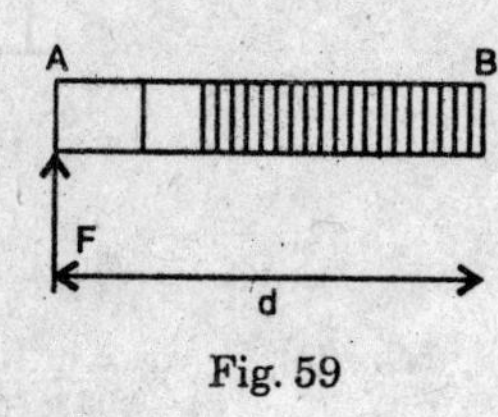

Fig. 59

51. (*d*) $\tau = (2mx^2)\alpha_1$

$\tau = (2 \times m \times 4x^2)\alpha_2$

Equating, $8mx^2\alpha_2 = 2mx^2\alpha_1$ or $\alpha_2 = \frac{\alpha_1}{4}$

52. (*d*) L is increased by a factor of $\frac{3}{2}$. E_k is increased by a factor of $\frac{9}{4}$.

Percentage increase = $\frac{\frac{9}{4}-1}{1} \times 100 = 125$

53. (*b*) Equating loss of gravitational potential energy with gain of kinetic energy, we get

$$mg\,\frac{l}{2}\sin\theta = \frac{1}{2}mv^2 + \frac{1}{2}I\omega^2$$

or $\frac{mgl\sin\theta}{2} = \frac{1}{2}mv^2 + \frac{1}{2}mR^2 \times \frac{v^2}{R^2}$

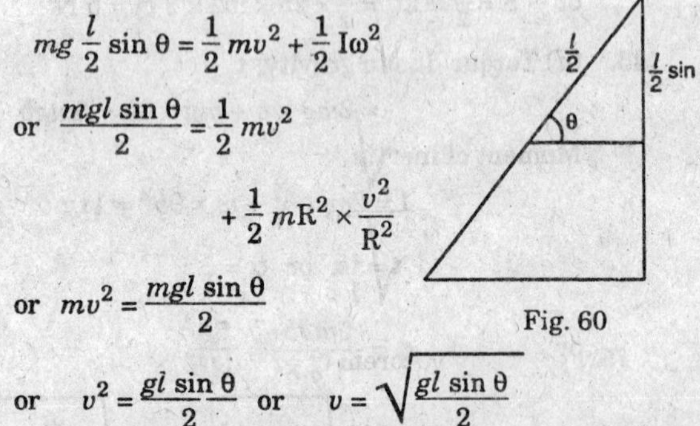

Fig. 60

or $mv^2 = \frac{mgl\sin\theta}{2}$

or $v^2 = \frac{gl\sin\theta}{2}$ or $v = \sqrt{\frac{gl\sin\theta}{2}}$

54. (*a*) If ρ is mass/area, then mass of square is $l^2\rho$. Again mass of circular plate is $\frac{\pi l^2\rho}{4}$. Clearly, the square plate is more massive than the circular plate.

55. (*b*) Instantaneous velocity of

$$P = \left[\frac{v}{r}\right]\sqrt{r^2\sin^2\theta + r^2(1+\cos\theta)^2}$$

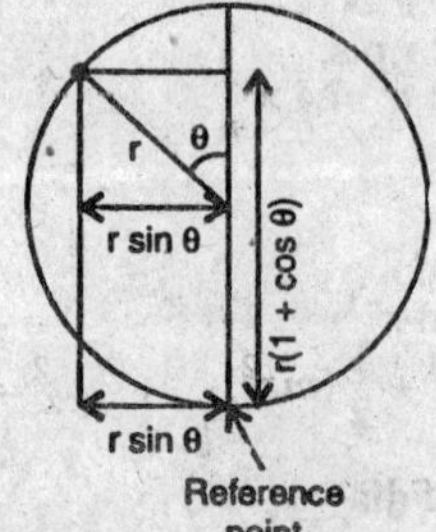

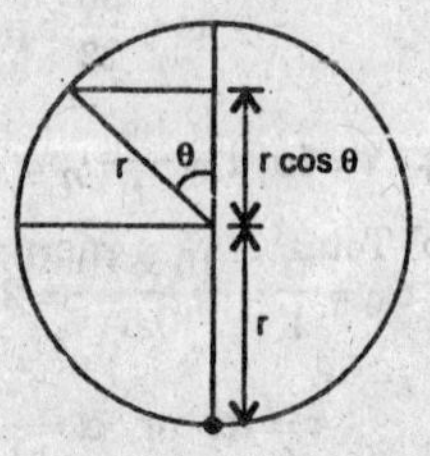

Fig. 61

$$= \frac{v}{r}\sqrt{\sin^2\theta + 1 + \cos^2\theta + 2\cos\theta}$$

$$= \frac{v}{r}r\sqrt{2+2\cos\theta} = 2v\cos\frac{\theta}{2}$$

56. (*c*) $a_c = \omega^2 R = 4\pi^2\nu^2 R$

$= 4 \times 9.87 \times 1 \times 1 \times 20 \times 10^3$ m s^{-2}

$= 7.9 \times 10^5$ m s$^{-2} \approx 8 \times 10^5$ m s^{-2}

57. (*b*) $S = ut + \frac{1}{2}at^2$

$u = 0$ and S is same for both the cases.

$\therefore \quad t \propto \frac{1}{\sqrt{a}}$

But $\quad a = \frac{g\sin\theta}{1+\frac{K^2}{R^2}}$

$\therefore \quad t \propto \sqrt{1+\frac{K^2}{R^2}}$

Now, $\frac{t_s}{t_d} = \frac{1+\frac{2}{5}}{1+\frac{1}{2}} = \sqrt{\frac{7}{5}\times\frac{2}{3}} = \sqrt{\frac{14}{15}}$

58. (*b*) $\tau = I\alpha$ or $\alpha = \frac{FR}{I} = \frac{2\times 9.8\times 0.1}{0.1}$ rad s^{-2} = 19.6 rad s^{-2}

59. (*a*) Since Iω is to remain constant,

$\therefore \quad I'\omega' = I\omega$

Clearly, $\quad I' = \frac{2}{5}I$ and $\omega' = \frac{5}{2}\omega$

New kinetic energy

$$= \frac{1}{2}\left(\frac{2}{5}I\right)\left(\frac{5}{2}\omega\right)^2 = \frac{1}{2}\times\frac{5}{2}I\omega^2 = 2.5\times\frac{1}{2}I\omega^2$$

60. (*b*) $I_1 = \frac{ml^2}{12}$; $I_2 = m\left[\frac{l}{2\pi}\right]^2 = \frac{ml^2}{4\pi^2}$

$$\frac{I_1}{I_2} = \frac{ml^2}{12}\times\frac{4\pi^2}{ml^2} = \frac{\pi^2}{3}$$

61. (*c*) $\frac{1}{2}I\omega^2 = \frac{1}{2}\left(\frac{1}{3}ml^2\right)\omega^2$

$$= \frac{1}{6}\times\frac{1}{2}\times 1\times 1\times 6\times 6\text{ J} = 3\text{ J}$$

62. (*b*) Speed is constant ; So, torque is zero. $F_1 \times 5 = F_2 \times 30$, With $F_2 = 4$ N, $F_1 = 24$ N.

63. (*a*) According to Kepler's third law,

$T^2 \propto r^3$ or $T \propto r^{3/2}$

$\omega = \frac{2\pi}{T}$ or $\omega \propto r^{-3/2}$

Now, $\quad L = mr^2\omega$ or $L \propto r^{2-3/2}$

or $\quad L \propto r^{1/2}$

64. (*d*) Work done, $W = \frac{1}{2}I\omega_2^2 - \frac{1}{2}I\omega_1^2$

$$= \frac{1}{2}I\times 4\pi^2[\nu_2^2 - \nu_1^2]$$

$$= \frac{1}{2}\times 30\times 4\pi^2\left[\frac{25}{36}-\frac{1}{36}\right]$$

$$= 60\pi^2\times\frac{24}{36}\text{ joule} = 40\pi^2\text{ joule}$$

65. (*b*) $\frac{1}{2}I\omega^2$ = Loss of gravitational potential energy

$$\frac{1}{2} \times \frac{ml^2}{3} \omega^2 = \frac{mgl}{2}(1 - \cos\theta)$$

or $$\frac{l\omega^2}{3} = g\left[1 - \left(1 - 2\sin^2\frac{\theta}{2}\right)\right]$$

or $$\frac{l\omega^2}{3} = 2g\sin^2\frac{\theta}{2} \quad \text{or} \quad \omega = \sqrt{\frac{6g}{l}}\sin\frac{\theta}{2}$$

> ☞ ***In this type of questions, think of the vertical distance moved by centre of gravity.***

66. (*d*) $$\tau = I\alpha = \frac{I\omega_0^2}{2\theta} = \frac{I \times 4\pi^2 v^2}{2 \times 2\pi \times 2\pi} = \frac{1}{2} I v^2$$

$$= \frac{1}{2} \times \frac{2}{5} MR^2 v^2 = \frac{1}{5} MR^2 v^2$$

$$= \frac{1}{5} \times 2000 \times 5 \times 5 \times 5 \times 5 \text{ dyne cm}$$

$$= 250 \times 10^3 \text{ dyne cm} = 2.5 \times 10^5 \text{ dyne cm}$$

67. (*c*) Using conservation of angular momentum,

$I\omega$ = constant or Iv = constant

$\therefore \; I \times 90 = [I + mr^2] \times 60$

or $3I = 2I + 2mr^2$ or $I = 2mr^2$

68. (*d*) Moments of inertia of rings I and II about

$$YY' = 2 \times \frac{3}{2} MR^2 = 3MR^2$$

Moment of inertia of ring III about $YY' = \frac{1}{2} MR^2$

Total moment of inertia

$$= 3MR^2 + \frac{1}{2} MR^2 = \frac{7}{2} MR^2$$

69. (*b*) $P = 10^5$ W, $\omega = 1800$ rev/min.

$$= \frac{1800}{60} = 30 \text{ rev/s} = 30 \times 2\pi \text{ rad/s} = 60\pi \text{ rad/s}$$

$$\tau = \frac{10^5 \text{ W}}{60\pi \text{ rad s}^{-1}} = \frac{10^4}{6\pi} \text{ Nm}$$

70. (*b*) Moment of inertia

$$= \frac{1}{2} \times 20 \times 0.25 \times 0.25 \text{ kg m}^2 = 0.625 \text{ kg m}^2$$

71. (*a*) $L = I\omega = 0.625 \times 100$ J s = 62.5 J s

72. $$a = \frac{g\sin\theta}{1 + \frac{K^2}{R^2}} = \frac{g\sin 30°}{1 + \frac{2}{5}}$$

or $$a = \frac{5}{7} g \times \frac{1}{2} = \frac{5g}{14}$$

73. (*c*) Total kinetic energy of ring

= Translational K.E. + Rotational K.E.

$$= \frac{1}{2}mv^2 + \frac{1}{2}I\omega^2 = \frac{1}{2}mv^2 + \frac{1}{2}(mr^2)\frac{v^2}{r^2}$$

$$= \frac{1}{2}mv^2 + \frac{1}{2}mv^2 = mv^2 = 1 \times 2 \times 2 \text{ J} = 4 \text{ J}$$

74. (*b*) $$E = \frac{1}{2}Mv^2\left[1 + \frac{K^2}{R^2}\right]$$

In the given problem, E is constant

$$\therefore \quad v \propto \frac{1}{\sqrt{1 + \frac{K^2}{R^2}}}$$

or $$\frac{v_d}{v_r} = \frac{\sqrt{1 + \frac{R^2}{R^2}}}{\sqrt{1 + \frac{\frac{1}{2}R^2}{R^2}}} = \frac{\sqrt{1+1}}{\sqrt{1 + \frac{1}{2}}} = \frac{\sqrt{2}}{\sqrt{\frac{3}{2}}} = \frac{2}{\sqrt{3}}$$

75. (*c*) Using theorem of parallel axes,

$$I = \frac{ML^2}{12} + M\left(\frac{L}{6}\right)^2$$

$$I = \frac{ML^2}{12} + \frac{ML^2}{36}$$

$$= \frac{4ML^2}{36} = \frac{ML^2}{9}$$

Fig. 62

76. (*a*) $$\frac{m_P}{m_o} = \frac{2\pi r\rho}{2\pi n r\rho} = \frac{1}{n} \quad \text{or} \quad m_P = \frac{1}{n} m_o$$

Again, $I_o = 4I_P$

$$m_o n^2 r^2 = 4m_P r^2 \quad \text{or} \quad m_o n^2 = 4 \times \frac{1}{n} m_o$$

or $n^3 = 4$ or $n = 4^{1/3}$

77. (*b*) Total kinetic energy of loop

$$= \frac{1}{2}mv^2 + \frac{1}{2}I\omega^2 = \frac{1}{2}mv^2 + \frac{1}{2}(mr^2)\frac{v^2}{r^2} = mv^2 = 8$$

Again, total kinetic energy of disc

$$= \frac{1}{2}mv^2 + \frac{1}{2}I\omega^2 = \frac{1}{2}mv^2 + \frac{1}{2} \times \frac{1}{2}mr^2 \times \frac{v^2}{r^2}$$

$$= \frac{1}{2}mv^2 + \frac{1}{4}mv^2 = \frac{3}{4}mv^2 = \frac{3}{4} \times 8 \text{ J} = 6 \text{ J}$$

78. (*b*) $$\frac{\text{Rotational K.E.}}{\text{Translational K.E.}} = \frac{\frac{1}{2}\left(\frac{2}{5}MR^2\right)\frac{v^2}{R^2}}{\frac{1}{2}Mv^2} = \frac{2}{5}$$

79. (*b*) Considering the free body diagram of centre of mass, we get

$$F_{ext.} = 6g - 2T$$

$$a_{C.M.} = \frac{F_{ext.}}{m_1 + m_2} = \frac{6g - 2T}{6} \quad ...(i)$$

Now, $\quad T = \frac{2 \times 4 \times 2}{4 + 2} g = \frac{8g}{3}$

From eqn. (*i*), $a_{C.M.} = g - \frac{1}{3}\left[\frac{8g}{3}\right]$

or $\quad a_{C.M.} = \frac{g}{9} = \frac{9}{9} \text{ m s}^{-2} = 1 \text{ m s}^{-2}$

2T

C.M.

6g

Fig. 63

A BETTER SOLUTION

$$M\vec{A} = m_1\vec{a_1} + m_2\vec{a_2} \quad ...(i)$$

| Acceleration of 4 kg mass |

= | Acceleration of 2 kg mass |

$$= \frac{4-2}{4+2} \times 9 = \frac{18}{6} \text{ m s}^{-2} = 3 \text{ m s}^{-2}$$

Note that the directions are opposite.

If $a_1 = 3 \text{ m s}^{-2}$, then $a_2 = -3 \text{ m s}^{-2}$

From eqn. (*i*), $(4 + 2)A = 4 \times 3 - 2 \times 3$

or $\quad A = 1 \text{ m s}^{-2}$

80. (*b*) $\dfrac{\frac{1}{2}I\omega^2}{\frac{1}{2}I\omega^2 + \frac{1}{2}mv^2} = \dfrac{I\omega^2}{I\omega^2 + mv^2} = \dfrac{\frac{2}{5}mr^2 \times \frac{v^2}{r^2}}{\frac{2}{5}mv^2 + mv^2} = \dfrac{2}{7}$

81. (*d*) $K_A = \dfrac{L_A^2}{2I_A} = \dfrac{25L_B^2 \times 4}{2 \times I_B}$

$$K_B = \frac{L_B^2}{2I_B}$$

Now, $\dfrac{K_A}{K_B} = 25 \times 4 = 100$

82. (*c*) $\quad mgh = \frac{1}{2}mv^2 + \frac{1}{2}I\omega^2$

$$mgh = \frac{1}{2}mv^2 + \frac{1}{2}\left(\frac{1}{2}mr^2\right)\frac{v^2}{r^2}$$

or $\quad mgh = \frac{3}{4}mv^2$ or $v = \sqrt{\frac{4gh}{3}}$

83. (*b*) $\left[\frac{1}{2}MR^2 + MR^2\right]\omega' = \left[\frac{1}{2}MR^2\right]\omega$

or $\quad \frac{3}{2}MR^2\omega' = \frac{1}{2}MR^2\omega$ or $\omega' = \frac{\omega}{3}$

84. (*c*) $\theta = \left(\frac{0 + 36}{2}\right)6$ radian = 108 radian.

85. (*c*) $\quad L = mvr$ or $v = \frac{L}{mr}$

$$F = \frac{mv^2}{r} = \frac{m}{r}\left[\frac{L}{mr}\right]^2 \text{ or } F = \frac{L^2}{mr^3}$$

86. (*b*) $\dfrac{40}{100} = \dfrac{\frac{1}{2}I\omega^2}{\frac{1}{2}mv^2} = \dfrac{\frac{Iv^2}{R^2}}{mv^2}$ or $I = \frac{2}{5}mR^2$

So, the body is solid sphere.

87. (*c*) Applying theorem of parallel axes,

$$I = \frac{MR^2}{2} + M\left(\frac{R}{4}\right)^2$$

$$I = \frac{MR^2}{2} + \frac{MR^2}{16}$$

$$I = \frac{8MR^2 + MR^2}{16} \text{ or } I = \frac{9MR^2}{16}$$

88. (*c*) $\quad E = \frac{L^2}{2I}, E \propto \frac{1}{I}$

$$\frac{E_1}{E_2} = \frac{I_2}{I_1}, \frac{I_2}{I_1} < 1, \frac{E_1}{E_2} < 1 \text{ or } E_1 < E_2$$

89. (*c*) $\quad \tau = I\alpha$

$$FR = I\frac{\omega - \omega_0}{t}$$

$$10 \times \frac{10}{100} = 5\frac{\Delta\omega}{10} \text{ or } \Delta\omega = 2 \text{ rad s}^{-1}$$

90. (*c*) Change in angular momentum = $I\Delta\omega$

$$= 5 \times 2 \text{ kg m}^2\text{ s}^{-1} = 10 \text{ kg m}^2\text{ s}^{-1}$$

91. (*d*) The acceleration of weight W is $\frac{2g}{3}$. It is independent of radius.

So, $v(= at)$ is also independent of radius.

92. (*a*) $\quad E_k = \frac{L^2}{2I}$

$$\log E_k = \log L^2 - \log(2I)$$

or $\quad \log E_k = 2\log L - \log(2I)$

Now, think of $y = mx - c$ to arrive at the result.

93. (*c*) We have to combine the tangential velocity due to rotational motion with linear translational velocity. So, the required velocity is $\sqrt{2}v_0$.

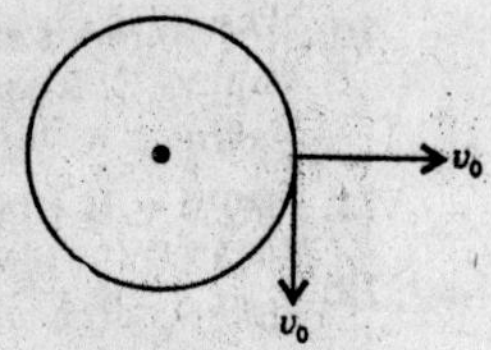

Fig. 64

94. (*c*) Moment of inertia about tangent = $\frac{2}{3}Mr^2 + Mr^2$

$$= \frac{5}{3}Mr^2 = MK^2, \quad K = \sqrt{\frac{5}{3}}\,r$$

95. (*c*) $\quad a = \dfrac{g\sin\theta}{1 + \frac{I}{MR^2}}$ or $a = \dfrac{g\sin\theta}{1 + \frac{MK^2}{MR^2}} = \dfrac{g\sin\theta}{1 + \frac{K^2}{R^2}}$

For a ring, $MK^2 = MR^2$ or $K^2 = R^2$

$\therefore \quad a = \frac{g \sin\theta}{2} = \frac{g}{2} \sin 30° = \frac{g}{4}$

96. (*a*) This is based on law of conservation of angular momentum.

$x\omega' = (2x)\omega$ or $\omega' = 2\omega$

97. (*d*) Rotational kinetic energy $= \frac{1}{2} I\omega^2$

$I = Mr^2$ and $\omega = \frac{v}{r}$

98. (*a*) The wheel will climb the step only if the clockwise moment of the applied force F is greater than the anticlockwise moment of the weight of the wheel. Moment of weight mg about point of contact, $\tau_1 = mg \times x$

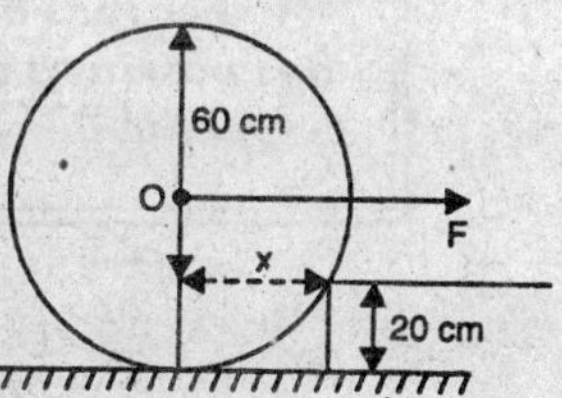

Fig. 65

Now, $x^2 = 60 \times 20$ (from geometry)

$\therefore \quad x = \sqrt{1200} = 20\sqrt{3}$ cm $= 0.2\sqrt{3}$ m

Moment of applied force F about point of contact,

$\tau_2 = F \times 0.2$ m

Now, $\tau_2 = \tau_1$

$F \times 0.2 = mg \times 0.2\sqrt{3} = 2 \times 9.8 \times 0.2\sqrt{3}$

On simplification, $F = 19.6 \times \sqrt{3}$ N

99. (*a*) $I = Mr^2, I' = (M + 2m)r^2$

$I'\omega' = I\omega$

or $\omega' = \frac{I}{I'}\omega = \frac{Mr^2}{(M + 2m)r^2}\omega$

or $\omega' = \frac{M}{M + 2m}\omega$

100. (*d*) In order to find the centre of gravity of the remaining portion of the disc, let us construct a situation which is equivalent to the given one. The 'disc with the hole' can be replaced by a 'solid disc' plus an extra disc whose size is the same as that of the hole and is located at the same place as the hole. The weight of the additional disc will be assumed to be negative *i.e.* acting vertically upwards.

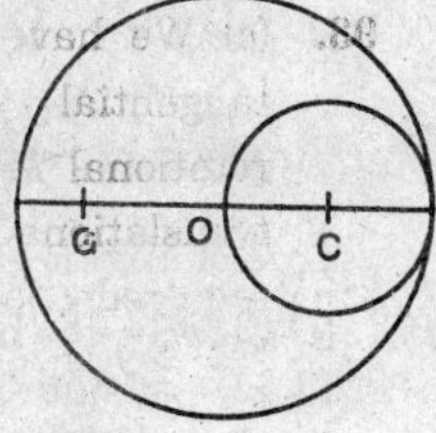

Fig. 66

Now, $\frac{CG}{OG} = \frac{\text{Weight of solid disc}}{\text{Weight of smaller disc}}$

$\frac{CG}{OG} = \frac{\pi (R^2)\rho g}{\pi\left(\frac{R}{2}\right)^2 \rho g} = \frac{4}{1}$

$CG = 4OG$ or $OG + OC = 4OG$

or $3OG = OC = \frac{R}{2}$ or $OG = \frac{R}{6}$

101. (*c*) $L = I\omega = \frac{2}{5}MR^2\omega$

Also, $V = \frac{4}{3}\pi R^3$ or $R^2 = \left[\frac{3V}{4\pi}\right]^{2/3}$

$\therefore \quad L = \frac{2}{5}M\left[\frac{3V}{4\pi}\right]^{2/3}\omega$

L and M do not change with temperature.

$\therefore \quad \omega \propto V^{-2/3}$

$\frac{d\omega}{\omega} \times 100 = -\frac{2}{3}\frac{dV}{V} \times 100$

So, the angular velocity decreases by 0.67%.

Aliter $V' = \frac{101}{100}V$

or $\frac{4}{3}\pi R'^3 = \frac{101}{100} \times \frac{4}{3}\pi R^3$

or $R'^3 = \left(\frac{101}{100}\right)R^3$ or $R' = \left(\frac{101}{100}\right)^{1/3} R$

Using law of conservation of angular momentum,

$\left(\frac{2}{5}MR^2\right)\omega = \frac{2}{5}M\left(\frac{101}{100}\right)^{2/3} R^2\omega'$

or $\omega' = \left(\frac{100}{101}\right)^{2/3}\omega = 0.9933\,\omega$

So, ω' is 99.33% of ω. So, the angular velocity decreases by 0.67%.

102. (*d*) $I = MK^2$

$\log I = \log M + 2\log K$

103. (*b*) Applying theorem of parallel axes,

$I = MR^2 + MR^2 = 2MR^2$

104. (*c*) $X_{C.M.} = \frac{1 \times 0 + 2 \times 2 + 3 \times 0 + 4 \times 2 + 5 \times 1}{1 + 2 + 3 + 4 + 5}$

$= \frac{4 + 8 + 5}{15} = \frac{17}{15} = 1.1$

This itself chooses option (*c*). You may try $Y_{C.M.}$ if you like.

105. (*c*) Double torque, Double angular acceleration, Double linear acceleration, Double distance.

106. (*b*) Two particles, each of mass m, are on the axis of rotation X′X.

Now, $\sin 60° = \frac{d}{l}$

or $d = l \sin 60°$

or $d = \frac{\sqrt{3}\, l}{2}$

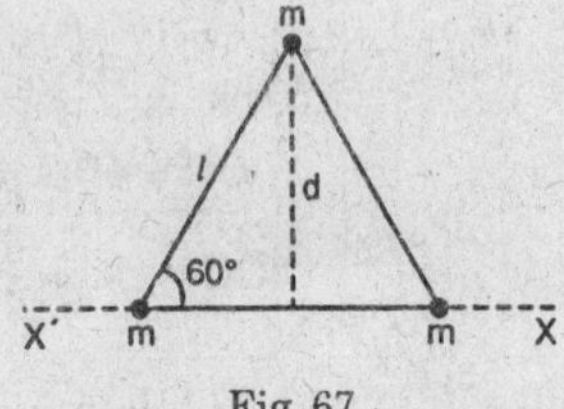

Fig. 67

$$I = md^2 \text{ or } I = \frac{3ml^2}{4}$$

107. (*b*) Mass of incomplete ring

$$= M - \frac{M}{2\pi} \times \frac{\pi}{6} = M - \frac{M}{12} = \frac{11M}{12}$$

Moment of inertia of incomplete ring

$$= \left(\frac{11M}{12}\right) R^2 = \frac{11}{12} MR^2$$

108. (*c*) Applying theorem of parallel axes,

M.I. about the given axis

$$= \frac{1}{2} MR^2 + MR^2 = \frac{3}{2} MR^2 = 6 \times \frac{1}{4} MR^2 = 6I$$

109. (*a*) $L = \sqrt{2IE_{kr}}$ or $L \propto \sqrt{E_{kr}}$

110. (*a*) Take a bisector line. A′B′ perpendicular to bisector line AB.

Moment of inertia about an axis perpendicular to the plane and passing through the point of intersection is

$$2 \times \frac{ML^2}{12} \text{ or } \frac{ML^2}{6}$$

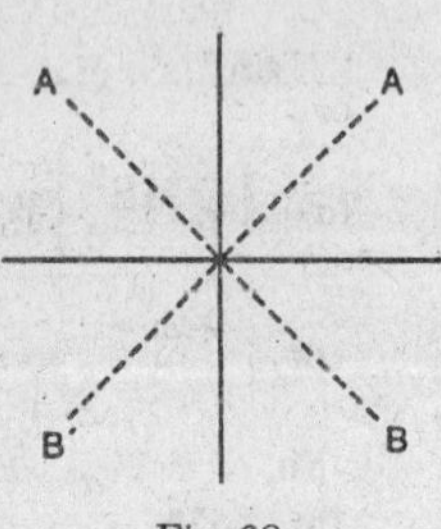

Fig. 68

Applying theorem of perpendicular axis, we get

$$\frac{ML^2}{6} = I_{AB} + I_{A'B'}$$

or $2I = \frac{ML^2}{6}$ or $I = \frac{ML^2}{12}$

111. (*d*) τ = force × lever arm of force

$\tau = mgl \sin \theta$.

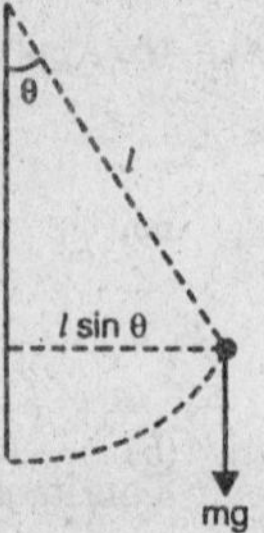

Fig. 69

112. (*c*) $\alpha = \frac{mgl \sin \theta}{ml^2} = \frac{g \sin \theta}{l}$

113. (*a*) M.I. of rod about z-axis is zero.

M.I. of remaining rods about z-axis

$$= \frac{1}{3} ml^2 + \frac{1}{3} ml^2$$

$$= \frac{2ml^2}{3}$$

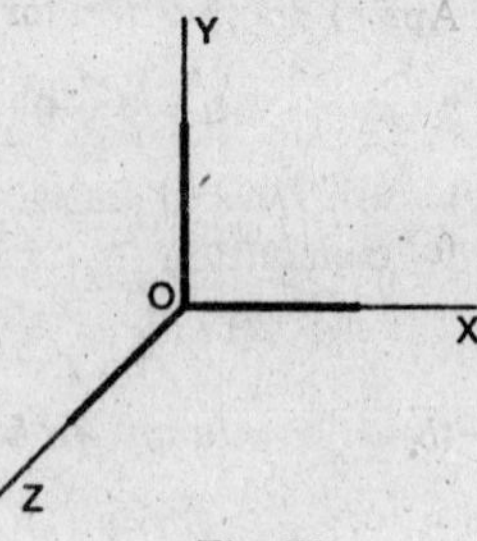

Fig. 70

> ☞ ***For rods along X and Y axes, Z-axis is an axis passing through one end of the rod and perpendicular to the length of the rod.***

114. (*d*) $t = \sqrt{\frac{2h}{g}}$

$$t = \sqrt{\frac{2 \times 2g}{g}}$$

$$= \sqrt{4} = 2 \text{ s}$$

Now $AB = 2v \times 2 = 4v$

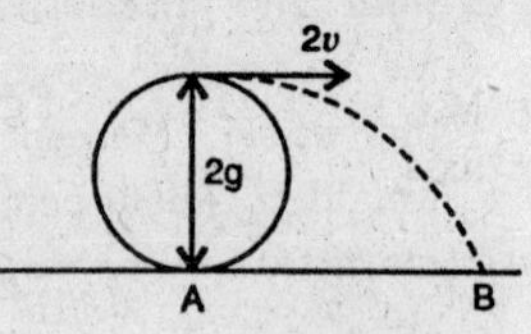

Fig. 71

115. (*d*) Using conservation of energy,

$$mgh = \frac{1}{2} mv^2 + \frac{1}{2} I\omega^2$$

$$mgh = \frac{1}{2} mv^2 + \frac{1}{2}\left(\frac{2}{5} mR^2\right)\frac{v^2}{R^2}$$

or $mgh = \frac{1}{2} mv^2 + \frac{1}{5} mv^2 = \frac{7}{10} mv^2$

or $v = \sqrt{\frac{10\, gh}{7}}$

116. (*c*) Applying conservation of angular momentum,

$$(5 + 5)\omega' = 5 \times 10 + 5 \times 20 = 150$$

or $\omega' = 15 \text{ rad s}^{-1}$

Initial kinetic energy

$$= \frac{1}{2} \times 5 \times 10 \times 10 + \frac{1}{2} \times 5 \times 20 \times 20 = 250 + 1000 = 1250 \text{ J}$$

Final kinetic energy $= \frac{1}{2} \times 10 \times 15 \times 15 = 1125$ J

Loss of kinetic energy = (1250 – 1125) J = 125 J

117. (*c*) Clearly, the block shall topple about its edge through O. The torque FL of the applied force is clockwise. The torque $\frac{mgL}{2}$ of the weight is anti-clockwise.

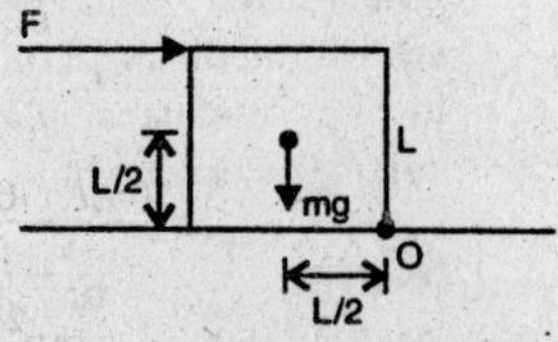

Fig. 72

Applying condition for rotational equilibrium,

$$-\text{FL} + mg\frac{\text{L}}{2} = 0 \quad \text{or} \quad \text{F} = \frac{mg}{2}$$

118. (*c*) $a = \dfrac{g \sin\theta}{1+\dfrac{2}{3}} = \dfrac{3}{5} g \sin\theta = \dfrac{3}{5} \times 10 \times \dfrac{2}{5} = \dfrac{12}{5} \text{ m s}^{-2}$

119. (*b*) $v = at = \dfrac{12}{5} \times 5 \text{ m s}^{-1} = 12 \text{ m s}^{-1}$

120. (*d*) $\dfrac{\frac{1}{2}\text{I}\omega^2}{\frac{1}{2}\text{M}v^2} = \dfrac{40}{60} = \dfrac{2}{3}$ or $\dfrac{\frac{\text{I}v^2}{\text{R}^2}}{\text{M}v^2} = \dfrac{2}{3}$ or $\text{I} = \dfrac{2}{3}\text{MR}^2$

Clearly, the body is a hollow sphere.

121. (*a*) M.I. of element of mass dm about the given axis

$= dm\text{R}^2$

M.I. of the wire about given axis

$$= \int_0^{\text{M}} dm\text{R}^2 = \text{MR}^2.$$

Fig. 73

122. (*b*) Torque due to the weight of the book $= \dfrac{\text{W}b}{2}$

This can be balanced only if the torque applied by the boy is equal in magnitude and opposite in direction. So, the torque applied by the person is $\dfrac{\text{W}b}{2}$ anticlockwise.

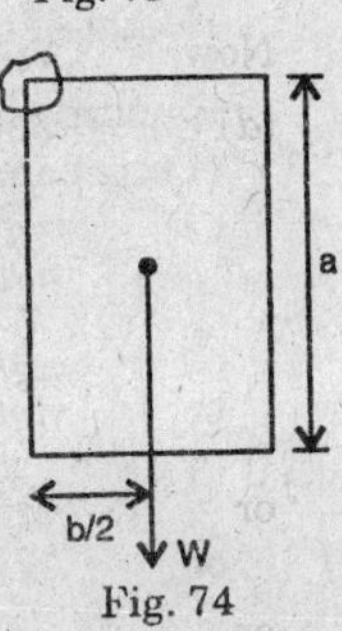

Fig. 74

123. (*b*) Since there is no slipping therefore the points of contact are instantaneous centres of rotation of the cylinder. The centre of the cylinder has to move through a distance l. So, the centre of mass of the string has to move through a distance l. So, the total length of the string passed through the hands of the man will be $2l$.

124. (*a*) $\text{X} = \dfrac{2m \times l + 2m \times 2l + m \times \dfrac{3l}{2} + 2m \times 0}{2m + 2m + m + 2m}$

or $\text{X} = \dfrac{2l + 4l + \dfrac{3l}{2}}{7} = \dfrac{15l}{14}$

$\text{Y} = \dfrac{2m \times 0 + 2m \times l + m \times 2l + 2m \times l}{7m}$

$= \dfrac{6ml}{7m} = \dfrac{6l}{7}$

125. (*b*) $\text{I} = \dfrac{7}{10}mv^2 = \dfrac{7}{10} \times 2 \times 0.1 \times 0.1 = 0.014 \text{ J}$

126. (*a*) The topmost block will be in equilibrium if its centre of gravity is at the edge of the block below it. So, the topmost block can have maximum equilibrium extension of $\dfrac{l}{2}$.

Fig. 75

127. (*b*) If G is the centre of gravity of blocks A and B, then

$\text{W}x = \text{W}\left(\dfrac{l}{2} - x\right)$

or $x = \dfrac{l}{2} - x$

or $2x = \dfrac{l}{2}$ or $x = \dfrac{l}{4}$

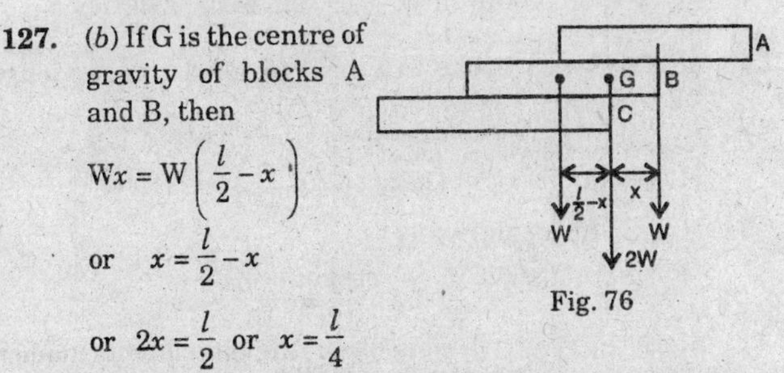

Fig. 76

128. (*c*) Let G represent the centre of gravity of the three blocks A, B and C.

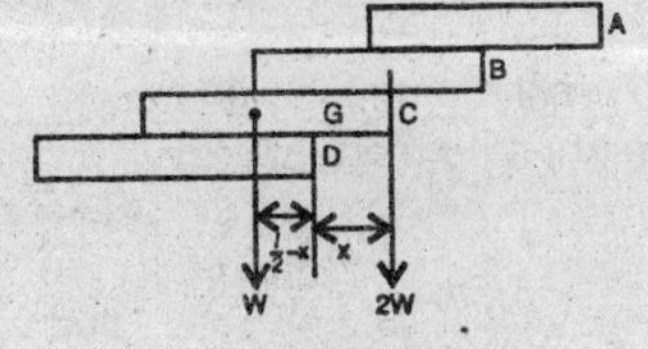

Fig. 77

Now, $2\text{W} \times x = \text{W}\left(\dfrac{l}{2} - x\right)$ or $2x = \dfrac{l}{2} - x$

or $3x = \dfrac{l}{2}$ or $x = \dfrac{l}{6}$

129. (*a*) Moment of inertia of system

$= m_1 \times 0^2 + m_2\left[\dfrac{a}{2}\right]^2 + m_3\left[\dfrac{a}{2}\right]^2 = \dfrac{a^2}{4}(m_2 + m_3)$

130. (*d*) $\text{X} = \dfrac{m \times \dfrac{l}{2} + 2m \times 0 + m \times \dfrac{l}{2}}{m + 2m + m}$

or $\text{X} = \dfrac{ml}{4m} = \dfrac{l}{4}$

$\text{Y} = \dfrac{m \times 0 + 2m \times l + m \times 2l}{m + 2m + m} = \dfrac{4ml}{4m} = l$

So, the x and y co-ordinates of the centre of mass are $\left(\dfrac{l}{4}, l\right)$

131. (*b*) $\text{P} = \vec{\tau} \cdot \vec{\omega}$

$\text{P} = (\hat{i} + 2\hat{j} + 3\hat{k}) \cdot (2\hat{i} + 3\hat{j} + 4\hat{k})$

$= 2(\hat{i} \cdot \hat{i}) + 6(\hat{j} \cdot \hat{j}) + 12(\hat{k} \cdot \hat{k})$

$= 2 + 6 + 12 = 20 \text{ W}.$

132. (c) $a = \dfrac{g \sin\theta}{1+\dfrac{K^2}{R^2}} = \dfrac{g\sin\theta}{1+\dfrac{2}{5}} = \dfrac{5}{7} g\sin\theta$

133. (a) $Mg\sin\theta - f = Ma = M\left(\dfrac{5}{7} g\sin\theta\right)$

or $f = Mg\sin\theta - \dfrac{5Mg}{7}\sin\theta$ or, $f = \dfrac{2}{7} mg\sin\theta$

134. (d) Treat each spoke as a thin rod of mass m and length 2R.

Moment of inertia of each spoke $= \dfrac{1}{12} m(2R)^2 = \dfrac{mR^2}{3}$

Moment of inertia of two spokes $= \dfrac{2mR^2}{3}$

Since the given wheel is like a ring therefore its moment of inertia is mR^2. Total moment of inertia, I

$$= \frac{2mR^2}{3} + mR^2 = \left(\frac{2}{3}+1\right)mR^2 = \frac{5}{3} mR^2$$

135. (d) $\dfrac{\text{Translational kinetic energy}}{\text{Translational kinetic energy} + \text{Rotational kinetic energy}}$

$$= \frac{\frac{1}{2}mv^2}{\frac{1}{2}mv^2 + \frac{1}{2}I\omega^2} = \frac{mv^2}{mv^2 + \frac{2}{5}mr^2 \times \frac{v^2}{r^2}} = \frac{5}{7}$$

136. (d) Taking moments of forces about point P, we get

$F \times 0.6 = 100 \times 0.8$

or $F = \dfrac{100 \times 0.8}{0.6}$ N

$= \dfrac{800}{6}$ N

$= \dfrac{400}{3}$ N

$= 133.3$ N

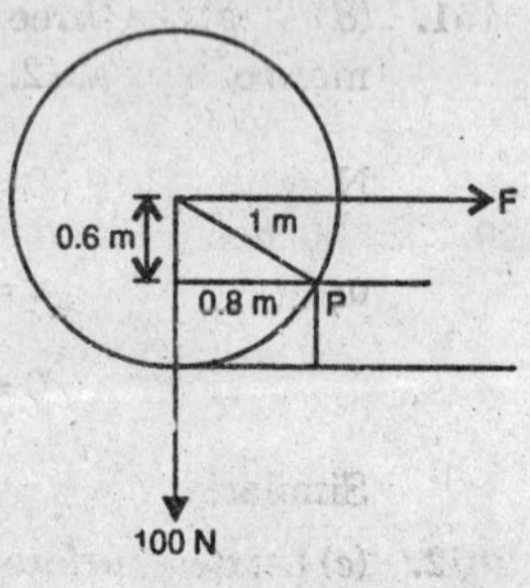

Fig. 78

137. (b) In one revolution, the angle described is 2π radian.

$\omega_0 = 0,\ \omega = ?,\ \theta = 2\pi$ radian

Using $\omega^2 - \omega_0^2 = 2\alpha\theta$, we get

$\omega^2 - 0^2 = 2\alpha \times 2\pi = 4\pi\alpha$ or $\omega = 2\sqrt{\pi\alpha}$

138. (a) Note that $\vec{r}$ and $\vec{F}$ are co-terminus.

So, angle between $\vec{r}$ and $\vec{F}$ is 110°.

Now, $\tau = rF\sin\theta$

$= 3 \times 50 \times \sin 110°$

$= 150 \sin(90° + 20°) = 150\cos 20°$

$= 150 \times 0.9$ N m $= 135$ N m

139. (c) $20 = 0 + \alpha \times 10$

or $\alpha = 2$ rad s^{-2}

$\tau = 5 \times 10^{-3} \times 2 = 0.01$ N m

140. (b) $\dfrac{\text{Rotational K.E.}}{\text{Translational K.E.}} = \dfrac{\frac{1}{2}I\omega^2}{\frac{1}{2}Mv^2} = \dfrac{I\omega^2}{Mv^2}$

$$= \frac{\frac{1}{2}MR^2 \times \frac{v^2}{R^2}}{Mv^2} = \frac{1}{2} : 1 = \frac{1}{2}$$

141. (a) $I = \dfrac{1}{2}MR^2,\ I \propto R^2$,

Percentage increase in moment of inertia $= 2 \times 1\% = 2\%$.

142. (d) $\vec{\tau} = \vec{r} \times \vec{F}$

$$\vec{\tau} = \begin{vmatrix} \hat{i} & \hat{j} & \hat{k} \\ 2 & -3 & 4 \\ 3 & 2 & 3 \end{vmatrix}$$

$= \hat{i}[-9-8] + \hat{j}[12-6] + \hat{k}[4+9] = -17\hat{i} + 6\hat{j} + 13\hat{k}$

143. (c) $l = 0.6$ m

Height of centre of mass, $h = \dfrac{l}{2}$ or $h = 0.3$ m

Equating loss of gravitational potential energy with gain of rotational energy, we get

$$mgh = \frac{1}{2}I\omega^2$$

or $mgh = \dfrac{1}{2}\left[\dfrac{ml^2}{3}\right]\left[\dfrac{v^2}{l^2}\right]$

or $2gh = \dfrac{v^2}{3}$ or $2g\left[\dfrac{l}{2}\right] = \dfrac{v^2}{3}$

or $v = \sqrt{3gl} = \sqrt{3 \times 10 \times 0.6} = \sqrt{18} = 3\sqrt{2}$ m s^{-1}

144. (c) Using condition for rotational equilibrium,

$F(r-h) = mgx$ ...(1)

But $r^2 = x^2 + (r-h)^2$

or $x^2 = r^2 - (r-h)^2$

or $x^2 = (2r-h)h$

or $x = \sqrt{(2r-h)h}$

From eqn. (1),

$$F = \frac{mg}{r-h}\sqrt{(2r-h)h}$$

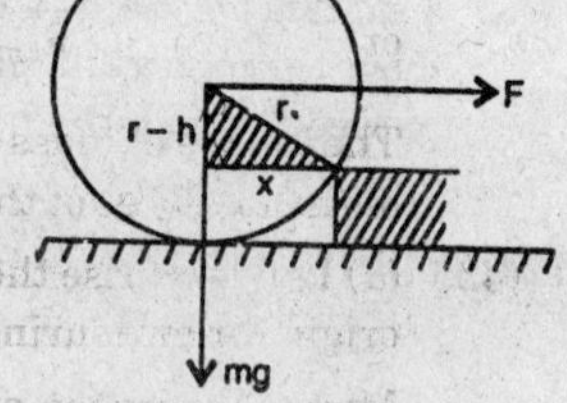

Fig. 79

145. (b) $E_k = \dfrac{1}{2}I\omega^2$. Also $L = I\omega$ or $I = \dfrac{L}{\omega}$

Now, $E_k = \dfrac{1}{2}\dfrac{L}{\omega}\omega^2 = \dfrac{L\omega}{2}$ or $L = \dfrac{2E_k}{\omega} = \dfrac{2E_k}{2\pi\nu} = \dfrac{E_k}{\pi\nu}$

Again, $L' = \dfrac{\frac{1}{2}E_k}{\pi(2v)} = \dfrac{1}{4}\dfrac{E_k}{\pi v} = \dfrac{1}{4}L$

146. (*b*) $\pi r = l$ or $r = \dfrac{l}{\pi}$

$$I = \frac{1}{2}mr^2$$

$$I = \frac{1}{2}m\frac{l^2}{\pi^2} = \frac{ml^2}{2\pi^2}$$

147. (*c*) Applying theorem of parallel axes, we get

$$I = \frac{1}{2}MR^2 + M\left[\frac{R}{2}\right]^2$$

$$I = \frac{3}{4}MR^2 \Rightarrow MK^2 = \frac{3}{4}MR^2$$

$$K = \frac{\sqrt{3}\,R}{2}$$

148. (*d*) The centre of mass of each sphere is at its geometrical centre. So, the centre of mass of the system is in fact the centre of mass of three equal point masses located at the vertices of an equilateral triangle ABC. Here A, B and C are the centres of three spheres. Each side of the triangle is 2R. In order to locate the centre of mass of the system, we have to choose one point, say B, as the origin. The position vector of the centre of mass with respect to the origin B is given by

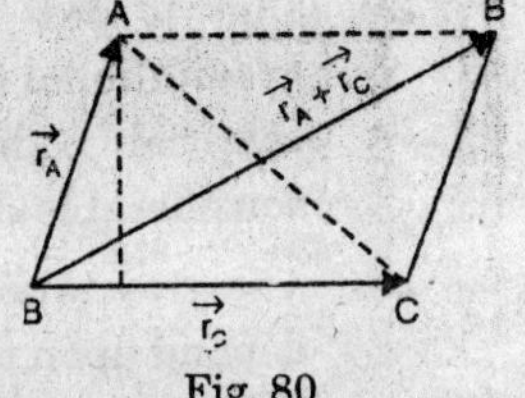

Fig. 80

$$\vec{R}_{cm} = \frac{M\vec{r}_A + M\vec{r}_C}{3M}$$

where M is the mass of each sphere.

or $$\vec{R}_{cm} = \frac{\vec{r}_A + \vec{r}_C}{3}$$

The centre of mass is located at the point of intersection of the medians of the triangle.

149. (*a*) Let us choose the nucleus of the hydrogen atom as the origin for measuring distance.

Mass of hydrogen atom, $m_1 = 1$ unit (say)

Since chlorine atom is 35.5 times as massive as hydrogen atom,

$\therefore$ mass of chlorine atom, $m_2 = 35.5$ unit (say)

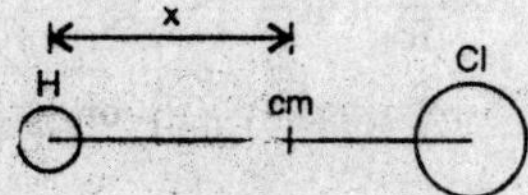

Fig. 81

Now, $x_1 = 0$ and $x_2 = 1.27$ Å $= 1.27 \times 10^{-10}$ m

Distance of centre of mass of HCl molecule from the origin is given by

$$X = \frac{m_1x_1 + m_2x_2}{m_1 + m_2}$$

$$= \frac{1 \times 0 + 35.5 \times 1.27 \times 10^{-10}}{1 + 35.5}\text{ m}$$

$$= \frac{35.5 \times 1.27}{36.5} \times 10^{-10}\text{ m.} = \frac{35.5 \times 1.27}{36.5}\text{ Å}$$

150. (*d*) If $\vec{a}$ is the acceleration of m_1, then $-\vec{a}$ is the acceleration of m_2.

We know that

$$(m_1 + m_2)\vec{A}_{c.m.} = m_1\vec{a}_1 + m_2\vec{a}_2$$

or $$(m_1 + m_2)\vec{A}_{c.m.} = m_1\vec{a} - m_2\vec{a}$$

or $$\vec{A}_{c.m.} = \frac{m_1 - m_2}{m_1 + m_2}\vec{a}$$

But $$\vec{a} = \frac{m_1 - m_2}{m_1 + m_2}\vec{g}$$ [connected-body problem]

$$\therefore \quad \vec{A}_{cm} = \frac{(m_1 - m_2)^2}{(m_1 + m_2)^2}\vec{g}$$

151. (*d*) Treat the three masses of 1 μg, 2 μg and 3 μg as a single mass of 6 μg at (2, 2, 2).

Now, $0 = \dfrac{4 \times x + 6 \times 2}{10}$ or $4x + 12 = 0$

or $x = -3$,

$0 = \dfrac{4 \times y + 6 \times 2}{10}$ or $y = -3$

Similarly, $z = -3$

152. (*e*) Let σ be surface mass density. The mass of the removed portion may be taken as negative. The given system may be regarded as two-particle system—one particle of mass 16σ at (0, 0) and second particle of mass '– 4σ' at (1, – 1).

Now, $x = \dfrac{16\sigma \times 0 - 4\sigma(1)}{12\sigma} = -\dfrac{1}{3}$ cm

and $y = \dfrac{16\sigma \times 0 - 4\sigma(-1)}{12\sigma} = \dfrac{1}{3}$ cm

$$r = \sqrt{\left(-\frac{1}{3}\right)^2 + \left(\frac{1}{3}\right)^2}\text{ cm} = \frac{\sqrt{2}}{3}\text{ cm}$$

GRAVITATION

Important Terms, Facts and Formulae

BASIC CONCEPTS

1. A uniform solid sphere of matter attracts a particle that is outside the sphere as if all the sphere's mass were concentrated at its centre.
2. Lasers have made it possible to measure distance up to 10^{-19} m. Again, by electronic devices, time can be measured accurately up to 10^{-9} s. Modern methods have made it possible to measure g with an accuracy of 1 part in 10^8.
3. Note that the gravitational field strength at a place is equal to the acceleration produced in a freely falling body at that place.
4. Gravitational potential is essentially a negative quantity.
5. Gravitational potential difference is defined as the work done to move a unit mass from one point to the other point in the gravitational field.
6. The surface of Earth is an equipotential surface because the gravitational potential at all points on the surface of Earth is the same.
7. The gravitational force between two bodies does not depend upon the nature of the medium between two bodies. So, **gravity screens are not possible.**
8. The value of G does not depend upon the nature, size or location of the bodies. It also does not depend upon the nature of the medium between the bodies.
9. In the motion of planets and satellites, the gravitational force provides the necessary centripetal force.
10. The value of acceleration due to gravity is maximum at the surface of Earth and becomes zero at the centre of the Earth.
11. There is no effect of rotational motion of Earth on the value of acceleration due to gravity at the poles.
12. The rotational motion of Earth has maximum effect on the value of acceleration due to gravity at the equator.
13. Since gravitational potential energy arises from attractive forces therefore it is taken as negative.
14. As we proceed away from the Earth, the gravitational potential energy goes on increasing. Its maximum value is zero at infinity.
15. The gravitational potential at a point is negative.
16. The gravitational potential increases as we move towards infinity. It becomes zero (maximum value) at infinity.
17. The orbital speed of a satellite does not depend upon the mass of the satellite.
18. The orbital speed of a satellite depends on the height of the satellite above the surface of Earth. Greater the height, smaller is the speed of the satellite.
19. The period of revolution of a satellite depends upon the height of the satellite. Greater the height, more is the period of revolution.
20. Due to low value of G, we do not experience gravitational force in daily life. However in the case of celestial bodies, the gravitational force becomes considerable due to large masses.
21. The gravitational forces between two bodies does not depend upon the presence or absence of other bodies.

IMPORTANT POINTS TO REMEMBER

1. When altitude $h << R$, then the variation of g with height is given by :

$$g_h = g\left(1 - \frac{2h}{R}\right)$$

2. If value of altitude h is not very small as compared with R, then the following equation holds good

$$\frac{g_h}{g} = \frac{R^2}{(R+h)^2}$$

3. The fractional decrease in the value of g with height $h(<< R)$ is $\frac{g - g_h}{g} = \frac{2h}{R}$

4. The percentage decrease in the value of g with height $h\ (<< R)$ is $\frac{2h}{R} \times 100$

5. If the Earth stops rotating about its axis *i.e.*, if $\omega = 0$, then the value of g will increase everywhere except at the poles. This increase will be maximum at the equator and will go on decreasing towards the poles.
6. If there is an increase in the angular velocity of Earth, then the value of g and hence the weight of a body will decrease at all places except at the poles.
7. The escape velocity does not depend on the direction in which a projectile is fired. However, it becomes easier to attain escape velocity if the projectile is fired in the direction the launch site is moving as Earth rotates about its axis.
8. In cgs system, the gravitational potential is measured in erg g^{-1}. In SI, it is measured in joule kg^{-1}.
9. The value of G is 6.67×10^{-8} dyne $cm^2\ g^{-2}$ in cgs system and 6.67×10^{-11} N $m^2\ kg^{-2}$ in SI.
10. The mass of Earth is 6×10^{24} kg.
11. The density of Earth is 5.5×10^3 kg m^{-3}.
12. At a height equal to the radius of the Earth, the value of acceleration due to gravity is $\frac{g}{4}$ where g is the value of acceleration due to gravity on the surface of Earth.
13. The universal constant G must not be confused with the acceleration due to gravity $\vec{g}$ which is the acceleration of a body arising from the Earth's gravitational pull on it. The constant G has the dimensions $\frac{L^3}{MT^2}$ and is a scalar. On the other hand, $\vec{g}$ has the dimensions $\frac{L}{T^2}$, is a vector and is neither universal nor constant.
14. The gravitational potential at a point on the surface of Earth is given by $V = -\frac{GM}{R}$ where M is the mass of Earth and R is the radius of Earth.
15. Change in gravitational potential

$$dV = \frac{\text{Change in gravitational potential energy } dU}{m}$$

$$= -\vec{E}\,.\,\vec{dr}$$

Here $\vec{E}$ represents gravitational field strength.
16. The period of revolution of a satellite revolving close to the surface of Earth is approximately 84 minutes.
17. If the orbital velocity of a satellite revolving close to the surface of Earth increases $\sqrt{2}$ times, the satellite would escape the gravitational field of Earth.
18. The polar radius of Earth is smaller than its equatorial radius by 21 km.
19. The variation of acceleration due to gravity with depth d is given by $g_d = g\left[1 - \frac{d}{R}\right]$
20. Percentage change in the value of acceleration due to gravity with depth $= \frac{d}{R} g \times 100$.
21. **Formula for period of revolution in terms of density.**

If a satellite is moving very close to the surface of Earth, then

$$T = 2\pi\sqrt{\frac{R^3}{GM}}$$

or $$T = 2\pi\sqrt{\frac{R^3}{G \times \frac{4}{3}\pi R^3 \rho}} \quad \text{or} \quad T = 2\pi\sqrt{\frac{3}{4\pi G\rho}}$$

22. **Formula for orbital velocity in terms of mean density of Earth.**

$$v_0 = \sqrt{\frac{GM}{R}} \quad \text{or} \quad v_0 = \sqrt{\frac{G}{R} \times \frac{4}{3}.\pi R^3 \rho}$$

or $$v_0 = R\sqrt{\frac{4}{3}\pi G\rho}$$

QUESTION BANK

Level I

Choose the correct answer from the four alternatives given in each of the following questions :

1. The distances of two satellites P and Q from Earth are in the ratio 3 : 1. The ratio of their total energy will be

(a) 3 : 1 (b) 1 : 3
(c) 1 : 1 (d) 1/3 : 1.

2. If R is the radius of the Earth and g the acceleration due to gravity on the Earth's surface, the mean density of the Earth is

(a) $\frac{4\pi G}{3gR}$ (b) $\frac{3\pi R}{4gG}$
(c) $\frac{3g}{4\pi RG}$ (d) $\frac{\pi Rg}{12G}$.

3. Due to a mass distribution, the gravitational field is $\frac{k}{x^3}$ along x-axis where k is a constant. If the gravitational potential is taken to be zero at infinity, then the gravitational potential at a distance x is

(a) $\frac{k}{x}$ (b) $\frac{k}{2x^2}$
(c) $\frac{k}{x^4}$ (d) $\frac{k}{x^6}$.

4. Two of Jupiter's moons have orbit radii which differ by a factor of 2. Their periods differ by

(a) 8 (b) 4
(c) 2.8 (d) 4.2.

5. Given : R = radius of Earth. What is the approximate distance of a geostationary satellite from the centre of Earth ? [Note that the satellite is above the equator.]

(a) 5R (b) 6R
(c) 7R (d) 8R.

6. Acceleration due to gravity at Earth's surface is 10 m s^{-2}. The value of acceleration due to gravity at the surface of a planet of mass $\frac{1}{5}$ th and radius $\frac{1}{2}$ of the earth is

(a) 4 m s^{-2} (b) 6 m s^{-2}
(c) 8 m s^{-2} (d) 12 m s^{-2}.

7. If radius of Earth is 4000 miles, then the linear velocity of a body at the equator is nearly

(a) 1000 mile/h (b) 2000 mile/h
(c) 3000 mile/h (d) 4000 mile/h.

8. Two identical copper spheres, each of radius R, are in contact with each other. The gravitational force of attraction between two copper spheres is proportional to

(a) R (b) 2R
(c) R^2 (d) R^4.

9. At sea level, a body will have minimum weight at

(a) pole (b) equator
(c) 42° south latitude (d) 37° north latitude.

10. The mean radius of the Earth is R. Its angular speed on its own axis is ω. If g is the value of acceleration due to gravity on the surface of Earth, then the cube of the radius of orbit of a geostationary satellite will be

(a) $\frac{R^2g}{\omega^2}$ (b) $\frac{R^2\omega^2}{g}$
(c) $\frac{R^2g}{\omega}$ (d) $\frac{Rg}{\omega^2}$.

11. Mass remaining constant, the radius of the Earth shrinks by 1%. The acceleration due to gravity on the earth's surface would

(a) increase by 2% (b) increase by 1%
(c) decrease by 1% (d) decrease by $\frac{1}{2}$ %.

12. A satellite of mass m is circulating around the Earth with constant angular velocity. If radius of the orbit is R_0 and mass of the Earth M, the angular momentum about the centre of the Earth is

(a) $m\sqrt{GMR_0}$ (b) $M\sqrt{GmR_0}$
(c) $m\sqrt{\frac{GM}{R_0}}$ (d) $M\sqrt{\frac{GM}{R_0}}$.

13. The time period of a satellite in a circular orbit of radius R is T. The radius of the orbit in which time period is 8T is

(a) 2R (b) 3R
(c) 4R (d) 5R.

14. A satellite is orbiting very close to a planet. Its time period depends only upon

(a) density of the planet (b) mass of the planet
(c) radius of the planet (d) mass of the satellite.

15. Escape velocity on a planet is v_e. If radius of the planet remains same and mass becomes 4 times, the escape velocity becomes

(a) $4v_e$ (b) $2v_e$

(c) v_e (d) $\frac{1}{2}v_e$.

16. The kinetic energy of a satellite in its orbit around the earth is E What should be the kinetic energy of the satellite so as to enable it to escape from the gravitational pull of the Earth ?

(a) 4E (b) 2E

(c) $\sqrt{2}$ E (d) E.

17. The Earth moves around the Sun in an elliptical orbit as shown in the Fig. 1. The ratio $\frac{OA}{OB} = x$. The ratio of the speed of the Earth at B and at A is nearly

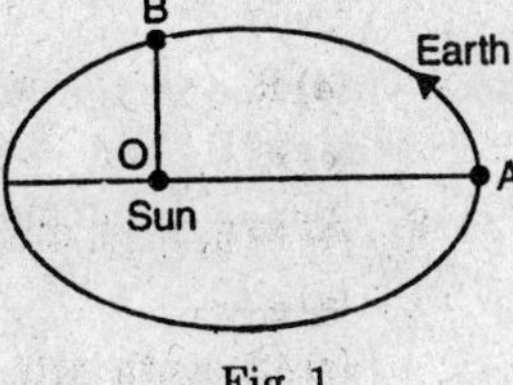

Fig. 1

(a) $\sqrt{x}$ (b) x

(c) $x\sqrt{x}$ (d) x^2.

18. A planet is revolving around the Sun in an elliptical orbit. Its closest distance from the Sun is $r_{min.}$. The farthest distance from the Sun is $r_{max.}$. If the orbital angular velocity of the planet when it is nearest to the Sun is ω, then the orbital angular velocity at the point when it is at the farthest distance from the Sun is

(a) $\sqrt{\frac{r_{min}}{r_{max}}}\,\omega$ (b) $\sqrt{\frac{r_{max}}{r_{min}}}\,\omega$

(c) $\frac{r_{max}^2}{r_{min}^2}\,\omega$ (d) $\frac{r_{min}^2\omega}{r_{max}^2}$.

19. A body weighs 700 g on the surface of the Earth. How much will it weigh on the surface of a planet whose mass is $\frac{1}{7}$ and radius is half that of the Earth ?

(a) 200 g wt (b) 400 g wt

(c) 50 g wt (d) 300 g wt.

20. If Earth describes an orbit round the Sun of double its present radius, the year on Earth will be of

(a) 365 days (b) $365 \times 2 \times \sqrt{2}$ days

(c) $\frac{365}{2}$ days (d) 365×4 days.

21. Two satellites revolve around the Earth at distances 3R and 6R from the centre of Earth. Their periods of revolutions will be in the ratio

(a) 1 : 2 (b) 2 : 1

(c) $1 : 2^{1.5}$ (d) $1 : 2^{0.67}$

22. Orbital velocity of satellite revolving in a circular path close to the planet

(a) is directly proportional to the density of planet

(b) is directly proportional to the square root of density of planet

(c) is directly proportional to square of density of planet

(d) is inversely proportional to square root of density of planet.

23. What is the SI unit of G ?

(a) $kg^{-1}\,m^3\,s^{-2}$ (b) $kg^3\,m\,s^{-2}$

(c) $kg\,m\,s^{-2}$ (d) $kg^2\,m\,s^{-2}$.

24. A body is broken into two parts of masses m_1 and m_2. These parts are then separated by a distance r. What is the value of m_1/m_2 so that the gravitational force has maximum possible value ?

(a) 1 : 1 (b) 1 : 2

(c) 2 : 1 (d) 4 : 3.

25. The binding energy of a system comprising of Earth and satellite is (Letters have usual meanings)

(a) $-\frac{GMm}{r}$ (b) $-\frac{GMm}{2r}$

(c) $\frac{GMm}{r}$ (d) $\frac{GMm}{2r}$.

26. Which graph correctly represents the variation of acceleration due to gravity with distance from the centre of the earth ?

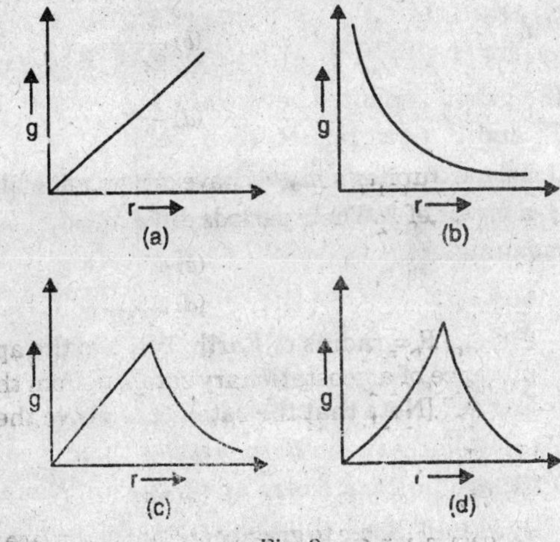

Fig. 2

27. The gravitational potential difference between a point at the surface of a certain planet and a point 10 m above the surface is 4 J kg^{-1}. Assume the gravitational field to be uniform. What is the work done in joule in moving a mass of 2 kg from the surface to a point 5 m above the surface ?

(a) 0.16 (b) 1.0

(c) 2.0 (d) 4.0.

28. In Q. 27, what is the gravitational field strength in the region under observation ?

(a) 0.40 N kg^{-1} (b) 2.5 N kg^{-1}

(c) 4 N kg^{-1} (d) 40 N kg^{-1}.

29. A geo-stationary satellite is orbiting the Earth at a height of 6 R above the surface of the Earth, R being the radius of the Earth. The time period of another satellite at a height of 2.5 R from the surface of Earth is

(a) $6\sqrt{2}$ hours (b) 6 hours

(c) $6\sqrt{3}$ hours (d) 10 hours.

30. One of the combination from the fundamental physical constants is hc/G. The unit of this expression is

(a) kg^2 (b) m^3

(c) s^{-1} (d) None of these.

31. A 2 kg body falls from infinity to the surface of earth. What will be the kinetic energy of the body on reaching the surface of Earth ? (Assume escape velocity to be 10^4 m s^{-1}.)

(a) 10^8 J (b) 10^{16} J

(c) 10^{20} J (d) 10^{24} J.

32. A satellite revolving around a planet has a time period T and orbital radius R. The mass of this planet is

(a) $4\pi^2R^3G^{-1}$ (b) $4\pi^2R^3G^{-1}T$

(c) $R^3G^{-1}T$ (d) $4\pi^2R^3G^{-1}T^{-2}$.

33. If a graph is plotted between T^2 and r^3 for a planet, then the slope of the graph will be : (Letters have usual meanings.)

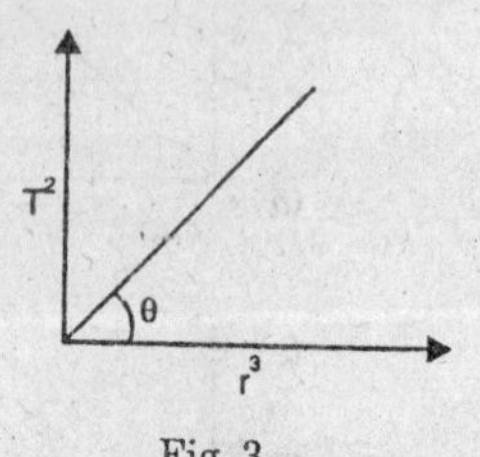

Fig. 3

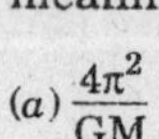

(a) $\frac{4\pi^2}{GM}$ (b) $\frac{GM}{4}$

(c) $4\pi GM$ (d) $\frac{GM}{4\pi}$.

34. Two satellites A and B, the ratio of their masses being 3 : 1, are in the circular orbits of radii r and $4r$. The ratio of total mechanical energy of A to that of B is

(a) 1/3 (b) 3

(c) 3/4 (d) 12.

35. A man weighs W on the surface of Earth. What is his weight at a height equal to R ?

(a) W (b) W/2

(c) W/4 (d) W/8.

36. What is the cube of the radius of the orbit of a geostationary satellite ? Given : R = mean radius of Earth, ω = angular speed of Earth about its own axis, g = value of acceleration due to gravity on the surface of Earth.

(a) R^2g/ω (b) R^2g/ω^2

(c) $R^2\omega^2/g$ (d) Rg/ω^2.

37. A body is projected from the surface of Earth with a velocity $2v_e$ where v_e is the escape velocity. The velocity of the body when it escapes the gravitational field of Earth is

(a) $\sqrt{2}\, v_e$ (b) $\sqrt{3}v_e$

(c) $\sqrt{7}v_e$ (d) $\sqrt{11}v_e$.

38. The ratio of the accelerations due to gravity on two planets P_1 and P_2 is k_1. The ratio of their respective radii is k_2. The ratio of their respective escape velocities is

(a) $\sqrt{k_1k_2}$ (b) $\sqrt{2k_1k_2}$

(c) $\sqrt{\frac{k_1}{k_2}}$ (d) $\sqrt{\frac{k_2}{k_1}}$.

39. If Earth describes an orbit round the Sun of double its present radius, the year on Earth will be of

(a) 182 days (b) 1032 days

(c) 730 days (d) None of these.

40. A satellite in a circular orbit of radius R has period 4 hours. A satellite of orbit radius 4R around the same planet will have period

(a) 8 h (b) 16 h

(c) 32 h (d) 64 h.

41. X and Y are two points at respective distances R and 2R from the centre of Earth, where R is greater than the radius of Earth. The gravitational potential at X is – 800 kJ kg^{-1}. When a 1 kg mass is taken from X to Y, the work done on the mass is

(a) 400 kJ (b) 200 kJ

(c) – 200 kJ (d) 80 kJ

(e) 800 kJ.

42. A satellite is revolving in a circular orbit of radius r around the Earth. The angular momentum of the satellite is proportional to

(a) r (b) $\sqrt{r}$

(c) $\frac{1}{\sqrt{r}}$ (d) $r^{3/2}$.

43. With what angular velocity the Earth should spin in order that a body lying at 45° latitude may become weightless ?

(a) $\sqrt{\frac{g}{R}}$ (b) $\sqrt{\frac{2g}{R}}$

(c) $2\sqrt{\frac{g}{R}}$ (d) None of these.

44. If the Earth suddenly shrinks to half of its present radius, then the acceleration due to gravity will be

(a) g (b) $2g$

(c) $4g$ (d) $8g$.

45. The orbital velocity of a satellite close to the surface of the Earth is v. If this satellite starts orbiting at an altitude of half the Earth's radius, the orbital velocity will be

(a) $\sqrt{\frac{2}{3}}\,v$ (b) $\sqrt{\frac{2}{5}}\,v$

(c) $\sqrt{\frac{2}{7}}\,v$ (d) $\sqrt{\frac{2}{9}}\,v$.

46. If different planets have different radii but same mean density, then acceleration due to gravity g on the surface of planet is related to radius R of planet as

(a) $g \propto R$ (b) $g \propto R^2$

(c) $g \propto \frac{1}{R}$ (d) $g \propto \frac{1}{R^2}$.

47. The gravitational potential energy of an isolated system of three particles, each of mass m, at the three corners of an equilateral triangle of side l is

(a) $-\frac{Gm^2}{l}$ (b) $-\frac{Gm^2}{2l}$

(c) $-\frac{2Gm^2}{l}$ (d) $-\frac{3Gm^2}{l}$.

48. Two satellites S and S′ revolve around the Earth at distances 3R and 6R from the centre of the Earth. The periods of revolution of S and S′ are in the ratio of

(a) 1 : 2 (b) 2 : 1

(c) $1 : (2)^{0.67}$ (d) $1 : (2)^{1.5}$.

49. The diameters of two planets are in the ratio of 4 : 1. Their mean densities are in the ratio 1 : 2. The ratio of accelerations due to gravity on the planets is

(a) 2 : 1 (b) 1 : 2

(c) 4 : 1 (d) 1 : 4.

50. The mass of a planet is 6 times that of earth. The radius is twice that of Earth. If v is the escape velocity from the surface of Earth, then the escape velocity from the surface of the planet will be

(a) v (b) $\sqrt{2}v$

(c) $\sqrt{3}v$ (d) $\sqrt{5}v$.

51. The K.E. of a satellite is 2 MJ. What is the total energy of the satellite ?

(a) – 2 MJ (b) – 1 MJ

(c) $-\frac{1}{2}$ MJ (d) – 4 MJ.

52. If the gravitational force between two objects were proportional to (1/R) [and not $1/R^2$] where R is the distance between them, then a particle in a circular path would have its orbital speed v proportional to

(a) $1/R^2$ (b) 1/R

(c) R (d) R°.

53. Two satellites A and B go around a planet P in circular orbits having radii 4R and R respectively. If the speed of the satellite A is 3V, the speed of satellite B would be

(a) 12V (b) 6V

(c) (3/2)V (d) (4/3)V.

54. The escape velocity from the surface of Earth is v_e. The escape velocity from the surface of a planet whose mass and radius are 3 times those of the Earth will be

(a) v_e (b) $3v_e$

(c) $9v_e$ (d) $27v_e$.

55. A body of mass m is approaching towards the centre of a hypothetical hollow planet of mass M and radius R. The velocity of the body when it passes, through the centre of the planet, through its diametrical chute is

(a) $\sqrt{\frac{2GM}{R}}$ (b) $\sqrt{\frac{GM}{R}}$

(c) $\sqrt{\frac{3}{5}\frac{GM}{R}}$ (d) $\sqrt{\frac{2}{5}\frac{GM}{R}}$.

56. Two particles of equal mass go around a circle of radius R under the action of their mutual gravitational force of attraction. The speed of each particle is : (M = mass of the particle)

(a) $\sqrt{\frac{GM}{R}}$ (b) $\frac{1}{2}\sqrt{\frac{GM}{R}}$

(c) $\frac{1}{3}\sqrt{\frac{GM}{R}}$ (d) $\frac{1}{4}\sqrt{\frac{GM}{R}}$.

QUESTION BANK

Level II

Choose the correct answer from the four alternatives given in each of the following questions :

1. The gravitational force of attraction between a uniform sphere of mass M and a uniform rod of length l and mass m oriented as shown is

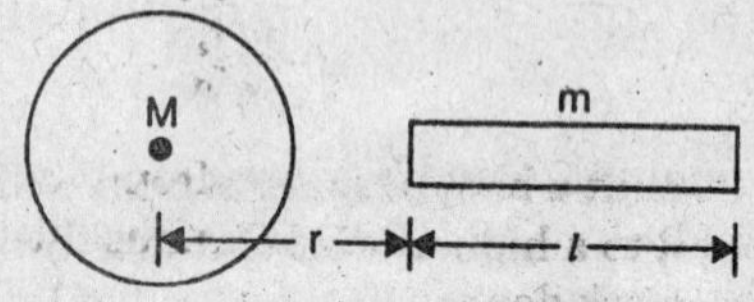

Fig. 4

(a) $\frac{GMm}{r(r+l)}$ (b) $\frac{GM}{r^2}$

(c) $Mmr^2 + l$ (d) $(r^2 + l)mM$.

2. Which of the following statements is wrong ?

(a) A ship moving from west to east, along the equator, shall have less weight as compared to when it is at rest at the equator.

(b) A ship moving from east to west, along the equator, shall have more weight as compared to when it is at rest.

(c) Earth has retained its atmosphere because the value of $\sqrt{\frac{3kT}{m}}$ for air molecules is larger than escape velocity.

(d) The time period of a simple pendulum of infinite length is the same as the time period of SHM of a ball in a tunnel along the diameter of earth.

3. Three masses, each equal to M, are placed at the three corners of a square of side a. The force of attraction on unit mass at the fourth corner will be

(a) $\frac{GM}{3a^2}$ (b) $\frac{GM}{a^2}\sqrt{3}$

(c) $\frac{3GM}{a^2}$ (d) $\frac{GM}{a^2}\left[\frac{1}{2}+\sqrt{2}\right]$.

4. If the mass of a planet is 10% less than that of the Earth and the radius 20% greater than that of the Earth, the acceleration due to gravity on the planet will be

(a) $\frac{5}{8}$ times that on the surface of the Earth

(b) $\frac{3}{4}$ times that on the surface of the Earth

(c) $\frac{1}{2}$ times that on the surface of the Earth

(d) $\frac{9}{10}$ times that on the surface of the Earth.

5. If the distance between the Earth and the Sun becomes half its present value, the number of days in a year would have been

(a) 64.5 (b) 129

(c) 182.5 (d) 730.

6. Which of the following correctly indicates the approximate effective values of g on various parts of a journey to the Moon ?

(All values are in m s^{-2})

	before take off from Earth	*one minute after lift-off*	*in Earth orbit*	*on the Moon*
(a)	9.80	9.80	0	1.6
(b)	9.80	0.98	0	1.6
(c)	9.80	0	0	9.8 × 6
(d)	9.80	7	0	1.6.

7. Consider a planet in some solar system which has a mass double the mass of the Earth and density equal to the average density of the Earth. An object weighing W on the Earth will weigh

(a) W (b) 2 W

(c) W/2 (d) $2^{1/3}$ W at the planet.

8. Acceleration due to gravity g and the mean density ρ of the Earth are related by the relation : [G is the gravitational constant and R_e is the radius of the Earth]

(a) $\frac{g}{G}\frac{4\pi}{3}R_e^3 = \rho$ (b) $\frac{g/G}{\frac{4\pi}{3}R_e} = \rho$

(c) $\frac{g}{G}\frac{4\pi}{3}R_e^2 = \rho$ (d) $\frac{g/G}{\frac{4\pi}{3}R_e^3} = \rho$.

9. A body is projected vertically from the surface of the Earth with a velocity equal to half of the escape velocity. Maximum height reached by the body is

(a) $\frac{5R}{3}$ (b) $\frac{R}{3}$

(c) $\frac{2R}{3}$ (d) $\frac{R}{2}$.

10. A sky laboratory of mass 2×10^3 kg is raised from a circular orbit of radius 2R to a circular orbit of radius 3R. The work done is approximately

(*a*) 1×10^{16} J (*b*) 2×10^{10} J
(*c*) 1×10^{6} J (*d*) 3×10^{10} J.

11. An artificial satellite is moving in a circular orbit around the Earth with a speed equal to half the magnitude of the escape velocity from the Earth. The height of the satellite above the surface of Earth is R. Suppose the satellite is stopped suddenly in its orbit and allowed to fall freely. On reaching Earth, its speed will be

(*a*) $\sqrt{gR}$ (*b*) $2\sqrt{gR}$
(*c*) $3\sqrt{gR}$ (*d*) $5\sqrt{gR}$.

12. A satellite of mass m is revolving at a height R above the surface of Earth. Here R is the radius of the Earth. The gravitational potential energy of this satellite is

(*a*) $-mgR$ (*b*) $-\frac{mgR}{2}$
(*c*) $-\frac{mgR}{3}$ (*d*) $-\frac{mgr}{7}$.

13. A spherical planet far out in space has a mass m_0 and diameter d_0. A particle of mass m falling freely near the surface of this planet will experience an acceleration

(*a*) $\frac{m_0}{{d_0}^2}$ (*b*) $\frac{4mm_0}{2}$
(*c*) $\frac{4Gm_0}{{d_0}^2}$ (*d*) $\frac{mm_0}{{d_0}^2}$.

14. The escape velocity for a projectile in the case of Earth is 11.2 km s^{-1}. A body is projected from the surface of the Earth with a velocity which is equal to twice the escape velocity. The velocity of the body when at infinite distance from the centre of the Earth is

(*a*) 11.2 km s^{-1} (*b*) $11.2\sqrt{3}$ km s^{-1}
(*c*) 22.4 km s^{-1} (*d*) $11.2\sqrt{2}$ km s^{-1}.

15. Masses of 200 g and 800 g are 12 cm apart. At which point from the 200 g mass, the intensity of the gravitational field due to the two masses would be zero ?

(*a*) 2 cm (*b*) 4 cm
(*c*) 6 cm (*d*) 8 cm.

16. A body is dropped from a height equal to the radius of the Earth. M is the mass of the Earth. R is the radius of the Earth. The velocity of the body on reaching the surface of Earth will be

(*a*) $\sqrt{\frac{GM}{R}}$ (*b*) $\sqrt{\frac{2GM}{R}}$
(*c*) $\sqrt{\frac{3GM}{R}}$ (*d*) $\sqrt{\frac{7GM}{R}}$

17. The time period of revolution of a satellite is T. The kinetic energy of the satellite is proportional to

(*a*) T (*b*) T^2
(*c*) T^3 (*d*) $T^{-2/3}$.

18. What should be the angular velocity of rotation of Earth about its own axis so that the weight of a body at the equator reduces to $\frac{3}{5}$ of its present value ? (Take R as the radius of the Earth)

(*a*) $\sqrt{\frac{g}{3R}}$ (*b*) $\sqrt{\frac{2g}{3R}}$
(*c*) $\sqrt{\frac{2g}{5R}}$ (*d*) $\sqrt{\frac{2g}{7R}}$.

19. In order to shift a body of mass m from a circular orbit of radius 3R to a higher orbit of radius 5R around the Earth, the work done is

(*a*) $\frac{3GMm}{5R}$ (*b*) $\frac{GMm}{2R}$
(*c*) $\frac{2}{15}\frac{GMm}{R}$ (*d*) $\frac{GMm}{5R}$.

20. Two equal masses m and m are hung from a balance whose scale pans differ in vertical height by h. If D is the mean density of Earth, then the error in weighing is

(*a*) Zero (*b*) $\frac{4}{3}\pi GD\ mh$
(*c*) $\frac{8}{3}\pi GD\ mh$ (*d*) $\frac{2}{3}\pi GD\ mh$.

21. Radius of the Earth is 6000 km. Weight of a body on the surface of Earth is 120 kg wt. If it is taken to a height of 2000 km above the surface of Earth, its weight is

(*a*) 60 kg wt. (*b*) 40 kg wt.
(*c*) 67.5 kg wt. (*d*) 47.5 kg wt.

22. Explorer 38, a radio-astronomy satellite of mass 200 kg, circles the Earth in an orbit of average radius $\frac{3R}{2}$ where R is the radius of the Earth. Assuming the gravitational pull on a mass of 1 kg at the earth's surface to be 10 N, calculate the pull on the satellite

(*a*) 889 N (*b*) 89 N
(*c*) 8889 N (*d*) 8.9 N.

23. A person brings a mass of 1 kg from infinity to a point A. Initially the mass was at rest but it moves at a speed of 2 m s^{-1} as it reaches A. The work done by the person on the mass is – 3 J. The potential at A is

(*a*) – 3 J/kg (*b*) – 2 J/kg
(*c*) – 5 J/kg (*d*) None of these.

24. A satellite of mass m moves around the Earth in a circular orbit with speed v. The potential energy of the satellite is

(a) mv^2 (b) $-mv^2$

(c) $\frac{3}{2}mv^2$ (d) $-\frac{3}{2}mv^2$.

25. Suppose the angular speed of Earth becomes 1.25×10^{-3} rad s^{-1}. If $g = 10$ m s^{-2}, then the weight of a body at the equator will become

(a) zero (b) double

(c) half (d) infinity.

26. A satellite is launched in a circular orbit of radius R and another satellite is launched in circular orbit of radius 1.01 R. The time period of second satellite is different from that of the first satellite by

(a) 1.5% increased (b) 1% decreased

(c) 1% increased (d) 1.5% decreased.

27. A satellite of mass m is revolving round the Earth at a height R above the surface of the Earth. If g is the gravitational intensity at the earth surface and R is the radius of the Earth, then the kinetic energy of the satellite will be

(a) $\frac{mgR}{4}$ (b) $\frac{mgR}{2}$

(c) mgR (d) $2mgR$.

28. A star suddenly shrinks and its density becomes 10^9 times the original value. The value of acceleration due to gravity on its surface will increase by a factor of

(a) 10^6 (b) 10^9

(c) 10^{19} (d) 10^{21}.

29. A uniform ring of mass m and radius r is placed directly above a uniform sphere of mass M and of equal radius. The centre of the ring is directly above the centre of the sphere at a distance $r\sqrt{3}$ as shown in the Fig. 5. The gravitational force exerted by the sphere on the ring will be

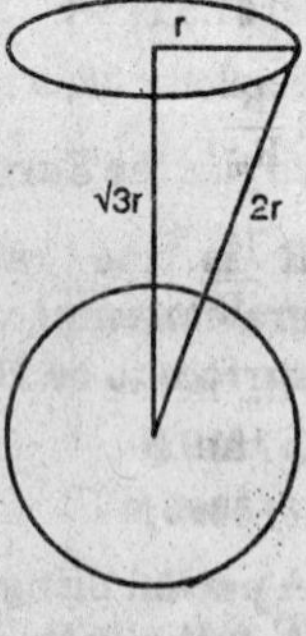

Fig. 5

(a) $\frac{GMm}{8r^2}$ (b) $\frac{GMm}{4r^2}$

(c) $\sqrt{3}\frac{GMm}{8r^2}$ (d) $\frac{GMm}{8r^3\sqrt{3}}$.

30. A tunnel is dug along a diameter of the Earth. If M_e and R_e are the mass and radius respectively of Earth, then the force on a particle of mass m placed in the tunnel at a distance r from the centre is

(a) $\frac{GM_em}{R_e^3}r$ (b) $\frac{GM_em}{R_e^3 r}$

(c) $\frac{GM_emR_e^3}{r}$ (d) $\frac{GM_em}{R_e^2}r$.

31. Escape velocity of a body from the surface of Earth is 11 km s^{-1}. Assume that the mass and radius of the Earth are 81 times and 4 times the mass and radius respectively of the Moon. Then, the escape velocity from the surface of Moon is

(a) 6.8 m s^{-1} (b) 2.44 km s^{-1}

(c) 3.97 km s^{-1} (d) 1118 km h^{-1}.

32. The velocity of a satellite in a parking orbit is nearly

(a) 3 km s^{-1} (b) 9.8 km s^{-1}

(c) 12.4 km s^{-1} (d) 36 km s^{-1}.

33. An object weighs 98 N at the north pole of Earth. What is its true weight in a geostationary satellite whose distance from the centre of Earth is 7R, where R is the radius of the Earth ?

(a) 1 N (b) 2 N

(c) 7 N (d) 9 N.

34. Two particles of equal mass m go round a circle of radius R under the action of their mutual gravitational attraction. The speed of each particle is

(a) $v = \frac{1}{2R}\sqrt{\frac{1}{Gm}}$ (b) $v = \sqrt{\frac{Gm}{2R}}$

(c) $v = \frac{1}{2}\sqrt{\frac{Gm}{R}}$ (d) $v = \sqrt{\frac{4Gm}{R}}$

35. If the Earth were assumed to have uniform density and spherical symmetry, then the value of g in m s^{-2} halfway towards the centre of Earth would be

(a) 0 (b) 1.25

(c) 5 (d) 10.

36. The acceleration due to gravity is reduced to a value one-fourth of that on the surface of Earth at a height h where h is equal to xR_e (radius of Earth). The value of x is

(a) 1 (b) $\frac{1}{2}$

(c) $\frac{1}{3}$ (d) $\frac{1}{4}$.

37. A satellite is revolving in an orbit close to the surface of Earth with kinetic energy K. When the kinetic energy is increased to nK, the satellite escapes. What is the minimum value of n ?

(a) 2 (b) 2.5

(c) 4 (d) 5.

38. If the gravitational force were proportional to $\frac{1}{r}$, then a particle in a circular orbit under such a force would have its original speed

(a) independent of r (b) $\propto \frac{1}{r}$

(c) $\propto \frac{1}{r^2}$ (d) $\propto r^2$.

39. The speed of Earth's rotation about its axis is ω. Its speed increases to x times to make the effective value of acceleration due to gravity zero at the equator. Then the value of x is

(a) 1 (b) 8.5

(c) 17 (d) 34.

40. Two bodies of masses 10^5 kg and 10^3 kg are separated by 3 m. At what distance from the smaller body the gravitational field strength will be zero ?

(a) $\frac{1}{11}$ m (b) $\frac{3}{11}$ m

(c) $\frac{5}{11}$ m (d) $\frac{7}{11}$ m.

41. A planet of mass 3×10^{26} kg moves around a star with a constant speed of 2×10^6 m s^{-1} in a circle of radius 1.5×10^{12} m. The gravitational force acting on the planet is

(a) 6.67×10^{22} dyne (b) 8×10^{27} dyne

(c) 8×10^{26} N (d) 6.67×10^{20} N.

42. Three particles of equal mass m are situated at the vertices of an equilateral triangle of side l. The work done in increasing the side of the triangle to $2l$ will be

(a) $\frac{3G^2m}{2l}$ (b) $\frac{Gm^2}{2l}$

(c) $\frac{3Gm^2}{2l}$ (d) $\frac{3Gm^2}{l}$.

43. The orbital velocity of a body at height h above the surface of Earth is 36% of that very near the surface of the Earth of radius R. If the escape velocity at the surface of Earth is 11.2 km s^{-1}, then its value at the height h will be

(a) 11.2 km s^{-1} (b) $\sqrt{\frac{h}{R}} \times 11.2$ km s^{-1}

(c) $\frac{9}{25} \times 11.2$ km s^{-1} (d) $\sqrt{\frac{R}{h}} \times 11.2$ km s^{-1}.

44. A uniform spherical shell gradually shrinks maintaining its shape. The gravitational potential at the centre

(a) increases (b) decreases

(c) remains constant (d) oscillates.

45. The orbital velocity of an artificial satellite in a circular orbit very close to Earth is v. The velocity of a geosynchronous satellite orbiting in a circular orbit at an altitude of 6R from Earth's surface will be

(a) $\frac{v}{\sqrt{7}}$ (b) $\frac{v}{\sqrt{6}}$

(c) v (d) $\sqrt{6}v$.

46. A projectile is fired vertically from the Earth with a velocity kv_e where v_e is the escape velocity and k is a constant less than unity. The maximum height to which projectile rises, as measured from the centre of Earth, is

(a) $\frac{R}{k}$ (b) $\frac{R}{k-1}$

(c) $\frac{R}{1-k^2}$ (d) $\frac{R}{1+k^2}$.

47. A satellite of mass m is revolving in a circular orbit radius r around the Earth. If M be the mass of Earth, then the angular momentum of the satellite

(a) $\sqrt{GMr}$ (b) $\sqrt{Gmr}$

(c) $\sqrt{GMm^2r}$ (d) $\sqrt{GM^2mr}$.

48. Assume that the time period of a satellite close to the surface of Earth is 80 minutes. What is the time period of another Earth satellite in an orbit of radius equal four times the Earth's radius ?

(a) 640 minutes (b) 660 minutes

(c) 680 minutes (d) 720 minutes.

49. If R_m is the radius of Moon's orbit around the Earth a_m the acceleration of Moon towards the Earth, R the radius of the Earth and g the value of acceleration due to gravity on the Earth, then the ratio $\frac{a_m}{g}$ is equal

(a) $\frac{R}{R_m}$ (b) $\left(\frac{R}{R_m}\right)^2$

(c) $\left(\frac{R_m}{R}\right)^2$ (d) $\left(\frac{R_m}{R}\right)^3$.

50. If R_m is the radius of Moon orbiting round the Earth, a_m the acceleration of Moon towards the centre Earth, and R_e the radius of Earth. Then a_m is equal (if g is acceleration due to gravity on the surface of Earth)

(a) $\left(\frac{R_e}{R_m}\right)g$ (b) $\left(\frac{R_m}{R_e}\right)g$

(c) $\left(\frac{R_m}{R_e}\right)^2 g$ (d) $\left(\frac{R_e}{R_m}\right)^2 g$.

51. The gravitational potential difference between the surface of a planet and a point 20 m above the surface, is 2 joule/kg. If the gravitational field is uniform, then the work done in carrying a 5 kg body to a height of 4 *m* above the surface is

(*a*) 2 J (*b*) 20 J
(*c*) 40 J (*d*) 10 J.

52. Two satellites A and B go around the Earth in circular orbits at heights R_A and R_B respectively from the surface of Earth. Assume the Earth to be a uniform sphere of radius R_E. The ratio of the magnitudes of the velocity, $\frac{v_A}{v_B}$ of the satellites is

(*a*) $\sqrt{\frac{R_B}{R_A}}$ (*b*) $\sqrt{\frac{R_B + R_E}{R_A + R_E}}$

(*c*) $\frac{R_B + R_E}{R_A + R_E}$ (*d*) $\left(\frac{R_A}{R_B}\right)^2$.

53. A communication satellite is put in an orbit around the Earth. What is the approximate time taken by a radiowave to go to the satellite and come back to the Earth ?

(*a*) $\frac{1}{4}$ s (*b*) $\frac{1}{8}$ s
(*c*) 1 s (*d*) 12 s.

54. The radius of Earth is 6400 km and g = 10 m s^{-2}. In order that a body of 5 kg weighs zero at the equator, the angular speed of Earth in radian/s is

(*a*) $\frac{1}{80}$ (*b*) $\frac{1}{400}$

(*c*) $\frac{1}{800}$ (*d*) $\frac{1}{1600}$.

55. A body weighs 63 N on the surface of the Earth. At a height *h* above the surface of Earth, its weight is 28 N while at a depth *h* below the surface of Earth, the weight is 31.5 N. The value of *h* is

(*a*) 0.4 R (*b*) 0.5 R
(*c*) 0.8 R (*d*) R.

56. A double star consists of two stars, one of mass M and the other 2M. Distance between them is *r*. The two stars revolve under their mutual gravitational force. Then

(*a*) the kinetic energy of the heavier star is 2 times the kinetic energy of the lighter star

(*b*) the two stars will revolve around the point of trisection of the straight line joining their mass-centres, nearer to the heavier star

(*c*) the two stars have equal periods of revolution

(*d*) both (*b*) and (*c*) are correct.

57. The escape velocity for a planet is 20 km s^{-1}. The potential energy of a particle of mass 1 milligram on the surface of planet will be

(*a*) 100 J (*b*) 200 J
(*c*) 400 J (*d*) 600 J.

58. A body is raised to a height *h* so that the force of gravity on the body is 1% of the value on the surface of Earth. If the radius of the Earth is 6400 km, then the value of h is

(*a*) 57600 km (*b*) 5999 km
(*c*) 9.8 km (*d*) 0.98 km.

59. The square of the angular velocity of a planet around the Sun is proportional to (where R is the radius of the orbit of the planet around the Sun.)

(*a*) R (*b*) R^2
(*c*) R^3 (*d*) R^{-3}

60. What is the shortest possible period of an Earth satellite in circular orbit ?

(*a*) $2\pi\sqrt{\frac{R}{g}}$ (*b*) zero
(*c*) 1 day (*d*) 1 year.

61. An artificial satellite is moving in a circular orbit around the Earth with a speed equal to half the magnitude of escape velocity from the Earth. The height of satellite above the surface of Earth is (where R = radius of Earth.)

(*a*) R (*b*) 2R
(*c*) 3R (*d*) 4R

62. The distances of Neptune and Saturn from the Sun are nearly 10^{13} m and 10^{12} m respectively. Their periodic times will be in the ratio

(*a*) 10 (*b*) 100
(*c*) $10\sqrt{10}$ (*d*) 1000.

63. A planet has twice the radius but the same mean density as compared to Earth. If v_e' is the escape velocity for the planet and v_e is the escape velocity for the Earth, then v_e'/v_e is equal to

(*a*) 1 : 1 (*b*) 2 : 1
(*c*) 1 : 2 (*d*) $1 : 2\sqrt{2}$.

64. A missile which missed its target went into orbit around the Earth at a mean radius 4 times as great as the parking orbit of a satellite. The period of the missile is

(*a*) 4 days (*b*) 8 days
(*c*) 16 days (*d*) 64 days.

65. The force of gravitational attraction between the Earth and a geostationary satellite of mass 1 quintal is nearly

(*a*) 10 N (*b*) 20 N
(*c*) 30 N (*d*) 40 N.

66. A mass m is raised from the surface of Earth to a point which is at a height nR from the surface of Earth. Change in potential energy is

(*a*) $\frac{mgR}{n}$ (*b*) $\frac{n}{n+1}(mgR)$
(*c*) $nmgR$ (*d*) $\frac{2mgR}{n}$.

67. A planet of mass m moves around the Sun of mass M in an elliptical orbit. The maximum and minimum distances of the planet from the Sun are r_1 and r_2 respectively. The time period of the planet is proportional to

(*a*) $(r_1 + r_2)$ (*b*) $(r_1 + r_2)^{1/2}$
(*c*) $(r_1 - r_2)^{3/2}$ (*d*) $(r_1 + r_2)^{3/2}$.

68. What amount of energy is required for making a 500 kg body escape from the Earth ? (Given : $g = 10$ m s^{-2} and radius of Earth $= 6 \times 10^6$ m)

(*a*) 3×10^9 J (*b*) 1.5×10^9 J
(*c*) 3×10^{10} J (*d*) 6×10^{10} J.

69. Two bodies of masses 100 kg and 1000 kg are separated by a distance of 1 m. What is the gravitational field intensity at the mid-point of the line joining them ?

(*a*) 2.4×10^{-6} N kg^{-1} (*b*) 2.4×10^{-7} N kg^{-1}
(*c*) 2.4×10^{-8} N kg^{-1} (*d*) 2.4×10^{-9} N kg^{-1}.

70. A body is orbiting around the Earth at a mean radius 16 times as great as the parking orbit of a satellite. The period of the body is

(*a*) 8 days (*b*) 16 days
(*c*) 64 days (*d*) 365 days.

71. The weight of an object at the equator is the same as its weight at a depth d below the pole. If ω is the rotational speed of Earth, then the value of d is

(*a*) $\frac{\omega R}{g}$ (*b*) $\frac{\omega R}{g^2}$
(*c*) $\frac{\omega^2 R^2}{g}$ (*d*) $\frac{3\omega^2 R^2}{g}$.

72. Earth is compressed to such a small-size sphere that the escape velocity becomes equal to the velocity of light. What is the approximate radius of this small-size sphere ? (Given : mass of Earth $= 6 \times 10^{24}$ kg)

(*a*) 4.5 mm (*b*) 9 mm
(*c*) 12 mm (*d*) 24 mm.

73. The mean radius of the Earth's orbit around the Sun is 1.5×10^{11} m. The mean radius of the orbit of Mercury around the Sun is 6×10^{10} m. The Mercury will rotate around the Sun in

(*a*) 1 year (*b*) nearly 4 years
(*c*) nearly $\frac{1}{4}$ years (*d*) 2.5 years.

74. In Q. 73, the ratio of the linear velocity of mercury in its orbit around the Sun to the linear velocity of Earth around the Sun is

(*a*) $\frac{\sqrt{10}}{2}$ (*b*) $\sqrt{\frac{2}{10}}$
(*c*) $2\sqrt{10}$ (*d*) $\frac{10}{2}$.

75. The ratio of the kinetic energy required to be given to a satellite so that it escapes the gravitational field of Earth to the kinetic energy required to put the satellite in a circular orbit just above the free surface of Earth is

(*a*) 1 (*b*) 2
(*c*) 3 (*d*) 9.

76. A satellite is circling around the Earth at a distance x from the centre of Earth. If the radius of the circular orbit is reduced by 1%, then the speed will increase by

(*a*) 0.5% (*b*) 1%
(*c*) 2% (*d*) 3%.

77. The increase in gravitational potential energy of an object of mass m raised from the surface of Earth to a height equal to the radius R of Earth is

(*a*) mgR (*b*) $\frac{mgR}{2}$
(*c*) $\frac{mgR}{3}$ (*d*) $\frac{mgR}{4}$.

78. The value of 'g' at a certain height h above the free surface of Earth is $\frac{x}{4}$ where x is the value of 'g' at the surface of Earth. The height h is

(*a*) R (*b*) 2R
(*c*) 3R (*d*) 4R.

79. What is the radius of a sphere with a density of ρ from the surface of which you could just barely throw a ball and have it never return ? Assume your best throw is v.

(*a*) zero (*b*) infinity
(*c*) $v\sqrt{\frac{3}{8\pi G\rho}}$ (*d*) $\rho v^2 + G$.

80. Two bodies of masses 10 kg and 100 kg are separated by a distance of 2 m. The gravitational potential at the mid-point of the line joining the two bodies is

(*a*) -7.3×10^{-7} J/kg (*b*) -7.3×10^{-8} J/kg
(*c*) -7.3×10^{-9} J/kg (*d*) -7.3×10^{-6} J/kg.

81. A planet is revolving in an elliptical orbit around the Sun. Its closest distance from the Sun is r_{min} and the farthest distance is r_{max}. If the velocity of the planet at the distance of closest approach is v_1 and that at the farthest distance from the Sun is v_2, then $v_1/v_2 =$

(a) $r_{max} \div r_{min}$ (b) $r_{min} \div r_{max}$

(c) $\frac{r_{min} + r_{max}}{r_{max} - r_{min}}$ (d) None of these.

82. Two satellites move in circular orbits around the Earth at distances 9000 km and 9010 km from the centre of Earth. If the nearer satellite has a period of revolution of 90 minutes around the Earth, then the difference in the periods of revolution around the Earth is

(a) 3 s (b) 6 s

(c) 9 s (d) 12 s.

83. The ratio of the radius of the Earth to that of moon is 10. The ratio of acceleration due to gravity on Earth and on Moon is 6. The ratio of the escape velocity from the surface of Earth to that from the Moon is

(a) 10 (b) 5

(c) nearly 8 (d) 1.66.

84. The eccentricity of Earth's orbit is 0.017. The ratio of the maximum speed in its orbit to its minimum speed is

(a) 1 (b) 1.017

(c) 1.034 (d) 1.051.

85. The gravitational potential energy of a body of mass m at a distance r from the centre of the Earth is U. What is the weight of the body at this distance ?

(a) U (b) Ur

(c) $\frac{U}{r}$ (d) $\frac{U}{2r}$.

86. The largest and the shortest distances of the Earth from the Sun are r_1 and r_2 respectively. Its distance from the Sun, when it is at the perpendicular to the major axis of the orbit drawn from the Sun is

(a) $\frac{r_1 + r_2}{2}$ (b) $\frac{r_1 r_2}{r_1 + r_2}$

(c) $\frac{2r_1 r_2}{r_1 + r_2}$ (d) $\frac{r_1 + r_2}{3}$.

87. What is the weight of a body at a distance $2r$ from the centre of the Earth if the gravitational potential energy of the body at a distance r from the centre of the Earth is U ?

(a) $\frac{U}{2r}$ (b) $\frac{U}{3r}$

(c) $\frac{U}{4r}$ (d) Ur.

88. Consider two point masses of 10^3 kg and 10^5 kg separated by 1 km. At a distance r from the lighter of the two masses, the gravitational potential due to the lighter mass is 10 times the gravitational potential due to the heavier mass. The value of r is

(a) $\frac{1}{1000}$ km (b) $\frac{1}{1001}$ km

(c) $\frac{1}{999}$ km (d) $\frac{1000}{1001}$ km.

89. Masses of 1 kg each are placed 1 m, 2 m, 4 m, 8 m, ... from a point P. The gravitational field intensity at P due to these masses is

(a) G (b) 2G

(c) 4G (d) $\frac{4G}{3}$.

90. If the Earth were to cease rotating about its own axis, the increase in the value of g at the equator is nearly

(a) 6.72 cm s^{-2} (b) 3.4 cm s^{-2}

(c) 1.68 cm s^{-2} (d) 0.84 cm s^{-2}.

91. If the value of acceleration due to gravity on the surface of Earth is 10 m s^{-2}, then a girl can jump safely from a height of 2 m. If she has to jump on a planet where acceleration due to gravity is 0.2 m s^{-2}, then the safe height will be

(a) 0.2 m (b) 10 m

(c) 2 m (d) 100 m.

92. Suppose Earth's orbital motion around the Sun is suddenly stopped. What time the Earth shall take to fall into the Sun ?

(a) 5 days (b) 27.3 days

(c) two months (d) 1 year.

93. Suppose Moon's orbital motion around the earth is suddenly stopped. What time the Moon shall take to fall on to the Earth ?

(a) 5 days (b) 27.3 days

(c) two months (d) 1 year.

94. Three uniform spheres, each having mass m and radius r, are kept in such a way that each touches the other two. The magnitude of the gravitational force on any sphere due to the other two is

(a) $\frac{Gm^2}{r^2}$ (b) $\frac{Gm^2}{4r^2}$

(c) $\sqrt{2}\,\frac{Gm^2}{4r^2}$ (d) $\sqrt{3}\,\frac{Gm^2}{4r^2}$.

95. The radius of the Earth shrinks by 1%, its mass remaining the same. The percentage change in the value of g is

(a) – 2% (b) + 2%

(c) – 3% (d) + 4%.

96. The escape velocity from a spherical planet is v_0. What is the escape velocity corresponding to another planet of twice the radius and half the mean density ?

(*a*) $\sqrt{2}v_0$ (*b*) $\frac{v_0}{\sqrt{2}}$

(*c*) $2v_0$ (*d*) $4v_0$.

97. A body of mass *m* rises to a height $h = \frac{R}{5}$ from the Earth's surface where R is Earth's radius. If *g* is acceleration due to gravity at the Earth's surface, the increase in potential energy is

(*a*) *mgh* (*b*) $\frac{4}{5}$ *mgh*

(*c*) $\frac{5}{6}$ *mgh* (*d*) $\frac{6}{7}$ *mgh*.

98. A particle of mass 1 kg is placed at a distance of 4 m from the centre and on the axis of a uniform ring of mass 5 kg and radius 3 m. The work done to increase the distance of the particle from 4 m to $3\sqrt{3}$ m is

(*a*) $\frac{G}{3}$ J (*b*) $\frac{G}{4}$ J

(*c*) $\frac{G}{5}$ J (*d*) $\frac{G}{6}$ J.

99. The Earth revolves around the Sun with an angular velocity of 2×10^{-7} rad s^{-1} in a circular orbit of radius 1.5×10^8 km. If mass of Earth is 6×10^{24} kg, then the force exerted by the Sun on Earth is

(*a*) zero (*b*) 36×10^{21} N

(*c*) 45×10^{23} N (*d*) 9.8 N.

100. A man weighs 80 kg on the surface of Earth of radius R. At what height above the surface of Earth his weight will be 40 kg ?

(*a*) $\frac{R}{2}$ (*b*) $\sqrt{2}$R

(*c*) $(\sqrt{2} - 1)$R (*d*) $(\sqrt{2} + 1)$R.

101. Masses of the Earth and the Mars are in the ratio 10 : 1 and their radii in the ratio 2 : 1. What will be the ratio of the weights of the same body on the surface of these planets ?

(*a*) $\frac{2}{5}$ (*b*) $\frac{5}{2}$

(*c*) $\frac{20}{1}$ (*d*) $\frac{1}{20}$.

102. A research satellite of mass 200 kg circles the Earth in an orbit of average radius 3R/2, where R is the radius of the Earth. Assuming the gravitational pull on a mass 1 kg on Earth's surface to be 10 N, the pull on the satellite will be

(*a*) 880 N (*b*) 889 N

(*c*) 890 N (*d*) 892 N.

103. What is the percentage change in the value of *g* as we shift from equator to pole on the surface of Earth ? (Given : Equatorial radius of Earth is greater than polar radius by 21 km and mean radius of Earth = 6300 km)

(*a*) 0.1% (*b*) 0.5%

(*c*) 0.67% (*d*) 10%.

104. A satellite moves in a circle around the Earth. The radius of this orbit is equal to one-half of the radius of the Moon's orbit. The satellite completes one revolution in

(*a*) $\frac{1}{2}$ lunar month (*b*) $2\sqrt{2}$ lunar month

(*c*) $2^{2/3}$ lunar month (*d*) $2^{-3/2}$ lunar month.

105. The Moon revolves around the Earth 13 times in one year. The ratio of Sun-Earth distance to Earth-Moon distance is 392. Then the ratio of the masses of Sun and Earth will be

(*a*) $\frac{393}{13}$ (*b*) $\frac{392}{13}$

(*c*) $\frac{392^2}{13^2}$ (*d*) $\frac{392^3}{13^2}$.

106. A particle is kept at rest at a distance R (Earth's radius) above the Earth's surface. The minimum speed with which it should be projected so that it does not return is

(*a*) $\sqrt{\frac{GM}{4R}}$ (*b*) $\sqrt{\frac{GM}{2R}}$

(*c*) $\sqrt{\frac{GM}{R}}$ (*d*) $\sqrt{\frac{2GM}{R}}$.

107. The kinetic energy needed to project a body of mass *m* from the Earth's surface to infinity is

(*a*) $\frac{1}{4}$ *mgR* (*b*) $\frac{1}{2}$ *mgR*

(*c*) *mgR* (*d*) 2*mgR*.

108. The weight of an object at the equator is the same as its weight at height *h* above the pole. If *w* is the rotational speed of Earth, then the value of *h* is

(*a*) $\frac{\omega^2R^2}{2g}$ (*b*) $\frac{\omega R}{2g}$

(*c*) $\frac{\omega}{2g}$ (*d*) $\frac{\omega^2R^2}{2g}$.

109. What would be the length of the day if the angular speed of Earth is increased, such that the bodies lying on the equator fly off ? (Take $\pi = 3$)

(a) 1.3h (b) 9h
(c) 12h (d) 16h.

110. Three particles, each of mass 10^{-2} kg, are brought from infinity to the vertices of an equilateral triangle of side 0.1 m. The work done is

(a) 2×10^{-9} J (b) 2×10^{-10} J
(c) 2×10^{-11} J (d) 2×10^{-13} J.

111. A diametrical tunnel is dug across the Earth. A ball is dropped into the tunnel from one side. The velocity of the ball when it reaches the centre of the Earth is ......

$\left(\text{Given : gravitational potential at the centre of Earth} = -\frac{3}{2}\frac{GM}{R}\right)$.

(a) $\sqrt{R}$ (b) $\sqrt{gR}$
(c) $\sqrt{2.5\,gR}$ (d) $\sqrt{7.1\,gR}$.

112. If the Earth is treated as a sphere of radius R and mass M, its angular momentum about the axis of its diurnal rotation with period T is

(a) $\frac{4\pi MR^2}{5T}$ (b) $\frac{2\pi MR^2}{T}$
(c) $\frac{MR^2T}{2\pi}$ (d) $\frac{\pi MR^2}{T}$.

113. Imagine a light planet revolving around a very massive star in a circular orbit of radius R with a period of revolution T. If the gravitational force of attraction between the planet and the star is proportional to $R^{-5/2}$, then $T^2 \propto$

(a) R^3 (b) $R^{7/2}$
(c) $R^{5/2}$ (d) $R^{3/2}$.

114. A space ship is released in a circular orbit near Earth's surface. How much additional velocity will have to be given to the ship in order to escape it out of this orbit ?

(a) 3.28 m s^{-1} (b) 3.3×10^3 m s^{-1}
(c) 3.28×10^7 m s^{-1} (d) 3.28×10^{-3} m s^{-1}.

115. The gravitational potential energy of a rocket of mass 100 kg at a distance of 10^7 m from the Earth's mass is 4×10^9 J. The weight of the rocket in newton at a distance of 10^9 m is

(a) 8×10^{-2} (b) 8×10^{-3}
(c) 4×10^{-3} (d) 4×10^{-2}.

116. The gravitational potential at height h above Earth's surface is -5.12×10^7 joule/kg and acceleration due to gravity at this point is 6.4 m s^{-2}. If radius of Earth is 6400 km, then the value of h is

(a) 4200 km (b) 1600 km
(c) 9800 km (d) 12400 km.

117. Consider two solid uniform spherical objects of the same density ρ. One has a radius R and the other of radius 2R. They are in outer space where the gravitational fields from other objects are negligible. If they are at rest with their surfaces touching, what is the contact force between the objects due to their gravitational attraction ?

(a) $G\pi^2R^4$ (b) $\frac{128}{81}G\pi^2R^4\rho^2$
(c) $\frac{128}{81}G\pi^2$ (d) $\frac{128}{87}\pi R^2G$.

118. Three particles, each of mass M, are situated at the vertices of an equilateral triangle of side length a. The only forces acting on the particles are their mutual gravitational forces. It is desired that each particle moves in a circle while maintaining the original mutual separation a. What is the net gravitational force on one of the particles ?

(a) $\frac{GM^2}{a^2}$ (b) $\sqrt{2}\frac{GM^2}{a^2}$
(c) $\sqrt{3}\frac{GM^2}{a^2}$ (d) $\frac{2GM^2}{a^2}$.

119. In Q. 118 the required initial velocity is

(a) $\sqrt{\frac{GM}{a}}$ (b) $\sqrt{\frac{2GM}{a}}$
(c) $\sqrt{\frac{3GM}{a}}$ (d) $\sqrt{\frac{4GM}{a}}$.

120. In Q. 118, the time period of revolution is

(a) $2\pi\sqrt{\frac{a}{3GM}}$ (b) $2\pi\sqrt{\frac{a^2}{3GM}}$
(c) $2\pi\sqrt{\frac{a^3}{GM}}$ (d) $2\pi\sqrt{\frac{a^3}{3GM}}$.

121. A body starts from rest from a point distant r_0 from the centre of the Earth. It reaches the surface of the Earth whose radius is R. The velocity acquired by the body is

(a) $2GM\sqrt{\frac{1}{R}-\frac{1}{r_0}}$ (b) $\sqrt{2GM\left(\frac{1}{R}-\frac{1}{r_0}\right)}$
(c) $GM\sqrt{\frac{1}{R}-\frac{1}{r_0}}$ (d) $\sqrt{GM\left(\frac{1}{R}-\frac{1}{R_0}\right)}$

122. Four particles, each of mass M, move along a circle of radius R under the action of their mutual gravitational attraction. The speed of each particle is

(a) $\frac{GM}{R}$ (b) $\sqrt{2\sqrt{2}\frac{GM}{R}}$
(c) $\sqrt{\frac{GM}{R}(2\sqrt{2}+1)}$ (d) $\sqrt{\frac{GM}{R}\left(\frac{2\sqrt{2}+1}{4}\right)}$.

123. If distance between Sun and Earth is 400 times the distance between Moon and Earth and gravitational pull of Sun on Earth is 170 times the gravitational pull of Earth on Moon, then the ratio of the mass of Sun to that of Moon is approximately equal to

(*a*) 2.7×10^7 (*b*) 4.6×10^9
(*c*) 6.8×10^4 (*d*) 7.4×10^4.

124. Given : R = radius of Earth and escape velocity from the surface of Earth = 11.2 km s^{-1}. What is the value of escape velocity at a height 3R from the surface of Earth ?

(*a*) 5.6 km s^{-1} (*b*) 9.8 km s^{-1}
(*c*) 10 km s^{-1} (*d*) 11.2 km s^{-1}.

125. The value of acceleration due to gravity at height *h* from Earth surface will become half its value on the surface if : (R = radius of Earth)

(*a*) $h = R$ (*b*) $h = 2R$
(*c*) $h = (\sqrt{2} - 1)R$ (*d*) $h = (\sqrt{2} + 1)R$.

126. A body is thrown up from the surface of Earth with a velocity equal to $\frac{3}{4}$ th of the escape velocity from the surface of Earth. The height attained by the body is

(*a*) $\frac{9R}{7}$ (*b*) $\frac{7R}{9}$
(*c*) $\frac{11R}{17}$ (*d*) $\frac{19R}{20}$.

127. An artificial satellite is orbiting at a height of 1800 km from the Earth's surface. The Earth's radius is 6300 km and g = 10 m s^{-2} on its surface. What is the radial acceleration of the satellite ?

(*a*) 6 m s^{-2} (*b*) 7 m s^{-2}
(*c*) 8 m s^{-2} (*d*) 9 m s^{-2}.

128. In Q. 127, what is the speed of the satellite ?

(*a*) 8 km s^{-1} (*b*) 7 km s^{-1}
(*c*) 6 km s^{-1} (*d*) 5 km s^{-1}.

129. The mass of the Earth is 6.00×10^{24} kg and that of the Moon is 7.40×10^{22} kg. The constant of gravitation $G = 6.67 \times 10^{-11}$ Nm^2 kg^{-2}. The potential energy of the system is -7.79×10^{28} J. The mean distance between the Earth and Moon is

(*a*) 3.80×10^8 metre (*b*) 3.37×10^6 metre
(*c*) 7.60×10^4 metre (*d*) 1.90×10^2 metre.

130. The value of g on the Earth's surface is 980 cm s^{-2}. Its value at a height of 64 km from the Earth's surface is (Radius of the Earth R = 6400 km)

(*a*) 960.4 cm s^{-2} (*b*) 984.9 cm s^{-2}
(*c*) 982.4 cm s^{-2} (*d*) 977.5 cm s^{-2}.

131. At what height above the surface of Earth the value of g is the same as that in a mine 10 km deep ?

(*a*) 10 km (*b*) 5 km
(*c*) 2.5 km (*d*) 1.25 km.

132. The period of revolution of planet A around the Sun is 8 times that of B. The distance of A from the Sun is n times greater than the distance of B from the Sun. The value of n is

(*a*) 4 (*b*) 8
(*c*) 12 (*d*) 16.

133. A satellite is moving very close to a planet of density 8×10^3 kg m^{-3}. If $G = 6.67 \times 10^{-11}$ Nm^2 kg^{-2}, then the time period of the satellite is nearly

(*a*) 420 s (*b*) 4200 s
(*c*) 1 hour (*d*) 1 day.

134. The orbital velocity at a height *h* above the surface of Earth is 90% of that near the surface of earth. If the escape velocity at the surface of Earth is v, then its value at height *h* will be

(*a*) 0.99 v (*b*) 0.90 v
(*c*) 0.81 v (*d*) 0.11 v.

135. If a body is to be projected vertically upwards from earth's surface to reach a height of 10R where R is the radius of Earth, then the velocity required to do so is

(*a*) $\sqrt{\frac{24}{11} gR}$ (*b*) $\sqrt{\frac{22}{11} gR}$
(*c*) $\sqrt{\frac{20}{11} gR}$ (*d*) $\sqrt{\frac{18}{11} gR}$.

136. A satellite is orbiting just above the surface of a planet of average density ρ with period T. If G is the universal gravitational constant, the quantity $T^2\rho$ is equal to

(*a*) $4\pi^2 G$ (*b*) $4\pi^2/G$
(*c*) $3\pi/G$ (*d*) $1/G$.

137. What is the total energy of a satellite of mass m revolving around the Earth in a circular orbit with speed v ? The total energy of the satellite is

(*a*) $\frac{1}{2} mv^2$ (*b*) $-\frac{1}{2} mv^2$
(*c*) $\frac{3}{4} mv^2$ (*d*) $\frac{5}{4} mv^2$
(*e*) $\frac{7}{4} mv^2$.

138. A thin wire of mass m and length l is bent to form a semi-circle. The gravitational potential at the centre of the semicircle is

(a) $-\frac{\pi Gm}{l}$ (b) $-\frac{2\pi Gm}{l}$

(c) $\frac{-3\pi Gm}{l}$ (d) zero.

139. If both the mass and radius of Earth decrease by 1%, the value of acceleration due to gravity will decrease by nearly

(a) 1% (b) 1.5%

(c) 2% (d) 2.5%.

140. In Q. 139, the escape velocity will

(a) be doubled (b) be halved

(c) be tripled (d) not change.

141. A planet has twice the density of Earth but the acceleration due to gravity on the surface is same as on the surface of Earth. Its radius in terms of the radius R of Earth is

(a) $\frac{R}{2}$ (b) $\frac{R}{4}$

(c) $\frac{R}{8}$ (d) $\frac{R}{64}$.

142. A satellite goes along an elliptical orbit around the Earth. The rate of change of area swept by the line joining earth and satellite is proportional to : (r is radius of orbit)

(a) r (b) $r^{1/2}$

(c) $r^{3/2}$ (d) $r^{7/2}$.

143. The gravitational potential at a height h above the surface of Earth is -5.12×10^7 J kg^{-1}. The value of acceleration due to gravity at this height is 6.4 m s^{-2}. If radius of Earth is 6400 km, then the value of h is

(a) 800 km (b) 1000 km

(c) 1600 km (d) 2000 km.

144. The escape velocity from the surface of Earth is 11.2 km s^{-1}. A body is projected with twice this velocity. What is the velocity at infinity ?

(a) 25.4 km s^{-1} (b) 19.4 km s^{-1}

(c) 9.8 km s^{-1} (d) 0.98 km s^{-1}.

145. Suppose the gravitational force varies inversely as the nth power of the distance. Then the time period of a planet in circular orbit of radius R around the Sun will be proportional to

(a) R^{-n} (b) R^{n}

(c) $R^{(n-1)/2}$ (d) $R^{(n+1)/2}$.

146. Consider a particle of mass m on the axis of a ring, of mass M and radius r, at a distance r from the centre of the ring. This particle, moving under the gravitational attraction of the ring, reaches the centre of the ring. The velocity of this particle at the centre of the ring will be

(a) $\sqrt{\frac{GM}{r}}$ (b) $\sqrt{\frac{2GM}{r}}$

(c) $\sqrt{\frac{2GM}{r(\sqrt{2}+1)}}$ (d) $\sqrt{\frac{2GM}{r}\left[1-\frac{1}{\sqrt{2}}\right]}$.

147. A tunnel is dug along a diameter of the planet. A particle is dropped into it at the surface. The particle reaches the centre of the planet with speed v. If v_e is the escape velocity from the surface of the planet, then

(a) $\sqrt{2}v = v_e$ (b) $v = v_e$

(c) $v_e = \sqrt{3}v$ (d) $v_e = \sqrt{5}v$.

148. A satellite revolving around the Earth loses some energy due to collision. What would be the effect on its velocity and distance from the centre of the Earth ?

(a) Velocity increases and distance decreases

(b) Both velocity and distance increase

(c) Both velocity and distance decrease

(d) Velocity decreases and distance increases.

149. The energy required to shift a satellite from orbital radius r to orbital radius $2r$ is E. What energy will be required to shift the satellite from orbital radius $2r$ to orbital radius $3r$?

(a) E (b) $\frac{E}{2}$

(c) $\frac{E}{3}$ (d) $\frac{E}{4}$.

150. The radius of Earth is h. If the acceleration due to gravity reduces to 1% of its value at the surface of Earth, then the altitude is

(a) $25h$ (b) $10h$

(c) $9h$ (d) $5h$.

151. If the distance of the Earth to the Moon is nearly 60 earth radii, then the gravitational field at the Moon due to Earth is, in N kg^{-1}, nearly

(a) 6 (b) 0.16

(c) 3.2×10^{-2} (d) 2.7×10^{-3}.

152. The gravitational potential energy of a body at a distance r from the centre of Earth is U. The weight of the body at that point is

(a) Ur (b) $\frac{U}{r}$

(c) Ur^2 (d) Ur^3.

153. Two identical satellites are moving around the Earth in circular orbits at heights 3R and R respectively where R is the radius of the Earth. The ratio of their kinetic energies is

(a) 2 : 1 (b) 1 : 2

(c) 3 : 1 (d) 2 : 3.

154. A boy jumps to a height of 2m on the surface of Earth. Suppose the same boy is now on a planet where when he jumps, he escapes from its gravitational pull. If the density of the planet were same as that of Earth, what would be the maximum radius of the planet ? (Take radius of earth as 6400 km)

(a) $6400 \times \sqrt{2}$ km (b) $3200 \times \sqrt{2}$ km

(c) $80\sqrt{2}$ km (d) $40\sqrt{2}$ km.

155. A solid sphere of mass m and radius r is placed inside a hollow thin spherical shell of mass M and radius R as shown in Fig. 6. A particle of mass m' is placed on the line joining the two centres at a distance x from the point of contact of the sphere and shell. If $2r < x < 2R$, then the net gravitational force on this particle is

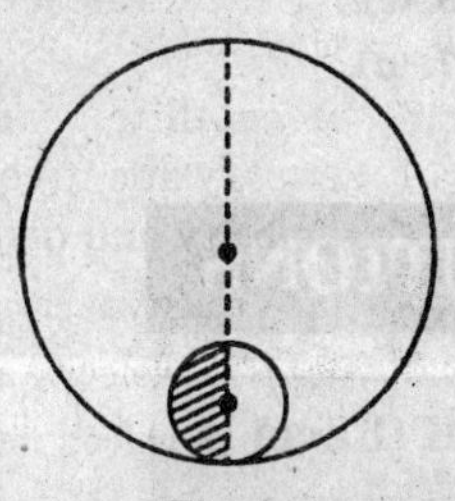

Fig. 6

(a) $\dfrac{Gmm'}{(x-r)^2}$ (b) $\dfrac{Gmm'}{2R-x}$

(c) $\dfrac{Gmm'}{R}$ (d) $\dfrac{Gmm'}{r}$.

156. The masses and radii of Earth and Moon are M_1, R_1 and M_2, R_2 respectively. Their centres are at a distance d apart. The minimum speed with which a particle of mass m should be projected from a point mid-way between the two centres so as to escape to infinity is

(a) $\sqrt{\dfrac{2G(M_1+M_2)}{d}}$ (b) $\sqrt{\dfrac{G(M_1+M_2)}{2d}}$

(c) $2\sqrt{\dfrac{G(M_1-M_2)}{d}}$ (d) $2\sqrt{\dfrac{G(M_1+M_2)}{d}}$.

157. A solid sphere of uniform density and radius R exerts a gravitational force of attraction F_1 on a particle placed at P. The distance of P from the centre of the sphere is 2R. A spherical cavity of radius $\dfrac{R}{2}$ is now made in the sphere. The sphere (with cavity) exerts a gravitational force F_2 on the same particle at P. The ratio $\dfrac{F_1}{F_2}$ is

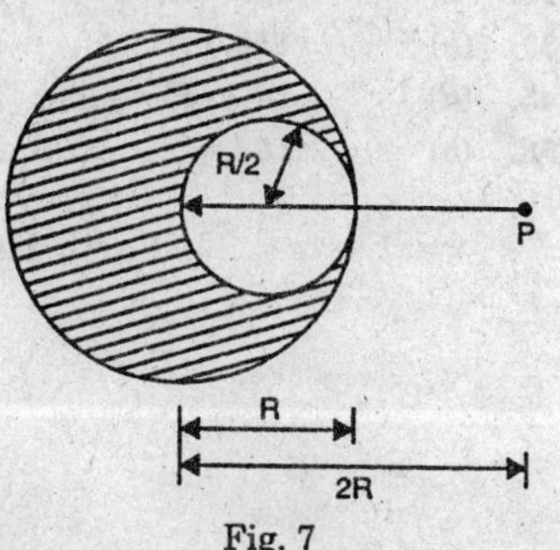

Fig. 7

(a) $\dfrac{9}{7}$ (b) $\dfrac{7}{9}$

(c) $\dfrac{1}{2}$ (d) $\dfrac{4}{3}$.

ANSWERS

QUESTION BANK (Level I)

1. (b)	2. (c)	3. (b)	4. (c)	5. (c)	6. (c)	7. (a)	8. (d)
9. (b)	10. (a)	11. (a)	12. (a)	13. (c)	14. (a)	15. (b)	16. (b)
17. (b)	18. (d)	19. (b)	20. (b)	21. (c)	22. (b)	23. (a)	24. (a)
25. (d)	26. (c)	27. (d)	28. (a)	29. (a)	30. (a)	31. (a)	32. (d)
33. (a)	34. (d)	35. (c)	36. (b)	37. (b)	38. (a)	39. (b)	40. (c)
41. (a)	42. (b)	43. (b)	44. (c)	45. (a)	46. (a)	47. (d)	48. (d)
49. (a)	50. (c)	51. (a)	52. (d)	53. (b)	54. (a)	55. (a)	56. (b).

QUESTION BANK (Level II)

1. (a)	2. (c)	3. (d)	4. (a)	5. (b)	6. (d)	7. (d)	8. (b)
9. (b)	10. (b)	11. (a)	12. (b)	13. (c)	14. (b)	15. (b)	16. (a)
17. (d)	18. (c)	19. (c)	20. (c)	21. (c)	22. (a)	23. (c)	24. (b)
25. (a)	26. (a)	27. (a)	28. (a)	29. (c)	30. (a)	31. (b)	32. (a)
33. (b)	34. (c)	35. (c)	36. (a)	37. (a)	38. (a)	39. (c)	40. (b)
41. (c)	42. (c)	43. (c)	44. (b)	45. (a)	46. (c)	47. (c)	48. (a)

49. (*b*)	50. (*d*)	51. (*a*)	52. (*b*)	53. (*a*)	54. (*c*)	55. (*b*)	56. (*d*)
57. (*b*)	58. (*a*)	59. (*d*)	60. (*a*)	61. (*a*)	62. (*c*)	63. (*b*)	64. (*b*)
65. (*b*)	66. (*b*)	67. (*d*)	68. (*c*)	69. (*b*)	70. (*c*)	71. (*c*)	72. (*b*)
73. (*c*)	74. (*a*)	75. (*b*)	76. (*a*)	77. (*b*)	78. (*a*)	79. (*c*)	80. (*c*)
81. (*a*)	82. (*c*)	83. (*c*)	84. (*c*)	85. (*c*)	86. (*c*)	87. (*c*)	88. (*b*)
89. (*d*)	90. (*b*)	91. (*d*)	92. (*c*)	93. (*a*)	94. (*d*)	95. (*b*)	96. (*a*)
97. (*c*)	98. (*d*)	99. (*b*)	100. (*c*)	101. (*b*)	102. (*b*)	103. (*c*)	104. (*d*)
105. (*d*)	106. (*c*)	107. (*c*)	108. (*a*)	109. (*a*)	110. (*d*)	111. (*b*)	112. (*a*)
113. (*b*)	114. (*b*)	115. (*d*)	116. (*b*)	117. (*b*)	118. (*c*)	119. (*a*)	120. (*d*)
121. (*b*)	122. (*d*)	123. (*a*)	124. (*a*)	125. (*c*)	126. (*a*)	127. (*a*)	128. (*b*)
129. (*a*)	130. (*a*)	131. (*b*)	132. (*a*)	133. (*b*)	134. (*b*)	135. (*c*)	136. (*c*)
137. (*b*)	138. (*a*)	139. (*a*)	140. (*d*)	141. (*a*)	142. (*b*)	143. (*c*)	144. (*b*)
145. (*d*)	146. (*d*)	147. (*a*)	148. (*a*)	149. (*c*)	150. (*c*)	151. (*d*)	152. (*b*)
153. (*b*)	154. (*c*)	155. (*a*)	156. (*d*)	157. (*a*).			

HINTS/SOLUTIONS

QUESTION BANK (Level I)

1. (*b*) $\frac{E_1}{E_2} = \frac{r_2}{r_1} = \frac{1}{3}$

2. (*c*) $g = \frac{GM}{R^2}$

or $g = \frac{G}{R^2} \times \frac{4}{3}\pi R^3 \rho$ or $\rho = \frac{3g}{4\pi RG}$

3. (*b*) $V = -\int_0^x E\, dx$

$$V = -\int_0^x \frac{k}{x^3}\, dx = -k\int_0^x x^{-3}\, dx = -k \left| \frac{x^{-3+1}}{-3+1} \right|_0^x = \frac{k}{2x^2}$$

4. (*c*) $\frac{T'}{T} = \left[\frac{R'}{R}\right]^{3/2}$

or $\frac{T'}{T} = (2)^{3/2} = 2\sqrt{2} = 2 \times 1.414 = 2.828$

5. (*c*) $\frac{42400}{6400} = 6.625 \approx 7$

6. (*c*) $g_p = \frac{GM_p}{R_p^{\ 2}} = G \times \frac{1}{5} M_e \times \frac{4}{R_e^{\ 2}} = \frac{4}{5} g = 8 \text{ m s}^{-2}$

7. (*a*) $v = R\omega = 4000 \times \frac{2\pi}{24} \approx 1000$ mile/h

8. (*d*) $F = G\frac{M \times M}{(2R)^2}$ or $F = G\frac{(\frac{4}{3}\pi R^3 \times \rho)^2}{(2R)^2}$ or $F \propto R^4$

9. (*b*) $g' = g - R\omega^2 \cos^2\phi$ or $mg' = mg - mR\omega^2 \cos^2\phi$

Out of the four given places, the value of ϕ is least at the equator. So, the term $mR\omega^2 \cos^2\phi$ will have maximum value at the equator. Thus, mg' will be least.

10. (*a*) $mr\omega^2 = \frac{GMm}{r^2}$ or $r^3 = \frac{GM}{\omega^2} = \frac{gR^2}{\omega^2}$

11. (*a*) $g = \frac{GM}{R^2}$

$\frac{\Delta g}{g} \times 100 = 2\frac{\Delta R}{R} \times 100 = 2 \times 1\% = 2\%$

12. (*a*) $L = mv_0 R_0$

$L = m\sqrt{\frac{GM}{R_0}}\, R_0$ or $L = m\sqrt{GMR_0}$

13. (*c*) $T^2 \propto R^3$ or $R \propto T^{2/3}$

R is increased by a factor of $[8]^{2/3}$ *i.e.* $[64]^{1/3}$ or 4. So, the new radius is 4R.

14. (*a*) $T = 2\pi\sqrt{\frac{R^3}{GM}}$

$T = 2\pi\sqrt{\frac{R^3}{G \times \frac{4}{3}\pi R^3 \rho}}$ or $T = 2\pi\sqrt{\frac{3}{4\pi G\rho}}$

Clearly, T depends upon density.

15. (*b*) $v_e = \sqrt{\frac{2GM}{R}}$

M is increased by a factor of 4. So, v_e is increased by a factor of 2.

16. (*b*) Kinetic energy required to escape $= \frac{1}{2} m v_e^2$

$$= \frac{1}{2} m \frac{2GM}{R} = \frac{GMm}{R}$$

Orbital kinetic energy $= \frac{1}{2} m \frac{GM}{R} = \frac{1}{2} \frac{GMm}{R}$

Clearly, kinetic energy required to escape is twice the orbital kinetic energy.

17. (*b*) Applying conservation of angular momentum,

$$m v_B \times OB = m v_A \times OA \quad \text{or} \quad \frac{v_B}{v_A} = \frac{OA}{OB} = x$$

[Note that $m v_B \times OB$ is only an approximate values because $\theta \neq 90°$].

18. (*d*) Applying conservation of angular momentum, we get

$$m r_{min}^2 \omega = m r_{max}^2 \omega' \quad \text{or} \quad \omega' = \frac{r_{min}^2 \omega}{r_{max}^2}$$

19. (*b*)

$$g' = G\left[\frac{M}{7}\right]\left[\frac{2}{R}\right]^2 \quad \text{or} \quad g' = \frac{4}{7} g$$

or $\quad mg' = \frac{4}{7} mg = \frac{4}{7} \times 700 \text{ g wt} = 400 \text{ g wt.}$

20. (*b*) $T \propto (r)^{3/2}$

Since r is doubled therefore T is increased by a factor of $[2]^{3/2}$ or $\sqrt{8}$ or $2\sqrt{2}$.

So, the new time period is $365 \times 2\sqrt{2}$ days.

21. (*c*) Since r is doubled therefore T is increased by a factor of $[2]^{3/2}$ *i.e.* $2^{1.5}$. So, the required ratio is $1 : 2^{1.5}$.

22. (*b*)

$$v_0 = \sqrt{\frac{GM}{R}}$$

$$v_0 = \sqrt{\frac{G}{R} \times \frac{4}{3} \pi R^3 \rho} \quad \text{or} \quad v_0 = \sqrt{\frac{4}{3} \pi G R^2 \rho}$$

or $\quad v_0 = R\sqrt{\frac{4}{3} \pi G \rho} \quad$ or $\quad v_0 \propto \sqrt{\rho}$

23. (*a*) $F = G\frac{m_1 m_2}{r^2}$ or $G = \frac{Fr^2}{m_1 m_2}$

The SI unit of G is $N\, m^2\, kg^{-2}$

or $1 \text{ kg} \times 1 \text{ m s}^{-2} \times \text{m}^2 \text{ kg}^{-2}$ or $\text{kg}^{-1} \text{ m}^3 \text{ s}^{-2}$

24. (*a*) $F = \frac{G m_1 m_2}{r^2}$

F is maximum when $m_1 m_2$ is maximum.

Now, $4 m_1 m_2 = \sqrt{(m_1 + m_2)^2 - (m_1 - m_2)^2}$

Clearly, $m_1 m_2$ will be maximum if

$$m_1 - m_2 = 0 \quad \text{or} \quad m_1 = m_2 \quad \text{or} \quad \frac{m_1}{m_2} = 1$$

25. (*d*) Binding energy

= – [Total energy].

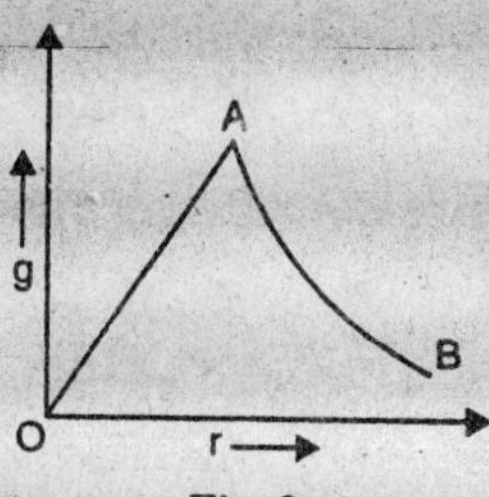

Fig. 8

26. (c)

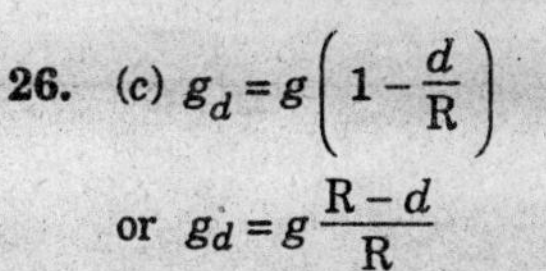

$$g_d = g\left(1 - \frac{d}{R}\right)$$

or $\quad g_d = g\frac{R - d}{R}$

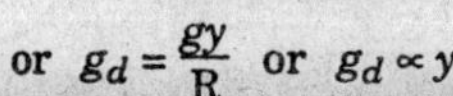

or $\quad g_d = \frac{gy}{R} \quad$ or $\quad g_d \propto y$

So, within the Earth, the acceleration due to gravity varies linearly, with the distance from the centre of the Earth. This explains the linear portion OA of the graph.

y = R – d

Fig. 9

27. (*d*) Work done is same whether we move a mass of 1 kg through 10 m or 2 kg through 5 m.

Since potential difference between two points separated by 10 m is 4 J kg^{-1} therefore the work done in moving 1 kg through 10 m is 4 J. Work done in moving 2 kg through 5 m will also be 4 J.

28. (*a*) $g = \left|\frac{dV}{dr}\right| = \frac{4}{10} = 0.4 \text{ N kg}^{-1}$

29. (*a*) In the first case, $r_1 = 7R$

In the second case, $r_2 = 3.5R$

Radius of orbit is halved. So, T becomes

$$\left[\frac{1}{2}\right]^{3/2} \quad \text{or} \quad \frac{1}{2\sqrt{2}}$$

New time period is $\frac{24}{2\sqrt{2}}$ or $6\sqrt{2}h$

30. (*a*) The SI unit of $\frac{hc}{G}$ is $\frac{(Js)(m\, s^{-1})}{Nm^2\, kg^{-2}}$ or kg^2

31. (*a*) Kinetic energy $= \frac{1}{2} m v_e^2$

$$= \frac{1}{2} \times 2 \times 10^4 \times 10^4 \text{ J} = 10^8 \text{ J}$$

32. (*d*) $\quad T = 2\pi\sqrt{\frac{R^3}{GM}}$

Squaring, $T^2 = 4\pi^2 \frac{R^3}{GM}$ or $M = 4\pi^2 R^3 G^{-1} T^{-2}$

33. (*a*) $\tan\theta = \frac{T^2}{r^3} = \frac{4\pi^2}{GM}$

34. (*d*) $\quad E = -\frac{GMm}{2r}$ or $E \propto \frac{m}{r}$

$$\therefore \quad \frac{E_1}{E_2} = \frac{3}{1} \times \frac{4}{1} = \frac{12}{1} = 12$$

35. (c) $\frac{W_h}{W} = \frac{mg_h}{mg} = \frac{g_h}{g} = \frac{R^2}{(R+R)^2}$ or $W_h = \frac{W}{4}$

36. (b) $v_o = \sqrt{\frac{GM}{r}} = \sqrt{\frac{gR^2}{r}}$ or $r\omega = \sqrt{\frac{gR^2}{r}}$

or $r^2\omega^2 = \frac{gR^2}{r}$ or $r^3 = \frac{gR^2}{\omega^2}$

37. (b) $v = \sqrt{n^2 - 1}\, v_e$ or $v = \sqrt{2^2 - 1}\, v_e$ or $v = \sqrt{3}\, v_e$

38. (a) $\frac{v_1}{v_2} = \sqrt{\frac{2g_1R_1}{2g_2R_2}} = \sqrt{k_1k_2}$

39. (b) $T' = (2^{3/2})\, T$
$= 2\sqrt{2} \times 365$ days
≈ 1032 days

40. (c) $T' = [4]^{3/2}T = 8T = 8 \times 4 = 32\, h$

41. (a) Work done/mass = Potential difference
$= [-400] - [-800] = 400$ kJ/kg.

> ☞ ***Note that gravitational potential is inversely proportional to distance.***

42. (b) Angular momentum $= mv_0r$
$= m\sqrt{\frac{GM}{r}}\, r = m\sqrt{GM} \times \sqrt{r}$

or Angular momentum $\propto \sqrt{r}$

43. (b) $0 = g - R\omega^2 \cos^2 45°$

$\frac{R\omega^2}{2} = g$ or $\omega = \sqrt{\frac{2g}{R}}$

44. (c) $g' = \frac{GM}{(R/2)^2} = 4\frac{GM}{R^2} = 4g$

45. (a) $v = \sqrt{\frac{GM}{R}}$, $v' = \sqrt{\frac{GM}{R + \frac{R}{2}}}$

or $v' = \sqrt{\frac{2}{3}\frac{GM}{R}} = \sqrt{\frac{2}{3}}\, v$

46. (a) $g = \frac{GM}{R^2} = \frac{G}{R^2} \times \frac{4}{3}\pi R^3\rho$ or $g = \frac{4}{3}\pi GR\rho$

Since ρ is given to be constant, ∴ $g \propto R$.

47. (d) Gravitational potential energy of a pair of particles $= -\frac{Gm^2}{l}$

Since we have three pairs, therefore the total gravitational potential energy is $-\frac{3Gm^2}{l}$

48. (d) $\frac{T}{T'} = \left[\frac{r}{r'}\right]^{3/2} = \left[\frac{3R}{6R}\right]^{1.5}$ or $\frac{T}{T'} = \frac{1}{2^{1.5}}$

49. (a) $g = \frac{2}{3}\pi GD\rho$

$\frac{g_1}{g_2} = \frac{D_1\rho_1}{D_2\rho_2} = \frac{4}{1} \times \frac{1}{2} = \frac{2}{1}$

50. (c) $v_p = \sqrt{\frac{2G(6M)}{2R}} = \sqrt{3 \times \frac{2GM}{R}} = \sqrt{3}v$

51. (a) Total energy of satellite = – Kinetic energy of satellite.

52. (d) $\frac{GMm}{R} = \frac{mv^2}{R}$

Clearly, $v \propto R^\circ$

53. (b) $3V = \sqrt{\frac{GM}{4R}}$ or $\sqrt{\frac{GM}{R}} = 6V$

54. (a) $v_e' = \sqrt{\frac{2G(3M)}{3R}}$

$v_e' = v_e$.

55. (a) $-\frac{GMm}{R} + \frac{1}{2}mv^2 = 0$ or $v = \sqrt{\frac{2GM}{R}}$

When the particle is inside the shell, the force is zero. So, the body would pass through the centre with the same velocity $\sqrt{\frac{2GM}{R}}$

56. (b) $\frac{GM^2}{(2R)^2} = \frac{Mv^2}{R}$ or $\frac{GM}{4R} = v^2$ or $v = \frac{1}{2}\sqrt{\frac{GM}{R}}$

QUESTION BANK (Level II)

1. Mass per unit length $= \frac{m}{l}$

Mass of section of length $dx = \frac{m}{l}\, dx$

Let dF be the gravitational force of attraction between this section and sphere.

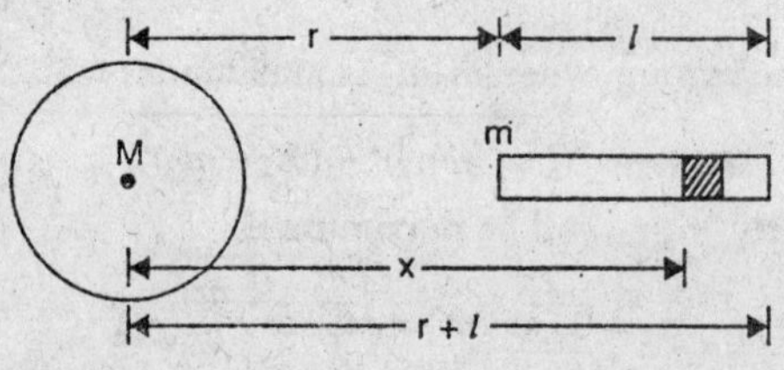

Fig. 10

Then, $dF = GM \dfrac{\left(\frac{m}{l}dx\right)}{x^2}$

$$F = \int_r^{r+l} \frac{GMm}{lx^2}dx = \frac{GMm}{l}\int_r^{r+l}\frac{1}{x^2}dx$$

$$= \frac{GMm}{l}\int_r^{r+l} x^{-2}dx = \frac{GMm}{l}\left[\frac{x^{-2+1}}{-2+1}\right]_r^{r+l}$$

$$= -\frac{GMm}{l}\left[x^{-1}\right]_r^{r+l} = -\frac{GMm}{l}\left[\frac{1}{x}\right]_r^{r+l} = \frac{GMm}{r(r+l)}$$

If $r >> l$, then $F = \dfrac{GmM}{r^2}$

2. (c) Let us discuss the correctness of the (a) option.

When the ship is travelling due east, its angular velocity will incre ase. This is because of earth's rotation about its axis from west to east along the equator. Due to increase in angular velocity, a component of the weight of the body has now to provide a greater centripetal force. So, there is a decrease in the apparent weight.

Let us now discuss the correctness of the (b) option.

When the ship is travelling due west, there is a decrease in the resultant velocity of the body. So, a smaller centripetal force is provided from the weight. So, there is an increase in the apparent weight.

Let us now discuss the correctness of the (c) option. For this, let us first find time period of a simple pendulum of infinite length.

Restoring force, F

$$= -mg\sin(\theta + \phi) = -mg[\theta + \phi]$$

$$= -mg\left[\frac{\xi}{l} + \frac{\xi}{R}\right]$$

$$= -mg\left[\frac{1}{l} + \frac{1}{R}\right]\xi$$

$$= -\frac{mg}{R}\left[1 + \frac{R}{l}\right]\xi$$

or $ma = -\dfrac{mg}{R}\left[1 + \dfrac{R}{l}\right]\xi$

or $a = -\dfrac{g}{R}\left[1 + \dfrac{R}{l}\right]\xi$

Fig. 11

Comparing with $a = -\omega^2\xi$, we get

$$\omega = \sqrt{\frac{g}{R}\left[1 + \frac{R}{l}\right]}$$

$$T = \frac{2\pi}{\omega} = 2\pi\sqrt{\frac{R}{g\left(1 + \frac{R}{l}\right)}}$$

When the pendulum is of infinite length, then $l \to \infty$

$\therefore \quad T = 2\pi\sqrt{\dfrac{R}{g}}$

Substituting values and simplifying, T = 84.6 min.

Let us now determine the time period of SHM of a ball in a tunnel along the diameter of Earth.

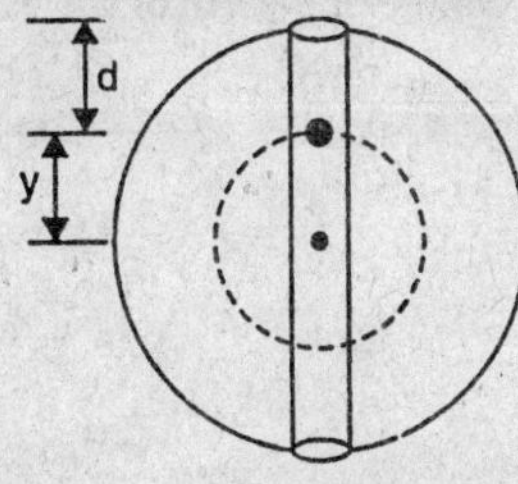

Fig. 12

$F = mg_d$

or $F = mg\left(1 - \dfrac{d}{R}\right)$

or $F = mg\dfrac{R - d}{R} = \dfrac{mgy}{R}$

or $\dfrac{F}{m} = \dfrac{g}{R}y$ or $a = \dfrac{g}{R}y$

Comparing with $a = \omega^2 y$, we get

$$\omega = \sqrt{\frac{g}{R}} \text{ or } \frac{2\pi}{T} = \sqrt{\frac{g}{R}} \text{ or } T = 2\pi\sqrt{\frac{R}{g}}$$

Substituting values, we get T = 84.6 min.

Let us now discuss the incorrectness of option (c).

If υ_{rms} is larger than escape velocity, then the molecule would escape the gravitational field of Earth.

3. (d) $F_1 = F_2 = \dfrac{GM}{a^2}$

Resultant of F_1 and F_2 is

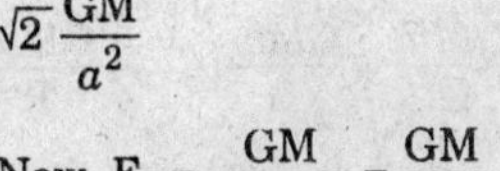

$\sqrt{2}\dfrac{GM}{a^2}$

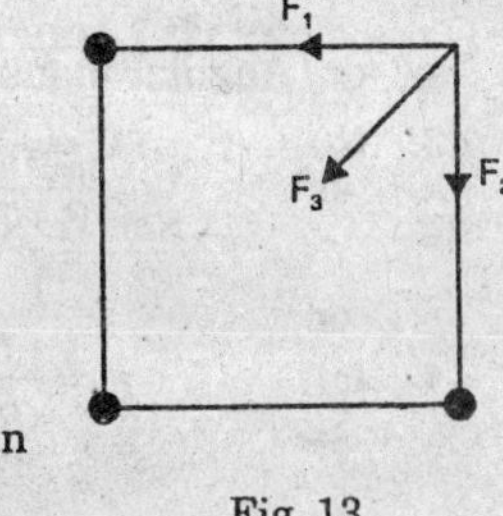

Fig. 13

Now, $F_3 = \dfrac{GM}{(\sqrt{2}a)^2} = \dfrac{GM}{2a^2}$

Now, $\dfrac{\sqrt{2}GM}{a^2}$ and $\dfrac{GM}{2a^2}$ act in the same direction. Their resultant is $\dfrac{\sqrt{2}GM}{a^2} + \dfrac{GM}{2a^2}$ or $\dfrac{GM}{a^2}\left[\sqrt{2} + \dfrac{1}{2}\right]$

4. (a) $g_p = \dfrac{G\left[M - \frac{10}{100}M\right]}{\left[R + \frac{20}{100}R\right]^2} = \dfrac{G \times 9M}{10} \times \dfrac{25}{36R^2} = \dfrac{5}{8}g$

5. (b) $T' = \left[\dfrac{1}{2}\right]^{3/2} \times 365$ days

$$= \frac{1}{2\sqrt{2}} \times 365 \text{ days} = \frac{182.5 \times \sqrt{2}}{\sqrt{2} \times \sqrt{2}} \text{ days}$$

$$= 91.25 \times 1.414 \text{ days} = 129 \text{ days}$$

6. (d) The effective value of acceleration due to gravity on the surface of Earth is 9.8 m s^{-2}.

After, 1 minute, the body will be at a certain height above the Earth. So, the value of g will be less than 9.8 m s^{-2}. When the satellite is orbiting the Earth, the effective value of acceleration due to gravity is zero.

The value of acceleration due to gravity on the Moon is one-sixth of the value of acceleration due to gravity on the Earth.

7. (*d*) $$M_p = 2M_e$$

$$\frac{4}{3}\pi R_p^3\rho = 2 \times \frac{4}{3}\pi R_e^3\rho$$

or $$R_p^3 = 2R_e^3 \text{ or } R_p = 2^{1/3} R_e$$

$$g_p = \frac{GM_p}{R_p^2} = \frac{G[2M_e]}{[2^{1/3}R_e]^2} = 2^{1-\frac{2}{3}}\frac{GM_e}{R_e^2} \text{ or } g_p = 2^{1/3} g_e$$

$$mg_p = 2^{1/3} mg_e = 2^{1/3}W.$$

8. (*b*) $$g = \frac{GM}{R_e^2} = \frac{G \times \frac{4}{3}\pi R_e^3\rho}{R_e^2} = G \times \frac{4}{3}\pi R_e\rho$$

or $$\rho = \frac{\frac{g}{G}}{\frac{4\pi}{3}R_e}$$

9. (*b*) $$-\frac{GMm}{R+h} + 0 = -\frac{GMm}{R} + \frac{1}{2}m\left[\frac{1}{2}\sqrt{\frac{2GM}{R}}\right]^2$$

$$-\frac{GMm}{R+h} + \frac{GMm}{R} = \frac{1}{8}m\frac{2GM}{R} = \frac{2GMm}{8R} = \frac{GMm}{4R}.$$

or $$-\frac{GMm}{R+h} = \frac{GMm}{4R} - \frac{GMm}{R} = -\frac{3}{4}\frac{GMm}{R}$$

or $$\frac{1}{R+h} = \frac{3}{4R} \text{ or } 3R + 3h = 4R$$

or $$3h = 4R - 3R = R \text{ or } h = \frac{R}{3}$$

10. (*b*) $$W = \left[-\frac{GMm}{3R}\right] - \left[-\frac{GMm}{2R}\right]$$

$$= \frac{GMm}{2R} - \frac{GMm}{3R} = \frac{1}{6}\frac{GMm}{R}$$

$$= \frac{1}{6}mgR = \frac{1}{6} \times 2 \times 10^3 \times 10 \times 6400 \times 10^3 \text{ J}$$

$$\approx 2 \times 10^{10} \text{ J}$$

11. (*a*) Using conservation of energy,

$$\frac{1}{2}mv^2 - \frac{GMm}{R} = 0 - \frac{GMm}{2R}$$

or $$\frac{1}{2}mv^2 = \frac{GMm}{R} - \frac{GMm}{2R}$$

or $$v^2 = 2\frac{GM}{R}\left[1 - \frac{1}{2}\right] \text{ or } v^2 = \frac{GM}{R}$$

But $$gR^2 = GM$$

∴ $$v = \sqrt{\frac{gR^2}{R}} \text{ or } v = \sqrt{gR}$$

12. (*b*) Potential energy $= -\frac{GMm}{2R} = -\frac{gR^2m}{2R} = -\frac{1}{2}mgR$.

13. (*c*) $g = \frac{Gm_0}{r_0^2}$ or $g = \frac{4Gm_0}{d_0^2}$ $[\because d_0 = 2r_0]$

14. (*b*) $$v = \sqrt{2^2 - 1}\, v_e$$

or $$v = \sqrt{3}v_e \text{ or } v = 11.2\sqrt{3} \text{ km s}^{-1}$$

15. (*b*) $$\frac{G \times 800}{(12-x)^2} = \frac{G \times 200}{x^2}$$

or $$\frac{4}{(12-x)^2} = \frac{1}{x^2}$$

or $$\frac{2}{12-x} = \frac{1}{x}$$

or $$12 - x = 2x \text{ or } 3x = 12 \text{ or } x = 4 \text{ cm}$$

Fig. 14

16. (*a*) Applying conservation of energy,

$$-\frac{GMm}{2R} + 0 = \frac{1}{2}mv^2 - \frac{GMm}{R}$$

or $$\frac{1}{2}mv^2 = \frac{GMm}{R} - \frac{GMm}{2R}$$

or $$\frac{1}{2}mv^2 = \frac{GMm}{2R} \text{ or } v = \sqrt{\frac{GM}{R}}$$

17. (*d*) $E_k = \frac{GMm}{2r}$ or $E_k \propto \frac{1}{r}$

Also, $T^2 \propto r^3$ or $r \propto T^{2/3}$

∴ $E_k \propto T^{-2/3}$

18. (*c*) $$mg' = mg - mR\omega^2\cos^2\phi$$

Now, $$\frac{3}{5}mg = mg - mR\omega^2 \text{ or } mR\omega^2 = mg - \frac{3}{5}mg$$

or $$mR\omega^2 = \frac{2}{5}mg \text{ or } \omega = \sqrt{\frac{2g}{5R}}$$

19. (*c*) $$W = \left[-\frac{GMm}{5R}\right] - \left[-\frac{GMm}{3R}\right]$$

or $$W = \frac{GMm}{3R} - \frac{GMm}{5R}$$

or $$W = \frac{GMm}{R}\left[\frac{1}{3} - \frac{1}{5}\right] = \frac{2}{15}\frac{GMm}{R}$$

20. (*c*) $$g_h = g\left[1 - \frac{2h}{R}\right]$$

$$g - g_h = \frac{2gh}{R} \text{ or } g - g_h = 2 \times \frac{GM}{R^3}h$$

$$g - g_h = \frac{2Gh}{R^3} \times \frac{4}{3}\pi R^3 D \text{ or } g - g_h = \frac{8}{3}\pi GDh$$

or $$mg - mg_h = \frac{8}{3}\pi GDhm$$

21. (*c*) $$\frac{mg_h}{mg} = \frac{R^2}{(R+h)^2}$$

or $$\frac{mg_h}{mg} = \frac{6000 \times 6000}{8000 \times 8000} \text{ or } \frac{mg_h}{mg} = \frac{9}{16}$$

or $$mg_h = \frac{9}{16} \times 120 \text{ kg wt} = 67.5 \text{ kg wt}$$

22. (*a*) $$\frac{g_h}{g} = \frac{R^2}{(R+h)^2} = \frac{R^2}{\left(\frac{3R}{2}\right)^2}$$

or $$\frac{g_h}{g} = \frac{4}{9} \text{ or } g_h = \frac{4}{9} g$$

Pull on the satellite $= mg_h = \frac{4}{9} mg$

$$= \frac{4}{9} \times 200 \times 10 \text{ N} \approx 889 \text{ N}$$

23. (*c*) If V is the required potential, then

$$\frac{1}{2} \times 1 \times 2 \times 2 + 1 \times V = -3$$

or $$V = -5 \text{ J/kg}$$

24. (*b*) $$U = -\frac{GMm}{r}$$

Kinetic energy $= \frac{GMm}{2r}$

$$U = (-2)\frac{GMm}{2r} = -2 \times \text{kinetic energy}$$

$$= -2 \times \frac{1}{2} mv^2 = -mv^2$$

25. (*a*) $$\omega = \frac{125}{10^5} \text{ rad s}^{-1} = \frac{1}{800} \text{ rad s}^{-1}$$

Now, $mg' = mg - mR\omega^2 \cos^2 \phi$

At the equator, $\cos \phi = 1$

$\therefore$ $mg' = mg - mR\omega^2$

$$mg' = mg - m \times 6400 \times 10^3 \times \frac{1}{800 \times 800}$$

or $$mg' = 10\,m - 10\,m = 0$$

26. (*a*) $$T \propto R^{3/2}$$

$$\frac{\Delta T}{T} = \frac{3}{2}\frac{\Delta R}{R}$$

or $$\frac{\Delta T}{T} \times 100 = \frac{3}{2}\frac{\Delta R}{R} \times 100$$

$$= \frac{3}{2} \times \frac{0.01R}{R} \times 100 = \frac{3}{2}\% = 1.5\%$$

Clearly, it is an increase.

27. (*a*) $$\frac{mv^2}{r} = \frac{GMm}{r^2} \text{ or } \frac{1}{2}mv^2 = \frac{GMm}{2r}$$

But $r = 2R$

$\therefore$ $$\frac{1}{2}mv^2 = \frac{GMm}{4R}$$

or $$E_k = \frac{gR^2m}{4R} \text{ or } E_k = \frac{mgR}{4}$$

28. (*a*) $$M = \frac{4}{3}\pi R^3 \rho$$

or $R^3\rho = \text{constant}$

or $$R^3 \propto \frac{1}{\rho} \text{ or } R \propto \frac{1}{\rho^{1/3}} \quad \text{...(1)}$$

Again, $$g = \frac{4}{3}\pi GR\rho$$

or $$g \propto R\rho \text{ or } g \propto \frac{\rho}{\rho^{1/3}} \quad \text{[from eq. (1)]}$$

or $$g \propto \rho^{2/3}$$

The required factor is $[10^9]^{2/3}$ *i.e.* 10^6

29. (*c*) $$dF = G\frac{Mdm}{4r^2}$$

$$F = \Sigma dF \cos\theta$$

or $$F = \Sigma \frac{GMdm}{4r^2} \cos\theta$$

or $$F = \frac{GM}{4r^2} \times \frac{\sqrt{3}r}{2r} \Sigma dm$$

or $$F = \frac{\sqrt{3}GMm}{8r^2}$$

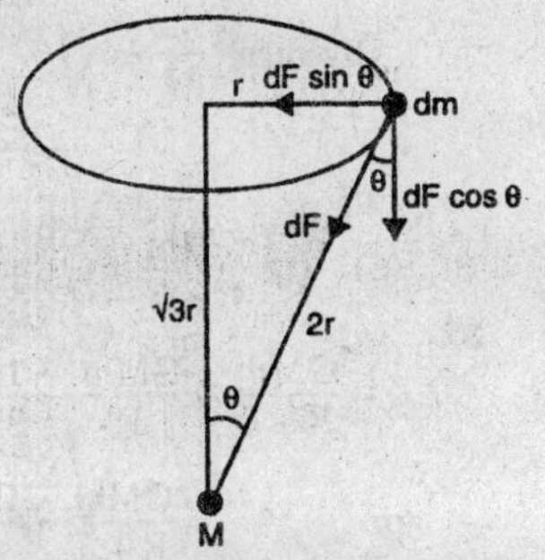

Fig. 15

Alternative Solution

The gravitational field due to the ring at a distance $\sqrt{3}r$ is given by

$$E = \frac{Gm(\sqrt{3}r)}{[r^2 + (\sqrt{3}r)^2]^{3/2}} \text{ or } E = \frac{\sqrt{3}Gm}{8r^2}$$

Required force is EM *i.e.* $\frac{\sqrt{3}GmM}{8r^2}$

> ☞ ***Follow the central argument in this solution.***
> ***The gravitational field at any point on the ring due to a sphere is equal to the field due to a single particle of mass M located at the centre of the sphere.***

30. (*a*) Let us first calculate the mass of inner solid sphere of radius *r*.

Mass of inner solid sphere, M′

$$= \frac{M_e}{\frac{4}{3}\pi R_e^3} \times \frac{4}{3}\pi r^3 = \frac{M_e}{R_e^3} r^3$$

Now, $g = \frac{GM_e r^3}{R_e^3} \times \frac{1}{r^2}$ or $g = \frac{GM_e r}{R_e^3}$

Force on particle of mass $m = mg = \frac{GM_e mr}{R_e^3}$

> ☞ ***Note that those layers of the sphere which have radii larger than r do not have any contribution to the gravitational force.***

31. (*b*) $11 = \sqrt{\frac{2G \times 81M}{4R}}$

$$v = \sqrt{\frac{2GM}{R}}$$

Dividing, $\frac{v}{11} = \sqrt{\frac{2GM}{R}} \times \sqrt{\frac{4R}{2G \times 81M}}$

or $\frac{v}{11} = \sqrt{\frac{4}{81}} = \frac{2}{9}$ or $v = \frac{22}{9}$ km s^{-1} = 2.44 km s^{-1}

32. (*a*) $v_0 = \sqrt{\frac{GM}{R+h}} = \sqrt{\frac{gR^2}{R+h}}$

$$v_0 \approx \sqrt{\frac{gR^2}{7R}} = \sqrt{\frac{1}{7}gR}$$

33. (*b*) $\frac{g_h}{g} = \frac{R^2}{(R+h)^2} = \frac{R^2}{(7R)^2} = \frac{1}{49}$

or $g_h = \frac{g}{49}$ or $mg_h = \frac{mg}{49}$

or $mg_h = \frac{98}{49}$ N = 2N

34. (*c*) The gravitational force of attraction provides the necessary centripetal force.

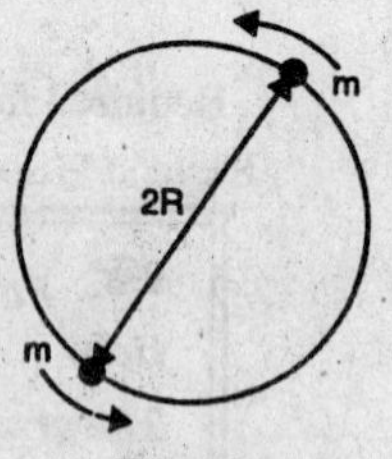

$\therefore \frac{mv^2}{R} = \frac{G(m)(m)}{(2R)^2}$

or $v^2 = \frac{Gm}{4R}$

or $v = \frac{1}{2}\sqrt{\frac{Gm}{R}}$

Fig. 16

35. (*c*) $g_d = g\left[1 - \frac{R/2}{R}\right]$ or $g_d = \frac{g}{2} = \frac{10 \text{ m s}^{-2}}{2} = 5 \text{ m s}^{-2}$

36. (*a*) $\frac{1}{4} = \frac{R^2}{(R + xR)^2}$

or $\frac{1}{2} = \frac{R}{R(1+x)}$ or $1 + x = 2$ or $x = 1$

37. (*a*) Orbital kinetic energy

$$= \frac{1}{2}mv_0^2 = \frac{1}{2}m\frac{GM}{R} = \frac{GMm}{2R}$$

Kinetic energy required to escape

$$= \frac{1}{2}mv_e^2 = \frac{1}{2}m\frac{2GM}{R} = 2\left[\frac{GMm}{2R}\right]$$

So, the value of n is 2.

38. (*a*) $\frac{GMm}{r^n} = \frac{mv^2}{r}$ or $v^2 r^n \propto r$ or $v^2 \propto r^{1-n}$

If $n = 1$, then $v^2 \propto r^0$

So, v is independent of r.

39. (*c*) $\omega = \frac{2\pi}{86400}$ rad s^{-1}

$$g' = g - R\omega^2 \cos^2\phi$$

For the equator, $\cos^2\phi = 1$

Also, $g' = 0$ (given)

$\therefore R\omega'^2 = g$ or $\omega' = \sqrt{\frac{g}{R}}$

Now, $\omega' = \sqrt{\frac{10}{6400 \times 1000}}$ rad s^{-1} or $\omega' = \frac{1}{800}$

Now $\frac{\omega'}{\omega} = \frac{1}{800} \times \frac{86400 \times 7}{2 \times 22} = 17.2 \approx 17$

40. (*b*) $\frac{G \times 10^3}{x^2} = \frac{G \times 10^5}{(3-x)^2}$

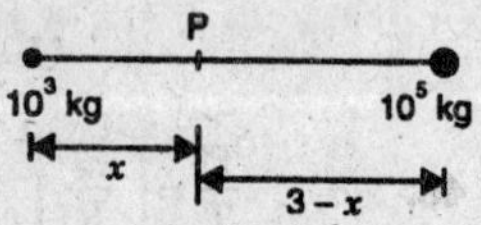

or $\frac{1}{x} = \frac{10}{3-x}$

or $10x = 3 - x$

or $11x = 3$ or $x = \frac{3}{11}$ m

Fig. 17

41. (*c*) $F = \frac{mv^2}{r} = \frac{3 \times 10^{26} \times (2 \times 10^6)^2}{1.5 \times 10^{12}}$ N $= 8 \times 10^{26}$ N

42. (*c*) Initial gravitational potential energy, $U_1 = -\frac{3Gm^2}{l}$

Final gravitational potential energy, $U_2 = -\frac{3Gm^2}{2l}$

Work done, W $= U_2 - U_1$

$$= \left[-\frac{3Gm^2}{2l}\right] - \left[-\frac{3Gm^2}{l}\right]$$

$$= \frac{3Gm^2}{l} - \frac{3Gm^2}{2l} = \frac{3Gm^2}{l}\left[1 - \frac{1}{2}\right] = \frac{3Gm^2}{2l}$$

43. (c) $v_e = \sqrt{2}v_0$

Clearly, if v_0 becomes 36%, v_e will also become 36%

$$\therefore \quad v'_e = \frac{36}{100} \times 11.2 \text{ km s}^{-1} = \frac{9}{25} \times 11.2 \text{ km s}^{-1}$$

44. (b) Gravitational potential at the centre of spherical shell $= -\frac{GM}{R}$

As the shell is shrinking, R decreases. $\frac{GM}{R}$ increases.

So, $-\frac{GM}{R}$ decreases.

45. (a) $v \propto \frac{1}{\sqrt{R}}$, $\quad v' \propto \frac{1}{\sqrt{7R}}$

$$\frac{v'}{v} = \frac{1}{\sqrt{7}} \text{ or } v' = \frac{v}{\sqrt{7}}$$

46. (c) Applying conservation of energy principle, we get

$$\frac{1}{2}mk^2v_e^2 - \frac{GMm}{R} = -\frac{GMm}{r}$$

or $\frac{1}{2}mk^2\frac{2GM}{R} - \frac{GMm}{R} = -\frac{GMm}{r}$

or $\frac{k^2}{R} - \frac{1}{R} = -\frac{1}{r}$ or $\frac{1}{r} = \frac{1}{R} - \frac{k^2}{R}$

or $\frac{1}{r} = \frac{1}{R}(1 - k^2)$ or $r = \frac{R}{1 - k^2}$

47. (c) $L = mvr$

$$L = m\sqrt{\frac{GM}{r}}\, r$$

$$L = \sqrt{\frac{GMm^2}{r}r^2} \text{ or } L = \sqrt{GMm^2r}$$

48. (a) $\frac{T'}{T} = \left[\frac{R'}{R}\right]^{3/2} = \left[\frac{4R}{R}\right]^{3/2} = 8$

or $T' = 8T = 8 \times 80$ minutes $= 640$ minutes.

49. (b) $a_m = \frac{gM}{R_m^2}, g = \frac{gM}{R^2}$

$$\frac{a_m}{g} = \frac{gM}{R_m^2} \times \frac{R^2}{gM} = \left[\frac{R}{R_m}\right]^2$$

50. (d) $\frac{a_m}{g} = \frac{GM_e/R_m^2}{GM_e/R_e^2}$ or $\frac{a_m}{g} = \frac{R_e^2}{R_m^2}$ or $a_m = \left[\frac{R_e}{R_m}\right]^2 g$

51. (a) Gravitational field strength, $g = -\frac{\Delta V}{\Delta r} = -\frac{-2}{20} = \frac{1}{10}$ m s^{-2}

Now, $\quad W = mgh = 5 \times \frac{1}{10} \times 4 \text{ J} = 2 \text{ J}$

52. (b) $\upsilon \propto \frac{1}{\sqrt{r}}$

$$\frac{\upsilon_A}{\upsilon_B} = \sqrt{\frac{R_B + R_E}{R_A + R_E}}$$

53. (a) $\text{Time} = \frac{72000 \times 1000}{3 \times 10^8} \text{ s} = \frac{72}{300} \approx \frac{1}{4} \text{ s}$

54. (c) $mg' = mg - mR\omega^2\cos^2\phi$

In the given problem, $mg' = 0$

At the equator, $\phi = 0°$, $\cos\phi = 1$

$\therefore \; mR\omega^2 = mg$

or $\quad \omega = \sqrt{\frac{g}{R}} = \sqrt{\frac{10}{6400 \times 1000}} \text{ rad s}^{-1} = \frac{1}{800} \text{ rad s}^{-1}$

55. (b) $mg_d = mg\left[1 - \frac{d}{R}\right]$

$$31.5 = 63\left[1 - \frac{d}{R}\right]$$

or $\quad 1 - \frac{d}{R} = \frac{31.5}{63} = \frac{1}{2}$ or $\frac{d}{R} = 1 - \frac{1}{2} = \frac{1}{2}$

or $\quad 2d = R$ or $d - \frac{R}{2} = 0.5 R$

Let us verify the answer

$$\frac{g_h}{g} = \frac{R^2}{(R + 0.5R)^2} \text{ or } \frac{28}{63} = \frac{R^2 \times 4}{9R^2} \text{ or } \frac{4}{9} = \frac{4}{9}$$

So, the answer is verified.

56. (d) Considering the origin of the co-ordinate system at $2m$, the position of the common centre of mass is given by

$$\frac{2m \times 0 + m \times r}{2m + m} = \frac{r}{3}$$

So, the centre of mass is at a distance of $\frac{r}{3}$ from $2m$. Also, it is at a distance of $\frac{2r}{3}$ from m. Thus, the stars revolve around the point of trisection of the straight line joining their mass-centres.

$$\text{Ratio of kinetic energies} = \frac{\frac{1}{2}[I\omega^2]_{2m}}{\frac{1}{2}[I\omega^2]_m} = \frac{2m\left(\frac{r}{3}\right)^2}{m\left(\frac{2r}{3}\right)^2} = \frac{2}{4} = \frac{1}{2}$$

57. (b) | Potential energy on the surface of planet | $= \frac{GMm}{R}$

$$= \frac{1}{2}m\frac{2GM}{R} = \frac{1}{2}mv_e^2$$

$$= \frac{1}{2} \times 1 \times 10^{-3} \times 10^{-3} \times (20 \times 10^3)^2 \text{ J} = 200 \text{ J}$$

58. (a) $\frac{mg'}{mg} = \frac{g'}{g} = \frac{R^2}{(R+h)^2}$

$$\frac{1}{100} = \frac{R^2}{(R+h)^2} \text{ or } \frac{1}{10} = \frac{R}{R+h} \text{ or } R + h = 10R$$

or $h = 9R = 9 \times 6400 \text{ km} = 57600 \text{ km}$

59. (*d*) $T^2 \propto R^3$

$$\frac{4\pi^2}{\omega^2} \propto R^3 \qquad \left[\because T = \frac{2\pi}{\omega}\right]$$

or $\omega^2 \propto R^{-3}$

60. (*a*) $T = 2\pi\sqrt{\frac{(R+h)^3}{GM}}$

T will be minimum if $h = 0$

$$\therefore \quad T_{min} = 2\pi\sqrt{\frac{R^3}{gR^2}} \text{ or } T_{min} = 2\pi\sqrt{\frac{R}{g}}$$

61. (*a*) $\frac{1}{2}\sqrt{\frac{2GM}{R}} = \sqrt{\frac{GM}{R+h}}$

$$\frac{1}{4} \times \frac{2GM}{R} = \frac{GM}{R+h}$$

$$2R = R + h \text{ or } h = R$$

62. (*c*) $\frac{T_n^2}{T_s^2} = \frac{R_n^3}{R_s^3}$ or $\frac{T_n}{T_s} = \left[\frac{10^{13}}{10^{12}}\right]^{3/2} = (\sqrt{10})^3 = 10\sqrt{10}$

63. (*b*) $v_e = R\sqrt{\frac{8}{3}\pi G\rho}$

$$v_e \propto R \text{ and } v_e' \propto 2R$$

$$\frac{v'_e}{v_e} = \frac{2}{1}$$

64. (*b*) $\frac{T'^2}{T^2} = \left[\frac{4R}{R}\right]^3 = 64$

or $\frac{T'}{T} = 8$ or $T' = 8T = 8 \times 1 \text{ day} = 8 \text{ days}$

65. (*b*) $\frac{g_h}{g} = \frac{R^2}{(7R)^2} = \frac{1}{49}$ or $g_h = \frac{g}{49} = \frac{9.8}{49} = 0.2 \text{ m s}^{-2}$

Weight $= mg_h$

$$= 100 \text{ kg} \times 0.2 \text{ m s}^{-2} = 20 \text{ N}$$

66. (*b*) Change in potential energy

$$= \left[-\frac{GMm}{nR+R}\right] - \left[-\frac{GMm}{R}\right] = \frac{GMm}{R} - \frac{GMm}{R(n+1)}$$

$$= \frac{GMm}{R}\left[1 - \frac{1}{n+1}\right] = \frac{GMm}{R}\frac{n}{n+1}$$

$$= \frac{n}{n+1}\frac{mgR^2}{R} = \frac{n}{n+1}mgR$$

67. (*d*) According to Kepler's third law, T^2 is proportional to cube of semi-major axis of the elliptical orbit.

Semi-major axis $= \frac{r_1 + r_2}{2}$

$$\therefore \quad T^2 \propto \left[\frac{r_1+r_2}{2}\right]^3 \text{ or } T \propto (r_1+r_2)^{3/2}$$

68. (*c*) Kinetic energy of escape

$$= \frac{1}{2}mv_e^2 = \frac{1}{2}m \times 2gR = mgR$$

$$= 500 \times 10 \times 6 \times 10^6 \text{ J} = 3 \times 10^{10} \text{ J}$$

69. (*b*) Gravitational field strength

$$= \frac{6.67 \times 10^{-11} \times 900 \times 4}{1} \text{ N kg}^{-1} = 2.4 \times 10^{-7} \text{ N kg}^{-1}$$

70. (*c*) $T \propto R^{3/2}$

Since R is increased by a factor of 16 therefore T is increased by a factor of $[16]^{3/2}$ or 4^3 *i.e.* 64.

71. (*c*) $mg - mR\omega^2 = mg\left(1 - \frac{d}{R}\right)$

or $-mR\omega^2 = -mg\frac{d}{R}$ or $R\omega^2 = \frac{gd}{R}$ or $d = \frac{R^2\omega^2}{g}$

72. (*b*) $v_e = \sqrt{\frac{2GM}{R}}$ or $v_e^2 = \frac{2GM}{R}$

or $R = \frac{2GM}{c^2} = \frac{2 \times \frac{20}{3} \times 10^{-11} \times 6 \times 10^{24}}{3 \times 10^8 \times 3 \times 10^8} \text{ m}$

$$= \frac{80}{9} \times 10^{-3} \text{ m} \approx 9 \text{ mm}$$

73. (*c*) $\frac{T_m}{T_e} = \left[\frac{R_m}{R_e}\right]^{3/2} = \left[\frac{6 \times 10^{10}}{1.5 \times 10^{11}}\right]^{3/2}$

$$= \left[\frac{6}{15}\right]^{3/2} = \frac{2}{5}\sqrt{\frac{2}{5}} = 0.4 \times \sqrt{0.4}$$

$$= 0.4 \times 0.63 \approx 0.25 = \frac{1}{4}$$

$$T_m = \frac{1}{4}T_e = \frac{1}{4} \text{ years}$$

An alternative solution

$$\frac{T_m}{T_e} = \left[\frac{2}{5}\right]^{3/2} = \left[\frac{8}{125}\right]^{1/2} \approx \sqrt{\frac{8}{128}} = \sqrt{\frac{1}{16}} = \frac{1}{4}$$

74. (*a*) $v = \sqrt{\frac{GM}{r}}$ or $v \propto \frac{1}{\sqrt{r}}$

$$\frac{v_m}{v_e} = \sqrt{\frac{r_e}{r_m}} = \sqrt{\frac{1.5 \times 10^{11}}{6 \times 10^{10}}}$$

$$= \sqrt{\frac{15}{6}} = \sqrt{\frac{5}{2}} = \sqrt{\frac{5 \times 2}{2 \times 2}} = \frac{\sqrt{10}}{2}$$

75. (*b*) $E_1 = \frac{1}{2} m v_e^2$ $= \frac{1}{2} m \times 2gR = mgR$ $E_2 = \frac{1}{2} m v_0^2 = \frac{1}{2} mgR$

$v_e = \sqrt{2gR}$
When orbit is close to Earth, $v_0 = \sqrt{gR}$

$\therefore \quad \frac{mgR}{\frac{1}{2} mgR} = 2$

76. (*a*) $v_0 = \sqrt{\frac{GM}{x}} = (GM)^{1/2} x^{-1/2}$

Both G and M are constants.

$\therefore \quad v_0 = kx^{-1/2}$ ...(1)

$\frac{\Delta v_0}{\Delta x} = k\left(-\frac{1}{2}\right) x^{-1/2-1}$

$\Delta v_0 = -\frac{k}{2} x^{-3/2} \Delta x$ ...(2)

Dividing (2) by (1),

$\frac{\Delta v_0}{v_0} = \frac{-\frac{k}{2} x^{-3/2} \Delta x}{kx^{-1/2}}$ or $\frac{\Delta v_0}{v_0} = -\frac{1}{2}\frac{\Delta x}{x}$

Let $v_0 = 100, x = 100$

Also, $\Delta x = 1$ (given)

$\therefore \quad \Delta v_0 = -\frac{1}{2} \times 1 = -\frac{1}{2} = -0.5$

Negative sign indicates that if radius is decreased, the velocity is increased.

Time-saving Tip

$v_0 = \sqrt{\frac{GM}{r}}\ ;\ \frac{\Delta v_0}{v_0} = \frac{1}{2}\frac{\Delta r}{r}$

77. (*b*) Increase in potential energy $= -\frac{GmM}{2R} - \left(-\frac{GmM}{R}\right)$

$= \frac{GmM}{2R} = \frac{GM}{R^2} \times \frac{mR}{2} = g \times \frac{1}{2} mR = \frac{1}{2} mgR$

78. (*a*) $x = \frac{GM}{R^2}$

Again, $\frac{x}{4} = \frac{GM}{(R+h)^2}$ or $x = GM\left(\frac{2}{R+h}\right)^2$

$\therefore \quad \frac{1}{R^2} = \left(\frac{2}{R+h}\right)^2$

or $\frac{1}{R} = \frac{2}{R+h}$ or $R + h = 2R$ or $h = R$

79. (*c*) $v = \sqrt{\frac{2GM}{R}}$ or $v = \sqrt{\frac{2G}{R} \times \frac{4}{3}\pi R^3 \rho}$

or $v = R\sqrt{\frac{8}{3}\pi G\rho}$ or $R = v\sqrt{\frac{3}{8\pi G\rho}}$

80. (*c*) Gravitational potential

$= -\frac{G \times 10}{1} - \frac{G \times 100}{1} = -110G$

$= -110 \times 6.67 \times 10^{-11}\ \text{J kg}^{-1} = -7.3 \times 10^{-9}\ \text{J kg}^{-1}$

81. (*a*) Conserving angular momentum,

$mv_1 r_{min} = mv_2 r_{max}$

or $\frac{v_1}{v_2} = \frac{r_{max}}{r_{min}}$

82. (*c*) $T^2 = kR^3$

$2T\Delta T = 3kR^2 \Delta R$

or $\frac{2T\Delta T}{T^2} = \frac{3kR^2\Delta R}{kR^3}$ or $\frac{\Delta T}{T} = \frac{3}{2}\frac{\Delta R}{R}$

or $\Delta T = \frac{3}{2}\frac{\Delta R}{R} T = \frac{3}{2} \times \frac{10}{9000} \times 90 \times 60\ \text{s} = 9\ \text{s}$

83. (*c*) For Moon, $v_m = \sqrt{2gR}$

For Earth, $v_e = \sqrt{2(6g)(10R)}$ or $v_e = \sqrt{120gR}$

$\frac{v_e}{v_m} = \frac{\sqrt{120gR}}{\sqrt{2gR}} = \sqrt{60} \approx 8$

84. (*c*) $\frac{v_{max}}{v_{min}} = \frac{r_{max}}{r_{min}}$

or $\frac{v_{max}}{v_{min}} = \frac{1+e}{1-e} = \frac{1+0.017}{1-0.017} = \frac{1.017}{0.983} = 1.034$

85. (*c*) $U = \frac{GMm}{r}$

or $U = \frac{GM}{r^2} \times mr$ or $U = g \times mr$

or $U = (mg)r$ or $mg = \frac{U}{r}$

86. (*c*) The general equation of a conic is

$\frac{1}{r} = \frac{1}{l}(1 + e\cos\theta)$

In the case of ellipse, the turning points are at $\theta = 0°$ and $\theta = 180°$

$\frac{1}{r_{max}} = \frac{1}{l}(1 + e\cos 180°)$ or $\frac{1}{r_{max}} = \frac{1}{l}(1 - e)$

Again, $\frac{1}{r_{min}} = \frac{1}{l}(1 + e\cos 0°)$ or $\frac{1}{r_{min}} = \frac{1}{l}(1 + e)$

Adding, $\frac{2}{l} = \frac{1}{r_{max}} + \frac{1}{r_{min}}$

Replacing, r_{max} by r_1 and r_{min} by r_2, we get

$\frac{2}{l} = \frac{1}{r_1} + \frac{1}{r_2}$ or $\frac{2}{l} = \frac{r_2 + r_1}{r_1 r_2}$ or $l = \frac{2r_1 r_2}{r_1 + r_2}$

87. (*c*) $W = \frac{GMm}{(2r)^2} = \frac{GMm}{4r^2} = \frac{1}{4r}\left[\frac{GMm}{r}\right] = \frac{U}{4r}$

88. (*b*) $\frac{G \times 10^3}{r} = 10\frac{G \times 10^5}{1-r}$

or $\frac{1}{r} = \frac{1000}{1-r}$

or $1 - r = 1000r$

or $1001r = 1$

or $r = \frac{1}{1001}$ km

Fig. 18

89. (*d*) $E = \frac{G \times 1}{1^2} + \frac{G \times 1}{2^2} + \frac{G \times 1}{4^2} + \ldots\ldots$

Fig. 19

$$E = G\left[1 + \frac{1}{4} + \frac{1}{16} + \ldots\ldots\right] \text{ or } E = G\frac{1}{1 - \frac{1}{4}} = \frac{4G}{3}$$

90. (*b*) $g' - g_e = g - [g - R\omega^2 \cos^2\phi] = R\omega^2\cos^2\phi$

$= R\omega^2$ [$\because$ $\phi = 0°$ at the equator]

$= \frac{4\pi^2 R}{T^2}$

$= 4 \times \frac{484}{49} \times \frac{6400 \times 1000}{86400 \times 86400}$ m s^{-2}

$= 0.03387$ m s^{-2} $= 3.4$ cm s^{-2}

91. (*d*) 'u' = 0, 'a' = 10 m s^{-2}, 'S' = 2 m, 'v' = ?

$v^2 - u^2 = 2aS$

$v^2 - 0^2 = 2 \times 10 \times 2 = 40$ or $v = \sqrt{40}$ m s^{-1}

In the second case, 'u' = 0, 'v' = $\sqrt{40}$ m s^{-1}

'a' = 0.2 m s^{-2}, 'S' = ?

$40 - 0^2 = 2 \times 0.2 \times S$

or $S = \frac{40 \times 10}{4}$ m or S = 100 m

92. (*c*) When the Earth's motion is suddenly stopped, it would fall into the Sun and (suppose) it comes back. If the effect of temperature of Sun is ignored, we can say that the Earth would continue to move along a strongly extended flat ellipse whose extreme points are located at the Earth's orbit and at the centre of the Sun.

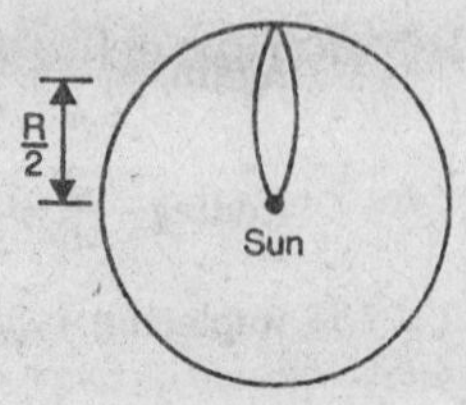

Fig. 20

The major axis is $\frac{R}{2}$.

Now, $\frac{T'^2}{T^2} = \left[\frac{R}{2}\right]^3\left[\frac{1}{R^3}\right]$ or $T'^2 = \frac{T^2}{8}$ or $T' = \frac{T}{2\sqrt{2}}$

Now, time required to fall into the Sun, $t = \frac{T'}{2}$

$$= \frac{T}{4\sqrt{2}} = \frac{365}{4\sqrt{2}} \approx 65 \text{ days}$$

So, the Earth would take slightly more than 2 months to fall into the Sun.

93. (*a*) $\frac{T'^2}{T^2} = \left[\frac{R}{2}\right]^3 \times \frac{1}{R^3} = \frac{1}{8}$

or $\frac{T'}{T} = \frac{1}{2\sqrt{2}}$ or $T' = \frac{27}{2\sqrt{2}}$

Required time, $t = \frac{T'}{2} = \frac{27}{4\sqrt{2}}$ days $\approx$ 5 days

So, if Moon's orbital motion around the Earth is stopped, the Moon shall take nearly five days to hit the Earth.

94. (*d*) The given system may be regarded as a system of three particles located at the three vertices of an equilateral triangle of side $2r$.

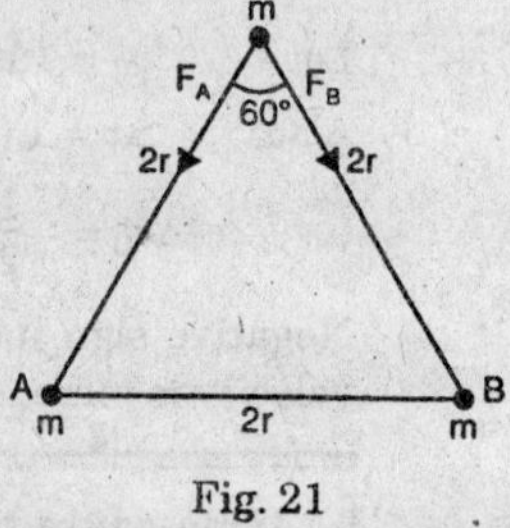

Fig. 21

Now, $F_A = F_B$

$$= \frac{Gm^2}{(2r)^2} = \frac{Gm^2}{4r^2}$$

F_A and F_B are inclined to each other at an angle of 60°. If F is the resultant of F_A and F_B, then

$$F = \sqrt{3} \times \frac{Gm^2}{4r^2}$$

95. (*b*) $g = \frac{GM}{R^2}$, $g' = \frac{GM}{\left(\frac{99R}{100}\right)^2}$

or $g' = \left[\frac{100}{99}\right]^2 \frac{GM}{R^2} = \left[\frac{100}{99}\right]^2 g$

Now, $\frac{g' - g}{g} \times 100 = \left[\frac{100 \times 100}{99 \times 99} - 1\right] \times 100$

$= [1.02 - 1] \times 100 = 2$

An alternative method

$$g = \frac{GM}{R^2}$$

$$\frac{\Delta g}{g} = -\frac{2\Delta R}{R}$$

$$\frac{\Delta g}{g} \times 100 = -2(-1)\% = 2\%$$

96. (*a*) $v_e = \sqrt{\frac{2GM}{R}}$

$v_e = \sqrt{\frac{2G}{R} \times \frac{4}{3}\pi R^3 \rho}$ or $v_e = R\sqrt{\frac{8}{3}\pi G\rho}$

or $v_e = R\sqrt{\rho}, v'_e = 2R\sqrt{\frac{\rho}{2}}$

or $v'_e = \sqrt{2}\, R\sqrt{\rho} = \sqrt{2} v_e$

Replace v_e by v_0.

97. (*c*) Increase in gravitational potential energy

$$= \left[-\frac{GMm}{R + \frac{R}{5}}\right] - \left[-\frac{GMm}{R}\right] = \frac{GMm}{R} - \frac{GMm \times 5}{6R}$$

$$= \frac{GMm}{R}\left[1 - \frac{5}{6}\right] = \frac{GMm}{R} \times \frac{1}{6}$$

$$= \frac{mgR^2}{6R} = \frac{1}{6} mgR = \frac{5}{6} mgh. \qquad [\because R = 5h]$$

98. (*d*) $U_1 = -\frac{G \times 5 \times 1}{5}$ J or $U_1 = -$ GJ

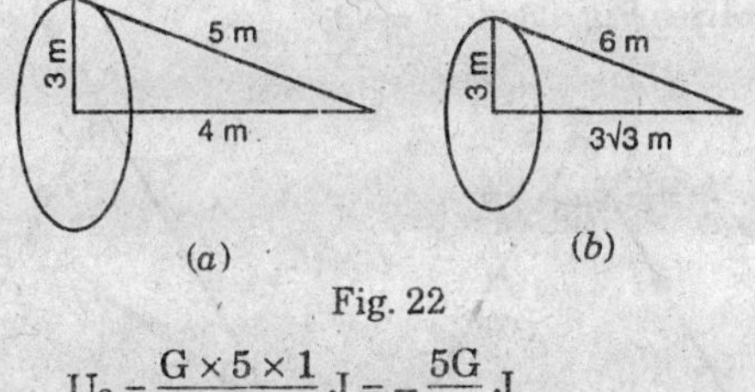

Fig. 22

$$U_2 = \frac{G \times 5 \times 1}{6} J = -\frac{5G}{6} J$$

$$\text{Work done} = \left[-\frac{5G}{6}\right] - [-G] = G - \frac{5G}{6} = \frac{G}{6}$$

> **☞ CAUTION**
>
> ***For determining gravitational potential due to a ring, you are not to directly use the distance between the observation point and the centre of the ring.***

99. (*b*) The required centripetal force is equal to $mr\omega^2$

$$mr\omega^2 = 6 \times 10^{24} \times 1.5 \times 10^8 \times 10^3 (2 \times 10^{-7})^2 \text{ N}$$

$$= 36 \times 10^{24+11-14} \text{ N} = 36 \times 10^{21} \text{ N}$$

100. (*c*) $\frac{40}{80} = \frac{R^2}{(R+h)^2}$

$\sqrt{2}R = R + h$ or $h = (\sqrt{2} - 1)R$

101. (*b*) $g_m = G \times \frac{M}{10} \times \left[\frac{2}{R}\right]^2 = \frac{4}{10} g_e$ or $\frac{g_e}{g_m} = \frac{10}{4} = \frac{5}{2}$

$\therefore \quad \frac{mg_e}{mg_m} = \frac{5}{2}$

102. (*b*) $\frac{g_h}{g} = \frac{R^2}{(R+h)^2}$ or $\frac{mg_h}{mg} = \frac{R^2}{\left(R + \frac{R}{2}\right)^2} = \frac{4}{9}$

or $mg_h = \frac{4}{9} mg = \frac{4}{9} \times 200 \times 10$

$$= \frac{8000}{9} \text{ N} = 888.88 \text{ N} \approx 889 \text{ N}$$

103. (*c*) $g = \frac{GM}{R^2}$

$$\frac{\Delta g}{g} = -2\frac{\Delta R}{R}$$

or $\frac{\Delta g}{g} \times 100 = -2\frac{\Delta R}{R} \times 100$

$$= -2\frac{(-21)}{6300} \times 100 = \frac{2}{3} = 0.67$$

104. (*d*) $\frac{T_s}{T_m} = \left[\frac{r_s}{r_m}\right]^{3/2} = \left[\frac{1}{2}\right]^{3/2}$ or $T_s = 2^{-3/2} \times T_m$

or $T_s = 2^{-3/2}$ lunar month

> ***Note that $T_m = 1$ lunar month.***

105. (*d*) $T_{es}^2 = \frac{4\pi^2 R_{es}^3}{GM_s}$

$$T_{me}^2 = \frac{4\pi^2 R_{me}^3}{GM_e}$$

Now, $\frac{T_{me}^2}{T_{es}^2} = \frac{R_{me}^3}{M_e} \times \frac{M_s}{R_{es}^3}$

$$\left[\frac{1}{13}\right]^2 = \left[\frac{1}{392}\right]^3 \frac{M_s}{M_e} \text{ or } \frac{M_s}{M_e} = \frac{392^3}{13^2}$$

106. (*c*) Using conservation of energy,

$$-\frac{GMm}{2R} + \frac{1}{2}mv^2 = 0$$

or $mv^2 = \frac{GMm}{R}$ or $v = \sqrt{\frac{GM}{R}}$

107. (*c*) $E_k - \frac{GMm}{R} = 0$ or $E_k = \frac{GMm}{R}$

or $E_k = \frac{gR^2 m}{R}$ or $E_k = mgR$

108. (*a*) $mg - m\omega^2 R = mg\left(1 - \frac{2h}{R}\right)$

or $-m\omega^2 R = -\frac{2mgh}{R}$ or $\omega^2 R = \frac{2gh}{R}$ or $h = \frac{\omega^2 R^2}{2g}$

109. (*a*) $mg' = mg - mR\omega^2 \cos^2 \phi$

At the equator, $\phi = 0°$, $\cos \phi = 1$

$$mg' = 0$$

$\therefore$ $mR\omega^2 = mg$ or $\omega = \sqrt{\frac{g}{R}}$

or $\frac{2\pi}{T} = \sqrt{\frac{g}{R}}$ or $T = 2\pi\sqrt{\frac{R}{g}}$

$$= 2 \times 3\sqrt{\frac{6400 \times 1000}{10}} \text{ s}$$

$$= \frac{6 \times 800}{3600} h = \frac{4800}{3600} h = \frac{4}{3} h = 1.33h$$

110. (*d*) $U_1 = 0$

$$U_2 = 3\left[-G\frac{10^{-2} \times 10^{-2}}{0.1}\right]$$

$$= -3 \times \frac{20}{3} \times 10^{-11} \times 10^{-3} = -2 \times 10^{-13} \text{ J}$$

$$\Delta U = 2 \times 10^{-13} \text{ J}$$

Work done $= 2 \times 10^{-13}$ J

111. (*b*) Gravitational potential at a point on the surface of Earth $= -\frac{GM}{R}$

If Earth is assumed to be a solid sphere, then the gravitational potential at the centre of Earth

$$= -\frac{3}{2}\frac{GM}{R}$$

Decrease in gravitational potential $= \frac{R}{2} \times \frac{GM}{R^2} = \frac{Rg}{2}$

Loss in potential energy $= \frac{Rg}{2} \times m$

Now, gain in kinetic energy = loss in potential energy

$\because$ $\frac{1}{2}mv^2 = \frac{1}{2}mgR$ or $v = \sqrt{gR}$

112. (*b*) Angular momentum = Iω

$$= \frac{2}{5}MR^2 \times \frac{2\pi}{T} = \frac{4\pi MR^2}{5T}$$

113. (*b*) $\frac{GMm}{R^2} \propto R^{-5/2}$ or $\frac{GMm}{R^2} = kR^{-5/2}$

or $mR\left[\frac{4\pi^2}{T^2}\right] = kR^{-5/2}$ or $T^2 \propto R^{7/2}$

114. (*b*) $v_0 = \sqrt{gR}$; $v_e = \sqrt{2gR}$

Required additional velocity

$$= v_e - v_0 = (\sqrt{2} - 1)\sqrt{gR}$$

$$= 0.414\sqrt{10 \times 6400 \times 10^3}$$

$$= 0.414 \times 8 \times 10^3 \text{ m s}^{-1} = 3.3 \times 10^3 \text{ m s}^{-1}$$

115. (*d*) $W = \frac{GMm}{r^2}$ and $U = \frac{GMm}{r'}$ [Negative sign neglected]

Now, $\frac{W}{U} = \frac{GMm}{r^2} \times \frac{r'}{GMm}$ or $\frac{W}{U} = \frac{r'}{r^2}$ or $W = \frac{Ur'}{r^2}$

or $W = \frac{4 \times 10^9 \times 10^7}{10^9 \times 10^9}$ N $= 4 \times 10^{-2}$ N

116. (*b*) $V = -\frac{GM}{R+h}$; $g = \frac{GM}{(R+h)^2}$

$$\frac{V}{g} = -(R+h) \Rightarrow \frac{-5.12 \times 10^7}{6.4} = -(R+h)$$

$$R + h = 0.8 \times 10^7 \text{ m} = 0.8 \times 10^4 \text{ km} = 8000 \text{ km}$$

$$h = (8000 - 6400) \text{ km} = 1600 \text{ km}$$

117. (*b*) $m_1 = \frac{4}{3}\pi R^3\rho$, $m_2 = \frac{4}{3}\pi(2R)^3\rho$, distance between the centres of two spherical objects, $r = 3R$.

$$F = \frac{Gm_1m_2}{r^2} = G\left(\frac{4}{3}\pi R^3\rho\right)\left(8 \times \frac{4}{3}\pi R^3\rho\right) \times \frac{1}{(3R)^2}$$

$$= \frac{128}{81}G\pi^2R^4\rho^2$$

118. (*c*) See Fig. 23. $F = \sqrt{3}\frac{GM^2}{a^2}$

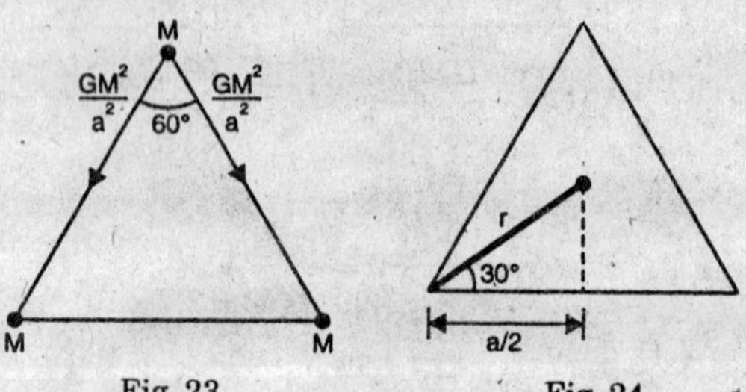

Fig. 23 Fig. 24

119. (*a*) See Fig. 24.

$\cos 30° = \frac{a/2}{r}$ or $r = \frac{a}{2\cos 30°}$ or $r = \frac{a}{2 \times \frac{\sqrt{3}}{2}} = \frac{a}{\sqrt{3}}$

Now $\frac{Mv^2}{r} = \frac{\sqrt{3}\,GM^2}{a^2}$

or $\frac{Mv^2 \times \sqrt{3}}{a} = \frac{\sqrt{3}\,GM^2}{a^2}$ or $v^2 = \frac{GM}{a}$ or $v = \sqrt{\frac{GM}{a}}$

120. (*d*) $T = \frac{2\pi r}{v} \Rightarrow T = \frac{2\pi a/\sqrt{3}}{\sqrt{\frac{GM}{a}}}$

$$T = \frac{2\pi a}{\sqrt{3}} \times \sqrt{\frac{a}{GM}} \text{ or } T = 2\pi\sqrt{\frac{a^3}{3GM}}$$

121. (*b*) $-\frac{GMm}{r_0} = \frac{1}{2}mv^2 - \frac{GMm}{R}$

or $\frac{1}{2}mv^2 = \frac{GMm}{R} - \frac{GMm}{r_0}$ or $\frac{v^2}{2} = \frac{GM}{R} - \frac{GM}{r_0}$

or $v^2 = 2GM\left[\frac{1}{R} - \frac{1}{r_0}\right]$ or $v = \sqrt{2GM\left(\frac{1}{R} - \frac{1}{r_0}\right)}$

122. (*d*) $F_{12} = \frac{GM^2}{2R^2}$; $F_{14} = \frac{GM^2}{2R^2}$

The resultant of these two forces is $\frac{\sqrt{2}\,GM^2}{2R^2}$

Now $F_{13} = \frac{GM^2}{4R^2}$

Fig. 25

Combined resultant of all the forces is

$$\frac{\sqrt{2}\,GM^2}{2R^2} + \frac{GM^2}{4R^2} \text{ or } \frac{GM^2}{R^2}\left[\frac{\sqrt{2}}{2} + \frac{1}{4}\right]$$

Equating this with centripetal force, we get

$$\frac{Mv^2}{R} = \frac{GM^2}{R^2}\left[\frac{2\sqrt{2}+1}{4}\right]$$

or $v^2 = \frac{GM}{R}\left[\frac{2\sqrt{2}+1}{4}\right]$ or $v = \sqrt{\frac{GM}{R}\left(\frac{2\sqrt{2}+1}{4}\right)}$

123. (*a*) $170F = \frac{GM_sM_e}{(400x)^2}$ or $F = \frac{GM_eM_m}{x^2}$

Dividing, $170 = \frac{M_s}{(400)^2} \times \frac{1}{M_m}$

or $\frac{M_s}{M_m} = 170 \times 400 \times 400 = 2.7 \times 10^7$

124. (*a*) $v_e = \sqrt{\frac{2GM}{r}}$

$$v_e = \sqrt{\frac{2GM}{R+3R}} = \sqrt{\frac{2GM}{4R}}$$

$$= \frac{1}{2}\sqrt{\frac{2GM}{R}} = \frac{1}{2} \times 11.2 \text{ km s}^{-1} = 5.6 \text{ km s}^{-1}$$

Formula for escape velocity of a body at a height r from the centre of the Earth

$$\frac{1}{2}mv_e^2 = W = \int_r^\infty \frac{GMm}{x^2}\,dx$$

or $\frac{v_e^2}{2} = GM\int_r^\infty x^{-2}\,dx = GM\left|\frac{x^{-1}}{-1}\right|_r^\infty$

or $\frac{v_e^2}{2} = -GM\left[\frac{1}{\infty} - \frac{1}{r}\right] = \frac{GM}{r}$ or $v_e = \sqrt{\frac{2GM}{r}}$.

125. (*c*) $\frac{1}{2} = \frac{R^2}{(R+h)^2}$ or $\frac{1}{\sqrt{2}} = \frac{R}{R+h}$

or $R + h = \sqrt{2}R$ or $h = (\sqrt{2} - 1)R$

126. (*a*) $\frac{1}{2}m\left[\frac{3}{4}\sqrt{\frac{2GM}{R}}\right]^2 - \frac{GMm}{R} = -\frac{GMm}{R+h}$

or $\frac{9 \times 2}{32R} - \frac{1}{R} = -\frac{1}{R+h}$ or $\frac{9}{16R} - \frac{1}{R} = -\frac{1}{R+h}$

or $\frac{7}{16R} = \frac{1}{R+h}$ or $7R + 7h = 16R$

or $7h = 9R$ or $h = \frac{9R}{7}$

127. (*a*) $\frac{a_c}{g} = \frac{R^2}{r^2}$ or $a_c = 10 \times \frac{6300 \times 6300}{8100 \times 8100}$ m s^{-2}

$$= \frac{10 \times 7 \times 7}{9 \times 9} \text{ m s}^{-2} \approx 6 \text{ m s}^{-2}$$

128. (*b*) $v = \sqrt{\frac{GM}{r}} = \sqrt{\frac{gR^2}{r}} = R\sqrt{\frac{g}{r}}$

$$= 6300 \times 10^3\sqrt{\frac{10}{8100 \times 10^3}} = \frac{6300 \times 10^3}{9 \times 10 \times 10}$$

$$= 7 \times 10^3 \text{ m s}^{-1} = 7 \text{ km s}^{-1}$$

129. (*a*) $-7.79 \times 10^{28} = -\frac{6.67 \times 10^{-11} \times 6 \times 10^{24} \times 7.4 \times 10^{22}}{r}$

or $r = \frac{6.67 \times 6 \times 7.4}{7.79} \times 10^7 = 3.80 \times 10^8$ m

130. (*a*) $\frac{g_h}{g} = \frac{6400 \times 6400}{6464 \times 6464}$ or $\frac{g_h}{g} = 0.98$

or $g_h = 0.98 \times 980$ cm s^{-2} = 960.4 cm s^{-2}.

131. (*d*) $g\left[1 - \frac{2h}{R}\right] = g\left[1 - \frac{d}{R}\right]$

$2h = d$ or $h = \frac{d}{2} = \frac{10 \text{ km}}{2} = 5$ km

132. (*a*) Using Kepler's third law,

$$\frac{T_A^2}{T_B^2} = \frac{R_A^3}{R_B^3}$$

or $\frac{(8T_B)^2}{T_B^2} = \left(\frac{R_A}{R_B}\right)^3$ $\quad | \because T_A = 8T_B$

or $\left(\frac{R_A}{R_B}\right)^3 = (4)^3$ or $\frac{R_A}{R_B} = 4$ or $R_A = 4R_B$

133. (*b*) $T = \sqrt{\frac{3\pi}{G\rho}} = \sqrt{\frac{3 \times 3.14}{6.67 \times 10^{-11} \times 8 \times 10^3}}$ s

≈ 4200 s.

134. (*b*) We know that $v_e = \sqrt{2} v_0$.

So, percentage change in escape velocity is the same as the percentage change in orbital velocity.

135. (*c*) $-\frac{GMm}{11R} = \frac{1}{2}mv^2 - \frac{GMm}{R}$

or $\frac{1}{2}mv^2 = \frac{GMm}{R} \times \frac{10}{11}$ or $\frac{v^2}{2} = \frac{10}{11}gR$

or $v^2 = \frac{20}{11}gR$ or $v = \sqrt{\frac{20}{11}gR}$

136. (*c*) $T = 2\pi\sqrt{\frac{R^3}{GM}}$

$T = 2\pi\sqrt{\frac{R^3}{G \times \frac{4}{3}\pi R^3 \rho}}$ or $T = 2\pi\sqrt{\frac{3}{4\pi G\rho}} = \sqrt{\frac{3\pi}{G\rho}}$

$T^2 = \frac{3\pi}{G\rho}$ or $T^2\rho = \frac{3\pi}{G}$

137. (*b*) Total energy $= \frac{1}{2}mv^2 - \frac{GMm}{r}$

But $\frac{mv^2}{r} = \frac{GMm}{r^2}$ or $mv^2 = \frac{GMm}{r}$

Total energy $= \frac{1}{2}mv^2 - mv^2 = -\frac{1}{2}mv^2$

138. (*a*) $V = -\frac{GM}{r}$, $\pi r = l$ or $r = \frac{l}{\pi}$

$\therefore$ $V = -\frac{\pi GM}{l}$

139. (*a*) $g' = G\frac{0.99M}{(0.99R)^2} = 1.01\frac{GM}{R^2} = 1.01\,g$

or $\frac{g'}{g} - 1 = 0.01$ or $\frac{g'-g}{g} \times 100 = 1$

140. (*d*) $v_e = \sqrt{\frac{2GM}{R}}$; Both M and R decrease by same percent.

141. (*a*) $g = \frac{GM}{R^2} = \frac{4}{3}\pi GR\rho \Rightarrow R_\rho = \text{constant}$

If density is doubled, radius is halved.

142. (*b*) $dA = \frac{1}{2}r^2 d\theta$

$\frac{dA}{dt} = \frac{1}{2}r^2\frac{d\theta}{dt} = \frac{1}{2}r^2\omega = \frac{1}{2}r^2\frac{v}{r}$

or $\frac{dA}{dt} = \frac{vr}{2} = \sqrt{\frac{GM}{r}} \times \frac{r}{2}$

or $\frac{dA}{dt} \propto \sqrt{r}$

Fig. 26

143. (*c*) $V = -\frac{GM}{r}$

$V = -\frac{GM}{R+h}$

$V = -\frac{gR^2}{R+h}$

$V = -\frac{g_h(R+h)^2}{R+h}$ $\quad \left[gR^2 = g_h(R+h)^2\right]$

$V = -g_h(R+h)$

$R + h = -\frac{V}{g_h} = -\frac{-5.12 \times 10^7}{6.4}$ m

$= 0.8 \times 10^7$ m $= 8000$ km

$h = (8000 - 6400)$ km $= 1600$ km

144. (*b*) $v = \sqrt{n^2 - 1}\, v_e$

$= \sqrt{3}\, v_e = 1.732 \times 11.2$ km s^{-1}

$= 19.4$ km s^{-1}

145. (*d*) $mR\omega^2 = \frac{GMm}{R^n}$

or $\omega^2 \propto \frac{1}{R^{n+1}}$ or $\frac{4\pi^2}{T^2} \propto \frac{1}{R^{n+1}}$

or $T^2 \propto R^{n+1}$ or $T \propto R^{\frac{n+1}{2}}$

146. (*d*) Gravitational potential at a distance r from the centre of the ring $= -\frac{GM}{\sqrt{2}\,r}$

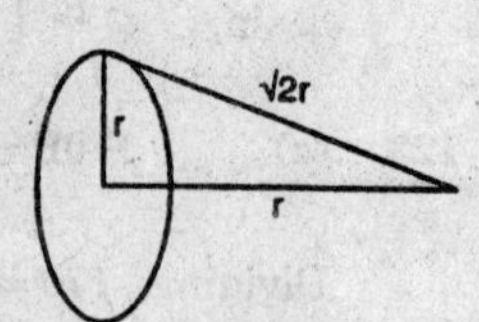

Fig. 27

Gravitational potential at the centre of the ring

$= -\frac{GM}{r}$

Decrease in gravitational potential energy

$= \left[-\frac{GMm}{\sqrt{2}\,r}\right] - \left[-\frac{GMm}{r}\right]$

$= \frac{GMm}{r} - \frac{GMm}{\sqrt{2}\,r} = \frac{GMm}{r}\left[1 - \frac{1}{\sqrt{2}}\right]$

Gain in kinetic energy $= \frac{1}{2}mv^2$

Equating, $\frac{v^2}{2} = \frac{GM}{r}\left[1 - \frac{1}{\sqrt{2}}\right]$

or $v = \sqrt{\frac{2GM}{r}\left[1 - \frac{1}{\sqrt{2}}\right]}$

147. (*a*) $\frac{1}{2}mv^2 = \left[-\frac{GMm}{R}\right] - \left[-\frac{3GMm}{2R}\right]$

or $\frac{1}{2}mv^2 = \frac{3GMm}{2R} - \frac{GMm}{R}$

or $\frac{v^2}{2} = \frac{3GM}{2R} - \frac{GM}{R}$

or $\frac{v^2}{2} = \frac{GM}{R} \times \frac{1}{2}$ or $v = \sqrt{\frac{GM}{R}}$

Also, $v_e = \sqrt{\frac{2GM}{R}} = \sqrt{2}v$

148. (*a*) Total energy $= -\frac{GMm}{2r}$

Total energy decreases.

So, $-\frac{GMm}{2r}$ decreases or $\frac{GMm}{2r}$ increases. So, r decreases.

Again, $\frac{mv^2}{r} = \frac{GMm}{r^2}$ or $v \propto \frac{1}{\sqrt{r}}$

Since r decreases therefore v increases.

149. (*c*) $E = \left[-\frac{GMm}{2r}\right] - \left[-\frac{GMm}{r}\right]$

or $E = \frac{GMm}{r} - \frac{GMm}{2r}$

or $E = \frac{GMm}{r}\left[1 - \frac{1}{2}\right]$ or $E = \frac{GMm}{2r}$

Again, $E' = \left[-\frac{GMm}{3r}\right] - \left[-\frac{GMm}{2r}\right]$

or $E' = \frac{GMm}{2r} - \frac{GMm}{3r}$ or $E' = \frac{GMm}{r}\left[\frac{1}{2} - \frac{1}{3}\right]$

or $E' = \frac{GMm}{6r} = \frac{1}{3}\left[\frac{GMm}{2r}\right]$ or $E' = \frac{E}{3}$

150. (*c*) $\frac{1}{100} = \frac{h^2}{(h+y)^2}$ or $\frac{h}{h+y} = \frac{1}{10}$

or $h + y = 10h$ or $y = 9h$

151. (*d*) $\frac{g_m}{g} = \frac{R^2}{(R+h)^2} = \frac{R^2}{(60R)^2}$

or $g_m = \frac{g}{3600}$ m s^{-2} $= \frac{9.8}{3600}$ m s^{-2} $= 2.7 \times 10^{-3}$ m s^{-2}

152. (*b*) $W = \frac{GMm}{r^2} = \frac{GMm}{r} \cdot \frac{1}{r}$

$W = \frac{U}{r}$

153. (*b*) $E_k = \frac{GMm}{2r}$

$\frac{E_{k1}}{E_{k2}} = \frac{r_2}{r_1} = \frac{R+R}{R+3R} = \frac{2R}{4R} = \frac{1}{2}$

154. (*c*) From the earth,

$0^2 - v^2 = -2g \times 2$ or $v^2 = 4g$

Again, from the planet,

$v^2 = 2g'R'$ or $4g = 2g'R'$ or $2g = g'R'$

But $g = \frac{GM}{R^2} = \frac{G}{R^2} \times \frac{4}{3}\pi R^3 \rho$ or $g = \frac{4}{3}\pi G R \rho$

$\therefore$ $2 \times \frac{4}{3}\pi G R \rho = \frac{4}{3}\pi G R' \rho' \times R'$

or $2R\rho = R'^2\rho'$ or $R'^2 = \frac{2R\rho}{\rho'}$

But $\rho = \rho'$ (given)

$\therefore$ $R' = \sqrt{2R}$

$= \sqrt{2 \times 6400}$ km $= 80\sqrt{2}$ km

155. (*a*) Two arguments favouring option (*a*) are as follows :

(*i*) Gravitational field intensity inside the shell is zero.

(*ii*) The mass of the solid sphere behaves as if concentrated at the centre.

156. (*d*) Total potential energy at mid-point is

$\left[-\frac{GM_1m}{d/2} - \frac{GM_2m}{d/2}\right]$

M₁ ——— M₂

Fig. 28

or $-\frac{2G}{d}(M_1 + M_2)m$

If v is the required escape velocity, then

$\frac{1}{2}mv^2 = \frac{2G}{d}(M_1 + M_2)m$

or $v = 2\sqrt{\frac{G(M_1 + M_2)}{d}}$

157. (*a*) $F_1 = \frac{GMm}{(2R)^2} = \frac{GMm}{4R^2}$

Now, mass of complete sphere, $M = \frac{4}{3}\pi R^3 \rho$

Mass removed $M' = \frac{4}{3}\pi\left[\frac{R}{2}\right]^3 \rho = \frac{1}{8}M$

Force due to hollow sphere F_2 = Force due to solid sphere – Force due to removed mass

$$F_2 = \frac{GMm}{4R^2} - \frac{G\left(\frac{M}{8}\right)m}{\left(\frac{3R}{2}\right)^2}$$

or $F_2 = \frac{GMm}{R^2}\left[\frac{1}{4} - \frac{4}{9 \times 8}\right]$ or $F_2 = \frac{GMm}{R^2}\left[\frac{1}{4} - \frac{1}{18}\right]$

or $F_2 = \frac{GMm}{R^2}\left(\frac{9-2}{36}\right)$ or $F_2 = \frac{7}{36}\frac{GMm}{R^2}$

Now, $\frac{F_1}{F_2} = \frac{GMm}{4R^2} \times \frac{36R^2}{7GMm}$ or $\frac{F_1}{F_2} = \frac{9}{7}$

8.1 ELASTICITY

Important Terms, Facts and Formulae

BASIC CONCEPTS

1. The term 'elasticity' is not relevant for the case of rigid bodies.
2. The stress required to cause actual fracture of a material is called the *breaking stress* or the *ultimate strength*.
3. The time delay in regaining the original form after the removal of the deforming force is called elastic relaxation time. Its value is negligibly small for gold, silver and phosphor bronze. For glass, its value is high. For quartz, its value is minimum.
4. The materials which have low elastic hysteresis have also low elastic relaxation time.
5. The addition of impurities (which is generally done to bind the crystals better together) affect the elastic properties of the metal according as they are themselves more elastic or plastic than the metal in question. Examples : carbon is added to iron and potassium to gold. In both these cases, the effect is strengthening of the elastic properties of both iron and gold.
6. Young's modulus is not defined for liquids and gases.
7. On increasing temperature, the value of Young's modulus decreases.

IMPORTANT POINTS TO REMEMBER

1. Maxwell was the first to define bulk modulus.
2. The nearest approach to a perfectly elastic body is quartz.
3. If the deforming force is inclined to the surface at an angle θ such that $\theta \neq 0°$ and $\theta \neq 90°$, then both tangential and normal stress are developed.
4. The maximum stress to which a wire can be subjected just before it breaks is called breaking stress or tensile strength.
5. The value of breaking stress depends on the nature of material.
6. Breaking force = Breaking stress × Cross-sectional area of the wire.
7. On hammering or rolling, elasticity increases.
8. By annealing, *i.e.*, heating and then cooling gradually, the elasticity is decreased.
9. When the deforming force is removed, the body takes some time to recover back the original shape or size.
10. Steel is more elastic than rubber.
11. Greater is the deforming force required to produce a given change in length, more is the value of Y.

QUESTION BANK

Level I

Choose the correct answer from the four alternatives given in each of the following questions :

1. The ratio of diameters of two wires of same material is $n : 1$. The length of each wire is 4 m. On applying the same load, the increase in length of thin wire will be $(n > 1)$

(*a*) n^2 times (*b*) n times

(*c*) $2n$ times (*d*) $(2n + 1)$ times.

2. Young's modulus of rubber is 104 N m^{-2} and area of cross- section is 2 cm^2. If force of 2×10^5 dyne is applied along its length, then its initial length l becomes

(*a*) $3l$ (*b*) $4l$

(*c*) $2l$ (*d*) None of the above.

3. A rope, made of steel, has a diameter of 5 cm. The breaking strength of the rope is 2×10^8 N. The breaking strength of a similar rope of double diameter is

(*a*) 2×10^8 N (*b*) 4×10^8 N

(*c*) 1×10^8 N (*d*) 8×10^8 N.

4. The potential energy U between the two molecules as a function of the distance x between them is shown in Fig. 1.

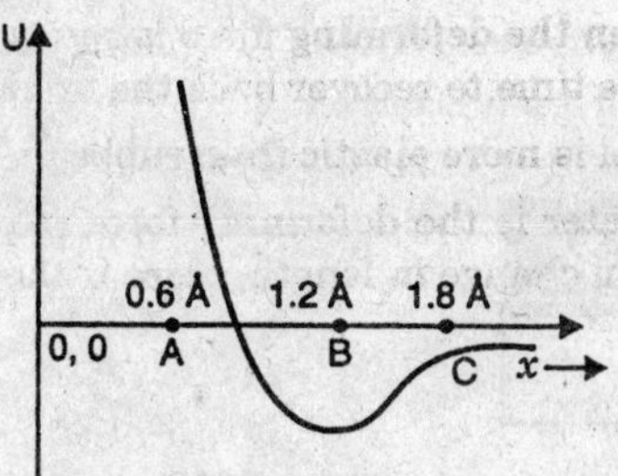

Fig. 1

A, B and C are points for which x = 0.6 Å, 1.2 Å and 1.8 Å. At A, B and C, force between the two molecules respectively is

(*a*) attractive, zero, repulsive

(*b*) zero, attractive, repulsive

(*c*) repulsive, zero, attractive

(*d*) zero, repulsive, attractive.

5. Bulk modulus of water is 2×10^9 N m^{-2}. The change in pressure required to increase the density of water by 0.1% is

(*a*) 2×10^9 N m^{-2} (*b*) 2×10^8 N m^{-2}

(*c*) 2×10^6 N m^{-2} (*d*) 2×10^4 N m^{-2}.

6. If a bar is made of copper whose coefficient of linear expansion is one and a half times that of iron, the ratio of the force developed in the copper bar to the iron bar of identical lengths and cross-sections, when heated through the same temperature range (Young's modulus for copper may be taken to be equal to that of iron) is

(*a*) 3/2 (*b*) 2/3

(*c*) 9/4 (*d*) 4/9.

7. A spring of constant K is cut into parts of length in the ratio 1 : 2. The spring constant of larger one is

(*a*) $\frac{K}{2}$ (*b*) $\frac{K}{3}$

(*c*) $\frac{2K}{3}$ (*d*) $\frac{3K}{2}$

8. A beam of metal supported at two ends is loaded at the centre. The depression at the centre is proportional to Y^n. The value of n is

(*a*) 1 (*b*) – 1

(*c*) 2 (*d*) 4.

9. When a certain weight is suspended from a long uniform wire, its length increases by one cm. If the same weight is suspended from another wire of the same material and length but having a diameter half of the first one, the increase in length will be

(*a*) 0.5 cm (*b*) 2 cm

(*c*) 4 cm (*d*) 8 cm.

10. A force F is needed to break a copper wire having radius R. The force needed to break a copper wire of same length and radius 2R will be

(*a*) F/2 (*b*) 2F

(*c*) 4F (*d*) F/4.

11. A string 1 mm in diameter breaks if the tension in it exceeds 80 N. The maximum tension that may be given to a similar string of diameter 2 mm is

(*a*) 40 N (*b*) 80 N

(*c*) 320 N (*d*) 1640 N.

12. A wire can sustain a weight of 10 kg before breaking. If the wire is cut into two equal parts, then each part can sustain a weight of

(*a*) 2.5 kg (*b*) 5 kg

(*c*) 10 kg (*d*) 15 kg.

13. A nylon rope 2 cm in diameter has a breaking strength of 1.5×10^5 N. The breaking strength of a similar rope 1 cm in diameter is

(*a*) 0.375×10^5 N (*b*) 2×10^5 N
(*c*) 6×10^5 N (*d*) 9×10^5 N.

14. On stretching a wire, the elastic energy stored per unit volume is [Letters have usual meanings.]

(*a*) $\frac{F\,\Delta l}{2Al}$ (*b*) $\frac{FA}{2l}$
(*c*) $\frac{Fl}{2A}$ (*d*) $\frac{Fl}{2}$.

15. In a wire stretched by hanging a weight from its end, the elastic potential energy per unit volume in terms of longitudinal strain σ and modulus of elasticity Y is

(*a*) $\frac{Y\sigma^2}{2}$ (*b*) $\frac{Y\sigma}{2}$
(*c*) $\frac{2Y\sigma^2}{2}$ (*d*) $\frac{Y^2\sigma}{2}$.

16. A cylinder of radius *r* is rigidly fixed at one end. It is twisted through a certain angle. The couple required is proportional to

(*a*) r^4 (*b*) r^2
(*c*) $\frac{1}{r^4}$ (*d*) $\frac{1}{r^2}$.

17. A tensile force of 2×10^3 N doubles the length of a rubber band of cross-sectional area 2×10^{-4} m^2. The Young's modulus of elasticity of the rubber band is

(*a*) 4×10^7 N m^{-2} (*b*) 2×10^7 N m^{-2}
(*c*) 10^7 N m^{-2} (*d*) 0.5×10^7 N m^{-2}.

18. Young's modulus = 21×10^{11} dyne cm^{-2}. It is equal to

(*a*) 210 G Pa (*b*) 2100 G Pa
(*c*) 21000 G Pa (*d*) 210000 G Pa.

19. A cable that can support a load of 800 N is cut into two equal parts. The maximum load that can be supported by either part is

(*a*) 100 N (*b*) 400 N
(*c*) 800 N (*d*) 1600 N.

20. The femur, which is the principal bone of the thigh, has a minimum diameter, in an adult male, of nearly 2.8 cm, corresponding to a cross-sectional area of 6×10^{-4} m^2. At what compressive load would it break ? The ultimate strength for bone in compression is 170×10^6 N m^{-2}.

(*a*) 10^2 N (nearly) (*b*) 10^3 N
(*c*) 10^4 N (*d*) 10^5 N (nearly).

21. A uniform wire, fixed at its upper end, hangs vertically and supports a weight at its lower end. If its radius is *r*, its length L and the Young's modulus for the material of the wire is E, the extension is

1. directly proportional to E
2. inversely proportional to *r*.
3. directly proportional to L

(*a*) if 1, 2, 3 are correct (*b*) if 1, 2 are correct
(*c*) if 2, 3 are correct (*d*) if 1 only is correct
(*e*) if 3 only is correct.

QUESTION BANK

Level II

Choose the correct answer from the four alternatives given in each of the following questions :

1. A 5 metre long wire is fixed to the ceiling. A weight of 10 kg is hung at the lower end and is 1 metre above the floor. The wire was elongated by 1 mm. The energy stored in the wire due to stretching is (g = 10 m s^{-2})

(*a*) Zero (*b*) 0.05 joule
(*c*) 100 joule (*d*) 500 joule.

2. The dimensions of four wires of the same material are given below. In which wire the increase in length will be maximum ?

(*a*) Length 100 cm, Diameter 1 mm
(*b*) Length 200 cm, Diameter 2 mm
(*c*) Length 300 cm, Diameter 3 mm
(*d*) Length 50 cm, Diameter 0.5 mm.

3. On increasing the length by 0.5 mm in a steel wire of length 2 m and area of cross-section 2 mm^2, the force required is [Y for steel = 2.2×10^{11} N/m^2]

(*a*) 1.1×10^5 N (*b*) 1.1×10^4 N
(*c*) 1.1×10^3 N (*d*) 1.1×10^2 N.

4. The breaking stress of a wire of length *l* and radius *r* is 5 kg wt m^{-2}. The wire of length 2*l* and radius 2*r* of the same material will have breaking stress in kg wt m^{-2}

(*a*) 5 (*b*) 10
(*c*) 20 (*d*) 80.

5. In order to twist one end of a wire, 2 m long and 4 mm in diameter, though 45°, the torque required is ($\eta = 5 \times 10^{10}$ N m^{-2})

(*a*) 0.49 N m (*b*) 3.49 N m
(*c*) 49 N m (*d*) 4.9×10^{10} N m.

6. In steel, the Young's modulus and the strain at the breaking point are 2×10^{11} N m^{-2} and 0.15 respectively. The stress at the break point for steel is
(*a*) 1.33×10^{11} N m^{-2} (*b*) 1.33×10^{-12} N m^{-2}
(*c*) 7.5×10^{-3} N m^{-2} (*d*) 3×10^{10} N m^{-2}.

7. If the Young's modulus of the material is 3 times its modulus of rigidity, then its volume elasticity will be
(*a*) Zero (*b*) Infinity
(*c*) 2×10^{10} N m^{-2} (*d*) 3×10^{10} N m^{-2}.

8. A steel wire of length 0.2 m and cross-sectional area 1 mm^2 is attached to two rigid supports at each end. The temperature during a day changes from 40°C to 20°C. The change of tension in the wire is :
[Y = 2×10^{11} N m^{-2}, coefficient of linear expansion = 1.1×10^{-5} (°C)$^{-1}$]
(*a*) 8 N (*b*) 16 N
(*c*) 44 N (*d*) 2.2×10^6 N.

9. The compressibility of water is 4×10^{-5} per unit atmospheric pressure. The decrease in volume of 100 cm^3 of water under a pressure of 100 atmosphere will be
(*a*) 0.4 cm^3 (*b*) 0.025 cm^3
(*c*) 4×10^{-5} cm^3 (*d*) 0.04 cm^3.

10. A 100 N force stretches the length of a hanging wire by 0.5 mm. The force required to stretch a wire, of the same material and length but having four times the diameter, by 0.5 mm is
(*a*) 100 N (*b*) 400 N
(*c*) 1200 N (*d*) 1600 N.

11. Two wires of the same material and length but diameters in the ratio 1 : 2 are stretched by the same force. The potential energy per unit volume for the two wires when stretched will be in the ratio
(*a*) 16 : 1 (*b*) 4 : 1
(*c*) 2 : 1 (*d*) 1 : 1.

12. If E_θ and E_ϕ denote the isothermal and adiabatic elasticities respectively of a gas, then $\frac{E_\theta}{E_\phi}$
(*a*) < 1 (*b*) > 1
(*c*) = 1 (*d*) = 3.2.

13. A rod of length l and radius r is joined to a rod of length $\frac{l}{2}$ and radius $\frac{r}{2}$ of same material. The free end of small rod is fixed to a rigid base. The free end of larger rod is given a twist of θ°. The twist angle at the joint will be
(*a*) $\frac{\theta}{2}$ (*b*) $\frac{\theta}{4}$
(*c*) $\frac{5\theta}{6}$ (*d*) $\frac{8\theta}{9}$.

14. Equal torsional torques act on two rods x and y having equal length. The diameter of rod y is twice the diameter of rod x. If θ_x and θ_y are the angles of twist, then $\frac{\theta_x}{\theta_y}$ =
(*a*) 1 (*b*) 2
(*c*) 4 (*d*) 16.

15. If the work done in stretching a wire by 1 mm is 2 J, the work necessary for stretching another wire of same material but with double radius of cross-section and half the length by 1 mm is
(*a*) $\frac{1}{4}$ J (*b*) 4 J
(*c*) 8 J (*d*) 16 J.

16. Two rods of different materials having coefficients of linear expansion α_1 and α_2 and Young's modulli, Y_1 and Y_2 respectively are fixed between two rigid massive walls. The rods are heated such that they undergo the same increase in temperature. There is no bending of rods. If $\frac{\alpha_1}{\alpha_2} = \frac{2}{3}$, then the thermal stresses developed in the two rods are equal, provided $\frac{Y_1}{Y_2}$ is equal to
(*a*) 2 : 3 (*b*) 1 : 1
(*c*) 3 :2 (*d*) 4 : 9.

17. Two wires of the same material and length are stretched by the same force. Their masses are in the ratio 3 : 2. Their elongations are in the ratio
(*a*) 3 : 2 (*b*) 9 : 4
(*c*) 2 : 3 (*d*) 4 : 9.

18. A slightly concal wire of length L and radii r_1 and r_2 is stretched by two forces F, F applied parallel to length in opposite directions and normal to end faces. If Y denotes the Young's modulus, then extension produced is
(*a*) $\frac{FL}{\pi r_1^2 Y}$ (*b*) $\frac{FL}{\pi r_1 Y}$
(*c*) $\frac{FLY}{\pi r_1 r_2}$ (*d*) $\frac{FL}{\pi r_1 r_2 Y}$.

19. The upper end of a wire of radius 4 mm and length 100 cm is clamped and its other end is twisted through an angle of 30°. Then angle of shear is
(*a*) 12° (*b*) 0.12°
(*c*) 1.2° (*d*) 0.012°.

20. Two wires of the same length and same material but radii in the ratio of 1 : 2 are stretched by unequal forces to produce equal elongation. The ratio of the two forces is

(*a*) 1 : 1 (*b*) 1 : 2
(*c*) 1 : 3 (*d*) 1 : 4.

21. You are given three wires A, B and C, of the same length and cross-section. They are each stretched by applying the same force to the ends. The wire A is stretched least and comes back to its original length when the stretching force is removed. The wire B is stretched more than A and also comes back to its original length when the stretching force is removed. The wire C is stretched most and remains stretched even when stretching force is removed. The greatest Young's modulus of elasticity is possessed by the material of wire

(*a*) A (*b*) B
(*c*) C (*d*) Data is not sufficient.

22. Two wires of the same material and same mass are stretched by the same force. Their lengths are in the ratio 2 : 3. Their elongations are in the ratio

(*a*) 3 : 2 (*b*) 2 :3
(*c*) 4 : 9 (*d*) 9 : 4.

23. A wire extends by 1 mm when a force is applied. Double the force is applied to another wire of same material and length but half the radius of cross-section. The elongation of the wire in mm will be

(*a*) 8 (*b*) 4
(*c*) 2 (*d*) 1.

24. When a weight of 5 kg is suspended from a copper wire of length 30 m and diameter 0.5 mm, the length of the wire increases by 2.4 cm. If the diameter is doubled, the extension produced is

(*a*) 1.2 cm (*b*) 0.6 cm
(*c*) 0.3 cm (*d*) 0.15 cm.

25. A cube of side 10 cm is subjected to a tangential force of 5×10^5 N at the upper face, keeping lower face fixed. The upper face is displaced by 0.001 radian relative to the lower face along the direction of tangential force. The shear modulus of the material of the cube is

(*a*) 5×10^{10} N m^{-2} (*b*) 5×10^{11} N m^{-2}
(*c*) 5×10^{12} N m^{-2} (*d*) 5×10^{13} N m^{-2}.

26. A helical spring extends 40 mm when stretched by a force of 10 N, and for tensions up to this value the extension is proportional to the stretching force. Two such springs are joined end-to-end and the double-length spring is stretched 40 mm beyond its natural length. The total strain energy, in J, stored in the double spring is

(*a*) 0.05 (*b*) 0.10
(*c*) 0.20 (*d*) 0.40
(*e*) 0.80.

27. The diameter of a brass wire is 0.6 mm and Y is 9×10^{10} N m^{-2}. The force which will increase its length by 0.2% is about

(*a*) 100 N (*b*) 51 N
(*c*) 25 N (*d*) None of these.

28. Forces of 100 N each are applied in opposite directions on the upper and lower faces of a cube of side 20 cm. The upper face is shifted parallel to itself by 0.25 cm. If the side of the cube were 10 cm, then the displacement would be

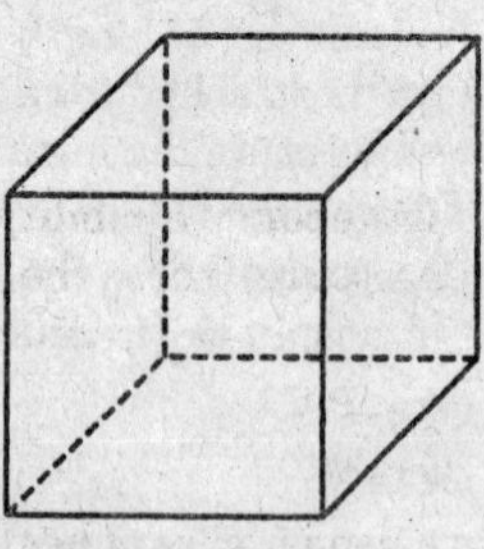

Fig. 2

(*a*) 0.25 cm (*b*) 0.5 cm
(*c*) 0.75 cm (*d*) 1 cm.

29. Two wires of the same material have lengths in the ratio 1 : 2 and their radii are in the ratio $1 : \sqrt{2}$. If they are stretched by applying equal forces, the increase in their lengths will be in the ratio

(*a*) $\sqrt{2} : 2$ (*b*) $2 : \sqrt{2}$
(*c*) 1 : 1 (*d*) 1 : 2.

30. A wire ($Y = 2 \times 10^{11}$ N m^{-2}) has length 1 m and cross-sectional area 1 mm^2. The work required to increase the length by 2 mm is

(*a*) 0.4 J (*b*) 4 J
(*c*) 40 J (*d*) 400 J.

31. The length of a wire is increased by 1 mm on the application of a given load. In a wire of the same material, but of length and radius twice that of the first, on the application of the same load, extension is

(*a*) 0.25 mm (*b*) 0.5 mm
(*c*) 2 mm (*d*) 4 mm.

32. A metal rod of Young's modulus 2×10^{10} N m^{-2} undergoes an elastic strain of 0.06%. The energy per unit volume stored in J m^{-3} is

(*a*) 3600 (*b*) 7200
(*c*) 10800 (*d*) 14400
(*e*) 17600.

33. An aluminium rod, Young's modulus 7.0×10^9 N m^{-2}, has a breaking strain of 0.2%. The minimum cross-sectional area of the rod in m^2 in order to support a load of 10^4 N is

(a) 1×10^{-2} (b) 1.4×10^{-3}
(c) 1.0×10^{-3} (d) 7.1×10^{-4}
(e) 1×10^{-4}.

34. A wire of length L and radius r is clamped rigidly at one end. When the other end of the wire is pulled by a force F, its length increases by l. Another wire of the same material of length 4 L and radius $4r$, is pulled by a force 4F. The increase in length will be

(a) $\frac{l}{2}$ (b) l
(c) $2l$ (d) $4l$.

35. If the compressibility of water is σ per unit atmospheric pressure, then the decrease in volume (V) due to atmospheric pressure P will be

(a) $\sigma P/V$ (b) σPV
(c) σ/PV (d) $\sigma V/P$.

36. A copper wire of negligible mass, 1 m length and cross-sectional area 10^{-6} m^2 is kept on a smooth horizontal table with one end fixed. A ball of mass 1 kg is attached to the other end. The wire and the ball are rotated with an angular velocity 20 rad s^{-1}. If the elongation in the wire is 10^{-3} m, then the Young's modulus is

(a) 4×10^{11} N m^{-2} (b) 6×10^{11} N m^{-2}
(c) 8×10^{11} N m^{-2} (d) 10×10^{11} N m^{-2}.

37. A piece of copper wire has twice the radius of a piece of steel wire. The Young's modulus for steel is twice that for the copper. One end of the copper wire is joined to one end of the steel wire so that both can be subjected to the same longitudinal force. By what fraction of its length will the steel have stretched when the length of the copper has increased by 1% ?

(a) 1% (b) 2%
(c) 2.5% (d) 3%.

38. For a given material, the Young's modulus is 2.4 times that of modulus of rigidity. Its Poisson's ratio is

(a) 0.1 (b) 0.2
(c) 0.3 (d) 0.4.

39. A 1 m long wire is stretched without tension at 30°C between two rigid supports. What strain will be produced in the wire if the temperature falls to 0°C ? (Given : $\alpha = 12 \times 10^{-6}$ K^{-1})

(a) 36×10^{-5} (b) 64×10^{-5}
(c) 0.78 (d) 0.32.

40. If 3×10^{-3} J of work is done in stretching a wire through 1 mm, the work done in stretching it through 3 mm is

(a) 9×10^{-3} J (b) 8×10^{-3} J
(c) 27×10^{-3} J (d) Data incomplete.

41. A spherical ball contracts in volume by 0.01% when subjected to a normal uniform pressure of 100 atmosphere. What is the bulk modulus of elasticity of the material of the ball ? Take 1 atmosphere = 10^6 dyne cm^{-2}.

(a) 10^9 dyne cm^{-2} (b) 10^{10} dyne cm^{-2}
(c) 10^{12} dyne cm^{-2} (d) 10^{14} dyne cm^{-2}.

42. The upper end of a wire 1 metre long and 2 mm radius is clamped. The lower end is twisted through an angle of 45°. The angle of shear is

(a) 0.09° (b) 0.9°
(c) 9° (d) 90°.

43. One end of uniform wire of length L and of weight W is attached rigidly to a point in the roof and a weight W_1 is suspended from its lower end. If s is the area of cross-section of the wire, the stress in the wire at a height (3L/4) from its lower end is

(a) $\frac{W_1}{s}$ (b) $\left[W_1 + \frac{W}{4}\right]s$
(c) $\left[W_1 + \frac{3W}{4}\right]/s$ (d) $\frac{W_1 + W}{s}$.

44. A one metre long steel wire of cross-sectional area 1 mm^2 is extended by 1 mm. If Y = 2×10^{11} N m^{-2}, then the work done is

(a) 0.1 J (b) 0.2 J
(c) 0.3 J (d) 0.4 J.

45. A cube is compressed at 0°C equally from all sides by an external pressure P. By what amount should its temperature be raised to bring it back to the size it had before the external pressure was applied ? (Given : K is bulk modulus of elasticity of the material of the cube and α is the coefficient of linear expansion.)

(a) $\frac{P}{K\alpha}$ (b) $\frac{P}{3K\alpha}$
(c) $\frac{3\pi\alpha}{P}$ (d) $\frac{K}{3P}$.

46. A 2 m long rod of radius 1 cm which is fixed from one end is given a twist of 0.8 radian. The shear strain developed will be

(a) 0.002 (b) 0.004
(c) 0.008 (d) 0.016.

47. A cylindrical tree has a breaking stress of 106 N m^{-2}. The maximum possible height of the tree is 5 m. The density of material of the tree is (Take g = 10 m s^{-2})

(a) 10^3 kg m^{-3} (b) 10^4 kg m^{-3}
(c) 2×10^4 kg m^{-3} (d) 1 kg m^{-3}.

48. A force F is required to break a wire of length l and radius r. What force is required to break a wire, of the same material, having twice the length and six times the radius ?

(a) F (b) 3 F
(c) 9 F (d) 36 F.

49. The Young's modulus of a material is 10^{11} N m^{-2} and its Poisson's ratio is 0.2. The modulus of rigidity of the material is

(a) 0.42×10^{11} N m^{-2} (b) 0.42×10^{14} N m^{-2}
(c) 0.42×10^{16} N m^{-2} (d) 0.42×10^{18} N m^{-2}.

50. When a 4 kg mass is hung vertically on a light spring that obeys Hooke's law, the spring stretches by 2 cm. The work required to be done by an external agent in stretching this spring by 5 cm will be

(a) 4.9 J (b) 2.45 J
(c) 0.495 J (d) 0.245 J.

51. A cube is shifted to a depth of 100 m in a lake. The change in volume is 0.1%. The bulk modulus of the material is nearly

(a) 10 Pa (b) 10^4 Pa
(c) 10^7 Pa (d) 10^9 Pa.

52. The pressure applied from all directions on a cube is P. How much its temperature should be raised to maintain the original volume ? The volume elasticity of the cube is β and the coefficient of volume expansion is α.

(a) $\frac{P}{\alpha\beta}$ (b) $\frac{P\alpha}{\beta}$
(c) $\frac{P\beta}{\alpha}$ (d) $\frac{\alpha\beta}{P}$.

53. A steel ring of radius r and cross-sectional area 'A' is fitted on to a wooden disc of radius R (R > r). If Young's modulus be E, then the force with which the steel ring is expanded is

(a) AE $\frac{R}{r}$ (b) AE $\left(\frac{R-r}{r}\right)$
(c) $\frac{E}{A}\left(\frac{R-r}{r}\right)$ (d) $\frac{Er}{AR}$.

54. A copper bar of length L and area of cross-section A is placed in a chamber at atmospheric pressure. If the chamber is evacuated, the percentage change in its volume will be (compressibility of copper is 8×10^{-12} m^2 N^{-1} and 1 atm = 10^5 N m^{-2})

(a) 8×10^{-7} (b) 8×10^{-5}
(c) 1.25×10^{-4} (d) 1.25×10^{-5}.

55. An elastic material of Young's modulus Y is subjected to a stress S. The elastic energy stored per unit volume of the material is

(a) $\frac{SY}{2}$ (b) $\frac{S^2}{2Y}$
(c) $\frac{S}{2Y}$ (d) $\frac{2S}{Y}$.

56. A wire is suspended vertically from a rigid support. When loaded with a steel weight in air, the wire extends by 16 cm. When the weight is completely immersed in water, the extension is reduced to 14 cm. The relative density of the material of the weight is

(a) 2 g cm^{-3} (b) 6 g cm^{-3}
(c) 8 g cm^{-3} (d) 16 g cm^{-3}.

57. The breaking stress for a substance is 10^6 N m^{-2}. What length of the wire of this substance should be suspended vertically so that the wire breaks under its own weight ? (Given : density of material of the wire = 4×10^3 kg m^{-3} and g = 10 m s^{-2})

(a) 10 m (b) 15 m
(c) 25 m (d) 34 m.

58. What stress would cause a wire to increase in length by one-tenth of one percent if the Young's modulus for the wire is 12×10^{10} N m^{-2} ? (Take g = 10 N kg^{-1})

(a) 12×10^3 N m^{-2} (b) 12×10^5 N m^{-2}
(c) 12×10^7 N m^{-2} (d) 12×10^9 N m^{-2}.

59. A cube of aluminium of sides 0.1 m is subjected to a shearing force of 100 N. The top face of the cube is displaced through 0.02 cm with respect to the bottom face. The shearing strain would be

(a) 0.02 (b) 0.1
(c) 0.005 (d) 0.002.

60. A small but heavy block of mass 10 kg is attached to a wire 0.3 m long. Its breaking stress is 4.8×10^7 N m^{-2}. The area of cross-section of the wire is 10^{-6} m^2. The maximum angular velocity with which the block can be rotated in the horizontal circle is

(a) 4 rad s^{-1} (b) 8 rad s^{-1}
(c) 10 rad s^{-1} (d) 32 rad s^{-1}.

61. Two rods A and B of the same material and length have their radii r_1 and r_2 respectively. When they are rigidly fixed at one end and twisted by the same couple applied at the other end, the ratio of the angle of twist at the end of A and the angle of twist at the end of B is

(a) $\frac{r_2^4}{r_1^4}$ (b) $\frac{r_1^4}{r_2^4}$
(c) $\frac{r_2^2}{r_1^2}$ (d) $\frac{r_1^2}{r_2^2}$.

62. What amount of work is done in increasing the length of a wire through unity ?

(a) $\frac{YL}{2A}$ (b) $\frac{YL^2}{2A}$

(c) $\frac{YA}{2L}$ (d) $\frac{YL}{A}$.

[Here the letters have usual meanings.]

63. Of the three materials, copper, glass and rubber, which two best illustrate the properties described in the table below when each is extended at room temperature ?

	Obeys Hooke's Law virtually up to its breaking point	*Tolerates a large strain while still exhibiting elastic behaviour*
(a)	copper	glass
(b)	copper	rubber
(c)	glass	copper
(d)	rubber	glass
(e)	glass	rubber

64. A long wire hangs vertically with its upper end clamped. A torque of 8 N m applied to the free end twists it through 45°. The potential energy of the twisted wire is

(a) π joule (b) $\frac{\pi}{2}$ joule

(c) $\frac{\pi}{4}$ joule (d) $\frac{\pi}{8}$ joule.

65. Two wires of the same material and radius but having lengths in the ratio of 1 : 2 are stretched with the same force. The ratio of the work done in two cases is

(a) 1 : 2 (b) 2 : 1

(c) 1 : 3 (d) 1 : 8.

66. When a rubber cord is stretched, the change in volume with respect to change in its linear dimensions is negligible. The Poisson's ratio for rubber is

(a) 1 (b) 0.25

(c) 0.5 (d) 0.75.

67. A wire whose cross-section is 4 mm^2 is stretched by 0.1 mm by a certain weight. How far will a wire of the same material and length stretch if its cross-sectional area is 8 mm^2 and the same weight is attached ?

(a) 0.1 mm (b) 0.05 mm

(c) 0.025 mm (d) 0.012 mm.

68. A wire is stretched 1 mm by a force of 1 kN. How far would a wire of the same material and length but of four times that diameter be stretched by the same force ?

(a) $\frac{1}{2}$ mm (b) $\frac{1}{4}$ mm

(c) $\frac{1}{8}$ mm (d) $\frac{1}{16}$ mm.

69. In Q. 68, the work done in the two wires is

(a) 0.5 J, 0.03 J (b) 0.25 J, 0 J

(c) 0.03 J, 0.25 J (d) 0 J, 0 J.

70. A substance breaks down by a stress of 10^6 N m^{-2}. If the density of the material of the wire is 3×10^3 kg m^{-3}, then the length of the wire of that substance which will break under its own weight when suspended vertically is nearly

(a) 3.4 m (b) 34 m

(c) 340 m (d) 3400 m.

71. Two bars A and B of circular cross-section and of same volume and made of the same material are subjected to tension. If the diameter of A is half that of B and if the force applied to both the rods is the same and it is in the elastic limit, the ratio of extension of A to that of B will be

(a) 16 (b) 8

(c) 4 (d) 2.

72. A ball falling in a lake of depth 200 m shows a decrease of 0.1% in its volume at the bottom. The bulk modulus of elasticity of the material of the ball is (Take g = 10 m s^2)

(a) 10^9 N m^{-2} (b) 2×10^9 N m^{-2}

(c) 3×10^9 N m^{-2} (d) 4×10^9 N m^{-2}.

73. The increase in length on stretching a wire is 0.05 %. If its Poisson's ratio is 0.4, the diameter is reduced by

(a) 0.01% (b) 0.02%

(c) 0.03% (d) 0.04%.

74. Two wires of equal cross-section but one made of steel and the other of copper are joined end to end. When the combination is kept under tension, the elongations in the two wires are found to be equal. What is the ratio of the lengths of the two wires ? (Given : Y_{steel} = 2.0×10^{11} N m^{-2} and Ycopper = 1.1×10^{11} N m^{-2})

(a) 2 : 11 (b) 11 : 2

(c) 20 : 11 (d) 11 : 20.

75. A massive stone pillar 20 m high and of uniform cross-section rests on a rigid base and supports vertical load of 5.0×10^5 N at its upper end. If the compressive stress in the pillar is not to exceed 1.6×10^6 N m^{-2}, what is the minimum cross-sectional area of the pillar ? Density of the stone = 2.5×10^3 kg m^{-3}. (Take g = 10 N kg^{-1})

(a) 0.15 m^2 (b) 0.25 m^2

(c) 0.35 m^2 (d) 0.45 m^2.

76. The following data were obtained when a wire was stretched within the elastic region

Force applied to wire 100 N.

Area of cross-section of wire 10^{-6} m^2.

Extension of wire 2×10^{-3} m.

Original length of wire 2 m.

Which of the following deductions can be correctly made from this data ?

1. The value of Young's Modulus is 10^{11} N m^{-2}.
2. The strain is 10^{-3}.
3. The energy stored in the wire when the load is applied is 10 J.

(*a*) 1, 2, 3 are correct (*b*) 1, 2 are correct
(*c*) 1 only (*d*) 3 only
(*e*) 3 only.

77. The Young's modulus of the material of a wire is 6×10^{12} N m^{-2} and there is no transverse strain in it, then its modulus of rigidity will be

(*a*) 3×10^{12} N m^{-2} (*b*) 2×10^{12} N m^{-2}
(*c*) 10^{12} N m^{-2} (*d*) None of the above.

78. The ratio of the adiabatic to isothermal elasticities of a triatomic gas is

(*a*) 3/4 (*b*) 4/3
(*c*) 1 (*d*) 5/3.

79. A uniform rod suffers a longitudinal strain of 2×10^{-3}. The Poisson's ratio of the material of the rod is 0.50. The percentage change in volume is

(*a*) zero (*b*) 0.1
(*c*) 0.2 (*d*) 0.6.

80. A uniform cylindrical wire is subjected to a longitudinal tensile stress of 5×10^7 N m^{-2}. The Young's modulus of the material of the wire is 2×10^{11} N m^{-2}. The volume change in the wire is 0.02%. The fractional change in the radius is

(*a*) 0.25×10^{-4} (*b*) 0.5×10^{-4}
(*c*) 1.0×10^{-4} (*d*) 1.5×10^{-4}.

81. A circular tube of mean radius 8 cm and thickness 0.04 cm is melted up and recast into a solid rod of the same length. The ratio of the torsional rigidities of the circular tube and the solid rod is

(*a*) $\dfrac{(8.02)^4 - (7.98)^4}{(0.8)^4}$ (*b*) $\dfrac{(8.02)^2 - (7.98)^2}{(0.8)^2}$

(*c*) $\dfrac{(0.8)^2}{(8.02)^4 - (7.98)^4}$ (*d*) $\dfrac{(0.8)^2}{(8.02)^3 - (7.98)^2}$.

82. A steel wire has length 2 m, radius 1 mm and $Y = 2 \times 10^{11}$ N m^{-2}. A 1 kg sphere is attached to one end of the wire and whirled in a vertical circle with an angular velocity of 2 revolutions per second. When the sphere is at the lowest point of the vertical circle, the elongation of the wire is nearly [Take $g = 10$ m s^{-2}]

(*a*) 1 mm (*b*) 2 mm
(*c*) 0.1 mm (*d*) 0.01 mm.

83. A 1 m long wire of cross-sectional area 10^{-6} m^2 is used to suspend a load of 840 N. Which of the following is incorrect ? (Given : $Y = 12 \times 10^{10}$ Pa, $\rho = 7000$ kg m^{-3}, $S = 420$ J kg^{-1} K^{-1})

(*a*) The stress developed in the wire is 840×10^6 N m^{-2}

(*b*) The strain is $\dfrac{840 \times 10^6}{12 \times 10^{10}}$

(*c*) Energy stored is 2.94 J

(*d*) If the wire snaps, heat will be produced and the temperature would rise by 1°C

(*e*) None of these.

84. Three wires P, Q and R of the same material and length have radii 0.1 cm, 0.2 cm and 0.3 cm respectively. Which wire has the highest value of Young's modulus of elasticity ?

(*a*) P (*b*) Q
(*c*) R (*d*) All have the same value.

85. The length of an elastic string is '*a*' metre when the tension is 4 N, and '*b*' metre when the tension is 5 N. The length in metre when the tension is 9 N is

(*a*) $5b - 4a$ (*b*) $4a - 5b$
(*c*) $a + b$ (*d*) $b - a$.

86. A copper wire 2 m long is stretched by 1 mm. If the energy stored in the stretched wire is converted to heat, calculate the rise in temperature of the wire. (Given : $Y = 12 \times 10^{11}$ dyne cm^{-2}, density of copper = 9 g cm^{-3} and specific heat of copper = 0.1 cal g^{-1} °C^{-1})

(*a*) 252°C (*b*) (1/252)°C
(*c*) 1000°C (*d*) 2000°C.

87. Find the extension produced in a copper wire of length 2 m and diameter 3 mm when a force of 30 N is applied. Young's modulus for copper = 1.1×10^{11} N m^{-2} and $g = 10$ N kg^{-1}.

(*a*) 0.02 mm (*b*) 0.04 mm
(*c*) 0.08 mm (*d*) 0.68 mm.

88. A spring is extended by 30 mm when a force of 1.5 N is applied to it. Calculate the energy stored in the spring when hanging vertically supporting a mass of 0.20 kg if the spring was unstretched before applying the mass. (Take $g = 10$ N kg^{-1})

(*a*) 0.01 J (*b*) 0.02 J
(*c*) 0.04 J (*d*) 0.08 J.

89. A steel wire is suspended vertically from a rigid support. When loaded with a weight in air, it extends by l_a. When the weight is completely immersed in water, the extension is reduced to l_w. Then the relative density of the material of the weight is

(a) $\frac{l_a}{l_a - l_w}$ (b) $\frac{l_a}{l_w}$

(c) $\frac{l_w}{l_a - l_w}$ (d) $\frac{l_w}{l_a}$.

90. Given the following values for an elastic material :

Young's modulus = 7×10^{10} N m^{-2} and Bulk modulus = 11×10^{10} N m^{-2}.

The Poisson's ratio of the material is

(a) 0.12 (b) 0.24

(c) 0.31 (d) 0.39.

91. A 1000 kg lift is tied with metallic wires of maximum safe stress of 1.4×10^8 N m^{-2}. If the maximum acceleration of the lift is 1.2 m s^{-2}, then the minimum diameter of the wire is

(a) 1 m (b) 0.1 m

(c) 0.01 m (d) 0.001 m.

92. The Young's modulus of brass and steel are 10×10^{10} N m^{-2} and 2×10^{11} N m^{-2} respectively. A brass wire and a steel wire of the same length are extended by 1 mm under the same force. The radii of the brass and steel wires are R_B and R_S respectively. Then

(a) $R_S = \sqrt{2}\, R_B$ (b) $R_S = \frac{R_B}{\sqrt{2}}$

(c) $R_S = 4R_B$ (d) $R_S = \frac{R_B}{4}$.

93. Two identical wires are suspended from the same rigid support but one is of copper and the other is of iron. Young's modulus of iron is thrice that of copper. The weights to be added on copper and iron wires so that the ends are on the same level must be in the ratio of

(a) 1 : 3 (b) 2 : 1

(c) 3 : 1 (d) 4 : 1.

94. A material has normal density ρ and bulk modulus K. The increase in the density of the material when it is subjected to an external pressure P from all sides is

(a) P/ρK (b) K/ρP

(c) ρP/K (d) ρK/P.

95. A rubber rope of length 8 m is hung from the ceiling of a room. What is the increase in length of the rope due to its own weight ? (Given : Young's modulus of elasticity of rubber = 5 × 106 N m^{-2} and density of rubber = 1.5×10^3 kg m^{-3}. Take g = 10 m s^{-2})

(a) 1.5 mm (b) 6 mm

(c) 24 mm (d) 96 mm.

96. When the tension in a metal wire is T_1, its length is l_1. When the tension is T_2, its length is l_2. The natural length of wire is

(a) $\frac{T_2}{T_1}(l_1 + l_2)$ (b) $T_1 l_1 + T_2 l_2$

(c) $\frac{l_1 T_2 - l_2 T_1}{T_2 - T_1}$ (d) $\frac{l_1 T_2 + l_2 T_1}{T_2 + T_1}$.

97. Two blocks of masses 1 kg and 2 kg are connected by a metal wire going over a smooth pulley as shown. The breaking stress of the metal is $\frac{40}{3\pi} \times 10^6$ N m^{-2}. If g = 10 m s^{-2}, then what should be the minimum radius of the wire used if it is not to break ?

(a) 0.5 mm (b) 1 mm

(c) 1.5 mm (d) 2 mm.

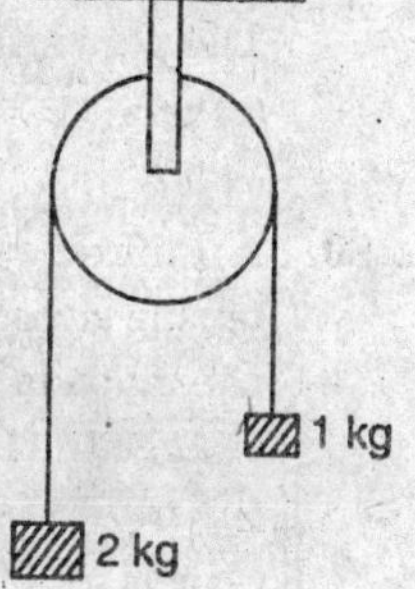

Fig. 3

ANSWERS

QUESTION BANK (Level I)

1. (a) **2.** (c) **3.** (d) **4.** (c) **5.** (c) **6.** (a) **7.** (d) **8.** (b)
9. (c) **10.** (c) **11.** (c) **12.** (c) **13.** (a) **14.** (a) **15.** (a) **16.** (a)
17. (c) **18.** (a) **19.** (c) **20.** (d) **21.** (e).

QUESTION BANK (Level II)

1. (b) **2.** (d) **3.** (d) **4.** (a) **5.** (a) **6.** (d) **7.** (b) **8.** (c)
9. (a) **10.** (d) **11.** (a) **12.** (a) **13.** (d) **14.** (d) **15.** (d) **16.** (c)
17. (c) **18.** (d) **19.** (b) **20.** (d) **21.** (a) **22.** (c) **23.** (a) **24.** (b)
25. (a) **26.** (b) **27.** (b) **28.** (b) **29.** (c) **30.** (a) **31.** (b) **32.** (a)
33. (d) **34.** (b) **35.** (b) **36.** (a) **37.** (b) **38.** (b) **39.** (a) **40.** (c)

41. (*c*)	**42.** (*a*)	**43.** (*c*)	**44.** (*a*)	**45.** (*b*)	**46.** (*b*)	**47.** (*c*)	**48.** (*d*)
49. (*a*)	**50.** (*b*)	**51.** (*d*)	**52.** (*a*)	**53.** (*b*)	**54.** (*b*)	**55.** (*b*)	**56.** (*c*)
57. (*c*)	**58.** (*c*)	**59.** (*d*)	**60.** (*a*)	**61.** (*a*)	**62.** (*c*)	**63.** (*e*)	**64.** (*a*)
65. (*a*)	**66.** (*c*)	**67.** (*b*)	**68.** (*d*)	**69.** (*a*)	**70.** (*b*)	**71.** (*a*)	**72.** (*b*)
73. (*b*)	**74.** (*c*)	**75.** (*d*)	**76.** (*b*)	**77.** (*a*)	**78.** (*b*)	**79.** (*a*)	**80.** (*a*)
81. (*a*)	**82.** (*a*)	**83.** (*e*)	**84.** (*d*)	**85.** (*a*)	**86.** (*b*)	**87.** (*c*)	**88.** (*c*)
89. (*a*)	**90.** (*d*)	**91.** (*c*)	**92.** (*b*)	**93.** (*a*)	**94.** (*c*)	**95.** (*d*)	**96.** (*c*)
97. (*b*).							

HINTS/SOLUTIONS

QUESTION BANK (Level I)

1. (*a*) $Y = \dfrac{\frac{F}{a}}{\frac{\Delta l}{l}} = \dfrac{Fl}{a\,\Delta l}$

or $Y = \dfrac{Fl \times 4}{\pi D^2 \times \Delta l}$ or $\Delta l \propto \dfrac{1}{D^2}$ or $\dfrac{\Delta l_2}{\Delta l_1} = \dfrac{D_1^2}{D_2^2} = \dfrac{n^2}{1}$

2. (*c*) $\Delta l = \dfrac{Fl}{aY}$

$\Delta l = \dfrac{2 \times 10^5 \times 10^{-5} \times l}{2 \times 10^{-4} \times 10^4}$ or $\Delta l = i$

So, new length is $2l$.

3. (*d*) Diameter doubled, cross-sectional area quadrupled ; Breaking strength ∝ Cross-sectional area.

> ☞ *In this problem, the term 'breaking strength' has been used in the sense of 'force'.*

4. (*c*) At A, P.E. is positive. So, force is repulsive. At C, P.E. is negative. So, force is attractive. At B, slope of graph is zero. $\left[F = -\dfrac{dU}{dr}\right]$

5. (*c*) The density would increase by 0.1% if the volume decreases by 0.1%.

$K = \dfrac{\Delta P}{\Delta V/V}$

$\Delta P = K\dfrac{\Delta V}{V} = 2 \times 10^9 \times \dfrac{0.1}{100} = 2 \times 10^6\ \text{N m}^{-2}$

6. (*a*) $F = YA\alpha t$; $\dfrac{F_{Cu}}{F_{Fe}} = \dfrac{\alpha_{Cu}}{\alpha_{Fe}} = \dfrac{3}{2}$

7. (*d*) $Y = \dfrac{Fl}{a\Delta l} = \left(\dfrac{F}{\Delta l}\right)\dfrac{l}{a} = \dfrac{Kl}{a}$; Kl = constant ;

$K \times 3 = K' \times 2$ or $K' = \dfrac{3K}{2}$

8. (*b*) $y = \dfrac{WL^3}{12Y\pi r^4}$ or $y \propto \dfrac{1}{Y}$

9. (*c*) $Y = \dfrac{Fl}{a\,\Delta l}$ or $\Delta l \propto \dfrac{1}{a}$; $\Delta l \propto \dfrac{1}{D^2}$.

$\dfrac{\Delta l_2}{\Delta l_1} = \dfrac{D_1^2}{D_2^2} = 4$ or $\Delta l_2 = 4\Delta l_1 = 4$ cm

10. (*c*) Breaking force ∝ cross-sectional area. Area is quadrupled.

11. (*c*) Breaking force = Breaking stress × area

$F \propto r^2$

When r is doubled, F increases four times.

12. (*c*) Maximum load is not related to length.

13. (*a*) Breaking strength = Breaking stress × $\dfrac{\pi D^2}{4}$

Breaking stress is unchanged.

D is halved. So, breaking strength becomes one-fourth *i.e.*, $\dfrac{1}{4} \times 1.5 \times 10^5$ N or 0.375×10^5 N.

14. (*a*) Energy density $= \dfrac{1}{2} \times$ stress × strain

$= \dfrac{1}{2} \times \dfrac{F}{A} \times \dfrac{\Delta l}{l} = \dfrac{F\Delta l}{2Al}$

15. (*a*) Energy density $= \dfrac{1}{2} \times$ stress × strain

$Y = \dfrac{\text{stress}}{\sigma}$ or stress $= Y\sigma$

∴ Energy density $= \dfrac{1}{2}Y\sigma \times \sigma = \dfrac{Y\sigma^2}{2}$

16. (*a*) $\tau = \frac{\pi \eta r^4}{2l} \theta \quad \therefore \quad \tau \propto r^4$

17. (*c*) Strain $= \frac{\Delta l}{l} = \frac{l}{l} = 1$

$\therefore$ $Y = \text{stress} = \frac{2 \times 10^3 \text{ N}}{2 \times 10^{-4} \text{ m}^2} = 10^7 \text{ N m}^{-2}$

18. (*a*) 21×10^{11} dyne cm^{-2}

$= 21 \times 10^{11} \times 10^{-5} \text{ N } (10^{-2} \text{ m})^{-2} = 21 \times 10^{10} \text{ N m}^{-2}$

$= 210 \times 10^9 \text{ N m}^{-2} = 210$ GPa

19. (*c*) Breaking stress does not depend upon the length of the cable.

20. (*d*) Force = Stress × Area

$= 170 \times 10^6 \times 6 \times 10^{-4} \text{ N} = 1.02 \times 10^5 \text{ N}$

21. (*e*) $E = \frac{FL}{\pi r^2 \Delta L}$ or $\Delta L = \frac{FL}{\pi r^2 E}$

Clearly, $\Delta L \propto L$.

QUESTION BANK (Level II)

1. (*b*) Energy stored $= \frac{1}{2} F \times \Delta l$

$= \frac{1}{2} \times 10 \times 10 \times 10^{-3} \text{ J} = 5 \times 10^{-2} \text{ J} = 0.05 \text{ J}$

2. (*d*) $Y = \frac{F/a}{\Delta l/l} = \frac{Fl}{a \, \Delta l}$ or $\Delta l \propto \frac{l}{D^2}$

(*a*) $\frac{100}{1^2} = 100$ (*b*) $\frac{200}{4} = 50$

(*c*) $\frac{300}{9} = 33.33$ (*d*) $\frac{50}{(1/2)^2} = 200$

3. (*d*) $Y = \frac{Fl}{A\Delta l}$ or $F = \frac{YA\,\Delta l}{l}$

or $F = \frac{2.2 \times 10^{11} \times 2 \times 10^{-6} \times 0.5 \times 10^{-3}}{2} \text{ N} = 1.1 \times 10^2 \text{ N}$

4. (*a*) For the same material, breaking stress is same.

5. (*a*) $\tau = \frac{22 \times 5 \times 10^{10} (2 \times 10^{-3})^4 \times 22 \times 45}{7 \times 7 \times 180 \times 2 \times 2} \text{ N m}$

$= 0.49$ N m

6. (*d*) Stress = Y × Strain

$= 2 \times 10^{11} \times 0.15 \text{ N m}^{-2} = 3 \times 10^{10} \text{ N m}^{-2}$

7. (*b*) $\frac{3}{\eta} + \frac{1}{K} = \frac{9}{Y}$

$\frac{1}{K} = \frac{9}{Y} - \frac{3}{\eta}$ or $\frac{1}{K} = \frac{9}{3\eta} - \frac{3}{\eta} = 0 \Rightarrow K = \infty$

8. (*c*) Tension $= YA\alpha t$

$= 2 \times 10^{11} \times 10^{-6} \times 1.1 \times 10^{-5} \times 20 \text{ N} = 44 \text{ N}$

9. (*c*) $K = \frac{\Delta P}{\Delta V/V}$ or $\frac{1}{K} = \frac{\Delta V}{V\Delta P}$

or $\Delta V = \frac{1}{K} V\Delta P$

$= 4 \times 10^{-5} \times 100 \times 100 \text{ cm}^3$

$= 4 \times 10^{-1} \text{ cm}^3 = 0.4 \text{ cm}^3$

10. (*d*) $Y = \frac{F \times 4 \times l}{\pi D^2 \, \Delta l}$

In the given problem, $F \propto D^2$. Since D is increased by a factor of 4 therefore F is increased by a factor of 16.

11. (*a*) Energy density $= \frac{1}{2}$ stress × strain

$= \frac{1}{2} \text{ stress} \times \frac{\text{stress}}{Y} = \frac{(\text{stress})^2}{2Y} \propto \frac{1}{D^4}$

Now, $\frac{u_A}{u_B} = \frac{D_B^4}{D_A^4} = [2]^4 = 16$

12. (*a*) Isothermal elasticity = P, Adiabatic elasticity = γP

$\therefore \quad \frac{E_\theta}{E_\phi} = \frac{1}{\gamma} \cdot \gamma > 1 \quad \therefore \quad \frac{E_\theta}{E_\phi} < 1$

13. (*d*) $\frac{\pi \eta r^4}{2l} (\theta - \theta') = \frac{\pi\eta \left(\frac{r}{2}\right)^4}{2\,(l/2)} \theta'$

or $\theta - \theta' = \frac{\theta'}{8}$

or $8\theta - 8\theta' = \theta'$

or $9\theta' = 8\theta$

or $\theta' = \frac{8\theta}{9}$

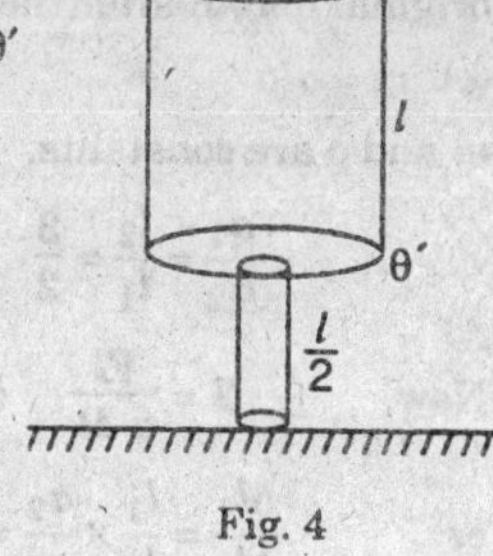

Fig. 4

14. (*d*) $\tau_x = \frac{\pi\eta r^4}{2l} \theta_x$ and $\tau_y = \frac{\pi\eta (2r)^4}{2l} \theta_y$

Since $\tau_x = \tau_y$,

$\therefore \quad \theta_x = 16\,\theta_y$ or $\frac{\theta_x}{\theta_y} = 16$

15. (*d*) $W = \frac{1}{2} F \Delta l$

$W = \frac{1}{2} \times \frac{Y\pi r^2 \Delta l}{l} \Delta l$

or $W = \frac{Y\pi r^2 \Delta l^2}{2l}$

$Y = \frac{Fl}{\pi r^2 \Delta l}$

$F = \frac{Y\pi r^2 \Delta l}{l}$

or $W \propto \frac{r^2}{l}$, $W' \propto \frac{(2r)^2 2}{l}$

$\frac{W'}{W} = 8$ or $W' = 8 \times 2$ J= 16 J

16. (*c*) Thermal stress = $Y\alpha t$

In the given problem,

$Y\alpha$ = constant

$\frac{Y_1}{Y_2} = \frac{\alpha_2}{\alpha_1} = \frac{3}{2}$

17. (*c*) $Y = \frac{Fl}{a\Delta l}$ or $\Delta l \propto \frac{1}{a}$

Again, $m = al\rho$ or $m \propto a$

$\therefore$ $\Delta l \propto \frac{1}{m}$

$\frac{\Delta l_1}{\Delta l_2} = \frac{m_2}{m_1} = \frac{2}{3}$

18. (*d*) Mean radius is $\sqrt{r_1 r_2}$.

19. (*b*) $\theta = \frac{r}{l}\phi = \frac{0.4}{100} \times 30° = 0.12°$

20. (*d*) $Y = \frac{Fl}{a\,\Delta l}$

In the given problem, Y, l and Δl are constants.

$\therefore$ $F \propto a$

or $F \propto \pi r^2$ or $F \propto r^2$ or $\frac{F_1}{F_2} = \frac{r_1^2}{r_2^2} = \frac{1}{4}$

21. (*a*) Note that A is stretched the least and recovers the original length after the removal of the deforming force.

22. (*c*) $m = al\rho$

m and ρ are constants.

$\therefore$ $\frac{a_1}{a_2} = \frac{l_2}{l_1} = \frac{3}{2}$

Now, $Y = \frac{Fl}{a\,\Delta l}$ or $\Delta l \propto \frac{l}{a}$

or $\frac{\Delta l_1}{\Delta l_2} = \frac{l_1}{l_2} \times \frac{a_2}{a_1} = \frac{2}{3} \times \frac{2}{3}$ or $\frac{\Delta l_1}{\Delta l_2} = \frac{4}{9}$

23. (*a*) $Y = \frac{Fl}{a\,\Delta l}$ or $\Delta l \propto \frac{F}{r^2}$

or $\frac{\Delta l_2}{\Delta l_1} = \frac{F_2}{F_1} \times \frac{r_1^2}{r_2^2}$

or $\frac{\Delta l_2}{\Delta l_1} = 2 \times 2 \times 2 = 8$

or $\Delta l_2 = 8\Delta l_1 = 8 \times 1$ mm = 8 mm

24. (*b*) $Y = \frac{Mg \times 4 \times l}{\pi D^2 \times \Delta l}$ or $\Delta l \propto \frac{1}{D^2}$

When D is doubled, Δl becomes one-fourth, *i.e.*, $\frac{1}{4} \times 2.4$ cm, *i.e.*, 0.6 cm.

25. (*a*) $\eta = \frac{F}{A\theta} = \frac{5 \times 10^5 \text{ N}}{100 \times 10^{-4} \times 0.001 \text{ m}^2} = 5 \times 10^{10} \text{ N m}^{-2}$

26. (*b*) $k = \frac{10 \text{ N}}{40 \times 10^{-3} \text{ m}} = \frac{1000}{4} \text{ N m}^{-1} = 250 \text{ N m}^{-1}$

Spring constant of combination $= \frac{250}{2} \text{ N m}^{-1} = 125 \text{ N m}^{-1}$

Energy $= \frac{1}{2} \times 125 \times (40 \times 10^{-3})^2 \text{ J} = 0.1$ J

27. (*b*) $F = \frac{Ya\,\Delta l}{l}$

$= 9 \times 10^{10} \times \frac{22}{7} \times \frac{(0.6 \times 10^{-3})^2}{4} \times \frac{0.2}{100} \text{ N} \approx 51$ N

28. (*b*) $\eta = \frac{Fl}{A\Delta l} = \frac{Fl}{l^2 \Delta l} = \frac{F}{l\,\Delta l}$ or $\Delta l \propto \frac{1}{l}$

If l is halved, then Δl is doubled.

29. (*c*) $Y = \frac{Fl}{\pi r^2 \Delta l}$ or $\Delta l = \frac{Fl}{\pi r^2 Y}$

$\Delta l \propto \frac{l}{r^2}$, $\Delta l' \propto \frac{2l}{(\sqrt{2}r)^2}$ or $\Delta l' \propto \frac{l}{r^2}$

$\therefore$ $\frac{\Delta l}{\Delta l'} = 1$

30. (*a*) Work done $= \frac{1}{2} F\Delta l$

$= \frac{1}{2} \frac{Y\,a\,\Delta l^2}{l}$ $\quad\left| \; Y = \frac{Fl}{a\,\Delta l} \right.$

$= \frac{2 \times 10^{11} \times 10^{-6} (2 \times 10^{-3})^2}{2 \times 1}$ J $\quad\left| \text{ or } F = \frac{Ya\Delta l}{l} \right.$

$= 4 \times 10^{-1}$ J = 0.4 J

31. (*b*) $Y = \frac{F/a}{\Delta l / l}$ or $Y = \frac{Fl}{a\,\Delta l}$ or $\Delta l = \frac{Fl}{\sigma Y} = \frac{Fl}{\pi r^2 Y}$

In the given problem, $\Delta l \propto \frac{l}{r^2}$; When both l and r are doubled, Δl is halved.

32. (*a*) Energy/volume $= \frac{1}{2} \times$ stress $\times$ strain

$= \frac{1}{2} Y \times \text{strain} \times \text{strain} = \frac{1}{2} Y \times \text{strain}^2$

$= \frac{1}{2} \times 2 \times 10^{10} \times 0.06 \times 10^{-2} \times 0.06 \times 10^{-2} = 3600 \text{ J m}^{-3}$

33. (*d*) $Y = \frac{F/a}{\text{Breaking strain}}$

or $a = \frac{F}{Y \times \text{Breaking strain}} = \frac{10^4 \times 100}{7 \times 10^9 \times 0.2}$

$= 0.71 \times 10^{-3} = 7.1 \times 10^{-4}$

34. (b) $Y = \frac{FL}{\pi r^2 l}$ or $l = \frac{FL}{\pi r^2 Y}$ or $l \propto \frac{FL}{r^2}$

Both F and L are quadrupled. Also, r is quadrupled. So, l remains unchanged.

35. (b) $K = \frac{P}{\frac{\Delta V}{V}}$ or $\frac{1}{K} = \frac{\Delta V/V}{P}$ or $\sigma = \frac{\Delta V}{PV}$ or $\Delta V = \sigma PV$

36. (a) $Y = \frac{Fl}{a\,\Delta l} = \frac{(ml\omega^2)l}{a\,\Delta l}$ or $Y = \frac{ml^2\omega^2}{a\,\Delta l}$

or $Y = \frac{1 \times 1 \times 1 \times 20 \times 20}{10^{-6} \times 10^{-3}} = 4 \times 10^{11}$ N m^{-2}

37. (b) $\Delta l = \frac{Fl}{\pi r^2 Y}$ or $\frac{\Delta l}{l} \propto \frac{1}{r^2 Y}$

$$\frac{\left(\frac{\Delta l}{l}\right)_s}{\left(\frac{\Delta l}{l}\right)_{Cu}} = \frac{r_{Cu}^2 Y_{Cu}}{r_s^2 Y_s} = \frac{4r_s^2 \frac{Y_s}{2}}{r_s^2 Y_s} = 2$$

$$\therefore \left(\frac{\Delta l}{l}\right)_s = 2\left(\frac{\Delta l}{l}\right)_{Cu}$$

38. (b) $\eta = \frac{Y}{2(1+\sigma)}$ or $\eta = \frac{2.4\eta}{2(1+\sigma)}$

or $1 + \sigma = 1.2$ or $\sigma = 0.2$

39. (a) Strain $= \frac{\Delta l}{l} = \frac{l\alpha t}{l} = \alpha t = 12 \times 10^{-6} \times 30 = 36 \times 10^{-5}$

40. (c) $W \propto \Delta l^2$

Since Δl is increased by a factor of 3, therefore, work is increased by a factor of 9.

> ☞ $W = \frac{1}{2} F\Delta l$
>
> $W = \frac{1}{2}\left[\frac{Ya\,\Delta l}{l}\right]\Delta l$ or $W = \frac{1}{2}\frac{Y a\,\Delta l^2}{l}$

41. (c) $\Delta P = 100$ atm $= 100 \times 10^6$ dyne cm^{-2} $= 10^8$ dyne cm^{-2}

$\frac{\Delta V}{V} = \frac{0.01}{100} = 10^{-4}$

$K = \frac{10^8}{10^{-4}}$ dyne cm^{-2} $= 10^{12}$ dyne cm^{-2}

42. (a) $\theta = \frac{r}{l}\phi = \frac{2 \times 10^{-3}}{1} \times 45° = 0.09°$

43. (c) Force = weight suspended + weight of $\frac{3L}{4}$ of wire

$= W_1 + \frac{3W}{4}$

stress $= \frac{\text{force}}{\text{area}}$

44. (a) $Y = \frac{FL}{A\Delta L}$ or $F = \frac{YA\,\Delta L}{L}$

Work dcne $= \frac{1}{2} F\Delta L$

$= \frac{1}{2}\frac{YA(\Delta L)^2}{L} = \frac{YA(\Delta L)^2}{2L}$

$= \frac{2 \times 10^{11} \times 10^{-6} \times 10^{-6}}{2 \times 1}$ J $= 0.1$ J

45. (b) $K = \frac{PV}{\Delta V} = \frac{PV}{\gamma VT} = \frac{P}{3\alpha T}$ or $T = \frac{P}{3K\alpha}$

46. (b) $\frac{1}{100} \times 0.8 = 2\phi$ or $\phi = \frac{0.4}{100} = 0.004$ radian

47. (c) $10^6 = h_{max} \times \rho \times g$

or $10^6 = 5 \times \rho \times 10$ or $\rho = \frac{10^5}{5}$ kg m^{-3}

or $\rho = 2 \times 10^4$ kg m^{-3}

48. (d) Breaking force does not depend upon length.

Breaking force = Breaking stress × Area of cross-section

For a given material, breaking stress is constant.

$\therefore$ Breaking force $\propto$ Area of cross-section

$$\frac{F_2}{F_1} = \frac{A_2}{A_1} = \frac{\pi(6r)^2}{\pi r^2} = 36$$

or $F_2 = 36\, F_1 = 36$ F

49. (a) $\eta = \frac{10^{11}}{2(1+0.2)} = \frac{10^{11}}{2.4} = 0.42 \times 10^{11}$ N m^{-2}

50. (b) $K = \frac{4 \times 9.8\text{ N}}{2 \times 10^{-2}\text{ m}}$ or $K = 19.6 \times 10^2$ N m^{-1}

Work done $= \frac{1}{2} \times 19.6 \times 10^2 \times (5 \times 10^{-2})^2$ J $= 2.45$ J

51. (d) 10 m column of water exerts nearly 1 atmosphere pressure. So, 100 m column of water exerts nearly 10 atmosphere pressure, *i.e.*, 10×10^5 Pa or 10^6 Pa.

Now, $K = \frac{(\Delta P)V}{\Delta V} = \frac{10^6 \times 100}{0.1}$ Pa $= 10^9$ Pa

52. (a) $\beta = \frac{PV}{v}$; $\alpha = \frac{v}{Vt}$ or $v = \alpha Vt$

$\therefore$ $\beta = \frac{PV}{\alpha Vt}$ or $t = \frac{P}{\alpha\beta}$

Note carefully the difference in Qs. 45 and 52.

53. (b) Change in length of ring $= 2\pi R - 2\pi r$

Longitudinal strain $= \frac{2\pi R - 2\pi r}{2\pi r} = \frac{R - r}{r}$

$E = \frac{F/A}{\frac{R-r}{r}}$ or $E = \frac{Fr}{A(R-r)}$ or $F = \frac{EA(R-r)}{r}$

54. (*b*) $\frac{1}{K} = \frac{\Delta V/V}{\Delta P}$ or $\frac{\Delta V}{V} = \Delta P\left[\frac{1}{K}\right]$

or $\frac{\Delta V}{V} \times 100 = 10^5 \times 8 \times 10^{-12} \times 100 = 8 \times 10^{-5}$

55. (*b*) Energy per unit volume $= \frac{1}{2} \times$ stress $\times$ strain

$= \frac{1}{2} \times \text{stress} \times \frac{\text{stress}}{Y}$ | $Y = \frac{\text{stress}}{\text{strain}}$

$= \frac{S^2}{2Y}$

56. (*c*) $Y = \frac{Fl}{a\Delta l}$; Y, l and a are constants

$\therefore \; F \propto \Delta l$

In the first case, $V\rho g \propto 16$

In the second case,

$(V\rho g - V \times 1 \times g) \propto 14$

Dividing, $\frac{Vg(\rho - 1)}{Vg\rho} = \frac{14}{16} = \frac{7}{8}$

or $\frac{\rho - 1}{\rho} = \frac{7}{8}$ or $8\rho - 7\rho = 8$ or $\rho = 8 \text{ g cm}^{-3}$

57. (*c*) Breaking stress $= \frac{\text{Maximum weight}}{\text{area of cross-section}}$

$10^6 = \frac{al\rho g}{a} = l\rho g$

or $l = \frac{10^6}{\rho g} = \frac{10^6}{4 \times 10^3 \times 10} \text{ m} = 25 \text{ m}$

58. (*c*) $\frac{\text{Stress}}{\frac{1}{10} \times \frac{1}{100}} = 12 \times 10^{10}$

or Stress $\times 1000 = 12 \times 10^{10}$

or Stress $= 12 \times 10^7 \text{ N m}^{-2}$

59. (*d*) Shearing strain $= \frac{0.02 \times 10^{-2}}{0.1} = 0.002$

60. (*a*) $mr\omega^2$ = Breaking stress × cross-sectional area

$10 \times 0.3\omega^2 = 4.8 \times 10^7 \times 10^{-6} = 48$

or $\omega^2 = \frac{48}{3} = 16$ or $\omega = 4 \text{ rad s}^{-1}$

61. (*a*) $\tau = \frac{\pi \eta r^4}{2l}\theta$

In the given problem, $r^4\theta$ = constant

$\therefore \; \frac{\theta_A}{\theta_B} = \frac{r_2^4}{r_1^4}$

62. (*c*) Work done $= \frac{1}{2}$ F × extension

$= \frac{1}{2} \times \frac{YA}{L} \times 1$ | $Y = \frac{F \times L}{A \times 1}$

$= \frac{YA}{2L}$ | $F = \frac{YA}{L}$

63. (*e*) Out of the given materials, copper has highest value of Young's modulus. It exhibits large plastic deformation between its yield point and fracture point. Glass is very hard but brittle material. It fractures almost immediately after the elastic stage ; little or no plastic deformation occurs. Rubber is capable of regaining its original shape after a very large deformation.

64. (*b*) $W = \frac{1}{2}\tau\phi = \frac{1}{2} \times 8 \times \frac{\pi}{4} = \pi$ joule.

> ***180° corresponds to π radian.***

65. (*a*) $W = \frac{F^2 l}{2aY}$ [Refer to Synopsis]

Y, a and F are constants.

$\therefore \; W \propto l$

66. (*c*) $V = \pi r^2 l$

$\frac{\Delta V}{V} = \frac{\Delta(\pi r^2 l)}{\pi r^2 l}$ or $\frac{\Delta V}{V} = \frac{r^2\Delta l + 2rl\Delta r}{r^2 l}$

$\frac{\Delta V}{V} = \frac{\Delta l}{l} + \frac{2\Delta r}{r}$

But $\frac{\Delta V}{V} = 0$ $\quad \therefore \; \frac{\Delta l}{l} = -\frac{2\Delta r}{r}$

Now, $\sigma = -\frac{\Delta r/r}{\frac{\Delta l}{l}} = -\frac{\Delta r/r}{-2\frac{\Delta r}{r}} = 0.5$

67. (*b*) $Y = \frac{Fl}{a\,\Delta l}$

Y, F and l are constants.

$\therefore \; \frac{\Delta l_2}{\Delta l_1} = \frac{a_1}{a_2} = \frac{4}{8} = \frac{1}{2}$

or $\Delta l_2 = \frac{\Delta l_1}{2} = \frac{0.1}{2}$ mm = 0.05 mm

68. (*d*) $Y = \frac{Fl}{a\,\Delta l}$

Y, l and F are constants.

$\therefore \; \Delta l \propto \frac{1}{D^2}$

$\frac{\Delta l_2}{\Delta l_1} = \frac{D_1^2}{D_2^2} = \frac{1}{16}$

$\therefore \; \Delta l_2 = \frac{1}{16}$ mm

69. (*a*) $W = \dfrac{F^2 l}{2\left(\dfrac{\pi D^2}{4}\right)Y}$

Y, l and F are constants.

$\therefore \quad W \propto \dfrac{1}{D^2}$

$\therefore \quad \dfrac{W_1}{W_2} = \dfrac{D_2^2}{D_1^2} = 16 \quad ...(1)$

Now, $W_1 = \dfrac{1}{2} \times 10^3 \times 1 \times 10^{-3} = 0.5$ J

$W_2 = \dfrac{1}{2} \times 10^3 \times \dfrac{10^{-3}}{16} = \dfrac{1}{32} = 0.03125$

Again, $\dfrac{W_1}{W_2} = \dfrac{0.5}{0.03125} = 16 \quad ...(2)$

Answer is confirmed by comparing (1) and (2).

70. (*b*) $10^6 = \dfrac{LAdg}{A}$

$\therefore \quad L = \dfrac{10^6}{3 \times 10^3 \times 9.8}$ m $= \dfrac{1000}{3 \times 9.8}$ m $= 34.01$ m

71. (*a*) $Y = \dfrac{FV}{a^2 \Delta l}$

$\Delta l \propto \dfrac{1}{a^2}$ or $\Delta l \propto \dfrac{1}{D^4}$

$\dfrac{\Delta l_A}{\Delta l_B} = \dfrac{D_B^4}{D_A^4} = \dfrac{1^4}{\left(\dfrac{1}{2}\right)^4} = 16$

> ☞ $Y = \dfrac{F}{a} \times \dfrac{l}{\Delta l}$
>
> *Now,* $V = al$ *or* $l = \dfrac{V}{a} \quad \therefore \quad Y = \dfrac{FV}{a^2 \Delta l}$

72. (*b*) $\Delta P = h\rho g = 200 \times 10^3 \times 10$ N m^{-2}

$= 2 \times 10^6$ N m^{-2}

$K = \dfrac{\Delta P}{\dfrac{\Delta V}{V}} = \dfrac{2 \times 10^6}{\dfrac{0.1}{100}}$

$= \dfrac{2 \times 10^8}{0.1}$ N m^{-2} $= 2 \times 10^9$ N m^{-2}

73. (*b*) $\sigma = \dfrac{\text{lateral strain}}{\text{longitudinal strain}}$

or lateral strain = $\sigma \times$ longitudinal strain

$= 0.4 \times \dfrac{0.05}{100} = \dfrac{0.02}{100}$

So, percentage reduction in diameter is 0.02

74. (*c*) $Y = \dfrac{\text{stress}}{\text{strain}}$ or strain $= \dfrac{\text{stress}}{Y}$ or $\dfrac{\Delta L}{L} = \dfrac{\text{stress}}{Y}$,

Since cross-sections are equal and same tension exists in both the wires, therefore, the stresses developed are equal. Also, ΔL is given to be the same for both the wires.

$\therefore \quad L \propto Y$

$\therefore \quad \dfrac{L_s}{L_{Cu}} = \dfrac{2 \times 10^{11}}{1.1 \times 10^{11}} = \dfrac{20}{11}$

75. (*d*) $\dfrac{20a \times 2.5 \times 10^3 \times 10 + 5 \times 10^5}{a} = 1.6 \times 10^6$

or $\quad 500a + 500 = 1600a$

or $\quad 1100a = 500$ or $a = \dfrac{5}{11}$ m^2 $= 0.45$ m^2

76. (*b*) Stress $= \dfrac{100 \text{ N}}{10^{-6} \text{ m}^2} = 10^8$ N m^{-2}

Strain $= \dfrac{2 \times 10^{-3}}{2} = 10^{-3}$

Young's Modulus $= \dfrac{10^8}{10^{-3}}$ N m^{-2} $= 10^{11}$ N m^{-2}

Energy stored $= \dfrac{1}{2} \times 100 \times 2 \times 10^{-3}$ J $= 10^{-1}$ J $= 0.1$ J

77. (*a*) $\eta = \dfrac{Y}{2(1+\sigma)}, \quad \sigma = 0$

$\therefore \quad \eta = \dfrac{Y}{2} = \dfrac{6 \times 10^{12}}{2}$ N m^{-2} $= 3 \times 10^{12}$ N m^{-2}

78. (*b*) $\dfrac{E_a}{E_i} = \dfrac{\gamma P}{P} = \gamma$

> ***For non-linear triatomic molecule,*** $\gamma = \dfrac{4}{3}$

79. (*a*) Poisson's ratio, $\sigma = \dfrac{\text{Lateral strain}}{\text{Longitudinal strain}}$

$= \dfrac{\dfrac{\Delta D}{D}}{\dfrac{\Delta L}{L}} = \dfrac{\dfrac{\Delta r}{r}}{\dfrac{\Delta L}{L}}$

or $\quad \dfrac{\Delta r}{r} = \sigma \times \dfrac{\Delta L}{L} = 0.5 \times 2 \times 10^{-3} = 10^{-3}$

Volume of rod, $V = \pi r^2 L$

$V + \Delta V = \pi(r - \Delta r)^2 (L + \Delta L)$

or $\quad V + \Delta V = (\pi r^2 + \pi \Delta r^2 - 2\pi r \Delta r)(L + \Delta L)$

Neglecting Δr^2 and $\Delta r \, \Delta L$, we get

$V + \Delta V = \pi r^2 L - 2\pi r L \, \Delta r + \pi r^2 \, \Delta L$

or $\quad \Delta V = \pi r^2 \Delta L - 2\pi r L \, \Delta r$

or $\frac{\Delta V}{V} = \frac{\pi r^2 \Delta L}{\pi r^2 L} - \frac{2\pi r L \Delta r}{\pi r^2 L}$

or $\frac{\Delta V}{V} = \frac{\Delta L}{L} - 2\frac{\Delta r}{r} = 2 \times 10^{-3} - 2 \times 10^{-3} = 0$

80. (*a*) $Y = \frac{\text{stress}}{\Delta L/L}$ or $\frac{\Delta L}{L} = \frac{\text{stress}}{Y} = \frac{5 \times 10^7}{2 \times 10^{11}} = 2.5 \times 10^{-4}$

Now, $V = \pi r^2 L$

$\frac{\Delta V}{V} = \frac{\pi \Delta (r^2 L)}{\pi r^2 L}$

or $\frac{\Delta V}{V} = \frac{r^2 \Delta L + L \times 2r\Delta r}{r^2 L}$

or $\frac{\Delta V}{V} = \frac{\Delta L}{L} + 2\frac{\Delta r}{r}$

or $2\frac{\Delta r}{r} = \frac{\Delta V}{V} - \frac{\Delta L}{L}$

or $2\frac{\Delta r}{r} = \frac{0.02}{100} - 2.5 \times 10^{-4}$

or $\frac{\Delta r}{r} = 1 \times 10^{-4} - \frac{2.5}{2} \times 10^{-4} = -0.25 \times 10^{-4}$

81. (*a*) Outer radius, $r_2 = \left(8 + \frac{0.04}{2}\right)\text{cm} = 8.02\text{ cm}$

Inner radius, $r_1 = \left(8 - \frac{0.04}{2}\right)\text{cm} = 7.98\text{ cm}$

$C_1 = \frac{\pi\eta\,(r_2^4 - r_1^4)}{2l}$; $C_2 = \frac{\pi\eta r^4}{2l}$

But $\pi\,[r_2^2 - r_1^2]l\rho = \pi r^2 l\rho$

or $r^2 = r_2^2 - r_1^2$

or $r^2 = (r_2 + r_1)(r_2 - r_1)$

or $r^2 = (8.02 + 7.98)(8.02 - 7.98)$

or $r^2 = 16 \times 0.04 = 0.64\text{ cm}$ or $r = 0.8\text{ cm}$

$\frac{C_1}{C_2} = \frac{r_2^4 - r_1^4}{r^4} = \frac{[8.02]^4 - [7.98]^4}{[0.8]^4}$

82. (*a*) $Y = \frac{[mg + ml\omega^2]l}{\pi r^2\,\Delta l}$

or $\Delta l = \frac{m[g + l\omega^2]l}{\pi r^2 Y}$

or $\Delta l = \frac{1\,[10 + 2 \times 4\pi^2 \times 4]2}{\pi\,(1 \times 10^{-3})^2 \times 2 \times 10^{11}}\text{ m}$

or $\Delta l = \frac{[20 + 64 \times 9.88]7}{2 \times 22 \times 10^5}\text{ m}$

$= \frac{4566.24}{44 \times 10^5} \times 10^3\text{ mm} \approx 1\text{ mm}$

83. (*e*) Stress $= \frac{840\text{ N}}{10^{-6}\text{m}^2} = 84 \times 10^7\text{ N m}^{-2}$

Strain $= \frac{\text{Stress}}{Y} = \frac{84 \times 10^7\text{ N m}^{-2}}{12 \times 10^{10}\text{ N m}^{-2}} = 7 \times 10^{-3}$

Volume $= 1 \times 10^{-6}\text{ m}^3 = 10^{-6}\text{ m}^3$

Energy $= \frac{1}{2} \times \text{Stress} \times \text{Strain} \times \text{Volume}$

$= \frac{1}{2} \times 84 \times 10^7 \times 7 \times 10^{-3} \times 1 \times 10^{-6}\text{ J} = 2.94\text{ J}$

Now, $\theta = \frac{2.94\text{ J}}{10^{-6} \times 7000 \times 420} = 1°\text{C}$

84. (*d*) Young's modulus depends upon the nature of material and not on the radii of the wires.

85. (*a*) $4 \propto (a - l)$ ...(1)

$5 \propto (b - l)$ ...(2)

$9 \propto (L - l)$ ...(3)

Dividing (1) by (2), we get

$\frac{4}{5} = \frac{a - l}{b - l}$

or $4b - 4l = 5a - 5l$ or $l = 5a - 4b$

Again, dividing (1) by (3),

$\frac{4}{9} = \frac{a - l}{L - l}$ or $4L - 4l = 9a - 9l$

or $4L = 9a - 5l$ or $4L = 9a - 5(5a - 4b)$

or $4L = 9a - 25a + 20b$

or $4L = 20b - 16a$ or $L = 5b - 4a$

86. (*b*) $E = \frac{1}{2}\frac{Ya\,\Delta l^2}{l}$

But $m = ald$ or $a = \frac{m}{ld}$

$\therefore$ $E = \frac{Ym\,\Delta l^2}{2l^2 d}$

E in calorie $= \frac{Ym\,\Delta l^2}{2l^2 d\,J}$

Now, $mS\theta = \frac{Ym\,\Delta l^2}{2l^2 dJ}$ or $\theta = \frac{Y\Delta l^2}{2l^2 d\,JS}$

or $\theta = \frac{12 \times 10^{11} \times 10^{-1} \times 10^{-3} \times 10^{-3}}{2 \times 2 \times 2 \times 9 \times 10^3 \times 4.2 \times 0.1 \times 10^3}$

$= \frac{12 \times 10^5}{72 \times 42 \times 10^5} = \frac{1}{252}°\text{C}$

87. (*c*) $\Delta l = \frac{4Fl}{\pi D^2 Y}$

$= \frac{4 \times 30 \times 2 \times 7}{22 \times (3 \times 10^{-3})^2 \times 1.1 \times 10^{11}}\text{ m}$

$= 7.7 \times 10^{-5}$ m = 0.077 mm

≈ 0.08 mm

88. (*c*) $K = \frac{1.5 \text{ N}}{30 \times 10^{-3} \text{ m}} 50 \text{ N m}^{-1}$

$l = \frac{0.20 \times 10}{50}$ m = 0.04 m

Energy stored $= \frac{1}{2} \times 0.20 \times 10 \times 0.04$ J = 0.04 J

89. (*a*) If V is the volume of load and ρ its relative density

$$Y = \frac{FL}{A\, l_a} = \frac{V\rho\, gL}{A l_a} \quad ...(1)$$

When the load is immersed in liquid,

$$Y = \frac{(V\rho g - V \times 1 \times g)L}{A l_w} \quad ...(2)$$

Equating, $\frac{\rho}{l_a} = \frac{\rho - 1}{l_w}$

or $\rho\,(l_w - l_a) = -l_a$ or $\rho = \frac{l_a}{l_a - l_w}$

90. (*d*) $K = \frac{Y}{3(1-2\sigma)}$ or $11 \times 10^{10} = \frac{7 \times 10^{10}}{3(1-2\sigma)}$

or $\frac{7}{33} = 1 - 2\sigma$ or $2\sigma = 1 - \frac{7}{33}$

or $\sigma = \frac{26}{66} = 0.39$

91. (*c*) T = 1000 [9.8 + 1.2] N = 11000 N

Maximum stress $= \frac{11000 \times 7}{22 \times r^2_{min.}}$

or $r^2_{min.} = \frac{11000 \times 7}{22 \times 1.4 \times 10^8} = \frac{50^2}{(10^4)^2}$

or $r_{min.} = 50 \times 10^{-4}$ m

$D_{min.} = 100 \times 10^{-4}$ m = 0.01 m

92. (*b*) $Y = \frac{F}{\pi R^2} \times \frac{l}{\Delta l}$

F, l and Δl are constants.

∴ $R^2 \propto \frac{1}{Y}$

$\frac{R_S^2}{R_B^2} = \frac{Y_B}{Y_S} = \frac{10^{11}}{2 \times 10^{11}} = \frac{1}{2}$

or $\frac{R_S}{R_B} = \frac{1}{\sqrt{2}}$ or $R_S = \frac{R_B}{\sqrt{2}}$

93. (*a*) $Y \propto F$

$\frac{F_{Cu}}{F_{Fe}} = \frac{Y_{Cu}}{Y_{Fe}} = \frac{1}{3}$

94. (*c*) $\frac{\rho'}{\rho} = \frac{\frac{M}{V - \Delta V}}{M/V} = \frac{V}{V - \Delta V}$

or $\frac{\rho'}{\rho} = V(V - \Delta V)^{-1}$

or $\frac{\rho'}{\rho} = 1 + \frac{\Delta V}{V}$ [Using Binomial Theorem]

or $\frac{\Delta V}{V} = \frac{\rho'}{\rho} - 1$ or $\frac{\Delta V}{V} = \frac{\rho' - \rho}{\rho}$

Again, $K = \frac{\Delta P}{\frac{\Delta V}{V}} = \frac{\rho\, \Delta P}{\rho' - \rho}$ or $\rho' - \rho = \frac{\rho\, \Delta P}{K}$

But $\Delta P = P$ (given)

∴ $\rho' - \rho = \frac{\rho P}{K}$

95. (*d*) $Y = \frac{Mg}{A} \times \frac{L/2}{\Delta L}$

(Length is taken as $\frac{L}{2}$ because weight acts at C.G.)

Now, $M = AL\rho$

[For the purpose of calculation of mass, the whole of geometrical length L is to be considered]

∴ $Y = \frac{AL\, \rho\, gL}{2A\, \Delta L}$

or $\Delta L = \frac{\rho g L^2}{2Y} = \frac{1.5 \times 10^3 \times 10 \times 8 \times 8}{2 \times 5 \times 10^6}$ m

$= 9.6 \times 10^{-2}$ m $= 9.6 \times 10^{-2} \times 10^3$ mm = 96 mm

96. (*c*) $Y = \frac{Fl}{a \Delta l}$

Y, l and a are constants

∴ $\frac{F}{\Delta l}$ = constant or $\Delta l \propto F$

Now, $l_1 - l \propto T_1$ and $l_2 - l \propto T_2$

Dividing, $\frac{l_1 - l}{l_2 - l} = \frac{T_1}{T_2}$

or $l_1 T_2 - l T_2 = l_2 T_1 - l T_1$ or $l\,(T_1 - T_2) = l_2 T_1 - l_1 T_2$

or $l = \frac{l_2 T_1 - l_1 T_2}{T_1 - T_2}$ or $l = \frac{l_1 T_2 - l_2 T_1}{T_2 - T_1}$

97. (*b*) $T = \frac{2 m_1 m_2}{m_1 + m_2} g = \frac{2 \times 1 \times 2}{1 + 2} \times 10 \text{ N} = \frac{40}{3}$ N

If r is the minimum radius, then

breaking stress $= \frac{\frac{40}{3}}{\pi r^2}$ or $\frac{40}{3\pi} \times 10^6 = \frac{40}{3\pi r^2}$

or $r^2 = \frac{1}{10^6}$ or $r = \frac{1}{10^3}$ m

or $r = \frac{1}{10^3} \times 10^3$ mm = 1 mm

8.2 FLUIDS AT REST

Important Terms, Facts and Formulae

BASIC CONCEPTS

1. When immiscible liquids of different densities are put in a vessel, the liquid of lowest density settles at the top. The liquid of highest density settles at the bottom. The interfaces are plane.
2. The pressure difference between the hydrostatic pressure and the atmospheric pressure is called gauge pressure.
3. If the weight of a body is less than the upthrust on the body, then the body floats with a part of it outside the liquid.
4. If the body floats completely immersed in a liquid, then the weight of the body is equal to the upthrust on the body. In this case, the apparent weight of the body is zero.
5. If the weight of a body is greater than the upthrust on the body, then the body sinks.

IMPORTANT POINTS TO REMEMBER

1. Relative density of a body

$$= \frac{\text{Weight of body in air}}{\text{Loss of weight of body in water}}$$

$$= \frac{\text{Weight of body in air}}{\text{Upthrust}}$$

$$= \frac{\text{Weight of body in air}}{\text{Weight of equal volume of water}}$$

$$= \frac{\text{Density of body}}{\text{Density of water}}$$

2. Consider a piece of ice which floats in water. When the ice melts completely, the level of water remains unchanged.
3. Consider a piece of ice which floats in a liquid whose density is greater than that of water. When the ice melts completely, the level of (liquid + water) rises.
4. Consider a piece of ice which floats in a liquid whose density is less than that of water. When the ice melts completely, the level of (liquid + water) falls.

QUESTION BANK

Level I

Choose the correct answer from the four alternatives given in each of the following questions :

1. The reading of a spring balance when a block is suspended from it in air is 60 N. This reading is changed to 40 N when the block is submerged in water. The specific gravity of the block is

(*a*) 1 (*b*) 2
(*c*) 3 (*d*) 4.

2. We have two different liquids A and B whose relative densities are 0.75 and 1.0 respectively. If we dip solid objects P and Q having relative densities 0.6 and 0.9 in these liquids, then

(*a*) P floats in A and Q sinks in B
(*b*) P sinks in A and Q floats in B
(*c*) P floats in B and Q sinks in A
(*d*) P sinks in B and Q floats in A.

3. In a hydraulic lift at a service station, the radii of the large and small pistons are in the ratio of 20 : 1. What weight placed on the small piston will be sufficient to lift a car of mass 1200 kg ?

(*a*) 3 kgf (*b*) 30 kgf
(*c*) 300 kgf (*d*) 3000 kgf.

4. A U-tube of uniform cross-section is partially filled with liquid I. Another liquid II which does not mix with liquid I is poured into one side. It is observed that the liquid levels in the two sides of the tube are the same, while the level of liquid I has risen by 2 cm. If the specific gravity of liquid I is 1.1, then the specific gravity of liquid II must be

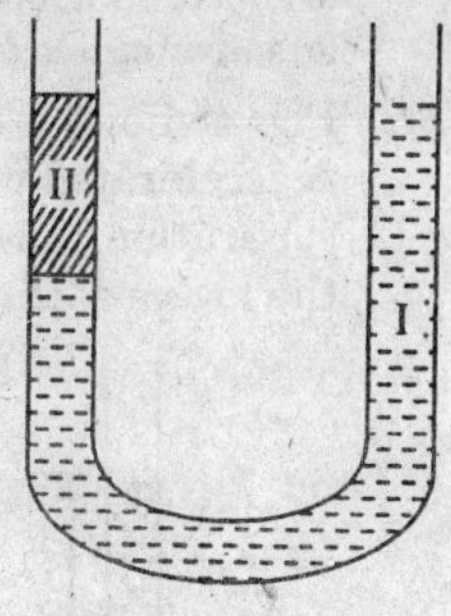

Fig. 1

(*a*) 1.0 (*b*) 1.1
(*c*) 1.05 (*d*) 1.12.

5. A wooden cylinder floats in water with half of its volume inside water. The density of wood is

(*a*) 125 kg m^{-3} (*b*) 0.8 kg m^{-3}
(*c*) 500 kg m^{-3} (*d*) 800 kg m^{-3}.

6. A cylindrical vessel containing a liquid is closed by a smooth piston of mass m. If A is the cross-sectional area of the piston and P_0 is the atmospheric pressure, then the pressure of the liquid just below the piston is

(*a*) P_0 (*b*) $P_0 + \frac{mg}{A}$
(*c*) $\frac{mg}{A}$ (*d*) Data is not sufficient.

7. In a hydraulic lift, the surface area of the input piston is 10 cm^2. The surface area of the output piston is 3000 cm^2. A 100 N force applied to the input piston raises the output piston by 1 m. The work done is

(*a*) 15 kJ (*b*) 30 kJ
(*c*) 45 kJ (*d*) 60 kJ.

8. A common hydrometer reads specific gravity of liquids. Compared to 1.6 on the stem, the mark 1.5 will be

(*a*) upward (*b*) downward
(*c*) in the same place
(*d*) may be upward or downward depending upon the hydrometer.

9. The liquids shown in the U-tube of uniform cross-section are water and mercury. If the difference of the heights of the mercury columns is 2 cm, then the height h of the water column is

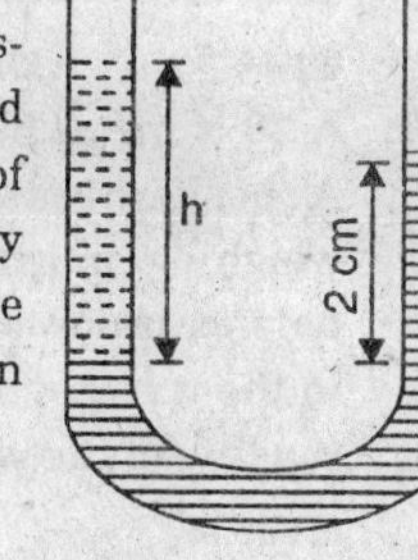

Fig. 2

(*a*) 6.8 cm (*b*) 13.6 cm
(*c*) 27.2 cm (*d*) 54 cm.

10. Given : σ is the compressibility of water, ρ is the density of water and K is the bulk modulus of water. What is the energy density of water at the bottom of a lake 'h' metre deep ?

(*a*) $\frac{1}{2}\sigma(h\rho g)^2$ (*b*) $\frac{1}{2}\sigma(h\rho g)$
(*c*) $\frac{1}{2}\frac{h\rho g}{\sigma}$ (*d*) $\frac{h\rho g}{\sigma}$.

11. A body weighs 160 g in air, 130 g in water and 136 g in oil. The specific gravity of oil is

(*a*) 0.2 (*b*) 0.6
(*c*) 0.7 (*d*) 0.8.

12. A diver is 10 m below the surface of water. The approximate pressure experienced by the diver is

(*a*) 10^5 Pa (*b*) 2×10^5 Pa

(*c*) 3×10^5 Pa (*d*) 4×10^5 Pa.

13. A penguin floats first in a fluid of density ρ_0, then in a fluid of density $0.95\rho_0$ and then in a fluid of density $1.1\rho_0$. Which of the following is correct ?

(*a*) maximum buoyant force in the fluid of density of $0.95\rho_0$

(*b*) maximum buoyant force in the case of fluid of density $1.1\rho_0$.

(*c*) maximum fluid is displaced in the case of density $1.1\rho_0$

(*d*) maximum fluid is displaced in the case of density $0.95\rho_0$.

14. Density of mercury = 13.6 g cm^{-3}, density of a liquid = 1.36 g cm^{-3}. If the height of mercury barometer is 76 cm, then the height of the liquid barometer is

(*a*) 76 cm (*b*) 7.6 cm

(*c*) 760 cm (*d*) 760 metre.

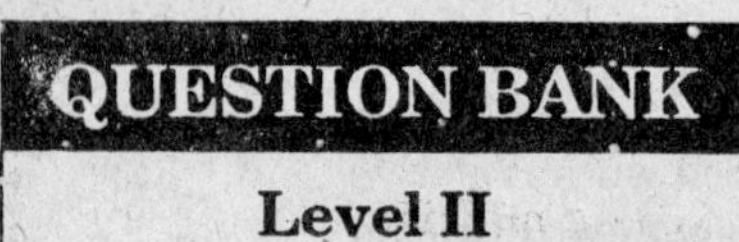

QUESTION BANK

Level II

Choose the correct answer from the four alternatives given in each of the following questions :

1. The fraction of a floating object of density d_0 above the surface of a liquid of density d will be

(*a*) $\frac{d_0}{d}$ (*b*) $\frac{d\,d_0}{d+d_0}$

(*c*) $\frac{d-d_0}{d}$ (*d*) $\frac{d_0}{d-d_0}$.

2. If a block of iron (density 5 g cm^{-3}) of size 5 cm × 5 cm × 5 cm was weighed whilst completely submerged in water, what would be the apparent weight ?

(*a*) 5 × 5 × 5 × 5 gf (*b*) 4 × 4 × 4 × 5 gf

(*c*) 3 × 5 × 5 × 5 gf (*d*) 4 × 5 × 5 × 5 gf.

3. A 20 N metal block is suspended by a spring balance. A beaker containing some water is placed on a weighing machine which reads 40 N. The spring balance is now lowered so that the block gets immersed in the water. The spring balance now reads 16 N. The reading of the weighing machine will be

(*a*) 36 N (*b*) 60 N

(*c*) 44 N (*d*) 56 N.

4. Two thirds of a beaker is filled with water. When placed on a balance, it weighs 1700 g. A body weighing 500 g and of volume 300 cm^3 is suspended such that it is completely immersed in water without touching the sides of the beaker. Now the beaker on the balance will weigh

(*a*) 1400 gf (*b*) 1200 gf

(*c*) 2200 gf (*d*) 2000 gf.

5. A piece of solid weighs 120 g in air, 80 g in water and 60 g in a liquid, then the relative density of the solid, and that of liquid are

(*a*) 3, 2 (*b*) 2, $\frac{4}{3}$

(*c*) 3, $\frac{3}{2}$ (*d*) 4, 3.

6. The three vessels shown in Fig. 3 have same base area. Equal volumes of a liquid are poured in the three vessels. The force on the base will be

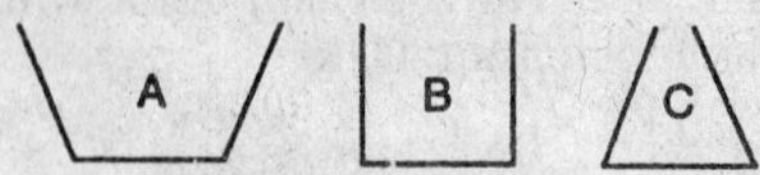

Fig. 3

(*a*) maximum in vessel A (*b*) maximum in vessel B

(*c*) maximum in vessel C (*d*) equal in all the vessels.

7. Fig. 4 shows a U-tube of uniform cross-sectional area A accelerated with acceleration a as shown. If d is the separation between the limbs, then the difference in the levels of the liquid in the U-tube is

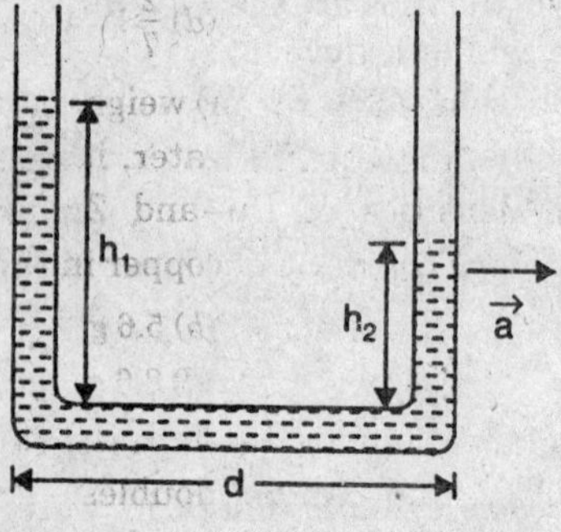

Fig. 4

(*a*) $\frac{ad}{g}$ (*b*) $\frac{g}{ad}$

(*c*) adg (*d*) $ad + g$.

8. A uniformly tapering vessel of height h whose lower and upper radii are r and R is completely filled with a liquid of density ρ. The force that acts on the base of the vessel due to the liquid is

(a) $\pi R^2 h\rho g$ (b) $\pi r^2 h\rho g$

(c) $\pi\left(\frac{R+r}{2}\right)^2 h\rho g$ (d) $\frac{1}{3}\pi(R^2 - r^2)h\rho g$.

9. A sphere of wax (relative density = 0.9) and volume = 18 cm^3 has some iron nails pierced into it. It floats on water just completely immersed. The mass of iron nails into it is

(a) 0.9 g (b) 1.8 g

(c) 2.7 g (d) 3.6 g.

10. A 0.5 kg block of brass (density : 8 × 103 kg m^{-3}) is suspended from a string. What is the tension in the string if the block is completely immersed in water ? (g = 10 m s^{-2})

(a) 5 N (b) $\frac{0.5}{8 \times 10^3}$ N

(c) $\frac{5}{8}$ N (d) $\left[5 - \frac{5}{8}\right]$ N.

11. A cubical box of wood of side 30 cm weighing 21.6 kg floats on water with two faces horizontal. The depth of immersion of the box is

(a) 12 cm (b) 24 cm

(c) 28 cm (d) 30 cm.

12. A cubical block of wood of specific gravity 0.5 and chunk of concrete of specific gravity 2.5 are fastened together. The ratio of the mass of wood to the mass of concrete which makes the combination to float with its entire volume submerged under water is

(a) $\frac{3}{5}$ (b) $\frac{4}{5}$

(c) $\frac{3}{7}$ (d) $\frac{2}{7}$.

13. A piece of brass (Cu and Zn) weighs 12.9 g in air. When completely immersed in water, it weighs 11.3 g. The relative densities of Cu and Zn are 8.9 and 7.1 respectively. The mass of copper in the alloy is

(a) 4.6 g (b) 5.6 g

(c) 7.6 g (d) 8.6 g.

14. When a large bubble rises from the bottom of a lake to the surface, its radius doubles. The atmospheric pressure is equal to that of a column of water of height H. The depth of the lake is

(a) 8 H (b) 7 H

(c) 2 H (d) H.

15. A steel block having an internal cavity weighs 234 g in air and 197 g in water. If the density of steel is 7.8 g cm^{-3}, then the volume of the cavity is

(a) 5 cm^3 (b) 7 cm^3

(c) 9 cm^3 (d) 11 cm^3.

16. A cubical block of wood 10 cm on a side floats at the interface between oil and water, as in Fig. 5, with its lower face 2 cm below the interface. The density of the oil is 0.6 g cm^{-3}. The mass of the block is

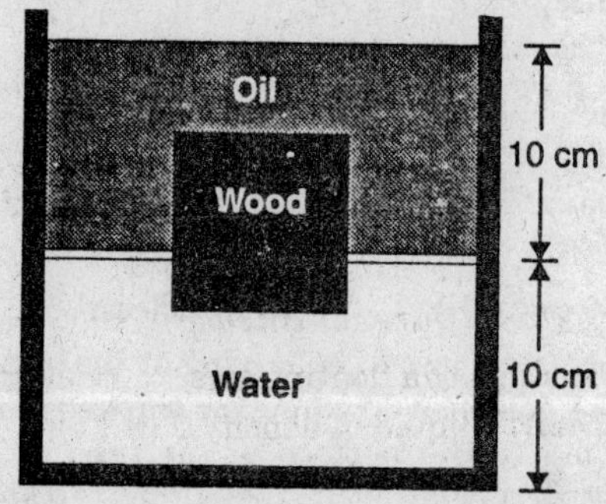

Fig. 5

(a) 340 g (b) 680 g

(c) 80 g (d) 10 g.

17. In Q. 16, the gauge pressure at the lower face of the block is

(a) 84 Pa (b) 384 Pa

(c) 484 Pa (d) 784 Pa.

18. A tube 1 cm^2 in cross-section is attached to the top of a vessel 1 cm high and of cross-section 100 cm^2. Water is poured into the system filling it to a depth of 100 cm above the bottom of the vessel as shown in Fig. 6. Take g = 10 m s^{-2}. Now,

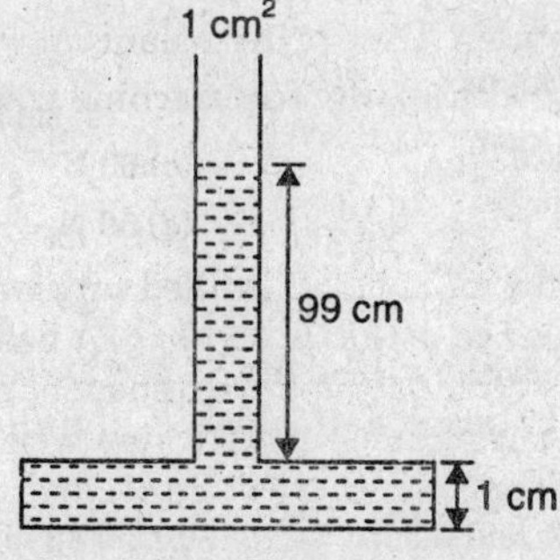

Fig. 6

(a) The force exerted by the water against the bottom of the vessel is 100 N.

(b) The weight of water in the system is 1.99 N.

(c) Both (a) and (b) are correct.

(d) Neither (a) nor (b) is correct.

19. The ratio of the diameters of an air bubble at the bottom and at the surface is 1 : 2. If 1 atmosphere is equal to pressure of 10 m long water column, then the depth of the lake is

(*a*) 70 m (*b*) 80 m

(*c*) 90 m (*d*) 140 m.

20. An open U-tube of uniform cross-section contains mercury. When 27.2 cm of water column is poured into one limb, then the difference in levels of liquids of two sides is

(*a*) 2.4 cm (*b*) 4.8 cm

(*c*) 25.2 cm (*d*) 1 cm.

21. A 700 g solid cube having an edge of length 10 cm floats in water. The volume of cube outside water is

(*a*) 50 cm^3 (*b*) 100 cm^3

(*c*) 300 cm^3 (*d*) 500 cm^3.

22. An inverted vessel (bell) lying at the bottom of a lake, 47.6 m deep has 50 cm^3 of air trapped in it. The bell is brought to the surface of the lake. If atmospheric pressure is 70 cm of Hg, the volume of trapped air when the vessel is brought to the surface of the lake is

(*a*) 100 cm^3 (*b*) 200 cm^3

(*c*) 300 cm^3 (*d*) 500 cm^3.

23. In Fig. 7, block A hangs by a cord from spring balance D and is submerged in a liquid C contained in a beaker B. The mass of the beaker is 1 kg. The mass of the liquid is 1.5 kg. Balance D reads 2.5 kg and balance E reads 7.5 kg. The volume of block A is 0.003 m^3. The mass per unit volume of the liquid is

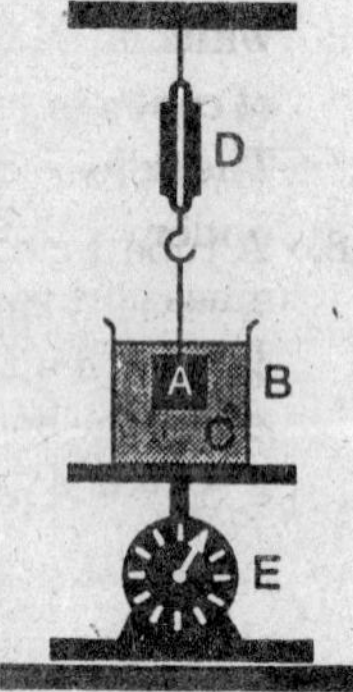

Fig. 7

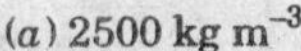
(*a*) 2500 kg m^{-3}

(*b*) 5000 kg m^{-3}

(*c*) 1 kg m^{-3}

(*d*) $\frac{5000}{3}$ kg m^{-3}.

24. In Q. 23, what will be the respective readings of balance D and balance E if block A is pulled up out of the liquid ?

(*a*) 7.5 kg and 2.5 kg (*b*) 2.5 kg and 7.5 kg

(*c*) 5 kg and 0 kg (*d*) 0 and 0.1 kg.

25. A body of density d and volume V floats with volume V_1 of its total volume V immersed in a liquid of density d_1 and the rest of the volume V_2 immersed in another liquid of density d_2 ($< d_1$). The volume V_1 immersed in liquid of density d_1 is

(*a*) $\left(\frac{d-d_2}{d_1-d_2}\right)V$ (*b*) $\left(\frac{d+d_2}{d_1+d_2}\right)V$

(*c*) $\left(\frac{d_1-d_2}{d_1}\right)V$ (*d*) $\frac{d_1}{d_2}$ V.

26. A mercury barometer reads 75 cm in vertical position. If the tube is inclined by 60° to the vertical, the length of the mercury in the tube will be

(*a*) 37.5 cm (*b*) 75 cm

(*c*) 112.5 cm (*d*) 150 cm.

27. The diameter of the piston of a hydraulic automobile is D metre. What pressure, in atmosphere is required to lift a car of mass m kg ?

(*a*) $\frac{4mg}{\pi D^2 \times 10^5}$ (*b*) $\frac{2mg}{\pi D^2}$

(*c*) $\frac{mg}{\pi D^2}$ (*d*) $10^5 \times \frac{4mg}{\pi D^2}$.

28. The volume of an air bubble rises to three times as it ascends from the bottom of the lake to its surface. Assuming the atmospheric pressure to be 75 cm of Hg and density of water to be $\frac{1}{10}$ th of the density of mercury, the depth of the lake is

(*a*) 5 m (*b*) 8 m

(*c*) 10 m (*d*) 15 m.

29. Water is filled in a flask upto a height of 20 cm. The bottom of the flask is circular and has an area of 1 m^2. If the atmospheric pressure is 1.01×10^5 Pa, the force exerted by water on the bottom is (Take g = 10 m s^{-2} and density of water = 1000 kg m^{-3})

(*a*) 1.03 N (*b*) 1.03×10^2 N

(*c*) 1.03×10^4 N (*d*) 1.03×10^5 N.

30. The surface area of a bubble increases four times when it rises from bottom to the top of a water tank. If the atmospheric pressure is 10 metre of water column, then the depth of water in the tank is

(*a*) 30 m (*b*) 40 m

(*c*) 70 m (*d*) 80 m.

31. Two solids A and B float in water. It is observed that A floats with half of its volume immersed and B floats with $\frac{2}{3}$ of its volume immersed. The ratio of densities of A and B is

(*a*) 2 : 3 (*b*) 3 : 2

(*c*) 3 : 4 (*d*) 4 : 3.

32. A wooden rod of uniform cross-section and of length 120 cm is hinged at the bottom of the tank which is filled with water to a height of 40 cm. In the equilibrium position, the rod makes an angle of 60° with the vertical. The centre of buoyancy is located on the rod at a distance (from the hinge) of

(a) 20 cm (b) 40 cm
(c) 60 cm (d) 75 cm.

33. A piece of lead weighs 100 g in water. If density of lead is 11 g cm^{-3}, then the weight of the piece of lead in air is

(a) 90 g (b) 100 g
(c) 101 g (d) 110 g.

34. A body weighs 150 g in air, 120 g in water and 130 g in a liquid. The density of liquid in g cm^{-3} is

(a) $\frac{2}{3}$ (b) $\frac{4}{5}$
(c) $\frac{13}{15}$ (d) $\frac{15}{13}$.

35. The weight of a wooden block of size $8 \times 8 \times 8$ cm is 307.2 g. When it is floating in water, the height of the wooden block above water level is

(a) 4.8 cm (b) 3.2 cm
(c) 4 cm (d) 6 cm.

36. When a loaded test tube floats vertically with $\frac{1}{3}$ and $\frac{1}{4}$ of the lengths inside two liquids, then the ratio of the densities of the two liquids is

(a) 3 : 4 (b) 4 : 3
(c) 9 : 16 (d) 16 : 9.

37. The density of ice is x gcm^{-3} and that of water is y gcm^{-3}. What is the change in volume when m g of ice melts ?

(a) $m(y - x)$ cm^3 (b) $\frac{y-x}{m}$ cm^3
(c) $m\,x\,y(x - y)$ cm^3 (d) $m\left(\frac{1}{y}-\frac{1}{x}\right)$cm^3.

38. An ice cube containing a piece of lead floats in water. What would be the effect on the level of water if the ice cube melts ?

(a) It would fall (b) It would rise
(c) It would remain the same (d) It would rise very high.

39. What volume of lead of specific gravity 11.3 must be placed on top of a 20 g block of cork of specific gravity 0.24 to cause the cork to be barely submerged in water ?

(a) 3.61 cm^3 (b) 4.61 cm^3
(c) 5.61 cm^3 (d) 6.61 cm^3.

40. An iceberg is floating partially immersed in sea water. The density of sea water is 1.03 g cm^{-3} and that of ice is 0.92 g cm^{-3}. The approximate percentage of total volume of iceberg above the level of sea water is

(a) 8 (b) 11
(c) 34 (d) 89.

41. Fig. 8 here shows the vertical cross-section of a vessel filled with a liquid of density ρ. The normal thrust per unit area on the walls of the vessel at point P, as shown, will be

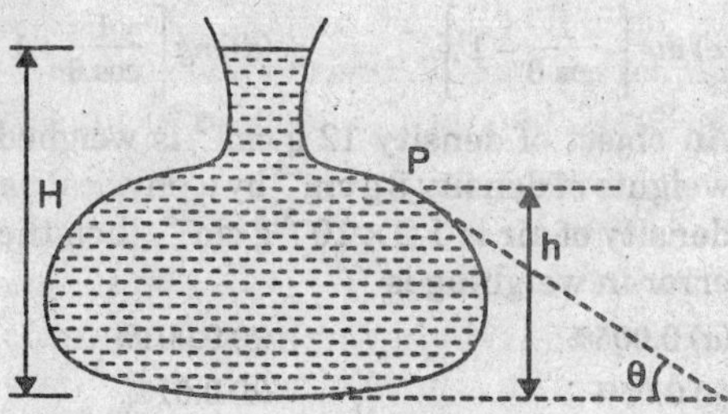

Fig. 8

(a) $h\rho g$ (b) $H\rho g$
(c) $(H - h)\rho g$ (d) $(H - h)\rho g \cos\theta$.

42. A piece of material weighing 50 g is coated with 6.3 g of wax of relative density 0.9. If the coated piece weighs 16.3 g in water, the density of the material is

(a) 1.515 (b) 2.3
(c) 4.8 (d) 6.2.

43. A concrete sphere of radius R has a cavity of radius r which is packed with saw-dust. The specific gravities of concrete and saw dust are 2.4 and 0.3 respectively. The sphere floats with its entire volume submerged under water. The ratio of mass of concrete and the mass of saw dust is

(a) 2 (b) 3
(c) 4 (d) 6.

44. A small ball of density r is immersed in a liquid of density σ (> ρ) to a depth h and released. The height above the surface of water upto which the ball will jump is

(a) $\left(\frac{\sigma}{\rho}-1\right)h$ (b) $\left(\frac{\rho}{\sigma}-1\right)h$
(c) $\left(\frac{\rho}{\sigma}+1\right)h$ (d) $\left(\frac{\sigma}{\rho}+1\right)h$.

45. A bowl of soup rests on a table in the dining car of a train. If the acceleration of the train is $\frac{g}{4}$ in the forward direction, what angle does the surface of the soup make with the horizontal ?

(a) $\tan^{-1}\left[\frac{1}{2}\right]$ (b) $\tan^{-1}\left[\frac{1}{3}\right]$
(c) $\tan^{-1}\left[\frac{1}{4}\right]$ (d) $\tan^{-1}\left[\frac{1}{5}\right]$.

46. A hollow cylinder of mass m made heavy at its bottom is floating vertically in water. It is tilted from its vertical position through an angle θ and is left. The restoring force acting on it is

(a) $mg \cos\theta$ (b) $\frac{mg}{\cos\theta}$

(c) $mg\left[\frac{1}{\cos\theta}-1\right]$ (d) $mg\left[\frac{1}{\cos\theta}+1\right]$.

47. An object of density 12 g cm^{-3} is weighed with brass weights of density 8 g cm^{-3} by a physical balance. If the density of air is 1.2×10^{-3} g cm^{-3}, then the percentage error in weighing is

(a) 0.005% (b) 0.001%

(c) 0.05% (d) 0.01%.

48. Assume the density of brass weights to be 8 g cm^{-3} and that of air to be 0.0012 g cm^{-3}. What fractional error arises from neglecting the buoyancy of air in weighing an object of density 3.4 g cm^{-3} on a beam balance ?

(a) 2×10^{-1} (b) 2×10^{-2}

(c) 2×10^{-3} (d) 2×10^{-4}.

49. A hollow spherical iron shell floats almost completely submerged in water. The outer diameter is 58.7 cm and the density of iron is 7.87 g cm^{-3}. What is the approximate inner diameter of the shell ?

(a) 56 cm (b) 57 cm

(c) 57.5 cm (d) 58 cm.

ANSWERS

QUESTION BANK (Level I)

1. (c)	2. (c)	3. (a)	4. (b)	5. (c)	6. (b)	7. (b)	8. (a)
9. (c)	10. (a)	11. (d)	12. (b)	13. (d)	14. (c).		

QUESTION BANK (Level II)

1. (c)	2. (d)	3. (c)	4. (d)	5. (c)	6. (c)	7. (a)	8. (b)
9. (b)	10. (d)	11. (b)	12. (a)	13. (c)	14. (b)	15. (b)	16. (b)
17. (d)	18. (c)	19. (a)	20. (c)	21. (c)	22. (c)	23. (d)	24. (a)
25. (a)	26. (d)	27. (a)	28. (d)	29. (d)	30. (c)	31. (c)	32. (b)
33. (d)	34. (a)	35. (b)	36. (a)	37. (d)	38. (a)	39. (c)	40. (b)
41. (c)	42. (a)	43. (c)	44. (a)	45. (c)	46. (c)	47. (a)	48. (d)
49. (a).							

HINTS/SOLUTIONS

QUESTION BANK (Level I)

1. (c) Specific gravity $= \frac{\text{Weight in air}}{\text{Loss of weight in water}}$

$$= \frac{60}{60-40} = 3$$

2. (c) The relative density of object P is less than the relative density of both the given liquids. So, P will float in both the liquids. The relative density of Q is higher than the relative density of liquid A. So, Q sinks in A.

3. (a) Equating pressures, $\frac{F}{\pi R^2} = \frac{1200}{\pi[20\,R]^2}$ or F = 3 kgf

4. (b) The weight of liquid II added must be same as that of 2 cm of liquid I, which is possible only if both have the same densities.

5. (c) $V\rho g = \frac{V}{2} \times 1000 \times g$ or $\rho = 500$ kg m^{-3}

6. (b) Since the piston is in equilibrium,

$$\therefore \quad PA = P_0A + mg \quad \text{or} \quad P = P_0 + \frac{mg}{A}$$

7. (b) Same pressure is transmitted to the output piston [Pascal's law]

$$\therefore \quad \frac{F}{3000} = \frac{100}{10} \quad \text{or} \quad F = 30{,}000 \text{ N}$$

Work done = 30 kN × 1 m = 30 kJ

8. (a) If specific gravity of liquid is less, then more volume of liquid is required to be displaced. This is possible if a greater portion of hydrometer is in the liquid.

9. (c) $h \times 1000 \times g = \frac{2}{100} \times 13600 \times g$

or $h = \frac{2 \times 13.6}{100}$ m = 27.2 cm

10. (a) Energy density, $u = \frac{1}{2}$ stress × strain

or $u = \frac{1}{2}$ stress × $\frac{\text{stress}}{\text{Bulk modulus}}$

or $u = \frac{1}{2} \times$ compressibility × (stress)^2

11. (d) Specific gravity of oil = $\frac{\text{Loss of weight in oil}}{\text{Loss of weight in water}}$

$= \frac{160 - 136}{160 - 130} = \frac{24}{30} = \frac{8}{10} = 0.8$

12. (b) 1 atmosphere ≈ 10^5 Pa

Also, $p = h\rho g$

$= 10 \times 1000 \times 10$ Pa $= 10^5$ Pa

So, total pressure is nearly 2×10^5 Pa.

13. (d) The buoyant force on the penguin is the same in all the cases. Maximum fluid is displaced in the case of least density.

14. (c) $h' \times 1.36 \times g = 76 \times 13.6 \times g$

or $h' = \frac{76 \times 13.6}{1.36}$ cm = 760 cm.

QUESTION BANK (Level II)

1. (c) Let v be the volume, of the object, above the surface of the liquid. If V be the total volume of the object, then

$(V - v)\, dg = V d_0 g$

or $\frac{V - v}{V} = \frac{d_0}{d}$ or $\frac{v}{V} = 1 - \frac{d_0}{d} = \frac{d - d_0}{d}$

2. (d) Upthrust = 5 × 5 × 5 × 1 gf

Weight = 5 × 5 × 5 × 5 gf

Apparent weight = 5 × 5 × 5(5 – 1) gf = 5 × 5 × 5 × 4 gf

3. (c) Upthrust = 4 N

Downthrust = 4 N

Weight = 40 N

Effective weight = (40 + 4) N = 44 N

4. (d) Downthrust by the body = 300 gf

Effective weight = (1700 + 300) gf = 2000 gf

5. (c) Relative density of solid = $\frac{120}{120 - 80} = 3$

Relative density of liquid = $\frac{120 - 60}{120 - 80} = \frac{60}{40} = \frac{3}{2}$

6. (c) Maximum height of liquid column in C.

7. (a) Mass of liquid in horizontal portion of U-tube = $Ad\rho$

Pseudo force on this mass = $Ad\rho a$

Force due to pressure difference in the two limbs

$= (h_1\rho g - h_2\rho g)A$

Equating, $(h_1 - h_2)\,\rho g A = Ad\rho a$

or $h_1 - h_2 = \frac{Ad\rho a}{\rho g A} = \frac{ad}{g}$

8. (b) Force = area of base × pressure due to liquid column of height $h = \pi r^2 \times h\rho g$

9. (b) Upthrust = 18 gf

Weight of wax = 18 × 0.9 gf = 16.2 gf

Weight of iron nails = (18 – 16.2) gf = 1.8 gf

10. (d) Volume of block = $\frac{0.5}{8 \times 10^3}$ m^3

Volume of displaced water = $\frac{0.5}{8 \times 10^3}$ m^3

Weight of displaced water = $\frac{0.5}{8 \times 10^3} \times 10^3 \times 10$ N $= \frac{5}{8}$ N

This is equal to upward buoyant force.

Tension in string is actual weight less the upward buoyant force. So, tension is $\left[5 - \frac{5}{8}\right]$ N.

Time-saving tip

The density of water is less than that of brass by a factor of 8. So, the weight of an amount of water of volume equal to that of the brass will be less than the weight of the brass by the same factor. So, weight of displaced water is $\frac{5}{8}$ N.

11. (b) If h be the depth of immersion, then

$30 \times 30 \times 10^{-4} \times h \times 10^3 \times g = 21.6\, g$

or $h = \frac{21.6}{90}$ m $= \frac{2160}{90}$ cm or $h = 24$ cm

12. (a) Let m_w and m_c be the masses of wood and concrete respectively. Then, $(m_w + m_c)\, g$

$= \left[\frac{m_w}{0.5} + \frac{m_c}{2.5}\right] \times 1 \times g$

or $m_w\left[1-\frac{1}{0.5}\right]=m_c\left[\frac{1}{2.5}-1\right]$

or $\frac{m_w}{m_c}=\frac{1.5}{2.5}\times\frac{0.5}{0.5}=\frac{3}{5}$

13. (*c*) Loss of weight = (12.9 – 11.3) gf = 1.6 gf

Weight of water displaced = 1.6 gf

If m is the mass of Cu, then

$$\left[\frac{m}{8.9}+\frac{12.9-m}{7.1}\right]g = 1.6\ g$$

or $7.1m + 12.9\times 8.9 - 8.9m = 1.6\times 8.9\times 7.1$

or $1.8m = 114.8 - 101.1$

or $m = 7.6$ gram

14. (*b*) $P_1V_1 = P_2V_2$

$P_1\times\frac{4}{3}\pi R^3 = H\times\frac{4}{3}\pi(2R)^3$ or $P_1 = 8H$

So, depth of lake = (8H – H) = 7H

15. (*b*) Weight of water displaced by steel block

= Loss of weight of block in water

= (234 – 197) gf = 37 gf

Volume of water displaced = 37 cm^3

This is the volume of the block with cavity.

Volume of block without cavity = $\frac{234}{7.8}$ cm^3 = 30 cm^3

Volume of cavity = (37 – 30) cm^3 = 7 cm^3

16. (*b*) $mg = [100\times 2\times 1 + 100\times 8\times 0.6]$ g

$\therefore$ $m = (200 + 480)$ g = 680 g

17. (*d*) P = Pressure due to 10 cm of oil column

+ Pressure due to 2 cm of water column

$= 10\times 0.6g + 2\times 1g$

$= 8g = 8\times 980$ dyne cm^{-2}

$= \frac{8\times 980}{10}$ N m^{-2} = 784 Pa

18. (*c*) $P = 100\ cm\times 1\ g\ cm^{-3}\times 1000\ cm\ s^{-2} = 10^5$ dyne cm^{-2}

$F = 10^5\times 100$ dyne = 100 N

Again, V = 199 cm^3

Weight = 199 × 1 × 1000 dyne = 1.99 N

19. (*a*) Diameter doubled. Radius doubled. Volume increases by a factor of 8.

$(h + 10)\ V = 10\times 8\ V$

or $h = (80 - 10)$ m = 70 m

20. (*c*) Considering the horizontal level AB,

$27.2\times 1\times g = 2h\times 13.6\times g$ or $h = 1$ cm

Difference in levels = (27.2 – 2 × 1) cm = 25.2 cm

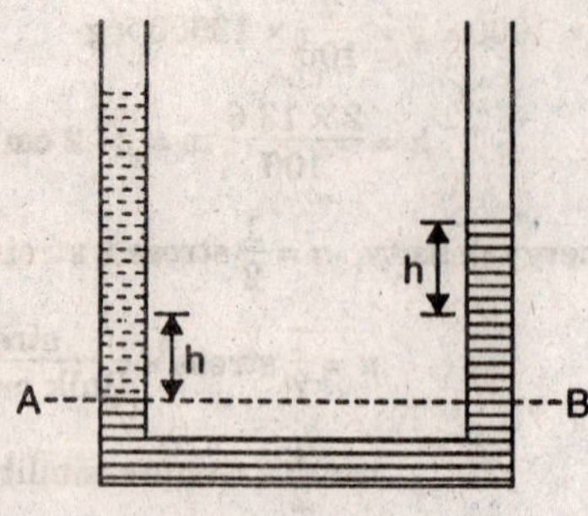

Fig. 9

21. (*c*) If V is the volume of the cube inside the water, then

$V\times 1000\times 10 = 0.7\times 10$

or $V = \frac{7}{10000}\ m^3 = 700\ cm^3$

Total volume of cube = $(10\ cm)^3 = 1000\ cm^3$

Volume outside the water = (1000 – 700) cm^3 = 300 cm^3

22. (*c*) $[47.6\times 10^2\times g + 70\times 13.6\times g]\ 50 = 70\times 13.6\times gV$

or $V = \frac{[4760+952]50}{952}\ cm^3 = \frac{5712\times 50}{952}\ cm^3 = 300\ cm^3$

23. (*d*) 7.5 g = 1 g + 1.5 g + Downward reaction force of block

or 5 g = | Downward reaction force |

= | Upthrust | = 0.003 $\rho_l\, g$

$\therefore$ $\rho_l = \frac{5}{0.003}$ kg m^{-3} = $\frac{5000}{3}$ kg m^{-3}

24. (*a*) When block A was in water, the upthrust was 5 g and the apparent weight was 2.5 g. So, total weight in air is 5*g* + 2.5*g*. So, the balance D would read 7.5 kg. Reading of balance E will be due to weights of beaker and liquid.

25. (*a*) $Vdg = V_1d_1g + V_2d_2g$ ...(*i*)

Also, $V = V_1 + V_2$

or $Vd_2g = V_1d_2g + V_2d_2g$ ...(*ii*)

(*i*) – (*ii*) gives

$Vg(d - d_2) = V_1g(d_1 - d_2)$ or $V_1 = \frac{d-d_2}{d_1-d_2}V$

26. (*d*) $\cos 60° = \frac{h}{l}$

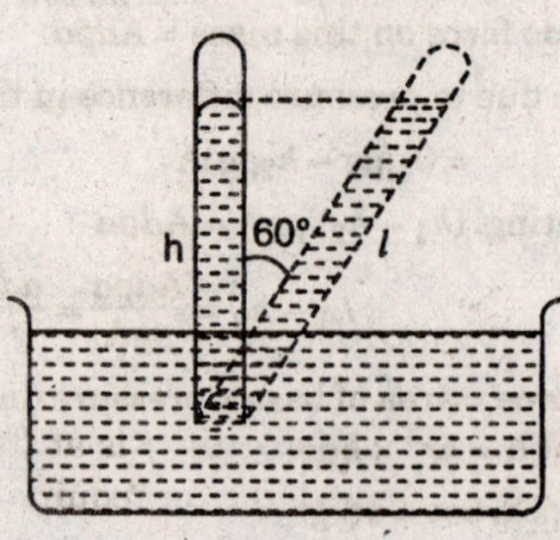

Fig. 10

or $l = \frac{h}{\cos 60°} = \frac{75}{1/2}$ cm = 150 cm

> ☞ *The vertical height of mercury level in a barometer does not change.*

27. (*a*) $P \times \frac{\pi D^2}{4} = mg$ or $P = \frac{4\,mg}{\pi D^2}$ pascal

or $P = \frac{4\,mg}{\pi D^2 \times 10^5}$ atmosphere

28. (*d*) $\left[h \times \frac{\rho}{10} \times g + 75\,\rho g\right] V = 75\,\rho g \times 3\,V$

or $\frac{h}{10} + 75 = 75 \times 3$

or $\frac{h}{10} = 75 \times 2$ cm or $h = 15$ m

29. (*d*) $P = P_0 + h\rho g$

$= 1.01 \times 10^5 + 0.20 \times 1000 \times 10$

$= 1.01 \times 10^5 + 0.02 \times 10^5 = 1.03 \times 10^5$ Pa

Force $= PA = 1.03 \times 10^5$ N

30. (*c*) Surface area increases four times. So, radius is doubled and volume becomes 8 times. If initial volume is V, then final volume is 8 V. Using PV = constant, we get

$10 \times 8\,V = (10 + h)V$ or $h = 70$ m

> ☞ *In such questions, it is understood that the temperature of water is uniform.*

31. (*c*) $V_A \rho_A = \frac{1}{2} V_A \times 1$ or $\rho_A = \frac{1}{2}$

Again, $V_B \rho_B = \frac{2}{3} V_B \times 1$ or $\rho_B = \frac{2}{3}$

$\therefore \frac{\rho_A}{\rho_B} = \frac{1}{2} \times \frac{3}{2} = \frac{3}{4}$

32. (*b*) OP is the portion of the rod immersed in water.

$\cos 60° = \frac{40}{OP}$

or $OP = \frac{40}{\cos 60°}$ cm

$= \frac{40}{1/2}$ cm = 80 cm

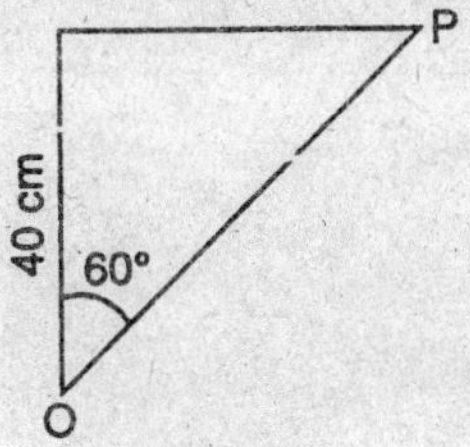

Fig. 11

The centre of buoyancy is at the centre of the immersed part of the rod. So, the required distance is 40 cm.

33. (*d*) $11 = \frac{w}{w - 100}$

$11w - w = 1100$ or $10w = 1100$ or $w = 110\,g$

34. (*a*) Specific gravity of body $= \frac{150}{30} = 5$

Volume of body $= \frac{150}{5} = 30$ cm^3

Now, $(150 - 130)\,g = 30 \times \rho_l \times g \Rightarrow \rho_l = \frac{20}{30} = \frac{2}{3}$

> ☞ **Time-saving solution**
>
> *Specific gravity of liquid*
>
> $= \frac{\text{Loss of weight of body in liquid}}{\text{Loss of weight of body in water}}$
>
> $= \frac{150 - 130}{150 - 120} = \frac{20}{30} = \frac{2}{3}$

35. (*b*) $307.2\,g = 8 \times 8 \times h \times 1 \times g$

$h = \frac{307.2}{64}$ cm = 4.8 cm

Outside height = (8.0 – 4.8) cm = 3.2 cm

36. (*a*) $A \times \frac{l}{3} \rho_1 g = A \times \frac{l}{4} \rho_2 g$ or $\frac{\rho_1}{\rho_2} = \frac{l}{4} \times \frac{3}{l} = \frac{3}{4}$

37. (*d*) Volume of ice $= \frac{m}{x}$; Volume of water $= \frac{m}{y}$

Change in volume $= \frac{m}{y} - \frac{m}{x} = m\left[\frac{1}{y} - \frac{1}{x}\right]$ cm^3

38. (*a*) Let M and m represent the masses of ice and lead respectively.

Now, $(M + m)g = V\rho_w g$

$\therefore V = \frac{M + m}{\rho_w}$

After melting of ice, volume of water and lead,

$V' = \frac{M}{\rho_w} + \frac{m}{\rho_l}$

Since $\rho_l > \rho_w$, $\therefore V' < V$

So, level of water goes down.

> ☞ **Commit to memory**
>
> *If density of impurity is greater than density of ice, then level of water would go down.*

39. (*c*) $V = \frac{20}{0.24}$ cm^3 $= \frac{2000}{24}$ cm^3 $= \frac{250}{3}$ cm^3

Now, $\frac{250}{3} = 20 + m$ or $m = \frac{190}{3}$ g

Again, $11.3 = \frac{190/3}{V}$

or $V = \frac{190}{3 \times 11.3}\text{ cm}^3 = 5.61\text{ cm}^3$

40. (*b*) $(V - \Delta V) \times 1.03 \times g = V \times 0.92 \times g$

or $\frac{V - \Delta V}{V} = \frac{0.92}{1.03}$

or $\frac{\Delta V}{V} = 1 - \frac{0.92}{1.03} = \frac{0.11}{1.03}$

or $\frac{\Delta V}{V} \times 100 = \frac{11}{1.03} \approx 11$

41. (*c*) Pressure is proportional to depth from the free surface. It is the same in all directions.

42. (*a*) Mass of material and wax = 56.3 g

Loss of weight in water = 56.3 g − 16.3 g = 40 g

Volume of water displaced = 40 cm^3

So, volume of material and wax is 40 cm^3

Now, $\frac{50}{\rho} + \frac{6.3}{0.9} = 40$ or $\frac{50}{\rho} = 33$ or $\rho = \frac{50}{33} = 1.515$

43. (*c*) $\frac{4}{3}\pi r^3 \times 0.3 \times g + \frac{4}{3}\pi[R^3 - r^3] \times 2.4\, g = \frac{4}{3}\pi R^3 \times 1 \times g$

or $0.3r^3 + 2.4R^3 - 2.4r^3 = R^3$ or $1.4R^3 = 2.1r^3$

or $\frac{R^3}{r^3} = \frac{3}{2}$ or $\frac{R^3 - r^3}{r^3} = \frac{1}{2}$

Now, $\frac{\text{mass of concrete}}{\text{mass of saw-dust}} = \frac{\frac{4}{3}\pi[R^3 - r^3]2.4}{\frac{4}{3}\pi r^3 \times 0.3} = \frac{1}{2} \times 8 = 4$

44. (*a*) Let V be the volume of the ball.

Net upward force = $V\sigma g - V\rho g$

Net upward acceleration, $a = \frac{V\sigma g - V\rho g}{V\rho} = \frac{(\sigma - \rho)g}{\rho}$

Velocity at the surface $= \sqrt{\frac{2(\sigma - \rho)gh}{\rho}}$

If h_a is the height in air to which the ball rises, then

$$0 - \frac{2(\sigma - \rho)gh}{\rho} = 2(-g)h_a$$

$$h_a = \frac{(\sigma - \rho)gh}{g\rho} = \left(\frac{\sigma}{\rho} - 1\right)h$$

45. (*c*) In the non-inertial reference frame of the train (and hence the soup bowl), any mass *m* experiences a fictitious force of magnitude $\frac{mg}{4}$ in a direction opposite to the

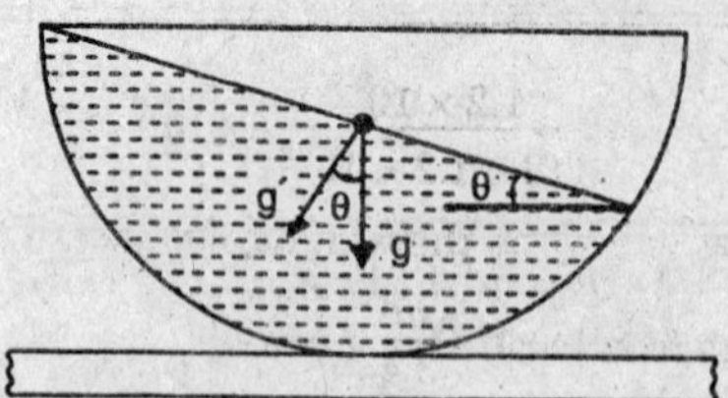

Fig. 12

horizontal forward acceleration of the train. When the soup is in equilibrium, its surface is perpendicular to effective value *g'* of acceleration due to gravity. The high side of the surface is towards the rear of the train, *i.e.*, in a direction opposite to the direction of acceleration of the train.

$$\tan\theta = \frac{g/4}{g} = \frac{1}{4}$$

or $\theta = \tan^{-1}\left[\frac{1}{4}\right]$

Fig. 13

46. (*c*) Let *l* be the length of the cylinder, when vertical, in water. Let A be the cross-sectional area of the cylinder. Equating weight of the cylinder with the upthrust, we get

$$mg = Al\rho g \text{ or } m = Al\rho$$

When the cylinder is tilted through an angle θ, length of cylinder in water $= \frac{l}{\cos\theta}$

Weight of water displaced $= \frac{l}{\cos\theta}A\rho g$

Restoring force $= \frac{lA\rho g}{\cos\theta} - lA\rho g$

$$= lA\rho g\left[\frac{1}{\cos\theta} - 1\right] = mg\left[\frac{1}{\cos\theta} - 1\right]$$

47. (*a*) When the beam is horizontal,

$$m_b - \frac{m_b}{8}\rho_a = m_o - \frac{m_o}{12}\rho_a \text{ or } \frac{m_b}{m_o} = \frac{1 - \frac{\rho_a}{12}}{1 - \frac{\rho_a}{8}}$$

or $\frac{m_b}{m_o} - 1 = \frac{1 - \frac{\rho_a}{12}}{1 - \frac{\rho_a}{8}} - 1 = \frac{\frac{\rho_a}{8} - \frac{\rho_a}{12}}{1 - \frac{\rho_a}{8}}$

$$= \frac{3\rho_a - 2\rho_a}{24} \times \frac{8}{8 - \rho_a} = \frac{\rho_a}{3(8 - \rho_a)}$$

or $\frac{m_b - m_o}{m_o} \times 100 = \frac{\rho_a}{3(8 - \rho_a)} \times 100$

$$= \frac{1.2 \times 10^{-3}}{3(8 - 1.2 \times 10^{-3})} \times 100 = \frac{0.12}{3 \times 7.9988} = 0.005$$

48. (*d*) Let m g wt be the weight of object in vacuum.

Volume of object $= \frac{m}{3.4}$

Weight of air displaced by object $= \frac{m}{3.4} \times 0.0012$

Volume of brass weight $= \frac{m}{8}$

Weight of air displaced by brass weights $= \frac{m}{8} \times 0.0012$

Error = difference in buoyancy

$$= 0.0012 \text{ m}\left[\frac{1}{3.4} - \frac{1}{8}\right]$$

$$\text{Fractional error} = \frac{0.0012 \text{ m}}{\text{m}}\left(\frac{1}{3.4} - \frac{1}{8}\right)$$

$$= \frac{0.0012 \times 4.6}{3.4 \times 8} = 2 \times 10^{-4}$$

49. (*a*) $\frac{4}{3} \times \frac{22}{7}\left[\frac{58.7}{2}\right]^3 \times 1 = \frac{4}{3} \times \frac{22}{7}\left[\left(\frac{58.7}{2}\right)^3 - R_i^{\ 3}\right] \times 7.87$

$$\frac{(58.7)^3}{8 \times 7.87} = \frac{(58.7)^3}{8} - R_i^{\ 3}$$

or $$R_i^{\ 3} = \frac{(58.7)^3}{8}\left[1 - \frac{1}{7.87}\right]$$

or $$R_i = \frac{58.7}{2}\left[\frac{6.87}{7.87}\right]^{1/3} = 28 \text{ cm}$$

$$D_i = 56 \text{ cm}$$

8.3 SURFACE TENSION

Important Terms, Facts and Formulae

BASIC CONCEPTS

1. Surface tension acts over the free surface of a liquid only and not within the interior of the liquid.
2. The free surface of a liquid behaves like a stretched elastic membrane. But there is one essential difference. The tension in an elastic membrane depends upon the amount of stretching. But the surface tension of a liquid remains constant.
3. While a wetting agent reduces angle of contact, a water-proofing agent increases angle of contact.

IMPORTANT POINTS TO REMEMBER

1. **If angle of contact is acute, $\left(0 \le \theta < \frac{\pi}{2}\right)$, then liquid wets solid.**
2. **If angle of contact is acute, then liquid rises in the capillary tube. The shape of liquid meniscus is concave upwards.**
3. **If angle of contact is obtuse, then liquid does not wet solid. The liquid falls in the capillary tube. The shape of liquid meniscus is convex upwards.**
4. **Water wets clean glass. Mercury wets zinc.**
5. **The surface tension of liquids decreases with rise in temperature.**
6. $h\rho g = \frac{2\sigma}{R}$
7. **The equation $2\pi r\,\sigma = \pi r^2\,h\rho g$ is to be applied only for vertical cylindrical tubes. This equation should not be used for capillary tubes of other shapes.**

QUESTION BANK

Level I

Choose the correct answer from the four alternatives given in each of the following questions :

1. When a capillary tube is dipped in water, water rises upto 8 cm in the tube. What happens when the tube is pushed down such that its end is only 5 cm above the outside water level ?

(*a*) The radius of the meniscus increases and therefore water does not overflow.

(*b*) The radius of the water meniscus decreases and therefore it does not overflow.

(*c*) The water forms a droplet on top of the tube but does not overflow.

(*d*) The water starts overflowing.

2. Energy needed in breaking a drop of radius R into n drops of radii r is given by

(*a*) $4\pi T(nr^2 - R^2)$ (*b*) $\frac{4}{3}\pi(nr^3 - R^2)$

(*c*) $4\pi T(R^2 - nr^2)$ (*d*) $4\pi T(nr^2 + R^2)$.

3. The surface tension of a liquid is 5 N m^{-1}. If a thin film is formed on a loop of area 0.02 m^2, then its surface energy will be

(*a*) 5×10^{-2} J (*b*) 2.5×10^{-2} J

(*c*) 2×10^{-1} J (*d*) 3×10^{-1} J.

4. Two soap bubbles, each with a radius r, coalesce in vacuum under isothermal conditions to form a bigger bubble of radius R. Then R is equal to

(*a*) $2^{-1/2} r$ (*b*) $2^{1/3} r$

(*c*) $2^{1/2} r$ (*d*) $2r$.

5. Two water drops, each of radius r coalesce to form a bigger drop of radius R. Then R is equal to

(*a*) $2^{-\frac{1}{2}} r$ (*b*) $2^{\frac{1}{3}} r$

(*c*) $2^{\frac{1}{3}} r$ (*d*) $2r$.

6. A disc of paper of radius R has a hole of radius r. It is floating on a liquid of surface tension T. The force of surface tension on the disc is

(*a*) $T.3\pi R$ (*b*) $T.2\pi(R + r)$

(*c*) $T.4\pi(R + r)$ (*d*) $T.2\pi(R - r)$.

7. If F_C and F_A denote cohesive and adhesive force on a liquid molecule near the surface of a solid, then the surface of liquid is convex if

(*a*) $F_A > \frac{F_C}{\sqrt{2}}$ (*b*) $F_A = \frac{F_C}{\sqrt{2}}$

(*c*) $F_A < \frac{F_C}{\sqrt{2}}$ (*d*) $F_A < F_C$.

8. Two soap bubbles, of radii 3 cm and 4 cm, coalesce in vacuum under isothermal conditions to form a bigger bubble of radius R. Then R is equal to

(*a*) 3 cm (*b*) 4 cm

(*c*) 5 cm (*d*) 7 cm.

9. Water rises to a height of 10 cm in a glass capillary tube. If the area of cross-section of the tube is reduced to one- fourth of the former value, the water will rise to

(*a*) 20 cm (*b*) 5 cm

(*c*) 2.5 cm (*d*) 40 cm.

10. A straw 6 cm long floats on water. The water film on one side has a surface tension of 50 dyne cm^{-1}. On the other side, camphor reduces the surface tension to 40 dyne cm^{-1}. The resultant force acting on the straw is

(*a*) $(50 \times 6 - 40 \times 6)$ dyne (*b*) 10 dyne

(*c*) $\left(\frac{50}{6} - \frac{40}{6}\right)$ dyne (*d*) 90 dyne.

11. Liquid in a capillary tube rises to a height h. The tube is filled with the liquid in the horizontal position. If the length of the tube is greater than $2h$, then the column of liquid which remains in the capillary tube when held in the vertical position is

(*a*) h (*b*) $\frac{h}{2}$

(*c*) $2h$ (*d*) zero.

12. What change in surface energy will be noticed when a drop of radius R splits up into 1000 droplets, each of radius 'r'. Surface tension is T.

(*a*) $4\pi R^2 T$ (*b*) $7\pi R^2 T$

(*c*) $16\pi R^2$ (*d*) $36\pi R^2 T$.

13. The lower end of a capillary tube is at a depth of 12 cm and the water rises 3 cm in it. The mouth pressure required to blow an air bubble at the lower end will be x cm of water column, where x is

(*a*) 12 (*b*) 15

(*c*) 3 (*d*) 9.

14. A film of soap solution is trapped between a vertical frame and a light wire *ab* of length 0.1 m. If $g = 10$ m s^{-2} and $\sigma = 25 \times 10^{-3}$ N m^{-1}, then the load W that should be suspended from the wire to keep it in equilibrium is

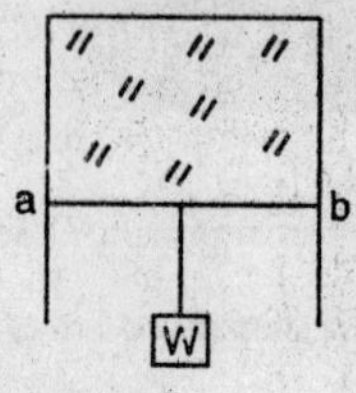

Fig. 1

(*a*) 0.2 g (*b*) 0.3 g
(*c*) 0.4 g (*d*) 0.5 g.

15. The addition of soap changes the surface tension of water to σ_1 and that of salt changes to σ_2. Then

(*a*) $\sigma_1 > \sigma_2$ (*b*) $\sigma_1 < \sigma_2$
(*c*) $\sigma_1 = \sigma_2$
(*d*) it is not possible to predict the above.

16. Water rises in a vertical glass capillary tube up to a length of 8 cm. If the tube is inclined at 45°, the length of water column in the tube will be

(*a*) 4 cm (*b*) 3 cm
(*c*) $8\sqrt{2}$ cm (*d*) $\frac{8}{\sqrt{2}}$ cm.

QUESTION BANK

Level II

Choose the correct answer from the four alternatives given in each of the following questions :

1. A number of droplets, each of radius *r*, combine to form a drop of radius R. If T is the surface tension, then the rise in temperature will be

(*a*) $\frac{2T}{r}$ (*b*) $\frac{3T}{R}$
(*c*) $2T\left[\frac{1}{r}-\frac{1}{R}\right]$ (*d*) $3T\left[\frac{1}{r}-\frac{1}{R}\right]$.

2. A wire ring of diameter 14 cm is gently lowered on to a liquid surface and then pulled up. When the film just breaks, the force required is 0.0616 N. The surface tension of the liquid is

(*a*) 70 N m^{-1} (*b*) 7 N m^{-1}
(*c*) 70 dyne cm^{-1} (*d*) None of these.

3. When two soap bubbles of radii r_1 and r_2 ($r_2 > r_1$) coalesce, the radius of curvature of common surface is

(*a*) $r_2 - r_1$ (*b*) $\frac{r_2 - r_1}{r_1 r_2}$
(*c*) $\frac{r_1 r_2}{r_2 - r_1}$ (*d*) $r_2 + r_1$.

4. The angle of contact between glass and water is 0° and water (surface tension 70 dyne cm^{-1}) rises in a glass capillary upto 6 cm. Another liquid of surface tension 140 dyne cm^{-1}, angle of contact 60° and relative density 2 will rise in the same capillary up to

(*a*) 12 cm (*b*) 24 cm
(*c*) 3 cm (*d*) 6 cm.

5. If work W is done in blowing a bubble of radius R from a soap solution, then the work done in blowing a bubble of radius 2R from the same solution is

(*a*) W/2 (*b*) 2W
(*c*) 4W (*d*) $2\frac{1}{3}$ W.

6. Water rises to a height of 2 cm in a capillary tube. If the tube is tilted 60° from the vertical, water will rise in the tube to a length

(*a*) 4.0 cm (*b*) 2.0 cm
(*c*) 1.0 cm (*d*) water will not rise at all.

7. A wire of mass 1 g is kept horizontally on the surface of water. The minimum length of the wire that does not break the surface film is (surface tension of water is 70 dyne cm^{-1})

(*a*) 3 cm (*b*) 4 cm
(*c*) 7 cm (*d*) 14 cm.

8. The surface tension of soap solution is 25×10^{-3} N m^{-1}. The excess pressure inside a soap bubble of diameter 1 cm is

(*a*) 10 Pa (*b*) 20 Pa
(*c*) 5 Pa (*d*) None of these.

9. Water rise in a capillary tube when its one end is dipped vertically in it, is 3 cm. If the surface tension of water is 75×10^{-3} N/m, then the diameter of capillary will be

(*a*) 0.1 mm (*b*) 0.5 mm
(*c*) 1.0 mm (*d*) 2.0 mm.

10. The surface tensions of two liquids are, respectively 30 and 60 dyne cm^{-1}. The liquids drop from the ends of two tubes of the same radius. The ratio of the weights of the two drops is

(a) 1 : 2 (b) 1 : 3
(c) 2 : 3 (d) 3 : 4.

11. Two glass plates are separated by water. If surface tension of water is 75 dyne per cm and the area of each plate wetted by water is 8 cm^2 and the distance between the plates is 0.12 mm, then the force applied to separate the two plates is

(a) 10^2 dyne (b) 10^4 dyne
(c) 10^5 dyne (d) 10^6 dyne.

12. Two spherical soap bubbles of radii r_1 and r_2 in vacuum coalesce under isothermal conditions. The resulting bubble has a radius R such that

(a) $R = r_1 + r_2$ (b) $R = (r_1 + r_2)^{1/2}$
(c) $R = \sqrt{r_1^2 + r_2^2}$ (d) $R = [r_1 r_2(r_1 + r_2)]$.

13. A ring is cut from a platinum tube 8.5 cm internal and 8.7 cm external diameter. It is supported horizontally from the pan of a balance, so that it comes in contact with the water in a glass vessel. If an extra 3.97 g.f. is required to pull it away from water, the surface tension of water is

(a) 72 dyne cm^{-1} (b) 70.80 dyne cm^{-1}
(c) 63.35 dyne cm^{-1} (d) 60 dyne cm^{-1}.

14. Two vertical glass plates 1 mm apart are dipped into water. How high will the water rise between the plates, if the surface tension of water is 70 dyne cm^{-1} ?

(a) 1. 43 cm (b) 1.63 cm
(c) 2.86 cm (d) 3.86 cm.

15. A hollow sphere has a small hole in it. On lowering the sphere in a tank of water, it is observed that water enters into the hollow sphere at a depth of 40 cm below the surface. Surface tension of water is 7×10^{-2} N m^{-1}. The diameter of the hole is

(a) $\frac{1}{28}$ mm (b) $\frac{1}{21}$ mm
(c) $\frac{1}{14}$ mm (d) $\frac{1}{7}$ mm.

16. A soap bubble has radius r. The work done in increasing its radius to three times its original radius, without any rise of temperature, is (Given : surface tension of soap solution is T)

(a) $12\pi r^2$T (b) $16\pi r^2$T
(c) $64\pi r^2$T (d) $48\pi r^2$T.

17. A soap film measures 10 cm × 6 cm. It is increased to 10 cm × 12 cm. If surface tension is 30×10^{-3} newton per metre, then the work done is

(a) 3.6×10^0 J (b) 3.6×10^{-2} J
(c) 3.6×10^{-4} J (d) 3.6×10^3 J.

18. The energy E supplied to a soap bubble of radius r and surface tension σ doubles the radius of the soap bubble under isothermal conditions. The value of E is

(a) $8\pi r^2\sigma$ (b) $12\pi r^2\sigma$
(c) $16\pi r^2\sigma$ (d) $24\pi r^2\sigma$.

19. A water drop of radius r is divided into eight equal droplets. What is the work done in the process ? Given : surface tension = σ.

(a) $2\pi r^2\sigma$ (b) $3\pi r^2\sigma$
(c) $4\pi r^2\sigma$ (d) $24\pi r^2\sigma$.

20. The excess pressure due to surface tension inside a spherical drop is 6 units. If eight such drops combine, then the excess pressure due to surface tension inside the larger drop is

(a) 3 units (b) 6 units
(c) 12 units (d) 48 units.

21. Work W is required to be done in forming a bubble of volume V from a given solution. What amount of work is required to be done to form a bubble of volume 2V ?

(a) W (b) 2 W
(c) $2^{1/3}$ W (d) $4^{1/3}$ W.

22. A square wire frame of length l is dipped in a solution. When the frame is taken out, a liquid film is formed. What is the force on the frame due to surface tension of the liquid ? Given : σ = surface tension of the liquid.

(a) $4\sigma l$ (b) $8\sigma l$
(c) $10\sigma l$ (d) $12\sigma l$.

23. A long cylindrical vessel has a small hole of diameter D at its bottom. This vessel can be lowered vertically in water to a depth h without any water entering the vessel. Given : A = surface tension, B = density and C = acceleration due to gravity. The value of h is

(a) $\frac{A}{DBC}$ (b) $\frac{2A}{DBC}$
(c) $\frac{3A}{DBC}$ (d) $\frac{4A}{DBC}$.

24. A capillary tube of radius r is immersed in a liquid. The liquid rises to a height h. The corresponding mass is m. What mass of water shall rise in the capillary if the radius of the tube is doubled ?

(a) m (b) $2m$
(c) $3m$ (d) $4m$.

25. The excess pressure inside the first soap bubble is three times that inside the second bubble. Then the ratio of the volumes of the first and second bubbles is

(a) 1 : 3 (b) 1 : 9
(c) 1 : 27 (d) 3 : 1.

26. A metal wire of density d floats, in a liquid of surface tension σ, in the horizontal position. The maximum radius of the wire so that it may not sink is

(a) $\sqrt{\frac{2\sigma}{\pi dg}}$ (b) $\sqrt{\frac{3\sigma}{\pi dg}}$

(c) $\sqrt{\frac{5\sigma}{\pi dg}}$ (d) $\sqrt{\frac{7\sigma}{\pi dg}}$.

27. Two soap bubbles, one of radius 50 mm and the other of radius 80 mm are brought in contact so that they have a common interface. The radius of curvature of common interface is

(a) 0.003 m (b) 0.133 m

(c) 1.2 m (d) 8.9 m.

28. The surface energy of a liquid drop is E. It is sprayed into 1000 equal droplets. Then its surface energy becomes

(a) 1000 E (b) 100 E

(c) 10 E (d) E.

29. In Q. 28, the work done in spraying is

(a) 999 E (b) 99 E

(c) 9 E (d) E.

30. The lower end of a capillary tube of radius r is placed vertically in water. With the rise h of water in the capillary tube, heat evolved is (Take ρ as the density of water.)

(a) $\frac{2\pi}{\rho g} \times h$ (b) $\frac{\pi}{\rho g} \times hr$

(c) zero (d) $\frac{\pi r^2 h^2 \rho g}{2}$.

31. Two mercury drops, each of radius r, merge to form a bigger drop. If σ is the surface tension of mercury, then the surface energy released is

(a) $8\pi r^2\sigma - 4 \times 2^{2/3}\pi r^2\sigma$ (b) zero

(c) negative (d) $4\pi r^2\sigma - 8 \times 2^{1/3} r^2\sigma$.

32. A long cylindrical glass vessel has a small hole of radius r at its bottom. The depth to which the vessel can be lowered vertically in a deep water bath, without water entering the vessel, is (σ = surface tension)

(a) $\frac{2\sigma}{rdg}$ (b) $\frac{3\sigma}{rdg}$

(c) $\frac{4\sigma}{rdg}$ (d) $\frac{\sigma}{rdg}$.

33. A soap film of surface tension 3×10^{-2} N m^{-1} formed in a rectangular frame can support a straw as shown in Fig. 2. If $g = 10$ m s^{-2}, the mass of the straw is

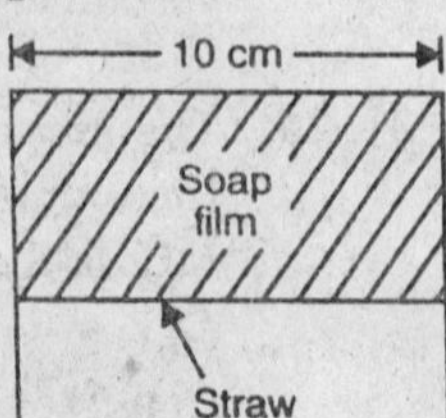

Fig. 2

(a) 0.006 g (b) 0.06 g

(c) 0.6 g (d) 6 g.

34. A cube with a mass m = 20 g wettable by water floats on the surface of water. Each face of the cube is a = 3 cm long. Surface tension of water is 70 dyne cm^{-1}. The distance of the lower face of the cube from the surface of water is : (g = 980 cm s^{-2})

(a) 2.3 cm (b) 4.6 cm

(c) 9.7 cm (d) 12.4 cm.

35. Consider a vertical tube of semi-circular cross-section dipped in a liquid. Assume that the wetting of the tube is complete. The forces of surface tension on the flat part and on the curved part of the tube are in the ratio

(a) 2 : π (b) 1 : π

(c) 3 : π (d) 2.7 : π.

36. A capillary tube is kept vertical with the lower end in water. The height of water raised in the capillary is 4 cm. If the length of the capillary tube is made 2 cm, then the angle made by the water surface in the capillary with the wall is

(a) 0° (b) 30°

(c) 60° (d) 90°.

37. A glass rod of radius r_1 is inserted symmetrically into a vertical capillary tube of radius r_2 such that their lower ends are at the same level. The arrangement is now dipped in water. The height to which water will rise into the tube will be (σ = surface tension of water, ρ = density of water)

(a) $\frac{2\sigma}{(r_2 - r_1)\rho g}$ (b) $\frac{\sigma}{(r_2 - r_1)\rho g}$

(c) $\frac{2\sigma}{(r_2 + r_1)\rho g}$ (d) $\frac{2\sigma}{(r_2^2 + r_1^2)\rho g}$.

38. A drop of liquid of density ρ is floating half-immersed in a liquid of density d. If σ is the surface tension, then the diameter of the drop of the liquid is

(a) $\sqrt{\frac{\sigma}{g(2\rho - d)}}$ (b) $\sqrt{\frac{2\sigma}{g(2\rho - d)}}$

(c) $\sqrt{\frac{6\sigma}{g(2\rho - d)}}$ (d) $\sqrt{\frac{12\sigma}{g(2\rho - d)}}$.

39. Work done in splitting a drop of water of 1 mm radius into 10^6 droplets is (Surface tension of water = 72×10^{-3} J/m^2)

(a) 9.58×10^{-5} J (b) 8.96×10^{-5} J

(c) 5.89×10^{-5} J (d) 5.98×10^{-6} J.

40. Water rises to a height of 10 cm in capillary tube and mercury falls to a depth of 3.1 cm in the same capillary tube. If the density of mercury is 13.6 and the angle of contact for mercury is 135°, the approximate ratio of surface tensions of water and mercury is

(*a*) 1 : 0.15 (*b*) 1 : 3

(*c*) 1 : 6 (*d*) 1.5 : 1.

41. A drop of water of volume V is pressed between the two glass plates so as to spread to an area A. If T is the surface tension, the normal force required to separate the glass plates is

(*a*) $\frac{TA^2}{V}$ (*b*) $\frac{2TA^2}{V}$

(*c*) $\frac{4TA^2}{V}$ (*d*) $\frac{TA^2}{2V}$.

42. A large number of droplets, each of radius *a*, coalesce to form a bigger drop of radius *b*. Assume that the energy released in the process is converted into the kinetic energy of the drop. The velocity of the drop is (σ = surface tension, ρ = density)

(*a*) $\left[\frac{\sigma}{\rho}\left(\frac{1}{a}-\frac{1}{b}\right)\right]^{1/2}$ (*b*) $\left[\frac{2\sigma}{\rho}\left(\frac{1}{a}-\frac{1}{b}\right)\right]^{1/2}$

(*c*) $\left[\frac{3\sigma}{\rho}\left(\frac{1}{a}-\frac{1}{b}\right)\right]^{1/2}$ (*d*) $\left[\frac{6\sigma}{\rho}\left(\frac{1}{a}-\frac{1}{b}\right)\right]^{1/2}$.

43. 125 water droplets, each of radius *r*, coalesce to form a single drop. The energy released raises the temperature of the drop. If σ represents surface tension, ρ represents density, S represents specific heat and J represents mechanical equivalent of heat, then the rise in temperature of the drop is

(*a*) $\frac{12\sigma}{5Jr\rho S}$ (*b*) $\frac{12\sigma\rho}{7JrS}$

(*c*) $\frac{2\sigma\rho}{SJr}$ (*d*) Zero.

44. Two vertical parallel glass plates are partially submerged in water. The distance between the plates is *d* and their width is *l*. Assume that the water between the plates does not reach the upper edges of the plates and that the wetting is complete. The water will rise to height (ρ = density of water and σ = surface tension of water)

(*a*) $\frac{2\sigma}{\rho gd}$ (*b*) $\frac{3\sigma}{\rho gd}$

(*c*) $\frac{4\sigma}{\rho gd}$ (*d*) $\frac{5\sigma}{\rho gd}$.

45. In Q. 44, the force of attraction between the plates is

(*a*) $\frac{l\rho g}{2}h^2$ (*b*) $\frac{l\rho g}{2}h$

(*c*) $\frac{l\rho g}{2}$ (*d*) $\frac{l\rho g}{2h}$.

[Here *h* represents the height to which the liquid rises.]

46. In Q. 45, if water is replaced by another liquid of surface tension 2σ, then the force F of attraction between two plates would become

(*a*) 2F (*b*) 4F

(*c*) 8F (*d*) 16F.

47. Two spherical soap bubbles coalesce. If V is the consequent change in volume of the contained air, S is the total change in surface area and σ is the surface tension of soap solution, then

(*a*) $P_0V + 4S\sigma = 0$ (*b*) $3P_0V + 4\sigma S = 0$

(*c*) $4P_0V + 3\sigma S = 0$ (*d*) $P_0V - \sigma S = 0$.

[Here P_0 represents atmospheric pressure.]

48. Two soap bubbles of radii *a* and *b* combine to form a single bubble of radius *c*. If P is the external pressure, then the surface tension of the soap solution is

(*a*) $\frac{P(c^3+a^3+b^3)}{4(a^2+b^2-c^2)}$ (*b*) $\frac{P(c^3-a^3-b^3)}{4(a^2+b^2-c^2)}$

(*c*) $Pc^3 - 4a^2 - 4b^2$ (*d*) $Pc^2 - 2a^2 + 3b^2$.

49. A thin square plate of side 5 cm is suspended vertically from a balance so that lower side just dips into water with side parallel to surface. When the plate is clean ($\theta = 0°$), it appears to weigh 0.044 N. But when the plate is greasy (θ = 180°), it appears to weigh 0.03 N. The surface tension of water is

(*a*) 35×10^{-2} N m^{-1} (*b*) 7.0×10^{-2} N m^{-1}

(*c*) 14.0×10^{-2} N m^{-1} (*d*) 1.08 N m^{-1}.

50. An ice cube of edge 1 cm melts in a gravity-free container. The approximate surface area of water formed is

(*a*) $[36\pi]^{1/2}$ cm^2 (*b*) $[36\pi]^{1/3}$ cm^2

(*c*) $[36\pi]^{2/3}$ m^2 (*d*) 36π m^2.

51. A loop of 6.28 cm long thread is put gently on a soap film in a wire loop. The film is pricked with a needle inside the soap film enclosed by the thread. If surface tension of soap solution is 0.030 N m^{-1}, then the tension in the thread is

(*a*) 1×10^{-4} N (*b*) 2×10^{-4} N

(*c*) 3×10^{-4} N (*d*) 4×10^{-4} N.

52. If the radii of two soap bubbles are R_1 and R_2, then the ratio of the respective masses of air in them will be (σ = surface tension and P_0 = atmospheric pressure)

(*a*) $\frac{R_1}{R_2}$ (*b*) $\frac{R_2}{R_1}$

(*c*) $\frac{P_0+\frac{4\sigma}{R_1}}{P_0+\frac{4\sigma}{R_2}}$ (*d*) $\frac{P_0+\frac{4\sigma}{R_1}}{P_0+\frac{4\sigma}{R_2}}\left[\frac{R_1}{R_2}\right]^3$.

ANSWERS

QUESTION BANK (Level I)

1. (*a*)	2. (*a*)	3. (*c*)	4. (*c*)	5. (*b*)	6. (*b*)	7. (*c*)	8. (*c*)
9. (*a*)	10. (*a*)	11. (*c*)	12. (*d*)	13. (*b*)	14. (*d*)	15. (*b*)	16. (*c*).

QUESTION BANK (Level II)

1. (*d*)	2. (*c*)	3. (*c*)	4. (*c*)	5. (*c*)	6. (*a*)	7. (*c*)	8. (*b*)
9. (*c*)	10. (*a*)	11. (*c*)	12. (*c*)	13. (*a*)	14. (*a*)	15. (*c*)	16. (*c*)
17. (*c*)	18. (*d*)	19. (*c*)	20. (*a*)	21. (*d*)	22. (*b*)	23. (*d*)	24. (*b*)
25. (*c*)	26. (*a*)	27. (*b*)	28. (*c*)	29. (*c*)	30. (*d*)	31. (*a*)	32. (*a*)
33. (*c*)	34. (*a*)	35. (*a*)	36. (*c*)	37. (*a*)	38. (*d*)	39. (*b*)	40. (*c*)
41. (*b*)	42. (*d*)	43. (*a*)	44. (*a*)	45. (*a*)	46. (*b*)	47. (*b*)	48. (*b*)
49. (*b*)	50. (*b*)	51. (*c*)	52. (*d*).				

HINTS/SOLUTIONS

QUESTION BANK (Level I)

1. (*a*) $h = \frac{2\sigma \cos \theta}{r\rho g}$ or $h = \frac{2\sigma \cos \theta}{(R \cos \theta)\rho g}$

or $h = \frac{2\sigma}{R\rho g}$ or $hR = \text{constant}$

If h decreases, the radius of curvature of the meniscus increases.

2. (*a*) Increase in surface area $= n \times 4\pi r^2 - 4\pi R^2$

Required energy is equal to the product of surface tension and increase in surface area.

3. (*c*) Effective area $= 2 \times 0.02 \text{ m}^2 = 0.04 \text{ m}^2$

Surface energy $= 5 \text{ N m}^{-1} \times 0.04 \text{ m}^2 = 2 \times 10^{-1}$ J

4. (*c*) Since conditions are isothermal, therefore, energy will be conserved.

$$2[2 \times 4\pi r^2 \sigma] = 2 \times 4\pi R^2 \sigma$$

or $R^2 = 2r^2$ or $R = 2^{1/2} r$

5. (*b*) $\frac{4}{3}\pi R^3 = 2 \times \frac{4}{3}\pi r^3$ or $R = 2^{1/3} r$

6. (*b*) Effective length $= 2\pi r + 2\pi R$

7. (*c*) $F_A < \frac{F_C}{\sqrt{2}}$ or $F_C > \sqrt{2}\, F_A$

Clearly, the cohesive force dominates. Now, think of mercury.

8. (*c*) $2 \times 4\pi \times 3^2 + 2 \times 4\pi \times 4^2 = 2 \times 4\pi R^2$

or $R^2 = 9 + 16 = 25$ or $R = 5$ cm

9. (*a*) Area reduced to one-fourth. Radius reduced to one-half. Since $h \propto \frac{1}{r}$, therefore, h is doubled.

10. (*a*) $F = (\sigma_1 - \sigma_2)l$

11. (*c*) $\frac{4\sigma}{R} = h'\rho g$

But $\cos\theta = \frac{r}{R}$ or $R = \frac{r}{\cos\theta}$

$\therefore \frac{4\sigma \cos\theta}{r} = h'\rho g$

or $h' = \frac{2\,(2\sigma\cos\theta)}{r\rho g}$

or $h' = 2h$

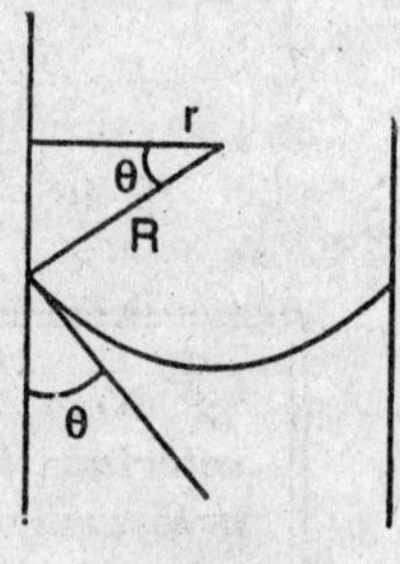

Fig. 3

12. (*d*) $1000 \times \frac{4}{3}\pi r^3 = \frac{4}{3}\pi R^3$ or $10r = R$

Increase in surface area

$= 1000 \times 4\pi r^2 - 4\pi R^2$

$= 1000 \times 4\pi \times \frac{R^2}{100} - 4\pi R^2 = 36\pi R^2$

Change in surface energy $= 36\pi R^2 T$

13. (*b*) Pressure due to 15 cm long liquid column needs to be balanced.

14. (*d*) $25 \times 10^{-3} \times 2 \times 0.1 = m \times 10$

or $m = \frac{5 \times 10^{-3}}{10}$ kg $= 0.5$ g

15. (*b*) When a substance is highly soluble in a liquid, the surface tension is increased.

Example : NaCl, $ZnSO_4$ etc. dissolved in water.

When the substance is sparingly soluble, the surface tension is reduced.

Example : Phenol or soap solution in water.

16. (c) $\cos 45° = \frac{8}{l}$

or $l = \frac{8}{\cos 45°}$

or $l = 8\sqrt{2}$ cm

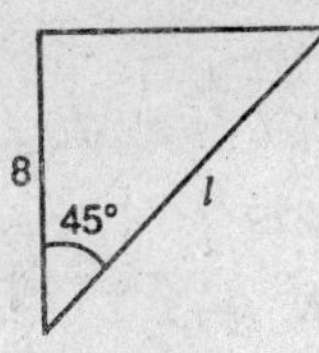

Fig. 4

QUESTION BANK (Level II)

1. (d) $\frac{4}{3}\pi R^3 = n \times \frac{4}{3}\pi r^3$ or $R^3 = nr^3$

Energy evolved = $[n \times 4\pi r^2 - 4\pi R^2]$ T

Now, $\theta = \frac{Q}{V\rho S} = \frac{Q}{V}$ [In cgs system, ρ = 1 and S = 1]

$\therefore \quad \theta = \frac{[n \times 4\pi r^2 - 4\pi R^2]T}{\frac{4}{3}\pi R^3}$

or $\theta = \frac{3T\,[nr^2 - R^2]}{R^3} = 3T\left[\frac{nr^2}{nr^3} - \frac{R^2}{R^3}\right] = 3T\left[\frac{1}{r} - \frac{1}{R}\right]$

2. (c) $2 \times 2\pi r \times \sigma = 0.0616 \times 10^5$

or $\sigma = \frac{6160 \times 7}{2 \times 22 \times 14}$ dyne cm^{-1} = 70 dyne cm^{-1}

3. (c) $\frac{4\sigma}{R} = \frac{4\sigma}{r_1} - \frac{4\sigma}{r_2}$ or $\frac{1}{R} = \frac{1}{r_1} - \frac{1}{r_2}$ or $R = \frac{r_1 r_2}{r_2 - r_1}$

> ☞ ***Note that the radius of curvature of the interface is determined by the difference of pressures of the two bubbles.***

4. (c) $\frac{h_2}{h_1} = \frac{\sigma_2 \cos\theta_2}{\rho_2} \times \frac{\rho_1}{\sigma_1 \cos\theta_1}$

or $\frac{h_2}{h_1} = \frac{140 \times \frac{1}{2}}{2} \times \frac{1}{70 \times 1}$ or $h_2 = \frac{h_1}{2} = \frac{6}{2}$ cm = 3 cm

5. (c) $W = 2 \times 4\pi R^2 \times \sigma$; R is increased by a factor of 2. So, W is increased by a factor of 4.

6. (a) $l \cos 60° = 2$ or $l = 2 \times 2$ cm = 4 cm

7. (c) $2l\sigma = 1 \times 980$ or $l = \frac{980}{2 \times 70}$ cm = 7 cm

8. (b) $p = \frac{4\sigma}{r} = \frac{8\sigma}{D} = \frac{8 \times 25 \times 10^{-3}}{1 \times 10^{-2}}$ N m^{-2}

$= 20$ N m^{-2} or 20 Pa

9. (c) $h = \frac{2\sigma\cos\theta}{r\rho g}$ or $r = \frac{2\sigma\cos\theta}{h\rho g}$

or $r = \frac{2 \times 75 \times 10^{-3} \times \cos 0°}{3 \times 10^{-2} \times 10^3 \times 10}$ m $= 5 \times 10^{-4}$ m

10. (a) Weight = $2\pi r\sigma$

$\frac{W_1}{W_2} = \frac{\sigma_1}{\sigma_2} = \frac{30}{60} = \frac{1}{2}$

11. (c) $F = \frac{2A\sigma}{t} = \frac{2 \times 8 \times 75 \times 10}{0.12}$ dyne $= 10^5$ dyne

12. (c) $2 \times 4\pi R^2\sigma = 2 \times 4\pi r_1^2\sigma + 2 \times 4\pi r_2^2\sigma$

or $R^2 = r_1^2 + r_2^2$ or $R = \sqrt{r_1^2 + r_2^2}$

13. (a) $\left[2\pi \times \frac{8.7}{2} + 2\pi \times \frac{8.5}{2}\right]\sigma = 3.97 \times 980$

or $\sigma = \frac{3.97 \times 980 \times 7}{22 \times 17.2}$ dyne cm^{-1} = 72 dyne cm^{-1}

14. (a) $h = \frac{2 \times 70}{10^{-1} \times 1 \times 980}$ cm = 1.43 cm

15. (c) $\frac{40}{100} \times 1000 \times 9.8 = \frac{2 \times 7 \times 10^{-2}}{R}$

or $R = \frac{14 \times 10^{-2} \times 100}{40 \times 1000 \times 9.8}$ m $= \frac{14 \times 1000}{40 \times 1000 \times 9.8}$ mm $= \frac{1}{28}$ mm

Diameter = 2R = $\frac{1}{14}$ mm

16. (c) $W = [2 \times 4\pi(3r)^2 - 2 \times 4\pi r^2]$ T

$W = 2 \times 4\pi r^2 \times 8T = 64\pi r^2$ T

17. (c) $W = 2[120 \times 10^{-4} - 60 \times 10^{-4}]\, 30 \times 10^{-3}$ J

$= 2 \times 60 \times 10^{-4} \times 30 \times 10^{-3}$ J $= 3.6 \times 10^{-4}$ J

18. (d) $E = [2 \times 4\pi\,(2r)^2 - 2 \times 4\pi r^2]\,\sigma$

$= 2 \times 4\pi r^2 \times 3 \times \sigma = 24\pi r^2\sigma$

19. (c) $\frac{4}{3}\pi r^3 = 8 \times \frac{4}{3}\pi r'^3$; $r' = \frac{r}{2}$

Increase in surface area = $8 \times 4\pi\left(\frac{r}{2}\right)^2 - 4\pi r^2 = 4\pi r^2$

Work done = $4\pi r^2 \times \sigma$

20. (a) $8 \times \frac{4}{3}\pi r^3 = \frac{4}{3}\pi R^3$; $R = 2r$

$\frac{2\sigma}{P} = \frac{2 \times 2\sigma}{p}$ or $P = \frac{p}{2} = \frac{6}{2}$ units = 3 units

21. (*d*) $V = \frac{4}{3}\pi R^3$; $2V = \frac{4}{3}\pi R'^3$; $2 \times \frac{4}{3}\pi R^3 = \frac{4}{3}\pi R'^3$; $R' = 2^{1/3} R$

$W' = 2 \times 4\pi [2^{1/3} R]^2 \sigma = 2^{2/3} \times 2 \times 4\pi R^2 \sigma = 4^{1/3} W$

22. (*b*) It may be noted that the soap film has two free surfaces. So, the effective length is $8l$.

23. (*d*) $h = \frac{2\sigma}{r\rho g} = \frac{2\sigma \times 2}{D\rho g} = \frac{4\sigma}{D\rho g} = \frac{4A}{DBC}$

24. (*b*) Mass of liquid which rises in the capillary,

$$m = (\pi r^2)\, h\rho = \pi r^2 \times \frac{2\sigma\cos\theta}{r\rho g} \times \rho \text{ or } m \propto r$$

Clearly, if the radius is doubled, the mass is doubled.

25. (*c*) $\frac{r_1}{r_2} = \frac{p_2}{p_1} = \frac{1}{3}$; $\quad \frac{V_1}{V_2} = \frac{r_1^3}{r_2^3} = \frac{1}{27}$

26. (*a*) $\sigma \times 2l = l \times \pi r^2 \times d \times g$

or $r^2 = \frac{2\sigma}{\pi d g}$ or $r = \sqrt{\frac{2\sigma}{\pi d g}}$

27. (*b*) $\left[P_0 + \frac{4\sigma}{R_2}\right] - \left[P_0 + \frac{4\sigma}{R_1}\right] = \frac{4\sigma}{R}$

or $\frac{1}{R} = \frac{1}{R_2} - \frac{1}{R_1}$

or $R = \frac{R_1R_2}{R_1 - R_2} = \frac{50 \times 80}{30}\text{ mm} = \frac{400}{3}\text{ mm}$

$= \frac{400}{3 \times 1000}\text{ m} = \frac{4}{30}\text{ m} = 0.133\text{ m}$

28. (*c*) Final surface energy $= 1000 \times 4\pi r^2\sigma$

Initial surface energy, $E = 4\pi R^2\sigma$

Again, $\frac{4}{3}\pi R^3 = 1000 \times \frac{4}{3}\pi r^3$ or $R = 10r$

Now, $\frac{1000 \times 4\pi r^2\sigma}{4\pi R^2\sigma} = \frac{10 \times 4\pi R^2\sigma}{E} = \frac{10E}{E} = 10$

29. (*c*) Work done $= 10E - E = 9E$.

31. (*a*) $2 \times \frac{4}{3}\pi r^3 = \frac{4}{3}\pi R^3$ or $R = 2^{1/3} r$

Final surface area $= 4\pi R^2 = 4\pi 2^{2/3} r^2$

Initial surface area $= 2 \times 4\pi r^2$

$\therefore$ Energy released $= [8\pi r^2 - 4 \times 2^{2/3}\pi r^2]\,\sigma$

32. (*a*) $hdg = \frac{2\sigma}{r}$ or $h = \frac{2\sigma}{rdg}$

33. (*c*) $m \times 10 = 2 \times 3 \times 10^{-2} \times \frac{10}{100}$

or $m = 6 \times 10^{-4}\text{ kg} = 6 \times 10^{-4} \times 10^3\text{ g} = 0.6\text{ g}$

34. (*a*) If y is the required distance, then

$a^2 y\rho g = mg + 4a\sigma$ or $y = \frac{mg \times 4a\sigma}{a^2\rho g}$

$= \frac{20 \times 980 + 4 \times 3 \times 70}{3 \times 3 \times 1 \times 980}\text{ cm} = 2.3\text{ cm}$

35. (*a*) Effective length of flat part $= 2r$

Effective length of curved part is πr.

$F = \sigma l$

So, required ratio is $2 : \pi$

36. (*c*) $4 \propto \cos 0°$; $2 \propto \cos\theta$

or $\frac{\cos\theta}{\cos 0°} = \frac{1}{2}$ or $\theta = 60°$

37. (*a*) Total upward force due to surface tension $= \sigma\,[2\pi r_1 + 2\pi r_2]$. This supports the weight of the liquid column of height h.

Weight of liquid column $= h\left[\pi r_2^2 - \pi r_1^2\right]\rho g$

Equating, $h\pi\,(r_2 + r_1)(r_2 - r_1)\,\rho g = 2\pi\sigma\,(r_1 + r_2)$

or $h(r_2 - r_1)\,\rho g = 2\sigma$ or $h = \frac{2\sigma}{(r_2 - r_1)\rho g}$

38. (*d*) $2\pi r\sigma + \frac{1}{2} \times \frac{4}{3}\pi r^3 dg = \frac{4}{3}\pi r^3 \rho g$

or $2\pi r\sigma = \frac{\pi r^3 g}{3}[4\rho - 2d]$ or $r^2 = \frac{3 \times 2\pi\sigma}{\pi g\,[4\rho - 2d]}$

or $r^2 = \frac{3\sigma}{g(2\rho - d)}$ or $r = \sqrt{\frac{3\sigma}{g(2\rho - d)}}$

Diameter $= 2r = \sqrt{\frac{12\sigma}{g(2\rho - d)}}$

39. (*b*) $10^6 r^3 = R^3$ or $r = \frac{R}{100}$ or $r = \frac{1 \times 10^{-3}}{100}\text{ m} = 10^{-5}\text{ m}$

$W = [10^6 \times 4\pi r^2 - 4\pi R^2]\sigma$

$= 4 \times \frac{22}{7}[10^6 \times 10^{-10} - 10^{-6}]\, 72 \times 10^{-3}\text{ J}$

$= \frac{88 \times 72}{7} \times 10^{-3} \times 10^{-6} \times 99\text{ J} = 8.96 \times 10^{-5}\text{ J}$

40. (*c*) $h = \frac{2\sigma\cos\theta}{r\rho g}$

or $h \propto \frac{\sigma\cos\theta}{\rho}$ or $\sigma \propto \frac{h\rho}{\cos\theta}$

$\frac{\sigma_w}{\sigma_m} = \frac{h_w\rho_w}{\cos\theta_w} \times \frac{\cos\theta_m}{h_m\rho_m}$

$= \frac{10 \times 1}{\cos 0°} \times \frac{\cos 135°}{-3.1 \times 13.6} = \frac{10 \times (-0.707)}{-3.1 \times 13.6} \approx \frac{1}{6}$

41. (*c*) Pressure difference

$= \frac{T}{R} = \frac{T}{t/2} = \frac{2T}{t}$

Force $= \frac{2T}{t}A$

But $V = At$ or $t = \frac{V}{A}$

$\therefore$ Force $= \frac{2TA^2}{V}$

t

Fig. 5

42. (d) Energy released = $[n \times 4\pi a^2 - 4\pi b^2]\ \sigma$

Now, $n \times \frac{4}{3}\pi a^3 = \frac{4}{3}\pi b^3$ or $n = \frac{b^3}{a^3}$

∴ Energy released

$$= \left[\frac{b^3}{a^3} \times 4\pi a^2 - 4\pi b^2\right]\sigma = 4\pi b^2\left[\frac{b}{a} - 1\right]\sigma$$

Now, $\frac{1}{2}\left(\frac{4}{3}\pi b^3\right)\rho v^2 = 4\pi b^2\left[\frac{b}{a} - 1\right]\sigma$

or $\frac{\rho}{6}v^2 = \left[\frac{1}{a} - \frac{1}{b}\right]\sigma$ or $v^2 = \frac{6\sigma}{\rho}\left[\frac{1}{a} - \frac{1}{b}\right]$

or $v = \left[\frac{6\sigma}{\rho}\left(\frac{1}{a} - \frac{1}{b}\right)\right]^{1/2}$

43. (a) $125 \times \frac{4}{3}\pi r^3 = \frac{4}{3}\pi R^3$ or $5r = R$

Energy released = $[125 \times 4\pi r^2 - 4\pi R^2]\ \sigma$

$= [125 \times 4\pi r^2 - 25 \times 4\pi r^2]\ \sigma = 400\ \pi r^2\sigma$

Now, $W = JQ = Jm\ S\theta$

or $400\pi r^2\sigma = J \times \frac{4}{3}\pi R^3 \rho\ S\theta$

or $\theta = \frac{3 \times 400\ \pi r^2\ \sigma}{J \times 4\pi \times 125 r^3 \rho S}$ or $\theta = \frac{12\sigma}{5Jr\rho S}$

44. (a) Total upward force due to surface tension = $2\sigma l$

Weight of lifted liquid = $(hld)\ \rho g$

Equating, $h = \frac{2\sigma}{\rho g d}$

45. (a) Increase in P.E. = $hld\ \rho g\left(\frac{h}{2}\right) = \frac{ld\rho g}{2}h^2$

If F is the force of attraction between the plates, then

$$Fd = \frac{ld\rho g}{2}h^2 \text{ or } F = \frac{l\rho g}{2}h^2$$

46. (b) $F \propto h^2$ But $h^2 \propto \sigma^2$ ∴ $F \propto \sigma^2$

If σ is doubled, then F will become 4F.

47. (b) Assume the conditions to be isothermal.

$PV = P_1V_1 + P_2V_2$

$$\left(P_0 + \frac{4\sigma}{R}\right)\left(\frac{4}{3}\pi R^3\right) = \left(P_0 + \frac{4\sigma}{R_1}\right)\left(\frac{4}{3}\pi R_1^3\right) + \left(P_0 + \frac{4\sigma}{R_2}\right)\left(\frac{4}{3}\pi R_2^3\right)$$

or $P_0\left[\frac{4}{3}\pi R^3 - \frac{4}{3}\pi R_1^3 - \frac{4}{3}\pi R_2^3\right]$

$= \frac{4\sigma}{3}[4\pi R_1^2 + 4\pi R_2^2 - 4\pi R^2]$

or $P_0V = -\frac{4\sigma}{3}S$ or $3P_0V + 4\sigma S = 0$

48. (b) Assuming isothermal conditions,

$$\left(P + \frac{4\sigma}{a}\right)\left(\frac{4}{3}\pi a^3\right) + \left(P + \frac{4\sigma}{b}\right)\left(\frac{4}{3}\pi b^3\right) = \left(P + \frac{4\sigma}{c}\right)\left(\frac{4}{3}\pi c^3\right)$$

or $P[a^3 + b^3 - c^3] = 4\sigma\ [c^2 - a^2 - b^2]$

or $\sigma = \frac{P(c^3 - a^3 - b^3)}{4(a^2 + b^2 - c^2)}$

49. (b) Difference in apparent weights is due to differences in forces of surface tension. Due to 180°, the force of surface tension in one case is opposite to the force of surface tension in the other case.

∴ $2 \times \sigma_w \times \frac{10}{100} = 0.044 - 0.03$

or $\sigma_w = \frac{0.014}{1} \times 5\ N\ m^{-1} = 0.07\ N\ m^{-1}$

50. (b) Water shall assume spherical shape.

$V = \frac{4}{3}\pi R^3$ or $1 = \frac{4}{3}\pi R^3$ or $R = \left[\frac{3}{4\pi}\right]^{1/3}$ cm

Surface area $= 4\pi R^2 = 4\pi\left[\frac{3}{4\pi}\right]^{2/3}$ cm^2

$= (4\pi)^{1/3} \times (9)^{1/3}$ cm^2 $= (36\pi)^{1/3}$ cm^2

51. (c) When the soap-film enclosed by the thread is pricked, the thread would take up a circular shape. Consider a small element of the thread. While horizontal components of tension get balanced, the vertical components get added up.

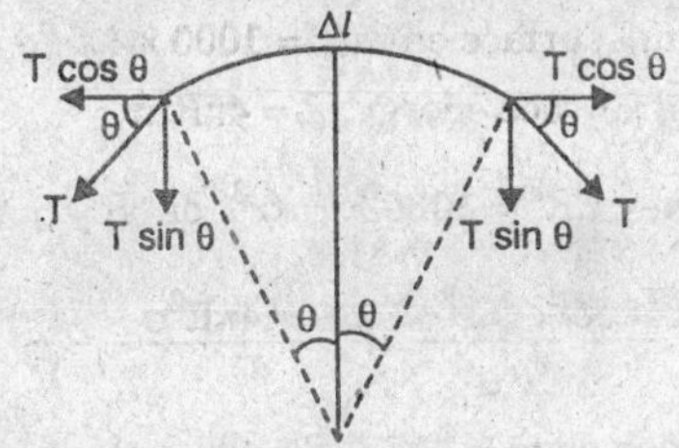

Fig. 6

or $2T \sin\theta = \sigma\ \Delta l$ or $2T\theta = \sigma(2\theta \times r)$

or $T = \sigma r = \sigma \times \frac{2\pi r}{2\pi}$ or $T = 0.030 \times \frac{6.28 \times 10^{-2}}{2\pi}$ N

$= 3 \times 10^{-4}$ N

52. (d) $M = \frac{4}{3}\pi R^3 \rho$

But $\rho \propto P$

∴ $M \propto PR^3$

8.4 FLUIDS IN MOTION

Important Terms, Facts and Formulae

BASIC CONCEPTS

1. Non-viscous and incompressible fluids are ideal fluids.
2. Stokes' law is valid under the following assumptions :
   (*i*) The viscous medium is homogeneous and of infinite extent.
   (*ii*) The spherical body is perfectly rigid and smooth.
   (*iii*) The medium is continuous, *i.e.*, the size of the moving body is much larger than the distance between the molecules of the medium.
   (*iv*) The body does not slip in the medium.
   (*v*) The velocity of the spherical body is less than critical velocity. So, the motion of the body through the fluid does not give rise to turbulent motion.
3. As water falls from a tap, the velocity of water goes on increasing. The cross-section of the water stream goes on decreasing [Equation of continuity.]
4. In wide and deep rivers, water has small velocity.
5. Bernoulli's equation is in accordance with the law of conservation of energy.

IMPORTANT POINTS TO REMEMBER

1. **1 decapoise** $= 1 \text{ N s m}^{-2} = 10^5 \times (10^2)^{-2}$ dyne s cm^{-2} **= 10 poise**
2. According to Stokes' law, $F = 6\pi\eta rv$
3. According to Poiseuille's equation,
$$V = \frac{\pi p r^4}{8\eta l}$$
where the letters have usual meanings.
4. For liquids of higher viscosity, Reynolds number is low.

QUESTION BANK

Level I

Choose the correct answer from the four alternatives given in each of the following questions :

1. A small spherical solid ball is dropped from a great height in a viscous liquid. Its journey in the liquid is best described, in the diagram given below, by

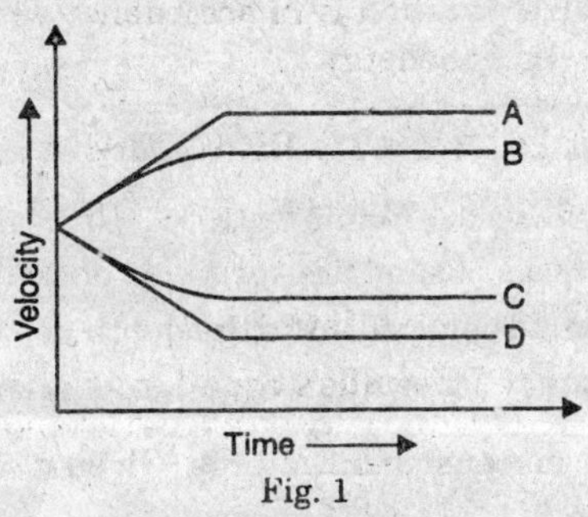

Fig. 1

(*a*) curve A (*b*) curve B
(*c*) curve C (*d*) curve D.

2. Water is flowing through a tube of non-uniform cross-section. If the radii of the tube at the entrance and exit are in the ratio 3 : 2, then the ratio of velocity of liquid entering and leaving the tube is

(*a*) 1 : 1 (*b*) 4 : 9
(*c*) 9 : 4 (*d*) 8 : 27.

3. 1 centipoise is equal to

(*a*) 1 kg m^{-1} s^{-1} (*b*) 1000 kg m^{-1} s^{-1}
(*c*) 0.1 kg m^{-1} s^{-1} (*d*) 0.001 kg m^{-1} s^{-1}.

4. A water tank of height 10 m, completely filled with water is placed on a level ground. It has two holes one at 3 m and the other at 7 m from its base. The water ejecting from

(*a*) both the holes will fall at the same spot
(*b*) upper hole will fall farther than that from the lower hole
(*c*) upper hole will fall closer than that from the lower hole
(*d*) more information is required.

5. A marble of mass x and diameter $2r$ is gently released in a tall cylinder containing honey. If the marble displaces mass y ($< x$) of the liquid, then the terminal velocity is proportional to

(*a*) $x + y$ (*b*) $x - y$
(*c*) $\frac{x+y}{r}$ (*d*) $\frac{x-y}{r}$.

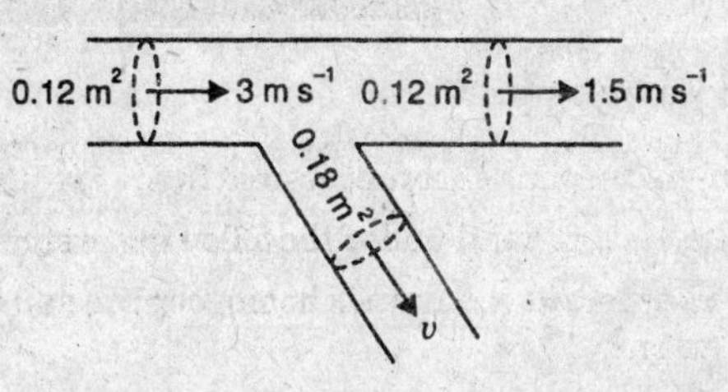

Fig. 2

6. An incompressible liquid travels as shown in Fig. 2. The speed of the fluid in the lower branch will be

(*a*) 1 m s^{-1} (*b*) 1.5 m s^{-1}
(*c*) 2.25 m s^{-1} (*d*) 3 m s^{-1}.

7. The rate of flow of a liquid through a capillary tube of radius r is V when the pressure difference across the two ends of the capillary is P. If pressure is increased by 3P and radius is reduced to $\frac{r}{2}$, then the rate of flow becomes

(*a*) $\frac{V}{8}$ (*b*) $\frac{V}{4}$
(*c*) $\frac{V}{2}$ (*d*) $\frac{V}{9}$.

8. A copper ball of radius r is moving with a uniform velocity v in the mustard oil. The dragging force acting on the ball is F. The dragging force on the copper ball of radius $2r$ moving with uniform velocity $2v$ in the mustard oil is

(*a*) F (*b*) 2F
(*c*) 4F (*d*) 8F.

9. A small steel ball of mass m and radius r is falling under gravity through a viscous liquid of coefficient of viscosity η. If g is the value of acceleration due to gravity, then the terminal velocity of the ball is proportional to

(*a*) $\frac{mg\eta}{r}$ (*b*) $mg\eta r$
(*c*) $\frac{mgr}{\eta}$ (*d*) $\frac{mg}{r\eta}$.

QUESTION BANK

Level II

Choose the correct answer from the four alternatives given in each of the following questions :

1. A cylinder containing water up to a height of 24 cm has a hole in its bottom. It is counterpoised in a balance. When water begins to flow out, the initial decrease in balancing weight is required to be 12 gf to keep the 'balance beam' horizontal. The cross-sectional area of the hole is

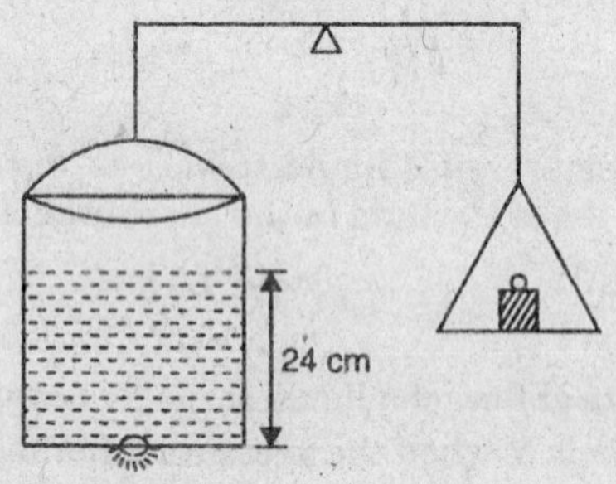

Fig. 3

(*a*) 0.12 cm^2 (*b*) 0.25 cm^2

(*c*) 0.50 cm^2 (*d*) 0.75 cm^2.

2. A small metal ball of diameter 4 mm and density 10.5 g cm^{-3} in dropped in glycerine of density 1.5 g cm^{-3}. The ball attains a terminal velocity of 8 cm s^{-1}. The coefficient of viscosity of glycerine is

(*a*) 4.9 poise (*b*) 9.8 poise

(*c*) 98 poise (*d*) 980 poise.

3. The velocity head of a stream of water is 40 cm. Then, the velocity of flow of water is

(*a*) 2.8 m s^{-1} (*b*) 5.6 m s^{-1}

(*c*) 7.2 m s^{-1} (*d*) 9.8 m s^{-1}.

4. A cylindrical tank has a hole of 1 cm^2 at its bottom. Water is flowing into the tank from a water reservoir above it. The maximum height upto which water can rise in the tank is 2.5 cm. The volume of water flowing into the tank per unit time is

(*a*) 20 cm^3 s^{-1} (*b*) 40 cm^3 s^{-1}

(*c*) 70 cm^3 s^{-1} (*d*) 90 cm^3 s^{-1}.

5. The vertical sections of the wing of a fan are shown in Fig. 4. Maximum upthrust is in

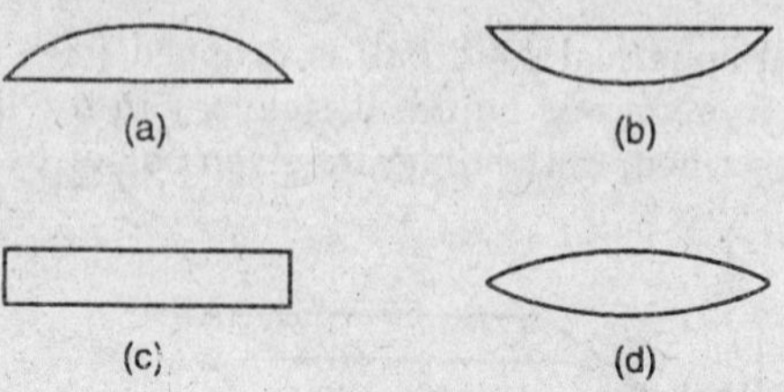

Fig. 4

6. Some particles of sand are sprinkled on to the surface of water in a beaker filled to the depth of $\frac{1}{2}$ units. The least time for which the grains of diameter 2 units remain in suspension in water is (Given $\eta = \frac{1}{4.5}$ units, density of sand = 2 units, density of water = 1 unit and $g = \frac{1}{2}$ units.)

(*a*) 1 s (*b*) $\frac{1}{2}$ s

(*c*) $\frac{1}{4}$ s (*d*) $\frac{1}{8}$ s.

7. *n* drops of water, each of radius 2 mm, fall through air at a terminal velocity of 8 cm s^{-1}. If they coalesce to form a single drop, then the terminal velocity of the combined drop is 32 cm s^{-1}. The value of *n* is

(*a*) 2 (*b*) 4

(*c*) 6 (*d*) 8.

8. In the streamlined flow of a liquid, the power due to pressure difference between the ends of the tube is

(*a*) equal to product of pressure and volume of liquid flowing per second

(*b*) equal to product of pressure and mass of liquid flowing per second

(*c*) equal to product of radius and coefficient of viscosity

(*d*) equal to product of pressure and velocity.

9. Two capillary tubes of same radius *r* but of lengths l_1 and l_2 are fitted in parallel to the bottom of a vessel. The length of a single tube that can replace the two tubes is

(*a*) $l_1 + l_2$ (*b*) $l_1 - l_2$

(*c*) $\frac{1}{l_1} + \frac{1}{l_2}$ (*d*) $\frac{l_1 l_2}{l_1 + l_2}$

(*e*) $\frac{1}{l_1 + l_2}$.

10. A tank has a hole at its bottom. The time needed to empty the tank from level h_1 to h_2 will be proportional to

(a) $h_1 - h_2$ (b) $h_1 + h_2$
(c) $\sqrt{h_1} - \sqrt{h_2}$ (d) $\sqrt{h_1} + \sqrt{h_2}$.

11. Two holes are made in the side of the tank such that the jets of water flowing out of them meet at the same point on the ground. If one hole is at a height of 3 cm above the bottom, then the distance of the other hole from the top surface of water is

(a) $\frac{3}{2}$ cm (b) 2×3 cm
(c) $\sqrt{3}$ cm (d) 3 cm.

12. A lawn sprinkler has 20 holes, each of cross-sectional area 2.0×10^{-2} cm^2 and is connected to a hose-pipe of cross-sectional area 2.4 cm^2. If the speed of water in the *hose-pipe* is 1.5 m s^{-1}, the speed of water as it emerges from the holes is

(a) 2.25 m s^{-1} (b) 4.5 m s^{-1}
(c) 9 m s^{-1} (d) 18 m s^{-1}.

13. A liquid is kept in a cylindrical vessel which is rotated along its axis. The liquid rises at the sides. If the radius of the vessel is r metre and the frequency of rotation is 2 revolutions per second, then the difference in the heights of the liquid at the centre of the vessel and at its sides is

(a) $\frac{8\pi^2 r^2}{g}$ (b) $\frac{8\pi r^2}{g}$
(c) $\frac{\pi r^2}{g}$ (d) $\frac{r^2}{g}$.

14. A cylindrical tank of height H is open at the top and has a radius of r. Water is filled in it upto a height of h. The time taken to empty the tank through a hole of radius r' in its bottom is

(a) $\sqrt{\frac{2h}{g}} \frac{r^2}{r'^2}$ (b) $\sqrt{\frac{2H}{g}} \frac{r^2}{r'^2}$
(c) $\sqrt{hH}$ (d) None of these.

15. A cylinder containing water up to a height of h cm has a hole of cross-section area a cm^2 at its bottom. It is counterpoised in a balance. When water just begins to flow out, the initial change in weight is

(a) $2gha$ (b) $\sqrt{2gha}$
(c) $\sqrt{gha}$ (d) None of these.

16. A liquid is poured into a vessel at rest with the hole in a wall closed by a valve. It is filled to a height H. The distance of hole from the top surface is h. With what horizontal acceleration should the vessel be moved so that the liquid does not come out when valve is opened ? (Given : l = length of base)

(a) $2gh$ (b) g
(c) $\frac{2gh}{l}$ (d) None of these.

17. A small steel ball falls through a syrup with a speed of 10 cm s^{-1}. If the steel ball is given an upward force equal to its effective weight, it would move upwards with a speed of

(a) 10 cm s^{-1} (b) 20 cm s^{-1}
(c) 5 cm s^{-1} (d) 0
(e) None of these.

18. A capillary tube is attached horizontally to a constant pressure head arrangement. If the radius of the capillary tube is increased by 10%, then the rate of flow of the liquid shall change nearly by

(a) + 10% (b) 46%
(c) – 10% (d) – 40%.

19. A hole is made at the bottom of a tank filled with water. If the total pressure at the bottom is 3 atmosphere, then velocity of efflux is

(a) 20 m s^{-1} (b) $10\sqrt{2}$ m s^{-1}
(c) $10\sqrt{6}$ m s^{-1} (d) $10\sqrt{5}$ m s^{-1}
(e) None of these.

20. A river 10 m deep is flowing at 5 m s^{-1}. The shearing stress between horizontal layers of the river is ($\eta = 10^{-3}$ SI units)

(a) 10^{-3} N m^{-2} (b) 0.8×10^{-3} N m^{-2}
(c) 0.5×10^{-3} N m^{-2} (d) 1 N m^{-2}.

21. 16 cm^3 of water flows per second through a capillary tube of radius r cm and of length l cm when connected to a pressure head of h cm of water. If a tube of the same length and radius $\frac{r}{2}$ cm is connected to the same pressure head, the quantity of water flowing the tube per second will be

(a) 1 cm^3 (b) 4 cm^3
(c) 8 cm^3 (d) 16 cm^3.

22. A liquid is flowing out of a horizontal tube with velocity v. The reaction force exerted by the liquid on the tube is proportional to

(a) v (b) v^2
(c) v^4 (d) v^8.

23. Two capillary tubes A and B are connected in parallel. A liquid flows through these tubes under the same pressure head. Both the tubes have the same length.

The radii of A and B are r and $\frac{r}{2}$ respectively. If the rate of flow of liquid through A is 8 cm^3 s^{-1}, then the rate of flow through the combination is

(a) 8 cm^3 s^{-1} (b) 8.5 cm^3 s^{-1}

(c) 9 cm^3 s^{-1} (d) 16 cm^3 s^{-1}.

24. A ball rises to the surface of a liquid with constant velocity. The density of the liquid is four times the density of the material of the ball. The frictional force of the liquid on the rising ball is greater than the weight of the ball by a factor of

(a) 2 (b) 3

(c) 4 (d) 6.

25. A liquid flows through two capillary tubes connected in series. Their lengths are L and 2L and radii r and $2r$ respectively. Then the pressure differences across the first and the second tube are in the ratio

(a) 8 (b) 4

(c) 2 (d) 1/8.

26. A tube of length L and radius R is joined to another tube of length $\frac{L}{3}$ and radius $\frac{R}{2}$. A fluid is flowing through this tube. If the pressure difference across the first tube is P, then the pressure difference across the second tube is

(a) $\frac{16P}{3}$ (b) $\frac{4P}{3}$

(c) P (d) $\frac{3P}{16}$.

27. Blood is flowing at the rate of 100 cm^3 s^{-1} in a capillary of cross-sectional area 0.25 m^2. The velocity of flow is

(a) 0.1 mm s^{-1} (b) 0.2 mm s^{-1}

(c) 0.3 mm s^{-1} (d) 0.4 mm s^{-1}.

28. A water tank placed on the floor has two small holes, pinched in the vertical wall, one above the other. The holes are 3.3 cm and 4.7 cm above the floor. If the jets of water issuing out from the holes hit the floor at the same point on the floor, then the height of water in the tank is

(a) 3 cm (b) 6 cm

(c) 8 cm (d) 9 cm.

29. A lead sphere of mass m falls in a viscous liquid with terminal velocity v_0. Another lead sphere of mass M falls through the same viscous liquid with terminal velocity $4v_0$. The ratio $\frac{M}{m}$ is

(a) 2 (b) 4

(c) 8 (d) 16.

30. The reading of a manometer fitted to a closed tap is 2.5×10^5 Pa. If the valve is opened, the reading of the manometer falls to 2×10^5 Pa. The velocity of water is

(a) 0.5 m s^{-1} (b) 1 m s^{-1}

(c) 10 m s^{-1} (d) 100 m s^{-1}.

31. Fig. 5 shows a liquid being pushed out of a tube by pressing a piston. The cross-sectional area of the piston is 1 cm^2 and that of the tube at the outlet is 20 mm^2. If the piston is pushed at a speed of 1 cm s^{-1}, then the velocity of the outgoing liquid is

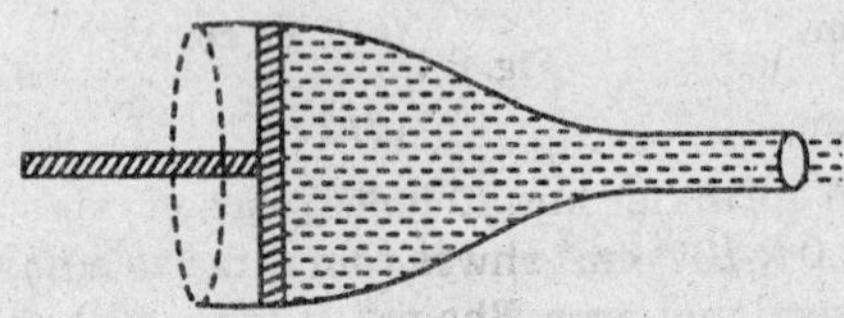

Fig. 5

(a) 2.5 cm s^{-1} (b) 5 cm s^{-1}

(c) 7.5 cm s^{-1} (d) 10 cm s^{-1}.

32. In a test experiment on a model aeroplane in a wind tunnel, the flow speed on the upper and lower surfaces of the wing are v_1 and v_2 respectively. If ρ be the density of air, then the upward lift is (Given : A = cross-sectional area of the wing)

(a) $\frac{1}{2}\rho A(v_1^2 - v_2^2)$ (b) $\frac{1}{2}\rho A(v_2^2 - v_1^2)$

(c) $Av_1 - Av_2$ (d) $v_1 - v_2$.

33. A tank containing a liquid is emptied through an orifice in 9.8 second when placed on the ground. If this tank is placed at a height of 9.8 metre above the ground, it would be emptied in

(a) 9.8 s (b) 1 s

(c) 9.8×9.8 s (d) $9.8 \times \frac{1}{2}$ s.

34. A flat plate is placed normal to a discharging jet of water. The jet discharges water at the rate of 0.2 m^3 s^{-1} and at a speed of 18 m s^{-1}. The force on the plate due to the jet, if the collision is inelastic is

(a) 6 N (b) 3 N

(c) 360 N (d) 3600 N.

35. Water contained in a tank flows through an orifice of radius 1 cm, under a constant pressure difference of 10 cm of water column. The rate of flow of water through the orifice is

(a) 4 cm^3 s^{-1} (b) 40 cm^3 s^{-1}

(c) 440 cm^3 s^{-1} (d) 4400 cm^3 s^{-1}.

36. Fig. 6 shows two holes in a wide tank containing a liquid commn. The water streams coming out of these holes strike the ground at the same point. The height of liquid column in the tank is

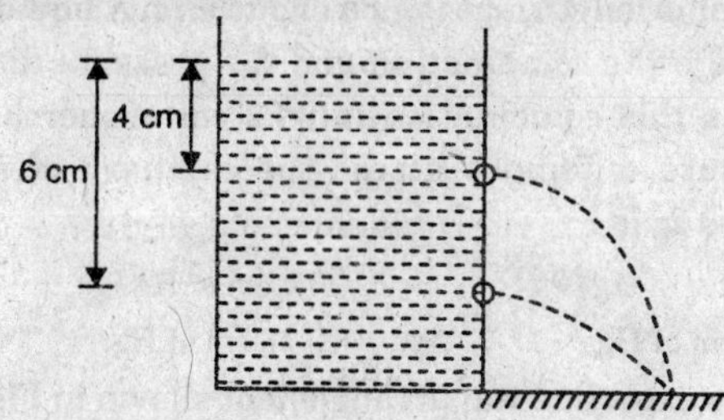

Fig. 6

(*a*) 10 cm (*b*) 8 cm
(*c*) 9.8 cm (*d*) 980 cm.

37. At Deoprayag in Garhwal (U.P.), river Alaknanda mixes with the river Bhagirathi and becomes river Ganga. Assume that Alaknanda, Bhagirathi and Ganga have widths 12 m, 8 m and 16 m respectively. Again, assume that the average speeds of water in Alaknanda and Bhagirathi are 20 km h^{-1} and 16 km h^{-1} respectively. Then, the average speed of water in the river Ganga is [Depth same in each case]

(*a*) 21 km h^{-1} (*b*) 23 km h^{-1}
(*c*) 25 km h^{-1} (*d*) 27 km h^{-1}.

38. The cross-sectional area of a large tank is 0.5 m^2. It has an opening near the bottom having cross-sectional area 1 cm^2. A load of 20 kg is applied on the water at the top. The velocity of water coming out of the opening, at the time when the height of water level is 50 cm above the bottom, is nearly (Take g = 10 m s^{-2})

(*a*) 1.3 m s^{-1} (*b*) 3.3 m s^{-1}
(*c*) 5.3 m s^{-1} (*d*) 7.3 m s^{-1}.

39. The gauge pressure of water in a tank is 2.88×10^5 pascal. If the density of water is 10^3 kg m^{-3}, then the velocity of efflux from the orifice in the tank, in m s^{-1}, is

(*a*) 12 (*b*) 144
(*c*) 28.8 (*d*) 24.

40. Coefficient of viscosity of water = 0.01 poise, density of water = 1 g cm^{-3}. Then the maximum velocity with which water can flow through a capillary tube of radius 0.05 cm, without turbulent flow setting in, is

(*a*) 2 cm s^{-1} (*b*) 20 cm s^{-1}
(*c*) 200 cm s^{-1} (*d*) 2000 cm s^{-1}.

41. A cylindrical tank stands on a frictionless surface. The seal over a circular hole of radius 0.5 cm in the wall of the tank ruptures when the level of water above the hole is 1 metre. The force that a person must apply on the cylinder to keep it from being set in motion is

(*a*) 1.54×10^{-3} N forward (*b*) 1.54×10^{-3} N backward
(*c*) 1.54 N forward (*d*) 1.54 N backward.

42. An air bubble of 1 cm radius is rising at a steady rate of 2.00 mm s^{-1} through a liquid of density 1.5 g cm^{-3}. Neglect density of air. If g = 1000 cm s^{-2}, then the coefficient of viscosity of the liquid is

(*a*) 0.166×10^3 poise (*b*) 1.66×10^3 poise
(*c*) 166×10^3 poise (*d*) 16.6×10^3 poise.

43. Three capillaries of lengths L, $\frac{L}{2}$ and $\frac{L}{3}$ are connected in series. Their radii are r, $\frac{r}{2}$ and $\frac{r}{3}$ respectively. Then if stream line flow is to be maintained and the pressure across the first capillary is P, then

(*a*) the pressure difference across the ends of the second capillary is 8 P
(*b*) the pressure difference across the third capillary is 43 P
(*c*) the pressure difference across the ends of the second capillary is 16 P
(*d*) the pressure difference across the third capillary is 59 P.

44. Air is streaming past a horizontal air plane wing such that its speed is 90 m s^{-1} at the lower surface and 120 m s^{-1} over the upper surface. If the wing is 10 m long and has an average width of 2 m, the difference of pressure on the two sides and the gross lift on the wing are [Density of air = 1.3 kg m^{-3}]

(*a*) 5 Pa, 900 N (*b*) 95 Pa, 900 N
(*c*) 4095 Pa, 900 N (*d*) 4095 Pa, 81900 N.

45. A large open tank has two holes in the wall. One is a square hole of side L at a depth y from the top and the other is a circular hole of radius R at a depth 4y from the top. When the tank is completely filled with water, the quantities of water flowing out per second from both holes are the same. Then, R =

(*a*) $\frac{L}{\sqrt{2\pi}}$ (*b*) $2\pi L$
(*c*) L (*d*) $\frac{L}{2\pi}$.

46. Two large tanks, open at the top, contain two liquids. A small hole is made in the side of each tank at the same depth h below the surface of the liquid. One hole has twice the cross-section of the other. The ratio of the densities of the liquids is [Mass flux of two liquids is same.]

(*a*) 1 : 2 (*b*) 4 : 3
(*c*) 3 : 4 (*d*) 1 : 21.

47. A spherical ball falls through a viscous medium with terminal velocity v. If this ball were replaced by another ball of the same mass but half the radius, then the terminal velocity will be [Neglect the effect of buoyancy.]

(a) v (b) $2v$
(c) $4v$ (d) $8v$.

48. A tank full of water has a small hole at its bottom. If t_1 and t_2 are the times taken to empty the first half and the remaining half of the tank respectively, then $\frac{t_1}{t_2} =$

(a) 1.732 (b) 0.928
(c) $2^{3.1}$ (d) 0.414.

49. There are two holes, each of cross-sectional area a, on the opposite sides of a wide rectangular tank containing a liquid of density ρ. When the liquid flows out of the holes, the net force on the tank is [h is the vertical distance between the two holes.]

(a) $2a\rho gh$ (b) $4a\rho gh$
(c) $0.5\ a\rho gh$ (d) ρgh.

50. There is a hole of area $\frac{1}{25}$ cm^2 in the bottom of a cylindrical vessel containing fluid up to height h. The liquid flows out in time t. If the liquid were filled in the vessel up to height $4h$, then it would flow out in time

(a) t (b) $2t$
(c) $4t$ (d) $\frac{t}{2}$.

51. A liquid of density 1200 kg m^{-3} is flowing steadily in a tube of varying cross-section. The cross-section at a point A is 1 cm^2 and that at B is 20 mm^2. The points A and B are in the same horizontal plane. The velocity of liquid at A is 10 cm s^{-1}. The velocity of liquid at B is

(a) 5 cm s^{-1} (b) 10 cm s^{-1}
(c) 25 cm s^{-1} (d) 50 cm s^{-1}.

52. In Q. 51, the pressure difference between A and B is

(a) 12 Pa (b) 36 Pa
(c) 144 Pa (d) 259 Pa.

53. Three capillaries of internal radii $2r$, $3r$ and $4r$, all of the same length, are joined end to end. A liquid passes through the combination and the pressure difference across this combination is 20.2 cm of mercury. The pressure difference across the capillary of internal radius $2r$ is

(a) 2 cm of Hg (b) 4 cm of Hg
(c) 8 cm of Hg (d) 16 cm of Hg.

54. In the experimental arrangement shown in Fig. 7, the areas of cross-section of the wide and narrow portions of the tube are 5 cm^2 and 2 cm^2 respectively. The rate of flow of water through the tube is 500 cm^3 s^{-1}. The difference of mercury levels in the U-tube is

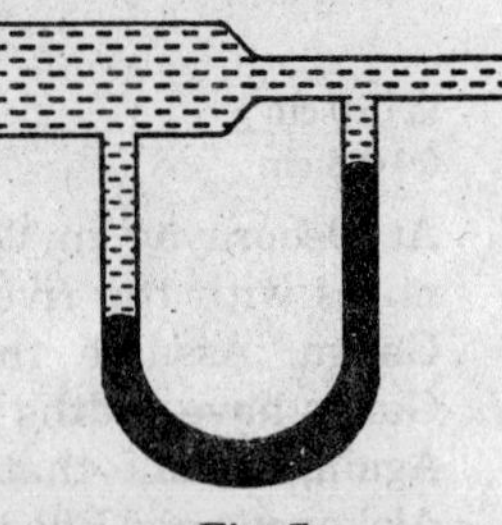
Fig. 7

(a) 0.97 cm (b) 1.97 cm
(c) 0.67 cm (d) 4.67 cm.

55. Fig. 8 shows how the stream of water emerging from a faucet "necks down" as it falls. The rate of water flow from the tap is

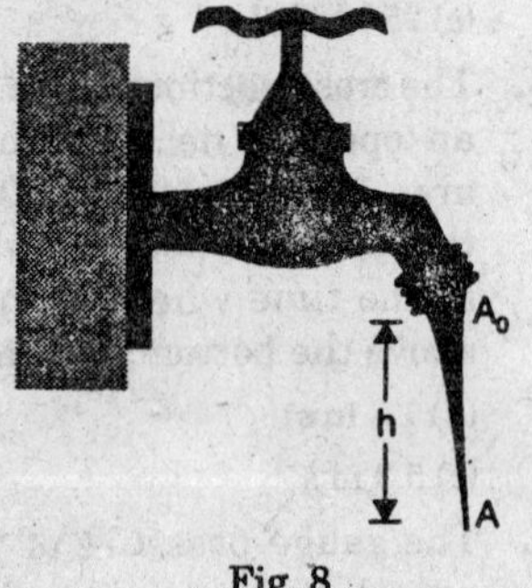

Fig. 8

(a) $\sqrt{2gh}$
(b) $\sqrt{gh}$
(c) $A_0\sqrt{\frac{2ghA^2}{A_0^2 - A^2}}$
(d) Data is incomplete.

ANSWERS

QUESTION BANK (Level I)

1. (c)	2. (b)	3. (d)	4. (a)	5. (d)	6. (a)	7. (b)	8. (c)
9. (d).							

QUESTION BANK (Level II)

1. (b)	2. (b)	3. (a)	4. (c)	5. (a)	6. (a)	7. (d)	8. (a)
9. (d)	10. (c)	11. (d)	12. (c)	13. (a)	14. (a)	15. (a)	16. (c)
17. (d)	18. (b)	19. (a)	20. (c)	21. (a)	22. (b)	23. (b)	24. (b)
25. (a)	26. (a)	27. (d)	28. (c)	29. (c)	30. (c)	31. (b)	32. (a)
33. (a)	34. (d)	35. (c)	36. (a)	37. (b)	38. (b)	39. (d)	40. (c)

41. (c) 42. (b) 43. (a) 44. (d) 45. (a) 46. (a) 47. (b) 48. (d)
49. (a) 50. (b) 51. (d) 52. (c) 53. (d) 54. (b) 55. (c).

HINTS/SOLUTIONS

QUESTION BANK (Level I)

1. (c) The ball acquires velocity due to gravity. In the viscous medium, this velocity goes on decreasing till it becomes constant.

> ☞ *Curve D is ruled out because it shows linear relationship between velocity and time in the beginning.*

2. (b) $\pi r^2 v$ = constant

$$\frac{v_1}{v_2} = \frac{r_2^2}{r_1^2} = \frac{2^2}{3^2} = \frac{4}{9}$$

3. (d) 1 centipoise

$= 10^{-2}$ g cm^{-1} s^{-1} $= 10^{-2} \times 10^{-3} \times 10^2$ kg m^{-1} s^{-1}

$= 10^{-3}$ kg m^{-1} s^{-1} $= 0.001$ kg m^{-1} s^{-1}

4. (a) Velocity of outcoming water from hole A $= v_1 = \sqrt{2gh}$

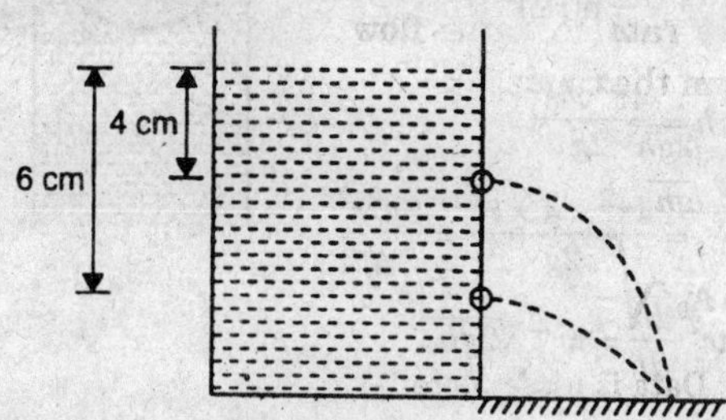

Fig. 9

Velocity of outcoming water from hole B

$= v_2 = \sqrt{2g(H_o - h)}$

Time of reaching the ground from hole A

$= t_1 = \sqrt{2(H_o - h)/g}$

Time of reaching the ground from hole B $= t_2 = \sqrt{2h/g}$

Obviously, Range on the ground for both is the same

$= R = v_1 t_1 = v_2 t_2 = 2g\sqrt{h(H_o - h)}$

5. (d) $v_0 = \dfrac{2}{9}\dfrac{r^2(\rho - \rho')g}{\eta}$

Now, $x = \dfrac{4}{3}\pi r^3 \rho$ or $\rho \propto \dfrac{x}{r^3}$

Similarly, $\rho' \propto \dfrac{y}{r^3}$ $\therefore$ $v_0 \propto \dfrac{x-y}{r}$

6. (a) $0.18v + 0.12 \times 1.5 = 0.12 \times 3$

or $0.18v = 0.12 \times 1.5$ or $v = \dfrac{0.12 \times 1.5}{0.18}$ m s^{-1} = 1 m s^{-1}

7. (b) $V = \dfrac{\pi P r^4}{8\eta l}$

$$V' = \frac{\pi (4P)\left(\frac{r}{2}\right)^4}{8\eta l}$$

$$\frac{V'}{V} = \frac{4}{16} = \frac{1}{4} \text{ or } V' = \frac{V}{4}$$

8. (c) $F = 6\pi\eta r v$

$F' = 6\pi\eta\,(2r)\,(2v) = 4F$

9. (d) Since the steel ball is given to be small therefore upthrust may be neglected.

Now, $6\pi\eta r v_0 = mg$ or $v_0 \propto \dfrac{mg}{\eta r}$

QUESTION BANK (Level II)

1. (b) Volume/s $= av$, Mass/s $= av\rho$

Momentum/s $= av^2\rho$

Force $= av^2\rho$

Upward reaction force $= av^2\rho$

Initial decrease in balancing weight $= av^2\rho$

$= 2a\rho gh$ $(\because v = \sqrt{2gh})$

Now, $12g = 2 \times a \times 1 \times g \times 24$

or $a = \dfrac{12}{2 \times 24} = \dfrac{1}{4}$ cm^2 = 0.25 cm^2

2. (b) $v_0 = \dfrac{2}{9}\dfrac{r^2(\rho - \rho')g}{\eta}$ or $\eta = \dfrac{2r^2(\rho - \rho')g}{9v_0}$

$= \dfrac{2 \times 0.2 \times 0.2 \times 9 \times 980}{9 \times 8}$ poise = 9.8 poise

3. (*a*) $\frac{v^2}{2g} = 40$ or $v = \sqrt{2 \times 980 \times 40}$ cm s^{-1}

$= 280$ cm s^{-1} $= 2.80$ m s^{-1}

4. (*c*) Volume/s $= av = a\sqrt{2gh}$

$= 1\sqrt{2 \times 980 \times 2.5}$ cm^3 s^{-1} $= 70$ cm^3 s^{-1}

5. (*a*) Air streams have to cover a larger distance above the surface as compared to the air streams below the surface. So, velocity above the surface is large and consequently the pressure is less. On the other hand, velocity below the surface is small and consequently the pressure is large. This explains maximum upthrust in Fig. 4 (*a*).

6. (*a*) $v_0 = \frac{2}{9}\frac{r^2(\rho - \rho')g}{\eta}$

$\frac{s}{t} = \frac{2}{9}\frac{r^2(\rho - \rho')g}{\eta}$ or $\frac{t}{s} = \frac{9\eta}{2r^2(\rho - \rho')g}$

or $t = \frac{9\eta s}{2r^2(\rho - \rho')g}$

or $t = 9 \times \frac{1}{4.5} \times \frac{1}{2} \times \frac{1}{2 \times 1 \times 1(2-1) \times \frac{1}{2}}$ s $= 1$ s

7. (*d*) $\frac{4}{3}\pi R^3 = n \times \frac{4}{3}\pi r^3$

or $R^3 = nr^3$ or $R = n^{1/3}r$ or $R = 2n^{1/3}$ mm

$v_0 \propto r^2$, $v_0' \propto R^2$

Now, $\frac{v_0'}{v_0} = \frac{R^2}{r^2} = \frac{4n^{2/3}}{4}$

or $\frac{32}{8} = n^{2/3}$ or $n^{2/3} = 4$

or $n = 4^{3/2} = \sqrt{64}$ or $n = 8$

8. (*a*) $W = PV$

Power $= \frac{W}{t} = \frac{PV}{t} = P\left[\frac{V}{t}\right]$

9. (*d*) $\frac{\pi p r^4}{8\eta L} = \frac{\pi p r^4}{8\eta l_1} + \frac{\pi p r^4}{8\eta l_2}$

or $\frac{1}{L} = \frac{1}{l_1} + \frac{1}{l_2}$ or $\frac{1}{L} = \frac{l_2 + l_1}{l_1 l_2}$ or $L = \frac{l_1 l_2}{l_1 + l_2}$

10. (*c*) Let $-\frac{dh}{dt}$ represent the rate of descent of water level. Let A and a represent the cross-sectional areas of the container and hole respectively.

Then, $-A\frac{dh}{dt} = a\sqrt{2gh}$

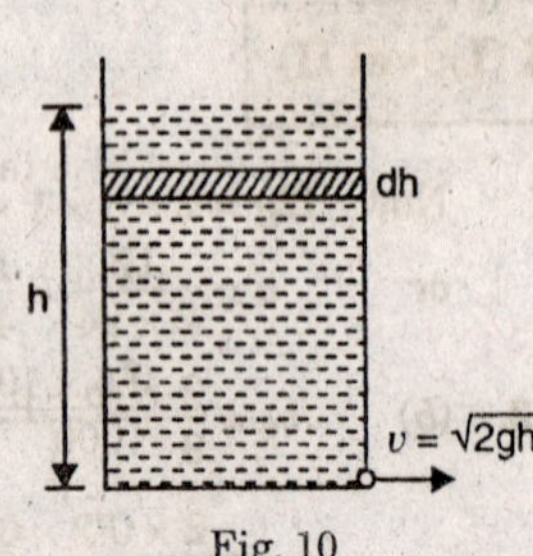

Fig. 10

or $dt = -k\frac{dh}{\sqrt{h}}$ [k is a constant]

or $\int_0^t dt = -k\int_{h_1}^{h_2}\frac{1}{\sqrt{h}}dh$ or $t = -k\left|\frac{h^{-\frac{1}{2}+1}}{-\frac{1}{2}+1}\right|_{h_1}^{h_2}$

or $t = -2k[\sqrt{h_2} - \sqrt{h_1}]$ or $t \propto (\sqrt{h_1} - \sqrt{h_2})$

11. (*d*) $x = \sqrt{2gh_1} \times \sqrt{\frac{2h_2}{g}}$

or $x = 2\sqrt{h_1 h_2}$

Now, imagine a hole at a depth h_2 below the free surface of the liquid. The height of this hole will be h_1. Clearly, x remains the same.

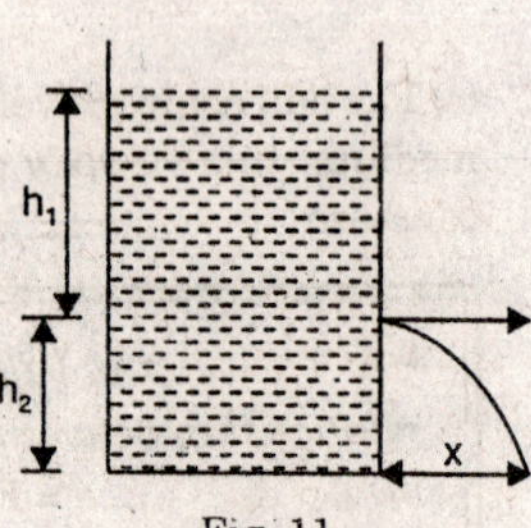

Fig. 11

12. (*c*) $2.4 \times 1.5 \times 10^{-4} = 20 \times 2 \times 10^{-2} \times v$

or $v = \frac{2.4 \times 1.5 \times 10^2}{40}$ m s^{-1}

$= \frac{24 \times 15}{40}$ m s^{-1} $= 9$ m s^{-1}

13. (*a*) $p_1 - p_2 = \frac{1}{2}\rho v^2$

or $h\rho g = \frac{1}{2}\rho(r\omega)^2$

or $h = \frac{4\pi^2 v^2 r^2}{2g}$

$= \frac{4\pi^2 \times 4 \times r^2}{2g} = \frac{8\pi^2 r^2}{g}$

Fig. 12

14. (*a*) $-\pi r^2\frac{dh}{dt} = \pi r'^2\sqrt{2gh}$

or $\frac{dh}{dt} = -\frac{r'^2}{r^2}\sqrt{2gh}$ or $dt = -\frac{r^2}{r'^2}\cdot\frac{1}{\sqrt{2gh}}dh$

$\int_0^t dt = -\frac{r^2}{r'^2\sqrt{2g}}\int_h^0 h^{-1/2}dh$

or $t = -\frac{r^2}{r'^2\sqrt{2g}} \times 2\left[\sqrt{h}\right]_h^0$ or $t = \sqrt{\frac{2h}{g}} \cdot \frac{r^2}{r'^2}$

15. (*a*) Upward thrust $= v\frac{dm}{dt}$

$= v(av)\rho = av^2\rho$

$= a \times 2gh$ [$\because \rho = 1$ g cm^{-3}]

$= 2gha$

16. (c) $$F = v\frac{dm}{dt}$$

$$F = v\,(av)\,\rho$$

This force acts on mass $al\rho$ of the liquid.

$$\therefore \text{ acceleration} = \frac{av^2\rho}{al\rho} = \frac{2gh}{l}$$

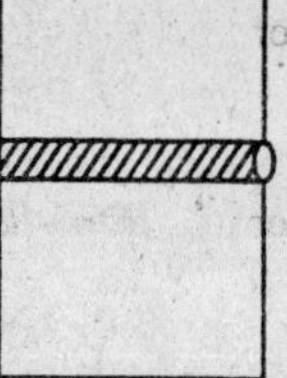

Fig. 13

> ☞ *Volume/s = av*
> *Mass/s = avρ*

17. (d) Initially, W = U + F, where W, U and F represent weight, upthrust and viscous force respectively.

Now, $W - U = F$

or given upward force = F

So, there will be no velocity.

18. (b) $V \propto r^4,\ V' \propto \left(r + \frac{1}{10}r\right)^4$

$$\frac{V'}{V} = \frac{11 \times 11 \times 11 \times 11}{10 \times 10 \times 10 \times 10} = 1.4641$$

or $\left(\frac{V'}{V} - 1\right) \times 100 = 46.4$

19. (a) $v = \sqrt{2gh}$

But $p = h\rho g$ or $\frac{p}{\rho} = gh$

$$\therefore \quad v = \sqrt{\frac{2p}{\rho}} = \sqrt{\frac{2 \times 2 \times 10^5}{10^3}}\ \text{m s}^{-1} = 20\ \text{m s}^{-1}$$

> ☞ *If total pressure is 3 atmosphere, then pressure due to liquid column is 2 atmosphere.*

20. (c) $\text{Shearing stress} = \frac{|\ \text{Viscous force}\ |}{\text{Area}}$

$$= \frac{\eta A \frac{dv}{dx}}{A} = \eta\frac{dv}{dx} = 10^{-3} \times \frac{5}{10}\ \text{N m}^{-2} = 0.5 \times 10^{-3}\ \text{N m}^{-2}$$

21. (a) $V \propto r^4,\quad V' \propto \frac{r^4}{16}$

or $\frac{V'}{V} = \frac{1}{16}$ or $V' = \frac{V}{16}$

or $V' = \frac{16}{16}\ \text{cm}^3 = 1\ \text{cm}^3$

22. (b) Volume/s $= av$

Mass/s $= av\rho$

Momentum/s $= av^2\rho$

Force $= av^2\rho$

23. (b) $V_A = 8\ \text{cm}^3\ \text{s}^{-1}$

$$V_B = \frac{\pi p\left(\frac{r}{2}\right)^4}{8\eta l} = \frac{1}{16} \times 8 = 0.5\ \text{cm}^3\ \text{s}^{-1}$$

$$V = V_A + V_B = 8 + 0.5 = 8.5\ \text{cm}^3\ \text{s}^{-1}$$

> ☞ *Note that the rates of flow have been added up.*

24. (b) $V\rho g + F = V(4\rho)\,g$

or $F = 3V\rho g = 3mg$ or $\frac{F}{mg} = 3$

25. (a) $\frac{\pi p_1 r^4}{8\eta L} = \frac{\pi p_2 (2r)^4}{8\eta(2L)}$ or $\frac{p_1}{p_2} = \frac{2^4}{2} = 2^3 = 8$

26. (a) $\frac{\pi P R^4}{8\eta L} = \frac{\pi P'\left(\frac{R}{2}\right)^4}{8\eta\frac{L}{3}}$ or $\frac{3P'}{16} = P$ or $P' = \frac{16P}{3}$

27. (d) $v = \frac{av}{a} = \frac{100 \times 10^{-6}}{0.25}\ \text{m s}^{-1}$

$$= 400 \times 10^{-3}\ \text{mm s}^{-1} = 0.4\ \text{mm s}^{-1}$$

28. (c) $R = 2\sqrt{h(H - h)}$

Now, $3.3\,[H - 3.3] = 4.7\,[H - 4.7]$

or $(4.7 - 3.3)\,H = 4.7 \times 4.7 - 3.3 \times 3.3$

or $$H = \frac{22.09 - 10.89}{1.4}\ \text{cm} = \frac{11.2}{1.4}\ \text{cm} = 8\ \text{cm}$$

> ☞ *If two holes at heights h_1 and h_2 are such that the liquid from these holes covers the same horizontal distance, then the height H of liquid column is such that $H = h_1 + h_2$.*

29. (c) Terminal velocity is quadrupled, $v_0 \propto r^2$

Clearly, radius is doubled.

Mass is increased by a factor of 8 ($M \propto r^3$).

30. (c) $2 \times 10^5 + \frac{1}{2} \times 1000 \times v^2 = 2.5 \times 10^5$

or $$v^2 = \frac{2 \times 0.5 \times 10^5}{1000} = 10^2\ \text{m}^2\ \text{s}^{-2}$$

or $v = 10\ \text{m s}^{-1}$

31. (*b*) $20 \times 10^{-2} \times v = 1 \times 1$

or $v = \frac{1}{20 \times 10^{-2}}$ cm s^{-1} = 5 cm s^{-1}

32. (*a*) $p_1 + \frac{1}{2}\rho v_1^2 = p_2 + \frac{1}{2}\rho v_2^2$

or $p_2 - p_1 = \frac{1}{2}\rho(v_1^2 - v_2^2)$

Upward lift $= \frac{1}{2}\rho A(v_1^2 - v_2^2)$

33. (*a*) Time to empty the tank does not depend upon the height of the bottom of tank above the ground.

34. (*d*) $F = av^2\rho = (av)v\rho$

$= 0.2 \times 18 \times 1000$ N = 3600 N

35. (c) Rate of flow $= av = \pi r^2 v$

$= \frac{22}{7} \times 1 \times 1 \times \sqrt{2 \times 980 \times 10}$ cm^3 s^{-1} = 440 cm^3 s^{-1}

36. (*a*) $4(H - 4) = 6(H - 6)$

or $2H = 36 - 16 = 20$ or $H = 10$ cm

37. (*b*) $12d \times 20 + 8d \times 16 = 16d \times v$

or $v = \frac{240 + 128}{16} = \frac{368}{16}$ km h^{-1} = 23 km h^{-1}

38. (*b*) $P_o + \frac{20 \times 10}{0.5} + 0.5 \times 1000 \times 10 + 0 = P_o + \frac{1}{2} \times 1000 \times v^2$

On simplification, $v \approx 3.3$ m s^{-1}

39. (*d*) $v = \sqrt{2gh}$

$= \sqrt{\frac{2h\rho g}{\rho}} = \sqrt{\frac{2p}{\rho}} = \sqrt{\frac{2 \times 2.88 \times 10^5}{1000}}$ m s^{-1}

$= \sqrt{2 \times 288}$ m s^{-1} = 24 m s^{-1}

40. (*c*) $v_c = \frac{2000 \times 0.01}{1 \times 2 \times 0.05}$ cm s^{-1} = 200 cm s^{-1}

41. (*c*) $F = av^2\rho$

$= \frac{22}{7} \times 0.5 \times 0.5 \times 2 \times 980 \times 100 \times 1$ dyne

$= 1.54 \times 10^5$ dyne = 1.54 N

42. (*b*) $v_0 = \frac{2}{9}\frac{r^2(\rho - \rho')g}{\eta}$

$\eta = \frac{2r^2(\rho - \rho')g}{9v_0} = \frac{2 \times 1 \times 1 \times 1.5 \times 1000}{9 \times 0.2}$ poise

$= 1.66 \times 10^3$ poise

43. (*a*) $\frac{p_1 r^4}{L} = \frac{p_2 \frac{r^4}{16}}{L/2}$ or $p_2 = P \times 8$

44. (*d*) $P_2 - P_1 = \frac{1}{2} \times 1.3\,[120^2 - 90^2] = 4095$ Pa

Lift = 4095 × 2 × 10 N = 81900 N

45. (*a*) $L^2\sqrt{2gy} = \pi R^2\sqrt{2g(4y)}$

or $L^2 = 2\pi R^2$ or $R = \frac{L}{\sqrt{2\pi}}$

46. (*a*) Mass flux $= av\rho = a\rho\sqrt{2gh}$

$\therefore\ 2a\rho_1\sqrt{2gh} = a\rho_2\sqrt{2gh}$

$\therefore\ \frac{\rho_1}{\rho_2} = \frac{1}{2}$

47. (*b*) $m = \frac{4}{3}\pi r^3\rho$

Keeping m constant, if r is halved, ρ will increased by a factor of 8.

Now, $v_0 \propto r^2\rho$

$v_0' \propto \frac{r^2}{4}(8\rho)$ or $v_0' \propto 2r^2\rho$

Dividing, $\frac{v_0'}{v_0} = 2$ or $v_0' = 2v_0$ or $v_0' = 2v$ $[\because v_0 = v]$

48. (*b*) $\frac{t_1}{t_2} = \frac{\sqrt{h} - \sqrt{\frac{h}{2}}}{\sqrt{\frac{h}{2}} - 0} = \frac{\sqrt{2} - 1}{1} = 0.414$

49. (*a*) $F = av^2\rho$

Net force

$= a\rho(v_2^2 - v_1^2)$

$= a\rho\,[2g(h + y) - 2gy]$

$= 2a\rho gh$

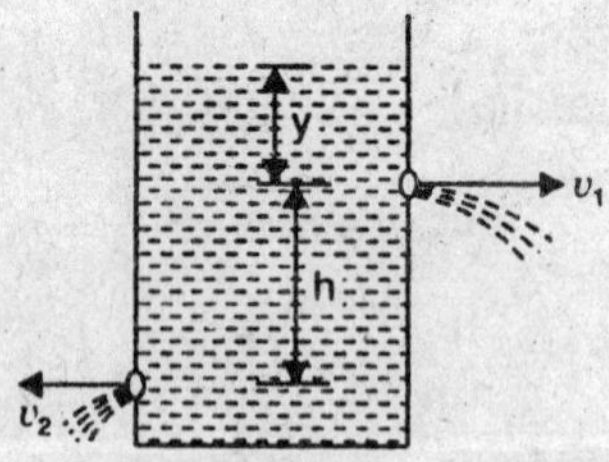

Fig. 14

50. (*b*) Let A and a be the cross-sectional areas of the vessel and hole respectively. Let h be the height of water in the vessel at time. Let $\left(-\frac{dh}{dt}\right)$ represent the rate of fall of level.

Then, $A\left(-\frac{dh}{dt}\right) = av = a\sqrt{2gh}$

or $-\frac{dh}{\sqrt{h}} = \frac{a\sqrt{2g}}{A}dt$

$-\int_h^0 \frac{1}{\sqrt{h}}dh = \frac{a\sqrt{2g}}{A}\int_0^t dt$ or $-(-2\sqrt{h}) = \frac{a\sqrt{2g}}{A}t$

or $t = \frac{A}{a}\frac{1}{\sqrt{2g}} \times 2\sqrt{h}$ or $t = \frac{A}{a}\sqrt{\frac{2h}{g}}$

Now, $t \propto \sqrt{h}$

When h is quadrupled, t is doubled.

51. (*d*) $20 \times 10^{-2} \times v = 10 \times 1$

or $v = \frac{1}{2 \times 10^{-2}}$ cm s^{-1} = 50 cm s^{-1}

52. (*c*) $P_A + \frac{1}{2} \times 1200 \times 10 \times 10 \times 10^{-4}$

$= P_B + \frac{1}{2} \times 1200 \times 50 \times 50 \times 10^{-4}$

or $P_A - P_B = 600 \times 10^{-4} \times 10^2 (25 - 1)$ N m^{-2}

= 144 N m^{-2} or 144 Pa

53. (*d*) $V = \frac{\pi p r^4}{8\eta l}$. In the given problem, pr^4 = constant

$\therefore$ $p_1 \times 16 = p_2 \times 81 = p_3 \times 256$

$\therefore$ $\frac{p_1}{p_2} = \frac{81}{16}, \frac{p_2}{p_3} = \frac{256}{81}$

or $\frac{p_1}{p_2} = \frac{81 \times 16}{16 \times 16}$

$\therefore$ $p_1 : p_2 : p_3 :: 81 \times 16 : 256 : 81$

or $p_1 : p_2 : p_3 :: 16 : 3.2 : 1$

or $p_1 = \frac{16}{16 + 3.2 + 1} \times 20.2$ cm of Hg = 16 cm of Hg

54. (*b*) $v_1 = \frac{500}{5}$ cm s^{-1} = 10^2 cm s^{-1}

$v_2 = \frac{500}{2}$ cm s^{-1} = 25×10^1 cm s^{-1}

$p_1 - p_2 = \frac{1}{2} \times 1\ [625 \times 10^2 - 100 \times 10^2]$

$h \times 13.6 \times 980 = 50 \times 525$

or $h = \frac{50 \times 525}{13.6 \times 980}$ cm = 1.97 cm

55. (*c*) $v^2 = v_0^2 + 2gh$

But $Av = A_0 v_0$ or $v = \frac{A_0 v_0}{A}$

$\therefore$ $\frac{A_0^2 v_0^2}{A^2} = v_0^2 + 2gh$

or $v_0^2 \left[\frac{A_0^2 - A^2}{A^2} \right] = 2gh$

8.5 KINETIC THEORY OF GASES

Important Terms, Facts and Formulae

BASIC CONCEPTS

1. Number of moles = $\dfrac{\text{mass of gas}}{\text{molecular mass of gas}}$

$$\mu = \frac{m}{M}$$

2. Intermolecular forces do not exist in an ideal gas. So, such a gas cannot be liquefied.
3. The values of k (Boltzmann constant) and R (Molar gas constant) do not depend upon the nature of the gas.
4. The value of r (gas constant for unit mass) varies with the nature of the gas.
5. $r = \dfrac{R}{M}$
6. At any given instant, each molecule may be moving along any six directions [$\pm x$, $\pm y$ and $\pm z$]. Thus, only one-sixth of the total molecules may be moving towards a wall.

IMPORTANT POINTS TO REMEMBER

1. For μ moles of ideal gas, $PV = \mu RT$
2. The value of R is 8.37 J/mol/°C or 2 cal/mol/°C
3. **Formulae for r.m.s. speed (C)**

   (*i*) $\sqrt{\dfrac{3PV}{M}}$ (for 1 mole)

   (*ii*) $\sqrt{\dfrac{3RT}{M}}$ (for 1 mole)

   (*iii*) $\sqrt{\dfrac{3kT}{m}}$ (for 1 molecule)

   (*iv*) $1.732\sqrt{\dfrac{kT}{m}}$ (for 1 molecule)

   (*v*) $\sqrt{\dfrac{3P}{\rho}}$ (for 1 mole)

4. **Formulae for most probable speed (C_{mp})**

   (*i*) $\sqrt{\dfrac{2PV}{M}}$ (for 1 mole)

   (*ii*) $\sqrt{\dfrac{2RT}{M}}$ (for 1 mole)

   (*iii*) $\sqrt{\dfrac{2kT}{m}}$ (for 1 molecule)

   (*iv*) $1.414\sqrt{\dfrac{kT}{m}}$ (for 1 molecule)

5. **Formulae for mean speed ($< C >$)**

   (*i*) $\sqrt{\dfrac{8RT}{\pi M}}$ (for 1 mole)

   (*ii*) $\sqrt{\dfrac{8PV}{\pi M}}$ (for 1 mole)

   (*iii*) $\sqrt{\dfrac{8kT}{\pi m}}$ (for 1 molecule)

   (*iv*) $1.6\sqrt{\dfrac{kT}{m}}$ (for 1 molecule)

6. $C : < C > : C_{mp} = 1.73 : 1.6 : 1.41$

   Clearly, C_{mp} is minimum, C is maximum and $< C >$ has intermediate value.
7. The pressure of a gas is equal to two-third of the kinetic energy per unit volume of the gas.
8. The temperature of an ideal gas is proportional to the translational kinetic energy of the molecules of the gas.

QUESTION BANK

Level I

Choose the correct answer from the four alternatives given in each of the following questions :

1. If the energy of a gas litre^{-1} is 300 joule, then its pressure will be

(*a*) 3×10^5 N m^{-2} (*b*) 6×10^5 N m^{-2}

(*c*) 10^5 N m^{-2} (*d*) 2×10^5 N m^{-2}.

2. A vessel of volume 0.1 m^3 containing air at a pressure of 76 cm of Hg is connected to an evacuated vessel of capacity 0.09 m^3. What is the resultant air pressure if temperature is assumed to remain constant ?

(*a*) 40 cm of Hg (*b*) 45 cm of Hg

(*c*) 52 cm of Hg (*d*) 76 cm of Hg.

3. By what percentage should the pressure of a given mass of a gas be increased so as to decrease its volume by 10% at a constant temperature ?

(*a*) 8.1% (*b*) 9.1%

(*c*) 10.1% (*d*) 11.1%.

4. The molecules of a given mass of gas have a root mean square velocity of 200 m s^{-1} at 27°C and 1.0×10^5 N m^{-2} pressure. When the temperature is 127°C and the pressure 0.5×10^5 N m^{-2}, the root mean square velocity in m s^{-1}, is

(*a*) $\frac{400}{\sqrt{3}}$ (*b*) $100\sqrt{2}$

(*c*) $\frac{100\sqrt{2}}{3}$ (*d*) $\frac{100}{3}$

(*e*) $\frac{50\sqrt{2}}{3}$.

5. If N is Avogadro's number, then the number of molecules in 6 g of hydrogen at NTP is

(*a*) 2 N (*b*) 3 N

(*c*) N (*d*) N/6.

6. Molecular hydrogen at one atmosphere and helium at two atmosphere occupy volume V each at the same temperature. The r.m.s. velocity of hydrogen molecules is x times the r.m.s. velocity of helium molecules. What is the value of x ?

(*a*) 1 (*b*) 2

(*c*) $\sqrt{2}$ (*d*) $\sqrt{3}$.

7. If the volume of air at 0°C and 10 atmospheric pressure is 10 litre, its volume, in litre, at normal temperature and pressure would be

(*a*) 1 (*b*) 10

(*c*) 100 (*d*) 1000.

8. What is the ratio of the total energy of all the molecules of one mole of O_2 to the total energy of all the molecules of two moles of helium at the same temperature ?

(*a*) 1 : 2 (*b*) 2 : 1

(*c*) 2 : 3 (*d*) 3 : 2.

9. It is required to double the pressure of a gas in a container at 27°C by heating it. To what temperature the gas should be raised

(*a*) 273°C (*b*) 373°C

(*c*) 327°C (*d*) 108°C.

10. Four molecules of a gas are having speeds of 1, 4, 8 and 16 m s^{-1}. The root mean square velocity of the gas molecules is

(*a*) 7.25 m s^{-1} (*b*) 52.56 m s^{-1}

(*c*) 84.25 m s^{-1} (*d*) 9.2 m s^{-1}.

11. Root mean square velocity of gas molecules is 300 m s^{-1}. The r.m.s. velocity of molecules of gas with twice the molecular weight and half the absolute temperature is

(*a*) 300 m s^{-1} (*b*) 600 m s^{-1}

(*c*) 75 m s^{-1} (*d*) 150 m s^{-1}.

12. One litre of gas A and two litre of gas B, both having same temperature 10°C and same pressure 2.5 Pa, will have the ratio of kinetic energies of their molecules as

(*a*) 1 : 1 (*b*) 1 : 2

(*c*) 1 : 4 (*d*) 4 : 1.

13. On the basis of kinetic theory of gases, the mean K.E. of 1 mole per degree of freedom is

(*a*) $\frac{1}{2}$ RT (*b*) $\frac{3}{2}$ RT

(*c*) $\frac{1}{2}kT$ (*d*) $\frac{3}{2}kT$.

14. The density of air at a pressure of 105 N m^{-2} is 1.2 kg m^{-3}. Under these conditions, the root mean square velocity of the air molecules in m s^{-1} is

(*a*) 500 (*b*) 1000

(*c*) 1500 (*d*) 3000

(*e*) 5000.

15. The mean translational kinetic energy of a perfect gas molecule at the temperature T K is (*a*) $\frac{1}{2}kT$ (*b*) kT

(*c*) $\frac{3}{2}kT$ (*d*) $2kT$.

16. A sample of a perfect gas occupies a volume V at a pressure P and absolute temperature T. The mass of each molecule is m. Which of the following expressions gives the number of molecules in the sample ?

(*a*) $\frac{PV}{m}$ (*b*) $\frac{k}{PVT}$

(*c*) $\frac{m}{k}$ (*d*) $\frac{PV}{kT}$.

17. A mixture of two gases A and B is in a container at a constant temperature. Gas A is diatomic and B is monoatomic. The ratio of molecular masses of A and B is 4. The ratio of the r.m.s. speed of A and B is

(*a*) 1 : 1 (*b*) $1 : \sqrt{2}$

(*c*) $\sqrt{2} : 1$ (*d*) 1 : 2.

18. What is the ratio of the r.m.s. speeds of the molecules of a gas at 200 K and 800 K ?

(*a*) 1 : 1 (*b*) 1 : 2

(*c*) 1 : 4 (*d*) 1 : 8.

19. Which of the following graphs best represents the relationship between absolute temperature of a gas and the average kinetic energy of the gas molecules ?

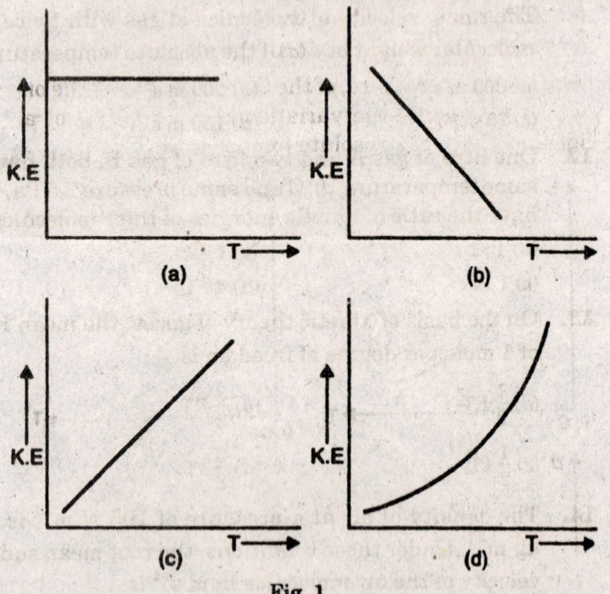

Fig. 1

20. In order to increase the volume of a gas to 3 times at constant pressure at 40°C, the final temperature should be

(*a*) 666°C (*b*) 777°C

(*c*) 555°C (*d*) 333°C.

21. At a certain temperature, the r.m.s. velocity for O_2 is 400 m s^{-1}. At the same temperature, the r.m.s. velocity for H_2 molecules will be

(*a*) 100 m s^{-1} (*b*) 25 m s^{-1}

(*c*) 1600 m s^{-1} (*d*) 6400 m s^{-1}.

22. A gas in a vessel is at the pressure P_0. If the masses of all the molecules be made half and their speeds be made double, then the resultant pressure will be

(*a*) $4P_0$ (*b*) $2P_0$

(*c*) P_0 (*d*) $P_0/2$.

23. The velocities of the three molecules are 3v, 4v and 5v. The r.m.s. velocity is

(*a*) 4v (*b*) $\frac{25}{3}v$

(*c*) $\frac{50}{3}v$ (*d*) $\sqrt{\frac{50}{3}}\,v$.

24. V versus T curves at constant pressures P_1 and P_2 for an ideal gas are shown in Fig. 2. Here,

(*a*) $P_1 > P_2$

(*b*) $P_1 < P_2$

(*c*) $P_1 = P_2$

(*d*) $P_1 \geq P_2$.

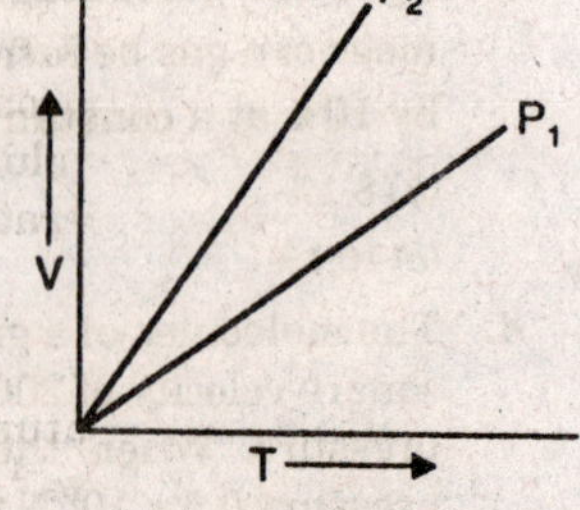

Fig. 2

25. The order of magnitude of the number of nitrogen molecules in an air bubble of diameter 2 mm under ordinary conditions is

(*a*) 10^5 (*b*) 10^9

(*c*) 10^{13} (*d*) 10^{17}.

26. The temperature of a gas is decreased from 273° to 0°C. The average kinetic energy of the molecules changes in the ratio

(*a*) 2 : 1 (*b*) 1 : 2

(*c*) 1 : 4 (*d*) 4 : 1.

27. Two molecules of a gas have speeds of 1 km s^{-1} and 9 km s^{-1} respectively. What is the root mean square speed of these two molecules ?

(*a*) 2 km s^{-1} (*b*) $\sqrt{3}$ km s^{-1}

(*c*) 4 km s^{-1} (*d*) $\sqrt{41}$ km s^{-1}

(*e*) $\sqrt{82}$ km s^{-1}.

28. The energy density $\frac{u}{V}$ of an ideal gas is related to its pressure P as

(*a*) $\frac{u}{V} = 3P$ (*b*) $\frac{u}{V} = \frac{3}{2}P$

(*c*) $\frac{u}{V} = \frac{1}{3}P$ (*d*) $\frac{u}{V} = \frac{2}{3}P$.

29. If the kinetic energy of a perfect gas is 60 joule and its volume is 3 litre, then its pressure will be

(*a*) 2×10^{-4} N m^{-2} (*b*) 4×10^{-4} N m^{-2}

(*c*) $4/3 \times 10^{4}$ N m^{-2} (*d*) $2/3 \times 10^{-4}$ N m^{-2}.

30. The temperature of an ideal gas is increased from 27°C to 927°C. The root mean square speed of its molecules becomes

(*a*) twice (*b*) half

(*c*) four times (*d*) one fourth.

QUESTION BANK

Level II

Choose the correct answer from the four alternatives given in each of the following questions :

1. At 30°C and atmospheric pressure, the volume of a given mass is 100 cm^3. Pressure remaining same, the volume will be doubled if the temperature upto which it should be heated is

(*a*) 15°C (*b*) 606°C

(*c*) 60°C (*d*) 333°C.

2. During an experiment, an ideal gas is found to obey an additional law VP^2 = constant. The gas is initially at temperature T and volume V. When it expands to a volume 2V, the temperature becomes

(*a*) T (*b*) 2T

(*c*) $T\sqrt{2}$ (*d*) T/2.

3. At a certain temperature, hydrogen molecules have r.m.s. velocity 2 kms^{-1}. The r.m.s. velocity of the oxygen molecules at the same temperature is

(*a*) 2 km s^{-1} (*b*) 8 km s^{-1}

(*c*) 0.5 km s^{-1} (*d*) 1 km s^{-1}.

4. How much should the pressure be increased in order to decrease the volume of a gas by 5% at a constant temperature ?

(*a*) 5% (*b*) 5.26%

(*c*) 10% (*d*) 4.26%.

5. We have a jar A filled with gas characterised by parameters P, V and T and another jar B filled with a gas with parameters 2P, V/4 and 2T, where the symbols have the usual meanings. The ratio of the number of molecules in jar A to those in jar B is

(*a*) 1 : 1 (*b*) 1 : 2

(*c*) 2 : 1 (*d*) 4 : 1.

6. The X-components of velocities of some particles (in km/s) came to 10, 8, 6, 4, 2, 0, – 2, – 4, – 6, – 8, – 10. The root mean square (r.m.s.) velocity calculated from these data (in km/s) is nearest to

(*a*) 0 (*b*) 4.6

(*c*) 5.4 (*d*) 6.3.

7. A balloon is filled with hydrogen at a given pressure at 20°C. What fraction of gas will escape out of the balloon if the temperature rises to 40°C at constant pressure ?

(*a*) 0.5 (*b*) 0.06

(*c*) 0.25 (*d*) 0.75.

8. At room temperature, the r.m.s. speed of the molecules of a certain diatomic gas is found to be 1930 m s^{-1}. The gas is

(*a*) H_2 (*b*) F_2

(*c*) O_2 (*d*) Cl_2.

9. At what temperature is the K.E. of a gas molecule half that of its value at 27°C ?

(*a*) 13.5°C (*b*) 150°C

(*c*) 150 K (*d*) – 123 K.

10. At a constant pressure, of the following graphs the one which represents the variation of the density of an ideal gas with the absolute temperature T, is

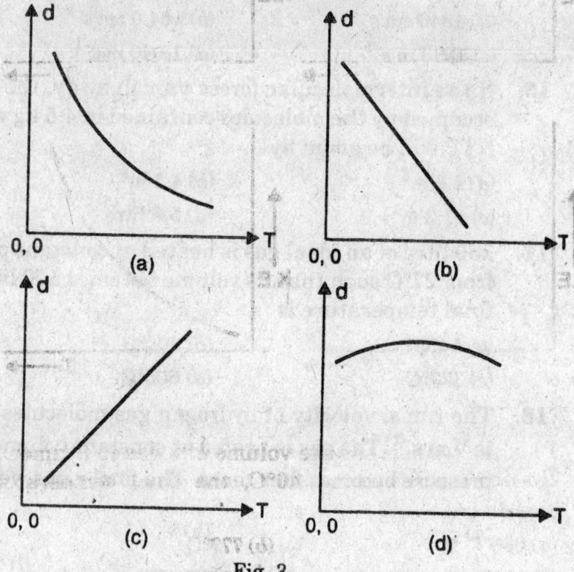

Fig. 3

(*a*) *a* (*b*) *b*
(*c*) *c* (*d*) *d*.

11. A container has an ideal gas at a pressure of 5 atmosphere and temperature 27°C. One-third of the mass of the gas from the container is removed and temperature is increased by 60°C. The pressure of gas in the enclosure is

(*a*) 5 atm. (*b*) 6 atm.
(*c*) 4 atm. (*d*) 3 atm.

12. At what temperature is the root mean square velocity of gaseous hydrogen molecules equal to that of oxygen molecules at 47°C ?

(*a*) 20 K (*b*) 80 K
(*c*) – 73 K (*d*) 3 K.

13. To what temperature should the hydrogen at 327°C be cooled at constant pressure so that the root mean square velocity of its molecules becomes half of its previous value ?

(*a*) – 123°C (*b*) 123°C
(*c*) – 100°C (*d*) 0°C.

14. Oxygen and hydrogen in two enclosures have same mass, volume and pressure. The ratio of the temperatures of the two gases is

(*a*) 1 : 4 (*b*) 4 : 1
(*c*) 16 : 1 (*d*) 1 : 16.

15. If density of hydrogen gas at NTP is 0.0000893 g cm^{-3}, then the root mean square velocity of the hydrogen gas at NTP is nearly

(*a*) 1840 cm s^{-1} (*b*) 184.0 cm s^{-1}
(*c*) 1833 m s^{-1} (*d*) 18.40 m s^{-1}.

16. If the intermolecular forces vanish away, the volume occupied by the molecules contained in 4.5 kg water at NTP will be given by

(*a*) 5.6 m^3 (*b*) 4.5 m^3
(*c*) 11.2 m^3 (*d*) 5.6 litre.

17. 250 litre of an ideal gas is heated at constant pressure from 27°C such that its volume becomes 500 litre. The final temperature is

(*a*) 54°C (*b*) 300°C
(*c*) 327°C (*d*) 600°C.

18. The r.m.s. velocity of hydrogen gas molecules at NTP is V m s^{-1}. The gas is heated at constant volume till the pressure becomes four times. The final r.m.s. velocity is

(*a*) $\frac{V}{2}$ (*b*) V
(*c*) 2V (*d*) 4V.

19. An ideal gas is at a temperature of 127°C. It is heated at constant pressure until its volume becomes 1.5 times. The new temperature of the gas is

(*a*) 300°C (*b*) 256°C
(*c*) 259°C (*d*) 327°C.

20. If the universal gas constant is 8.3 joule/mole/K and the Avogadro's number 6×10^{23}, then mean kinetic energy of the oxygen molecules at 327°C will be

(*a*) 415×10^{-23} joule (*b*) 2490×10^{-22} joule
(*c*) 1245×10^{-23} joule (*d*) 830×10^{-22} joule.

21. The pressure of a gas in a closed vessel is increased by 0.4% when heated by 1°C. The initial temperature of the gas is

(*a*) 25°C (*b*) 250°C
(*c*) 250 K (*d*) 2500 K.

22. In a given diatomic gas, sound travels at a speed of 300 m s^{-1}. The root mean square speed of the molecules of the gas is

(*a*) 4 m s^{-1} (*b*) 40 m s^{-1}
(*c*) 438 m s^{-1} (*d*) 43900 m s^{-1}.

23. The root mean square velocity of the molecules in a sample of helium is 5/7th that of the molecules in a sample of hydrogen. If the temperature of hydrogen sample is 0°C, then the temperature of the helium sample is about

(*a*) 0°C (*b*) 0 K
(*c*) 273°C (*d*) 100°C.

24. The volume of a vessel A is twice the volume of another vessel B, and both of them are filled with the same gas. If the gas in A is at twice the temperature and twice the pressure in comparison to the gas in B, then the ratio of the gas molecules in A and B is

(*a*) 1 : 4 (*b*) 4 : 1
(*c*) 2 : 1 (*d*) 1 : 2.

25. An open mouthed bottle contains a gas at 60°C. To what temperature should the gas be heated so that $\frac{1}{4}$ th mass of the gas may leave it ?

(*a*) 143°C (*b*) 171°C
(*c*) 444°C (*d*) 416°C.

26. A mixture of helium and hydrogen gases is filled in a vessel at 30°C. The r.m.s. velocities of helium and hydrogen molecules are in the ratio

(*a*) 1 : 2 (*b*) 2 : 1
(*c*) 1 : $\sqrt{2}$ (*d*) $\sqrt{2}$: 1.

27. The ratio of the vapour densities of two gases at the same temperature is 8 : 9. The ratio of the r.m.s. velocities of their molecules is

(a) 8 : 9 (b) 9 : 8
(c) $\sqrt{9} : \sqrt{8}$ (d) $\sqrt{8} : \sqrt{9}$.

28. Two perfect gases at absolute temperatures T_1 and T_2 are mixed. There is no loss of energy. The masses of the molecules are m_1 and m_2. The number of molecules in the gases are n_1 and n_2. The temperature of the mixture is

(a) $T_1 + T_2$ (b) $\frac{T_1 + T_2}{2}$
(c) $\frac{n_1T_1 + n_2T_2}{n_1 + n_2}$ (d) $\frac{n_1T_1 - n_2T_2}{n_1 + n_2}$.

29. A vessel is filled with a gas at a pressure of 76 cm of mercury at a certain temperature. The mass of the gas is increased by 50% by introducing more gas in the vessel at the same temperature. The resultant pressure, in cm of Hg, is

(a) 76 (b) 152
(c) 114 (d) 1117.

30. At what temperature, r.m.s. velocity of O_2 molecules will be 1/3 of H_2 molecules at – 3°C ?

(a) 90 K (b) 1167°C
(c) – 3°C (d) 207°C.

31. When the temperature of a gas is raised from 27°C to 90°C, the percentage increase in the r.m.s. velocity of the molecules will be

(a) 10% (b) 15%
(c) 20% (d) 17.5%.

32. An assembly of smoke particles in air at NTP is under consideration. If the mass of each particle is 5×10^{-17} kg, then the r.m.s. speed is (Given : $k = 1.38 \times 10^{-23}$ J K^{-1})

(a) 1.5 cm s^{-1} (b) 1.5 mm s^{-1}
(c) 1.5 m s^{-1} (d) 1.5 km s^{-1}.

33. The average kinetic energy of translatory motion of hydrogen molecules at 27°C is 6.2×10^{-21} J. The average kinetic energy at 127°C will be

(a) 30×10^{-21} J (b) 8.27×10^{-21} J
(c) 1.2×10^{-21} J (d) 3.1×10^{-21} J.

34. Four molecules of a gas have speeds 1, 2, 3 and 4 km s^{-1} respectively. The value of r.m.s. speed of the molecules is (in km s^{-1})

(a) $\sqrt{\frac{15}{2}}$ (b) $2\sqrt{15}$
(c) $\frac{\sqrt{15}}{2}$ (d) None of these.

35. At a pressure of 24×10^5 dyne cm^{-2}, the volume of O_2 is 10 litre and mass is 20 g. The r.m.s. velocity will be

(a) 800 m s^{-1} (b) 400 m s^{-1}
(c) 600 m s^{-1} (d) Data is incomplete.

36. A vessel contains a gas at a pressure of 10^{-10} dyne cm^{-2} and at a temperature of 27°C. If $k = 1.38 \times 10^{-16}$ erg/molecule/kelvin, then the number of molecules per cubic centimetre volume of the gas is

(a) $\frac{10}{3 \times 1.38}$ (b) $\frac{10^2}{3 \times 1.38}$
(c) $\frac{10^3}{3 \times 1.38}$ (d) $\frac{10^4}{3 \times 1.38}$.

37. The mean kinetic energy per mole of an ideal gas at 0°C is approximately

(a) 3.4 J (b) 3.4×10^2 J
(c) 3.4×10^3 J (d) 207 J.

38. The temperature at which the root mean square velocity of the gas molecules would become twice of its value at 0°C is

(a) 819°C (b) 1092°C
(c) 1100°C (d) 1400°C.

39. The root mean square velocity of gas molecules at 0°C is $\sqrt{20} \times 10^2$ m s^{-1}. The density of the gas at NTP is (Given : 1 atm. pressure = 10^5 N m^{-2})

(a) 1 kg m^{-3} (b) 1.5 kg m^{-3}
(c) 2 kg m^{-3} (d) 2.5 kg m^{-3}.

40. The average kinetic energy of hydrogen molecules at NTP is

(a) $\frac{3}{2} \times 1.38 \times 10^{-23} \times 273$ J (b) 3×1.38 J
(c) 1.38 J (d) 273 J.

41. At 0°C, the value of the density of a fixed mass of an ideal gas divided by its pressure is x. At 100°C, this quotient is

(a) $\frac{100}{273} x$ (b) $\frac{273}{100} x$
(c) $\frac{273}{373} x$ (d) $\frac{373}{273} x$
(e) x.

42. A gas is filled in a cylinder. Its temperature is increased by 20% on Kelvin scale and volume is reduced by 10%. How much percentage of the gas has to leak for pressure to remain constant ?

(a) 30% (b) 40%
(c) 15% (d) 25%.

43. A thin bulb containing air at normal atmospheric pressure at 41°C is sealed. The bulb is heated to 198°C and the bulb bursts. The bursting pressure is

(a) 1 atm (b) 1.5 atm
(c) 2 atm (d) 2.5 atm.

44. A flask containing air at 27°C at atmospheric pressure is corked up. A pressure of 2.5 atm inside the flask would force the cork out. The temperature at which it will happen is

(*a*) 67.5°C (*b*) 577°C
(*c*) 670°C (*d*) 750 K.

45. Fig. 4 shows graphs of pressure vs density for an ideal gas at two temperatures T_1 and T_2. Which of the following is correct ?

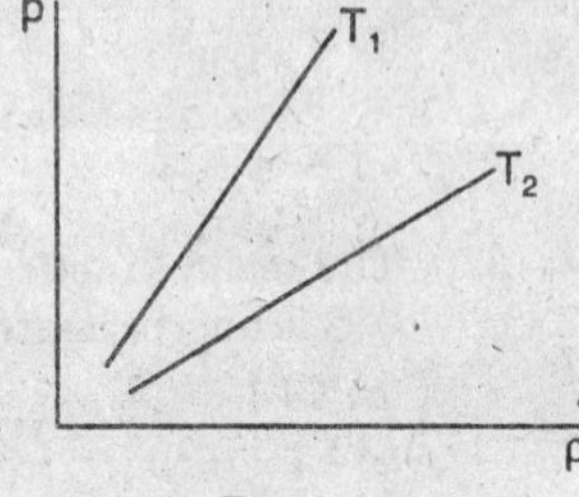

Fig. 4

(*a*) $T_1 > T_2$
(*b*) $T_1 = T_2$
(*c*) $T_1 < T_2$
(*d*) any of the three is possible.

46. At identical temperatures, the r.m.s. speed of hydrogen molecules is 4 times that for oxygen molecules. In a mixture of these in mass ratio $H_2 : O_2 = 1 : 8$, the r.m.s. speed of all molecules is *n* times the r.m.s. speed for O_2 molecules, where *n* is

(*a*) 3 (*b*) 4/3
(*c*) $(8/3)^{1/2}$ (*d*) $(11)^{1/2}$.

47. The molecular weights of O_2 and N_2 are 32 and 28 respectively. At 15°C, the pressure of 1 g O_2 will be the same as that of 1 g N_2 in the same bottle at the temperature

(*a*) – 21°C (*b*) 13°C
(*c*) 15°C (*d*) 56.4°C.

48. The temperature, at which r.m.s. velocity of helium molecules is equal to the r.m.s. velocity of hydrogen molecules at NTP, is

(*a*) 100 K (*b*) 300 K
(*c*) 502 K (*d*) 546 K.

49. If $P = 10^6\ kT$, then the number of molecules per unit volume of the gas is [The letters have usual meanings.]

(*a*) 1 (*b*) 10^2
(*c*) 10^3 (*d*) 10^6.

50. A sample of an ideal gas occupies volume V at a pressure P and absolute temperature T. The mass of each molecule is *m*. Which of the following expressions gives the density of the gas ?

(*a*) P/kT (*b*) Pm/kT
(*c*) m/kT (*d*) P/kTV.

51. A container of volume 1.5 litre contains helium gas at a temperature of 27°C and 80 cm Hg pressure. The number of moles of the helium gas in the container is

(*a*) 0.64 (*b*) 0.064
(*c*) 6.4 (*d*) 64.

52. 16 g of oxygen and 14 g of nitrogen are mixed in an enclosure of volume 5000 cm^3 at a temperature of 300 K. The resulting pressure of the mixture is nearly

(*a*) 5×10^5 Pa (*b*) 4×10^5 Pa
(*c*) 3×10^5 Pa (*d*) 1.2×10^5 Pa.

53. The temperature at which the r.m.s. velocity of oxygen molecules is equal to that of the nitrogen molecules at 100°C is

(*a*) 42.63 K (*b*) 426.3 K
(*c*) 4263 K (*d*) 4.263 K.

54. The gas in a vessel is subjected to a pressure of 20 atm at a temperature of 27°C. The pressure of the gas in the vessel after one half the mass of the gas is released from the vessel and the temperature of the remainder is raised by 50°C is

(*a*) 11.67 atm (*b*) 17 atm
(*b*) 8.5 atm (*d*) 10.8 atm.

55. A flask is filled with 4 g of a gas at 12°C and then heated to 50°C. Owing to the escape of the gas, the pressure in the flask is the same at the beginning as well as at the end of the experiment. Then the mass of the gas that has escaped is nearly

(*a*) 4 g (*b*) 3.53 g
(*c*) 0.47 g (*d*) 3.06 g.

ANSWERS

QUESTION BANK (Level I)

1. (*d*) **2.** (*a*) **3.** (*d*) **4.** (*a*) **5.** (*b*) **6.** (*c*) **7.** (*c*) **8.** (*a*)
9. (*c*) **10.** (*d*) **11.** (*d*) **12.** (*a*) **13.** (*a*) **14.** (*a*) **15.** (*c*) **16.** (*d*)
17. (*d*) **18.** (*b*) **19.** (*c*) **20.** (*a*) **21.** (*c*) **22.** (*b*) **23.** (*d*) **24.** (*a*)
25. (*d*) **26.** (*a*) **27.** (*d*) **28.** (*b*) **29.** (*c*) **30.** (*a*).

QUESTION BANK (Level II)

1. (*d*)	2. (*c*)	3. (*c*)	4. (*b*)	5. (*d*)	6. (*d*)	7. (*b*)	8. (*a*)
9. (*c*)	10. (*a*)	11. (*c*)	12. (*a*)	13. (*a*)	14. (*c*)	15. (*c*)	16. (*a*)
17. (*c*)	18. (*c*)	19. (*d*)	20. (*c*)	21. (*c*)	22. (*c*)	23. (*a*)	24. (*c*)
25. (*b*)	26. (*c*)	27. (*c*)	28. (*c*)	29. (*c*)	30. (*d*)	31. (*a*)	32. (*a*)
33. (*b*)	34. (*a*)	35. (*c*)	36. (*d*)	37. (*c*)	38. (*a*)	39. (*b*)	40. (*a*)
41. (*c*)	42. (*d*)	43. (*b*)	44. (*d*)	45. (*a*)	46. (*d*)	47. (*a*)	48. (*d*)
49. (*d*)	50. (*b*)	51. (*b*)	52. (*a*)	53. (*b*)	54. (*a*)	55. (*c*).	

HINTS/SOLUTIONS

QUESTION BANK (Level I)

1. (*d*) 1 litre = 1000 cm^3 = 10^{-3} m^3

Energy/m^3 = 300×10^3 J

$$P = \frac{2}{3} \times 300 \times 10^3 \text{ N m}^{-2} = 2 \times 10^5 \text{ N m}^{-2}$$

2. (*a*) $P' = \frac{76 \times 0.10}{0.19}$ cm of Hg = 40 cm of Hg

3. (*d*) $PV = P' \times \frac{90V}{100}, \frac{P'}{P} = \frac{10}{9}$

$$\frac{P' - P}{P} \times 100 = \left(\frac{10}{9} - 1\right) \times 100 = \frac{100}{9} = 11.1$$

4. (*a*) $\frac{C_2}{C_1} = \sqrt{\frac{400}{300}} = \frac{2}{\sqrt{3}}$

$$C_2 = \frac{2}{\sqrt{3}} \times 200 \text{ m s}^{-1} = \frac{400}{\sqrt{3}} \text{ m s}^{-1}$$

5. (*b*) 3 moles of H_2 are given.

6. (*c*) $C_{H_2} = xC_{He}$

$\sqrt{M_{He}} = x\sqrt{M_{H_2}} \Rightarrow \sqrt{4} = x\sqrt{2}$

or $x = \frac{2}{\sqrt{2}}$ or $x = \sqrt{2}$

7. (*c*) Pressure is reduced by a factor of 10. Volume is increased by a factor of 10.

8. (*a*) $\frac{E_1}{E_2} = \frac{N \times \frac{3}{2}kT}{2N \times \frac{3}{2}kT} = \frac{1}{2}$

9. (*c*) $T_1 = 300$ K, $P \propto T$

So, $T_2 = 600$ K or 327°C

10. (*d*) $C = \sqrt{\frac{1^2 + 4^2 + 8^2 + 16^2}{4}}$ m s^{-1}

$$= \sqrt{\frac{1 + 16 + 64 + 256}{4}} \text{ m s}^{-1} = \sqrt{84.25} \text{ m s}^{-1} = 9.2 \text{ m s}^{-1}$$

11. (*d*) $C' = \sqrt{\frac{3R\left(\frac{T}{2}\right)}{2M}} = \frac{1}{2}C = \frac{300}{2} \text{ m s}^{-1} = 150 \text{ m s}^{-1}$

12. (*a*) KE/molecule is constant at a given temperature.

13. (*a*) $\frac{1}{2}kT \times N = \frac{1}{2}RT$

14. (*a*) $C = \sqrt{\frac{3P}{\rho}} = \sqrt{\frac{3 \times 10^5}{1.2}} = \frac{1}{2} \times 10^3 \text{ m s}^{-1} = 500 \text{ m s}^{-1}$

15. (*c*) $\frac{1}{2}kT \times 3$

16. (*d*) $PV = \mu RT$, $PV = \frac{n}{N} \times kNT$ or $n = \frac{PV}{kT}$

17. (*d*) $\frac{C_A}{C_B} = \sqrt{\frac{M_B}{M_A}} = \sqrt{\frac{1}{4}} = \frac{1}{2}$

18. (*b*) $\frac{C_1}{C_2} = \sqrt{\frac{200}{800}} = \frac{1}{2}$

19. (*c*) KE $\propto$ T

20. (*a*) $\frac{T'}{T} = \frac{3V}{V} = 3$

$T' = 3T = 3[40 + 273]$ K $= 3 \times 313$ K $= 939$ K

$t = (939 - 273)$°C = 666°C

21. (*c*) $\frac{C_{H_2}}{C_{O_2}} = \sqrt{\frac{32}{2}} = 4$

or $C_{H_2} = 4 \times C_{O_2} = 4 \times 400 \text{ m s}^{-1} = 1600 \text{ m s}^{-1}$

22. (*b*) $P' = \frac{1}{3}\frac{\left(\frac{m}{2}\right)n}{V}(2C)^2 = 2P_0$

23. (*d*) $C = \sqrt{\frac{3^2+4^2+5^2}{3}}\, v = \sqrt{\frac{50}{3}}\, v$

24. (*a*) $PV = \mu RT$ or $P = \frac{\mu RT}{V}$ or $P = \frac{\mu R}{V/T}$

or $P = \frac{\mu R}{\tan\theta}$, where $\tan\theta$ is the slope of the graph.

So, more the slope, lesser will be the pressure.

25. (*d*) Molar volume $= 2 \times 10^4\ cm^3$; Bubble volume $= 4 \times 10^{-3}\ cm^3$. That gives 2×10^{-7} mole. With $N = 6 \times 10^{23}$, we reach 10^{17}.

26. (*a*) The temperature changes from 2×273 K to 273 K. Since temperature is halved, therefore, kinetic energy is halved.

27. (*d*) $C = \sqrt{\frac{1^2+9^2}{2}}\ km\ s^{-1} = \sqrt{\frac{82}{2}}\ km\ s^{-1} = \sqrt{41}\ km\ s^{-1}$

28. (*b*) $P = \frac{2}{3}\frac{u}{V}$ or $\frac{u}{V} = \frac{3}{2}P$

29. (*c*) $P = \frac{2}{3} \times \frac{60}{3 \times 10^{-3}}\ N\ m^{-2}$

$= \frac{40}{3} \times 10^3\ N\ m^{-2} = \frac{4}{3} \times 10^4\ N\ m^{-2}$

30. (*a*) $T_1 = 27 + 273 = 300$ K

$T_2 = 927 + 273 = 1200$ K

Temperature is increased by a factor of 4. So, C is increased by a factor of 2.

QUESTION BANK (Level II)

1. (*d*) $\frac{T_2}{T_1} = \frac{V_2}{V_1} = 2$ or $T_2 = 2[273 + 30] = 606°K$

or $t_2 = (606 - 273)°K = 333°C$

2. (*c*) $VP^2 = \text{constant}$

or $V\left[\frac{RT}{V}\right]^2 = \text{constant}$ or $\frac{T^2}{V} = \text{constant}$

or $\frac{T'^2}{T^2} = \frac{V'}{V} = \frac{2V}{V} = 2$ or $T' = \sqrt{2}\,T$

3. (*c*) $\frac{C_{O_2}}{C_{H_2}} = \sqrt{\frac{M_{H_2}}{M_{O_2}}} = \sqrt{\frac{2}{32}}$

or $C_{O_2} = \frac{C_{H_2}}{4} = \frac{2}{4}$ or $C_{O_2} = 0.5\ km\ s^{-1}$

4. (*b*) $\frac{P'}{P} = \frac{V}{V - \frac{5}{100}V}$ or $\frac{P'}{P} = \frac{100}{95}$

or $\left(\frac{P'}{P} - 1\right) \times 100 = \left(\frac{100}{95} - 1\right) \times 100 = \frac{500}{95} = 5.26$

5. (*d*) $nRT = PV$ or $n = \frac{PV}{RT}$

Again, $n' = \frac{(2P)\left(\frac{V}{4}\right)}{R(2T)}$ or $n' = \frac{n}{4}$ or $\frac{n}{n'} = \frac{4}{1}$

6. (*d*) $C = \sqrt{\frac{c_1^2 + c_2^2 + \ldots\ldots + c_n^2}{n}}$

7. (*b*) $PV = \mu RT \Rightarrow \mu T = \text{constant}$

$\therefore\ \mu'[273 + 40] = \mu[273 + 20]$

or $\frac{\mu'}{\mu} = \frac{293}{313} = 0.94$

Fraction escaped $= 1 - 0.94 = 0.06$

8. (*a*) $C = \sqrt{\frac{3RT}{M}}$

or $M = \frac{3RT}{C^2} = \frac{3 \times 8.4 \times 300}{1930 \times 1930} = 2 \times 10^{-3}$ kg

M = 2*g*, which is the molecular weight of hydrogen.

9. (*c*) $\frac{T}{300} = \frac{1}{2}$ or $T = 150$ K

10. (*a*) $P = \frac{1}{3}dC^2$

or $dC^2 = \text{constant}$ or $dT = \text{constant}$

11. (*c*) $PV = \mu RT$

$P \propto \mu T$

$\frac{P_2}{P_1} = \frac{\mu_2 T_2}{\mu_1 T_1} = \frac{2}{3} \times \frac{87 + 273}{27 + 273}$

or $\frac{P_2}{P_1} = \frac{2}{3} \times \frac{360}{300} = 0.8$

or $P_2 = 0.8 \times 5 = 4$ atm.

12. (*a*) $C = \sqrt{\frac{3RT}{M}}$

For the given data, $T \propto M$

$\frac{T_{H_2}}{T_{O_2}} = \frac{2}{32} = \frac{1}{16}$ or $T_{H_2} = \frac{273 + 47}{16}$ K = 20 K

13. (*a*) $\sqrt{\frac{T}{273 + 327}} = \frac{1}{2}$ or $\frac{T}{600} = \frac{1}{4}$

or $T = \frac{600}{4}$ K = 150 K

or $t = (150 - 273)°C$ or $t = -123°C$

14. (c) $PV = \mu RT$ or $PV = \frac{m}{M}RT$

For the given data, $\frac{T}{M}$ = constant

$$\therefore \quad \frac{T_{O_2}}{T_{H_2}} = \frac{M_{O_2}}{M_{H_2}} = \frac{32}{2} = 16$$

15. (c) $C = \sqrt{\frac{3P}{\rho}}$ or $C = \sqrt{\frac{3 \times 10^5}{893 \times 10^{-4}}}$ m s^{-1}

$$= \sqrt{\frac{30 \times 10^8}{893}} \text{ m s}^{-1} = 10^4 \times 0.1833 \text{ m s}^{-1} = 1833 \text{ m s}^{-1}$$

16. (a) Due to vanishing of intermolecular forces, the liquid would get converted into vapour.

We know that 18×10^{-3} kg of water vapour occupies a volume of 22.4×10^{-3} m^3 at NTP.

So, volume occupied by 4.5 kg is $\frac{22.4 \times 10^{-3}}{18 \times 10^{-3}} \times 4.5$ m^3,

i.e., 5.6 m^3

17. (c) V doubled, T doubled

$T_1 = (27 + 273)$ K = 300 K

$T_2 = 2 \times 300$ K = 600 K

$t_2 = (600 - 273)$°C = 327°C

18. (c) $P \propto T$. P is increased by a factor of 4. Same with T. $\sqrt{T}$ increased by a factor of 2. So, C is increased by a factor of 2.

19. (d) $\frac{T'}{400} = \frac{1.5V}{V} = 1.5$

or $T' = 1.5 \times 400$ K = 600 K

$t' = (600 - 273)$°C = 327°C

20. (c) $E = \frac{3}{2}kT = \frac{3}{2}\frac{R}{N}T = \frac{3 \times 8.3 \times 600}{2 \times 6 \times 10^{23}}$ J

21. (c) $\frac{T+1}{T} = \frac{P + \frac{0.4}{100}P}{P}$ or $1 + \frac{1}{T} = 1 + \frac{0.4}{100}$

or $T = \frac{100}{0.4} = \frac{1000}{4}$ K $\Rightarrow$ T = 250 K

22. (c) $v = \sqrt{\frac{\gamma P}{\rho}}$ and $C = \sqrt{\frac{3P}{\rho}}$

$$\frac{C}{v} = \sqrt{\frac{3}{\gamma}} = \sqrt{\frac{3}{1.4}} \text{ or } C = 300\sqrt{\frac{3}{1.4}} \text{ m s}^{-1}$$

$= 300 \times 1.46$ m s^{-1} = 438 m s^{-1}

23. (a) $(C_{He})_t = \frac{5}{7}(C_{H_2})_{0°C}$

$$\frac{273 + t}{4} = \left(\frac{5}{7}\right)^2 \frac{273}{2} \text{ or } 273 + t = \frac{25 \times 273 \times 4}{49 \times 2}$$

or $t = \frac{100}{98} \times 273 - 273 \approx 0$°C

24. (c) $P = \frac{1}{3}\frac{mn}{V}C^2$ or $n = \frac{3PV}{mC^2}$

$\frac{n_1}{n_2} = \frac{2 \times 2}{2} = \frac{2}{1}$ $[\because C^2 \propto T]$

25. (b) μT = constant

$\frac{3}{4}[273 + t] = 1[273 + 60]$

or $273 + t = \frac{4}{3} \times 333 = 444$

or $t = (444 - 273)$°C = 171°C

26. (c) $\frac{C_{He}}{C_{H_2}} = \sqrt{\frac{M_{H_2}}{M_{He}}} = \sqrt{\frac{2}{4}} = \frac{1}{\sqrt{2}}$

27. (c) $\frac{C_1}{C_2} = \sqrt{\frac{\rho_2}{\rho_1}} = \sqrt{\frac{9}{8}}$

28. (c) $\frac{3}{2}kT_1 \times n_1 + \frac{3}{2}kT_2 \times n_2 = (n_1 + n_2)\frac{3}{2}kT$

or $T = \frac{n_1T_1 + n_2T_2}{n_1 + n_2}$

29. (c) $P \propto m$

Since m is increased by a factor of $\frac{3}{2}$, therefore, P will increase by a factor of $\frac{3}{2}$.

$\therefore$ New pressure $= \frac{3}{2} \times 76$ cm of Hg = 114 cm of Hg

30. (d) $(C_{O_2})_t = \frac{1}{3}(C_{H_2})_{-3}$

$$\sqrt{\frac{273 + t}{32}} = \frac{1}{3}\sqrt{\frac{273 - 3}{2}}$$

or $\frac{273 + t}{32} = \frac{1}{9} \times \frac{270}{2}$

or $273 + t = \frac{270 \times 32}{18}$

or $273 + t = 480$

or $t = 480 - 273 = 207$°C

31. (a) $\frac{C'}{C} = \sqrt{\frac{273 + 90}{273 + 27}} = \sqrt{\frac{363}{300}}$

or $\frac{C'}{C} = \sqrt{1.21} = 1.1$ or $\frac{C'}{C} - 1 = 0.1$

or $\frac{C' - C}{C} \times 100 = 0.1 \times 100 = 10$

32. (a) $\frac{1}{2}mC^2 = \frac{3}{2}kT, C = \sqrt{\frac{3kT}{m}}$

$= \sqrt{\frac{3 \times 1.38 \times 10^{-23} \times 273}{5 \times 10^{-17}}}$

$= 15 \times 10^{-3} \text{ m s}^{-1} = 1.5 \text{ cm s}^{-1}$

33. (b) $T_1 = 27 + 273 = 300$ K

$T_2 = 127 + 273 = 400$ K

$\frac{E_2}{E_1} = \frac{T_2}{T_1}$ or $E_2 = \frac{4}{3} \times 6.2 \times 10^{-21} \text{ J} = 8.27 \times 10^{-21}$ J

34. (a) $C = \sqrt{\frac{1^2 + 2^2 + 3^2 + 4^2}{4}} = \sqrt{\frac{1 + 4 + 9 + 16}{4}}$

$= \sqrt{\frac{15}{2}} \text{ km s}^{-1}$

35. (c) $C = \sqrt{\frac{3PV}{M}} = \sqrt{\frac{3 \times 24 \times 10^5 \times 10 \times 1000}{20}} \text{ cm s}^{-1}$

$= 6 \times 10^4 \text{ cm s}^{-1} = 600 \text{ m s}^{-1}$

36. (d) $PV = nkT$ or $\frac{n}{V} = \frac{P}{kT}$

or $\frac{n}{V} = \frac{10^{-10}}{1.38 \times 10^{-16} \times 300} = \frac{10^4}{1.38 \times 3}$

37. (c) Mean kinetic energy per mole

$= \frac{3}{2}RT = \frac{3}{2} \times 8.32 \times 273 \text{ J} = 3.4 \times 10^3$ J

38. (a) $\frac{C_t}{C_0} = \sqrt{\frac{273 + t}{273}}$

or $4 \times 273 - 273 = t$ or $t = 3 \times 273 = 819$°C

39. (b) $C = \sqrt{\frac{3P}{\rho}}$ or $\rho = \frac{3P}{C^2} = \frac{3 \times 10^5}{20 \times 10^4} = 1.5 \text{ kg m}^{-3}$

40. (a) $E_k = \frac{3}{2}kT$

> ☞ *At low temperature, we have to consider only the translational degrees of freedom.*

41. (c) The quotient is density/pressure. Since density $= \frac{\text{mass}}{\text{volume}}$, x can be expressed as $\frac{\text{mass}}{\text{volume} \times \text{pressure}}$. Mass is fixed but pressure times volume is proportional to the absolute temperature ; thus the quotient is inversely proportional to the absolute temperature.

42. (d) $V \propto \mu T$, $\frac{90}{100} V \propto \mu' \frac{6T}{5}$

$\frac{\mu'}{\mu} \times \frac{6}{5} = \frac{9}{10}$ or $\frac{\mu'}{\mu} = \frac{9}{10} \times \frac{5}{6} = \frac{3}{4}$

Percentage of gas leaked out

$= \frac{\mu - \mu'}{\mu} \times 100 = \frac{4 - 3}{4} \times 100 = 25$

43. (b) $\frac{P'}{P} = \frac{198 + 273}{41 + 273} = \frac{471}{314} = 1.5$

or $P' = 1.5 \times 1 = 1.5$ atmosphere

44. (d) $\frac{T_2}{T_1} = \frac{2.5}{1}$ or $\frac{T_2}{300} = 2.5$

or $T_2 = 2.5 \times 300 \text{ K} = 750$ K

45. (a) For a constant value of density, pressure is more at T_1.

$\therefore \quad T_1 > T_2 \quad [\because P \propto T]$

46. (d) Molecule number ratio is $H_2 : O_2 = \frac{2}{3} : \frac{1}{3}$. That gives $(C_{rms})^2 = 16\left(\frac{2}{3}\right) + 1\left(\frac{1}{3}\right)$ times the value for O_2.

47. (a) $PV = \mu RT = \frac{m}{M} RT$, $\frac{T}{M}$ = constant.

48. (d) $C = \sqrt{\frac{3RT}{M}}$ or $T \propto M$

$\therefore \quad \frac{T'}{T} = \frac{4}{2} = 2$ or $T' = 2T$

or $T' = 2 \times 273 \text{ K} = 546$ K

49. (d) $PV = \mu RT = \frac{n}{N} kNT = nkT$

or $\frac{n}{V} = \frac{P}{kT} = \frac{10^6 kT}{kT} = 10^6$

50. (b) $P = \frac{1}{3}\rho C^2$

Also, $\frac{1}{2}mC^2 = \frac{3}{2}kT$ or $C^2 = \frac{3kT}{m}$

Now, $\rho = \frac{3P}{C^2} = \frac{3Pm}{3kT} = \frac{Pm}{kT}$

51. (b) $PV = \mu RT$

$\mu = \frac{PV}{RT} = \frac{80 \times 10^{-2} \times 13600 \times 9.8 \times 1.5 \times 10^{-3}}{8.3 \times 300} = 0.064$

52. (a) $P = \frac{m}{M} \frac{RT}{V}$

Using Dalton's law of partial pressures

$P = \left[\frac{16}{32} + \frac{14}{28}\right] \frac{8.3 \times 300}{5000 \times 10^{-6}} \text{ N m}^{-2}$

$= \frac{8.3 \times 300 \times 10^3}{5} \text{ N m}^{-2}$

$= 4.98 \times 10^5 \text{ N m}^{-2} \approx 5 \times 10^5 \text{ N m}^{-2}$ (Pa)

53. (*b*) $\left(C_{O_2}\right)_t = \left(C_{N_2}\right)_{100°C}$

$$\sqrt{\frac{3RT}{32}} = \sqrt{\frac{3R(273+100)}{28}}$$

or $\frac{T}{32} = \frac{373}{28}$ or $T = \frac{32}{28} \times 373\ K = 426.3\ K$

54. (*a*) $PV = \mu RT$ or $PV = \frac{m}{M} RT$ or $P \propto mT$

Now, $20 \propto m \times 300$ and $P' \propto \frac{m}{2} \times 350$

$\therefore \quad \frac{P'}{20} = \frac{350}{2 \times 300} = 0.583$

or $P' = 20 \times 0.583$ atm $= 11.67$ atm

55. (*c*) $PV = \mu RT$ or $PV = \frac{m}{M} RT$

For the given data, mT = constant

$4(273 + 12) = m(273 + 50)$ or $m = \frac{4 \times 285}{323} = 3.53$ g

Mass of gas escaped = (4 – 3.53) g = 0.47 g.

9 I. HEAT AND THERMODYNAMICS II. TRANSFERENCE OF HEAT

Important Terms, Facts and Formulae

IMPORTANT POINTS TO REMEMBER

1. $\frac{C}{100} = \frac{F - 32}{180} = \frac{R}{80}$

2. (*i*) In linear expansion,

$$l_t = l_0(1 + \alpha t)$$

(*ii*) In superficial expansion,

$$S_t = S_0(1 + \beta t)$$

(*iii*) In cubical expansion,

$$V_t = V_0(1 + \gamma t)$$

3. Relationship between the three coefficients of expansion of a body

$$\frac{\alpha}{1} = \frac{\beta}{2} = \frac{\gamma}{3}$$

or $\beta = 2\alpha, \ \gamma = 3\alpha$

4. Specific heat of a material at any temperature is defined by

$$c = \frac{dQ}{m dT}$$

5. Molar specific heat is given by

$$c = \frac{1}{n}\frac{dQ}{dT}$$

6. Two specific heats of a gas are defined as

$$c_v = Mc_v, \ \ c_p = Mc_p$$

7. Number of moles, $n = \frac{m}{M}$

8. Ratio of specific heats, $\frac{c_p}{c_v} = \gamma$

9. Thermal capacity = Mc

QUESTION BANK

Level I

Choose the correct answer from the four alternatives given in each of the following questions :

1. The temperature of a gas increases by 10°C. What is the equivalent increase on the kelvin scale ?

(*a*) 10 K (*b*) 0 K
(*c*) 100 K (*d*) 273.90 K.

2. What is 273.00 K on the Celsius scale of temperature ?

(*a*) – 0.15°C (*b*) 0.00°C
(*c*) 0.15°C (*d*) 273.15°C
(*e*) 546.15°C.

3. What thermodynamic temperature is equivalent to 501.85°C ?

(*a*) 775.01 K (*b*) 228.85 K
(*c*) 775.00 K (*d*) 228.70 K
(*e*) 774.85 K.

4. The thermo e.m.fs. generated in a thermocouple thermometer are 3 mV at the triple point temperature of water and 4.5 mV when the junction is kept in a hot liquid. The temperature of the liquid is

(*a*) 309.74 K (*b*) 409.74 K
(*c*) 509.74 K (*d*) 609.74 K.

5. The table lists the approximate range, accuracy and response time of different types of thermometers.

Ther-mometer	*Range / K*	*Accuracy*	*Response time*
(*a*)	3—1750	very good	long
(*b*)	30—1750	average	short
(*c*)	75—1550	good	long
(*d*)	230—630	poor	medium

Which set of properties belongs to a thermocouple ?

6. Which one of the following thermometers would be most suitable for monitoring the temperature of gases in a factory chimney if the temperature can vary over a range of 200 K in a minute ?

(*a*) alcohol in glass (*b*) mercury in glass
(*c*) constant-volume gas (*d*) thermoelectric
(*e*) platinum resistance.

7. Which combination of thermometers would be most appropriate for measuring the following three temperatures ?

	1	*2*	*3*
	temperature at various positions in a flame	*boiling point of sulphur (717K)*	*boiling point of liquid nitrogen (80 K)*
(*a*)	liquid-in-glass	resistance	thermocouple
(*b*)	resistance	liquid-in-glass	thermocouple
(*c*)	resistance	thermocouple	liquid-in-glass
(*d*)	thermocouple	liquid-in-glass	resistance
(*e*)	thermocouple	resistance	resistance.

8. A temperature may be measured by using a constant volume gas thermometer and measuring the pressure of the gas both at the triple point of water and at the unknown temperature. Which one of the following procedures is necessary if two gas thermometers, using different real gases, are to agree ?

(*a*) Use gases of the same density in both thermometers
(*b*) Use gases of the same specific heat capacity in both thermometers.
(*c*) Take readings at very low pressures in both thermometers..
(*d*) Use the same mass of gas in both thermometers.
(*e*) Use the same volume of gas in both thermometers.

9. Latent heat of vaporisation is the energy required to

(*a*) separate the molecules of the liquid
(*b*) force back the atmosphere to make space for the vapour
(*c*) increase the average molecular speed in the liquid phase to that in the gas phase
(*d*) separate the molecules and to force back the atmosphere
(*e*) separate the molecules and to increase their average molecular speed to that in the gas phase.

10. A small quantity, mass m, of water at a temperature θ (in°C) is poured on to a large mass M of ice which is at its melting point.

If c is the specific heat capacity of water and L the specific latent heat of fusion of ice, then the mass of ice melted is given by

(*a*) $\frac{ML}{mc\theta}$ (*b*) $\frac{Mc\theta}{L}$
(*c*) $\frac{mc\theta}{ML}$ (*d*) $\frac{mc\theta}{L}$.

11. A thermodynamics system goes from states (*i*) P_1, V to $2P_1$, V (*ii*) P, V to P, 2V. Then work done in the two cases is

(a) Zero, Zero (b) Zero, PV
(c) PV_1, Zero (d) PV_1, P_1V_1.

12. In the case of diatomic gas, the ratio of energy used for expansion and the heat supplied at constant pressure is

(a) 2/5 (b) 3/7
(c) 2/7 (d) 5/7.

13. Thermometers M_1 and M_2 are placed inside an evacuated enclosure X with opaque walls maintained at temperature T. The thermometers are identical except that the bulb of M_1 is blackened. If T_1 and T_2 are the temperatures indicated by M_1 and M_2 respectively after thermal equilibrium has been established, then

(a) $T_2 > T_1 > T$ (b) $T_1 > T > T_2$
(c) $T_1 = T_2 = T$ (d) $T_1 = T > T_2$
(e) $T_1 > T = T_2$.

14. Two thermometers X and Y are placed inside an enclosure whose walls are maintained at a steady temperature T. When equilibrium is established, X and Y will give the same reading T

(a) under all circumstances
(b) only if they absorb radiation at equal rates
(c) only if both are perfect absorbers
(d) only if they have equal thermal capacities
(e) only if the enclosure is evacuated.

15. The molar heat capacity of an ideal gas at constant pressure is greater than that at constant volume because

(a) work has to be done against intermolecular forces as the gas expands
(b) work has to be done against external pressure as the gas expands
(c) the molecules gain rotational kinetic energy as the gas expands
(d) when the molecules are farther apart, they collide less frequently and the average molecular energy decreases
(e) the molecules move faster when heat is supplied at constant pressure than when supplied at constant volume.

16. The molar heat capacity of chromium at room temperature is 24 J K^{-1} mol^{-1} and its relative atomic mass is 52. What is the approximate specific heat capacity of chromium ?

(a) 0.46 J K^{-1} kg^{-1} (b) 1.2 JK^{-1} kg^{-1}
(c) 2.2 JK^{-1} kg^{-1} (d) 460 JK^{-1} kg^{-1}
(e) 1200 JK^{-1} kg^{-1}.

17. A fixed mass of an ideal gas undergoes the change represented by $\overrightarrow{XYZX}$ below (Fig. 1).

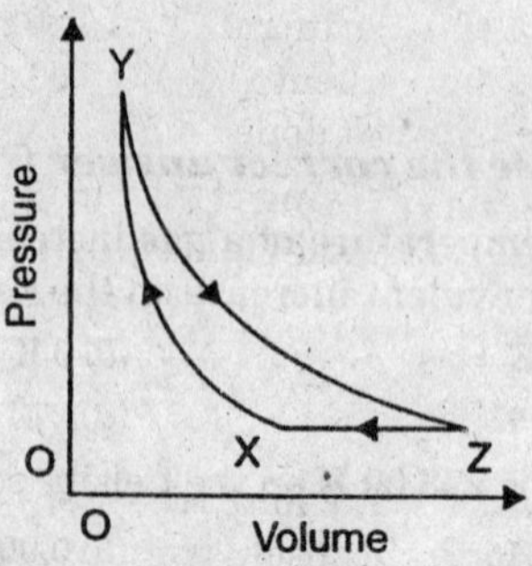

Fig. 1

Which one of the following sets could describe this of changes ?

	XY	YZ	ZX
(a)	isothermal expansion	adiabatic compression	compression at constant pressure
(b)	adiabatic expansion	isothermal compression	pressure reduction constant volume
(c)	isothermal compression	adiabatic expansion	compression at constant pressure
(d)	adiabatic compression	isothermal expansion	pressure reduction constant volume
(e)	adiabatic compression	isothermal expansion	compression at constant pressure

18. With the usual notation, the first law of thermodynamics applied to one mole of an ideal gas can be written in the following form :

$$C_v \Delta T = \Delta Q - p\Delta V$$

In a change for which Boyle's law is obeyed, which of the following would necessarily be zero ?

(a) ΔQ (b) C_v
(c) ΔT (d) p
(e) ΔV.

19. The first law of thermodynamics can be stated in the form $\Delta U = \Delta Q + \Delta W$

ΔU is the increase in the internal energy of the system ;
ΔQ is the heat supplied to the system ;
ΔW is the external work done on the system.

Which of the quantities ΔU, ΔQ and ΔW are necessarily zero when the system is an ideal gas that undergoes a change at constant temperature ?

(a) ΔU only (b) none of ΔU, ΔQ, ΔW
(c) ΔQ only (d) all of ΔU, ΔQ, ΔW
(e) ΔW only.

20. The first law of thermodynamics may be written

$$\Delta U = Q + W$$

where ΔU is the increase in internal energy of the system, Q is the heat transfer to the system and W is the external work done on the system.

Which of the following is correct for the case of isothermal expansion of an ideal gas ?

(*a*) $W > 0$ (*b*) $W = 0$

(*c*) $\Delta U = 0$ (*d*) $\Delta U > 0$

(*e*) $Q = 0$.

21. If K, σ and T stand for thermal conductivity, electrical conductivity and absolute temperature of a metal, then

(*a*) $K + \sigma = T$ (*b*) $K - \sigma = T$

(*c*) $\frac{K}{\sigma} \propto T$ (*d*) $\frac{K}{\sigma} = T^2$.

22. If wavelengths of maximum intensity of radiations emitted by Sun and Moon are 0.5×10^{-6} metre and 10^{-4} metre respectively, the ratio of their temperatures is

(*a*) $\frac{1}{100}$ (*b*) $\frac{1}{200}$

(*c*) 100 (*d*) 200.

23. Two rods of the same length and areas of cross-section A_1 and A_2 have their ends at the same temperature. K_1 and K_2 are the thermal conductivities of the two rods. The rate of flow of heat is same in both the rods if

(*a*) $\frac{A_1}{A_2} = \frac{K_1}{K_2}$ (*b*) $\frac{A_1}{A_2} = \frac{K_2}{K_1}$

(*c*) $A_1A_2 = K_1K_2$ (*d*) placed in hot oil.

24. The rate of radiation of a black body at 0°C is E J s^{-1}. The rate of radiation of this black body at 273°C will be

(*a*) 16E (*b*) 8E

(*c*) 4E (*d*) E.

25. On heating one end of a rod, the temperature of the whole rod will be uniform when

(*a*) $K = 1$ (*b*) $K = 0$

(*c*) $K = 100$ (*d*) $K = \infty$.

26. The thermal resistance of two blocks connected in series will be, if their separate thermal resistances are 2 and 3 is

(*a*) 1 (*b*) 6

(*c*) 5 (*d*) 1.5.

27. If three rods of lengths l_1, l_2 and l_3 coated with wax and thermal conductivities k_1, k_2 and k_3 are dipped in a metal box containing hot water at 100°C in Ingen Hausz experiment, then

(*a*) $k_1 : k_2 : k_3 :: l_1 : l_2 : l_3$

(*b*) $k_1 : k_2 : k_3 :: l_1^2 : l_2^2 : l_3^2$

(*c*) $k_1^2 : k_2^2 : k_3^2 :: l_1 : l_2 : l_3$

(*d*) $\sqrt{k_1} : \sqrt{k_2} : \sqrt{k_3} :: l_1^2 : l_2^2 : l_3^2$.

28. A container contains hot water at 100°C. If in time T_1 temperature falls to 80°C and in time T_2 temperature falls to 60°C from 80°C, then

(*a*) $T_1 = T_2$ (*b*) $T_1 > T_2$

(*c*) $T_1 < T_2$ (*d*) None of these.

29. The top of a lake is frozen. Air in contact with the surface of the lake is at – 5°C. Then the maximum temperature of the water in contact with the lower surface of ice will be

(*a*) 0°C (*b*) 4°C

(*c*) – 7.5°C (*d*) – 15°C.

30. The wax melts up to lengths 105 cm and 60 cm for two identical rods of different metals in Ingen Hausz experiment. The ratio of thermal conductivities of the two metals is

(*a*) 7 : 4 (*b*) 4 : 7

(*c*) 49 : 16 (*d*) 16 : 49.

31. If *r*, *a* and *t* represent the reflection coefficient, absorption coefficient and transmission coefficient respectively, then for a perfectly black body

(*a*) $r = 0, a = 1, t = 0$ (*b*) $r = 1, a = 0$ and $t = 0$

(*c*) $r = 0, a = 0, t = 1$ (*d*) $r = 0, a = 0$ and $t = 0$.

32. If l is the length, A is the area of cross-section and K the thermal conductivity, then the thermal resistance of the block is given by

(*a*) $\frac{Kl}{A}$ (*b*) $\frac{l}{KA}$

(*c*) $\frac{AK}{l}$ (*d*) $\frac{A}{Kl}$.

33. If Wien's constant $b = 0.3$ cm K, then the temperature of the Sun having maximum intensity of radiation at 5000 Å wavelength is

(*a*) 5000 K (*b*) 6000 K

(*c*) 4000 K (*d*) 7000 K.

34. The maximum wavelength of radiation emitted at 2000 K is 4 μm. What will be the maximum wavelength of radiation emitted at 2400 K ?

(*a*) 3.33 μm (*b*) 0.66 μm

(*c*) 1 μm (*d*) 1 m.

35. Which of the following is the v_m – T graph for a perfectly black body

(a) A

(b) B

(c) C

(d) D.

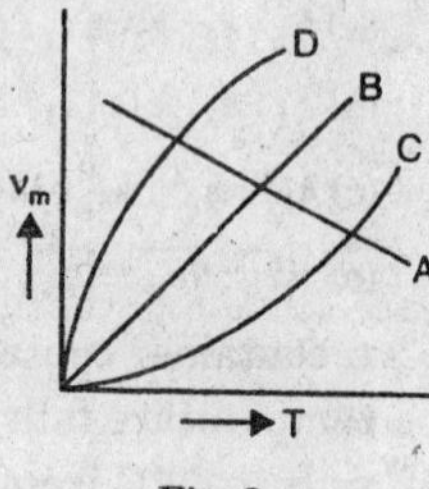

Fig. 2

36. Two metal spheres of different radii are in thermal contact in vacuum as shown in Fig. 3.

The spheres are at the same temperature.

Which statement must be correct ?

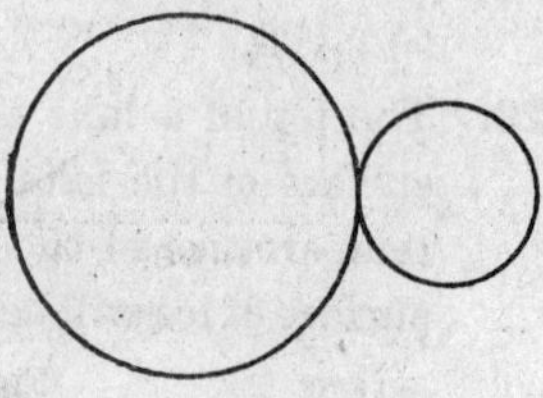
Fig. 3

(a) Each sphere has the same internal energy.

(b) There is no net transfer of thermal energy between the spheres.

(c) Both spheres radiate electromagnetic energy at the same rate.

(d) The larger sphere has a greater mean internal energy per atom than the smaller sphere.

37. Experiments show that even the best thermal insulators conduct some thermal energy, whereas the best electrical insulators conduct virtually no current. Which of the following statements explains this difference ?

(a) In all materials, thermal energy is transmitted by molecular vibration as well as electron diffusion, but molecular vibration plays no part in electrical conduction

(b) During thermal conduction, the free electrons collide with atoms of the material, but during electrical conduction they do not

(c) Unless a material is at absolute zero, the free electrons will always have some random thermal motion.

(d) Electron motion is random in thermal conduction, but in a constant direction in electrical conduction.

(e) Electrical conduction requires a potential difference, but it is impossible to have a potential difference across an insulator.

38. A body X at an original temperature 100°C and another body Y at an original temperature 0°C are placed in an evacuated enclosure, the walls of which are maintained at 10°C. Which one of the following statements is consistent with Prevost's Theory ?

(a) X emits but does not absorb heat.

(b) Y absorbs but does no emit heat.

(c) The final temperature of the bodies will be the mean of their initial temperatures (i.e. 50°C).

(d) X receives heat only from Y.

(e) The walls of the enclosure radiate heat to both X and Y.

39. Which one of the following graphs best represents the way in which the total power P radiated by a black body depends upon the thermodynamic temperature T of the body ?

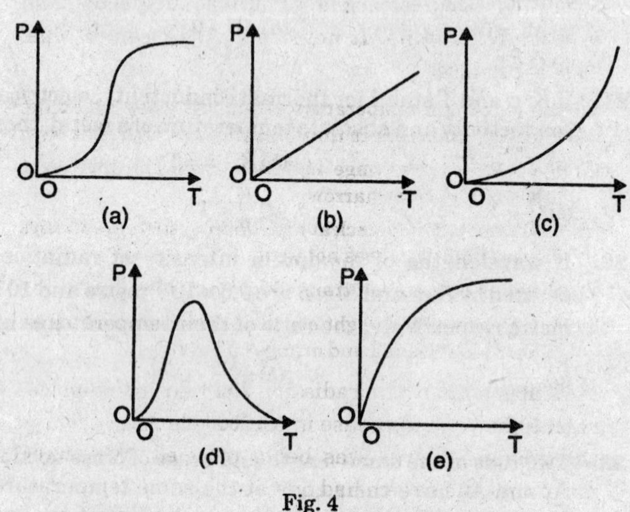

Fig. 4

40. The ends of well-lagged copper bar of length L are held at steady temperatures θ_1 and θ_2 ($\theta_1 > \theta_2$).

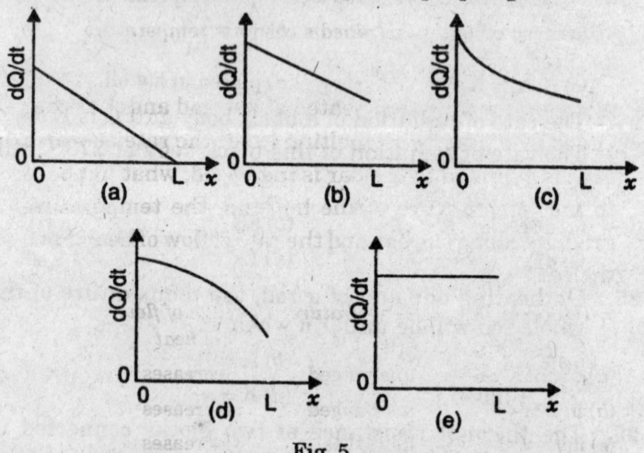

Fig. 5

Which graph best represents the relationship between dQ/dt, the rate of flow of heat, and x, the distance from the hot end when all temperatures in the bar are steady ?

41. Only a very small fraction of the energy supplied to a domestic light bulb is emitted as light. This is because (*a*) the filament is not a black body

(*b*) the filament surface area is too small to emit light efficiently

(*c*) most of the energy is given out at longer wavelengths

(*d*) most of the energy is given out at shorter wavelengths

(*e*) energy is absorbed by the gas in the bulb.

42. When the filament of a light bulb is heated slowly, its colour changes from red to orange to yellow and eventually to white. It never appears green or blue. This is because

(*a*) the maximum temperature attainable gives maximum intensity of emission in the ultraviolet part of the spectrum

(*b*) the temperature range in which green and blue light predominate is too narrow

(*c*) the filament is not black at high temperatures and Wien's displacement law does not apply

(*d*) the glass bulb cannot transmit green or blue light

(*e*) any green and blue light emitted is always accompanied by more intense red and orange emission.

43. Measurements of the radiation from an astronomical source showed a decrease in the wavelength at which the greatest energy was being received. This could mean that the source had

(*a*) increased in temperature

(*b*) decreased in temperature

(*c*) expanded but maintained a constant temperature

(*d*) contracted but maintained a constant temperature

(*e*) moved further away.

44. A lagged copper bar is heated at one end and the other end is in contact with melting ice. If the rate at which heat is supplied to the bar is increased, what happens to the temperature of the hot end, the temperature gradient along the bar and the rate of flow of heat down the bar ?

Temperature of hot end	*Temperature gradient*	*Rate of flow of heat*
(*a*) unchanged	unchanged	increases
(*b*) increases	unchanged	increases
(*c*) increases	increases	increases
(*d*) increases	increases	unchanged
(*e*) unchanged	increases	unchanged.

45. Which one of the sketches below correctly compares the distribution of radiated energy of a 'grey' body with that of a black body at the same temperature ? (A 'grey' body is one that has an emissivity which is less than unity and which is independent of wavelength.)

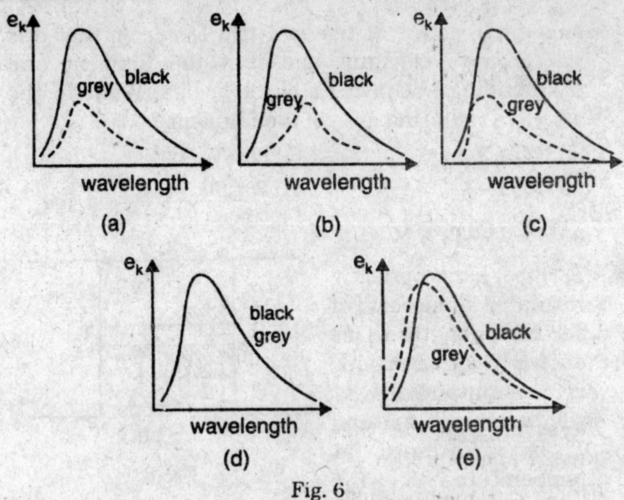

Fig. 6

46. The ends of a metal bar of constant cross-sectional area are maintained at temperatures T_1 and T_2 which are both higher than the temperature of the surroundings. If the bar is unlagged, which one of the following sketches best represents the variation of temperature with distance along the bar ?

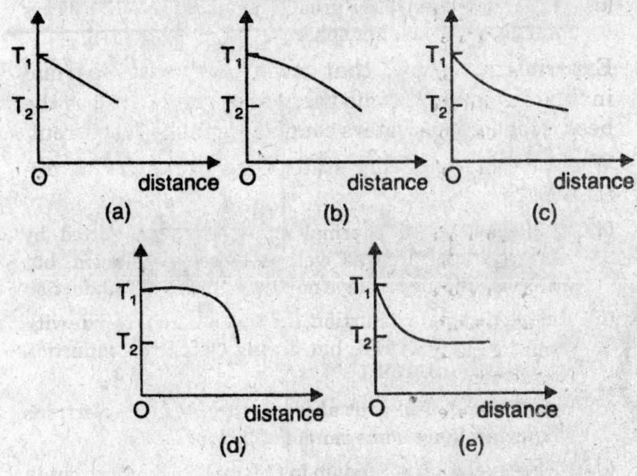

Fig. 7

47. The thermal conductivity of copper decreases significantly as the temperature is increased. This is because, at higher temperatures,

(*a*) ionic forces between copper atoms decrease

(*b*) lattice vibrations travel more slowly

(*c*) fewer conduction electrons are present

(*d*) the distance between copper atoms increases

(*e*) the rate of collision between conduction electrons and copper atoms increases.

48. The filament of a lamp is maintained at temperature T by dissipation of electrical power P. The surroundings are at temperature T_0. Given that σ is the Stefan's constant and that the filament and surroundings behave as black bodies. What is the effective radiating area of the filament ?

(*a*) $P/(T - T_0)$ (*b*) $P/\sigma (T - T_0)^4$

(*c*) $P/\sigma (T - T_0)$ (*d*) $P/\sigma T^4$

(*e*) $P/\sigma (T^4 - T_0^4)$.

49. A copper bar is perfectly insulated along part l of its total length L, as shown in Fig. 8. End X is maintained at temperature T_1 and end Y is maintained at temperature T_2 ($T_1 > T_2$).

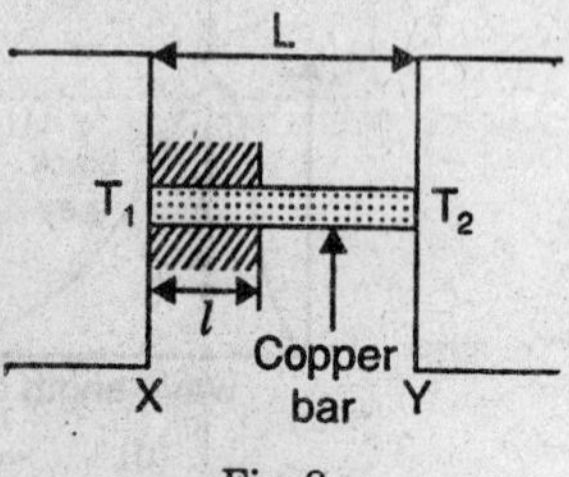

Fig. 8

Which one of the following sketches shows the steady-state variation of temperature T with distance x measured from X along the bar ?

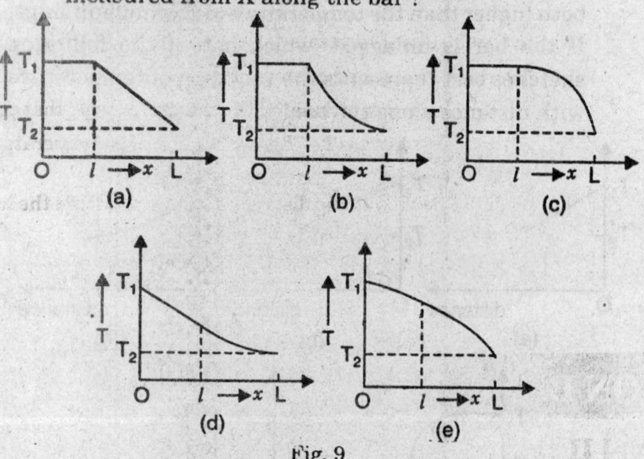

Fig. 9

50. The distribution of the heat energy radiated by the human body, shown in Fig. 10, is very similar to that of a black body at the same temperature.

Which one of the following is the most likely value for x ?

Fig. 10

(*a*) 9 nm (*b*) 90 nm

(*c*) 9000 nm (*d*) 90 mm

(*e*) 9000 mm.

51. The ends of an unlagged rod are in contact with reservoirs maintained at 100°C and at 0° C through thin end-pieces of a material of lower thermal conductivity than the rod (Fig. 11)

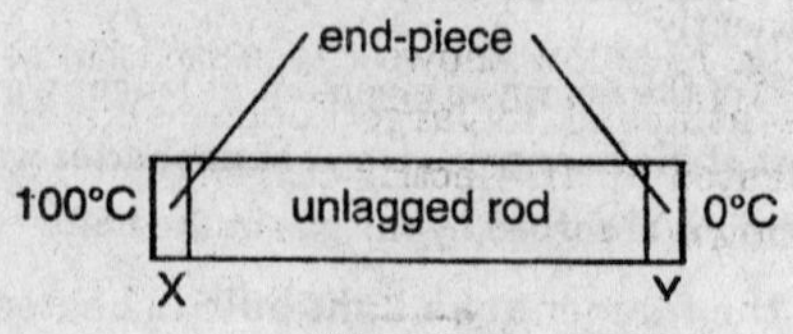

Fig. 11

Which one of the following graphs best represents the variation of temperature with distance between the faces of the reservoirs ?

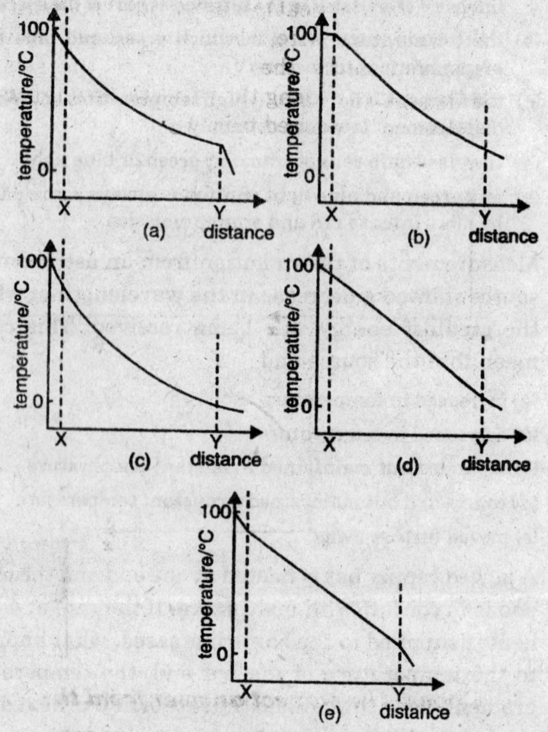

Fig. 12

52. A certain bolometer detects thermal radiation by the change of resistance of a metal foil when exposed to the radiation. In order to have a bolometer of high sensitivity and rapid response, it is desirable that the foil should have

(*a*) **a polished surface, a high heat capacity and a low temperature coefficient of resistance**

(*b*) **a blackened surface, a high heat capacity and a high temperature coefficient of resistance**

(*c*) **a blackened surface, a low heat capacity and a high temperature coefficient of resistance**

(*d*) **a polished surface, a low heat capacity and a low temperature coefficient of resistance** (*e*) a blackened surface, a low heat capacity and a low temperature coefficient of resistance.

53. Fig. 13 below shows a thin resistance wire MN which is attached to large copper supports and which is heated by an electric current flowing in the direction shown.

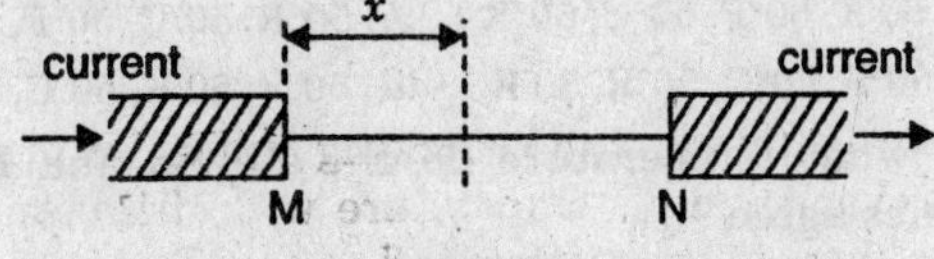

Fig. 13

Assuming that heat loss occurs predominantly through the ends of the wire, which one of the following graphs best represents the variation of the steady state temperature θ along the wire MN as a function of the distance x measured from M ?

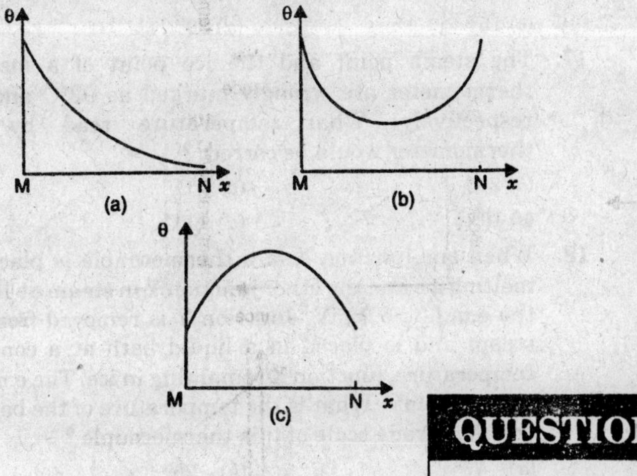

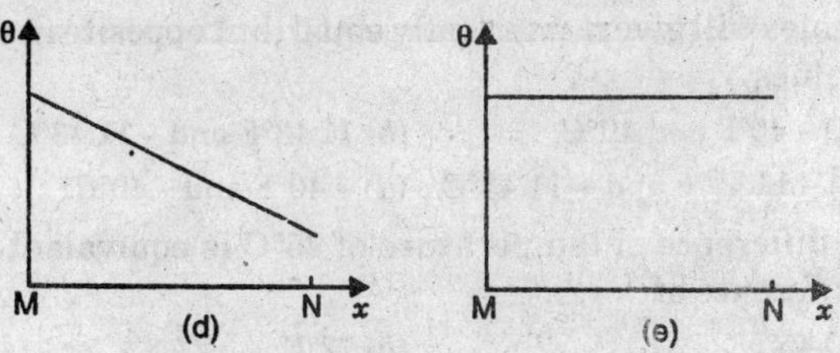

Fig. 14

76. At low temperatures, the thermal conductivity of a particular metal is proportional to its thermodynamic temperature T. The ends of a well-lagged cylindrical rod of the metal are maintained at low temperatures T_1 and T_2 ($T_1 > T_2$). The rate of flow of heat across unit cross-sectional area of the rod at different points along it is

(*a*) the same at all points

(*b*) proportional to the temperature at that point

(*c*) inversely proportional to the temperature at that point

(*d*) proportional to $(T - T_2)$, where T is the temperature at that point

(*e*) proportional to the distance of that point from the hotter end.

QUESTION BANK

Level II

Choose the correct answer from the four alternatives given in each of the following questions :

1. The readings of air thermometer at 0°C and 100°C are 50 cm and 75 cm of mercury column respectively. The temperature at which its reading is 80 cm of mercury column is

(*a*) 105°C (*b*) 110°C

(*c*) 115°C (*d*) 120°C.

2. When a platinum resistance thermometer is put in contact with ice, steam and a liquid, the resistances of platinum wire recorded are 2.56 ohm, 3.56 ohm and 5.06 ohm respectively. The temperature of the liquid is

(*a*) 100°C (*b*) 250°C

(*c*) 40°C (*d*) 25°C.

3. Two thermometers x and y have fundamental intervals of 80° and 120°. When immersed in ice, they show the readings of 20° and 30°. If y measures the temperature of a body as 120°, the reading of x is

(*a*) 55° (*b*) 65°

(*c*) 75° (*d*) 80°.

4. Which of the following temperatures is the highest

(*a*) 100 K (*b*) – 13°F

(*c*) – 20°C (*d*) – 23°C.

5. A faulty thermometer has its fixed points marked 5° and 95°. This thermometer reads the temperature of a body as 59°. Then correct temperature on Celsius scale is

(a) 59°　　(b) 48.6°

(c) 60°　　(d) 58°.

6. At what temperature, the Fahrenheit and Celsius scales will give numerically equal (but opposite in sign) values

(a) – 40°F and 40°C　　(b) 11.43°F and – 11.43°C

(c) – 11.43°F and + 11.43°C　　(d) + 40°F and – 40°C.

7. A difference of temperature of 25°C is equivalent to a difference of

(a) 45°F　　(b) 72°F

(c) 32°F　　(d) 25°F.

8. The lower and upper fixed points of a faulty thermometer are 5° and 99° respectively. If the reading of the thermometer is 52°, the temperature on the Fahrenheit scale is

(a) 132°F　　(b) 122°F

(c) 154°F　　(d) 151°F.

9. A constant volume gas thermometer shows pressure reading of 50 cm and 90 cm of mercury at 0°C and 100°C respectively. When the pressure reading is 60 cm of mercury, the temperature is

(a) 25°C　　(b) 40°C

(c) 15°C　　(d) 12.5°C.

10. The temperature of a body on Kelvin scale is found to be x K. When it is measured by a Fahrenheit thermometer, it is found to be x°F. Then x is

(a) 301.25　　(b) 574.25

(c) 313　　(d) 40.

11. A thermometer has wrong calibration (of course at equal distances and the capillary is of uniform diameter). It reads the melting point of ice as – 10°C. It reads 60°C in place of 50°. The temperature of boiling point of water on this scale is

(a) 100°C　　(b) 130°C

(c) 110°C　　(d) 120°C.

12. A fixed mass of an ideal gas is maintained at constant volume. The pressure of the gas at the triple point of water is p_{tr}. What is the thermodynamic temperature of the gas when its pressure is p ?

(a) $273.16\left(\frac{p_{tr}}{p}\right)$ K　　(b) $273.16\left(\frac{p}{p_{tr}}\right)$ K

(c) $273.16\left(\frac{p-p_{tr}}{p_{tr}}\right)$ K　　(d) $273.16\left(\frac{p-p_{tr}}{p}\right)$ K

(e) $273.16\left(\frac{p+p_{tr}}{p_{tr}}\right)$ K.

13. A graph is plotted between the temperature of a copper cube in °C versus °F. The sine of the angle made by the graph with °F axis is

(a) $\frac{2}{\sqrt{106}}$　　(b) $\frac{3}{\sqrt{106}}$

(c) $\frac{4}{\sqrt{106}}$　　(d) $\frac{5}{\sqrt{106}}$.

14. Given : 50°R, 50°F, 50°C and 50 K. Which of the following represents the increasing order of these temperatures ?

(a) 50 K, 50°F, 50°C, 50°R　　(b) 50°R, 50°C, 50°F, 50 K

(c) 50°F, 50°C, 50°R, 50 K　　(d) 50°F, 50 K, 50°C, 50°R.

15. At what temperature do the Kelvin and Reaumer scales agree ?

(a) 0°　　(b) 3°

(c) – 9°　　(d) They never agree.

16. If the reading of the Reaumer scale is numerically less than that of Celsius scale by 3, then the reading of the Celsius scale is

(a) 5°　　(b) 10°

(c) 15°　　(d) 20°.

17. The steam point and the ice point of a mercury thermometer are wrongly marked as 92°C and 2°C respectively. What temperature read by this thermometer would be correct ?

(a) 2°C　　(b) 5°C

(c) 10°C　　(d) 20°C.

18. When one junction X of a thermocouple is placed in melting ice and the other junction Y in steam at 100°C, the e.m.f. is 6.0 mV. Junction Y is removed from the steam and is placed in a liquid bath at a constant temperature, junction X remaining in ice. The e.m.f. is now – 1.5 mV. What is the temperature of the bath on the centigrade scale of this thermocouple ?

(a) – 75°　　(b) – 25°

(c) 25°　　(d) 75°

(e) 125°.

19. Out of the four given temperatures 40°R, 49°C, 113°F and 320 K, the highest temperature is

(a) 40°R　　(b) 49°C

(c) 113°F　　(d) 320 K.

20. The upper fixed point and lower fixed point of a thermometer are wrongly marked as 96°C and – 2°C respectively. The reading of this thermometer is (correct thermometer reads 50°C)

(a) 37°C　　(b) 40°C

(c) 42°C　　(d) 47°C.

21. On a hypothetical scale X, the ice point is 40° and the steam point is 120°. For another scale Y, the ice point and steam point are – 30° and 130° respectively. If X reads 50°, then Y would read

(*a*) – 5° (*b*) – 8°
(*c*) – 10° (*d*) – 12°.

22. A Fahrenheit thermometer reads 113°F while a faulty Celsius thermometer reads 44°C. The correction required to be applied to the Celsius thermometer is
(*a*) – 1°C (*b*) + 1°C
(*c*) + 2°C (*d*) – 2°C.

23. When a Celsius thermometer reads 90°C, a faulty Fahrenheit thermometer reads 190°F. The correction to be made in the latter scale is
(*a*) + 2°F (*b*) – 2°F
(*c*) – 4°F (*d*) + 4°F.

24. The upper and lower fixed points of a faulty mercury thermometer are 210°F and 34°F respectively. What temperature read by this thermometer would be correct ?
(*a*) 22°F (*b*) 80°F
(*c*) 100°F (*d*) 122°F.

25. A metal cube of length 10.0 mm at 0°C (273 K) is heated to 200°C (473 K). Given : its coefficient of linear expansion is 2×10^{-5} K^{-1}. The percent change of its volume is
(*a*) 0.1 (*b*) 0.2
(*c*) 0.4 (*d*) 1.2.

26. The fractional change in the period of a pendulum clock is
(*a*) $\frac{1}{2}\alpha(t_2 - t_1)^2$ (*b*) $2\alpha(t_2 - t_1)$
(*c*) $\frac{1}{2}\alpha(t_2 - t_1)$ (*d*) $2\alpha(t_2 - t_1)^2$.

27. The density of a substance at 0°C is 10 g cm^{-3} and at 100°C, its density is 9.7 g cm^{-3}. The coefficient of linear expansion of substance is
(*a*) 0.0001/°C (*b*) 0.001/°C
(*c*) 0.00001/°C (*d*) 0.1/°C.

28. A sphere made of iron is rotating about its diameter as axis. $\alpha = 1 \times 10^{-5}$ °C^{-1}. If the temperature rises by 100°C, the percentage increase in its moment of inertia is
(*a*) 0.1% (*b*) 0.2%
(*c*) 0.5% (*d*) 0.002%.

29. A solid occupies 1000 cm^3 at 20°C. Its volume becomes 1016.2 cm^3 at 320°C. Coefficient of linear expansion of the material is
(*a*) 18×10^{-6} °C^{-1} (*b*) 16×10^{-6} °C^{-1}
(*c*) 12×10^{-6} °C^{-1} (*d*) 36×10^{-6} °C^{-1}.

30. A clock with a metal pendulum beating seconds keeps correct time at 0°C. If it loses 10 second a day at 20°C, the coefficient of linear expansion of metal of pendulum is
(*a*) $\frac{1}{43200}$ °C^{-1} (*b*) $\frac{1}{86400}$ °C^{-1}
(*c*) $\frac{1}{64800}$ °C^{-1} (*d*) 16×10^{-6} °C^{-1}.

31. A clock which keeps correct time at 25°C has a pendulum made of a metal. The temperature falls to 0°C. If the coefficient of linear expansion of the metal is 1.9×10^{-5} per °C, then number of second the clock gains per day is
(*a*) 10.25 s (*b*) 20.52 s
(*c*) 30.75 s (*d*) 41 s.

32. Two rods, one hollow and the other solid, made of the same material have the same length of 20 cm and radius of 2 cm. When their temperature is increased through the same amount of 50°C, their expansion ratio $E_h : E_s$ will be
(*a*) 1 : 8 (*b*) 1 : 4
(*c*) 1 : 1 (*d*) 1 : 2.

33. Two rods of lengths l_1 and l_2 are made of materials whose coefficients of linear expansion are α_1 and α_2 respectively. If the difference between the two lengths is independent of temperature, then
(*a*) $\frac{l_1}{l_2} = \frac{\alpha_2}{\alpha_1}$ (*b*) $l_1^2\alpha_2 = l_2^2\alpha_1$
(*c*) $\frac{l_1}{l_2} = \frac{\alpha_1}{\alpha_2}$ (*d*) $\alpha_2^2 l_1 = \alpha_1^2 l_2$.

34. The loss in weight of a solid when immersed in a liquid at 0°C is W_0 and at t°C is W. If the coefficients of volume expansion of the solid and the liquid be γ_s and γ_l respectively, then
(*a*) $W = W_0 [(\gamma_s - \gamma_l)t]$ (*b*) $W = W_0 [1 + (\gamma_s - \gamma_l)t]$
(*c*) $W = \frac{W_0 t}{\gamma_l - \gamma_s}$ (*d*) $W = W_0 [1 - (\gamma_s - \gamma_l)t]$.

35. A grid iron pendulum consists of 5 iron rods and 4 brass rods. What will be the length of each brass rod if the length of each iron rod is 1 m ? ($\alpha_{Fe} = 12 \times 10^{-6}$/°C, $\alpha_{brass} = 20 \times 10^{-6}$/°C)
(*a*) 0.7 m (*b*) 0.8 m
(*c*) 0.9 m (*d*) 1.0 m.

36. The coefficient of linear expansion of iron is 0.000011 °K. An iron rod is 10 metre long at 27°C. The length of the rod will be decreased by 1.1 mm when the temperature of the rod changes to
(*a*) 0°C (*b*) 10°C
(*c*) 17°C (*d*) 20°C.

37. A uniform metal rod of 2 mm^2 cross-section is heated from 0°C to 20°C. The coefficient of linear expansion of the rod is 12×10^{-6} per °C, Y = 10^{11} N/m^2. The energy stored per unit volume of the rod is

(*a*) 1440 J/m^3 (*b*) 1500 J/m^3

(*c*) 2880 J/m^3 (*d*) 5760 J/m^3.

38. A bimetal made of copper and iron strips welded together is straight at room temperature. It is held vertically in the hand so that iron strip is towards the left hand and copper strip is towards the right hand side. This bimetal is then heated by flame. The bimetal strip will

(*a*) remain straight (*b*) bend towards right

(*c*) bend towards left (*d*) No change.

39. A sphere of diameter 7 cm and mass 266.5 g floats in a bath of liquid. As the temperature is raised, the sphere is about to sink at 35°C. If the density of liquid is 1.527 g cm^{-3} at 0°C, the coefficient of cubical expansion of the liquid will be (neglect the expansion of sphere)

(*a*) 8.486×10^{-3}/°C (*b*) 8.48×10^{-4}/°C

(*c*) 8.486×10^{-5}/°C (*d*) 8.486×10^{-6}/°C.

40. The real coefficient of volume expansion of glycerine is 0.000597 per °C and the linear coefficient of expansion of glass is 0.000009 per °C. Then the apparent volume coefficient of expansion of glycerine in glass will be

(*a*) 0.000606 per °C (*b*) 0.000588 per °C

(*c*) 0.00057 per °C (*d*) 0.00027 per °C.

41. 10 g of ice at – 20°C is added to 10 g of water at 50°C. The amount of ice in the mixture at resulting temperature is (Specific heat of ice = 0.5 cal g^{-1} °C^{-1} and latent heat of ice = 80 cal g^{-1})

(*a*) 10 g (*b*) 5 g

(*c*) 0 g (*d*) 20 g.

42. 80 g of water at 30°C is poured on a large block of ice at 0°C. The mass of ice that melts is

(*a*) 30 g (*b*) 80 g

(*c*) 150 g (*d*) 1600 g.

43. 10 g of ice cubes at 0°C is released in a tumbler containing water (water equivalent 55 g) at 40°C. Assuming that negligible heat is taken from surroundings, the temperature of water in the tumbler becomes (L = 80 cal g^{-1})

(*a*) 31°C (*b*) 21.5°C

(*c*) 19°C (*d*) 15°C.

44. 5 kg of water at 10°C is added to 10 kg of water at 40°C. Neglecting heat capacity of vessel and other losses, the equilibrium temperature will be

(*a*) 30°C (*b*) 25°C

(*c*) 35°C (*d*) 33°C.

45. A beaker contains 200 g of water. The heat capacity of beaker is equal to that of 20 g of water. The initial temperature of water in the beaker is 20°C. If 440 g of hot water at 92°C is poured in, the final temperature, neglecting radiation loss, will be

(*a*) 58°C (*b*) 68°C

(*c*) 73°C (*d*) 78°C.

46. 10 g of ice at – 20°C is added to 10 g of water at 50°C. Then the resulting temperature is (Specific heat of ice = 0.5 cal g^{-1} °C^{-1} and latent heat of ice = 80 cal g^{-1})

(*a*) – 20°C (*b*) 15°C

(*c*) 0°C (*d*) 50°C.

47. 100 g of ice is mixed with 100 g of water at 100°C. The final temperature of the mixture is

(*a*) 10°C (*b*) 20°C

(*c*) 30°C (*d*) 40°C.

48. 0.93 watt-hour of energy is supplied to a block of ice weighing 10 g. It is found that

(*a*) half of the block melts

(*b*) the entire block melts and the water attains a temperature of 4°C

(*c*) the entire block just melts

(*d*) the block remains unchanged.

49. If there are no heat losses, the heat released by the condensation of x g of steam at 100°C into water at 100°C can be used to convert y g of ice at 0°C into water at 100°C. Then the ratio $y : x$ is nearly

(*a*) 1 : 1 (*b*) 2 : 1

(*c*) 3 : 1 (*d*) 2.5 : 1.

50. Hailstone at 0°C falls from a height of 1 km on an insulating surface converting whole of its kinetic energy into heat. What part of it will melt ? (Given : $g = 10$ m s^{-2})

(*a*) $\frac{1}{33}$ (*b*) $\frac{1}{8}$

(*c*) $\frac{1}{33} \times 10^{-4}$ (*d*) All of it will melt.

51. 5 g of water at 30°C and 5 g of ice at – 20°C are mixed together in a calorimeter. Neglect the water equivalent of the calorimeter. The final temperature of the mixture is (Specific heat of ice = 0.5 cal g^{-1} °C^{-1})

(*a*) 0°C (*b*) – 20°C

(*c*) – 10°C (*d*) + 1.2°C.

52. In Q. 51, the amount of ice melted is

(*a*) 0 g (*b*) 0.25 g

(*c*) 0.50 g (*d*) 1.25 g.

53. One gram of ice at 0°C is added to 5 gram of water at 10°C. If the latent heat of ice be 80 cal/g, then the final temperature of the mixture is

(*a*) 5°C (*b*) 0°C
(*c*) – 5°C (*d*) None of these.

54. Steam at 100°C is passed into 1.1 kg of water contained in a calorimeter of water equivalent 0.02 kg at 15°C, till the temperature of the calorimeter and its contents rises to 80°C. The mass of steam condensed (in kg) is (Take latent heat of steam = 540 cal g^{-1})

(*a*) 0.130 (*b*) 0.065
(*c*) 0.260 (*d*) 0.135.

55. The temperature of equal masses of three different liquids A, B and C are 12°C, 19°C and 28°C respectively. The temperature when A and B are mixed is 16°C and when B and C are mixed is 23°C. The temperature when A and C are mixed is

(*a*) 18.2°C (*b*) 20.3°C
(*c*) 22.2°C (*d*) 24.2°C.

56. One gram of ice is mixed with one gram of steam. After thermal equilibrium, the temperature of the mixture is

(*a*) 0°C (*b*) 100°C
(*c*) 55°C (*d*) 80°C.

57. An airgun pellet, mass m and specific heat capacity c, hits a steel plate at speed v. During the impact, 50% of the pellet's kinetic energy is converted to thermal energy in the pellet. What is the rise in temperature of the pellet ?

(*a*) $\frac{v^2}{2c}$ (*b*) $\frac{mv^2}{2c}$
(*c*) $\frac{v^2}{4c}$ (*d*) $\frac{mv^2}{4c}$.

58. Three liquids with masses m_1, m_2, m_3 are thoroughly mixed. If their specific heats are s_1, s_2, s_3 and their temperatures are $\theta_1, \theta_2, \theta_3$ respectively, then the temperature of the mixture is

(*a*) $\frac{s_1\theta_1 + s_2\theta_2 + s_3\theta_3}{m_1s_1 + m_2s_2 + m_3s_3}$ (*b*) $\frac{m_1s_1\theta_1 + m_2s_2\theta_2 + m_3s_3\theta_3}{m_1s_1 + m_2s_2 + m_3s_3}$

(*c*) $\frac{m_1s_1\theta_1 + m_2s_2\theta_2 + m_3s_3\theta_3}{m_1\theta_1 + m_2\theta_2 + m_3\theta_3}$ (*d*) $\frac{m_1\theta_1 + m_2\theta_2 + m_3\theta_3}{s_1\theta_1 + s_2\theta_2 + s_3\theta_3}$.

59. An electric kettle contains 1.5 kg of water at 100°C and is powered by a 2.0 kW electric element. If the thermostat of the kettle fails to operate, approximately how long will the kettle take to boil dry ? (Take the specific latent heat of vaporisation of water as 2000 kJ kg^{-1}.)

(*a*) 500 s (*b*) 1000 s
(*c*) 1500 s (*d*) 3000 s.

60. Heat required to convert 1 g of ice at 0°C into steam at 100°C is (Latent heat of steam = 536 cal g^{-1})

(*a*) 100 cal (*b*) 0.01 kcal
(*c*) 716 cal (*d*) 1 kcal.

61. Which one of the following would raise the temperature of 20 g of water at 30°C most, when added to water ? (Specific heat of copper is 0.1 cal g^{-1} °C^{-1})

(*a*) 20 g of water at 40°C (*b*) 10 g of water at 27°C
(*c*) 1 kg of copper at 25°C (*d*) 4 g of water at 80°C.

62. An electric heater rated as 2 kW is used to heat 200 kg of water from 10°C to 70°C. Assuming no heat losses, the time taken is

(*a*) 25.2 s (*b*) 6×10^3 s
(*c*) 25.2×10^3 s (*d*) 25.2×10^6 s.

63. Cooling water enters the heat exchanger in the turbine hall of a nuclear power station at 6°C and leaves at 14°C. The specific heat capacity of water is 4200 J kg^{-1} K^{-1}. If the rate of heat removal by the water is 6.72×10^9 J per minute, what is the rate of water flow ?

(*a*) $\frac{6.72 \times 10^9}{4200 \times 8}$ kg s^{-1} (*b*) $\frac{6.72 \times 10^9 \times 60}{4200 \times 8}$ kg s^{-1}

(*c*) $\frac{6.72 \times 10^9}{4200 \times 8 \times 60}$ kg s^{-1} (*d*) $\frac{4200 \times 8}{6.72 \times 10^9 \times 60}$ kg s^{-1}

(*e*) $\frac{4200 \times 8 \times 60}{6.72 \times 10^9}$ kg s^{-1}.

64. One kilogram of ice at 0°C is mixed with one kilogram of water at 10°C. The resulting temperature will be

(*a*) between 0°C and 10°C (*b*) less than 0°C
(*c*) greater than 10°C (*d*) equal to 0°C.

65. A metal ball-bearing of specific heat capacity c, moving with speed v, is brought to rest. All its kinetic energy is converted into thermal energy which it absorbs, causing a temperature rise $\Delta\theta$. What was the value of v ?

(*a*) $\frac{1}{2} c\, \Delta\theta$ (*b*) $2c\, \Delta\theta$
(*c*) $\sqrt{c\, \Delta\theta}$ (*d*) $\sqrt{2c\, \Delta\theta}$.

66. In a heating experiment, it was noted that the temperature of liquid in a beaker rose at 4.0 K per minute just before it began to boil, and that 40 minute later, all the liquid had boiled away.

For this liquid, what is the numerical ratio :

$$\frac{\text{specific heat capacity}}{\text{specific latent heat of vaporisation}} ?$$

(a) $\frac{1}{10}$ (b) $\frac{1}{40}$

(c) $\frac{1}{160}$ (d) $\frac{1}{640}$

(e) $\frac{1}{2400}$.

67. 5 g of steam at 100°C is passed into 6 g of ice at 0°C. If the latent heats of steam and ice are 540 cal/g and 80 cal/g, then the final temperature is

(a) 0°C (b) 50°C

(c) 30°C (d) 100°C.

68. In the Q. 67, the mixture contains

(a) 11 g of water (b) 8 g of water

(c) 6 g of water (d) 5 g of water.

69. 6 kg of ice at – 10°C is added to 6 kg of water at 0°C. The temperature of the resultant mixture is

(a) – 10°C (b) 0°C

(c) – 2.5°C (d) – 3.3°C.

70. 540 g of ice at 0°C is mixed with 540 g of water at 80°C. The final temperature of mixture is

(a) 0°C (b) 40°C

(c) 80°C (d) less than 0°C.

71. A waterfall is 84 m high. Assuming that half the kinetic energy of falling water get converted to heat, the rise in temperature of water is

(a) 0.098°C (b) 0.98°C

(c) 98°C (d) 0.0098°C.

72. In an adiabatic process, the pressure is increased by $\frac{2}{3}$%. If $\gamma = \frac{3}{2}$, then the volume decreases by nearly

(a) $\frac{4}{9}$ % (b) $\frac{2}{3}$ %

(c) 1% (d) $\frac{9}{4}$ %.

73. A gas at NTP is slowly compressed to one-fourth of its original volume. Then final pressure is

(a) 4 atmosphere (b) 0.75 atmosphere

(c) 8 atmosphere (d) 0.25 atmosphere.

74. The pressure and density of a diatomic gas $\left(\gamma = \frac{7}{5}\right)$ change adiabatically from (P, d) to (P′, d'). If $\frac{d'}{d} = 32$, then $\frac{P'}{P}$ should be

(a) $\frac{1}{128}$ (b) 32

(c) 128 (d) None of these.

75. A gas at NTP is suddenly compressed to one-fourth of its original volume. If γ is supposed to be 3/2, then final pressure is

(a) 4 atmosphere (b) 0.75 atmosphere

(c) 8 atmosphere (d) 0.25 atmosphere.

76. When a thermodynamic system is taken from state A to state B via path ACB (Fig. 15), 100 cal is given to the system and 60 cal worth work is done. Along the path ADB, the work done is worth 20 cal ; the heat flowing into the system in this case would be

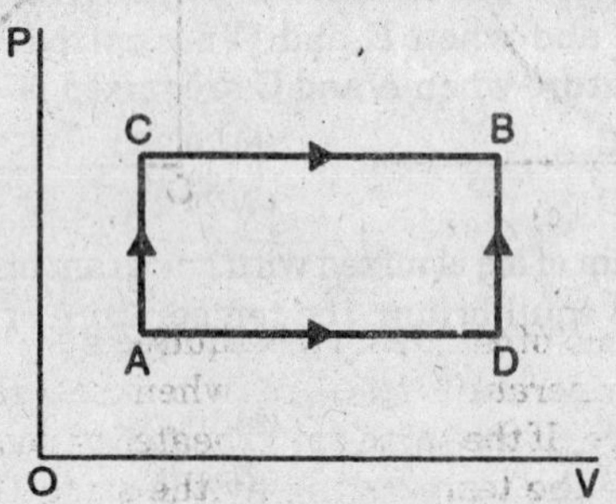

Fig. 15

(a) 120 cal (b) 40 cal

(c) 140 cal (d) 60 cal.

77. If one mole of a monoatomic gas $\left(\gamma = \frac{5}{3}\right)$ is mixed with one mole of a diatomic gas $\left(\gamma = \frac{7}{5}\right)$, the value of γ for the mixture is

(a) 1.40 (b) 1.50

(c) 1.53 (d) 3.07.

78. A monoatomic ideal gas expands at constant pressure, with heat Q supplied. The fraction of Q which goes as work done by the gas is

(a) 1 (b) $\frac{2}{3}$

(c) $\frac{3}{5}$ (d) $\frac{2}{5}$.

79. Which one of graphs below best illustrates the relationship between internal energy U of an ideal gas and temperature T of the gas in K ?

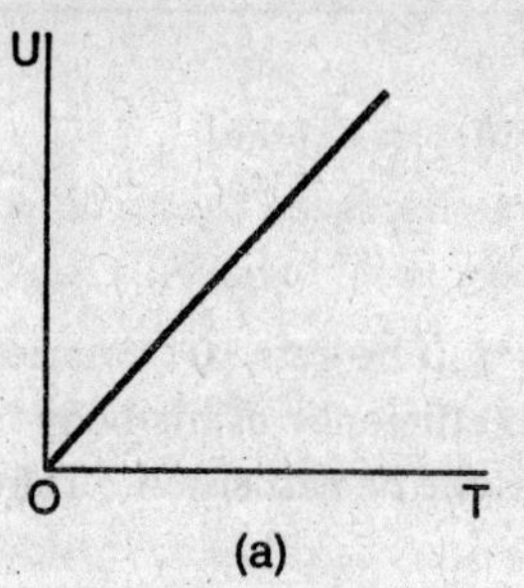

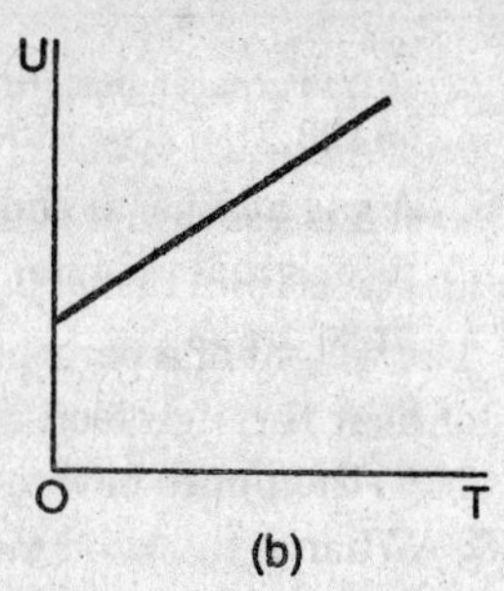

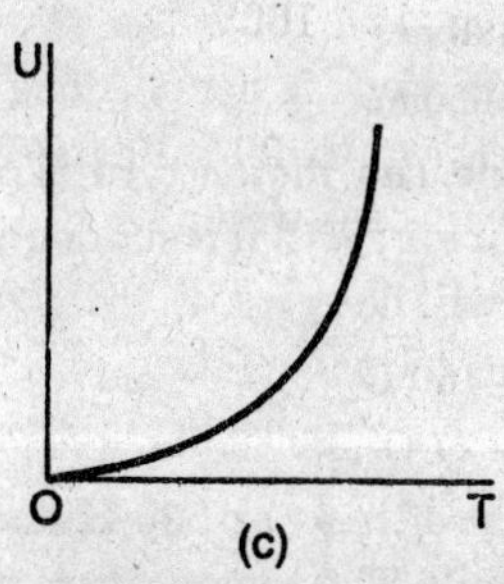

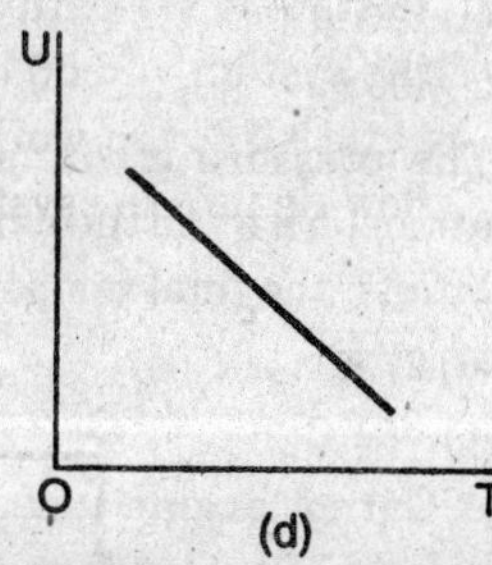

Fig. 16

80. One mole of an ideal gas requires 207 J heat to raise the temperature by 10 K when heated at constant pressure. If the same gas is heated at constant volume to raise the temperature by the same 10 K, then the heat required is

(*a*) 198.7 J (*b*) 215.3 J
(*c*) 124 J (*d*) 24 J.

81. 70 calories of heat are required to raise the temperature of 2 moles of an ideal gas at constant pressure from 30°C to 35°C. The amount of heat required in calories to raise the temperature of same gas through the same range (30°C to 35°C) at constant volume is

(*a*) 30 (*b*) 50
(*c*) 70 (*d*) 90.

82. Certain perfect gas is found to obey $PV^{3/2}$ = constant during an adiabatic process. If such a gas at initial temperature T is adiabatically compressed to half the initial volume, its final temperature will be

(*a*) $\sqrt{2}$ T (*b*) 2T
(*c*) $2\sqrt{2}$ T (*d*) 4T.

83. What is the approximate coefficient of performance of a Carnot refrigerator working between 30°C and 0°C ?

(*a*) 0 (*b*) 1
(*c*) 9 (*d*) 10.

84. An ideal gas heat engine operates Carnot cycle between 227°C and 127°C. It absorbs 6×10^4 calories at the higher temperature. The quantity of heat converted into work is equal to

(*a*) 4.8×10^4 cal (*b*) 3.5×10^4 cal
(*c*) 1.6×10^4 cal (*d*) 1.2×10^4 cal.

85. An ideal gas at 27°C is compressed adiabatically to $\frac{8}{27}$ of its original volume. If $\gamma = \frac{5}{3}$, then the rise in temperature is

(*a*) 450°C (*b*) 375°C
(*c*) 225°C (*d*) 405°C.

86. One mole of an ideal gas at temperature T was cooled isochorically till the gas pressure fell from P to $\frac{P}{n}$. Then, by an isobaric process, the gas was restored to the initial temperature. The net amount of heat absorbed by the gas in the process is

(*a*) nRT (*b*) $\frac{RT}{n}$
(*c*) RT $(1 - n^{-1})$ (*d*) RT $(n - 1)$.

87. Two identical containers A and B with frictionless pistons contain the same ideal gas at the same temperature and same volume V. The mass of the gas in A is m_A and that in B is m_B. The gas in each cylinder is now allowed to expand isothermally to the same final volume 2V. The changes in the pressure in A and B are found to be ΔP and 1.5 ΔP respectively. Then

(*a*) $4m_A = 9m_B$ (*b*) $2m_A = 3m_B$
(*c*) $3m_A = 2m_B$ (*d*) $9m_A = 4m_B$.

88. When an ideal diatomic gas is heated at constant pressure, the fraction of the heat energy supplied which increases the internal energy of the gas is

(*a*) $\frac{2}{5}$ (*b*) $\frac{3}{5}$
(*c*) $\frac{3}{7}$ (*d*) $\frac{5}{7}$.

89. If one mole of a monoatomic gas $\left(\gamma = \frac{5}{3}\right)$ is mixed with one mole of a triatomic gas $\left(\gamma = \frac{4}{3}\right)$, the value of γ for the mixture is

(*a*) 1.40 (*b*) 1.44
(*c*) 1.53 (*d*) 3.07.

90. If f be the number of degrees of freedom of a gas molecule, then the ratio of the specific heats $\frac{C_p}{C_v}$ is given by

(*a*) $1 + \frac{1}{f}$ (*b*) $1 - \frac{1}{f}$
(*c*) $1 - \frac{2}{f}$ (*d*) $1 + \frac{2}{f}$.

91. For a gas, $\gamma = 1.286$. What is the number of degrees of freedom of the molecules of this gas ?

(a) 3 (b) 5

(c) 6 (d) 7.

92. A Carnot engine is working between 127°C and 27°C. The increase in efficiency will be maximum when the temperature of

(a) the source is increased by 50°C

(b) the sink is decreased by 50°C

(c) source is increased by 25°C and that of sink is decreased by 25°C

(d) both source and sink are decreased by 25°C each.

93. From what height should a piece of ice fall so that it melts completely ? (Given : one-quarter of the heat produced is absorbed by ice and latent heat is L)

(a) $\frac{L}{g}$ (b) $\frac{2L}{g}$

(c) $\frac{3L}{g}$ (d) $\frac{4L}{g}$.

94. The P-V graph for a thermodynamical system is shown in Fig. 17. The work done by the system in the process A to B is

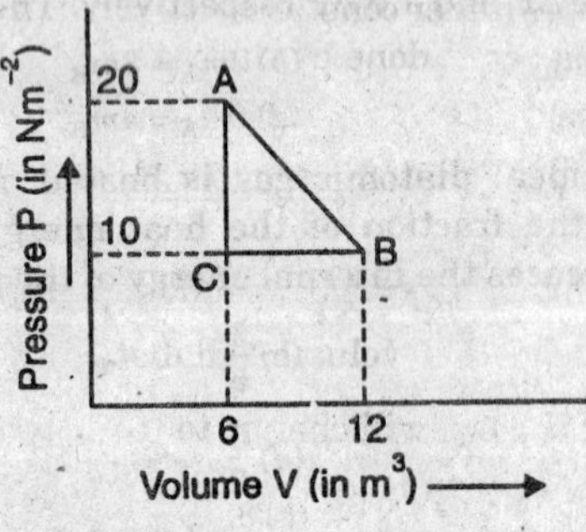

Fig. 17

(a) 90 J (b) 60 J

(c) 0 J (d) 30 J.

95. In Q. 94, the work done in the process B to C is

(a) – 90 J (b) – 60 J

(c) 0 J (d) – 30 J.

96. In Q. 94, the work done in the complete cycle ABCA is

(a) 90 J (b) 60 J

(c) 0 J (d) 30 J.

97. Two samples A and B of a gas which are initially at the same temperature and pressure are compressed from volume V to $\frac{V}{2}$ (A isothermally and B adiabatically). The final pressure

(a) of A is greater than that of B

(b) of A is equal to that of B

(c) of A is less than that of B

(d) of A is twice that of B.

98. The weight of a person is 60 kg. If he gets 10^5 calories of heat through food and the efficiency of his body is 28%, then upto how much height he can climb ? Take $g = 10 \text{ m s}^{-2}$

(a) 100 m (b) 196 m

(c) 400 m (d) 1000 m.

99. The pressure of the air inside the motor tyre is 2 atmosphere and the temperature is 27°C. If it suddenly bursts, the final temperature will be ($\gamma = 1.4$)

(a) 27 K (b) – 150°C

(c) – 81°C (d) – 27°C.

100. A Carnot engine takes 300 cal of heat at 500 K and rejects 150 cal of heat to the sink. The temperature of sink is

(a) 1000 K (b) 250 K

(c) 750 K (d) 125 K.

101. 5 mol of oxygen is heated at constant volume from 10°C to 20°C. Given : $C_p = 8$ cal/mol/°C and R = 8.36 J/mol/°C. The amount of heat consumed by oxygen is

(a) 100 cal (b) 200 cal

(c) 300 cal (d) 400 cal.

102. In Q. 101, the change in internal energy is

(a) 100 cal (b) 200 cal

(c) 300 cal (d) 400 cal.

103. A gas at pressure P is adiabatically compressed so that its density becomes twice that of initial value. Given that the ratio of specific heats at constant pressure and constant volume is 7/5. What will be the final pressure of the gas ?

(a) P (b) 2 P

(c) 2.6 P (d) $\frac{7P}{5}$.

104. A Carnot engine working between 300 K and 600 K has a work output of 800 J per cycle. The amount of heat energy supplied from the source to the engine in each cycle is

(a) 800 J (b) 3200 J

(c) 1600 J (d) 6400 J.

105. In a thermodynamic process, the pressure of a fixed mass of gas is changed in such a manner that the gas molecules give out 30 J of heat and 10 J of work is done on the gas. If the initial internal energy of the gas was 40 J, then the final internal energy will be

(*a*) 0 (*b*) 80 J
(*c*) 20 J (*d*) – 20 J.

106. **140 cal of heat is required to raise the temperature of 2 mol of a diatomic gas at constant pressure from 0°C to 10°C. How much heat will be required to heat it from 0°C to 10°C at constant volume ?**

(*a*) 60 cal (*b*) 100 cal
(*c*) 140 cal (*d*) 180 cal.

107. **A system absorbs 1000 cal of heat and does 1675 J of external work. If J = 4.18 J/cal, then the change in internal energy of the system is**

(*a*) + 1675 J (*b*) + 2505 J
(*c*) + 1000 J (*d*) + 650 J.

108. **If R = universal gas constant, then the amount of heat needed to raise the temperature of 2 mol of an ideal monoatomic gas from 273 K to 373 K when no work is done, is**

(*a*) 100 R (*b*) 150 R
(*c*) 300 R (*d*) 500 R.

109. **During the adiabatic expansion of 2 mol of a gas, the internal energy was found to have decreased by 100 J. The work done by the gas in this process is**

(*a*) zero (*b*) – 100 J
(*c*) 200 J (*d*) 100 J.

110. **An ideal gas with pressure P is compressed isothermally till the mean separation between molecules of the gas is reduced to half of the original value. The final pressure will be**

(*a*) P (*b*) 2P
(*c*) 8P (*d*) 4P.

111. **The height of a waterfall is 84 m. Assuming that the entire kinetic energy of falling water is converted into heat, the rise in temperature of the water will be (g = 10 m s^{-2}, J = 4.2 joule/cal)**

(*a*) 0.2°C (*b*) 1.960°C
(*c*) 0.96°C (*d*) 0.0196°C.

112. **A gas is expanded adiabatically at an initial temperature of 300 K so that its volume is doubled. The final temperature of the gas is (γ = 1.40)**

(*a*) 227.3 K (*b*) 500.30 K
(*c*) 454.76 K (*d*) – 47°C.

113. **One mole of O_2 gas having a volume equal to 22.4 litre at 0°C and 1 atmospheric pressure is compressed isothermally so that its volume reduces to 11.2 litre. The work done in this process is nearly**

(*a*) 1672.5 J (*b*) 1728 J
(*c*) – 1728 J (*d*) – 1570 J.

114. **A freezer has coefficient of performance 5. When 3.6 × 106 J work is done on the freezer, what mass of water at 0°C is converted into ice cubes at 0°C ?**

(*a*) ≈ 5 kg (*b*) ≈ 3.6 kg
(*c*) ≈ 54 kg (*d*) ≈ 107 kg.

115. **Two absolute scales A and B have triple points of water defined to be 200 A and 350 B. The ratio $T_A : T_B$ is**

(*a*) 7 : 4 (*b*) 4 : 7
(*c*) $\frac{4}{7.23}$ (*d*) Not defined.

116. **An ideal gas is taken through series of changes represented in Fig. 18. The net work done by the gas at the end of the cycle is equal to**

(*a*) P_1V_1
(*b*) $3P_1V_1$
(*c*) $6P_1V_1$
(*d*) $12P_1V_1$.

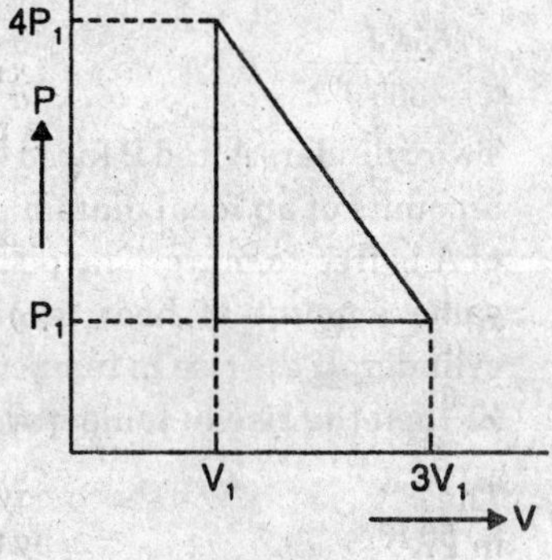

Fig. 18

117. **A gas expands under constant pressure P from volume V_1 to V_2. The work done by the gas is**

(*a*) $P(V_2 - V_1)$ (*b*) $P(V_1 - V_2)$
(*c*) $P(V_1^\gamma - V_2^\gamma)$ (*d*) $P\frac{V_1V_2}{V_2 - V_1}$.

118. **A monoatomic gas (γ = 5/3) is suddenly compressed to $\frac{1}{8}$ th of its original volume adiabatically. Then the pressure of the gas will change to**

(*a*) $\frac{24}{5}$ (*b*) 8
(*c*) $\frac{40}{3}$ (*d*) 32 times its initial pressure.

119. **Four curves A, B, C and D are drawn in the Fig. 19 for a given amount of gas. The curves which represent adiabatic and isothermal changes are**

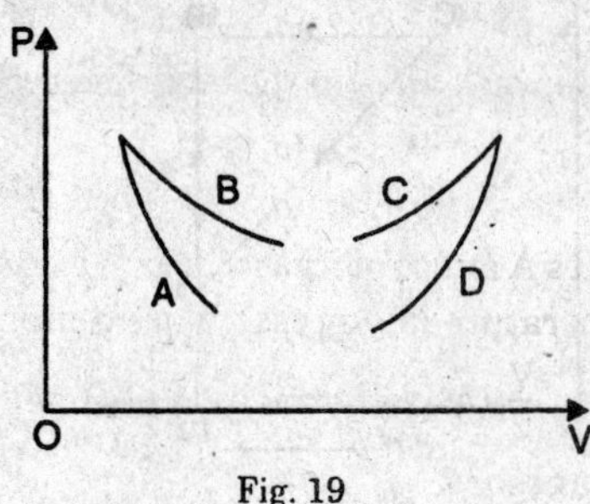

Fig. 19

(*a*) C and D respectively (*b*) D and C respectively
(*c*) A and B respectively (*d*) B and A respectively.

120. With respect to the figures of Q. 119, no heat exchange between the gas and the surroundings will take place, if the gas is taken along curve

(*a*) A (*b*) B
(*c*) C (*d*) D.

121. A system is given 300 calorie of heat and it does 600 joule of work. How much does the internal energy of the system increase in this process ? (J = 4.18 J cal^{-1})

(*a*) 654 J (*b*) 156.5 J
(*c*) – 300 J (*d*) – 528.2 J.

122. Two cylinders A and B fitted with pistons contain equal amounts of an ideal diatomic gas at 300 K. The piston of A is free to move. The piston of B is held fixed. The same amount of heat is given to the gas in each cylinder. If the rise in temperature of the gas in A is 30 K, then the rise in temperature of the gas in B is

(*a*) 30 K (*b*) 18 K
(*c*) 50 K (*d*) 42 K.

123. A polyatomic gas $\left(\gamma = \frac{4}{3}\right)$ is compressed to $\frac{1}{8}$ th of its volume adiabatically. If its initial pressure is P_0, its new pressure will be

(*a*) $8P_0$ (*b*) $16P_0$
(*c*) $6P_0$ (*d*) $2P_0$.

124. Work done by air when it expands from 50 litre to 150 litre at a constant pressure of 2 atmosphere is

(*a*) 2×10^4 J (*b*) 2×100 J
(*c*) $2 \times 10^5 \times 100$ J (*d*) $2 \times 10^{-5} \times 100$ J.

125. An ideal gas is taken through the cycle $A \rightarrow B \rightarrow C \rightarrow A$, as shown in the Fig. 20. If the net heat supplied to the gas in the cycle is 5 J, the work done by the gas in the process $C \rightarrow A$ is

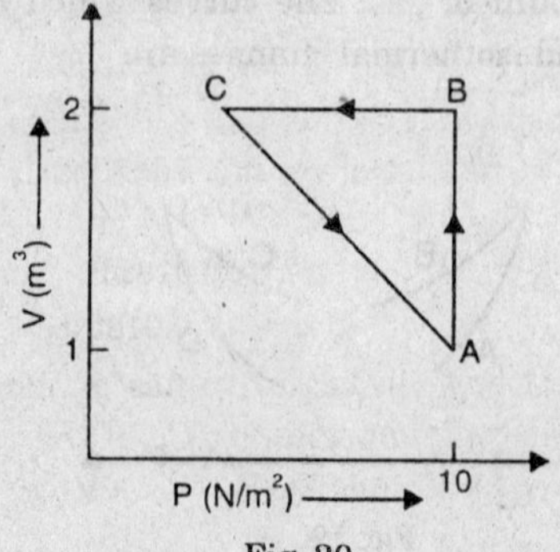

Fig. 20

(*a*) – 5 J (*b*) – 10 J
(*c*) – 15 J (*d*) – 20 J.

126. A gas is compressed at a constant pressure of 50 N m^{-2} from a volume of 10 m^3 to a volume of 4 m^3. Energy of 100 J is then added to the gas by heating. Its internal energy is

(*a*) increased by 400 J (*b*) increased by 200 J
(*c*) increased by 100 J (*d*) decreased by 200 J.

127. During the adiabatic expansion of 2 mol of a gas, the internal energy of the gas is found to decrease by 2 J. The work done during the process on the gas will be equal to

(*a*) 1 J (*b*) – 1 J
(*c*) 2 J (*d*) – 2 J.

128. When a charged particle passes through a liquid-hydrogen bubble chamber, ions are formed along its track. If the pressure is suddenly dropped when the particle traverses the chamber so that the liquid hydrogen reaches its boiling point, boiling occurs preferentially around the ions and the track can be observed. The graph below (Fig. 21) shows a pressure-temperature diagram for hydrogen.

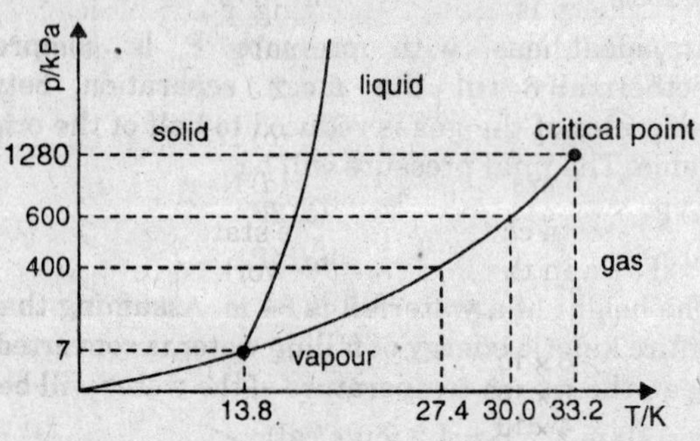

Fig. 21

The pressure in certain bubble chamber is dropped from 600 kPa to 400 kPa. For boiling to occur during the decrease in pressure, the temperature must be maintained in the range

(*a*) 13.8 K to 27.4 K (*b*) 13.8 K to 30.0 K
(*c*) 27.4 K to 30.0 K (*d*) 27.4 K to 33.2 K
(*e*) 30.0 K to 33.2 K.

129. A vessel containing 5 litre of a gas at 0.8 m pressure is connected to an evacuated vessel of volume 3 litre. The resultant pressure inside will be (assuming whole system to be isolated)

(*a*) 4/3 m (*b*) 0.5 m
(*c*) 2.0 m (*d*) 3/4 m.

130. If the amount of heat given to a system be 35 J and the amount of work done by the system be – 15 J, then the change in the internal energy of the system is

(*a*) – 50 J (*b*) 20 J
(*c*) 30 J (*d*) 50 J.

131. An ideal monoatomic gas is taken round the cycle ABCDA as shown in following P-V diagram. The work done during the cycle is

(*a*) PV
(*b*) 2PV
(*c*) 4PV
(*d*) zero.

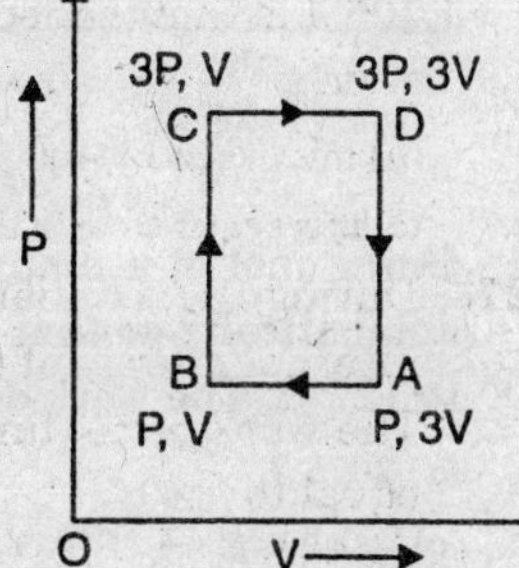

Fig. 22

132. A piece of lead of specific heat 0.03 cal/g/°C falls from a height of 20 m. If on reaching the ground, 50% of its energy is converted into heat, then rise of its temperature will be ($J = 4.2$ J cal^{-1} and $g = 10$ m s^{-2})

(*a*) 12.5°C (*b*) 21.8°C
(*c*) 10.4°C (*d*) 0.79°C.

133. A lead bullet at 27°C just melts when stopped by an obstacle. Assuming that 25% of heat is absorbed by the obstacle, then the velocity of the bullet at the time of striking is nearly (Melting point of lead = 327°C, specific heat of lead = 0.03 cal/g/°C, latent heat of fusion of lead = 6 cal/g and J = 4.2 J cal^{-1})

(*a*) 410 m s^{-1} (*b*) 1230 m s^{-1}
(*c*) 307.5 m s^{-1} (*d*) None of these.

134. A system changes from the state (P_1, V_1) to (P_2, V_2) as shown in the Fig. 23. The work done by the system is

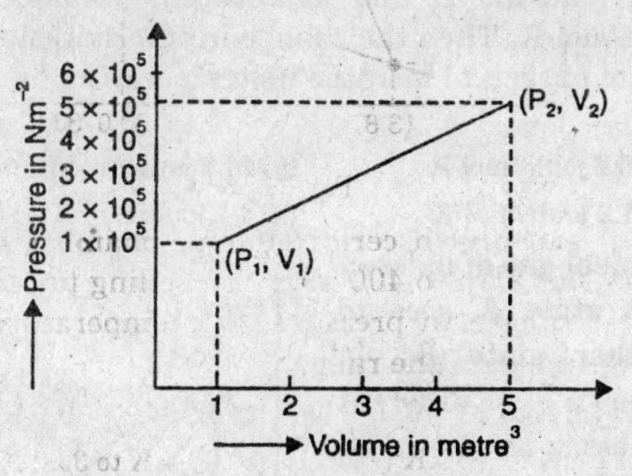

Fig. 23

(*a*) 7.5×10^5 J (*b*) 7.5×10^5 erg
(*c*) 12×10^5 J (*d*) 6×10^5 J.

135. A monoatomic ideal gas initially at 17°C is suddenly compressed to one-eighth of its original volume. The temperature after compression is

(*a*) 17°C (*b*) 16°C
(*c*) 887°C (*d*) None of these.

136. For an isothermal expansion of a perfect gas, the value of $\frac{\Delta P}{P}$ is equal to

(*a*) $-\gamma^{1/2}\frac{\Delta V}{V}$ (*b*) $-\frac{\Delta V}{V}$
(*c*) $-\gamma\frac{\Delta V}{V}$ (*d*) $-\gamma^2\frac{\Delta V}{V}$.

137. The temperature of 5 mol of a gas which was held at constant volume was changed from 100°C to 120°C. The change in internal energy was found to be 80 J. The specific heat capacity of the gas at constant volume will be equal to

(*a*) 8 J K^{-1} (*b*) 0.8 J K^{-1}
(*c*) 4.0 J K^{-1} (*d*) 0.4 J K^{-1}.

138. Heat energy absorbed by a system in going through a cyclic process shown in Fig. 24, is

(*a*) 10^7 π J
(*b*) 10^4 π J
(*c*) 10^2 π J
(*d*) 10^{-3} π J.

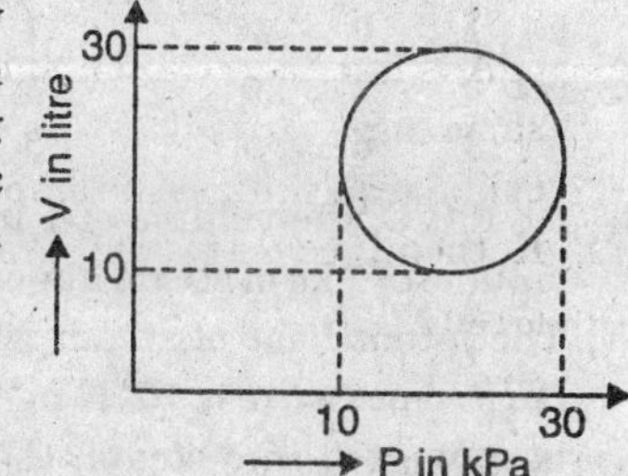

Fig. 24

139. A container of volume 1 m^3 is divided into two equal compartments by a partition. One of these compartments contains an ideal gas at 300 K. The other compartment is vacuum. The whole system is thermally isolated from its surroundings. The partition is removed and the gas expands to occupy the whole volume of the container. Its temperature now would be

(*a*) 300 K (*b*) 239 K
(*c*) 200 K (*d*) 100 K.

140. In an isothermal reversible expansion, if the volume of 96 g of oxygen at 27°C is increased from 70 litre to 140 litre, then the work done by the gas will be

(*a*) 300 R $\log_{10} 2$ (*b*) 81 R $\log_e 2$
(*c*) 900 R $\log_{10} 2$ (*d*) 2.3 × 900 R $\log_{10} 2$.

141. When 1 g of water at 0°C and 1×10^5 N m^{-2} pressure is converted into ice of volume 1.091 cm^3, the external work done will be

(*a*) 0.0091 J (*b*) 0.0182 J
(*c*) – 0.0091 J (*d*) – 0.0182 J.

142. The table below gives the values of the molar heat capacity at constant volume, C_v of the diatomic gas hydrogen at 50 K and 500 K.

T (in K)	C_v *(in J K^{-1} mol^{-1})*
50	12.4
500	20.6

You may assume that the gas remains dia atomic at 500 K and that the molar gas constant R is 8.3 J K^{-1} mol^{-1}. Which one of the following responses correctly gives the number of active translational (T), rotational (R) and vibrational (V) degrees of freedom at these temperatures ?

	at 50 K			*at 500 K*		
	T	*R*	*V*	*T*	*R*	*V*
(*a*)	1	0	0	1	1	0
(*b*)	3	0	0	3	2	0
(*c*)	1	0	2	1	2	2
(*d*)	3	2	0	3	2	2
(*e*)	3	0	0	3	0	2

143. At 100°C, the volume of 1 kg of water is 10^{-3} m^3 and volume of 1 kg of steam at normal pressure is 1.671 m^3. The latent heat of steam is 2.3×10^6 J kg^{-1} and the normal pressure is 10^5 N m^{-2}. If 5 kg of water at 100°C is converted into steam, the increase in the internal energy of water in this process will be

(*a*) 8.35×10^5 J (*b*) 10.66×10^6 J
(*c*) 11.5×10^6 J (*d*) Zero.

144. A fixed mass of an ideal gas slowly absorbs 1000 J of heat and as a result expands slowly, at a constant pressure of 2.0×10^4 Pa, from a volume of 0.050 m^3 to a volume of 0.075 m^3. What is the effect on the internal energy of the gas ?

(*a*) It decreases by 1000 J. (*b*) It decreases by 500 J.
(*c*) It is unchanged. (*d*) It increases by 500 J.
(*e*) It increases by 1000 J.

145. Which of the following graphs correctly represents the variation of $\beta = -(dV/dP)/V$ with P for an ideal gas at constant temperature ?

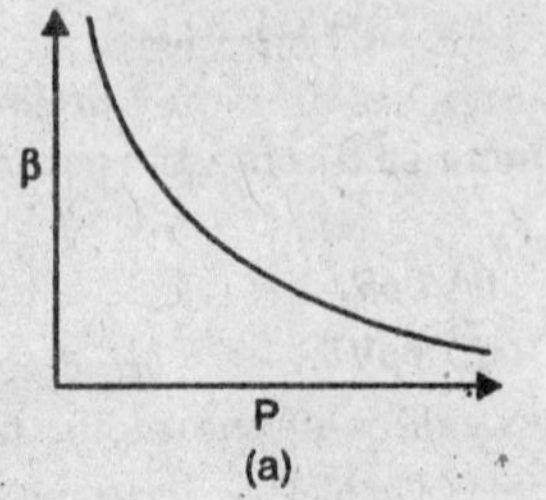

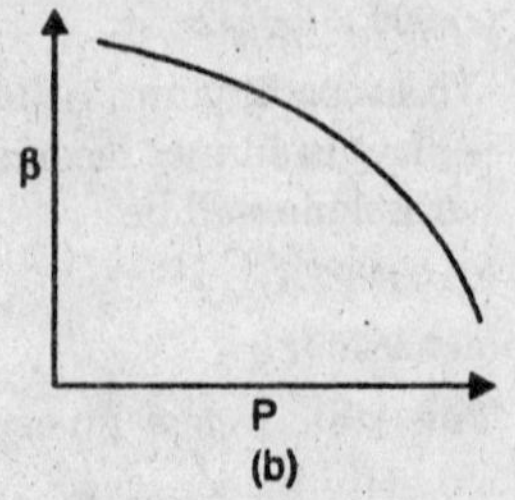

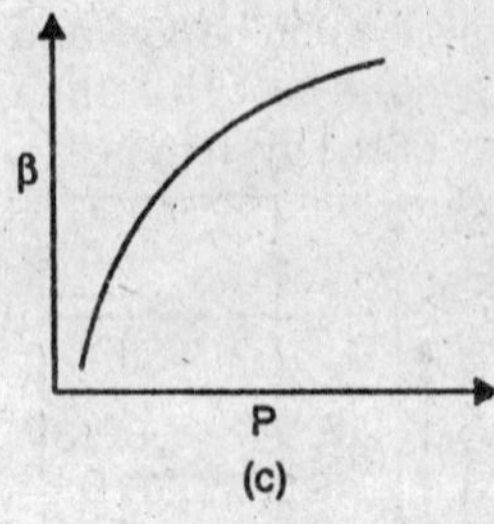

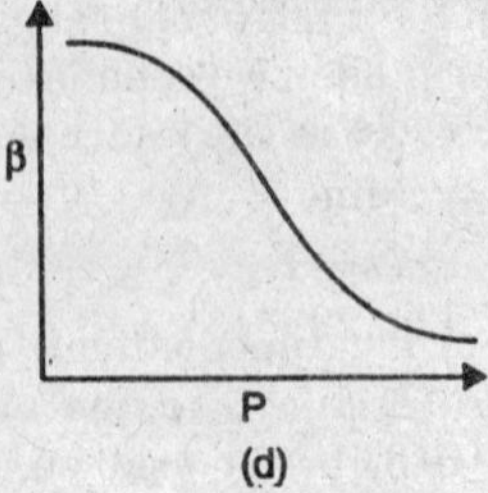

Fig. 25

146. One g mol of a diatomic gas ($\gamma = 1.4$) is compressed adiabatically so that its temperature rises from 27°C to 127°C. The work done will be (R = 8.3 J mol^{-1} K^{-1})

(*a*) 2075 J (*b*) 207.5 J
(*c*) 207.5 erg (*d*) None of these.

147. A sample of an ideal gas initially having internal energy U_1 is allowed to expand adiabatically performing external work W. Heat Q is then supplied to it, keeping the volume constant at its new value, until the pressure rises to its original value. The internal energy is then U_2. (See Fig. 26)

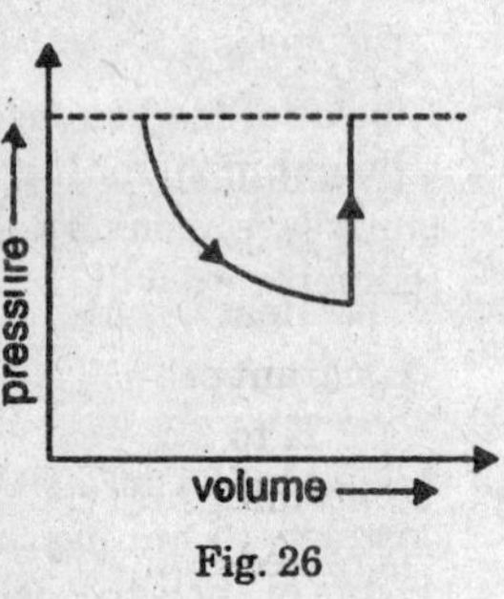

Fig. 26

The increase in internal energy, U_2—U_1 is equal to

(*a*) zero (*b*) W
(*c*) Q (*d*) Q – W
(*e*) W – Q.

148. The specific heat at constant pressure for oxygen is 29.5 joule/mol-K and at constant volume is 21.2 joule/mol-K. Then the change in internal energy of 1 mol of oxygen to increase its temperature by 1°C will be

(*a*) 30.2 joule/mol-K (*b*) 21.2 joule/mol-K
(*c*) 11.2 joule/mol-K (*d*) 1.2 joule/mol-K.

149. /n ideal gas of mass *m* in a state A goes to another state B via three different processes as shown in Fig. 27. If Q_1, Q_2 and Q_3 denote the heat absorbed by the gas along the three paths, then

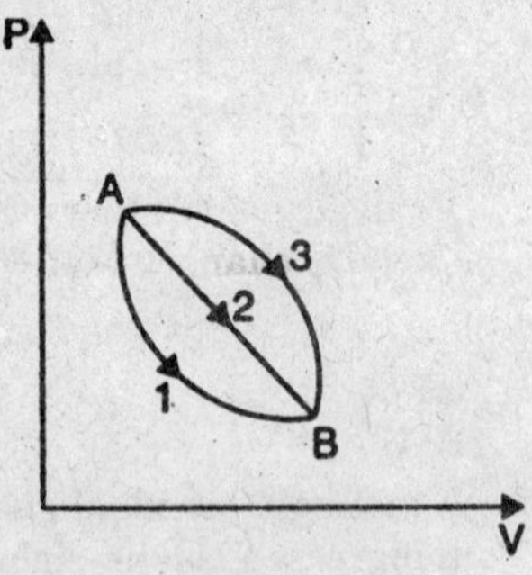

Fig. 27

(*a*) $Q_1 < Q_2 < Q_3$ (*b*) $Q_1 < Q_2 = Q_3$
(*c*) $Q_1 = Q_2 > Q_3$ (*d*) $Q_1 > Q_2 > Q_3$.

150. A gas undergoes the cycle of pressure and volume changes $W \to X \to Y \to Z \to W$ shown in Fig. 28.

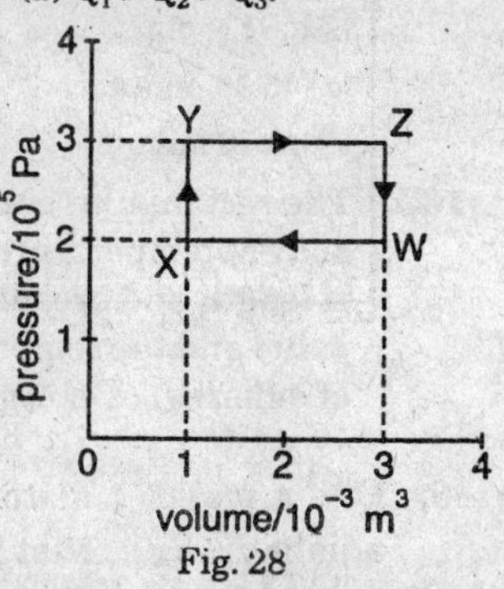

Fig. 28

What is the net work done by the gas ?

(*a*) – 600 J (*b*) 200 J
(*c*) – 200 J (*d*) 600 J
(*e*) 0 J.

151. The value of the gas constant (R) calculated from the perfect gas equation is 8.32 J mol^{-1} K^{-1}, whereas its value calculated from the knowledge of C_p and C_v of the gas is 1.98 cal mol^{-1} K^{-1}. From this data, the value of J is

(*a*) 4.16 J/cal (*b*) 4.18 J/cal
(*c*) 4.20 J/cal (*d*) 4.22 J/cal.

152. The heat absorbed by a mole of monoatomic gas at constant pressure, when its temperature is raised from 273 K to 373 K is Q. The heat absorbed by a mole of diatomic gas at constant volume, when its temperature is raised from 273 K to 323 K is Q′. The ratio of Q to Q′ is

(*a*) 1 (*b*) 1.25
(*b*) 1.75 (*d*) 2.

153. A fixed mass of ideal gas undergoes changes of pressure and volume starting at L, as shown in Fig. 29.

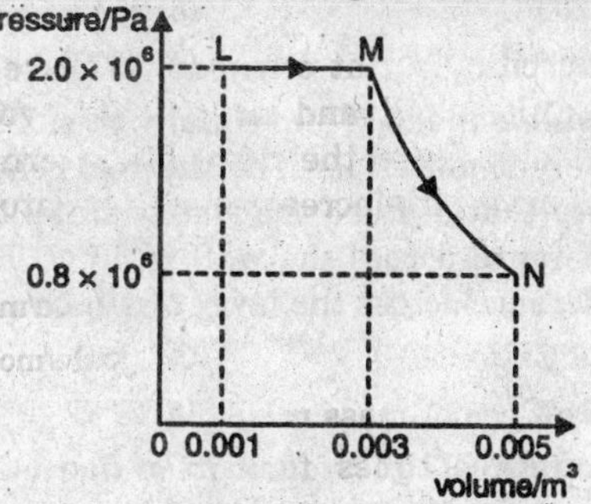

Fig. 29

Which graph shows how temperature (measured in kelvin) changes with volume ?

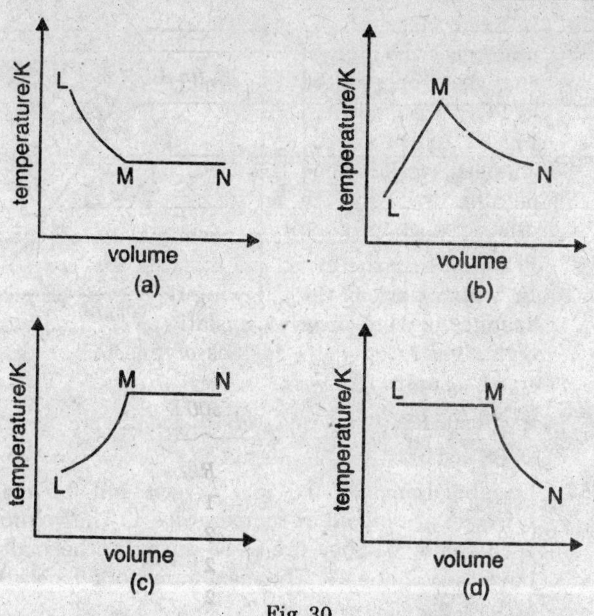

Fig. 30

154. A mass of an ideal gas of volume V at pressure *p* undergoes the cycle of changes shown in the graph

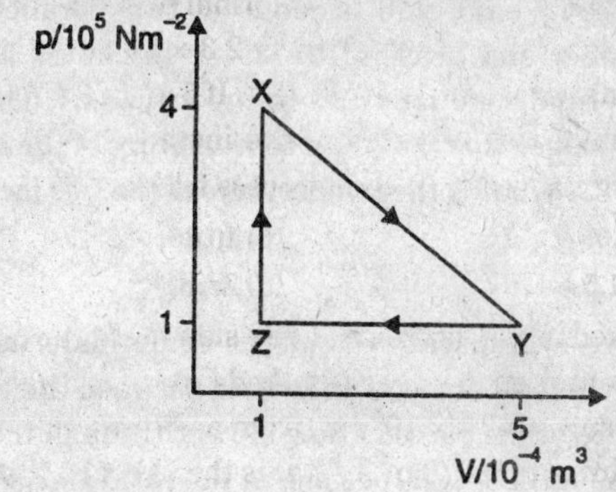

Fig. 31

At which point is the gas coolest and hottest ?

	coolest	*hottest*
(*a*)	X	Y
(*b*)	Y	X
(*c*)	Y	Z
(*d*)	Z	X
(*e*)	Z	Y

155. Air is injected from a cylinder of compressed air into balloon of volume V, causing its diameter to double. What is the work done against the pressure *p* of the atmosphere ?

(*a*) *p*V (*b*) 7 *p*V
(*c*) 3 *p*V (*d*) 8 *p*V
(*e*) 4 *p*V.

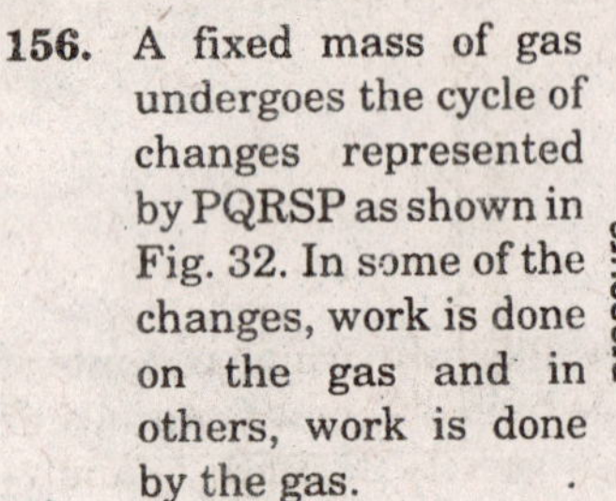

156. A fixed mass of gas undergoes the cycle of changes represented by PQRSP as shown in Fig. 32. In some of the changes, work is done on the gas and in others, work is done by the gas.

In which pair of the changes is work done on the gas ?

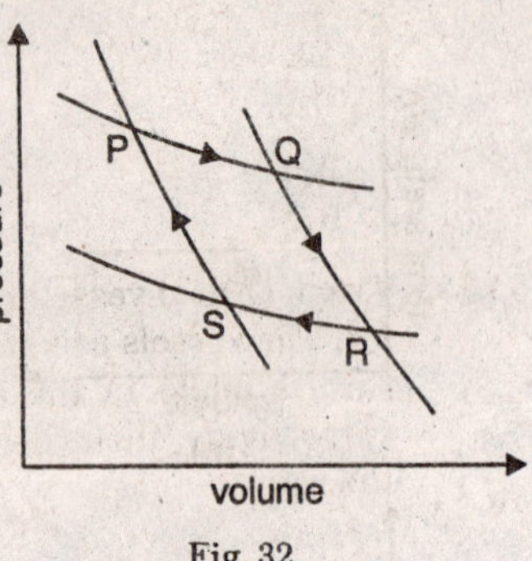

Fig. 32

(*a*) PQ and RS (*b*) PQ and QR

(*c*) QR and RS (*d*) QR and SP

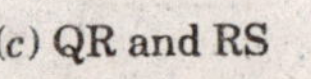

(*e*) RS and SP.

157. Equal volumes of 0.5 mol oxygen and 0.5 mol of nitrogen are mixed at temperature T. The volume of each gas is V. They are to be mixed isothermally to have the volume 2V. The maximum work done will be

(*a*) RT $\log_e$ 2 joule (*b*) RT $\log_e$ 4 joule

(*c*) $\frac{1}{2}$ RT $\log_e$ 4 joule (*d*) $\frac{1}{2}$ RT $\log_e$ 1 joule.

158. A compound slab is made of two parallel plates of copper and brass of the same thickness and having thermal conductivities in the ratio 4 : 1. The free face of copper is at 0°C. The temperature of the interface is 20°C. What is the temperature of the free face of brass ?

(*a*) 0°C (*b*) 20°C

(*c*) 40°C (*d*) 100°C.

159. Two cylindrical rods of the same material have the same temperature difference between their ends. The ratio of the rates of flow of heat through them is 1 : 8. The radii of the rods are in the ratio 1 : 2. What is the ratio of their lengths ?

(*a*) 2 : 1 (*b*) 4 : 1

(*c*) 1 : 8 (*d*) 1 : 32.

160. A cylindrical rod with one end in a steam chamber and the other end in ice results in melting of 0.1 g of ice per second. If the rod is replaced by another rod with half the length and double the radius of the first and if the thermal conductivity of material of the second rod is 0.25 times that of first, the rate at which ice melts in g s^{-1} will be

(*a*) 0.1 (*b*) 0.2

(*c*) 1.6 (*d*) 3.2.

161. There is ice formation on a tank of water of thickness 10 cm. How much time it will take to have a layer of 0.1 cm below it ? The outer temperature is – 5°C, the thermal conductivity of ice is 0.005 cal cm^{-1} s^{-1} °C^{-1} and latent heat of ice is 80 cal/g and the density of ice is 0.91 g cm^{-3}.

(*a*) 46.39 minute (*b*) 47.63 minute

(*c*) 48.78 minute (*d*) 49.31 minute.

162. The rectangular surface of area 8 cm × 4 cm of a black body at temperature 127°C emits energy E per second. If length and breadth are reduced to half of the initial value and the temperature is raised to 327°C, the rate of emission of energy becomes

(*a*) $\frac{3}{8}$ E (*b*) $\frac{81}{16}$ E

(*c*) $\frac{9}{16}$ E (*d*) $\frac{81}{64}$ E.

163. The emissivity and surface area of tungsten filament of an electric bulb are 0.35 and 0.25 × 10^{-4} $metre^2$ respectively. The operating temperature of filament is 3000 K. If $\sigma = 5.67 \times 10^{-8}$ watt $metre^{-2}$ K^{-4}, then power of bulb is approximately

(*a*) 40 watt (*b*) 143 watt

(*c*) 3000 watt (*d*) 1050 watt.

164. Ice starts forming in a lake with water at 0°C when the atmospheric temperature is – 10°C. If time taken for 1 cm of ice to be formed is 7 hour, the time taken for the thickness of ice to change from 1 cm to 2 cm is

(*a*) 3.5 hour (*b*) 7 hour

(*c*) 14 hour (*d*) 21 hour.

165. An object is at temperature of 400°C. At what approximate temperature would it radiate energy twice as first ? The temperature of surroundings may be assumed to be negligible.

(*a*) 200°C (*b*) 200 K

(*c*) 800°C (*d*) 800 K.

166. A wall is made of two equally thick layers A and B of different materials. The thermal conductivity of A is twice that of B. In the steady state, the temperature difference across the wall is 36°C. The temperature difference across the layer A will be

(*a*) 6°C (*b*) 12°C

(*c*) 18°C (*d*) 24°C.

167. The ratio of the coefficients of thermal conductivity of two different materials is 5 : 3. If the thermal resistance of the rods of the same thickness of these materials is same, then the ratio of the lengths of these rods is

(*a*) $\frac{3}{5}$ (*b*) $\frac{5}{3}$

(*c*) $\frac{2}{7}$ (*d*) $\frac{7}{2}$.

168. Which of the following cylindrical rods of the same metal has the highest rate of flow of heat ? The rods have equal difference of temperature between their ends.

(*a*) $l = 2$ m, $r = 1$ cm (*b*) $l = 4$ m, $r = 2$ cm
(*c*) $l = 2$ m, $r = 2$ cm (*d*) $l = 2$ m, $r = 4$ cm.

169. In Q. 168, the lowest rate of flow of heat is for

(*a*) $l = 2$ m, $r = 1$ cm (*b*) $l = 4$ m, $r = 2$ cm
(*c*) $l = 2$ m, $r = 2$ cm (*d*) $l = 2$ m, $r = 4$ cm.

170. Three rods made of the same material and having the same cross-section have been joined as shown in the Fig. 33. Each rod is of the same length. The left and right ends are kept at 0°C and 90°C respectively. The temperature of the junction of the three rods will be

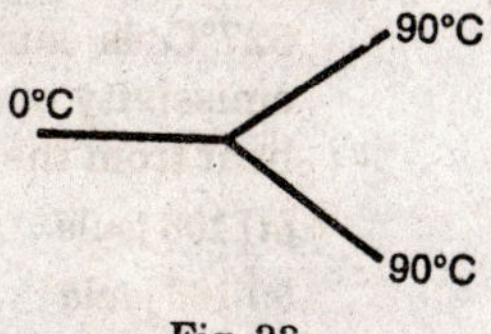

Fig. 33

(*a*) 45°C (*b*) 60°C
(*c*) 30°C (*d*) 20°C.

171. The intensity of radiation emitted by the Sun has its maximum value at a wavelength of 510 nm and that emitted by the North Star has the maximum value at 350 nm. If these Stars behave like black bodies, then the ratio of the surface temperatures of the Sun and the North Star is

(*a*) 1.46 (*b*) 0.69
(*c*) 1.21 (*d*) 0.83.

172. A cylinder of radius R made of a material of thermal conductivity K_1 is surrounded by a cylindrical shell of inner radius R and outer radius 2R made of a material of thermal conductivity K_2. The two ends of the combined system are maintained at two different temperatures. There is no loss of heat across the cylindrical surface and the system is in steady state. The effective thermal conductivity is

(*a*) $K_1 + K_2$ (*b*) $\frac{K_1K_2}{K_1 + K_2}$
(*c*) $\frac{K_1 + 3K_2}{4}$ (*d*) $\frac{3K_1 + K_2}{4}$.

173. Two solid spheres of radii R_1 and R_2 are made of same material and have similar surface. The spheres are raised to the same temperature and then allowed to cool under identical conditions. Assuming spheres to be perfect conductors of heat, their initial rates of loss of heat are

(*a*) $\frac{R_1^2}{R_2^2}$ (*b*) $\frac{R_1}{R_2}$
(*c*) $\frac{R_2}{R_1}$ (*d*) $\frac{R_2^2}{R_1^2}$.

174. In Q. 173, the initial rates of cooling are

(*a*) $\frac{R_1^2}{R_2^2}$ (*b*) $\frac{R_1}{R_2}$
(*c*) $\frac{R_2}{R_1}$ (*d*) $\frac{R_2^2}{R_1^2}$.

175. Two identical vessels are filled with equal amounts of ice. The vessels are made from different materials. If the ice melts in the two vessels in times t_1 and t_2 respectively, then their thermal conductivities are in the ratio

(*a*) $\frac{t_1}{t_2}$ (*b*) $\frac{t_2}{t_1}$
(*c*) $t_2^2 : t_1^2$ (*d*) $t_1^2 : t_2^2$.

176. Two identical rods of a metal are welded as shown in Fig. 34 (*a*). 20 cal of heat flows through them in 4 minute. If the rods are welded as shown in Fig. 34 (*b*), then the same amount of heat will flow in

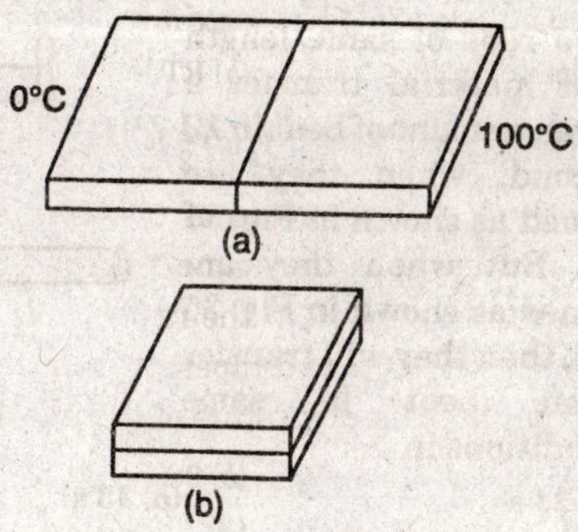

Fig. 34

(*a*) 1 minute (*b*) 2 minute
(*c*) 4 minute (*d*) 16 minute.

177. A small hole is made in a hollow enclosure whose walls are maintained at a temperature of 1000 K. The amount of energy being emitted per square metre per second is

(*a*) 567 J (*b*) 5670 J
(*c*) 56700 J (*d*) 567000 J.

178. The ends of the two rods of different materials with their lengths, diameters of cross-section and thermal conductivities all in the ratio 1 : 2 are maintained at the same temperature difference. The rate of flow of heat in the shorter rod is 1 cal s^{-1}. What is the rate of flow of heat in the larger rod ?

(*a*) 1 cal s^{-1} (*b*) 4 cal s^{-1}
(*c*) 8 cal s^{-1} (*d*) 16 cal s^{-1}.

179. The heat is flowing through two cylindrical rods of same material. The diameters of the rods are in the ratio 1 : 2 and their lengths are in the ratio 2 : 1. If the temperature difference between their ends is the same, the ratio of rates of flow of heat through them will be

(a) 1 : 1 (b) 2 : 1
(c) 1 : 4 (d) 1 : 8.

180. The temperature gradient in a rod of 0.5 m length is 80°C/m. If the temperature of hotter end of the rod is 30°C, then the temperature of the cooler end is

(a) 40°C (b) – 10°C
(c) 10°C (d) 0°C.

181. A body at 300°C radiates 10 J cm^{-2} s^{-1}. If Sun radiates 10^5 J cm^{-2} s^{-1}, then its temperature is

(a) 3000°C (b) 5457°C
(c) 300×10^4 °C (d) 5730°C.

182. The ratio of energy of radiation emitted by a black body at 27°C and 927°C is

(a) 1 : 4 (b) 1 : 16
(c) 1 : 64 (d) 1 : 256.

183. Two rods of same length and material transfer a given amount of heat in 12 second, when they are joined as shown in Fig. 35 (i). But when they are joined as shown in Fig. 35 (ii), then they will transfer same heat in same conditions in

l
(i)
l l
(ii)
Fig. 35

(a) 24 s (b) 13 s
(c) 15 s (d) 48 s.

184. The energy emitted per second by a black body at 27°C is 10 J. If the temperature of the black body is increased to 327°C, the energy emitted per second will be

(a) 20 J (b) 40 J
(c) 80 J (d) 160 J.

185. The dimensional formula of thermal resistance is

(a) $[M^{-1}L^{-2}T^3K]$ (b) $[ML^2T^{-2}K^{-1}]$
(c) $[ML^2T^{-3}K]$ (d) $[ML^2T^{-2}K^{-2}]$.

186. Two vessels of different materials are similar in size in every respect. The same quantity of ice filled in them gets melted in 20 minute and 30 minute. The ratio of their thermal conductivities will be

(a) 1.5 (b) 1
(c) $\frac{2}{3}$ (d) 4.

187. The temperature of a body is increased by 50%. The amount of radiation emitted by it would be nearly

(a) 50% (b) 225%
(c) 250% (d) 400%.

188. The temperature of a body is increased from 27°C to 127°C. The radiation emitted by it increases by a factor of

(a) $\frac{256}{81}$ (b) $\frac{15}{9}$
(c) $\frac{4}{5}$ (d) $\frac{12}{27}$.

189. A metal ball of surface area 200 cm^2 and temperature 527°C is surrounded by a vessel at 27°C. If the emissivity of the metal is 0.4, then the rate of loss of heat from the ball is nearly ($\sigma = 5.67 \times 10^{-8}$ J/m^2-s-K^4)

(a) 108 joule (b) 168 joule
(c) 186 joule (d) 192 joule.

190. A spherical black body with a radius of 12 cm radiates 450 W power at 500 K. If the radius were halved and the temperature doubled, the power radiated in watt would be

(a) 225 (b) 450
(c) 900 (d) 1800.

191. Wires A and B have identical lengths and have circular cross-sections. The radius of A is twice the radius of B i.e. $R_A = 2R_B$. For a given temperature difference between the two ends, both wires conduct heat at the same rate. The relation between the thermal conductivities is given by

(a) $K_A = 4K_B$ (b) $K_A = 2K_B$
(c) $K_A = K_B/2$ (d) $K_A = K_B/4$.

192. The ratio of thermal conductivities of two rods of different material is 5 : 4. The two rods of same area of cross-section and same thermal resistance will have the lengths in the ratio

(a) 4 : 5 (b) 9 : 1
(c) 1 : 9 (d) 5 : 4.

193. A compound slab consists of two parallel plates of different materials of same thickness and having thermal conductivities k_1 and k_2. What is the equivalent thermal conductivity of the slab ?

(a) $k_1 + k_2$ (b) $k_1 k_2$
(c) $\frac{k_1 + k_2}{k_1 k_2}$ (d) $\frac{2k_1 k_2}{k_1 + k_2}$.

194. A cylinder of radius R made of material of thermal conductivity K_1 is surrounded by a cylindrical shell of inner radius R and outer radius 3R made of a material of thermal conductivity K_2. The two ends of the combined system are maintained at two different temperatures. What is the effective thermal conductivity of the system ?

(a) $K_1 + K_2$ (b) $\frac{K_1 + 8K_2}{9}$

(c) $\frac{K_1K_2}{K_1 + K_2}$ (d) $\frac{8K_1 + K_2}{9}$.

195. The lengths of the two rods made up of the same metal and having the same area of cross-section are 0.6 m and 0.8 m respectively. The temperatures of the ends of first rod are 90°C and 60°C and that for the ends of the other rod are 150°C and 110°C. For which rod, the rate of conduction will be greater ?

(a) First (b) Second

(c) Same for both (d) None of these.

196. In a steady state of thermal conduction, temperature of the ends A and B of a 20 cm long rod are 100°C and 0°C respectively. What will be the temperature of the rod at a point at a distance of 6 cm from the end A of the rod ?

(a) – 30°C (b) 70°C

(c) 5°C (d) None of these.

197. A black metal foil is warmed by radiation from a small sphere at temperature T and at a distance *d*. It is found that the power received by the foil is 'P'. If both the temperature and the distance are doubled, the power received by the foil will be

(a) 16P (b) 4P

(c) 2P (d) P.

198. If the coefficient of conductivity of aluminium is 0.5 cal $cm^{-1}\ s^{-1}\ °C^{-1}$, then in order to conduct 10 cal $s^{-1}\ cm^{-2}$ in the steady state, the temperature gradient in aluminium must be

(a) 5°C/cm (b) 10°C/cm

(c) 20°C/cm (d) 10.5°C/cm.

199. Wien's constant is 2892×10^{-6} SI unit and the value of λ_m for Moon is 14.46 micron. The surface temperature of Moon is

(a) 100 K (b) 300 K

(c) 400 K (d) 200 K.

200. Which of the following cylindrical rods, of the same material, will conduct most heat, when their ends are maintained at the same steady temperature ?

(a) Length 1 m ; radius 1 cm

(b) Length 2 m ; radius 1 cm

(c) Length 2 m ; radius 2 cm

(d) Length 1 m ; radius 2 cm.

201. The coefficients of thermal conductivity of copper, mercury and glass are respectively K_c, K_m and K_g such that $K_c > K_m > K_g$. If the same quantity of heat is to flow per second per unit area of each and corresponding temperature gradients are X_c, X_m and X_g, then

(a) $X_c = X_m = X_g$ (b) $X_c > X_m > X_g$

(c) $X_c < X_m < X_g$ (d) $X_m < X_c < X_g$.

202. A heat flux of 4000 J s^{-1} is to be passed through a copper rod of length 10 cm and area of cross-section 100 cm^2. The thermal conductivity of copper is 400 W/m°C. The two ends of this rod must be kept at a temperature difference of

(a) 1°C (b) 10°C

(c) 100°C (d) 1000°C.

203. In designing a method for measuring the thermal conductivity of polystyrene, care must be taken to choose a specimen of appropriate dimensions as well as to decide whether or not the specimen requires lagging.

Which of the following would be the correct choice ?

	cross-sectional area of specimen	*thickness of specimen*	*lagging required*
(a)	small	thin	no
(b)	small	thick	yes
(c)	large	thin	no
(d)	large	thick	yes

204. The graph shows how temperature varies with distance along a well-insulated metal rod which is conducting thermal energy at a steady rate. The slope of this graph is the temperature gradient.

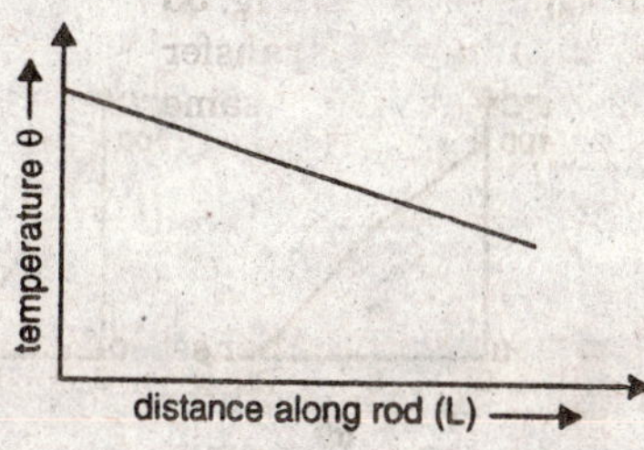

Fig. 36

There is an analogy between electrical conduction and thermal energy conduction.

If an equivalent electrical-graph were to be drawn, which electrical quantity, when plotted against distance along the rod, would have the slope shown ?

	Electrical quantity	*Slope*
(a)	charge	current
(b)	potential	potential gradient
(c)	potential difference	resistance
(d)	current	current gradient.

205. The temperature of a liquid drops from 365 K to 361 K in 2 minute. The time during which temperature of the liquid drops from 344 K to 342 K is (Temperature of room is 293 K.)

(a) 84 s (b) 72 s
(c) 66 s (d) 60 s.

206. A composite rod of uniform cross-section has copper and aluminium sections of the same length in good thermal contact. The ends of the rod, which is well-lagged, are maintained at 100°C and at 0°C as shown in the diagram (Fig. 37). The thermal conductivity of copper is twice that of aluminium.

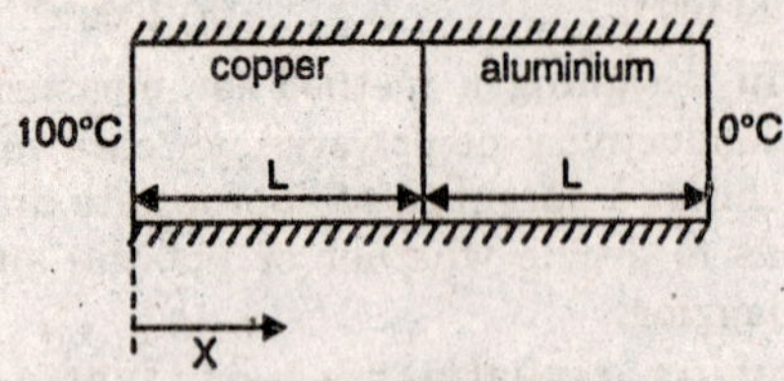

Fig. 37

Which one of the following graphs represents the variation of temperature t with distance x along the rod in the steady state ?

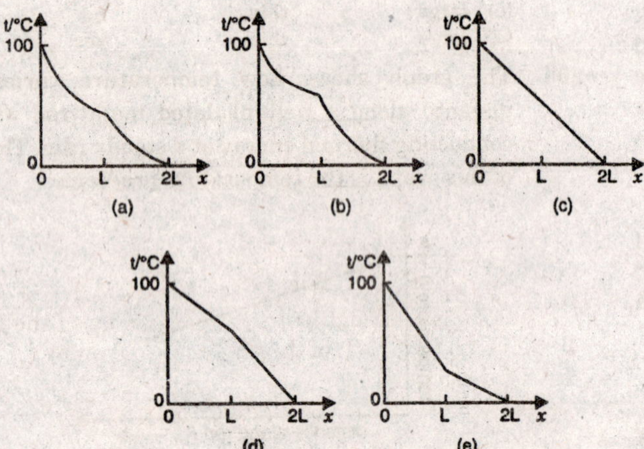

Fig. 38

207. Consider two hot bodies B_1 and B_2 which have temperatures 100°C and 80°C respectively at $t = 0$. The temperature of the surroundings is 40°C. The ratio of the respective rates of cooling R_1 and R_2 of these two bodies at $t = 0$ will be

(a) $R_1 : R_2 = 3 : 2$ (b) $R_1 : R_2 = 5 : 4$
(c) $R_1 : R_2 = 2 : 3$ (d) $R_1 : R_2 = 4 : 5$.

208. The temperature of a perfect black body is 727°C and its area is 0.1 m^2. If Stefan's constant is 5.67×10^{-8} watt/m^2-s-K^4, then heat radiated by it in 1 minute is

(a) 8100 cal (b) 81000 cal
(c) 810 cal (d) 81 cal.

209. The temperature of a piece of metal is raised from 27°C to 51.2°C. The rate at which the metal radiates energy increases nearly

(a) 2 times (b) 4 times
(c) 4.46 times (d) 1.36 times.

210. A parallel-sided slab is made of two different materials. The upper half of the slab is made of material X, of thermal conductivity λ ; the lower half is made of material Y, of thermal conductivity 2λ.

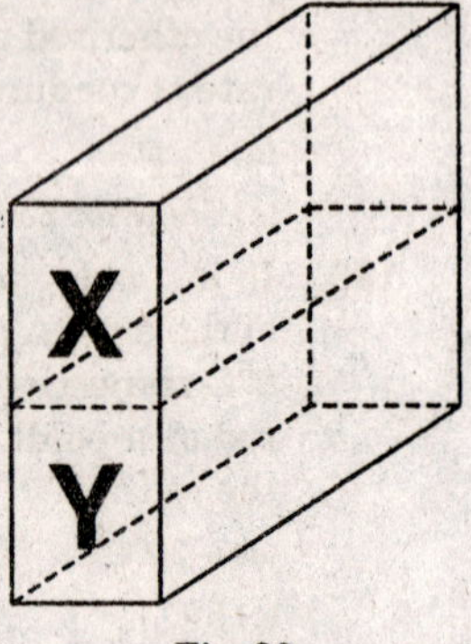

Fig. 39

In the steady state, the left hand face of the composite slab is at a higher, uniform temperature than the right-hand face, and the flow of heat through the slab is parallel to its shortest sides.

What fraction of the total heat flow through the slab passes through material X ?

(a) $\frac{1}{4}$ (b) $\frac{1}{3}$
(c) $\frac{1}{2}$ (d) $\frac{2}{3}$
(e) $\frac{3}{4}$.

211. Metal rods X and Y of identical cross-sectional area, have lengths 60 cm and 30 cm respectively. They are made of metals of thermal conductivities λ_X and λ_Y. They are well-lagged and joined end-to-end as shown in the diagram. One end of X is maintained at 100°C and the opposite end of Y is maintained at 0°C. When steady conditions have been reached, the temperature of the junction is found to be 25°C.

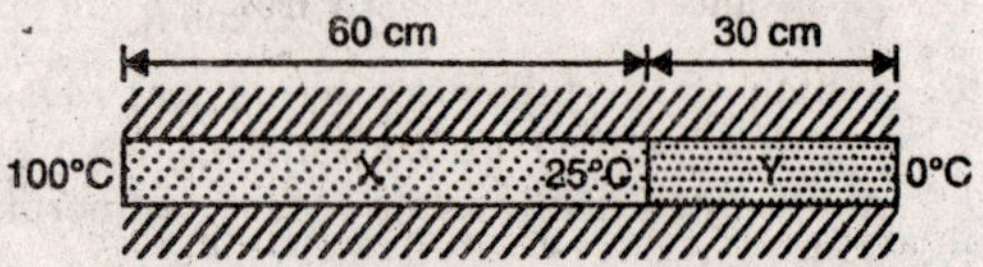

Fig. 40

What is the value of $\frac{\lambda_X}{\lambda_Y}$?

(a) $\frac{1}{6}$ (b) $\frac{2}{3}$
(c) $\frac{25}{24}$ (d) $\frac{3}{2}$
(e) 6.

212. The diagram below shows rods of the same size of two different materials P and Q placed end to end in thermal contact and heavily lagged at their sides. The outer ends of P and Q are kept at 0°C and 100°C, respectively. The thermal conductivity of P is four times that of Q.

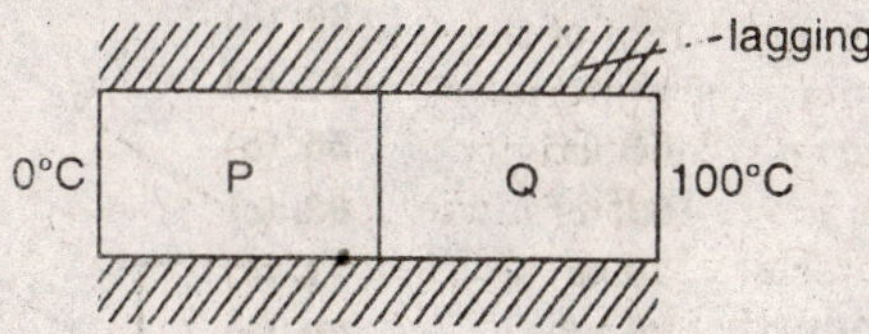

Fig. 41

What is the steady-state temperature of the interface ?

(*a*) 20°C (*b*) 75°C
(*c*) 25°C (*d*) 80°C
(*e*) 50°C.

213. PQ is a fully-lagged metal bar, containing a section XY of a material of lower thermal conductivity. The thermal conductivities of the two materials are independent of temperature. Ends P and Q are maintained at different temperatures.

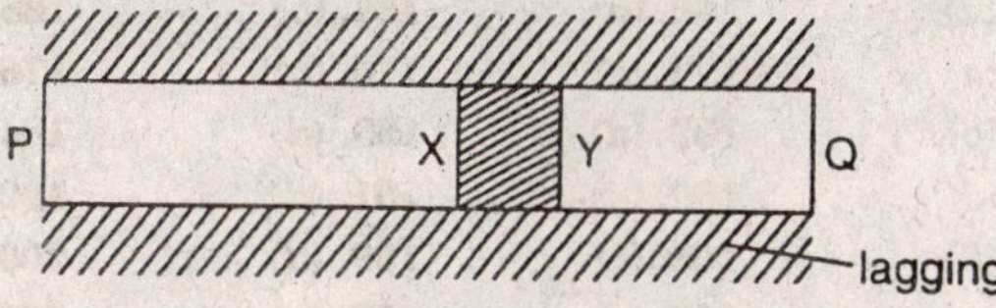

Fig. 42

In the steady state, the temperature difference across XY would be independent of

(*a*) the temperature difference between P and Q
(*b*) the metal of which the bar is made
(*c*) the thickness of the section XY
(*d*) the material of the section XY
(*e*) the distance of the section XY from the end P.

214. Radiation from a black body at the thermodynamic temperature T_1 is measured by a small detector at distance d_1 from it. When the temperature is increased to T_2 and the distance to d_2, the power received by the detector is unchanged. What is the ratio d_2/d_1 ?

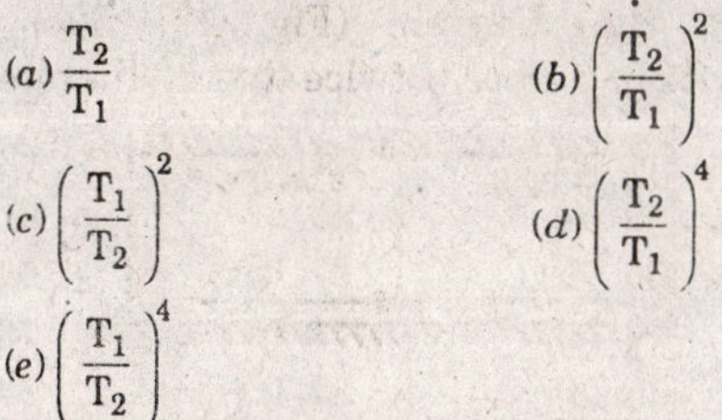

(*a*) $\frac{T_2}{T_1}$ (*b*) $\left(\frac{T_2}{T_1}\right)^2$
(*c*) $\left(\frac{T_1}{T_2}\right)^2$ (*d*) $\left(\frac{T_2}{T_1}\right)^4$
(*e*) $\left(\frac{T_1}{T_2}\right)^4$

215. Two bars of equal length and the same cross-sectional area but of different thermal conductivities, k_1 and k_2, are joined end to end as shown in Fig. 43. One end of the composite bar is maintained at temperature T_h whereas the opposite end is held at T_c.

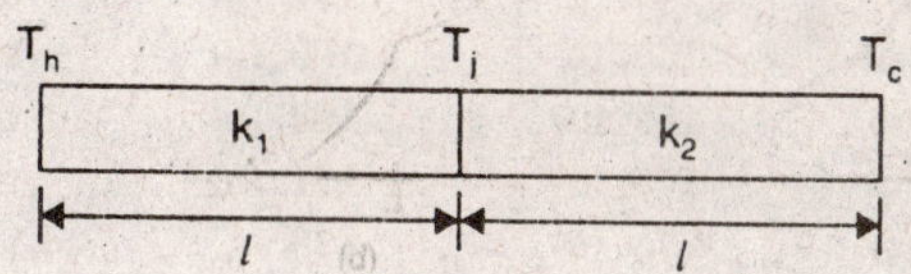

Fig. 43

If there are no heat losses from the sides of the bars, the temperature T_j of the junction is given by

(*a*) $\frac{k_2}{k_1}\frac{(T_h + T_c)}{2}$ (*b*) $\frac{k_2}{k_1 + k_2}(T_h + T_c)$
(*c*) $\frac{k_1 + k_2}{2}\frac{(T_h + T_c)}{2}$ (*d*) $\frac{1}{k_1 + k_2}(k_1 T_h + k_2 T_c)$
(*e*) $\frac{1}{2}(k_1 T_h + k_2 T_c)$.

ANSWERS

QUESTION BANK (Level I)

1. (*a*)	**2.** (*a*)	**3.** (*c*)	**4.** (*b*)	**5.** (*b*)	**6.** (*d*)	**7.** (*e*)	**8.** (*c*)
9. (*d*)	**10.** (*d*)	**11.** (*b*)	**12.** (*c*)	**13.** (*c*)	**14.** (*a*)	**15.** (*b*)	**16.** (*d*)
17. (*e*)	**18.** (*c*)	**19.** (*a*)	**20.** (*c*)	**21.** (*c*)	**22.** (*d*)	**23.** (*b*)	**24.** (*a*)
25. (*d*)	**26.** (*c*)	**27.** (*b*)	**28.** (*c*)	**29.** (*a*)	**30.** (*c*)	**31.** (*a*)	**32.** (*b*)
33. (*b*)	**34.** (*a*)	**35.** (*b*)	**36.** (*b*)	**37.** (*a*)	**38.** (*e*)	**39.** (*c*)	**40.** (*e*)
41. (*c*)	**42.** (*e*)	**43.** (*a*)	**44.** (*c*)	**45.** (*a*)	**46.** (*c*)	**47.** (*c*)	**48.** (*c*)
49. (*d*)	**50.** (*c*)	**51.** (*a*)	**52.** (*c*)	**53.** (*c*)	**54.** (*a*).		

QUESTION BANK (Level II)

1. (*d*)	**2.** (*b*)	**3.** (*d*)	**4.** (*c*)	**5.** (*c*)	**6.** (*b*)	**7.** (*a*)	**8.** (*b*)
9. (*a*)	**10.** (*b*)	**11.** (*b*)	**12.** (*b*)	**13.** (*d*)	**14.** (*a*)	**15.** (*d*)	**16.** (*c*)
17. (*d*)	**18.** (*b*)	**19.** (*a*)	**20.** (*d*)	**21.** (*c*)	**22.** (*b*)	**23.** (*d*)	**24.** (*d*)
25. (*d*)	**26.** (*c*)	**27.** (*a*)	**28.** (*b*)	**29.** (*a*)	**30.** (*b*)	**31.** (*b*)	**32.** (*c*)
33. (*a*)	**34.** (*b*)	**35.** (*c*)	**36.** (*c*)	**37.** (*c*)	**38.** (*c*)	**39.** (*b*)	**40.** (*c*)
41. (*b*)	**42.** (*a*)	**43.** (*b*)	**44.** (*a*)	**45.** (*b*)	**46.** (*c*)	**47.** (*a*)	**48.** (*c*)
49. (*c*)	**50.** (*a*)	**51.** (*a*)	**52.** (*d*)	**53.** (*b*)	**54.** (*a*)	**55.** (*b*)	**56.** (*b*)
57. (*c*)	**58.** (*b*)	**59.** (*c*)	**60.** (*c*)	**61.** (*d*)	**62.** (*c*)	**63.** (*c*)	**64.** (*d*)
65. (*d*)	**66.** (*c*)	**67.** (*d*)	**68.** (*b*)	**69.** (*b*)	**70.** (*a*)	**71.** (*a*)	**72.** (*a*)
73. (*a*)	**74.** (*c*)	**75.** (*c*)	**76.** (*d*)	**77.** (*b*)	**78.** (*d*)	**79.** (*a*)	**80.** (*c*)
81. (*b*)	**82.** (*a*)	**83.** (*c*)	**84.** (*d*)	**85.** (*b*)	**86.** (*c*)	**87.** (*c*)	**88.** (*d*)
89. (*b*)	**90.** (*d*)	**91.** (*d*)	**92.** (*b*)	**93.** (*d*)	**94.** (*a*)	**95.** (*b*)	**96.** (*d*)
97. (*c*)	**98.** (*b*)	**99.** (*d*)	**100.** (*b*)	**101.** (*c*)	**102.** (*c*)	**103.** (*c*)	**104.** (*c*)
105. (*c*)	**106.** (*b*)	**107.** (*b*)	**108.** (*c*)	**109.** (*d*)	**110.** (*c*)	**111.** (*a*)	**112.** (*a*)
113. (*d*)	**114.** (*c*)	**115.** (*b*)	**116.** (*b*)	**117.** (*a*)	**118.** (*d*)	**119.** (*c*)	**120.** (*a*)
121. (*a*)	**122.** (*d*)	**123.** (*b*)	**124.** (*a*)	**125.** (*a*)	**126.** (*a*)	**127.** (*d*)	**128.** (*c*)
129. (*b*)	**130.** (*d*)	**131.** (*c*)	**132.** (*d*)	**133.** (*a*)	**134.** (*c*)	**135.** (*c*)	**136.** (*b*)
137. (*b*)	**138.** (*c*)	**139.** (*b*)	**140.** (*d*)	**141.** (*a*)	**142.** (*b*)	**143.** (*b*)	**144.** (*d*)
145. (*a*)	**146.** (*a*)	**147.** (*d*)	**148.** (*b*)	**149.** (*a*)	**150.** (*b*)	**151.** (*c*)	**152.** (*d*)
153. (*b*)	**154.** (*e*)	**155.** (*b*)	**156.** (*e*)	**157.** (*a*)	**158.** (*d*)	**159.** (*a*)	**160.** (*b*)
161. (*c*)	**162.** (*d*)	**163.** (*a*)	**164.** (*d*)	**165.** (*d*)	**166.** (*b*)	**167.** (*b*)	**168.** (*d*)
169. (*a*)	**170.** (*b*)	**171.** (*b*)	**172.** (*c*)	**173.** (*a*)	**174.** (*c*)	**175.** (*b*)	**176.** (*a*)
177. (*c*)	**178.** (*b*)	**179.** (*d*)	**180.** (*b*)	**181.** (*b*)	**182.** (*d*)	**183.** (*d*)	**184.** (*d*)
185. (*a*)	**186.** (*a*)	**187.** (*d*)	**188.** (*a*)	**189.** (*c*)	**190.** (*d*)	**191.** (*d*)	**192.** (*d*)
193. (*d*)	**194.** (*b*)	**195.** (*c*)	**196.** (*b*)	**197.** (*b*)	**198.** (*c*)	**199.** (*d*)	**200.** (*d*)
201. (*c*)	**202.** (*c*)	**203.** (*c*)	**204.** (*b*)	**205.** (*a*)	**206.** (*d*)	**207.** (*a*)	**208.** (*b*)
209. (*d*)	**210.** (*b*)	**211.** (*b*)	**212.** (*a*)	**213.** (*e*)	**214.** (*b*)	**215.** (*d*).	

HINTS/SOLUTIONS

QUESTION BANK (Level I)

1. (*a*) $\Delta C = \Delta F$

2. (*a*) The Celsius temperature, symbol θ, is defined by $\theta = T - 273.15$ where T is the thermodynamic temperature.

Thus, T = 273.00 K is equivalent to

$$\theta = 273.00 - 273.15 = -0.15°C.$$

3. (*c*) The celsius temperature, symbol θ is defined by $\theta = T - 273.15$ where T is the thermodynamic temperature.

Thus, $\theta = 501.85°C$ is equivalent to $T = \theta + 273.15$

$$= 501.85 + 273.15 \text{ or } 775.00 \text{ K}$$

4. (*b*) $T = 273.16 \times \frac{4.5}{3} \text{ K} = 409.74 \text{ K}$

5. (*b*) Types of thermometers corresponding to the accuracy and response time are listed below.

	Range / K	*Accuracy*	*Response time*	*Thermometer*
a	3—1750	very good	long	constant-volume gas
b	30—1750	average	short	thermocouple
c	75—1550	good	long	platinum resistance
d	230—630	poor	medium	mercury in glass

6. (*d*) Both alcohol-in-glass and mercury-in-glass thermometers give direct reading but not accurate. The quick fluctuation of temperatures makes the reading of temperatures very difficult and is far from the accuracy we want.

Constant-volume gas thermometer is a large awkward instrument, demanding much skill and time, and is thus useless for measuring changing temperatures. Platinum resistance thermometer has wide temperature range and is best for small steady temperature differences. It can be used to measure steady temperature with very high accuracy.

Except for gas thermometer, thermoelectric thermometer is more accurate but is quicker in action and hence is most suitable for this application.

7. (*e*) Thermocouple is fairly accurate and is suitable for measurement of rapid changing temperature. Resistance thermometer, on the other hand, is very accurate for measurement of slow changing temperature and it is capable of measuring temperature from 20 K to 1400 K.

Thus thermocouple is most suitable for case 1 and resistance thermometer is most suitable for both cases 2 and 3.

8. (*c*) At very low pressures, all real gases behave as ideal gas and they give same temperature readings. Thus, readings should be taken at very low pressures in both thermometers.

9. (*d*) By definition, the specific latent heat of vaporisation of a liquid is the heat required to convert unit mass of it, at its boiling point, into vapour at the same temperature. It is expressed in joule per kilogram (J kg^{-1}), or, with high values, in kJ kg^{-1}. The only statement that describes correctly this definition is the option (*d*).

10. (*d*) By the principle of mixtures, heat given out by water = heat absorbed by the melted ice

$$\Rightarrow mc(\theta - 0) = ML \Rightarrow M = \frac{mc\theta}{L}$$

11. (*b*) In first case, there is no change in volume.

So work done is zero.

In the second case, W = P(2V – V) = PV

12. Energy used for expansion = $PdV = RdT$

Heat supplied at constant pressure = $C_p\,dT = \frac{7R}{2}dT$

Ratio is $\frac{R\,dT}{\frac{7R}{2}dT}$ *i.e.* $\frac{2}{7}$

13. (*c*) By the Zeroth law of Thermodynamics, if two bodies are in thermal equilibrium, they have the same temperature. We thus have $T_1 = T = T_2$

14. (*a*) By Zeroth law of thermodynamics, if two bodies are in thermal equilibrium with each other, then they have the same temperature. On the other hand, if both are in thermal equilibrium with the third body, then all three of them have the same temperature as well. Now, thermometers X and Y are in thermal equilibrium with the enclosure whose walls are maintained at steady temperature T. By the Zeroth law of thermodynamics, X and Y must have the same temperature T under all circumstances.

15. (*b*) Molar heat capacity of an ideal gas at constant volume is the heat required to warm one mole of gas through one degree when its volume is kept constant.

By First law of thermodynamics, since no external work is done by the gas, all the heat supplied goes wholly to increase the internal energy U. That is, for one mole of gas warmed through 1 K,

$$U = C_v$$

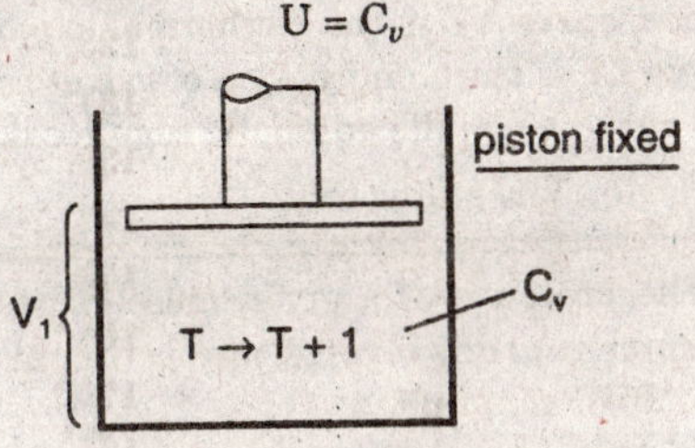

Fig. 44

On the other hand, molar heat capacity of an ideal gas at constant pressure is the heat required to warm one mole of it by one degree when its pressure is kept constant. Since the temperature rise of the gas is 1 K, the internal energy of the gas is increased by C_v,

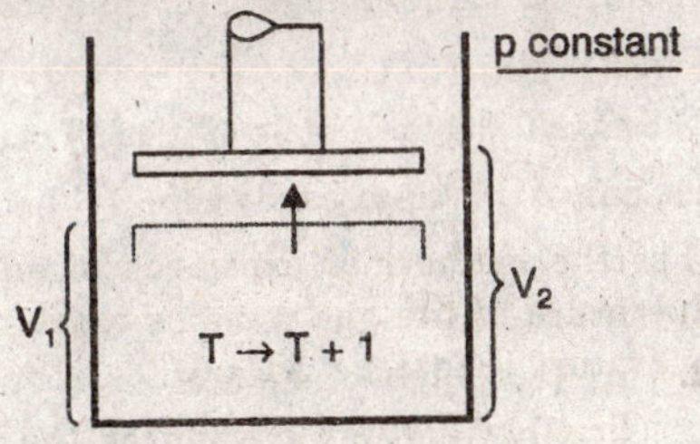

Fig. 45

By the First law of thermodynamics, $\Delta Q = \Delta U + p\Delta V$. The total amount of heat required to warm the gas at constant pressure is therefore,

$$C_p = C_v + p(V_2 - V_1)$$

where $\begin{cases} pV_1 = RT_1 \\ pV_2 = R(T_1 + 1) \end{cases}$

$$\Rightarrow \quad C_p = C_v + R$$

Thus, C_p is greater than C_v because work has to be done $[p(V_2 - V_1) = R]$ against external pressure as the gas expands.

16. (*d*) The molecular mass of chromium is 52 g.

Given its molar heat capacity is 24 J K^{-1} mol^{-1} *i.e.* per 52 g. Its specific heat capacity (1 kg) is given by

$\frac{1000}{52}(24)$ or $461.5 \text{ J K}^{-1} \text{kg}^{-1}$

To the nearest tens, it may be written as 460 JK^{-1} kg^{-1}.

17. (*e*) There are three different processes here, namely the *isothermal change, adiabatic change* and *change at constant pressure.*

The curve for adiabatic change is steeper than that for isothermal change, therefore curve XY is the one for adiabatic change and YZ has to be the one for isothermal change.

Hence, curve YZ is the isothermal expansion of the gas. Curve ZY is the compression of gas at constant pressure and curve XY is the adiabatic compression of the gas.

18. (*c*) Boyle's law states that

> The pressure of a given mass of gas, *at constant temperature*, is inversely proportional to its volume, or PV = constant.

Since temperature is kept constant, DT = 0 in the equation for the first law of thermodynamics.

19. (*a*) When the system of an ideal gas undergoes an isothermal change, the temperature is constant. Since the internal energy of the gas is dependent only on its temperature T, the change of internal energy is thus zero as given below.

$\Delta U = C_v \Delta T = 0$ for a mole of it as $\Delta T = 0$.

Thus, only ΔU is necessarily zero.

0. (*c*) For the isothermal expansion of an ideal gas, the temperature of the gas remains constant or change in temperature is zero, *i.e.* $\Delta T = 0$.

In the first law of thermodynamics, the internal energy ΔU of the gas is dependent only on the thermodynamic temperature T and is given by

$\Delta U = C_v \Delta T$ for a mole of it.

Thus, $\Delta U = 0$ since $\Delta T = 0$.

21. (*c*) In the year 1853, Wiedmann and Franz established a relationship between thermal conductivity and electrical conductivity.

$$\frac{K}{\sigma} \propto T$$

22. (*d*) $\frac{T_S}{T_M} = \frac{\lambda_M}{\lambda_S} = \frac{10^{-4}}{0.5 \times 10^{-6}}$ or $\frac{T_S}{T_M} = \frac{1000}{5} = 200$

23. (*b*) $Q = \frac{KA(\theta_1 - \theta_2)t}{d}$ or $\frac{Q}{t} = \frac{KA(\theta_1 - \theta_2)}{d}$

In the given problem, $\frac{Q}{t}$, $(\theta_1 - \theta_2)$ and d are constants.

$\therefore$ KA = constant

$\therefore$ $\frac{A_1}{A_2} = \frac{K_2}{K_1}$

24. (*a*) $T_1 = 273$ K, $T_2 = 546$ K

$\frac{E_2}{E_1} = \frac{(2T)^4}{T^4} = 16$ or $E_2 = 16E_1 = 16E$

25. (*d*) $\frac{d\theta}{dx} \propto \frac{1}{K}$. When $K = \infty$, then $\frac{d\theta}{dx}$ is zero. So, θ becomes independent of x.

26. (*c*) Thermal resistances are to be added up.

27. (*b*) Ingen Hausz experiment helps to compare thermal conductivities of different materials.

28. (*c*) In the first case, the temperature difference is greater as compared to the second case. So, the rate of loss of heat is quicker.

29. (*a*) Note that the water in contact with the lower surface of ice is about to be converted into ice.

30. (*c*) $\frac{K_1}{K_2} = \frac{l_1^2}{l_2^2} = \frac{105 \times 105}{60 \times 60} = \frac{7 \times 7}{4 \times 4} = \frac{49}{16}$

31. (*a*) Perfectly black body absorbs all radiations falling on it.

32. (*b*) $Q = \frac{KA(\theta_1 - \theta_2)t}{l}$ or $\frac{Q}{t} = \frac{\theta_1 - \theta_2}{l/KA}$

Compare with $I = \frac{V}{R}$

So, thermal resistance $= \frac{l}{KA}$

33. (*b*) $5000 \times 10^{-10}\, T = 0.3 \times 10^{-2}$

or $T = \frac{0.3 \times 10^{-2}}{5000 \times 10^{-10}}$ K

$= \frac{3 \times 10^4}{5}$ K $= \frac{30{,}000}{5}$ K = 6000 K

34. (*a*) $\frac{\lambda_m'}{\lambda_m} = \frac{T}{T'} = \frac{2000}{2400} = \frac{5}{6}$

$\lambda_m' = \frac{5}{6} \times 4\ \mu m = \frac{20}{6}\ \mu m = \frac{10}{3}\ \mu m = 3.33\ \mu m$

35. (*b*) $\lambda_m T$ = constant

or $T \propto \frac{1}{\lambda_m}$ or $T \propto \nu_m$

36. (*b*) Thermal conduction in metal from one point to another occurs if there is temperature difference between these two points as described in the following equation,

$$\frac{Q}{t} = KA\frac{(\theta_2 - \theta_1)}{d}$$

Since the two spheres are in thermal equilibrium (same temperature), there is no net transfer of thermal energy between them as $\frac{Q}{t} = 0$.

37. (*a*) Thermal conduction can be achieved by lattice vibration mechanism. On the other hand, electrical insulators contain no free conduction electron.

38. (*e*) Prevost's theory predicts that all bodies absorb and radiate simultaneously to their environment. This precludes *a* and *b*. The rates of absorption and of radiation depend on the temperature difference with the environment.

Body X is hot and radiates heat faster than it can absorb. Body Y is cold and absorbs heat faster than it can radiate.

When both are at 10°C, they radiate and absorb at equal rates. This precludes *c*. X receives heat from Y and the enclosure at all times. This precludes *d*. Radiation and absorption occur between all three. This includes *e*.

39. (*c*) The total power radiated by a black body of area A at temperature T K is given by $P = A\sigma T^4$

where σ = Stefan's constant = 5.7×10^{-8} W m^{-2} K^{-4},

which is best represented in graph (*c*).

40. (*e*) At steady state condition, the rate of heat flow along the copper bar is a constant given by

$$\frac{dQ}{dt} = -KA\frac{\theta_1 - \theta_2}{L}$$

where K = thermal conductivity of copper bar,

A = cross-sectional area of copper bar,

L = length of bar,

and θ_1, θ_2 = temperatures at ends of bar.

Thus, it is a horizontal line as represented in graph *e*.

41. (*c*) When a light bulb is heated up, it radiates energy in all directions and at all wavelengths. Because the bulb is relatively cool, the wavelengths radiated are mostly very long ones. These are invisible. Hence, only small fraction of the energy is emitted as light.

42. (*e*) Most bodies are coloured. They transmit or reflect some wavelengths *better* than others. Similarly, for the filament of the light bulb, when it is heated, it turns the heat into the random kinetic energy of its molecules and then radiates energy at *all wavelengths*. The colour changes from red to orange to yellow and eventually to white. It never appears green or blue not because it cannot transmit green or blue light, but because of the fact that any green and blue light emitted is always accompanied by more intense red and orange emission.

43. (*a*) By Planck's quantum theory of radiation, the wavelength λ_m of the peak of the intensity (greatest energy) curve is related to the temperature T (in K) by the equation

$$\lambda_m T = \text{constant}.$$

Since the wavelength at which the greatest energy was being received is decreased, the temperature of the source

$$T = \frac{\text{constant}}{\lambda_m} \text{ has increased.}$$

44. (*c*) The rate of heat flow along a lagged copper bar is given by $\frac{Q}{t} = KA\frac{\theta_2 - \theta_1}{l}$

where K = thermal conductivity of copper bar,

A = cross-sectional area of the bar,

θ_1 = temperature of melting ice,

θ_2 = temperature of hot end of the copper bar,

l = length of the copper bar.

Since the heat Q supplied to the bar is increased, rate of heat flow $\frac{Q}{t}$ increases. The temperature gradient $\frac{\theta_2 - \theta_1}{l}$ also increases since K and A remain constant. The hot end temperature θ_2 increases as well since l and θ_1 remain unchanged.

45. (*a*) A black body is a good radiator and absorber of almost all the radiation that falls upon it. At the same temperature and wavelength of radiation, a black body thus radiates more energy than a *'grey'* body. The emissivity of the black body is almost unity whereas that of the *'grey'* body is much less.

The peak height of the curve of radiated energy of either a black body or a *'grey'* body is a function of temperature T.

At the same temperature, the radiation peak occurs at λ_m, the same wavelength for both the black body and the *'grey'* body.

Thus, graph *'a'* correctly compares the distribution of radiated energy of a *'grey'* body with that of a black body at the same temperature.

46. (*c*) If a metal bar is well lagged with a bad conductor, its temperature will fall uniformly from the hot to the cold end as shown in Fig. 46.

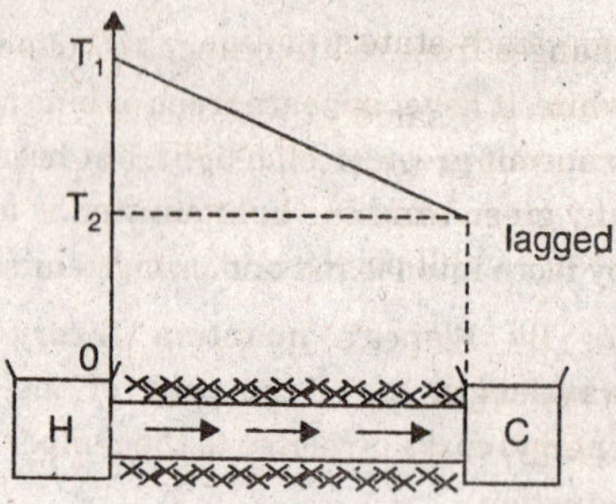

Fig. 46

In the unlagged bar, heat escapes from its sides, by convection to the surrounding air. Thus, the temperature falls more rapidly at the beginning than the well-lagged case. If the temperatures are still maintained at T_1 and T_2 at the ends, the temperature variation is then given by

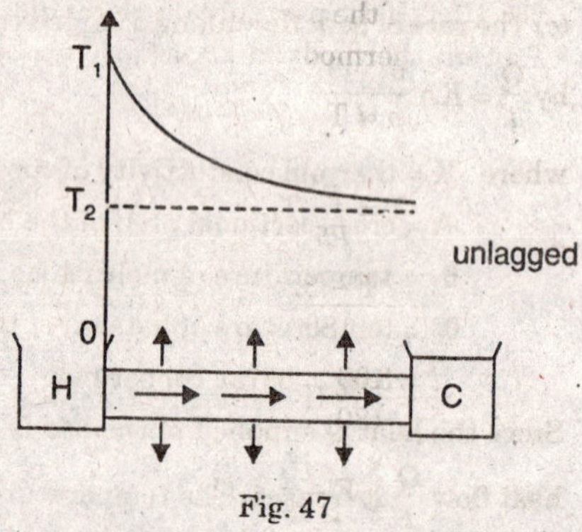

Fig. 47

47. (*e*) When the temperature of copper increases, the amplitude of vibrations of the lattice atoms at the lattice sites in copper is also increased. Thus, conduction electrons, which is the means of transport heat in copper, collide more often with the copper atoms. The mean time between collision is decreased and on the average, the mean speed of the electrons at which the heat is transported is thus reduced. As a result, the thermal conductivity of copper decreases with temperature rise of copper.

Note. For copper (metal), thermal conductivity K is analogous to electrical conductivity σ. Since σ decreases with temperature rise for the same reason, and so does K.

48. (*d*) The energy per second P_1 radiated from the black body of effective radiating area A at temperature T is given by

$$P_1 = A\,\sigma T^4 \qquad ...(1)$$

where σ is the Stefan's constant.

The heat P_0 absorbed from the surroundings at temperature T_0 is given by

$$P_0 = A\,\sigma T_0^4 \qquad ...(2)$$

In this case, from Prevost's theory, electrical power from source, P = net energy per second radiated by the black body.

$$P = P_1 - P_0 = A\,\sigma(T^4 - T_0^4)$$

$$\therefore \quad A = \frac{P}{\sigma(T^4 - T_0^4)}$$

49. (*d*) For the length *l* from X, the copper bar is well lagged so that negligible heat is lost through its sides. Thus, by the equation of rate of heat flow, the temperature variation along the bar is uniform. Beyond this length of *l*, the copper bar is unlagged and hence heat escapes from its sides. Temperature drop will be more rapid and the temperature curve is expected to concave towards the horizontal axis of distance. Thus, the temperature curve for the copper bar is best represented in graph *d*.

50. (*c*) From Wien's displacement law,

$$\lambda_m T = 2.9 \times 10^{-3}\ mK$$

Human body temperature, T = 273 + 37 = 310 K

$$\therefore \quad \lambda_m = \frac{2.9 \times 10^{-3}}{310} = 9355 \text{ nm or about } 9000 \text{ nm.}$$

51. (*a*) From the equation of rate of heat flow, we have

$$\frac{Q}{t} = -KA\frac{d\theta}{dt}$$

where

K is the thermal conductivity of the metal,

A is the cross-sectional area of the metal rod,

$\frac{d\theta}{dt}$ is the temperature gradient along the rod.

The end-pieces conduct poorly. So, their temperature gradients are steep. The unlagged rod loses heat along the sides. So, the temperature gradient curves downward.

52. (*c*) The bolometer should have a blackened surface so that it would better absorb the incident radiation. It should also have a low heat capacity in order that a small amount of radiation would result in a large temperature rise. It should also have a high temperature coefficient of resistance so that it would respond rapidly to the temperature rise when thermal radiation is detected.

53. (*c*) The heat dissipation per second in the thin wire due to current flow is given by $P = I^2R$

where R is the resistance in the wire.

Since the heat loss occurs predominantly through the ends of the wire therefore the thermal conductivity at the ends is higher than that in the centre of the wire. Thus, heat escapes much easier at ends than at the centre and thus temperature rise at the ends is less than the points at the centre of wire. The only graph that best represents this temperature variation is given in graph *c*.

54. (*a*) The rate of flow of heat across unit cross-sectional area of the rod is given by $\frac{Q}{tA} = K\frac{T_1 - T_2}{l}$

where K is the thermal conductivity of the metal and l is the length of the metal rod.

In steady state situation, $\frac{Q}{tA}$ is a constant quantity and is thus the same at all points along the metal rod. It is analogous to the electric current I flowing through the metal rod and by Kirchhoff's law, it should be a constant value at all points along the rod.

QUESTION BANK (Level II)

1. (*d*) $t = \frac{80-50}{75-50} \times 100$ or $t = \frac{30}{25} \times 100 = 120°C$

☞ ***For air thermometers***

$$t = \frac{P_t - P_0}{P_{100} - P_0} \times 100$$

2. (*b*) $t = \frac{R_t - R_0}{R_{100} - R_0} \times 100 = \frac{5.06 - 2.56}{3.56 - 2.56} \times 100$

$= \frac{2.5}{1} \times 100 = 250°C$

3. (*d*) $\frac{x-20}{80} = \frac{120-30}{120} = \frac{90}{120} = \frac{3}{4}$

or $\frac{x-20}{20} = 3$ or $x = 20 + 60 = 80$

4. (*c*) $-13°$ F is $(13 + 32)°$ below ice point on F scale.

5. (*c*) $\frac{\text{correct reading} - 0}{100 - 0}$

$= \frac{\text{Faulty reading} - \text{Faulty lower fixed point}}{\text{Faulty upper fixed point} - \text{Faulty lower fixed point}}$

or correct reading $= \frac{59-5}{95-5} \times 100 = 60°$

6. (*d*) $\frac{F-32}{9} = \frac{C}{5} \Rightarrow \frac{x-32}{9} = \frac{-x}{5} \Rightarrow x = 11.43$

$\therefore$ 11.43°F = – 11.43°C

7. (*a*) $\Delta F = \frac{9}{5}\Delta C = \frac{9}{5} \times 25 = 45°F$

8. (*b*) $\frac{52-5}{99-5} = \frac{x-32}{212-32}$ or $\frac{47}{94} = \frac{x-32}{180}$

or $x - 32 = 90$ or $x = 122$

9. (*a*) $t = \frac{P_t - P_0}{P_{100} - P_0} \times 100 = \frac{60-50}{90-50} \times 100 = \frac{10}{40} \times 100 = 25°C$

10. (*b*) $\frac{x-273}{100} = \frac{x-32}{180}$ or $\frac{x-273}{5} = \frac{x-32}{9}$

or $4x = +273 \times 9 - 5 \times 32$

or $4x = 2457 - 160 = 2297$

or $x = 574.25$

11. (*b*) $\frac{60-(-10)}{x-(-10)} = \frac{50}{100}$ or $\frac{70}{x+10} = \frac{1}{2}$

or $x + 10 = 140$ or $x = 130°C$

12. (*b*) Using the constant-volume gas thermometer, the gas pressure p_{tr} is measured at the triple point of water, 273.16 K. If the pressure is p at an unknown temperature T on the thermodynamic scale, then by gas laws,

$$p \propto T \text{ or } \frac{T}{T_{tr}} = \frac{p}{p_{tr}}$$

$$\therefore \quad T = \frac{p}{p_{tr}} T_{tr} = 273.16\left(\frac{p}{p_{tr}}\right) K$$

13. (*d*) $\frac{C}{100} = \frac{F-32}{180}$

or $C = \frac{100}{180}(F - 32)$

or $C = \frac{5}{9}F - \frac{5}{9} \times 32$

If follows from here that

$\tan\theta = \frac{5}{9}$

$\sin\theta = \frac{5}{\sqrt{106}}$

√106 5 θ 9

Fig. 48

14. (*a*) (*i*) $\frac{50-32}{180} = \frac{K-273}{100}$ or $K = 273 + \frac{5}{9} \times 18$ or $K = 283$

(*ii*) $\frac{50-0}{100-0} = \frac{K-273}{100}$ or $K = 273 + 50 = 323$

(*iii*) $\frac{50-0}{80-0} = \frac{K-273}{100}$ or $K = 273 + \frac{500}{8}$

$= 273 + 62.5 = 335.5$

15. (*d*) $\frac{K-273}{100} = \frac{R-0}{80-0}$

Suppose $K = R = x$

Then $\frac{x-273}{5} = \frac{x}{4}$

or $5x - 4x = -273 \times 4 = -1092$

or $x = -1092$

This is absurd, being out of range.

16. (*c*) $\frac{(C-3)-0}{80-0} = \frac{C}{100}$ or $\frac{C-3}{4} = \frac{C}{5}$

or $5C - 4C = 15$ or $C = 15$

17. (*d*) $\frac{x-2}{92-2} = \frac{x-0}{100-0}$ or $\frac{x-2}{90} = \frac{x}{100}$

or $9x = 10x - 20$ or $x = 20$

18. (*b*) Using the thermocouple, the Celsius temperature θ of the liquid can be calculated from $\theta = \frac{E_\theta}{E_{100}}\ 100°C$

where E_{100} is the measured e.m.f. when one junction X is in melting ice and the other junction Y is in steam at 100°C.

E_θ is the measured emf when the junction X is in melting ice and the other junction Y is in the liquid at θ °C.

Given $E_\theta = -1.5$ mV ; $E_{100} = 6.0$ mV

Therefore, $\theta = \frac{-1.5}{6.0} \times 100°C = -25°C$

19. (*a*) $\frac{C}{100} = \frac{40}{80}$ or $C = 50$

(*c*) $\frac{C}{100} = \frac{113-32}{180} = \frac{81}{180} = \frac{9}{20}$ or $C = 45$

(*d*) $\frac{C}{100} = \frac{320-273}{100}$ or $C = 47$

20. (*d*) $\frac{\theta-(-2)}{96-(-2)} = \frac{50-0}{100-0}$

or $\frac{\theta+2}{98} = \frac{1}{2}$ or $\theta + 2 = 49$ or $\theta = 47°C$

21. (*c*) $\frac{\theta-(-30)}{130-(-30)} = \frac{50-40}{120-40} = \frac{1}{8}$

or $\theta + 30 = \frac{1}{8} \times 160 = 20$ or $\theta = 20 - 30 = -10$

22. (*b*) If C be the correct reading, then

$$\frac{C}{5} = \frac{F-32}{9} = \frac{113-32}{9} = \frac{81}{9} = 9 \text{ or } C = 45$$

Correction is (45 – 44) *i.e.* + 1°C

23. (*d*) If F be the correct reading, then

$$\frac{F-32}{9} = \frac{90}{5} \text{ or } F - 32 = \frac{810}{5} = 162$$

or $F = 32 + 162 = 194$

Correction is (194 – 190) or + 4°F

24. (*d*) $\frac{\theta-32}{180} = \frac{\theta-34}{210-34} = \frac{\theta-34}{176}$ or $\frac{\theta-32}{45} = \frac{\theta-34}{44}$

or $45\,\theta - 45 \times 34 = 44\,\theta - 44 \times 32$

or $\theta = 45 \times 34 - 44 \times 32 = 1530 - 1408 = 122$

25. (*d*) Percentage change in volume $= \gamma t \times 100$

$= 3\,\alpha t \times 100 = 300\,\alpha t$

$= 300 \times 2 \times 10^{-5} \times 200 = 1.2$

26. (*c*) $T = 2\pi\sqrt{\frac{l}{g}}$

$$T' = 2\pi\sqrt{\frac{l[1+\alpha(t_2-t_1)]}{g}}$$

$$T' = T\left[1 + \frac{1}{2}\alpha\,(t_2 - t_1)\right]$$

$$\frac{T'}{T} - 1 = \frac{1}{2}\alpha\,(t_2 - t_1)$$

27. (*a*) $d_0 = d_t(1 + 3\,\alpha t)$

$10 = 9.7(1 + 3\alpha \times 100)$

or $\left[\frac{10}{9.7} - 1\right]\frac{1}{300} = \alpha$ or $\alpha = \frac{0.3}{9.7 \times 300} = 0.0001/°C$

28. (*b*) Percentage increase in radius

$= \alpha t \times 100 = 10^{-5} \times 100 \times 100 = 0.1$

Percentage increase in $(\text{radius})^2 = 2 \times 0.1 = 0.2$

This is the same as percentage increase in moment of inertia. $\left[I = \frac{2}{5}MR^2\right]$

29. (*a*) $1016.2 = 1000(1 + \gamma \times 300)$

or $\gamma = \left(\frac{1016.2}{1000} - 1\right)\frac{1}{300} = 5.4 \times 10^{-5} = 54 \times 10^{-6}$

$\alpha = \frac{\gamma}{3} = 18 \times 10^{-6}\ °C^{-1}$

30. (*b*) Time lost in one day $= \frac{1}{2}\alpha\, t \times 86400$

or $10 = \frac{1}{2} \times \alpha \times 20 \times 86400$ or $\alpha = \frac{1}{86400}\ °C^{-1}$

31. (*b*) $\frac{T_0}{T_{25}} = \sqrt{\frac{l_0}{l_{25}}}$

or $\frac{T_0}{T_{25}} = \sqrt{\frac{l_0}{l_0\,(1+25\alpha)}}$ or $\frac{T_0}{T_{25}} = 1 - \frac{1}{2} \times 25\alpha$

or $\frac{T_0}{T_{25}} - 1 = -\frac{25\alpha}{2}$ or $\frac{T_{25} - T_0}{T_{25}} = \frac{1}{2}\alpha \times 25$

$\therefore$ gain in 1s $= \frac{1}{2}\alpha \times 25$

$\therefore$ gain per day $= \frac{1}{2} \times 1.9 \times 10^{-5} \times 25 \times 86400$ s $= 20.52$ s

32. (*c*) Expansion does not depend upon air cavity.

33. (*a*) $\Delta l_1 = l_1\,\alpha_1\,\Delta t$

$\Delta l_2 = l_2\,\alpha_2\,\Delta t$

$\Delta l_1 - \Delta l_2 = 0$

$\therefore$ $l_1\,\alpha_1\,\Delta t = l_2\,\alpha_2\,\Delta t$ or $\frac{l_1}{l_2} = \frac{\alpha_2}{\alpha_1}$

34. (*b*) Loss in weight = Weight of liquid displaced

$$\frac{W}{W_0} = \frac{V_t d_t}{V_0 d_0}$$

But $\frac{V_t}{V_0} = 1 + \gamma_s t$ and $\frac{d_0}{d_t} = 1 + \gamma_l t$

$\therefore$ $\frac{W}{W_0} = \frac{1+\gamma_s t}{1+\gamma_l t}$ or $\frac{W}{W_0} = (1+\gamma_s t)(1-\gamma_l t)$

or $\frac{W}{W_0} = 1 - \gamma_l t + \gamma_s t$ [Neglecting $\gamma_s \gamma_l$]

or $\frac{W}{W_0} = 1 + (\gamma_s - \gamma_l)t$ or $W = W_0[1 + (\gamma_s - \gamma_l)t]$

35. (*c*) There are 5 iron rods and 4 brass rods in the pendulum. The number of iron rods in one half of the pendulum including the central rod is 3. The number of brass rods is 2.

Since the pendulum is compensated,

$\therefore$ $\Delta l_{brass} = \Delta l_{iron}$

or $2l_2\alpha_2 t = 3l_1\alpha_1 t$ or $l_2 = \frac{3l_1\alpha_1}{2\alpha_2}$

$= \frac{3 \times 1 \times 12 \times 10^{-6}}{2 \times 20 \times 10^{-6}}$ m = 0.9 m

36. (*c*) $\Delta l = l_0\alpha(27 - \theta)$

or $1.1 \times 10^{-3} = 10 \times 11 \times 10^{-6}(27 - \theta)$

or $10^{-4} = 10^{-5}(27 - \theta)$ or $27 - \theta = 10$

or $\theta = (27 - 10)°C = 17°C$

37. (*c*) Energy per unit volume

$= \frac{1}{2} \times \text{stress} \times \text{strain} = \frac{1}{2}(Y\alpha t)(\alpha t) = \frac{1}{2}Y\alpha^2 t^2$

$= \frac{10^{11} \times 144 \times 10^{-12} \times 400}{2}$ J m^{-3}

$= 288 \times 10$ J m^{-3} = 2880 J m^{-3}

38. (*c*) Since Cu has a larger value of α therefore Cu strip has to suffer larger expansion.

39. (*b*) Weight of sphere is equal to weight of liquid displaced at 35°C.

$\therefore$ $mg = \frac{4}{3}\pi r^3 d_t g$ or $\frac{4}{3} \times \frac{22}{7} \times (3.5)^3 d_t = 266.5$

or $d_t = \frac{3 \times 7 \times 266.5}{4 \times 22 \times (3.5)^3}$ or $d_t = 1.483$ g cm^{-3}

Now $d_0 = d_t(1 + \gamma t) \Rightarrow 1.527 = 1.483(1 + \gamma \times 35)$

or $\left[\frac{1.527}{1.483} - 1\right] \times \frac{1}{35} = \gamma$

or $\gamma = 8.477 \times 10^{-4}$ /°C

40. (*c*) $\gamma_a = \gamma_r - \gamma_g = 0.000597 - 0.000027 = 0.00057$/°C

41. (*b*) Suppose the temperature of water falls to 0°C.

Heat lost by water = $10 \times 1 \times (50 - 0) = 500$ cal

If x g of ice melts, then $500 = 10 \times 0.5 \times 20 + x \times 80$

or $500 = 100 + 80x$ or $x = \frac{400}{80}$ g = 5 g

$\therefore$ amount of ice in mixture = (10 – 5) g = 5 g

42. (*a*) $xL = 80 \times 1 \times (30 - 0)$

or $x = \frac{80 \times 30}{80}$ g = 30 g

43. (*b*) Heat gained by ice = Heat lost by water

$10 \times 80 + 10 \times 1 \times \theta = 55 \times 1 \times (40 - \theta)$

or $800 + 10\theta = 55 \times 40 - 55\theta$

or $65\theta = 2200 - 800$ or $\theta = \frac{1400}{65} = 21.5°C$

44. (*a*) Mass ratio 1 : 2 ; hence ΔT ratio 2 : 1. This fits 30°C

Aliter. $5 \times 1 \times (\theta - 10) = 10 \times 1 \times (40 - \theta)$

or $\theta - 10 = 80 - 2\theta$ or $3\theta = 90$ or $\theta = 30°C$

45. (*b*) Heat capacity of cold : hot = 1 : 2

So, final t is $\frac{1}{3}$ of $(2 \times 92 + 1 \times 20)$ *i.e.*, 68°C

46. (*c*) Refer to solution of Q. 41. Since the whole of the ice has not melted therefore the temperature of the mixture is 0°C.

47. (*a*) $100 \times 80 + 100 \times 1 \times \theta = 100 \times 1 \times (100 - \theta)$

or $80 + \theta = 100 - \theta$ or $2\theta = 20$ or $\theta = 10°C$

48. (*c*) 0.93 J s^{-1} × 3600 s = $x \times 4.2 \times 80$

or $x = \frac{0.93 \times 3600}{4.2 \times 80} = 9.96$ g ≈ 10 g

So, the entire block just melts.

49. (*c*) $y \times 80 + y \times 1 \times 100 = x \times 540$

or $y \times 180 = x \times 540$ or $\frac{y}{x} = \frac{540}{180} = 3$

50. (*a*) $x\,m \times 80 \times 1000 = \frac{m \times 10 \times 1000}{4.2}$

or $x = \frac{10 \times 1000}{4.2 \times 80 \times 1000} = \frac{1}{33.6} \approx \frac{1}{33}$

51-52. $5 \times 1 \times (30 - 0) = 5 \times 0.5 \times 20 + x \times 80$

$150 - 50 = 80x$ or $x = \frac{100}{80} = \frac{5}{4}$ g = 1.25 g

Since whole of ice is not melted therefore the temperature of the mixture is 0°C.

53. (*b*) Suppose temperature of water falls to 0°C.

Then $x \times 80 = 5 \times 1 \times 10$ or $x = \frac{50}{80}$ g $= \frac{5}{8}$ g

Since whole of ice is not melted therefore the temperature of the mixture is 0°C.

54. (*a*) $x \times 540 + x \times 1(100 - 80) = 1.12 \times 1000 \times (80 - 15)$

or $560x = 72800$

or $x = \frac{7280}{56}\ \text{g} = \frac{7280}{56 \times 1000}\ \text{kg} = 0.13\ \text{kg}$

55. (*b*) When A and B are mixed,

$mS_B\,(19 - 16) = mS_A\,(16 - 12)$

$3S_B = 4S_A$ ...(1)

When B and C are mixed,

$mS_B\,(23 - 19) = mS_C\,(28 - 23)$

or $4S_B = 5S_C$ ...(2)

When A and C are mixed,

$mS_A\,(T - 12) = mS_C\,(28 - T)$

or $\frac{S_A}{S_C} = \frac{28 - T}{T - 12}$ ...(3)

Dividing (1) by (2),

$\frac{4S_A}{5S_C} = \frac{3S_B}{4S_B} = \frac{3}{4}$ or $\frac{S_A}{S_C} = \frac{3}{4} \times \frac{5}{4} = \frac{15}{16}$

From eqn. (3), $\frac{15}{16} = \frac{28 - T}{T - 12}$

$15T + 16T = 16 \times 28 + 15 \times 12$

or $31T = 448 + 180 = 628$

or $T = \frac{628}{31} = 20.3°$

56. (*b*) Heat required to melt 1 g of ice at 0°C into 1 g of water at 0°C is 80 cal.

Heat required to convert 1 g of water at 0°C into 1 g of water at 100°C = $1 \times 1 \times 100 = 100$ cal

Heat required to condense 1 g of steam $= 1 \times 540$ cal $= 540$ cal.

Clearly, whole of steam is not condensed. So, temperature of the mixture is 100°C.

57. (*c*) Thermal energy in the pellet is

$Q = \frac{1}{2}$(kinetic energy of the pellet)

$= \frac{1}{2}\left(\frac{1}{2}mv^2\right) = \frac{1}{4}mv^2$

The rise in temperature of the pellet, θ, is given by

$mc\theta = Q \Rightarrow mc\theta = \frac{1}{4}mv^2 \Rightarrow \theta = \frac{v^2}{4c}$.

58. (*b*) Total heat lost by the three liquids, when their temperatures fall to 0°C $= m_1s_1\theta_1 + m_2s_2\theta_2 + m_3s_3\theta_3$.

If *t* is the final temperature, then the total heat gained when the temperature increases from 0°C to *t*°C

$= m_1s_1t + m_2s_2t + m_3s_3t$

$\therefore \quad t = \frac{m_1s_1\theta_1 + m_2s_2\theta_2 + m_3s_3\theta_3}{m_1s_1 + m_2s_2 + m_3s_3}$

59. (*c*) If *t* is the time taken to boil the water dry, then

$Pt = Q = mL$

where P = electric power supplied = 2.0 kW

Q = heat energy absorbed by water

m = mass of water = 1.5 kg

L = specific latent heat of vaporisation of water = 2000 kJ kg^{-1}

$\Rightarrow (2.0 \times 10^3)t = (1.5)(2000 \times 10^3) \Rightarrow t = 1500$ s

60. (*c*) $Q = 1 \times 80 + 1 \times 1 \times 100 + 1 \times 536$

$= 80 + 100 + 536 = 716$ cal.

61. (*d*) (*a*) $20 \times 1 \times (40 - \theta) = 20 \times 1 \times (\theta - 30)$

$2\theta = 70$ or $\theta = 35°C$

(*b*) $10 \times 1 \times (\theta - 27) = 20 \times 1 \times (30 - \theta)$

or $3\theta = 87$ or $\theta = 29°$

(*c*) $1 \times 0.1 \times (\theta - 25) = 20 \times (30 - \theta)$

or $\theta - 25 = 200 \times 30 - 200\,\theta$

or $201\theta = 6000 + 25$ or $\theta = \frac{6025}{201} \approx 30°$

(*d*) $4 \times 1 \times (80 - \theta) = 20 \times 1(\theta - 30)$

or $80 - \theta = 5\theta - 150$

or $6\theta = 80 + 150 = 230$ or $\theta = \frac{230}{6} = 38.3°$

62. (*c*) $2000t = 200 \times 1000 \times 4.2 \times 60$

or $t = 100 \times 4.2 \times 60\ \text{s} = 25.2 \times 10^3$ s

63. (*c*) The rate of heat absorption by the water is

$\frac{Q}{t} = \frac{mc(\theta_0 - \theta_i)}{t} = \left(\frac{m}{t}\right)c(\theta_0 - \theta_i)$

where $\frac{Q}{t}$ = heat removed per second $= \frac{6.72 \times 10^9}{60}$ J s^{-1}

c = specific heat capacity of water = 4200 J kg^{-1} K^{-1}

θ_i = temperature of water entering the heat exchanger = 6°C

θ_0 = temperature of water leaving the heat exchanger = 14°C

$\frac{m}{t}$ = rate of water flow in kg s^{-1}

$\Rightarrow \quad \frac{m}{t} = \frac{Q}{t}\frac{1}{c(\theta_0 - \theta_i)}$

$= \frac{6.72 \times 10^9}{60 \times 4200 \times (14 - 6)} = \frac{6.72 \times 10^9}{4200 \times 8 \times 60}$ kg s^{-1}

64. (*d*) Assume that the temperature of water falls to 0°C.

Now, $x \times 80 = 1000 \times 1 \times 10$

or $x = \frac{1000 \times 10}{80}\ \text{g} = \frac{1}{8}$ kg

Since whole of ice is not melted therefore the resulting temperature will be 0°.

65. (*d*) Loss in kinetic energy $= \frac{1}{2}mv^2$

Gain in thermal energy $= mc(\Delta\theta)$

By conservation of energy,

$$mc(\Delta\theta) = \frac{1}{2}mv^2 \Rightarrow v = \sqrt{2c(\Delta\theta)}$$

66. (*c*) The rate of heat supply P remains constant throughout the experiment.

If m is the mass of the liquid,

c is its specific heat capacity,

L is its specific latent heat of vaporisation,

then, just before the liquid starts to boil,

$$P = \frac{Q}{t} = mc\frac{\Delta\theta}{\Delta t} = 4mc$$

where Q is the heat absorbed, $\frac{\Delta\theta}{\Delta t} = 4\ K\ min^{-1}$

During next 40 minute,

$$P = \frac{Q}{t} = mL\left(\frac{1}{40}\right) = \frac{mL}{40}$$

$$\therefore \quad 4mc = \frac{mL}{40} \text{ or } \frac{c}{L} = \frac{1}{4(40)} = \frac{1}{160}$$

67-68. Suppose the whole of ice is melted and the temperature of the resulting water is increased from 0°C to 100°C.

Then $m \times 540 = 6 \times 80 + 6 \times 1 \times 100$

or $m = \frac{480 + 600}{540} g$ or $m = 2g$

Since whole of steam has not condensed therefore the temperature of the mixture is 100°C.

Amount of water in the mixture = (2 + 6) g = 8 g.

69. (*b*) $Q = 6 \times 0.5 \times 10 = 30$ cal

Again, $30 = 6 \times 1000 \times \theta$

$$\frac{30}{6000} = \theta \text{ or } \theta \approx 0°$$

70. (*a*) $540 \times 80 + 540\,\theta = 540(80 - \theta)$

or $80 + \theta = 80 - \theta$

or $2\theta = 0$ or $\theta = 0°$

71. (*a*) $\frac{1}{2} \times m \times 9.8 \times 84 \times \frac{1}{J} = Q = m \times 1000 \times \theta$

or $\theta = \frac{9.8 \times 84}{2 \times 1000 \times 4.2} = 0.098°C$

72. (*a*) $PV^{3/2} = K$; $\log P + \frac{3}{2}\log V = \log K$

$$\frac{\Delta P}{P} + \frac{3}{2}\frac{\Delta V}{V} = 0 ; \frac{\Delta V}{V} = -\frac{2}{3}\frac{\Delta P}{P}$$

or $\frac{\Delta V}{V} = \left(-\frac{2}{3}\right)\left(\frac{2}{3}\right) = -\frac{4}{9}$

73. (*a*) $P'V' = PV$ or $P' = \frac{PV}{V'} = \frac{PV \times 4}{V} = 4P$

But P = 1 atmosphere $\therefore$ P′ = 4 atmosphere

74. (*c*) $PV^\gamma = P\frac{m^\gamma}{d^\gamma}, \frac{P}{d^\gamma} = \text{constant} ; \frac{P'}{P} = \left(\frac{d'}{d}\right)^\gamma = (32)^{7/5}$

$= [(32)^{1/5}]^7 = 2^7 = 128$

75. (*c*) $P'V'^\gamma = PV^\gamma ; P' = P\left[\frac{V}{V'}\right]^\gamma = P\left[\frac{V}{V/4}\right]^{3/2} = 8P$

76. (*d*) If ΔW is the work done by the gas, then $\Delta U = \Delta Q - \Delta W$. That leads to $100 - 60 = -20 + x$ (Since U_{AC} **is unique**). That leads to $x = 60$ cal.

77. (*b*) Let U_1 and U_2 denote the internal energies **per gram** mole of monoatomic and diatomic gas **molecules** respectively.

Then, $U_1 = \frac{3}{2}RT$ and $U_2 = \frac{5}{2}RT$

Now, $U = \frac{U_1 + U_2}{2} = 2RT ; C_v = \frac{dU}{dT} = 2R$;

$$C_p = C_v + R = 3R ; \gamma = \frac{C_p}{C_v} = \frac{3R}{2R} = 1.5$$

78. (*d*) $Q = \mu C_p\,\Delta T$ and $W = P\,\Delta V = \mu R\,\Delta T$.

For monoatomic gas, $C_p = \frac{5R}{2}$. That gives $\frac{W}{Q}$ as $\frac{2}{5}$.

79. (*a*) For an ideal gas, the internal energy is the total **kinetic** energy of the random motion of the molecules.

80. (*c*) $\Delta Q = \mu C_p\,\Delta T ; 207 = 1 \times C_p \times 10$ or $C_p = 20.7$ J/mol/K

But $C_v = C_p - R = 20.7 - 8.3 = 12.4$ J/mol/K

$\Delta Q = \mu C_v \Delta T = 1 \times 12.4 \times 10 = 124$ J

81. (*b*) $70 = 2 \times C_p \times 5 ; C_p = 7$ cal/g mol/°C

$C_v = C_p - R = 7 - 2 = 5$ cal/g mol/°C

$\Delta Q = \mu C_v \Delta T = 2 \times 5 \times 5 = 50$ cal

82. (*a*) Eliminate P between the given equation and PV = RT.

So, T varies as $\frac{1}{\sqrt{V}}$.

83. (*c*) Coefficient of performance,

$$\beta = \frac{T_2}{T_1 - T_2} = \frac{273 + 0}{(273 + 30) - 273} = \frac{273}{30} \approx 9$$

84. (*d*) $\eta = \frac{W}{Q_1} = 1 - \frac{T_2}{T_1}$ or $\frac{W}{6 \times 10^4} = 1 - \frac{273 + 127}{273 + 227}$

or $\frac{W}{6 \times 10^4} = 1 - \frac{4}{5} = \frac{1}{5}$ or $W = 1.2 \times 10^4$ cal

85. (*b*) $\frac{T_2}{T_1} = \left(\frac{27}{8}\right)^{\frac{5}{3}-1} = \left(\frac{27}{8}\right)^{2/3} = \left[\left(\frac{27}{8}\right)^{1/3}\right]^2 = \frac{9}{4}$;

Now calculate $T_2 - T_1$

86. (*c*) Since temperature remains unchanged therefore $U_f = U_i$. So, $\Delta Q = \Delta W$. In the first step which is isochoric, $\Delta W = 0$. In second step, pressure $= \frac{P}{n}$. Volume V is increased from V to nV.

$\therefore \quad W = \frac{P}{n}(nV - V) = PV\left(\frac{n-1}{n}\right) = RT(1 - n^{-1})$

87. (*c*) $\Delta P_A V + P_A \Delta V = \frac{m_A}{M} \Delta T\, R = 0 \Rightarrow \Delta P_A V = -P_A \Delta V$

$\Delta P_B V + P_B \Delta V = \frac{m_B}{M} \Delta T\, R = 0 \Rightarrow \Delta P_B V = -P_B \Delta V$

$\therefore \quad \frac{\Delta P_A}{\Delta P_B} = \frac{P_A}{P_B} = \frac{m_A}{m_B}$ or $\frac{\Delta P}{1.5\Delta P} = \frac{m_A}{m_B}$

or $\quad 2m_B = 3m_A$

88. (*d*) $\Delta U = \mu C_v \Delta T$; $\Delta Q = \mu C_p \Delta T$; $\frac{\Delta U}{\Delta Q} = \frac{C_v}{C_p} = \frac{1}{\gamma} = \frac{1}{7/5} = \frac{5}{7}$

89. (*b*) $U = \frac{\frac{3}{2}RT + \frac{6}{2}RT}{2} = \frac{9}{4}RT$; $C_v = \frac{dU}{dT} = \frac{9}{4}R$

$C_p = C_v + R = \frac{13}{4}R$; $\frac{C_p}{C_v} = \frac{13}{9} = 1.44$

90. (*d*) $U = \frac{f}{2}RT$; $C_v = \frac{dU}{dT} = \frac{f}{2}R$

Now, $C_p = C_v + R = \frac{f}{2}R + R = \left(\frac{f}{2} + 1\right)R$

$\gamma = \frac{C_p}{C_v} = \frac{\left(\frac{f}{2}+1\right)R}{\frac{f}{2}R} = 1 + \frac{2}{f}$

91. (*d*) $1 + \frac{2}{n} = 1.286$ or $\frac{2}{n} = 0.286$ or $n = \frac{2}{0.286} = 7$

92. (*b*) $\frac{T_1 - T_2}{T_1}$ is maximum is case of (*b*).

93. (*d*) $\frac{1}{4}(mgh) = mL$ or $h = \frac{4L}{g}$

☞ *In this problem, it is understood that L is in SI units.*

94. (*a*) W_{AB} = Area of trapezium between AB and V-axis

$= \frac{1}{2}(20 + 10)6 \text{ J} = 90 \text{ J}$

95. (*b*) W_{BC} = Area of rectangle between BC and V-axis

$= -10 \times 6 \text{ J} = -60 \text{ J}$

96. (*d*) Work done in complete cycle = Area of ΔACB

$= \frac{1}{2} \times 6 \times 10 \text{ J} = 30 \text{ J}$

97. (*c*) $(P')_{iso} = P\left[\frac{V}{V/2}\right] = 2P$

$(P')_{adia} = 2^\gamma P$

$\gamma > 1$

$\therefore \quad 2^\gamma P > 2P$

$\therefore \quad (P)'_{adia} > (P')_{iso}$ or $(P')_{iso} < (P')_{adia}$

98. (*b*) $60 \times 10 \times h = \frac{4.2 \times 10^5 \times 28}{100}$

or $\quad h = \frac{4.2 \times 10^5 \times 28}{100 \times 600} \text{ m} = 196 \text{ m}$

99. (*d*) Use $\left[\frac{T_1}{T_2}\right]^\gamma = \left[\frac{P_1}{P_2}\right]^{\gamma - 1}$

100. (*b*) $\frac{Q_2}{Q_1} = \frac{T_2}{T_1}$

$\frac{150}{300} = \frac{T_2}{500}$ or $T_2 = 250 \text{ K}$

101. (*c*) $C_v = C_p - R = 6$ cal/mole/°C

$Q = \mu C_v \Delta T$

$= 5 \times 6 \times 10 \text{ cal} = 300 \text{ cal}$

102. (*c*) $dQ = dU + PdV$.

Heating is done at constant volume.

$\therefore \quad dV = 0$

$\therefore \quad dU = dQ = 300$ cal

103. (*c*) $\frac{P_2}{P_1} = \left(\frac{\rho_2}{\rho_1}\right)^\gamma = 2^{7/5} = [2^7]^{1/5} = [2^5 \times 2^2]^{1/5}$

$= 2 \times 4^{1/5}$

$= 2 \times 1.3195 = 2.6$

$y = 4^{1/5}$

$\log y = \frac{1}{5} \log 4$

$\log y = 0.1204$

$y = 1.3195$

104. (*c*) $\frac{W}{Q_1} = 1 - \frac{T_2}{T_1}$

$\frac{800}{Q_1} = 1 - \frac{300}{600} = 1 - \frac{1}{2} = \frac{1}{2}$ or $Q_1 = 1600 \text{ J}$

105. (*c*) Decrease in internal energy is (30 – 10) J *i.e.* 20 J.

$\therefore$ Final internal energy = 40 J – 20 J = 20 J.

106. (*b*) $C_p = \frac{140}{2 \times 10}$ cal/mol/°C = 7 cal/mol/°C

$C_v = C_p - R = 7 - 2 = 5$ cal/mol/°C

Required quantity of heat = 2 × 5 × 10 cal = 100 cal

107. (*b*) Increase in internal energy = 1000 × 4.18 cal = 4180 J

Decrease in internal energy = 1675 J

Net increase = 2505 J

108. (*c*) $dQ = dU + dW$

$dQ = \mu C_v dT + 0$

$dQ = 2 \times \frac{3R}{2} \times 100 = 300\ R$

109. (*d*) $dU = -100$ J

$dQ = dU + dW$

But $dQ = 0$ (adiabatic process)

$\therefore$ $dW = -dU = +100$ J

110. (*c*) $(P')\left(\frac{V}{8}\right) = PV$ or $P' = 8P$

111. (*a*) $\frac{mgh}{J} = mS\theta$ or $\theta = \frac{gh}{JS} = \frac{10 \times 84}{4.2 \times 1000} = 0.2$°C

112. (*a*) $TV^{\gamma-1}$ = constant

$T[2V]^{\gamma-1} = 300[V]^{\gamma-1}$

or $T = \left[\frac{1}{2}\right]^{\gamma-1} \times 300$ or $T = \frac{300}{2^{0.4}} = 227.3$ K

113. (*d*) $W = 2.3026\ RT \log_{10} \frac{V_2}{V_1}$

$= 2.3026 \times 8.3 \times 273 \log_{10} 0.5$

$= -5217.46 \times 0.3 \approx -1570$ J

114. (*c*) $Q_2 = \beta W = \frac{5 \times 3.6 \times 10^6}{4.2}$ cal

Now, $m = \frac{Q_2}{L} = \frac{5 \times 3.6 \times 10^6}{4.2 \times 80 \times 1000}$ kg ≈ 54 kg

115. (*b*) $T_A = \frac{P}{P_{triple}} \times 200$ and $T_B = \frac{P}{P_{triple}} \times 350$

$\therefore$ $\frac{T_A}{T_B} = \frac{200}{350} = \frac{4}{7}$

116. (*b*) Work done $= \frac{1}{2} \times 2V_1 \times 3P_1 = 3P_1V_1$

117. (*a*) Use $dW = P\,dV$

118. (*d*) $\frac{P_2}{P_1} = \left(\frac{V_1}{V_2}\right)^\gamma = (8)^{5/3}$ or $P_2 = 2^5 P_1 = 32P_1$

119. (*c*) Adiabatic is steeper than isothermal.

120. (*c*) No exchange of heat is allowed in an adiabatic process.

121. (*a*) $dQ = dU + dW$

or $dU = dQ - dW$

$= 300 \times 4.18 - 600 = 1254 - 600 = 654$ J

122. (*d*) $Q = mS\theta$

In the given problem, $\theta \propto \frac{1}{S}$

$\frac{\Delta T_v}{\Delta T_p} = \frac{C_p}{C_v} = \gamma = \frac{14}{10} = \frac{7}{5}$

or $\Delta T_v = \frac{7}{5} \times \Delta T_p = \frac{7}{5} \times 30 = 42$ K

123. (*b*) $P\left[\frac{V}{8}\right]^{4/3} = P_0[V]^{4/3}$ or $P = P_0[8]^{4/3}$ or $P = 16P_0$

124. (*a*) $W = 2 \times 10^5\ [150 - 50] \times 10^3 \times 10^{-6}\ J = 2 \times 10^4$ J

125. (*a*) For cyclic process,

$Q_{cyclic} = W_{AB} + W_{BC} + W_{CA}$

$5 = 10 + 0 + W_{CA}$

or $W_{CA} = -5$ J

126. (*a*) $dQ = dU + dW$

or $dU = dQ - dW = 100 + 50 \times 6 = 400$ J

127. (*d*) $dQ = 0, dU + dW = 0$

$dW = -dU = -(-2) = 2$ J

This is work done by the gas.

So, work done on the gas is – 2 J

128. (*c*) For boiling to occur during the decrease in pressure from 600 kPa to 400 kPa, the temperature must be maintained in such a way that the state of the hydrogen falls on the liquid-vapour line in the diagram. Thus the temperature range is 27.4 K to 30.0 K.

129. (*b*) $5 \times 0.8 = (5 + 3)P$ or $P = \frac{5 \times 0.8}{8}$ m = 0.5 m

130. (*d*) $35 = dU - 15$ or $dU = (35 + 15)$ J = 50 J

131. (*d*) Work done = 2 V × 2 P = 4 PV

132. (*d*) $\frac{1}{2} \times \frac{m \times 10 \times 20}{4.2} = m \times 0.03 \times 1000 \times \theta$

or $\theta = \frac{200}{2 \times 4.2 \times 0.03 \times 1000} = 0.79$°C

133. (*a*) $\frac{3}{4} \times \frac{1}{2} mv^2 \times \frac{1}{J} = mS\theta + mL$

or $\frac{3v^2}{8J} = S\theta + L$ or $v^2 = \frac{8[S\theta + L]J}{3}$

or $v^2 = \frac{8[0.03 \times 300 + 6] \times 1000 \times 4.2}{3}$

or $v^2 = \frac{8 \times 15 \times 1000 \times 4.2}{3} = 168000$

or $v = 409.88 \approx 410$ m s^{-1}

134. (*c*) $W = \frac{1}{2}\ [1 \times 10^5 + 5 \times 10^5]\ 4$ J

$= 2 \times 6 \times 10^5\ J = 12 \times 10^5$ J

135. (c) $TV^{\gamma-1}$ = constant

$$(273 + 17)\, V^{\frac{5}{3}-1} = T\left[\frac{V}{8}\right]^{\frac{5}{3}-1}$$

or $290 \times 8^{2/3} = T$

or $T = 290 \times 4 \text{ K} = 1160 \text{ K}$

or $t = (1160 - 273)°C = 887°C$

136. (b) PV = constant

$$P\,\Delta V + V\,\Delta P = 0$$

or $V\Delta P = -P\Delta V$ or $\frac{\Delta P}{P} = -\frac{\Delta V}{V}$

137. (b) $dU = \mu C_v dT$

$$80 = 5 \times C_v \times 20 \text{ or } C_v = \frac{80}{100} = 0.8 \text{ J K}^{-1}$$

138. (c) Area $= \frac{\pi D^2}{4} = \frac{\pi \times 20 \times 20}{4} \text{ J} = 100\,\pi\text{J}$

139. (b) $TV^{\gamma-1}$ = constant

140. (d) $W = \mu RT \log_e \frac{V_2}{V_1}$

141. (a) $W = 10^5\,[1.091 - 1] \times 10^{-6} \text{ J} = \frac{0.091}{10} \text{ J} = 0.0091 \text{ J}$

142. (b) At 50 K, the number of degrees of freedom is given by

$$C_v = \frac{f}{2}R \Rightarrow 12.4 = \frac{f}{2}(8.3) \Rightarrow f = 3$$

∴ number of active translational degrees of freedom = 3.

number of active rotational degrees of freedom = 0.

number of active vibrational degrees of freedom = 0.

At 500 K, the number of degrees of freedom is now

$$C_v = \frac{f}{2}R \Rightarrow 20.6 = \frac{f}{2}(8.3) \Rightarrow 5$$

∴ number of active translational degrees of freedom = 3

number of active rotational degrees of freedom = 2

number of active vibrational degrees of freedom = 0

143. (b) $dQ = 5 \times 2.3 \times 10^6 \text{ J} = 11.5 \times 10^6 \text{ J}$

$dW = 5 \times 10^5\,(1.671 - 10^{-3})$

Now, use $dQ = dU + dW$

144. (d) The work done by the gas ΔW is given by

$$\Delta W = p\Delta V$$
$$= (2.0 \times 10^4)(0.075 - 0.050)$$
$$= (2.0 \times 10^4)(0.025) = 500 \text{ J}$$

The heat supplied to the gas ΔQ is given by

$$\Delta Q = 1000 \text{ J}$$

By first law of thermodynamics, the increase in internal energy is

$$\Delta Q = \Delta U + \Delta W$$

$\Rightarrow$ $1000 \text{ J} = \Delta U + 500 \text{ J}$ or $\Delta U = 1000 - 500 = 500 \text{ J}$

145. (a) Since temperature is constant,

∴ PV = constant

Differentiating,

$$PdV + VdP = 0$$

or $PdV = -VdP$ or $P = -\frac{VdP}{dV}$

or $P = \frac{1}{-\left(\frac{dV}{dP}\right)/V}$ or $P = \frac{1}{\beta}$

So, graph between P and β is a hyperbola.

146. (a) Work done on the gas $= \frac{R}{\gamma - 1}(T_2 - T_1)$

147. (d) Since the gas expands adiabatically and performs external work W, applying the First law of thermodynamics, the decrease in internal energy is given by

$$\Delta Q = \Delta U + \Delta W \Rightarrow 0 = \Delta U + (W)$$

or $\Delta U = -W$ ...(1)

Next, heat Q is supplied to the gas while keeping the volume constant. All the heat supplied is used to increase the gas internal energy.

That is $\Delta Q = \Delta U + \Delta W = \Delta U + p\Delta V$

$\Rightarrow$ $Q = \Delta U + 0$ ($\Delta V = 0$)

or $\Delta U = Q$ ...(2)

Thus, total increase in internal energy of gas is

$$U_2 - U_1 = \Delta U \text{ (in (1))} + \Delta U \text{ (in (2))}$$
$$= Q - W$$

148. (b) $\Delta U = \mu C_v dT$

149. (a) $dQ = dU + dW$

Since dU depends upon initial and final conditions therefore dU is same for all three curves.

$dW_3 > dW_2 > dW_1 \Rightarrow dQ_3 > dQ_2 > dQ_1$

$\Rightarrow Q_1 < Q_2 < Q_3$

150. (b) The net work done by the gas is positive and is given by the enclosed area defined by WXYZ.

Thus, the net work done is

$$\Delta W = (3 \times 10^5 - 2 \times 10^5)(3 \times 10^{-3} - 1 \times 10^{-3})$$
$$= (1 \times 10^5)(2 \times 10^{-3}) = 2 \times 10^2 \text{ or } 200 \text{ J}$$

151. (b) $C_p - C_v = \frac{R}{J}$

152. (d) $Q = \frac{5}{2}R \times 100;\ Q' = \frac{5}{2}R \times 50$

153. (c) Applying the equation of state,

$$\frac{2.0 \times 10^6 \times 0.003}{T_M} = \frac{2.0 \times 10^6 \times 0.001}{T_L} \Rightarrow T_M = 3T_L$$

Again, $\frac{0.8 \times 10^6 \times 0.005}{T_N} = \frac{2.0 \times 10^6 \times 0.003}{T_M}$

$\Rightarrow \quad T_N = 0.67T_M$ or $T_M = 1.5T_N$

154. (*e*) For a mole of an ideal gas, the equation of state is

$pV = RT$

or $\quad T = \frac{pV}{R}$ which is proportional to the product pV

At X, $pV = (4 \times 10^5)(1 \times 10^{-4}) = 40$ N m

At Y, $pV = (1 \times 10^5)(5 \times 10^{-4}) = 50$ N m

At Z, $pV = (1 \times 10^5)(1 \times 10^{-4}) = 10$ N m

Thus, T is maximum at Y since pV is the highest and T is minimum at Z since pV is the smallest.

155. (*b*) The new volume of the balloon is now

$V_2 = \frac{4}{3}\pi(2r)^3$ since its radius is doubled from r to $2r$

$\Rightarrow V_2 = \frac{4}{3}\pi(8r^3) = 8\left(\frac{4}{3}\pi r^3\right) = 8$ V

where V_1 = initial volume of the balloon $= \frac{4}{3}\pi r^3 = V$.

Thus, the work done in expanding the balloon against the atmosphere pressure p is given by

$\Delta W = p\Delta V = p(V_2 - V_1) = p(8V - V) = 7pV$

156. (*e*) From the graph, when the gas expands from P to Q and Q to R, work is done by the gas. On the other hand, when the gas moves from R to S and then from S to P, it is compressed. Thus, work is done on gas when it is compressed from state R to state S and then to state P.

157. (*a*) $W = 0.5 \text{ RT} \log_e \frac{2V}{V} + 0.5 \text{ RT} \log_e \frac{2V}{V}$

158. (*d*) $\quad \frac{Q}{t} = \frac{4KA\,(20-0)}{d} = \frac{KA\,(\theta - 20)}{d}$

or $\quad 80 = \theta - 20$ or $\theta = 100°C$

159. (*a*) $\quad \frac{Q}{t} = \frac{K\pi r^2\,(\theta_1 - \theta_2)}{l_1}$

$\frac{8Q}{t} = \frac{K\pi(2r)^2\,(\theta_1 - \theta_2)}{l_2}$

Dividing, $8 = \frac{l_1}{l_2} \times 4$ or $\frac{l_1}{l_2} = \frac{2}{1}$

160. (*b*) $mL = \frac{KA\,(\theta_1 - \theta_2)t}{d}$ or $\frac{m}{t} = \frac{KA(\theta_1 - \theta_2)}{L\,d}$

Now, $0.1 = \frac{K\pi r^2 \times 100}{Ll}$

and $\quad \frac{m}{t} = \frac{0.25K\,\pi(2r)^2 \times 100}{L \times l/2}$

or $\quad \frac{m}{t} = \frac{1}{4} \times 2 \times 4 \times 0.1 \text{ g s}^{-1} = 0.2 \text{ g s}^{-1}$

161. (*c*) $\quad mL = \frac{KA(\theta_1 - \theta_2)t}{d}$

or $A \times \Delta x \times \rho \times L = \frac{KA(\theta_1 - \theta_2)t}{d}$

or $\quad t = \frac{\Delta x \rho L d}{K(\theta_1 - \theta_2)}$

$= \frac{0.1 \times 0.91 \times 80\left[\frac{10 + 10.1}{2}\right]}{0.005 \times 5}$ s

$= 2926.56$ s $= 48.78$ minute

162. (*d*) $E = \sigma \times \text{area} \times T^4$; T increases by a factor $\frac{3}{2}$

Area increases by a factor $\frac{1}{4}$

163. (*a*) $\frac{Q}{t} = Ae\sigma\, T^4$

164. (*d*) We know that $Q = \frac{KA(\theta_1 - \theta_2)t}{d}$

For the given problem, $t \propto d$

Average thickness changes from $\frac{0+1}{2}$ to $\frac{1+2}{2}$

i.e., from $\frac{1}{2}$ cm to $\frac{3}{2}$ cm *i.e.*, d increases by a factor of 3

t increases by a factor of 3.

165. (*d*) $\frac{2Q}{Q} = \left[\frac{T_2}{673}\right]^4$

166. (*b*) $K\,\Delta\theta$ = constant

$\therefore \quad \frac{\Delta\theta_A}{\Delta\theta_B} = \frac{K}{2K} = \frac{1}{2}$

$\therefore \; \Delta\theta_A = \left(\frac{1}{2+1}\right)36 = 12°C$

☞ ***Note carefully the simple way to solve this problem.***

167. (*b*) Thermal resistance $= \frac{1}{K}\frac{l}{A}$

For the given data, $l \propto K$

$\therefore \quad \frac{l_1}{l_2} = \frac{K_1}{K_2} = \frac{5}{3}$

168-169. $\frac{Q}{t} = \frac{K\pi r^2(\theta_1 - \theta_2)}{l}$ or $\frac{Q}{t} \propto \frac{r^2}{l}$

(*a*) $\frac{1 \times 1}{2} = \frac{1}{2}$ $\qquad$ (*b*) $\frac{2 \times 2}{4} = 1$

(*c*) $\frac{2 \times 2}{2} = 2$ $\qquad$ (*d*) $\frac{4 \times 4}{2} = 8$

Clearly, the highest rate of cooling corresponds to (*d*) and the lowest rate of cooling corresponds to (*a*).

> ☞ ***Do not bother to have the same units of r and l in this particular problem.***

170. (*b*) $\frac{KA(\theta - 0)}{l} = \frac{KA(90° - \theta)}{l} + \frac{KA(90° - \theta)}{l}$

∴ $\theta = 2(90° - \theta)$

or $3\theta = 180°$ or $\theta = \frac{180°}{3} = 60°$

171. (*b*) Use $\lambda_m T$ = constant

172. (*c*) $Q = \frac{K\pi R^2 (\theta_1 - \theta_2)t}{d}$

In the given problem, KR^2 = constant

∴ $K'[2R]^2 = K_1R^2 + K_2[(2R)^2 - R^2]$

or $4K' = K_1 + 3K_2$ or $K' = \frac{K_1 + 3K_2}{4}$

173. (*a*) $\frac{dQ}{dt} = Ae\sigma T^4 = 4\pi R^2 e\sigma T^4$

Clearly, $\frac{dQ}{dt} \propto R^2$

174. (*c*) $\frac{d}{dt}(mS\theta) \propto R^2$ or $\frac{d\theta}{dt} \propto \frac{R^2}{R^3}$ or $\frac{d\theta}{dt} \propto \frac{1}{R}$

175. (*b*) $mL = \frac{KA(\theta_1 - \theta_2)t}{d}$

In the given problem, Kt = constant

∴ $\frac{K_1}{K_2} = \frac{t_2}{t_1}$

176. (*a*) $t \propto \frac{d}{A}$, $t' \propto \frac{d/2}{2A}$

Dividing, $\frac{t'}{t} = \frac{1}{4}$ or $t' = \frac{t}{4} = \frac{4 \text{ min.}}{4} = 1$ min.

177. (*c*) $E = \sigma T^4 = 5.67 \times 10^{-8} \times 10^{12} = 5.67 \times 10^4$ J

178. (*b*) $\left(\frac{Q}{t}\right)_1 = \frac{KA(\theta_1 - \theta_2)}{d} = 1 \text{ cal s}^{-1}$

$\left(\frac{Q}{t}\right)_2 = \frac{2K(4A)(\theta_1 - \theta_2)}{2d} = \frac{4KA(\theta_1 - \theta_2)}{d} = 4 \text{ cal s}^{-1}$

179. (*d*) $\frac{dQ}{dt} \propto \frac{A}{d}$

180. (*b*) $\frac{[illegible] - \theta}{[illegible].5} = 80$

181. (*b*) [illegible] $= \frac{(573)^4 \times 10^5}{10}$

182. (*b*) $E \propto T^4$

183. (*d*) $t \propto \frac{l}{A}$, $t' \propto \frac{2l}{A/2}$

184. (*d*) $\frac{E_2}{E_1} = \left[\frac{T_2}{T_1}\right]^4$

185. (*a*) $\frac{Q}{t} = \frac{KA(\theta_1 - \theta_2)}{l}$ or $\frac{Q}{t} = \frac{(\theta_1 - \theta_2)}{\frac{l}{KA}}$

or Thermal resistance $= \frac{(\theta_1 - \theta_2)t}{Q}$

So, dimensional formula is $[M^{-1}L^{-2}T^2][KT]$

or $[M^{-1}L^{-2}T^3K]$

186. (*a*) $mL = \frac{KA(\theta_1 - \theta_2)t}{d}$

In the given problem, Kt = constant

∴ $K \propto \frac{1}{t}$ or $\frac{K_1}{K_2} = \frac{t_2}{t_1} = \frac{30}{20} = 1.5$

187. (*d*) $T' = \frac{3}{2}T$; $E' = \frac{81}{16}E$

$\left(\frac{E'}{E} - 1\right) \times 100 = \left(\frac{81}{16} - 1\right) \times 100$

$= \frac{81 - 16}{16} \times 100 = 406.25 \approx 400$

188. (*a*) $T_1 = 300$ K, $T_2 = 400$ K

Required factor is $\left[\frac{400}{300}\right]^4$ or $\frac{256}{81}$

189. (*c*) $R = \sigma e A T^4$

190. (*d*) $\frac{P_2}{P_1} = \frac{r_2^2}{r_1^2} \times \frac{T_2^4}{T_1^4}$

191. (*d*) $\frac{K_A}{K_B} = \frac{A_B}{A_A}$

192. (*d*) $\frac{d_1}{d_2} = \frac{K_1A_1}{K_2A_2}$

193. (*d*) $\frac{Q}{t} = \frac{KA(\theta_1 - \theta_2)}{d}$ or $(\theta_1 - \theta_2) \propto \frac{d}{K}$

Now, $\theta_1 - \theta_0 \propto \frac{d}{k_1}$

$\theta_0 - \theta_2 \propto \frac{d}{k_2}$

Adding $\theta_1 - \theta_2 \propto d\left[\frac{1}{k_1} + \frac{1}{k_2}\right]$

Also, $\theta_1 - \theta_2 \propto (2d)\left(\frac{1}{k}\right)$

∴ $\frac{2}{k} = \frac{1}{k_1} + \frac{1}{k_2}$ or $k = \frac{2k_1k_2}{k_2 + k_1}$

194. (*b*) $Q = \frac{K\pi R^2 (\theta_1 - \theta_2)t}{d}$

For the given problem, KR^2 = constant

$\therefore \quad K[3R]^2 = K_1R^2 + K_2[9R^2 - R^2]$

or $\quad 9K = K_1 + 8K_2$ or $K = \frac{K_1 + 8K_2}{9}$

195. (*c*) In both the cases, the rate of flow of heat is 50 KA.

196. (*b*) Temperature gradient $= \frac{100}{20} = 5°\text{C cm}^{-1}$

Fall of temperature over 6 cm distance = 30°C

Temperature at a distance of 6 cm

$= (100 - 30)°\text{C} = 70°\text{C}$

197. (*b*) $\frac{P_2}{P_1} = \left[\frac{T_2}{T_1}\right]^4 \left[\frac{d_1}{d_2}\right]^2$

198. (*c*) $\frac{d\theta}{dx} = \frac{1}{K}\frac{dQ}{Adt}$

199. (*d*) $\lambda_m T = 2892 \times 10^{-6}$

200. (*d*) $\frac{dQ}{dt} \propto \frac{A}{d}$

201. (*c*) Use $\frac{d\theta}{dt} \propto \frac{1}{K}$

202. (*c*) $\frac{dQ}{dt} = \frac{KA\Delta\theta}{d}$ or $\Delta\theta = \frac{d}{KA} \cdot \frac{dQ}{dt}$

$= \frac{10}{100} \times \frac{4000}{400 \times 100 \times 10^{-4}} = 100°\text{C}$

203. (*c*) The rate of heat flow through polystyrene $\frac{dQ}{dt}$ is related to its cross-sectional area A, its thickness *d* by the equation $\frac{dQ}{dt} = -KA\frac{\theta_2 - \theta_1}{d}$

where K = thermal conductivity of the polystyrene,

θ_2, θ_1 = temperatures at surface of polystyrene.

To have appreciable heat flow through the polystyrene where thermal conductivity of material is small, we could choose wide (big area A) and thin (small *d*, therefore high temperature gradient $\frac{\theta_2 - \theta_1}{l}$) specimen.

Since the specimen is relatively thin and its thermal conductivity is small, lagging is not necessary in this case.

204. (*b*) Since in metal, the heat conduction and current flow is by means of free electrons, there will be an analogue in plotting the graph of temperature variation versus distance along the rod, with the graph of voltage/potential variation versus distance along the rod as can readily be seen in the equations listed below.

For electrical conduction : $I = \frac{V}{R} \Rightarrow \frac{q}{t} = \sigma A\left(\frac{V}{L}\right)$

For thermal energy conduction : $\frac{Q}{t} = \lambda A\left(\frac{\Delta\theta}{\Delta L}\right)$

The slope of this graph is the potential gradient $\left(\frac{V}{L}\right)$.

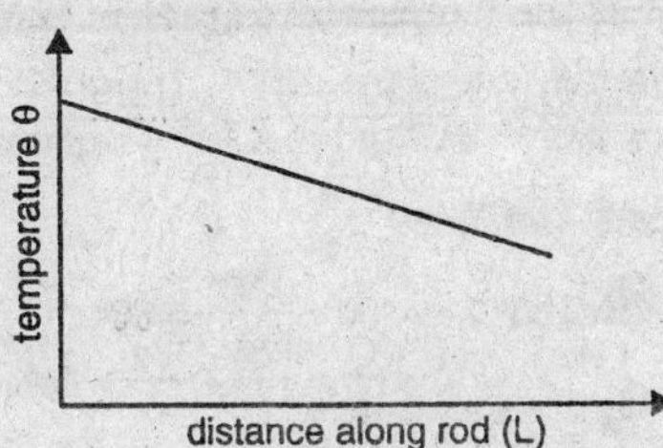

Fig. 49

205. (*a*) $\frac{365 - 361}{2} = k\left[\frac{365 + 361}{2} - 293\right]$

or $\quad 2 = k[70] \quad$ ...(1)

Again, $\frac{344 - 342}{t} = k\left[\frac{344 + 342}{2} - 293\right]$

or $\quad \frac{2}{t} = k[50] \quad$ ...(2)

Dividing (1) by (2),

$t = \frac{70}{50} = \frac{7}{5}\text{ min} = \frac{7}{5} \times 60\text{ s} = 84\text{ s}$

206. (*d*) Suppose K = thermal conductivity of aluminium

then $\quad$ 2K = thermal conductivity of copper.

Let $\quad$ A = cross-sectional area of the composite rod

T = temperature in °C at the copper-aluminium interface.

Then the rate of heat flow $\frac{Q}{t}$ is given by

$\frac{Q}{t} = (2K)(A)\frac{100 - T}{L} = (K)(A)\frac{T - 0}{L}$

$\Rightarrow 2(100 - T) = T$ or $T = \frac{200}{3} \approx 67°\text{C}$

which is best represented in graph *d*.

Note that the temperature gradient in the copper bar should be half that in the aluminium since the rate of heat flow is the same and the thermal conductivity of copper is twice that of aluminium. This is evident in the graph of temperature variation in graph *d*.

207. (*a*) $\frac{R_1}{R_2} = \frac{100 - 40}{80 - 40} = \frac{60}{40} = \frac{3}{2}$

208. (*b*) Use $Q = A\sigma T^4$

This will give heat radiated in one second.

209. (*d*) $E \propto T^4$

210. (*b*) Rate of heat flow, $\frac{Q}{t}$, through a parallel-sided slab of cross-sectional area A and thermal conductivity λ is given by $\frac{Q}{t} = -\lambda A \frac{d\theta}{dx}$

where $\frac{d\theta}{dx}$ is the temperature gradient across the slab.

Now $\lambda_Y = 2\lambda_X = 2\lambda$, and $-A\frac{d\theta}{dx}$ = constant = c for both materials X and Y.

$$\text{Hence} \begin{cases} \frac{Q_X}{t} = c\lambda \\ \frac{Q_Y}{t} = 2c\lambda \end{cases} \text{ and } \frac{Q}{t} = \frac{Q_X}{t} + \frac{Q_Y}{t} = 3c\lambda.$$

$\therefore$ fraction of total flow through the slab passes through material X $= \frac{Q_X}{Q} = \frac{c\lambda}{3c\lambda} = \frac{1}{3}$

211. (*b*) The steady state rate of heat flow is given by

$$\frac{dQ}{dt} = \lambda_X A \frac{100-25}{0.60} = \lambda_Y A \frac{25}{0.30}$$

where A is the cross-sectional area of the rods X and Y

Solving the equation, we have

$$\lambda_X \frac{75}{0.60} = \lambda_Y \frac{25}{0.30}$$

$$\Rightarrow \quad \frac{\lambda_X}{\lambda_Y} = \left(\frac{25}{0.30}\right)\left(\frac{0.60}{75}\right) = \frac{2}{3}$$

212. (*a*) The rate of heat flow at steady state condition is given by

$$\frac{Q}{t} = KA\frac{100-\theta}{L} = (4K)A\frac{\theta-0}{L}$$

where K = thermal conductivity of Q

4K = thermal conductivity of P

L = length of rods P and Q

A = cross-sectional area of the rods

θ = temperature of the interface.

$\Rightarrow \quad 100 - \theta = 4\theta \Rightarrow 5\theta = 100°C \;\therefore\; \theta = 20°C$

213. (*e*) At steady state condition, the rate of heat flow is a constant given by

$$\frac{Q}{t} = K_1A\frac{\theta_P - \theta_X}{L_{PX}} = K_2A\frac{\theta_X - \theta_Y}{L_{XY}} = K_1A\frac{\theta_Y - \theta_Q}{L_{YQ}}$$

where K_1 = thermal conductivity of the metal bar

K_2 = thermal conductivity of material XY

A = cross-sectional area of the metal bar.

$\theta_P, \theta_X, \theta_Y, \theta_Q$ are temperatures at locations P, X, Y and Q respectively.

Since $\frac{Q}{t}$ is constant and so is $\frac{\theta_X - \theta_Y}{L_{XY}}$ as both K_2 and A are independent of temperature. Hence, the temperature difference $(\theta_X - \theta_Y)$ across XY is independent of the distance of the section XY from the end P if the length L_{XY} is fixed.

214. (*b*) The intensity I or power per unit area of radiation received by the detector is the same for both cases. This may be represented by the equations below :

$$\begin{cases} I = \frac{P_1}{4\pi d_1^2} = \frac{A\sigma T_1^4}{4\pi d_1^2} \\ I = \frac{P_2}{4\pi d_2^2} = \frac{A\sigma T_2^4}{4\pi d_2^2} \end{cases}$$

where A is the area of the black body and σ is the Stefan's constant.

Equating the two equations, we have

$$\frac{T_1^4}{d_1^2} = \frac{T_2^4}{d_2^2} \Rightarrow \left(\frac{d_2}{d_1}\right)^2 = \left(\frac{T_2}{T_1}\right)^4 \text{ or } \frac{d_2}{d_1} = \left(\frac{T_2}{T_1}\right)^2$$

215. (*d*) The rate of heat flow through the bars is given by

$$\frac{Q}{t} = k_1A\frac{T_h - T_j}{l} = k_2A\frac{T_j - T_c}{l}$$

where A is the cross-sectional area of the two bars.

Hence, $\quad k_1(T_h - T_j) = k_2(T_j - T_c)$

$\Rightarrow \quad k_1T_h + k_2T_c = (k_1 + k_2)T_j$

or $\quad T_j = \frac{1}{k_1 + k_2}(k_1T_h + k_2T_c)$